Spectroscopic Data of Steroid Glycosides: Stigmastanes, Furostanes, Spirtostanes

Volume 2

Volume 1
SPECTROSCOPIC DATA OF STEROID GLYCOSIDES:
CHOLESTANES, ERGOSTANES, WITHANOLIDES,
STIGMASTANE
Edited by Viqar Uddin Ahmad and Anwer Basha

Volume 2
SPECTROSCOPIC DATA OF STEROID GLYCOSIDES:
STIGMASTANES, FUROSTANES, SPIRTOSTANES
Edited by Viqar Uddin Ahmad and Anwer Basha

Volume 3
SPECTROSCOPIC DATA OF STEROID GLYCOSIDES:
SPIROSTANES, BUFANOLIDES, CARDENOLIDES
Edited by Viqar Uddin Ahmad and Anwer Basha

Volume 4
SPECTROSCOPIC DATA OF STEROID GLYCOSIDES:
CARDENOLIDES AND PREGNANES
Edited by Viqar Uddin Ahmad and Anwer Basha

Volume 5
SPECTROSCOPIC DATA OF STEROID GLYCOSIDES:
PREGNANES, ANDROSTANES, AND MISCELLANEOUS
Edited by Viqar Uddin Ahmad and Anwer Basha

Volume 6
SPECTROSCOPIC DATA OF STEROID GLYCOSIDES:
MISCELLANEOUS STEROIDS AND INDEXES
Edited by Viqar Uddin Ahmad and Anwer Basha

Spectroscopic Data of Steroid Glycosides: Stigmastanes, Furostanes, Spirtostanes

Volume 2

Edited by

Viqar Uddin Ahmad

HEJ Research Institute of Chemistry
University of Karachi
Pakistan

and

Anwer Basha

Abbott Laboratories
Abbott Park, Illinois
USA

 Springer

Viqar Uddin Ahmad
HEJ Research Institute of Chemistry
International Center for Chemical Sciences
University of Karachi
Karachi 75270
Pakistan
vuahmad@cyber.net.pk

Anwer Basha
41 Heron Road
Lake Forest, IL 60045
USA
anwerbasha@hotmail.com

Library of Congress Control Number: 2006922481

ISBN-10: 0-387-31160-2
ISBN-13: 978-0387-31160-9
Set ISBN: 0-387-34348-2

9 8 7 6 5 4 3 2 1

springer.com

PREFACE

The present volumes reproduce the spectroscopic data of naturally occurring steroidal glycosides as far as they are available in the chemical literature published until the end of 2004. Steroids have the basic skeleton of cyclopentanoperhydrophenanthrene. Generally they do not have methyl groups attached to C-4 and thus differ from triterpenes. Many of the steroidal glycosides, or saponins, have interesting biological activities and constitute the active principles of the natural drugs. The cardiac glycosides (cardenolides) included in the present work act as life-saving medicines in certain ailments.

Not included in this work are the glycosides of steroidal alkaloids. However, the compounds which contain a nitrogen atom in the sugar or in the ester moiety (e.g. nicotinoyl moiety) are included.

The steroidal glycosides are arranged according to the class of their aglycones (steroidal parts). Within each class increasing molecular weight is taken as the basis for this arrangement. If the compounds of the same class have the same molecular weight, then the glycosides with lesser number of carbon atoms come earlier than those with more carbon atoms. Finally, if all these factors are the same, then the compounds are arranged in alphabetical order.

The chemical shifts in the proton nuclear magnetic resonance (PMR) spectral data are arranged according to the increasing δ (ppm) values. Each signal represents one proton unless indicated otherwise. The small alphabets used as superscript in PMR and ^{13}C-NMR (CMR)-spectral data mean that the assignments are ambiguous and may be reversed with signals having the same superscripts. The signals masked by solvent peaks or by other signals of the compound are marked by an asterisk.

Compounds can be easily located in this book with the help of the four indexes at the end of the last volume. The trivial names of the compounds given by the original authors are used as the heading of the compound. If no trivial name has been given, then the name of the plant from which the glycoside has been isolated followed by the word "saponin" or "glycoside" and then the numerical order are used as the main heading. For the subheading, the name of the aglycone (trivial names if available) followed by names of the sugars are used with clear indication of glycosidic linkages and branching of the sugar chain if present.

I am very grateful to Ms. Judy Watson of Chemical Abstract Service who has helped me greatly in finding the registry numbers of several compounds. This work would not have been possible without the help of literature surveyors Dr. Akbar Ali, Dr. Hidayat M. Khan, Dr. M. Athar Abbasi, Mr. Touseef Ali Khan, Mr. Umair Quyyum Khan, Miss. Humera Zaheer, Miss. Rukhsana Kausor, Miss Husna Qamar, Miss. Fouzia Shamim,

Ms. Zeenat Siddiqui, Muhammad Zubair, Afsar Khan, and Shazia Yasmeen to whom my sincere thanks are due. The whole book has been typed, composed, and structures drawn by Mr. Rafat Ali, Mr. Shabbir Ahmed, and Tariq Ilyas and I wish to express my sincere thanks to them.

ABBREVIATIONS

Aco	Acofrose
Afr	Acrofriose
Agl	Aglycone
All	Allose
Alt	Altrose
Ang	Angeloyl
Ant	Antirose
Ara	Arabinose
Boi	Boivinose
Ben	Benzoyl
Can	Canarose
Cin	Cinnamoyl
CMR	^{13}C-Nuclear Magnetic Resonance
Cym	Cymarose
DAC	4-Deoxy-4-aminocymarose
DMC	4-Deoxy-4-methylaminocymarose
Dal	6-Deoxyallose
Ddg	Dideoxygulopyranoside
Def	2-Deoxyfucose
Dex	6-Deoxy-D-glycero-L-threo-4-hexosulose
DHMP	2,3-Dihydroxy-3-methylpentanoyl
DMB	Dimethoxybenzoyl
DMC	4-Deoxy-4-methylaminocymarose
Dil	Digitalose
Din	Diginose
Dix	Digitoxose
Dma	Deoxymethylallose
DMP	3,4-Dimethyl-2(E)-pentenoyl
DMX	Dimethylxylose
EI	Electron ionization
ESI	Electro-spray ionization
F	Furanosyl
FAB	Fast Atom Bombardment
FD	Field desorption
Fuc	Fucose
Gal	Galactose

Glc	Glucose
Glum	6-Deoxy-α-L-glucopyranoside
Gum	Gulomethylose
HMB	Hydroxymethoxybenzoyl
HMG	Hydroxymethylglutaroyl
HR	High resolution
Ike	Ikemoyl (3,4-dimethyl-2-pentenoyl)
LD	Laser Desorption
Meb	2-Methylbutanoyl
MeXyl	Methylxylose
MGl	Methylglucose
Neg	Negative
Nic	Nicotinoyl
Ole	Oleandrose
Oli	Olivose
PMB	*Para* -methoxybenzoyl
Pos	Positive
PMR	Proton Magnetic Resonance
Qui	Quinovose
Rha	Rhamnose
Sar	Sarmentose
Tam	Talomethylose
Tar	Triacetylarabinose
The	Thevetose
TMB	Trimethoxybenzoyl
TOF	Time of flight
Xyl	Xylose

CONTENTS

COLEBRIN E

7β-Hydroxyclerosterol 3β-O-β-D-(6'-O-margaroyl)-glucopyranoside

Source : *Clerodendron colebrookianum* Walp.
(Verbenaceae)
Mol. Formula : $C_{52}H_{90}O_8$
Mol. Wt. : 842
$[\alpha]_D^{24}$: -42.2° (c=0.450, CHCl₃)
Registry No. : [329315-72-2]

IR (film) **:** 3421 (br), 2927, 2854, 1738, 1466, 1377, 1175, 1081, 1020, 888, 722 cm⁻¹.

PMR (CDCl₃, 400 MHz) **:** δ 0.63 (s, 3xH-18), 0.77 (t, *J*=7.3 Hz, 3xH-29), 0.83 (t, *J*=7.3 Hz, 3xH-17 of Mar), 0.88 (d, *J*=6.5 Hz, 3xH-21), 0.98 (s, 3xH-19), 1.14-1.37 (m, 28H, H-3 to H-16 of Mar), 1.55 (s, 3xH-27), 2.32 (t, *J*=7.6 Hz, H-2 of Mar), 3.36 (m, H-2, H-5 of Glc), 3.47 (m, H-3α), 3.53 (m, H-3, H-4 of Glc), 4.07 (br d, *J*=8.6 Hz, H-7α), 4.31 (br s, 2xH-6 of Glc), 4.36 (d, *J*=7.4 Hz, H-1 of Glc), 4.61 (br d, *J*=2.0 Hz, H-26A), 4.70 (br d, *J*=2.0 Hz, H-26B), 5.59 (br s, H-6).

CMR (CDCl₃, 100.6 MHz) **:** δ C-1) 36.9 (2) 31.9 (3) 79.2 (4) 38.6 (5) 145.2 (6) 122.3 (7) 86.3 (8) 34.7 (9) 48.8 (10) 36.7 (11) 21.1 (12) 39.6 (13) 42.9 (14) 56.1 (15) 28.2 (16) 29.3 (17) 55.7 (18) 11.8 (19) 18.7 (20) 35.5 (21) 18.7 (22) 33.7 (23) 24.9 (24) 49.5 (25) 147.6 (26) 111.3 (27) 17.9 (28) 26.5 (29) 12.0 **Glc** (1) 101.5 (2) 70.3 (3) 76.3 (4) 73.6 (5) 73.9 (6) 63.4 **Mar** (1) 174.3 (2) 34.3 (3) 31.9 (4-14) 29.7 (15) 24.9 (16) 22.7 (17) 14.0.

Mass (FAB, Positive ion) **:** *m/z* (rel.intens.) 842 [M]⁺ (8), 427 (38), 409 (100), 393 (60), 269 (16), 255 (10), 239 (25), 161 (28), 159 (30), 145 (32).

Mass (FAB, Positive ion, H.R.) **:** *m/z* 843.6617 [(M+H)⁺, calcd. for 843.6655].

Reference

1.   H. Yang, J. Wang, A.-J. Hou, Y.-P. Guo, Z.-W. Lin and H.-D. Sun, *Fitoterapia*, **71**, 641 (2000).

VERNONIA COLORATA GLYCOSIDE 4

(5α,21S,22R,23S,24ξ,25S,28S)-3β,21,24-Trihydroxy-21,23:22,28:26,28-triepoxy-5α-stigmasta-8(9),14(15)-diene
3-O-[β-D-galactopyranosyl-(1→2)-(6-O-acetyl)-β-D-glucopyranoside]

Source : *Vernonia colorata* (Willd.) Drake
(Asteraceae)
Mol. Formula : $C_{43}H_{64}O_{17}$
Mol. Wt. : 852
[α]$_D^{25}$: +61° (c=0.1, MeOH)
Registry No. : [678185-29-0]

UV (MeOH) : λ_{max} 236, 242, 250 nm.

PMR (CD₃OD, 600 MHz) : δ 1.95 (s, OCOCH₃), 3.30 (t, J=9.0 Hz, H-4 of Glc), 3.36 (m, H-5 of Glc), 3.38 (dd, J=3.0, 9.0 Hz, H-3 of Gal), 3.42 (dd, J=7.5, 9.0 Hz, H-2 of Glc), 3.58 (t, J=9.0 Hz, H-3 of Glc), 3.60 (m, H-5 of Gal), 3.65 (dd, J=7.0, 9.0 Hz, H-2 of Gal), 3.73 (dd, J=2.5, 12.0 Hz, H-6A of Gal), 3.78 (dd, J=4.5, 12 Hz, H-6B of Gal), 3.88 (m, H-4 of Gal), 3.98 (dd, J=4.5, 12.0 Hz, H-6A of Glc), 4.28 (dd, J=3.5, 12.0 Hz, H-6B of Glc), 4.58 (d, J=7.5 Hz, H-1 of Glc), 4.61 (d, J=7.5 Hz, H-1 of Gal).

CMR (CD₃OD, 150 MHz) : δ **Glc** C-1) 105.0 (2) 83.5 (3) 77.4 (4) 71.2 (5) 77.0 (6) 64.2 **Gal** (1) 102.9 (2) 73.5 (3) 74.8 (4) 69.8 (5) 77.0 (6) 61.4 (OCOCH₃) 172.0 (OCOCH₃) 21.8.

Mass (FAB, Negative ion) : m/z 851 [M-H]⁻, 809 [M-H-COCH₂]⁻, 647 [M-H-Glc-COCH₂]⁻.

Reference

1.   G. Gioffi, R. Sanogo, D. Diallo, G. Romussi and N.D. Tommasi, *J. Nat. Prod.*, **67**, 389 (2004).

ACODONTASTEROSIDE C

(5α)-Stigmast-24(28)-(*E*)-ene-3β,4β,6α,8,15β,29-hexol 29-O-[(2-O-methyl-β-D-xylopyranosyl-(1→2)-
β-D-xylpoyranoside]-6-hydrogen sulfate

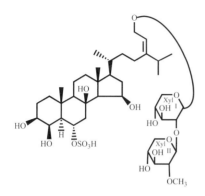

Source : *Acodontaster conspicuus* (Odontasteridae, Starfish)
Mol. Formula : $C_{40}H_{68}O_{17}S$
Mol. Wt. : 852
[α]$_D$: -11.1° (c=1.0, MeOH)
Registry No. : [195061-78-0]

PMR (CD$_3$OD, 500 MHz) : δ 1.02 (d, *J*=7.0 Hz, 3xH-21), 1.06 (d, *J*=7.0 Hz, 3xH-27), 1.07 (d, *J*=7.0, 3xH-26), 1.26 (s, 3xH-26), 1.31 (s, 3xH-18), 2.21 (m, H-20), 2.30 (m, H-25), 2.40 (m, H-16), 2.74 (dd, *J*=12.2, 4.0 Hz, H-7), 2.95 (dd, *J*=9.7, 7.3 Hz, H-2 of Xyl II), 3.19 (dd, *J*=10.5, 7.9 Hz, H-5A of Xyl II), 3.20 (t, *J*=10.5 Hz, H-5B of Xyl I), 3.26 (t, *J*=9.0 Hz, H-3 of Xyl I), 3.42 (t, *J*=9.7 Hz, H-3 of Xyl II), 3.48 (H-2 of Xyl I), 3.50 (m, H-3), 3.52 (m, H-4 of Xyl II), 3.52 (H-4 of Xyl I), 3.64 (s, OCH$_3$ of Xyl II), 3.89 (dd, *J*=10.5, 4.7 Hz, H-5A of Xyl II), 3.91 (dd, *J*=10.5, 4.7 Hz, H-5B of Xyl II), 4.19 (dd, *J*=10.5, 6.8 Hz, 2xH-29), 4.32 (dd, *J*=10.5, 7.0 Hz, 2xH-29), 4.33 (br s, H-4),4.37 (d, *J*=6.9 Hz, H-1 of Xyl I), 4.44 (t, *J*=5.5 Hz, H-15), 4.74 (d, *J*=7.3 Hz, H-1 of Xyl II), 4.95 (dt, H-6).

CMR (CD$_3$OD, 125 MHz) : δ C-1) 39.5 (2) 26.5 (3) 72.7 (4) 68.9 (5) 56.0 (6) 74.5 (7) 47.6 (8) 77.5 (9) 57.6 (10) 38.7 (11) 19.1 (12) 43.7 (13) 44.2 (14) 62.7 (15) 71.1 (16) 42.6 (17) 58.1 (18) 16.7 (19) 17.0 (20) 36.7 (21) 19.0 (22) 36.9 (23) 27.2 (24) 152.5 (25) 35.9 (26) 22.5 (27) 22.5 (28) 119.2 (29) 66.5 **Xyl I** (1) 103.4 (2) 81.1 (3) 76.8 (4) 71.0 (5) 66.3 **Xyl II** (1) 104.4 (2) 84.8 (3) 77.5 (4) 70.9 (5) 66.5 (OCH$_3$) 60.7.

Mass (FAB, Negative ion) : *m/z* 851 [MSO$_3$]$^-$.

Reference

1.   S.D. Marino, M. Iorizzi, F. Zollo, L. Minale, C.D. Amsler, B.J. Baker and J.B. Mc Clintock, *J. Nat. Prod.*, **60**, 959 (1997).

ACODONTASTEROSIDE B

(5α,24*R*)-Stigmastane-3β,4β,6α,8,15β,29-hexol 29-O-[(2-O-methyl)-β-D-xylopyranosyl-(1→2)-β-D-xylopyranoside]-6-hydrogen sulfate

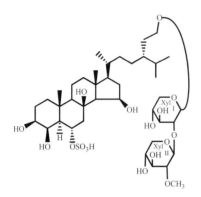

Source : *Acodontaster conspicuus* (Odontasteridae, Starfish)
Mol. Formula : $C_{40}H_{70}O_{17}S$
Mol. Wt. : 854
[α]$_D$: +10.0° (c=1.0, MeOH)
Registry No. : [195061-76-8]

PMR (CD$_3$OD, 500 MHz) : δ 0.87 (d, *J*=6.8 Hz, 3xH-27), 0.90 (d, *J*=6.8 Hz, 3xH-26), 0.97 (d, *J*=7.0 Hz, 3xH-21), 1.26 (s, 3xH-19), 1.31 (s, 3xH-18), 1.83 (m, H-20), 2.40 (m, H-16), 2.74 (dd, *J*=12.2, 4.0 Hz, H-7), 2.95 (dd, *J*=9.7, 7.3 Hz, H-2 of Xyl II), 3.19 (dd, *J*=10.5, 7.9 Hz, H-5A of Xyl II), 3.20 (t, *J*=10.5 Hz, H-5 of Xyl I), 3.26 (t, *J*=9.0 Hz, H-3 of Xyl I), 3.42 (t, *J*=9.7 Hz, H-4 of Xyl II), 3.48* (H-2 of Xyl I), 3.50 (m, H-3), 3.52* (m, H-4 of Xyl I and Xyl II), 3.64 (s, OC*H$_3$* of Xyl II), 3.89 (dd, *J*=10.5 ,4.7 Hz, H-5B of Xyl I), 3.91 (dd, *J*=10.5, 4.7 Hz, H-5B of Xyl II), 4.33 (br s, H-4), 4.44 (t, *J*=5.5 Hz, H-15), 4.37 (d, *J*=6.9 Hz, H-1 of Xyl I), 4.74 (d, *J*=7.3 Hz, H-1 of Xyl II), 4.95 (dt, H-6).
* overlapped signal.

CMR (CD$_3$OD, 125 MHz) : δ C-1) 39.5 (2) 26.5 (3) 72.7 (4) 68.9 (5) 56.0 (6) 74.5 (7) 47.6 (8) 77.5 (9) 57.6 (10) 38.7 (11) 19.1 (12) 43.7 (13) 44.2 (14) 62.7 (15) 71.1 (16) 42.6 (17) 58.1 (18) 16.7 (19) 17.0 (20) 36.6 (21) 19.0 (22) 34.8 (23) 281.1 (24) 41.9 (25) 30.5 (26) 19.1 (27) 20.0 (28) 31.7 (28) 69.5 **Xyl I** (1) 103.4 (2) 81.1 (3) 76.8 (4) 71.0 (5) 66.3 **Xyl II** (1) 104.4 (2) 84.8 (3) 77.5 (4) 70.9 (5) 66.5 (OC*H$_3$*) 60.7.

Mass (FAB, Negative ion) : *m/z* 853 [M-H]⁻, 707 [853-MeXyl]⁻, 757 [707-Xyl]⁻.

Reference

1.   S.D. Marino, M. Iorizzi, F. Zollo, L. Minale, C.D. Amsler, B.J. Baker and J.B. Mc Clintock, *J. Nat. Prod.*, **60**, 959 (1997).

CHRYSOLINA GEMINATA SAPONIN 1

16-Acetyloxy-3β,20,25-trihydroxy-5β,24ξ–stigmastan-29-oic acid γ-lactone
3-O-[β-D-glucopyranosyl-(1→2)-β-D-glucopyranoside]

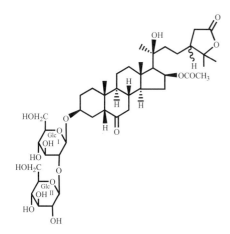

Source : *Chrysolina geminata* (beetle, Chrysomelidae)
Mol. Formula : $C_{43}H_{68}O_{17}$
Mol. Wt. : 856
[α]$_{579}$: -5.7° (c=0.35, CH$_3$OH)
Registry No. : [129562-47-6]

IR (KBr) : 3550 (OH), 1760, 1730 (C=O), 1695 (C-O) cm^{-1}.

PMR (CD$_3$OD, 250 MHz) : δ 0.87 (s, 3xH-19), 1.11 (s, 3xH-18), 1.28 (s, 3xH-26), 1.33 (s, 3xH-21), 1.44 (s, 3xH-27), 2.07 (s, OCOCH_3), 2.38 (dd, J=7.8, 1.7 Hz, H-28A), 2.56 (dd, J=4.0, 12.0 Hz, H-5), 2.62 (dd, J=11.3, 17.0 Hz, H-28B), 4.08 (br s, H-3), 4.44 (d, J=7.5 Hz, H-1 of Glc), 4.65 (d, J=7.5 Hz, H-1 of Glc II), 5.39 (dt, J=4.0, 6.2 Hz, H-16).

CMR (CD$_3$OD, 62.5 MHz) : δ C-1) 30.7^a (2) 26.8^b (3) 76.5 (4) 30.2^a (5) 55.9^c (6) 218.4 (7) 39.2 (8) 36.1^d (9) 45.3 (10) 37.4^d (11) 22.2 (12) 40.8 (13) 43.3 (14) 56.1^c (15) 35.6 (16) 77.9 (17) 63.3 (18) 15.6 (19) 25.2 (20) 76.2 (21) 27.1^b (22) 47.1 (23) 24.2 (24) 45.0 (25) 88.9 (26) 22.2 (27) 29.0 (28) 41.2 (29) 178.7 (CH$_3$COO) 21.6 (CH$_3$COO) 172.1 **Glc I** (1) 101.7 (2) 81.9 (3) 78.5^e (4) 72.0^g (5) 78.5^e (6) 63.2^f **Glc II** (1) 105.1 (2) 74.2 (3) 78.0^e (4) 71.7^f (5) 78.1 (6) 63.0^g.

Mass (FAB, Positive ion) : m/z 857 [M+H]$^+$, 533 [M+H-2x162]$^+$, 515 [M+H-2x162-18]$^+$.

Mass (FAB, Negative ion) : m/z 855 [M-H]$^-$.

Reference

1.   T. Randoux, J.C. Braekman, D. Daloze, J.M. Pasteels and R. Riccio, *Tetrahedron*, **46**, 3879 (1990).

CHRYSOLINA HYPERICI SAPONIN 7

28-Acetyloxy-3β,20,25-trihydroxy-5α,24ξ–stigmastane-3,16-dione
3-O-[β-D-glucopyranosyl-(1→2)-β-D-glucopyranoside]

Source : *Chrysolina hyperici* (beetle, Chrysomelidae)
Mol. Formula : $C_{43}H_{70}O_{17}$
Mol. Wt. : 858
Registry No. : [129562-49-8]

PMR (CD$_3$OD, 250 MHz) : δ 0.80 (s, 3xH-19), 1.00 (s, 3xH-18), 1.22 (s, 3xH-26), 1.22 (s, 3xH-27), 1.25 (d, *J*=6.6 Hz 3xH-29), 1.29 (s, 3xH-21), 2.0 (s, CH$_3$COO), 2.23 (m, H-5), 3.69 (dd, *J*=5.0, 12.5 Hz, H-6A of Glc II), 3.71 (dd, *J*=5.0, 12.5 Hz, H-6A of Glc I), 3.84 (dd, *J*=2.5, 12.5 Hz, H-6B of Glc II), 3.87 (dd, *J*=2.5, 12.5 Hz, H-6B of Glc I), 4.53 (d, *J*=8.0 Hz, H-1 of Glc I).

Reference

1.   T. Randoux, J.C. Braekman, D. Daloze and J.M. Pasteels, *Tetrahedron*, **46**, 3879 (1990).

SOPHOROSIDE 1

28-Acetyloxy-3β,16β,20,25-Tetrahydroxy-5β,24ξ-Stigmastane-6-one
3-O-[β-D-glucopyranosyl-(1→2)-β-D-glucopyranoside]

Source : *Chrysolina quadrigemina* Suffrian (Chrysomelidae)

Mol. Formula : $C_{43}H_{72}O_{17}$

Mol. Wt. : 860

Registry No. : [140381-59-5]

IR : 1730 (C=O), 1705 (C-O), 1274 cm^{-1}.

PMR (CD$_3$OD, 250 MHz) : δ 0.87 (s, 3xH-19), 1.14 (s, 3xH-18), 1.24 (s, 3xH-26), 1.25 (d, *J*=6.5 Hz, 3xH-29), 1.29 (s, 3xH-27), 1.31 (s, 3xH-21), 2.01 (s, OCOC*H*$_3$), 2.56 (dd, *J*=12.3, 4.2 Hz, H-5), 3.68 (dd, *J*=12.5, 5.0 Hz, H-6A of Glc I), 3.68 (dd, *J*=12.5, 5.0 Hz, H-6A of Glc II), 3.85 (dd, *J*=12.5, 2.0 Hz, H-6B of Glc I), 3.85 (dd, *J*=12.5, 2.0 Hz, H-6B of Glc II), 4.08 (br s, H-3), 4.43 (d, *J*=7.5 Hz, H-1 of Glc I), 4.66 (d, *J*=7.5 Hz, H-1 of Glc I), 5.32 (dq, *J*=6.5, 3.0 Hz, H-28).

Mass (FAB, Positive ion) : *m/z* 883 [M-Na]$^+$, 682 [M+H-OH-Glc]$^+$, 664 [M+H-OH-Glc-H$_2$O]$^+$, 520 [M+H-OH-2xGlc]$^+$, 502 [M+H-OH-2xGlc-H$_2$O]$^+$.

Mass (Negative ion) : *m/z* 859 [M-H]$^-$.

Reference

1.   D. Daloze, J.C. Braekman, A. Delbrassine and J.M. Pasteels, *J. Nat. Prod.*, **54**, 1553 (1991).

SOPHOROSIDE 2

28-Acetyloxy-3β,16β,20,25-Tetrahydroxy-5β,24ξ-Stigmastane-6-one
3-O-[β-D-glucopyranosyl-(1→2)-β-D-glucopyranoside]

Source : *Chrysolina quadrigemina* Suffrian (Chrysomelidae)
Mol. Formula : $C_{43}H_{72}O_{17}$
Mol. Wt. : 860
Registry No. : [140381-60-8]

PMR (CD₃OD, 250 MHz) : δ 0.77 (s, 3xH-19), 4.53 (d, *J*=7.7 Hz, H-1 of Glc I), 4.57 (dd, *J*=7.7 Hz, H-1 of Glc II).

Reference

1.   D. Daloze, J.C. Braekman, A. Delbrassine and J.M. Pasteels, *J. Nat. Prod.*, **54**, 1553 (1991).

ECHINASTEROSIDE F

(24R,5α)-3β,6β,8β,15α,16β,29-Hexahydroxystigmast-4-ene-15-hydrogen sulfate
3-O-[(2-O-methyl)-β-D-xylopyranoside]-29-O-β-D-xylopyranoside monosodium salt

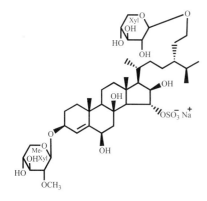

Source : *Echinaster brasiliensis* Muller and Trochel (Echinasteridae)
Mol. Formula : $C_{40}H_{67}O_{17}SNa$
Mol. Wt. : 874
[α]$_D$: -1.2°
Registry No. : [156398-66-2]

PMR (CD$_3$OD) : δ 0.87 (d, J=7.0 Hz, 3xH-26), 0.90 (d, J=7.0 Hz, 3xH-27), 0.96 (d, J=7.0 Hz, 3xH-21), 1.22 (s, 3xH-18), 1.40 (s, 3xH-19), 2.69 (dd, J=15.0, 3.0 Hz, H-7), 2.84 (dd, H-2 of MeXyl), 3.18 (dd, J=7.5, 9.0 Hz, H-2 of Xyl), 3.20 (dd, H-5A of MeXyl), 3.24 (dd, J=11.5, 9.5 Hz, H-5A of Xyl), 3.30* (H-3 of MeXyl), 3.36* (H-3 of Xyl), 3.50 (m, H-4 of MeXyl), 3.51 (m, H-4 of Xyl), 3.62 (s, OCH_3), 3.85 (dd, H-5B of MeXyl), 3.87 (dd, J=11.5, 5.0 Hz, H-5B of Xyl), 4.23 (d, J=7.5 Hz, H-1 of Xyl), 4.27 (br t, H-3), 4.33 (t, J=3.0 Hz, H-6), 4.38 (dd, J=7.5, 2.5 Hz, H-16), 4.46 (d, H-1 of MeXyl), 4.80 (dd, J=11.0, 2.5 Hz, H-15), 5.67 (br s, H-4). * masked by solvent signal.

CMR (CD$_3$OD) : δ C-20) 39.7 (2) 28.0 (3) 77.5 (4) 126.3 (5) 148.5 (6) 76.0 (7) 45.0 (8) 76.4 (9) 57.6 (10) 37.2 (11) 19.6 (12) 42.8 (13) 43.6 (14) 61.8 (15) 87.5 (16) 80.2 (17) 60.7 (18) 16.9 (19) 22.7 (20) 31.8 (21) 18.5 (22) 34.9 (23) 28.9 (24) 42.1 (25) 30.7 (26) 19.1 (27) 20.1 (28) 30.6 (29) 70.0 **MeXyl** (1) 104.9 (2) 85.0 (3) 77.5 (4) 71.4 (5) 66.8 (OCH_3) 61.2 **Xyl** (1) 105.2 (2) 75.0 (3) 77.9 (4) 71.5 (5) 67.0.

Mass (FAB, Negative ion) : *m/z* 851 [MSO$_3$]⁻.

Reference

1.   M. Iorizzi, F.de. Riccardis, L. Minale and R. Riccio, *J. Nat. Prod.*, **56**, 2149 (1993).

NARDOA TUBERCULATA SAPONIN 1

(5α,24R)-Stigmast-22(E)-ene-3β,4β,6α,8,15β,16β,29-heptol-6-hydrogen sulfate
29-O-[(2-O-methyl)-β-D-xylopyranosyl-(1→2)-β-D-xylopyranoside]

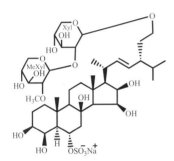

Source : *Nardoa tuberculata* (Starfish, Ophidiasteridae)
Mol. Formula : $C_{40}H_{67}O_{18}SNa$
Mol. Wt. : 890
M.P. : 192-193°C
$[\alpha]_D^{20}$: -8.5°
Registry No. : [151507-64-1]

PMR (CD₃OD, 250 MHz) : δ 0.88 (d, *J*=7.5 Hz, 3xH-26), 0.93 (d, *J*=7.0 Hz, 3xH-27), 1.10 (d, *J*=6.5 Hz, 3xH-21), 1.27 (s, 3xH-19), 1.30 (s, 3xH-18), 2.76 (dd, *J*=4.5, 12.0 Hz, H-7), 2.97 (dd, *J*=8.7, 7.0 Hz, H-2 of MeXyl), 3.20 (dd, *J*=11.9, 9.5 Hz, H-5 of Xyl), 3.21 (dd, *J*=7.0, 9.0 Hz, H-2 of Xyl), 3.2-3.6* (H-3 of Xyl), 3.2-3.6* (H-3 of MeXyl), 3.2-3.6* (H-5 of MeXyl), 3.5 (m, H-3), 3.65 (br m, 2xH-29), 3.65 (s, OC*H₃*), 3.86 (dd, *J*=8.0, 4.8 Hz, H-5 of MeXyl), 3.92 (dd, *J*=11.9, 4.8, H-5 of Xyl), 4.17 (t, *J*=7.0 Hz, H-16), 4.34 (br s, H-4), 4.41 (d, *J*=7.0 Hz, H-1 of Xyl), 4.43 (dd, *J*=7.0, 12.0 Hz, H-15), 4.75 (d, *J*=7.0 Hz, H-1 of MeXyl), 4.90 (m, H-6), 5.23 (dd, *J*=15.0, 9.0 Hz, H-23), 5.58 (dd, *J*=15.0, 7.5 Hz, H-22). * overlapped signals.

CMR (CD₃OD, 62.9 MHz) : δ C-1) 39.4 (2) 26.5 (3) 73.3 (4) 68.9 (5) 56.1 (6) 74.4 (7) 47.8 (8) 77.2 (9) 58.2 (10) 38.8 (11) 18.8 (12) 43.3 (13) 44.7 (14) 61.2 (15) 71.2 (16) 72.9 (17) 63.8 (18) 18.0 (19) 16.9 (20) 34.8 (21) 20.7 (22) 139.8 (23) 130.3 (24) 49.8 (25) 33.5 (26) 19.6 (27) 21.3 (28) 33.9 (29) 69.6 **Xyl** (1) 103.6 (2) 81.5 (3) 76.8 (4) 71.2 (5) 66.5 **MeXyl** (1) 104.3 (2) 84.7 (3) 77.7 (4) 71.2 (5) 66.5 (OC*H₃*) 60.7.

Mass (FAB, Negative ion) : *m/z* 867 [M-Na]⁻.

Reference

1.   I. Bruno, L. Minale, R. Riccio, L. Cariello, T. Higa and J. Tanaka, *J. Nat. Prod.*, **56**, 1057 (1993).

HALITYLOSIDE A 6-O-SULFATE

5α-Stigmastane-3β,4β,6α,8β,15β,16β,29-heptol 29-O-[(2-O-methyl)-β-D-xylopyranosyl-(1→2)-β-D-xylopyranoside]-6-hydrogen sulfate monosodium salt

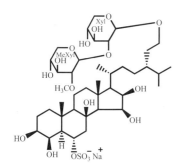

Source : *Nardoa tuberculata* Perrier (Ophidiasteridae)
Mol. Formula : $C_{40}H_{69}O_{18}SNa$
Mol. Wt. : 892
[α]$_D$: +3.7° (MeOH)
Registry No.: [151507-62-9]

PMR (CD$_3$OD, 250 MHz) : δ 0.88 (d, *J*=7.5 Hz, 3xH-26), 0.91 (d, *J*=7.5 3xHz, H-27), 0.98 (d, *J*=6.5 Hz, 3xH-21), 1.27 (s, 3xH-18, 3xH-19), 2.76 (dd, *J*=4.5, 12.0 Hz, H-7), 2.97 (dd, *J*=8.7, 7.0 Hz, H-2 of MeXyl), 3.20 (dd, *J*=11.9, 9.5 Hz, H-5A of Xyl), 3.21 (dd, *J*=7.0, 0.0 Hz, H-2 of Xyl), 3.50 (m, H-3), 3.65 (s, OC*H$_3$*), 3.86 (dd, *J*=8.0, 4.8 Hz, H-5B of Xyl), 3.92 (dd, *J*=11.9, 4.8 Hz, H-5B of Xyl), 4.24 (t, *J*=7.0 Hz, H-16), 4.34 (br s, H-4), 4.39 (m, H-15), 4.41 (d, *J*=7.0 Hz, H-1 of Xyl), 4.75 (d, *J*=7.0 Hz, H-1 of MeXyl), 4.90 (m, H-6).

CMR (CD$_3$OD, 62.9) : δ C-1) 39.5 (2) 26.6 (3) 73.0 (4) 69.0 (5) 56.2 (6) 74.4 (7) 47.8 (8) 77.2 (9) 58.3 (10) 38.8 (11) 18.9 (12) 43.5(13) 44.7 (14) 61.1 (15) 71.3 (16) 72.9 (17) 63.1 (18) 17.9 (19) 16.9 (20) 31.4 (21) 18.7 (22) 34.8 (23) 28.7 (24) 42.3 (25) 30.7 (26) 19.2 (27) 20.0 (28) 31.9 (29) 69.7 **Xyl** (1) 103.6 (2) 81.5 (3) 76.8 (4) 71.2 (5) 66.5 **MeXyl** (1) 104.3 (2) 84.7 (3) 77.7 (4) 71.2 (5) 66.5 (OC*H$_3$*) 60.7.

Mass (FAB, Negative ion) : *m/z* 869 [M-Na]⁻.

Reference

1.   I. Bruno, L. Minale, R. Riccio, L. Cariello, T. Higa and J. Tanaka, *J. Nat. Prod.*, **56**, 1057 (1993).

ASPARAGUS ADSCENDENS SAPONIN 1

Stigmasterol 3-O-[β-D-(2'-tetracosyl)-xylopyranoside]

Source : *Asparagus adscendens* Roxb. (Liliaceae)
Mol. Formula : $C_{58}H_{102}O_6$
Mol. Wt. : 894
M.P. : 140-142°C (decomp.)
[α]$_D$: -47° (CHCl$_3$)
Registry No. : [131870-90-1]

IR (KBr) : 3420, 2920, 2850, 1740, 1470, 1380, 1370, 1180, 1090, 1060, 1020, 840, 805, 725 cm^{-1}.

PMR (CDCl$_3$, 80 MHz) : δ 0.69 (s, 3xH-18), 0.82 (d, *J*=6.0 Hz, 3xH-26 and 3xH-27), 0.84 (t, *J*=6.0 Hz, 3xH-29), 0.90 (t, *J*=6.0 Hz, C*H*$_3$ of acid moiety), 1.00 (s, 3xH-19), 1.04 (d, *J*=3.0 Hz, 3xH-21), 1.25 (br s, CH$_2$), 2.35 (t, *J*=6.0 Hz, COC*H*$_2$ of acid moiety), 3.30-4.65 (5xH of Xyl), 4.38 (d, *J*=8.0 Hz, H-1 of Xyl), 5.07 (m, H-22, H-23), 5.35 (m, H-6).

CMR (CDCl$_3$, 25 MHz) : δ C-1) 37.2 (2) 29.8 (3) 79.0 (4) 38.6 (5) 140.2 (6) 121.8 (7) 31.5 (8) 31.5 (9) 50.0 (10) 36.2 (11) 21.0 (12) 39.7 (13) 42.0 (14) 56.4 (15) 24.3 (16) 28.7 (17) 55.9 (18) 12.0 (19) 19.4 (20) 40.3 (21) 21.1 (22) 138.1 (23) 129.0 (24) 51.0 (25) 31.9 (26) 21.2 (27) 18.9 (28) 25.2 (29) 12.3 **Xyl** (1) 101.0 (2) 78.0 (3) 74.8 (4) 72.8 (5) 63.5 **Acid moiety** (1) 13.9 (2) 22.2 (3) 31.9 (4) 34.6 (5) 27.4 (C-6 to 21) 29.0 (C-22) 22.5 (C-23) 35.8 (C-24) 173.8.

Mass (E.I.) : *m/z* (rel.intens.) 412 [M]$^+$ of Agl (9), 397 (20), 394 (28), 379 (1), 368 [(M-43)$^+$, 19], 368 [(C$_{24}$H$_{48}$O$_2$)$^+$, 12] 327 (1) 301 (4) 273 [(M-sc)$^+$, 4], 231 [(M-sc-42), 3], 213 (18) 185 (11) 158 (12) 150 (6) 145 (16) 139 (27) 133 (13) 111 (4) 105 (20) 85 (12) 55 (85) 43 (100). sc = side chain.

Reference

1.   M. Tandon, Y.N. Shukla and R.S. Thakur, *Phytochemistry*, **29**, 2957 (1990).

AMARANTHUS SPINOSUS SAPONIN 1

α-Spinosterol 3-O-[β-D-glucopyranosyl-(1→2)-β-D-glucopyranosyl-(1→2)-β-D-glucopyranoside]

Source : *Amaranthus spinosus* L. (Amaranthaceae)
Mol. Formula : $C_{47}H_{78}O_{16}$
Mol. Wt. : 898
M.P. : 304°C (decomp.)
[α]$_D^{28}$: -51.1° (C_5H_5N)
Registry No. : [75422-87-6]

Reference

1.   N. Banerji, *J. Indian Chem. Soc.*, **57**, 417 (1980).

HALITYLOSIDE H 6-O-SULFATE

(5α,24R)-Stigmastane-3β,4β,6α,8,15β,16β,29-heptol-6-O-hydrogen sulfate 2,4-di-O-methylxylopyranosyl-(1→2)-α-L-arabinopyranoside]

Source : *Halityle regularis* Fisher (Oreasteridea, Starfish)
Mol. Formula : $C_{41}H_{71}O_{18}SNa$
Mol. Wt. : 906
Registry No. : [102052-32-4]

PMR (CD$_3$OD, 500 MHz) : δ 1.27 (s, 3xH-19), 2.73 (dd, H-7), 2.90 (dd, *J*=9.0, 7.6 Hz, H-2 of DMX), 3.14 (t, *J*=10.6 Hz, H-5A of DMX), 3.20 (m, H-4 of DMX), 3.42 (t, *J*=9.0 Hz, H-3 of DMX), 3.50 (s, OC*H*$_3$ of DMX), 3.61 (s, OC*H*$_3$ of DMX), 3.65 (dd, *J*=12.5, 4.8 Hz, H-5A Ara), 3.79 (dd, *J*=12.5, 3.0 Hz, H-5B of Ara), 3.98 (m, H-4 of Ara), 4.02 (m, H-3 of Ara), 4.02 (dd, *J*=10.6, 4.0 Hz, H-5B of DMX), 4.08 (d, *J*=4.0 Hz, H-2 of Ara), 4.33 (br s, H-4), 4.44 (d, *J*=7.6, Hz, H-1 of DMX), 4.95 (ddd, H-6), 5.11 (br s, H-1 of Ara).

Mass (FAB, Positive ion) : *m/z* 929 [M+Na]$^+$.

Reference

1.   M. Iorizzi, L. Minale, R. Riccio, M. Debray, and J.L. Menou, *J. Nat. Prod.*, **49,** 67 (1986).

4"-O-METHYLHALITYLOSIDE A 6-O-SULFATE

5α-Stigmastane-3β,4β,6α,8β,15β,16β,29-heptol 29-O-[(2,4-di-O-methyl)-β-D-xylopyranosyl-(1→2)-β-D-xylopyranoside]-6-hydrogen sulfate monosodium salt

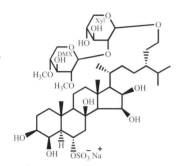

Source : *Nardoa tuberculata* Perrier (Ophidiasteridae)
Mol. Formula : $C_{41}H_{71}O_{18}SNa$
Mol. Wt. : 906
[α]$_D$: -5° (MeOH)
Registry No. : [151507-63-0]

PMR (CD$_3$OD, 250 MHz) : δ 0.88 (d, *J*=7.5 Hz, 3xH-26), 0.91 (d, *J*=7.5 Hz, 3xH-27), 0.98 (d, *J*=6.5 Hz, 3xH-21), 1.27 (s, 3xH-18, 3xH-19), 2.76 (dd, *J*=4.5, 12.0 Hz, H-7), 2.97 (dd, *J*=7.5, 8.7 Hz, H-2 of DMX), 3.16 (t, *J*=8.5 Hz, H-5A of DMX), 3.20 (m, H-4 of DMX), 3.26 (dd, *J*=12.5, 7.5 Hz, H-5A of Xyl), 3.50 (m, H-3), 3.50 (s, OC*H$_3$*), 3.62 (s, OC*H$_3$*), 3.90 (dd, *J*=12.5, 5.0 Hz, H-5 of Xyl), 4.02 (dd, *J*=8.5, 3.5 Hz, H-5B of DMX), 4.24 (t, *J*=7.0 Hz, H-16), 4.34 (br s, H-4), 4.39 (m, H-15), 4.42 (d, *J*=7.0 Hz, H-1 of Xyl), 4.75 (d, *J*=7.5 Hz, H-1 of DMX), 4.90 (m, H-6).

Mass (FAB, Negative ion) : *m/z* 883 [M-Na]⁻, 723, 591.

Reference

1.   I. Bruno, L. Minale, R. Riccio, L. Cariello, T. Higa and J. Tanaka, *J. Nat. Prod.*, **56**, 1057 (1993).

MONILOSIDE I

(24*R*)-5α-Stigmastane-3β,4β,6α,8,15β,16β,29-heptaol 29-O-[(3-O-methyl)-β-xylopyranosyl-(1→4)-β-xylopyranosyl-(1→4)-β-xylopyranoside]

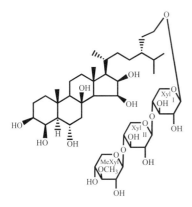

Source : *Fromia monilis* Perrier (Ophidiasteridae), Starfish
Mol. Formula : $C_{45}H_{78}O_9$
Mol. Wt. : 922
[α]$_D$: -3.5° (c=1.0 or 0.1, MeOH)
Registry No. : [147362-21-8]

PMR (CD₃OD, 500 MHz) : δ 3.08 (t, *J*=10.0 Hz, H-3 of MeXyl), 3.84-3.90, 4.01, 4.08 (dd, *J*=11.5, 5.0 Hz, H-5 of Xyl). The chemical shifts of aglycone moiety are virtually identical to those of Moniloside E (q.v.)

Mass (FAB, Negative ion) : *m/z* 921 [M-H]⁻, 775 [M-H-MeXyl]⁻, and at *m/z* 643 [M-H-MeXyl-Xyl]⁻, 511 [M-H-MeXyl-2xXyl]⁻.

Biological Activity : The compound showed marked anti-Herpes activity at a dose of 10 μg/ml as well as strong cytotoxicity on KB cells, 100% at 10 μg/ml.

Reference

1.   A. Casapullo, E. Finamore, L. Minale, F. Zollo, J.B. Carré, C. Debitus, D. Laurent, A. Folgore and F. Galdiero, *J. Nat. Prod.*, **56**, 105 (1993).

PISTIA STRATIOTES SAPONIN 1

Sitosterol-3-O-[2,4-di-O-acetyl-6-O-stearyl-β-D-glucopyranoside]

Source : *Pistia stratiotes* L. (Araceae)
Mol. Formula : $C_{57}H_{98}O_9$
Mol. Wt. : 926
Registry No. : [135757-63-0]

PMR (CDCl$_3$, 400.135 MHz) : δ 1.26 (br s), 2.36 (t, J=6.5 Hz), 2.12 (s, OCOCH$_3$), 2.14 (s, OCOCH$_3$), 3.61 (m, H-5 of Glc), 3.70 (m, H-3 of Glc), 4.13 (dd, *J*=2.7, 12.1 Hz, H-6A of Glc), 4.21 (dd, *J*=5.4, 12.1 Hz, H-6B of Glc), 4.53 (d, *J*=7.8 Hz, H-1 of Glc), 4.80 (dd, *J*=7.8, 8.1 Hz, H-2 of Glc), 4.91 (t, *J*=9.4 Hz, H-4 of Glc).

CMR (CDCl$_3$, 100.61 MHz) : δ **Glc** C-1) 99.4 (2) 74.4 (3) 79.9 (4) 71.3 (5) 71.7 (6) 62.3.

Reference

1.   M.D. Greca, A. Molinaro, P. Monaco and L. Previtera, *Phytochemistry*, **30,** 2422 (1991).

MONILOSIDE H

(24R)-5α-Stigmast-22(E)-ene-3β,4β,6α,8,15β,16β,29-heptaol 29-O-[(3-O-methyl)-β-xyloyranosyl-(1→4)-(3-O-methyl)-β-xylopyranosyl-(1→4)-β-xylopyranoside]

Source : *Fromia monilis* Perrier (Ophidiasteridae), Starfish
Mol. Formula : $C_{46}H_{78}O_{19}$
Mol. Wt. : 934
[α]$_D$: +23.5° (MeOH)
Registry No. : [147385-58-8]

PMR (CD$_3$OD, 500 MHz) : δ 3.08 (t, *J*=9.0 Hz, H-3), 3.88 (dd, *J*=5.0, 11.0 Hz, H-5eq), 4.03 and 4.09 (H-5eq), 4.21, 4.35, 4.39 (each d, *J*=7.5 Hz, anomeric H), 5.35 (dd, *J*=15.7 Hz, olefinic H), 5.48 (dd, *J*=15.7 Hz olefinic H). Chemical shifts of aglycone moiety are virtually identical with those of Moniloside F (q.v.).

CMR (CD$_3$OD, 62.9 MHz) : δ C-1) 39.7 (2) 26.2 (3) 73.7 (4) 69.2 (5) 57.4 (6) 64.8 (7) u.m (8) 77.2 (9) 58.6 (10) 38.2 (11) 19.0 (12) 43.6 (13) 44.6 (14) 61.4 (15) 71.3 (16) 72.9 (17) 63.1 (18) 17.9 (19) 16.9 (20) 31.5 (21) 18.5 (22) 34.8 (23) 28.8 (24) 42.6 (25) 30.8 (26) 19.1 (27) 20.0 (28) 31.9 (29) 70.1 **Xyl** (1) 105.0 (2) 74.8 (3) 75.9 (4) 78.2 (5) 64.4 **MeXyl I** (1) 103.6 (2) 74.3 (3) 85.1 (4) 76.6 (5) 64.4 (OCH$_3$) 60.6 **MeXyl II** (1) 103.8 (2) 73.8 (3) 87.2 (4) 70.8 (5) 66.9 (OCH$_3$) 60.6.

Mass (FAB) : *m/z* 933 [M-H]⁻, 777 [M-H-MeXyl]⁻, 641 [M-H-2xMeXyl]⁻, 509 [M-H-2xMeXyl-Xyl]⁻.

Biological Activity : The compound showed marked anti-Herpes activity at a dose of 10 mg/ml as well as strong cytotoxicity on KB cells, 100% at 10 μg/ml.

Reference

1. A. Casapullo, E. Finamore, L. Minale, F. Zollo, J.B. Carré, C. Debitus, D. Laurent, A. Folgore and F. Galdiero, *J. Nat. Prod.*, **56**, 105 (1993).

MONILOSIDE G

(24*R*)-5α-Stigmastane-3β,4β,6α,8,15β,16β,29-heptaol 29-O-[(3-O-methyl)-
β-xylopyranosyl-(1→4)-(3-O-methyl)-β-xylopyranosyl-(1→4)-β-xylopyranoside]

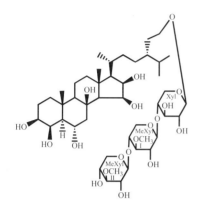

Source : *Fromia monilis* Perrier (Ophidiasteridae),
Starfish
Mol. Formula : $C_{46}H_{80}O_{19}$
Mol. Wt.: 936
[α]$_D$: -17.3° (MeOH)
Registry No. : [147362-20-7]

PMR (CD$_3$OD, 500 MHz) : δ 3.08 (t, *J*=9.0 Hz, H-3), 3.88 (dd, *J*=5.0, 11.0 Hz, H-5eq), 4.03 and 4.09 (H-5eq), 4.21, 4.35, 4.39 (each d, *J*=7.5 Hz, anomeric H). The chemical shifts of aglycone moiety are virtually identical to those of Maniloside E (q.v.).

CMR (CD$_3$OD, 62.9 MHz) : δ C-1) 39.7 (2) 26.2 (3) 73.7 (4) 69.2 (5) 57.4 (6) 64.8 (7) u.m (8) 77.2 (9) 58.6 (10) 38.2 (11) 19.0 (12) 43.6 (13) 44.6 (14) 61.4 (15) 71.3 (16) 72.9 (17) 63.1 (18) 17.9 (19) 16.9 (20) 31.5 (21) 18.5 (22) 34.8 (23) 28.8 (24) 42.6 (25) 30.8 (26) 19.1 (27) 20.0 (28) 31.9 (29) 70.1 **Xyl** (1) 105.0 (2) 74.8 (3) 75.9 (4) 78.2 (5) 64.4 **MeXyl I** (1) 103.6 (2) 74.3 (3) 85.1 (4) 76.6 (5) 64.4 (OCH$_3$) 60.6 **MeXyl II** (1) 103.8 (2) 73.8 (3) 87.2 (4) 70.8 (5) 66.9 (OCH$_3$) 60.6.

Mass (FAB, Negative ion) : *m/z* 935 [M-H]⁻, 789 [M-H-MeXyl]⁻, 643 [M-H-2xMeXyl]⁻, 511 [M-H-2xMeXyl-Xyl]⁻.

Biological Activity : The compound showed marked anti-Herpes activity at a dose of 10 mg/ml as well as strong cytotoxicity on KB cells, 100% at 10 μg/ml.

References

1.   P.A. Gorin and M. Magureck, *Can. J. Chem.*, **53**, 1212 (1975).

2.   M. Lorizzi, L. Minale, R. Riccio and T. Yasumoto, *J. Nat. Prod.*, **55**, 866 (1992).

3.   A. Casapullo, E. Finmore, L. Minale, F. Zollo, J.B. Carré, C. Debitus, D. Laurent, A. Folgore and F. Galdiero, *J. Nat. Prod.*, **56**, 105 (1993).

CHRYSOLINA BRUNSVICENSIS SAPONIN 1

16β,25,28ξ–Trisacetyloxy-3β,20-dihydroxy-5β,24ξ-stigmastan-6-one
3-O-[β-D-glucopyranosyl-(1→2)-β-D-glucopyranoside]

Source : *Chrysolina brunsvicensis* (beetle, Chrysomelidae)
Mol. Formula : $C_{47}H_{76}O_{19}$
Mol. Wt. : 944
Registry No. : [129562-45-4]

IR (KBr) : 3350 (OH), 1730, 1700 (C=O), 1245 (C-O) cm^{-1}.

PMR (CD$_3$OD, 250 MHz) : δ 0.87 (s, 3xH-19), 1.12 (s, 3xH-18), 1.26 (d, J=6.5 Hz, 3xH-29), 1.33 (s, 3xH-21), 1.48 (s, 3xH-26), 1.53 (s, 3xH-27), 1.96 (s, OCOCH_3), 1.99 (s, OCOCH_3), 2.07 (s, OCOCH_3), 2.56 (dd, J=4.2, 12.3 Hz, H-5), 4.08 (br s, H-3), 4.42 (d, J=7.3 Hz, H-1 of Glc I), 4.65 (d, J=7.3 Hz, H-1 of Glc II), 5.29 (dq, J=2.9, 6.5 Hz, H-28A), 5.39 (dt, J=4.0, 6.5 Hz, H-16).

Mass (FAB, Positive ion) : *m/z* 945 [M+H]$^+$, 886 [M+H-59]$^+$, 857 [M+H-88]$^+$, 797 [M+H-88-60]$^+$, 603 [M+H-2x162-18]$^+$.

Mass (FAB, Negative ion) : *m/z* 943 [M-H]$^-$, 888 [M-H-60]$^-$, 855 [M-H-88]$^-$, 813 [M-H-88-42]$^-$, 781 [M-H-Glc]$^-$, 347 [M-H-2xGlc-side chain].

Reference

1.   T. Randoux, J.C. Braekman, D. Daloze, J.M. Pasteels and R. Riccio, *Tetrahedron*, **46**, 3879 (1990).

CHRYSOLINA GEMINATA SAPONIN 2

16β,25,28ξ–Trisacetyloxy-3β,20-dihydroxy-5α,24ξ-stigmastan-6-one
3-O-[β-D-glucopyranosyl-(1→2)-β-D-glucopyranoside]

Source : *Chrysolina geminata* (beetle, Chrysomelidae)
Mol. Formula : $C_{47}H_{76}O_{19}$
Mol. Wt. : 944
Registry No. : [129562-48-7]

PMR (CD$_3$OD, 250 MHz) : δ 0.77 (s, 3xH-19), 1.12 (s, 3xH-18), 1.26 (d, *J*=6.0 Hz, 3xH-29), 1.33 (s, 3xH-21), 1.48 (s, 3xH-26), 1.51 (s, 3xH-27), 1.96 (s, OCOC*H*$_3$), 1.99 (s, OCOC*H*$_3$), 2.06 (s, OCOC*H*$_3$), 4.51 (d, *J*=7.5 Hz, H-1 of Glc I), 4.59 (d, *J*=7.3 Hz, H-1 of Glc II), 5.28 (d, *J*=3.0, 5.0 Hz, H-28A), 5.39 (m, H-16).

Reference

1.   T. Randoux, J.C. Braekman, D. Daloze, J.M. Pasteels and R. Riccio, *Tetrahedron*, **46**, 3879 (1990).

AJUGASALICIOSIDE D

3R,5β,16S,17R,20S,22R,24S,25S)-22,25-Epoxy-3a,16,27-trihydroxystigmast-7-ene
3-O-[β-D-glucopyranoside-(1→2)-β-D-glucopyranoside]-27-O-β-D-glucopyranoside

Source : *Ajuga salicifolia* (L.) Schreber. (Lamiaceae)
Mol. Formula : $C_{47}H_{78}O_{19}$
Mol. Wt. : 946
$[\alpha]_D^{20}$: -16.6° (c=0.1, MeOH/CH$_2$Cl$_2$)
Registry No. : [478015-49-5]

UV (MeOH) : 214 (log ε, 2.4) nm.

PMR (DMSO-d_6, 500 MHz) : δ 0.68 (s, 3xH-18), 0.72 (s, 3xH-19), 0.85 (d, J=6.0 Hz, 3xH-21), 0.87 (t, J=7.2 Hz, 3xH-29), 0.98 (s, 3xH-26), 0.98* (m, 2xH-1A), 1.00* (m, H-17), 1.08 (m, 2xH-28A), 1.15* (m, 2xH-12A), 1.20* (m, 2xH-4A), 1.26* (m, H-5), 1.31* (m, 2xH-15A), 1.32* (m, 2xH-2A), 1.38* (m, 2xH-11A), 1.41 (m, 2xH-28B), 1.52* (m, 2xH-11B), 1.56* (m, 2xH-23A), 1.57* (m, H-9), 1.60* (m, H-14), 1.65* (m, 2xH-6A), 1.72* (m, 2xH-1B), 1.74* (m, 2xH-6B), 1.77* (m, 2xH-4B), 1.80* (m, 2xH-2B, H-24), 1.87 (m, 2xH-12B), 1.88 (m, 2xH-23B), 2.01* (m, 2xH-15B, H-20), 2.95* (m, H-2 of Glc III), 2.98* (m, H-2 of Glc II), 3.02* (m, H-4 of Glc II), 3.05* (m, H-5 of Glc II, H-5 of Glc III), 3.08* (m, H-4, H-5 of Glc I), 3.10* (m, H-3 of Glc III), 3.12* (m, H-2 of Glc I), 3.13* (m, H-4 of Glc III), 3.15* (m, H-3 of Glc II), 3.31 (d, J=10.1 Hz, 2xH-27A), 3.34* (m, H-3 of Glc I), 3.40* (m, 2xH-6A of Glc II, 2xH-6A of Glc III), 3.50* (m, 2xH-6A of Glc I), 3.51 (m, H-3), 3.63* (m, 2xH-6B of Glc I, 2xH-6B of Glc II, 2xH-6B of Glc III), 3.69 (2xH-27B), 4.06 (m, H-22), 4.14 (d, J=7.8 Hz, H-1 of Glc III), 4.17* (m, H-16), 4.31 (d, J=7.8 Hz, H-1 of Glc II), 4.37 (d, J=7.8 Hz, H-1 of Glc I), 5.09 (m, H-7). * overlapped signals.

CMR (DMSO-d_6, 75.5 MHz) : δ C-1) 36.5 (2) 29.1 (3) 77.6 (4) 33.9 (5) 40.3 (6) 29.2 (7) 117.4 (8) 138.6 (9) 48.8 (10) 33.9 (11) 21.0 (12) 39.8 (13) 43.4 (14) 52.1 (15) 35.1 (16) 70.4 (17) 59.3 (18) 12.9 (19) 12.9 (20) 34.5 (21) 13.5 (22) 78.8 (23) 23.7 (24) 44.9 (25) 83.5 (26) 17.5 (27) 74.5 (28) 23.1 (29) 13.3 **Glc I** (1) 100.0 (2) 83.0 (3) 75.9 (4) 69.9 (5) 76.5 (6) 60.8 **Glc II** (1) 104.5 (2) 75.3 (3) 75.9 (4) 70.0 (5) 77.2 (6) 60.9 **Glc III** (1) 103.2 (2) 73.4 (3) 76.8 (4) 69.7 (5) 76.8 (6) 61.1.

Mass (MALDI, Positive ion, H.R.) : *m/z* 969.5025 [(M+Na)⁺, calcd. for 969.5035].

Biological Activity : The compound showed weak cytotoxicity against Jurkat t cells with IC_{50} value of 8µM.

Reference

1.   P. Akbay, J. Gertsch, I. Calis, J. Heilmanm, O. Zerbe and O. Sticher, *Helv. Chim. Acta*, **85**, 1930 (2002).

SITOINDOSIDE III
Sitosterol 3-O-[(6-O-palmitoyl)-β-D-glucopyranosyl-(1→6)-β-D-glucopyranoside]

Source : *Musa paradisiaca* L. (Musaceae)
Mol. Formula : $C_{57}H_{100}O_{12}$
Mol. Wt. : 976
$[\alpha]_D^{20}$: -34.8° (MeOH)
Registry No. : [98941-83-4]

IR (Nujol) : 3400 (br, OH), 1730 (ester>CO), 1600 (br, sugar moiety) cm^{-1}.

Mass : *m/z* 976 $[M]^+$.

Trimethylsilyl ether :

PMR (CDCl₃, 100 MHz) : δ 0.66-1.0 (7xmethyl groups), 1.2-1.35 (methylenes), 3.7-3.9 (glucosyl protons), 4.95 (d, *J*=7.0 Hz, glucosyl anomeric H), 5.3 (m, H-6).

Reference

1.   S. Ghosal, *Phytochemistry*, **24**, 1807 (1985).

SITOINDOSIDE IV

Sitosterol 3-[2"-O-palmitoyl-myoinosityl-(1"→6')-β-D-glucopyranoside]

Source : *Musa paradisiaca* L. (Musaceae)
Mol. Formula : $C_{57}H_{100}O_{12}$
Mol. Wt. : 976
$[\alpha]_D$ **:** -9.8° (MeOH)
Registry No. : [98941-86-7]

IR (Nujol) : 3400 (br), 1735, 1600, 1040, 835 cm^{-1}.

Hepta-acetate : M.P. : 146-148°C

IR (KBr) : 1735, 1728, 1605, 1205 cm^{-1}.

PMR (CDCl$_3$, 100 MHz) : δ 0.68-1.0 (7xCH_3), 1.33-1.2 (methylenes), 2.0-2.1 (21H, OCOCH_3), 3.6-3.9 (H-5, 2xH-6 of Glc), 5.3 (11H, m).

Reference

1.   S. Ghosal, *Phytochemistry*, **24**, 1807 (1985).

MYCALOSIDE I

Stigmast-7,24(28)-*E*-dien-3β,15β,29-triol 3-O-{β-D-galactopyranosyl-(1→2)-β-D-arabinopyranosyl-(1→3)-[β-D-galactopyranosyl-(1→4)]-β-D-glucopyranoside}

Source : *Mycale laxissima* (Sponge)
Mol. Formula : $C_{52}H_{86}O_{22}$
Mol. Wt. : 1062
M.P. : 222-225°C
[α]$_D^{25}$: -6.5° (c=0.7, MeOH)
Registry No. : [593280-58-1]

PMR (C_5D_5N-CD_3OD, 500 MHz) : δ 0.63 (s, 3xH-18), 0.66 (s, 3xH-19), 0.90 (m, H-1ax), 0.99 (d, *J*=6.5 Hz, H-21A), 1.05 (d, *J*=7.0 Hz, 3xH-26, 3xH-27), 1.12 (m, H-5), 1.25 (m, H-4ax, H-22A), 1.28 (m, H-12ax), 1.38 (m, H-20), 1.46 (m, H-2ax), 1.47 (m, H-11), 1.56 (m, H-17, H-22), 1.57 (m, H-9), 1.62 (m, H-6A), 1.66 (m, H-1eq), 1.68 (m, H-6), 1.84 (m, H-4eq), 1.91 (m, H-2eq), 1.95 (m, H-12eq), 2.01 (m, H-23A), 2.02 (m, H-16A), 2.17 (m, H-17), 2.19 (m, H-15), 2.21 (m, H-23B), 2.28 (sept, *J*=7.0 Hz, H-25), 3.74 (m, H-3), 4.46 (m, H-15), 4.56 (d, *J*=6.5 Hz, 2xH-29), 5.72 (t, *J*=6.5 Hz, H-28), 5.80 (m, H-7).

For signals of carbohydrte moiety see Mycaloside B.

CMR (C_5D_5N, 95:5) : δ C-1) 37.2 (2) 29.8 (3) 77.9 (4) 34.5 (5) 39.9 (6) 29.8 (7) 118.5 (8) 136.7 (9) 49.6 (10) 34.4 (11) 21.6 (12) 40.0 (13) 44.1 (14) 62.5 (15) 69.6 (16) 40.6 (17) 53.5 (18) 13.2 (19) 13.0 (20) 36.6 (21) 18.7 (22) 36.3 (23) 26.5 (24) 147.1 (25) 34.8 (26) 22.1 (27) 22.0 (28) 123.7 (29) 58.8.

For signals of carbohydrate moiety see Mycaloside B.

Mass (MALDI, Positive ion, H.R.) : *m/z* 1085.5481 [(M+Na)⁺, calcd. for 1085.5509], 1101.52 [M+K]⁺.

Reference

1.   A.S. Antonov, S.S. Afiyatullov, A.I. Kalinovsky, L.P. Ponomarenko, P.S. Dmitrenok, D.L. Aminin, I.G. Agafonova and V.A. Stonik, *J. Nat. Prod.*, **66**, 1082 (2003).

AJUGASALICIOSIDE H

(5α,20*S*,22*R*,24*S*,25*S*)-22,25-Epoxy-3β,16β,27-Trihydroxystigmast-7-ene 3-O-[β-D-glucopyranosyl-(1→2)-
β-D-glucopyranoside]-27-O-β-D-glucopyranosyl-(1→2)-β-D-glucopyranoside]

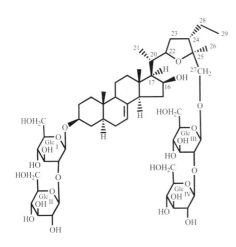

Source : *Ajuga salicifolia* (L.) Schreber (Lamiaceae)
Mol. Formula : $C_{53}H_{88}O_{24}$
Mol. Wt. : 1108
M.P. : 154°C
[α]$_D^{20}$: -13.2° (c=0.1, MeOH)
Registry No. : [503623-38-9]

UV (MeOH) : λ_{max} 206 (log ε, 2.68) nm.

PMR (CD$_3$OD, 500 MHz) : δ 0.78 (s, 3xH-18), 0.83 (s, 3xH-19), 0.98 (t, *J*=7.2 Hz, 3xH-29), 0.99 (d, *J*=6.2 Hz, 3xH-21), 1.04 (s, 3xH-26), 1.11 (m, H-1A), 1.19 (m, H-17), 1.22* (H-12A), 1.22* (m, H-28A), 1.22 (m, H-28A), 1.29 (m, H-6B), 1.35* (m, H-4A), 1.38* (m, H-5), 1.45* (m, H-15A), 1.47* (m, H-28B), 1.48* (m, H-2A), 1.52 (m, H-11A), 1.60* (m H-11B), 1.67 (m, H-9), 1.73 (m, H-14B), 1.80* (m, H-6), 1.85 (m, H-1B), 1.88* (m, H-4B), 1.92 (m, H-2B), 1.98 (m, 2xH-23), 2.07* (m, H-12), 2.08* (m, H-20), 2.09 (m, H-15), 2.18 (m, H-24), 3.22* (m, H-2 of Glc II, H-2 of Glc IV), 3.25* (m, H-4 of Glc II, H-5 of Glc IV), 3.27* (m, H-5 of Glc II), 3.28* (H-3 of Glc IV), 3.30* (m, H-4 of Glc I), 3.35* (m, H-4 of Glc III, H-4 of Glc IV), 3.37* (m, H-3 of Glc III), 3.40 (dd, *J*=7.4, 9.1 Hz, H-2 of Glc I), 3.50 (d, *J*=10.3 Hz, H-27A), 3.55* (m, H-3 and H-5 of Glc I), 3.56* (m, H-2, H-3, H-5 of Glc III), 3.65* (m, H-6 of Glc IV), 3.71* (m, H-6B of Glc I, H-6B of Glc II, H-6B of Glc III), 3.72* (m, H-3), 3.83* (m, H-6B of Glc II), 3.84 (m H-22), 3.85& (m, H-6B of Glc I), H-6B of Glc III), 3.88 (d, *J*=10.3 Hz, H-27B), 4.41 (d, *J*=7.5 Hz, H-1 of Glc III), 4.45 (m, H-16), 4.54 (d, *J*=7.8 Hz, H-1 of Glc I), 4.58 (d, *J*=7.8 Hz H-1 of Glc II), 4.66 (d, *J*=7.8 Hz, H-1 of Glc IV), 5.22 (m, H-9).

CMR (CD$_3$OD, 125 MHz) : δ C-1) 38.2 (2) 30.4 (3) 80.1 (4) 35.3 (5) 41.5 (6) 30.8 (7) 119.1 (8) 139.8 (9) 50.7 (10) 35.4 (11) 22.5 (12) 41.2 (13) 45.0 (14) 53.6 (15) 35.1 (16) 73.6 (17) 61.9 (18) 13.5 (19) 13.5 (20) 37.4 (21) 16.2 (22) 83.0 (23) 37.1 (24) 44.2 (25) 85.6 (26) 17.6 (27) 74.6 (28) 24.5 (29) 13.8 **Glc I** (1) 101.4 (2) 83.0 (3) 77.8 (4) 71.5 (5) 77.8 (6) 62.7 **Glc II** (1) 105.2 (2) 76.1 (3) 77.6 (4) 71.7 (5) 78.4 (6) 63.0 **Glc III** (1) 102.6 (2) 81.5 (3) 78.2 (4) 71.2 (5) 77.8 (6) 62.7 **Glc IV** (1) 104.7 (2) 76.1 (3) 78.3 (4) 71.4 (5) 77.8 (6) 62.5.

Mass (MALDI, Positive ion, H.R.) : *m/z* 1131.5552 [(M+Na)$^+$, requires 1131.5563].

Reference

1.   P. Akbay, I. Calis, J. Heilmann and O. Sticher, *J. Nat. Prod.*, **66**, 461 (2003).

PECTINIOSIDE C

(20*S*,24*S*)-6α,20-Dihydroxy-3β-sulfoxy-5α-stigmast-9(11)-en-23-one 6-O-{β-D-fucopyranosyl-(1→3)-β-D-fucopyranosyl-(1→2)-β-D-glucopyranosyl-(1→4)-[β-D-quinovopyranosyl-(1→2)]-β-D-xylopyranosyl-(1→3)-β-D-quinovopyranoside} monosodium salt

Source : *Asterina pectinifera* Muller and Trochel (Asteroidea)
Mol. Formula : $C_{64}H_{105}O_{32}SNa$
Mol. Wt. : 1440
M.P. : 230-235°C[1]
[α]$_D$: +5.38° (c=0.275, MeOH)[1]
Registry No. : [113322-00-2]

IR (KBr)[1] : 3400 (OH), 1690 (C=O), 1240, 1210 (sulfate) cm^{-1}.

CMR (C$_5$D$_5$N, 67.5 MHz)[1] : δ C-1) 36.0 (2) 29.4 (3) 77.7 (4) 30.8 (5) 49.3 (6) 79.9 (7) 41.7 (8) 35.4 (9) 145.5 (10) 38.4 (11) 116.7 (12) 42.5 (13) 41.6 (14) 54.1 (15) 23.3 (16) 25.2 (17) 58.8 (18) 13.8 (19) 19.5 (20) 73.6 (21) 27.2 (22) 55.4 (23) 215.8 (24) 61.3 (25) 29.3 (26) 21.2 (27) 19.5 (28) 21.2 (29) 12.1 **Qui I** (1) 105.1 (2) 74.2 (3) 90.3 (4) 74.6 (5) 72.0 (6) 17.9 **Xyl** (1) 104.5 (2) 82.7 (3) 75.6 (4) 77.9^a (5) 64.4 **Qui II** (1) 105.3 (2) 75.6 (3) 76.6 (4) 76.2 (5) 73.5 (6) 18.5 **Glc** (1) 101.8 (2) 84.3 (3) 77.8^a (4) 71.0 (5) 78.4 (6) 62.2 **Fuc I** (1) 106.5 (2) 71.8 (3) 85.1 (4) 72.7 (5) 72.1 (6) 17.2 **Fuc II** (1) 106.7 (2) 71.8 (3) 75.1 (4) 72.6 (5) 72.1 (6) 17.3.

Mass (FAB, glycol, Negative ion)[1] : m/z 1417 [M-Na]$^-$, 1271 [M-Na-146]$^-$, 1125 [M-Na-292]$^-$, 963 [M-Na-454]$^-$, 539 [M-Na-878]$^-$, 439, 411, 393, 265, 233.

Stereochemistry of the side chain.[2]

References

1. M.A. Dubois, Y. Noguchi, R. Higuchi and T. Komori, *Liebigs Ann. Chem.*, 495 (1988).

2. M. Honda, T. Igarashi and T. Komori, *Liebigs Ann. Chem.*, 547 (1990).

FUROSTANE

Basic skeleton

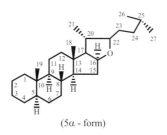

(5α - form)

TERRESTRININ A
(25*S*)-Furostan-4,20 (2)-dien-26-ol-3,12-dione 26-O-β-D-glucopyranoside

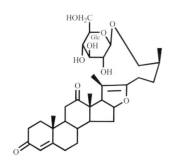

Source : *Tribulus terrestris* Linn. (Zygophyllaceae)
Mol. Formula : $C_{33}H_{48}O_9$
Mol. Wt. : 588
M.P. : 134-136°C
[α]$_D^{20}$: +56.5° (c=0.56, CHCl$_3$)
Registry No. : [681277-16-7]

IR (KBr) : 3412, 2918, 1707, 1670, 1450, 1379, 1269, 1238, 1076, 1040, 752 cm^{-1}.

PMR (CDCl$_3$, 400 MHz) : δ 0.93 (d, *J*=6.6 Hz, 3xH-27), 0.98 (s, 3xH-18), 1.28 (s, 3xH-19), 1.58 (s, 3xH-21), 4.28 (d, *J*=7.6 Hz, H-1 of Glc), 4.71 (m, H-16), 5.78 (s, H-4).

CMR (CDCl$_3$, 100 MHz) : δ C-1) 35.3 (2) 32.4 (3) 198.8 (4) 124.8 (5) 168.4 (6) 33.7 (7) 31.3 (8) 34.2 (9) 54.6 (10) 38.7 (11) 37.3 (12) 212.2 (13) 56.9 (14) 53.3 (15) 30.7 (16) 82.5 (17) 55.6 (18) 13.9 (19) 16.9 (20) 103.4 (21) 11.2 (22) 152.3 (23) 23.2 (24) 33.4 (25) 32.9 (26) 75.1 (27) 16.8 **Glc** (1) 103.0 (2) 73.6 (3) 76.3 (4) 69.7 (5) 75.4 (6) 61.7.

Mass (E.I.) : *m/z* 588 [M]$^+$, 480, 426, 181, 109, 69.

726

Mass (E.I., H.R.) : *m/z* 588.3293 [(M)$^+$, requires 588.3298].

Reference

1.   J.W. Huang, C.-H. Tan, S.-H. Jiang and D.-Y. Zhu, *J. Asian Nat. Prod. Res.*, **5**, 285 (2003).

KRYPTOGENIN 26-O-β-D-GLUCOPYRANOSIDE
Kryptogenin 26-O-β-D-glucopyranoside

Source : *Dioscorea olfersiana* Klotzsch ex Griseb.(Dioscoreaceae)
Mol. Formula : $C_{33}H_{52}O_9$
Mol. Wt. : 592
M.P. : 230-235°C
Registry No. : [157142-70-6]

CMR (CDCl$_3$+CD$_3$OD, 50 MHz) : δ C-1) 37.0 (2) 30.9 (3) 70.9 (4) 36.7 (5) 140.9 (6) 120.5 (7) 31.4 (8) 30.7 (9) 49.4 (10) 36.4 (11) 20.3 (12) 39.3 (13) 41.5 (14) 51.0 (15) 38.4 (16) 218.8 (17) 66.0 (18) 15.2 (19) 19.1 (20) 43.2 (21) 12.7 (22) 215.0 (23) 37.0 (24) 26.0 (25) 34.8 (26) 74.8 (27) 16.2 **Glc** (1) 102.9 (2) 73.4 (3) 76.1 (4) 69.7 (5) 75.5 (6) 61.5.

Mass (E.I.) : *m/z* (rel.intens.) 592 [(M)$^+$, 7], 574 (40), 413 (90), 412 (62), 395 (58), 381 (15), 356 (28), 343 (38), 316 (18) 213 (18), 180 (43), 145 (31), 133 (29), 115 (95), 69 (100).

Note : The compound was isolated admixed with kryptogenin 3-O-β-D-glucopyranoside.

Reference

1.   M. Haraguchi, A.P.Z. D. Santos, M.C. M.Young and E.P. Chu, *Phytochemistry*, **36**, 1005 (1994).

TRILLIUM KAMTSCHATICUM SAPONIN Tk

Kryptogenin 3-O-β-D-glucopyranoside

Source : *Trillium kamtschaticum* Pall.[1] (Liliaceae),
Dioscorea olfersiana Klotzsch ex Griseb.[2] (Dioscoreaceae)
Mol. Formula : $C_{33}H_{52}O_9$
Mol. Wt. : 592
M.P. : 238-241°C (decomp)[1]
[α]$_D$: -129.4° (c=0.51, EtOH)[1]
Registry No. : [55916-53-5]

IR (KBr)[1] : 3600-3200 (OH), 1740, 1740 (C=O) cm^{-1}. No spiroketal absorption.

CMR $(C_5D_5N)^2$: δ C-1) 37.4 (2) 30.1 (3) 78.2 (4) 39.2 (5) 141.0 (6) 121.2 (7) 31.9 (8) 30.9 (9) 49.9 (10) 37.0 (11) 20.7 (12) 40.3 (13) 41.6 (14) 51.1 (15) 38.7 (16) 217.3 (17) 66.3 (18) 15.6 (19) 19.6 (20) 43.7 (21) 12.9 (22) 213.2 (23) 37.1 (24) 27.6 (25) 36.1 (26) 67.4 (27) 17.2 **Glc** (1) 102.5 (2) 75.2 (3) 78.4 (4) 71.7 (5) 78.4 (6) 62.8.

Mass (E.I.) : *m/z* (rel.intens.) 592 [M]$^+$ (7), 574 (40), 413 (90), 412 (62), 395 (58), 381 (15), 356 (28), 343 (38), 316 (18) 213 (18), 180 (43), 145 (31), 133 (29), 115 (95), 69 (100).

Mass (FD)^2 : *m/z* (rel.intens.) 4631 [M+K]$^+$, 615 [(M+Na)$^+$, 100], 575 [(M+H-H$_2$O)$^+$, 75].

References

1. T. Nohara, K. Miyahara and T. Kawasaki, *Chem. Pharm. Bull*, **23**, 872 (1975).

2. M. Haraguchi, A.P.Z.D. Santos, M.C. M. Young and E.P. Chu, *Phytochemistry*, **36**, 1005 (1994).

3. H.R. Schulten, T. Komori and T. Kawasaki, *Tetrahedron*, **33**, 2595 (1977).

TUMAQUENONE
(25R)-Furost-4-en-22,26-diol-3-one 26-O-β-D-glucopyranoside

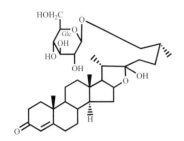

Source : *Solanum nudum* (Solanaceae)
Mol. Formula : $C_{33}H_{52}O_9$
Mol. Wt. : 592
[α]$_D$: +2.7° (c=1.0, MeOH)
Registry No. : [214130-69-5]

IR (film) : 3420, 2927, 1682, 1213 cm^{-1}.

PMR (CDCl$_3$, 300 MHz) : δ 0.70 (s, 3xH-18), 0.77 (d, *J*=7.0 Hz, 3xH-27), 0.80 (H-9), 0.87 (d, *J*=7.0 Hz, 3xH-21), 0.92 (H-7A), 1.01 (H-14), 1.04 (H-12A), 1.06 (s, 3xH-19), 1.18 (H-15A), 1.20 (H-24A), 1.41 (H-11A, H-11B), 1.53 (H-24B), 1.58 (H-1A, H-1B), 1.62 (H-8, H-23A), 1.64 (H-25), 1.66 (H-12B), 1.67 (H-17), 1.73 (H-7B), 1.86 (H-15B), 1.90 (H-23B), 1.92 (H-20), 2.16 (H-6A), 2.23 (H-2), 2.32 (H-6B), 3.18 (H-2 of Glc), 3.19 (H-5 of Glc), 3.22 (H-26 A), 3.32 (H-3, H-4 of Glc), 3.64 (H-26B), 3.66 (H-6A of Glc), 3.70 (H-6B Glc), 4.12 (d, *J*=8.0 Hz, H-1 of Glc), 4.28 (dd, *J*=7.0, 14.0 Hz, H-16), 5.60 (s, H-4).

CMR (CDCl$_3$, 75.0 MHz) : δ C-1) 35.4 (2) 33.5 (3) 200.6 (4) 123.3 (5) 172.6 (6) 32.6 (7) 31.8 (8) 34.9 (9) 53.5 (10) 38.5 (11) 20.5 (12) 39.2 (13) 40.4 (14) 55.2 (15) 31.3 (16) 80.7 (17) 62.2 (18) 16.0 (19) 17.0 (20) 39.7 (21) 15.0 (22) 110.4 (23) 35.2 (24) 26.8 (25) 32.3 (26) 75.0 (27) 16.5 **Glc** (1) 103.0 (2) 73.4 (3) 75.6 (4) 69.8 (5) 76.1 (6) 61.4.

Mass (FAB, Positive ion, H.R.) : *m/z* 615.3528 [(M+Na)$^+$, calcd. for 615.3509].

Biological Activity : The compound displayed antimalarial activity *in vitro* against *Plasmodium falciparum* a chloroquine-resistant FCB-1 strain, with IC$_{50}$ value of 16 μM or 9.54 μg/ml.

Reference

1. J. Saez, W. Cardona, D, Espinal S, Blair, J. Mesa, M. Bocar and A. Jossang, *Tetrahedron*, **54**, 10771 (1998).

FUNKIOSIDE B
(25R)-Furost-5-en-3β,22α,26-triol 26-O-[β-D-glucopyranoside]

Source : *Funkia ovata* Spr.[1] Syn. *Hosta caerulea*
(Liliaceae), *Melilotus tauricus* (Bieb.) Ser.[2] (Fabaceae)
Mol. Formula : $C_{33}H_{54}O_9$
Mol. Wt. : 594
M.P. : 258-266°C[1]
[α]$_D$: -135.0° (c=0.75, MeOH)[1]
Registry No. : [60433-64-9]

IR (KBr)[2] **:** 910 cm^{-1}.

References

1.   P.K. Kintya, N.E. Mashchenko, N.I. Kononova and G.V. Lazur'evskii, *Khim. Prir. Soedin.*, **12**, 267 (1976); *Chem. Nat. Comp.*, **12**, 241 (1976).

2.   G.V. Khodakov, A.S. Shashkov, P.K. Kintya and Yu. A. Akimov, *Khim. Prir. Soedin.*, **30**, 766 (1994); *Chem. Nat. Comp.*, **30**, 713 (1994).

ASPARAGOSIDE B
(25S,5β)-3β,22α,26-Trihydroxyfurostane 26-O-β-D-glucopyranoside

Source : *Asparagus officinalis* L. (Liliaceae)
Mol. Formula : $C_{33}H_{56}O_9$
Mol. Wt. : 596
M.P. : 152-155°C
[α]$_D^{20}$: -81° (c=1.0, MeOH)
Registry No. : [60237-69-6]

Reference

1. G.M. Goryanu, V.V. Krokhmalyuk and P.K. Kintya, *Khim. Prir. Soedin*, 400 (1976), *Chem. Nat. Comp.*, **12**, 353 (1976).

TORVOSIDE G

(25S)-26-Hydroxy-22α-methoxy-5α-furostan-3-one 26-O-[β-D-glucopyranoside]

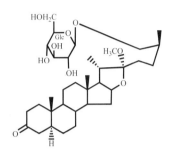

Source : *Solanum torvum* Swartz (Solanaceae)
Mol. Formula : $C_{34}H_{56}O_9$
Mol. Wt. : 608
$[\alpha]_D^{27}$ **:** -47.4° (c=0.10, MeOH)
Registry No. : [184777-22-8]

PMR (C_5D_5N, 400 MHz) **:** δ 0.84 (s, 3xH-18), 0.89 (s, 3xH-19), 1.07 (d, *J*=6.6 Hz, 3xH-27), 1.19 (d, *J*=6.6 Hz, 3xH-21), 3.28 (s, 3xOC*H₃*), 4.87 (d, *J*=8.1 Hz, H-1 of Glc).

CMR (C_5D_5N, 100 MHz) **:** δ C-1) 38.6 (2) 38.3 (3) 210.3 (4) 44.8 (5) 46.6 (6) 28.9 (7) 32.0 (8) 35.0 (9) 53.8 (10) 35.8 (11) 21.4 (12) 39.9 (13) 41.1 (14) 56.1 (15) 32.0 (16) 81.3 (17) 64.3 (18) 16.5 (19) 11.3 (20) 40.5 (21) 16.3 (22) 112.6 (23) 29.9 (24) 28.2 (25) 34.3 (26) 74.9 (27) 17.5 (OCH₃) 47.3 **Glc** (1) 105.1 (2) 75.2 (3) 78.6 (4) 71.7 (5) 78.5 (6) 62.9.

Mass (FAB, Positive ion) **:** *m/z* 577 [M-OMe]⁺, 415 [M-OMe-Glc]⁺.

Reference

1. S. Yahara, T. Yamashita, N. Nozawa (nee Fujimura) and T. Nohara, *Phytochemistry*, **43**, 1069 (1996).

POLYGONATUM ODORATUM GLYCOSIDE 1

(25*R* and *S*) Furost-5-en-12-en-3β,22ξ,26-triol 26-O-β-D-glucopyranoside]

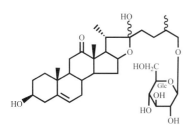

Source : *Polygonatum odoratum* (Mill.) Druce (Liliaceae).
The compound was isolated as a mixture of *R* and *S* epimers
Mol. Formula : $C_{33}H_{52}O_{10}$
Mol. Wt. : 608
M.P. : 142-143°C
$[\alpha]_D^{17}$: -0.024 (c=0.11, MeOH)
Registry No. : [681431-96-9]

IR : 3425 (br, OH), 1707 (C=O) cm⁻¹.

PMR (CD₃OD, 500 MHz) : δ 0.89 and 0.90 (each d, *J*=6.6 Hz, total 3H, 27-CH₃ of *R* and *S*-epimzer), 1.01 (d, *J*=6.5 Hz, 3xH-21), 1.08 (s, 3xH-18), 1.09 (s, 3xH-19), 2.58 (m, H-14), 3.71 (m, H-26A), 4.17 (d, *J*=8.0 Hz, H-1 of Glc), 4.26 (ddd, *J*=5.5, 7.0, 8.5 Hz, H-16), 5.35 (m, H-6).

CMR (C₅D₅N, 125 MHz) : δ C-1) 38.1ᵃ (2) 32.1ᵇ (3) 72.1 (4) 42.8 (5) 142.2 (6) 122.1 (7) 31.3ᵇ (8) 32.1 (9) 53.9 (10) 38.4 (11) 38.3ᵃ (12) 215.8 (13) 56.4 (14) 57.3 (15) 32.6ᵇ (16) 80.9 (17) 56.0, 56.1 (18) 16.4 (19) 19.3 (20) 41.6, 41.8 (21) 14.7, 14.8 (22) 114.0 (23) 32.06, 32.12 (24) 28.87, 28.93 (25) 35.0, 35.1 (26) 75.8, 76.0 (27) 17.2, 17.4 **Glc** (1) 104.6 (2) 75.2 (3) 78.1 (4) 71.7 (5) 78.0 (6) 62.8.

The double chemical shifts of some carbons is due to the presence of 25*R* and *S* epimers.

Mass (FAB, Positive ion) : *m/z* 631.3487 [(M+Na)⁺, calcd. for 631.3458], 591.3536 [(M-H₂O+H)⁺, calcd. for 591.3534].

Reference

1.   H.L. Qin, Z.H. Li and P. Wang, *Chin. Chem. Lett.*, **14**, 1259 (2003).

ALLIUM AMPELOPRASUM SAPONIN 2

(25*R*)-22α-Methoxy-5α-furostane-2α,3β,6β,26-tetrol-26-O-β-D-glucopyranoside

Source : *Allium ampeloprasum* (Liliaceae)
Mol. Formula : $C_{34}H_{58}O_{11}$
Mol. Wt. : 642
$[\alpha]_D^{26}$: -58.0° (c=0.10, MeOH)
Registry No. : [256642-46-3]

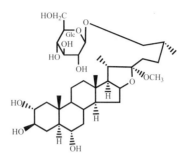

IR (KBr) **:** 3380 (OH), 2930, 2870 (C–H), 1050 cm⁻¹.

PMR (C₅D₅N, 400 MHz) **:** δ 0.84 (s, 3xH-18), 1.01 (d, *J*=6.7 Hz, 3xH-27),1.19 (d, *J*=6.9 Hz, 3xH-21), 1.41 (s, 3xH-19), 3.26 (s, OC*H₃*), 4.85 (d, *J*=7.8 Hz, H-1 of Glc).

CMR (C₅D₅N, 100 MHz) **:** δ C-1) 48.0 (2) 73.2 (3) 77.3 (4) 35.0 (5) 48.5 (6) 70.4 (7) 40.9 (8) 30.0 (9) 54.8 (10) 37.7 (11) 21.4 (12) 40.1 (13) 41.3 (14) 56.3 (15) 32.2 (16) 81.4 (17) 64.3 (18) 16.3 (19) 17.5 (20) 40.5 (21) 16.5 (22) 112.7 (23) 30.7 (24) 28.2 (25) 34.2 (26) 75.2 (27) 17.2 **Glc** (1) 104.9 (2) 75.2 (3) 78.5 (4) 71.8 (5) 78.6 (6) 62.9 (OCH₃) 47.3.

Mass (FAB, Negative ion) **:** *m/z* 641 [M-H]⁻.

Reference

1.   Y. Mimaki, M. Kuroda and Y. Sashida, *Natural Medicine*, **53**, 134 (1999).

AGAMENOSIDE J
(5α,22*S*,23*S*,25*R*,26*S*)-23,26-Epoxyfurostane-3β,22,26-triol 26-O-β-D-glucopyranoside

Source : *Agave americana* L. (Agavaceae)
Mol. Formula : C₃₃H₅₄O₁₀
Mol. Wt. : 610
[α]$_D^{21}$ **:** -37.1° (c=0.08, Pyridine)
Registry No. : [738584-21-9]

PMR (C$_5$D$_5$N, 400 MHz) : δ 0.57 (H-9), 0.74 (H-7α), 0.78 (s, 3xH-19), 0.80 (d, *J*=7.4 Hz, 3xH-27), 0.93 (s, 3xH-18), 0.94 (H-1α), 1.04 (H-14), 1.08 (H-5), 1.10 (H-12α), 1.20 (H-6), 1.24 (H-11α), 1.25 (d, *J*=5.2 Hz, 3xH-21), 1.47 (H-11β), 1.48 (H-8), 1.51 (H-2β), 1.54 (H-4β), 1.59 (H-24β), 1.60 (H-7β), 1.64 (H-1β), 1.67 (H-15β), 1.76 (H-4α, H-12β), 1.87 (dd, *J*=5.8, 7.4 Hz, H-17), 1.97 (H-15α), 2.64 (H-2α), 2.15 (dd, *J*=5.2, 6.4 Hz, H-20), 2.44 (H-25), 2.55 (H-24α), 3.84 (H-5 of Glc), 4.02 (H-2 of Glc), 4.15 (H-4 of Glc), 4.16 (H-3 of Glc), 4.32 (H-6A of Glc), 4.47 (H-6B of Glc), 4.58 (dd, *J*=.6, 6.4 Hz, H-23), 5.29 (d, *J*=8.0 Hz, H-1 of Glc), 5.47 (br s, *J*=1.9 Hz, H-26), 5.93 (s, H-22).

CMR (C$_5$D$_5$N, 125 MHz) : δ C-1) 37.6 (2) 32.5 (3) 70.7 (4) 39.3 (5) 45.3 (6) 29.1 (7) 32.5 (8) 35.5 (9) 54.7 (10) 35.9 (11) 21.4 (12) 40.2 (13) 41.3 (14) 56.7 (15) 32.2 (16) 81.5 (17) 64.9 (18) 16.6 (19) 12.6 (20) 39.3 (21) 16.6 (22) 109.1 (23) 85.2 (24) 32.5 (25) 40.3 (26) 107.3 (27) 16.8 **Glc** (1) 100.7 (2) 74.7 (3) 78.8 (4) 71.6 (5) 78.9 (6) 62.6

Mass (FAB, Negative ion) : *m/z* 609 [M-H]⁻, 447 [M-H-Glc]⁻.

Mass (FAB, Negative io n, H.R.) : *m/z* 609.3590 [(M-H)⁻, calcd. for 609.3639].

Reference

1.   J.M. Jin, Y.-J. Zhang and C.R. Yang, *Chem. Pharm. Bull.*, **52**, 654 (2004).

NOLINOFUROSIDE A
(25*S*)-Furost-5-ene-1β,3β,22α,26-tetraol 26-*O*-β-D-glucopyranoside

Source : *Nolina microcarpa* S. Wats. (Dracaenacea)
Mol. Formula : C$_{33}$H$_{54}$O$_{10}$
Mol. Wt. : 610
[α]$_D^{20}$: -56.0 ± 2° (c=1.0, Pyridine)
Registry No. : [145645-63-2]

IR (KBr) : 3200-3600 (OH), 915 (br weak) cm⁻¹.

PMR (C$_5$D$_5$N, 250/300 MHz) : δ 0.86 (s, 3xH-18), 0.94 (d, *J*$_{27,25}$=6.5 Hz, 3xH-27), 1.20 (d, *J*$_{21,20}$=7.0 Hz, 3xH-21), 1.23 (s, 3xH-19), 3.42 (dd, *J*$_{26,25}$=7.0 Hz, H-26A) 3.68 (dd, *J*$_{1ax,2ax}$=11.5 Hz, *J*$_{1ax,2ax}$=4.3 Hz, H-1ax), 3.80 (m, H-3), 3.85 (m, H-5 of Glc), 3.92 (dd, *J*$_{2,3}$=8.8 Hz, H-2 of Glc), 3.99 (dd, *J*$_{26B,25}$=5.0 Hz, *J*$_{26A,26B}$=9.0 Hz, H-26B), 4.06 (t, *J*$_{4,5}$=8.8 Hz, H-4 of Glc), 4.16 (t, *J*$_{3,4}$=8.8, H-3 of Glc), 4.22 (dd, *J*$_{6A,5}$=5.5 Hz, *J*$_{6A,6B}$=11.5 Hz, H-6A of Glc), 4.43 (dd, *J*$_{6B,5}$=2.5 Hz, H-6B), 4.70 (d, *J*$_{1,2}$=7.5 Hz, H-1 of Glc), 4.87 (m, H-16), 5.48 (br d, *J*$_{6,7}$=5.5 Hz, H-6).

CMR (C$_5$D$_5$N, 62.5/75 MHz) : δ C-1) 78.19 (2) 48.98 (3) 68.19 (4) 43.64 (5) 140.42 (6) 124.53 (7) 32.39 (8) 33.03 (9) 51.45 (10) 43.64 (11) 24.26 (12) 40.62 (13) 40.62 (14) 56.97 (15) 32.76 (16) 81.19 (17) 64.15 (18) 16.81 (19) 13.99 (20) 40.78 (21) 16.49 (22) 110.78 (23) 37.24 (24) 28.38 (25) 34.49 (26) 75.47 (27) 17.51 **Glc** (1) 105.21 (2) 75.28 (3) 78.53 (4) 71.78 (5) 78.67 (6) 62.88.

Mass : m/z 592 [M-H$_2$O]$^+$.

Reference

1.   G.V. Shevchuk, Yu.S. Vollerner, A.S. Shashkov and Y.Ya. Chirva, *Khim. Prir. Soedin.*, **27**, 672 (1991); *Chem. Nat. Comp.*, **27**, 592 (1991).

PROTOYONOGENIN
(25*R*)-5β-Furostan-2β,3α,22ξ,26-tetraol 26-O-β-D-glucopyranoside

Source : *Dioscorea tokoro* Makino (Dioscoreaceae)
Mol. Formula : C$_{33}$H$_{56}$O$_{10}$
Mol. Wt. : 612
Registry No. : [87585-29-3]

Isolated admixed with protoneoyonogenin (q.v.).

CMR (C$_5$D$_5$N, 25.16 MHz) : δ C-1) 44.8 (2) 71.3 (3) 77.0 (4) 35.6 (5) 42.4 (6) 26.9 (7) 26.9 (8) 35.7 (9) 42.4 (10) 37.2 (11) 21.1 (12) 40.2 (13) 41.2 (14) 56.2 (15) 32.3 (16) 81.2 (17) 63.9 (18) 16.7 (19) 23.6 (20) 40.6 (21) 16.4 (22) 110.7 (23) 37.2 (37.1) (24) 28.3 (25) 34.2 (34.3) (26) 75.3 (27) 17.4 **Glc** (1) 104.8 (105.0) (2) 75.1 (3) 78.5 11 (4) 71.7 (5) 78.4 (6) 62.8.

Reference

1.   A. Uomori, S. Seo, K. Tori and Y. Tomita, *Phytochemistry*, **22**, 203 (1983).

PROTONEOYONOGENIN
(25S)-5β-Furostan-2β,3α,22ξ,26-tetraol 26-O-β-D-glucopyranoside

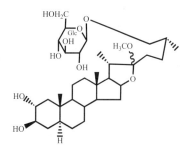

Source : *Dioscorea tokoro* Makino (Dioscoreaceae)
Mol. Formula : $C_{33}H_{56}O_{10}$
Mol. Wt. : 612
Registry No. : [87562-60-5]

Isolated admixed with protoyonogenin (q.v.).

Reference

1.   A. Uomori, S. Seo, K. Tori and Y. Tomita, *Phytochemistry*, **22**, 203 (1983).

HOSTA PLANTAGINEA SAPONIN 2
(25R)-22-O-methyl-5α-furostane-2α,3β,22 ,26-tetrol 26-O-β-O-glucopyranoside

Source : *Hosta plantaginea* Lam. var. *japonica* (Liliaceae)
Mol. Formula : $C_{34}H_{58}O_{10}$
Mol. Wt. : 626
Registry No. : [186960-65-6]

CMR (C_5D_5N, 100/125 MHz) : δ C-1) 46.5 (2) 73.0 (3) 76.7 (4) 37.2 (5) 45.2 (6) 28.3 (7) 32.1[a] (8) 34.6 (9) 54.6 (10) 37.5 (11) 21.4 (12) 40.0 (13) 41.1 (14) 56.3 (15) 32.3[a] (16) 81.3 (17) 64.3 (18) 16.3 (19) 13.7 (20) 40.5 (21) 16.5 (22)

112.6 (23) 30.8 (24) 28.2 (25) 34.2 (26) 75.2 (27) 17.1 (OCH_3) 47.2 **Glc** (1) 105.0 (2) 75.2 (3) 78.5[b] (4) 71.7 (5) 78.6[b] (6) 62.9.

Reference

1.   Y. Mimaki, A. Kameyama, M. Kuroda, Y. Sashida, T. Hirano, K. Oka, K. Koike and T. Nikaido, *Phytochemistry*, **44**, 305 (1997).

ASPIDISTRA SICHUANENSIS SAPONIN 1
22ξ-Methoxyfurostane-1β,3β,4β,5β,26-pentol 26-O-β-D-glucopyranoside

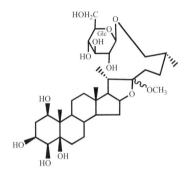

Source : *Aspidistra leshanensis* K. Lang et Z.Y. Zhu
Mol. Formula : $C_{34}H_{58}O_{12}$
Mol. Wt. : 658
M.P. : 217-220°C
Registry No. : [166822-28-2]

IR (KBr) : 3100 (OH), 1080 (OH) cm⁻¹.

CMR (50 MHz) : δ C-1) 72.03 (2) 34.83 (3) 71.80 (4) 68.11 (5) 77.89 (6) 30.79 (7) 28.99 (8) 34.58 (9) 45.59 (10) 47.36 (11) 21.62 (12) 40.42 (13) 40.53 (14) 56.12 (15) 32.14 (16) 81.36 (17) 64.51 (18) 16.43 (19) 13.80 (20) 40.93 (21) 16.24 (22) 112.38 (23) 31.57 (24) 28.78 (25) 34.48 (26) 75.09 (27) 17.11 (OCH_3) 47.28 **Glc** (1) 104.86 (2) 74.76 (3) 78.39 (4) 71.68 (5) 78.52 (6) 62.80.

Mass (E.I.) : *m/z* 464 [M-MeOH-Glc]⁺, 446, 404, 391, 373, 303, 181, 163, 145, 127, 121, 109, 73, 57.

Reference

1.   M. Chen, *Tianran Chanwu Yanjiu Yu Kaifa,* (*Nat. Prod. Res. Dev.*), **7**, 19 (1994).

TUPSTROSIDE G
(25S)-22α-Methoxyfurostane-1α,2β,3α,5α,26-pentol 26-O-β-D-glucopyranoside

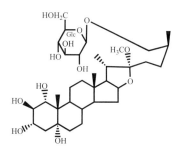

Source : *Aspidistra elatior* Blume (Liliaceae)
Mol. Formula : $C_{34}H_{58}O_{12}$
Mol. Wt. : 658
$[\alpha]_D^{23}$: -38.1° (c=0.3, C_5D_5N)
Registry No. : [288255-69-6]

IR (KBr) : 3399 (br, OH), 2936 (C–H), 1453, 1102, 1068, 907 cm^{-1}.

PMR (400/500 MHz) : δ 0.80 (s, 3xH-18), 1.03 (d, J=6.7 Hz, 3xH-27), 1.14 (d, J=6.9 Hz, 3xH-21), 1.56 (s, 3xH-19), 3.25 (s, OC*H₃* of H-22), 4.32 (br s, H-1), 4.37 (dd, J=11.8, 5.4 Hz, H-26A), 4.48 (m, H-16), 4.55 (dd, J=11.8, 2.0 Hz, H-26B), 4.65 (br s, H-3), 4.82 (d, J=7.7 Hz, H-1 of Glc).

CMR (100/125 MHz) : δ C-1) 78.02 (2) 68.23 (3) 71.69 (4) 39.32 (5) 75.02 (6) 36.09 (7) 28.98 (8) 34.97 (9) 45.72 (10) 45.72 (11) 21.75 (12) 40.01 (13) 41.06 (14) 56.26 (15) 32.30 (16) 81.39 (17) 64.41 (18) 16.53 (19) 13.90 (20) 40.62 (21) 16.35 (22) 112.80 (23) 31.06 (24) 28.30 (25) 34.53 (26) 74.88 (27) 17.65 (OCH₃) 47.48 **Glc** (1) 105.10 (2) 75.26 (3) 78.46 (4) 71.93 (5) 78.67 (6) 62.99.

Mass (FAB) : *m/z* 657 [M-H]⁻, 495 [M-Glc-H]⁻.

Reference

1.   Q.X. Yang and C.R. Yang, *Yunnan Zhiwu Yanjiu* (*Acta Botanica Yunnanica*), **22**, 109 (2000).

WATTOSIDE B

22ξ-Methoxy-25(R,S)-furost-1β,3β,4β,5β,26β-pentol-26-O-β-D-glucopyranoside

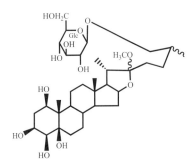

Source : *Tupistra wattii* Hook. f. (Liliaceae)
Mol. Formula : $C_{34}H_{58}O_{12}$
Mol. Wt. : 658
$[\alpha]_D^{18}$ **:** -18.7° (c=0.68, MeOH)
Registry No. : [177910-43-9]

The glucoside was isolated as a mixture of 25R and 25S isomers.

PMR (C_5D_5N, 400 MHz) **:** δ 0.83 (s, 3xH-18), 0.97, 1.02 (d, *J*=6.1 Hz, 3xH-27), 1.15, 1.17 (d, *J*=6.8 Hz, H-21), 1.58 (s, 3xH-19), 2.24 (m, H-20), 4.84, 4.86 (d, *J*=7.7 Hz, H-1 of Glc), 4.97 (m, H-16).

CMR (C_5D_5N, 100 MHz) **:** δ C-1) 73.9 (2) 33.5 (3) 71.2 (4) 68.2 (5) 78.6 (6) 30.5 (7) 28.6 (8) 34.9 (9) 45.5 (10) 46.0 (11) 21.6 (12) 40.2 (13) 41.2 (14) 56.4 (15) 32.6 (16) 81.5 (17) 64.0 (18) 17.0 (19) 13.9 (20) 40.9 (21) 16.4, 17.0 (22) 110.0, 112.6 (OCH₃) 47.5 (23) 30.9 (24) 28.3 (25) 34.4, 34.6 (26) 75.4 (27) 17.3, 17.6 **Glc** (1) 104.9 (2) 75.3 (3) 78.4 (4) 71.5 (5) 78.4 (6) 63.0.

Mass (FAB, Negative ion) **:** *m/z* 657 [M-H]⁻, 481 [M-H-Glc]⁻.

Reference

1.   F. Yang, P. Shen, Y.-F. Wang, R.-Y. Zhang and C.-R. Yang, *Acta Botanica Yunnanica*, **23**, 373 (2001).

TUPILOSIDE H

(25S)-22α-Methoxyfurostane-1β,2β,3β,4β,5β,26-hexol 26-O-β-D-glucopyranoside

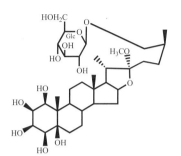

Source : *Aspidistra elatior* Blume (Liliaceae)
Mol. Formula : $C_{34}H_{58}O_{13}$
Mol. Wt. : 674
[α]$_D^{23}$: -31.5° (c=0.2, C_5D_5N)
Registry No. : [288255-54-9]

IR (KBr) **:** 3397 (br, OH), 2951 (C–H), 1453, 1102, 1067 cm^{-1}.

PMR (400/500 MHz) **:** δ 0.81 (s, 3xH-18), 1.03 (d, J=6.6 Hz, 3xH-27), 1.14 (d, J=6.9 Hz, 3xH-21), 1.57 (s, 3xH-19), 3.24 (s, OCH_3), 4.84 (d, J=7.5 Hz, H-1 of Glc).

CMR (100/125 MHz) **:** δ C-1) 77.96 (2) 68.28 (3) 75.66 (4) 67.46 (5) 78.23 (6) 30.42 (7) 28.44 (8) 34.49 (9) 45.43 (10) 45.12 (11) 21.63 (12) 39.97 (13) 41.02 (14) 56.18 (15) 32.20 (16) 81.32 (17) 64.37 (18) 16.47 (19) 13.76 (20) 40.57 (21) 16.28 (22) 112.72 (23) 31.02 (24) 28.25 (25) 34.87 (26) 74.97 (27) 17.57 (OCH_3) 47.37 **Glc** (1) 105.09 (2) 75.24 (3) 78.46 (4) 71.87 (5) 78.68 (6) 62.47.

Mass (FAB) : *m/z* 673 [M-H]$^-$, 511 [M-Glc-H]$^-$.

Reference

1.   Q.X. Yang and C.R. Yang, *Yunnan Zhiwu Yanjiu* (*Acta Botanica Yunnanica*), **22**, 109 (2000).

WATTOSIDE D
22ξ-Methoxy-25(*S*)-furost-1β,2β,3β,4β,5β,26β-hexol 26-O-β-D-glucopyranoside

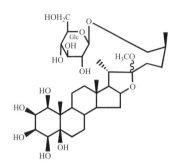

Source : *Tupistra wattii* Hook. f. (Liliaceae)
Mol. Formula : $C_{34}H_{58}O_{13}$
Mol. Wt. : 674
$[\alpha]_D^{20}$: -45.7° (c=0.26, MeOH)
Registry No. : [177910-41-7]

PMR (C₅D₅N, 400 MHz) **:** δ 0.90 (s, 3xH-18), 0.99 (d, *J*=6.6 Hz, 3xH-27), 1.30 (d, *J*=6.8 Hz, 3xH-21), 1.63 (s, 3xH-19), 4.80 (d, *J*=7.8 Hz, H-1 of Glc).

CMR (C₅D₅N, 100 MHz) **:** δ C-1) 78.3 (2) 67.5 (3) 75.7 (4) 68.3 (5) 78.1 (6) 30.4 (7) 28.5 (8) 35.0 (9) 45.5 (10) 45.2 (11) 21.7 (12) 40.2 (13) 41.1 (14) 56.3 (15) 32.5 (16) 81.3 (17) 64.0 (18) 16.8 (19) 13.9 (20) 40.8 (21) 16.4 (22) 112.7 (23) 30.8 (24) 28.4 (25) 34.5 (26) 75.2 (27) 17.5 (OCH₃) 47.4 **Glc** (1) 105.1 (2) 75.2 (3) 78.4 (4) 71.8 (5) 78.6 (6) 63.0.

Mass (FAB, Negative ion) **:** *m/z* 673 [M-H]⁻, 511 [M-Glc-H]⁻.

Reference

1. F. Yang, P. Shen, Y.-F. Wang, R.-Y. Zhang and C.-R. Yang, *Acta Botanica Yunnanica*, **23**, 373 (2001).

RUSCUS ACULEATUS SAPONIN 1
(25S)-Furost-5-en-1β,3β,22ξ,26-tetrol-1-sulphate 26-O-β-D-glucopyranoside

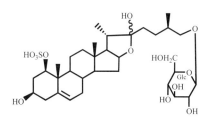

Source : *Ruscus aculeatus* L. (Liliaceae)
Mol. Formula : $C_{33}H_{54}O_{13}S$
Mol. Wt. : 690
$[\alpha]_D^{24}$ **: -86°** (c=0.01, MeOH)
Registry No. : [177747-12-5]

PMR (CD₃OD, 200.13 /300.13 MHz) : δ 0.89 (s, 3xH-18), 1.00 (d, *J*=6.6 Hz, 3xH-27), 1.05 (d, *J*=6.9 Hz, 3xH-21), 1.15 (s, 3xH-19), 4.29 (d, *J*=7.7 Hz, H-1 of Glc), 5.65 (d, *J*=5.1 Hz, H-6).

CMR (CD₃OD, 50.2/75.4 MHz) : δ C-1) 85.7 (2) 36.9 (3) 68.8 (4) 41.5 (5) 138.8 (6) 126.7 (7) 34.0 (8) 32.7 (9) 50.6 (10) 43.7 (11) 24.3 (12) 39.0 (13) 43.0 (14) 57.5 (15) 33.1 (16) 82.3 (17) 63.9 (18) 15.9 (19) 17.4 (20) 41.3 (21) 14.8 (22) 111.0 (23) 30.7 (24) 28.5 (25) 34.9 (26) 75.9 (27) 17.0 **Glc** (1) 104.6 (2) 75.1 (3) 78.1 (4) 71.6 (5) 77.8 (6) 62.7.

Mass (FAB, Positive ion) : *m/z* 713 [M+Na]⁺.

Reference

1. A.O.-Ali, D. Guillaume, R. Belle, B. David and R. Anton, *Phytochemistry*, **42**, 895 (1996).

NOLINOFUROSIDE G

(25S)-Furost-5,20(22)-diene-1β,3β-diol 1-sulfate 26-O-β-D-glucopyranoside monosodium salt

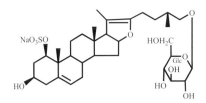

Source : *Nolina microcarpa* S. Wats. (Dracaenaceae)
Mol. Formula : $C_{33}H_{51}O_{12}SNa$
Mol. Wt. : 694
[α]$_D^{20}$: -38.0 ± 2° (c=1.01, Pyridine)
Registry No. : [144721-24-4]

IR (KBr) : 3200-3600 (OH), 1200-1300 (ester bond)cm^{-1}.

PMR (C_5D_5N) : δ 0.69 (s, 3xH-18), 0.99 (d, $J_{27,25}$=7.0 Hz, 3xH-27), 1.28 (s, 3xH-19), 1.53 (s, 3xH-21), 3.45 (m, H-26A), 3.93 (m, H-26B), 3.90 (m, H-3), 4.00 (m, H-2 and H-5 of Glc), 4.22 (m, H-3 of Glc), 4.22 (m, H-4 of Glc), 4.35 (dd, $J_{6A,6B}$=12.0 Hz, $J_{6A,5}$=5.0 Hz, H-6A of Glc), 4.51 ($J_{6B,5}$=2.5 Hz, H-6B of Glc), 4.77 (m, H-1), 4.79 (d, $J_{1,2}$=7.5 Hz, H-1 of Glc), 4.80 (m, H-16), 5.57 (br d, $J_{6,7}$=5.0 Hz, H-6).

CMR (C_5D_5N) : δ C-1) 85.17 (2) 39.67 (3) 67.97 (4) 43.53 (5) 138.78 (6) 125.57 (7) 32.41 (8) 33.78 (9) 49.96 (10) 43.18 (11) 23.91 (12) 40.57 (13) 43.18 (14) 55.00 (15) 32.86 (16) 84.57 (17) 64.84 (18) 14.45 (19) 14.78 (20) 103.73 (21) 11.76 (22) 152.47 (23) 23.72 (24) 31.49 (25) 34.74 (26) 75.16 (27) 17.24 **Glc** (1) 105.07 (2) 75.16 (3) 78.26 (4) 71.95 (5) 78.53 (6) 62.98.

Mass (L.S.I.) : *m/z* 717 [M+Na]$^+$, 597 [M+Na-NaHSO$_4$]$^+$, 695 [M+H]$^+$.

Reference

1. G.V. Shevchuk, Yu.S. Vollerner, A.S. Shashkov, M.B. Gorovits and Y.Ya. Chirva, *Khim. Prir. Soedin.*, **27**, 801 (1991); *Chem. Nat. Comp.*, **27**, 706 (1991).

ALLIUM VINEALE SAPONIN 2

Nuatigenin 3-O-[α-L-rhamnopyranosyl-(1→2)-β-D-glucopyranoside]

Source : *Allium vineale* (Liliaceae)
Mol. Formula : $C_{39}H_{62}O_{13}$
Mol. Wt. : 738
[α]$_D$: -39.6° (c=0.17, Pyridine)
Registry No. : [113561-09-4]

PMR ($C_5D_5N-D_2O$, 400 MHz) : δ 0.80 (s, CH_3), 1.04 (s, CH_3), 1.10 (d, *J*=6.9 Hz, CH_3), 1.37 (CH_3), 1.78 (d, *J*=6.2 Hz, 3xH-6 of Rha), 4.17 (t, *J*=9.4 Hz, H-4 of Glc), 3.85* (H-5 of Glc), 4.30* (H-2 and H-3 of Glc, H-6A of Glc), 4.37 (t, *J*=9.2 Hz, H-4 of Rha), 4.51 (d, *J*=11.4 Hz, H-6B of Glc), 4.63 (dd, *J*=9.2, 3.2 Hz, H-3 of Rha), 4.82 (br, H-2 of Rha), 5.00* (H-5 of Rha), 5.03 (d, *J*=7.3 Hz, H-1 of Glc), 5.29 (d, *J*=3.3 Hz, H-6), 6.38 (H-1 of Rha). * overlapped signal.

CMR ($C_5D_5N-D_2O$, 100 MHz) : δ C-1) 37.5 (2) 30.2 (3) 78.3 (4) 40.5 (5) 140.8 (6) 120.2 (7) 32.2 (8) 31.6 (9) 50.2 (10) 37.1 (11) 21.1 (12) 38.9 (13) 39.8 (14) 56.5 (15) 32.3 (16) 80.8 (17) 62.6 (18) 16.1 (19) 19.4 (20) 38.4 (21) 15.2 (22) 121.7 (23) 32.6 (24) 33.8 (25) 85.6 (26) 70.1 (27) 24.1 **Glc** (1) 100.3 (2) 79.7 (3) 77.9 (4) 71.8 (5) 77.8 (6) 62.5 **Rha** (1) 102.1 (2) 72.6 (3) 72.9 (4) 74.2 (5) 69.5 (6) 18.7.

Mass (FAB, Negative ion, H.R.) : *m/z* 737.41165 [(M-1)⁻, calcd. 737.41122].

Biological Activity : It showed molluscicidal at 50 ppm in <24h.

Reference

1.   S. Chen and J.K. Snyder, *J. Org. Chem.*, **54**, 3679 (1989).

CORDYLINE STRICTA SAPONIN 7
(25S) 5α-Furost-20 (20)-ene 1β,3α,26-triol 3,26-*bis*-O-β-D-glucopyranoside

Source : *Cordyline stricta* (Agavaceae)
Mol. Formula : $C_{39}H_{64}O_{14}$
Mol. Wt. : 756
$[\alpha]_D^{27}$: -10.0° (c=0.1, MeOH)
Registry No. : [194143-97-0]

IR (KBr) : 3400 (OH), 2920 (CH), 1445, 1370, 1155, 1075, 1030, 940, 895 cm^{-1}.

PMR (C$_5$D$_5$N, 400 MHz) : δ 0.76 (s, 3xH-18), 1.03 (d, J=6.6 Hz, 3xH-27), 1.06 (s, 3xH-19), 1.59 (s, 3xH-21), 4.82 (d, J=7.7 Hz, H-1 of Glc II), 4.96 (d, J=7.7 Hz, H-1 of Glc I).

CMR (C$_5$D$_5$N, 100 MHz) : δ C-1) 73.7 (2) 37.1 (3) 73.8 (4) 35.1 (5) 39.2 (6) 28.7 (7) 32.7 (8) 35.8 (9) 55.1 (10) 42.6 (11) 25.0 (12) 40.6 (13) 43.4 (14) 55.1 (15) 34.7 (16) 84.4 (17) 64.9 (18) 14.6 (19) 6.5 (20) 103.7 (21) 11.8 (22) 152.3 (23) 31.4 (24) 23.6 (25) 33.7 (26) 75.2 (27) 17.2 (OCH$_3$) 47.3 **Glc I** (1) 102.6 (2) 75.3 (3) 78.7 (4) 71.7 (5) 78.4 (6) 62.9 **Glc II** (1) 105.2 (2) 75.2 (3) 78.6 (4) 71.7 (5) 78.5 (6) 62.9.

Mass (FAB, Negative ion) : *m/z* 756 [M]$^-$.

Reference

1.   Y. Mimaki, Y. Takaashi, M. Kuroda and Y. Sashida, *Phytochemistry*, **45**, 1229 (1997).

NOLINOFUROSIDE C

(25S)-Furost-5-ene-1β,3β,22α,26-tetraol 1-O-β-D-fucopyranoside-26-O-β-D-glucopyranoside

Source : *Nolina microcarpa* S. Wats. (Dracaenacea)
Mol. Formula : $C_{39}H_{64}O_{14}$
Mol. Wt. : 756
$[\alpha]_D^{20}$ **:** -38.0 ± 2° (c=1.2, Pyridine)
Registry No. : [145603-90-3]

IR (KBr) **:** 3200-3600 (OH), 915 (br, weak) cm^{-1}.

PMR (C_5D_5N, 250/300 MHz) **:** δ 0.84 (s, 3xH-18), 0.97 (d, $J_{27,25}$=6.5 Hz, 3xH-27), 1.20 (s, 3xH-19), 1.20 (d, $J_{21,20}$=6.5 Hz, 3xH-21),1.52 (d, $J_{6,5}$=6.5 Hz, 3xH-6 of Fuc), 3.42 (dd, $J_{26,25}$=6.5 Hz, H-26a), 3.68 (dq, $J_{5,6}$=6.5 Hz, H-5 of Fuc), 3.77 (dd, $J_{1a,2a}$=11.5 Hz, $J_{1a,2e}$=4.0 Hz, H-3A), 3.86 (m, H-3), 3.88 (m, H-5 of Glc), 3.96 (dd, $J_{2,3}$=9.0 Hz, H-2 of Glc), 4.00 (dd, $J_{4,5}$=1.5 Hz, H-4 of Fuc), 4.02 (dd, $J_{26B,25}$=6.0 Hz, $J_{26A,26B}$=9.5 Hz, H-26B), 4.04 (dd, $J_{3,4}$=3.5 Hz, H-3 of Fuc), 4.15 (t, $J_{4,5}$=9.0 Hz, H-4 of Glc), 4.21 (t, $J_{3,4}$=9.0 Hz, H-3 of Glc), 4.28 (dd, $J_{2,3}$=9.0 Hz, H-2 of Fuc), 4.30 (dd, $J_{6,5}$=5.5 Hz, $J_{6A,6B}$=11.5 Hz, H-6A of Glc), 4.48 (dd, $J_{6,5}$=2.5, H-6B of Glc), 4.69 (d, $J_{1,2}$=7.5 Hz, H-1 of Fuc), 4.74 (d, $J_{1,2}$=7.5 Hz, H-1 of Glc), 4.89 (m, H-16), 5.52 (br d, $J_{6,7}$=5.0 Hz, H-6 of Glc).

CMR (C_5D_5N, 62.5/75 MHz) **:** δ C-1) 83.98 (2) 38.11 (3) 68.21 (4) 43.81 (5) 139.72 (6) 124.85 (7) 32.09 (8) 33.10 (9) 50.60 (10) 42.94 (11) 23.88 (12) 40.63 (13) 40.63 (14) 57.09 (15) 32.75 (16) 81.25 (17) 64.04 (18) 16.41 (19) 14.91 (20) 40.80 (21) 14.91 (22) 110.82 (23) 37.17 (24) 28.34 (25) 34.48 (26) 75.46 (27) 17.07 **Fuc** (1) 102.53 (2) 72.25 (3) 75.46 (4) 72.59 (5) 71.30 (6) 17.46 **Glc** (1) 105.17 (2) 75.26 (30 78.48 (4) 71.78 (5) 78.63 (6) 62.88.

Mass : *m/z* 738 [M-H_2O]⁺.

Reference

1. G.V. Shevchuk, Yu.S. Vollerner, A.S. Shaskov and Y.Ya. Chirva, *Khim. Prir. Soedin.*, **27, 67**2 (1991); *Chem. Nat. Comp.*, **27**, 592 (1991).

ANEMARRHENASAPONIN-I

3β,15α,22ξ-Trihydroxy-5β-furostane 3-O-[β-D-glucopyranosyl-(1→2)-β-D-galactopyranoside]

Source : *Anemarrhena asphodeloides* Bunge (Liliaceae)
Mol. Formula : $C_{39}H_{66}O_{14}$
Mol. Wt. : 758
M.P. : 202-204°C[1]
[α]$_D$: -41.7° (c=1.08, C_5D_5N)[1]
Registry No. : [163047-21-0]

PMR (C_5D_5N, 500 MHz)[1] : δ 0.89 (d, *J*=7.3 Hz, 3xH-26), 0.90 (d, *J*=7.3 Hz, 3xH-27), 0.97 (s, 3xH-18), 1.02 (s, 3xH-19), 1.33 (d, *J*=6.8 Hz, 3xH-21), 2.00-3.00 (H-17), 3.78 (m, H-5 of Glc), 3.79-4.00 (2xH-6 of Glc), 3.98 (dd, *J*=6.7, 6.7 Hz, H-5 of Gal), 4.01 (dd, *J*=8.7, 7.7 Hz, H-2 of Glc), 4.13 (dd, *J*=9.1, 8.7 Hz, H-3 of Glc), 4.20-4.36 (2xH-6 of Gal), 4.22 (dd, *J*=8.6, 3.1 Hz, H-3 of Gal), 4.25 (dd, *J*=9.4, 9.1 Hz, H-4 of Glc), 4.26 (br s, H-23), 4.52 (d, *J*=3.1 Hz, H-4 of Gal), 4.60 (dd, *J*=8.6, 7.6 Hz, H-2 of Gal), 4.85 (d, *J*=7.6 Hz, H-1 of Gal), 5.05 (H-15), 5.05 (dd, *J*=8.8, 3.6 Hz, H-16), 5.21 (d, *J*=7.7 Hz, H-1 of Glc).

CMR (C_5D_5N, 125 MHz) : δ C-1) 30.6[a] (2) 27.1[b] (3) 75.4 (4) 31.0[a] (5) 36.3 (6) 26.9[b] (7) 26.7[b] (8) 36.9 (9) 40.2 (10) 35.2 (11) 21.1 (12) 37.6 (13) 41.2 (14) 60.8 (15) 78.0 (16) 91.2 (17) 61.3 (18) 18.0 (19) 24.1 (20) 40.6 (21) 16.5 (22) 110.3 (23) 35.6 (24) 24.1 (25) 28.1 (26) 22.8 (27) 22.7 **Gal** (1) 102.2 (2) 81.6 (3) 75.0 (4) 69.7 (5) 76.4 (6) 62.0 **Glc** (1) 105.9 (2) 76.7 (3) 77.8 (4) 71.5 (5) 78.2 (6) 62.6.

Mass (FAB, Positive ion)[1] : *m/z* 781 [M+Na]$^+$.

Reference

1.   S. Saito, S. Nagase and K. Ichinose, *Chem. Pharm. Bull.*, **42**, 2342 (1994).

ANEMARRHENASAPONIN-II

3β,15β,22ξ-Trihydroxy-5β-furostane 3-O-[β-D-glucopyranosyl-(1→2)-β-D-galactopyranoside]

Source : *Anemarrhena asphodeloides* Bunge (Liliaceae)
Mol. Formula : $C_{39}H_{66}O_{14}$
Mol. Wt. : 758
M.P. : 174-176°C
$[\alpha]_D$: -39.1° (c=0.95, C_5D_5N)
Registry No. : [163047-22-1]

PMR (C_5D_5N, 500 MHz) : δ 0.92 (s, 3xH-18), 0.92 (d, *J*=7.3 Hz, 3xH-26 and 3xH-27), 1.02 (s, 3xH-19), 1.33 (d, *J*=6.8 Hz, 3xH-21), 2.00-3.00* (H-17), 3.81 (m, H-5 of Glc), 3.99 (dd, *J*=6.7, 6.7 Hz, H-5 of Gal), 4.00-4.35* (H-15), 4.05 (dd, *J*=8.9, 7.6 Hz, H-2 of Glc), 4.16 (dd, *J*=9.2, 8.9 Hz, H-3 of Glc), 4.23 (dd, *J*=9.1, 3.0 Hz, H-3 of Gal), 4.25-4.45 (2xH-6 of Glc), 4.28 (br s, H-3), 4.28 (dd, *J*=9.2, 9.2 Hz, H-4 of Glc), 4.35-4.45 (2xH-6 of Gal), 4.54 (d, *J*=3.0 Hz, H-4 of Gal), 4.63 (dd, *J*=9.1, 7.6 Hz, H-2 of Gal), 4.87 (d, *J*=7.6 Hz, H-1 of Gal), 5.07 (dd, *J*=7.8, 2.7 Hz, H-16), 5.24 (d, *J*=7.6 Hz, H-1 of Glc). * overlapped signals.

CMR (C_5D_5N, 125 MHz) : δ C-1) 30.6[a] (2) 27.1[b] (3) 75.4 (4) 31.0[a] (5) 36.2 (6) 26.9[b] (7) 26.7[b] (8) 36.9 (9) 40.2 (10) 35.4 (11) 21.1 (12) 36.9 (13) 41.2 (14) 60.5 (15) 79.0 (16) 92.4 (17) 61.5 (18) 17.7 (19) 24.1 (20) 40.3 (21) 15.9 (22) 110.3 (23) 35.3 (24) 24.1 (25) 28.7 (26) 22.8[c] (27) 22.7[c] **Gal** (1) 102.4 (2) 81.7 (3) 75.1 (4) 69.7 (5) 76.6 (6) 62.1 **Glc** (1) 106.0 (2) 76.9 (3) 77.9 (4) 71.6 (5) 78.3 (6) 62.7.

Mass (FAB, Positive ion) : *m/z* 781 [M+Na]⁺.

Reference

1.   S. Saito, S. Nagase and K. Ichinose, *Chem. Pharm. Bull.,* **42**, 2342 (1994).

FILICINOSIDE B

22(*S*)-5β-Furostan-3β,22α,26-triol 3-O-β-D-glucopyranosyl-26-O-β-D-glucopyranoside

Source : *Asparagus filicinus* Buch.-Ham. (Liliaceae)
Mol. Formula : $C_{39}H_{66}O_{14}$
Mol. Wt. : 758
M.P. : 186-193°C
[α]$_D^{20}$: -46.7° (C_5D_5N)
Registry No. : [156857-60-2]

Reference

1.   S.C. Sharma and N.K. Thakur, *Phytochemistry*, **36**, 469 (1994).

PETUNIOSIDE I

(25R)-5α-Furostan-3β,22α,26-triol 3-O-β-D-galactopyranoside-26-O-β-D-glucopyranoside

Source : *Petunia hybrida* L. (Solanaceae)
Mol. Formula : $C_{39}H_{66}O_{14}$
Mol. Wt. : 758
M.P. : 176°C
$[α]_D^{20}$ **:** -68° (c=1.0, CH_3OH)
Registry No. : [174693-28-8]

IR : 3300, 3020, 968, 920 < 900, 856 cm^{-1}.

Reference

1.   S.A. Shvets, A.M. Naibi and P.K. Kintya, *Khim. Prir. Soedin.*, **31**, 247 (1995), *Chem. Nat. Comp.*, **31**, 203 (1995).

PROTOTOKORONIN

(5α,25R)-Furost-1β,2β,3α,22ξ,26-pentol 1-O-α-arabinopyranoside-26-O-β-D-glucopyranoside

Source : Cell cultures of *Dioscorea tokoro* Makino[1,2] (Dioscoreaceae)
Mol. Formula : $C_{38}H_{64}O_{15}$
Mol. Wt. : 760
M.P. : 177-180°C
$[α]_D$ **:** -3.8° (MeOH)
Registry No. : [52989-08-9]

PMR (C$_5$D$_5$N, 90.0 MHz, 80°C)[2] : δ 0.88 (s, 3xH-18), 0.99 (d, *J*=7.0 Hz, 3xH-27), 1.26 (d, *J*=7.0 Hz, 3xH-21), 1.35 (s, 3xH-19), 1.95 (H-25), 3.60 (H-26A), 4.03 (H-26B).

CMR (C$_5$D$_5$N, 25.16/50.18 MHz, 30°C)[2] : δ C-1) 88.9 (2) 74.9 (3) 71.6 (4) 34.9 (5) 36.5 (6) 26.2 (7) 26.5 (8) 35.7 (9) 42.4 (10) 41.7 (11) 21.1 (12) 40.3 (13) 41.1 (14) 56.3 (15) 32.3 (16) 81.1 (17) 63.9 (18) 16.6 (19) 19.1 (20) 40.7 (21) 16.1 (22) 110.8 (23) 36.9 (24) 28.3 (25) 34.2 (26) 75.2 (27) 17.4 **Ara** (1) 107.6 (2) 73.8 (3) 75.0 (4) 69.6 (5) 67.3 **Glc** (1) 104.6 (2) 75.0 (3) 78.4 (4) 72.0 (5) 77.9 (6) 63.0.

References

1. Y. Tomita and A. Uomori, *Phytochemistry*, **13**, 729 (1974).

2. S. Seo, A. Uomori, Y. Yoshimura and K. Tori., *J. Chem. Soc. Perkin Trans. I*, 869 (1984).

PROTONEOTOKORIN
(25*S*)-Furost-1β,2β,3α,22ξ,26-pentol 1-O-α-L-arabinopyranoside-26-O-β-D-glucopyranoside

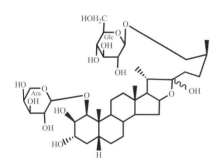

Source : Cell cultures of *Dioscorea tokoro* Makino (Dioscoreaceae)
Mol. Formula : C$_{38}$H$_{64}$O$_{15}$
Mol. Wt. : 760
Registry No. : [87562-59-2]

Reference

1. S. Seo, A. Uomori, Y. Yoshimura and K. Tori., *J. Chem. Soc. Perkin Trans. I*, 869 (1984).

REINECKIA CARNEA SAPONIN R$_S$-2

(25R)-22ξ-Methoxy-5β-furostane-1β,3β,4β,5β-26-pentaol 4-O-sulfate 26-O-β-D-glucopyranoside (sodium salt)

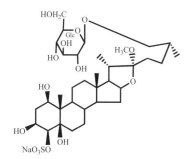

Source : *Reineckia carnea* Kunth (Liliaceae)
Mol. Formula : C$_{34}$H$_{57}$O$_{15}$SNa
Mol. Wt. : 760
M.P. : 243-247°C (decomp.)
[α]$_D^{36}$: -32.6 (c=0.95, MeOH)
Registry No. : [108382-80-5]

IR (KBr) : 3500-3000 (OH), 1250, 865 (S-O) cm^{-1}.

PMR (C$_5$D$_5$N, 80 MHz) : δ 0.75 (s, 3xH-18), 1.02 (d, *J*=7.0 Hz, 3xH-21), 1.03 (d, *J*=7.0 Hz, 3xH-27), 1.50 (s, 3xH-19), 3.25 (s, OC*H*$_3$-22), 4.75 (d, *J*=6.0 Hz, anomeric H).

CMR (20 MHz) : C-1) 73.8 (2) 34.3 (3) 68.2 (4) 75.1^a (5) 78.2 (6) 30.6 (7) 28.3 (8) 35.0 (9) 45.8 (10) 46.0 (11) 21.5 (12) 40.5 (13) 40.8 (14) 55.7 (15) 32.3 (16) 81.1 (17) 64.2 (18) 16.3 (19) 13.7 (20) 40.8 (21) 16.4 (22) 112.8 (23) 34.3 (24) 28.3 (25) 31.2 (26) 72.0 (27) 17.2 (OCH$_3$) 47.3 **Glc** (1) 104.8 (2) 75.4 (3) 78.4 (4) 71.7 (5) 78.4 (6) 62.8.

Reference

1.   K. Iwagoe, T. Konishi and S. Kiyosawa, *Yakugaku Zasshi*, **107**, 140 (1987).

DOWNEYOSIDE A

3β,20(*R*),23(*S*)-Trihydroxy-24(*R*)-methyl-6α-sulfooxy-5α-furost-9(11)-ene
3-O-β-D-glucuronopyranoside disodium salt

Source : *Henricia downeyae* (Echinasteridae, Starfish)
Mol. Formula : $C_{34}H_{52}O_{14}SNa_2$
Mol. Wt. : 762
[α]$_D$: -13.3° (MeOH)
Registry No. : [170894-36-7]

IR (FT) : 1700 (COONa), 1240 (OSO$_3$Na) cm^{-1}.

PMR (CD$_3$OD, 500 MHz) : δ 0.89 (d, *J*=6.8 Hz, 3xH-27), 0.94 (d, *J*=7.2 Hz, 3xH-28), 0.95 (d, *J*=6.8 Hz, 3xH-26), 1.07 (m, H-7A), 1.08 (s, 3xH-19), 1.18 (s, 3xH-18), 1.23 (H-14), 1.26 (m, H-5), 1.32 (m, H-4A), 1.41 (H-16A), 1.41 (s, 3xH-21), 1.50 (m, H-1A), 1.51 (m, H-24), 1.65 (H-2A), 1.75 (m, H-1B), 2.01 (dd, *J*=5.2, 10.0 Hz, H-12A), 2.02 (m, H-25), 2.07 (m, H-2B), 2.24 (m, H-8 and H-16B), 2.28 (dd, *J*=5.2, 10.0 Hz, H-12B), 2.40 (H-4B), 2.60 (m, H-7B), 3.19 (t, *J*=7.8, 8.0 Hz, H-2 of GlcUA), 3.45 (H-3 and H-4 of GlcUA), 3.60 (H-5 of GlcUA), 3.64 (dd, *J*=9.0, 4.6 Hz, H-23), 3.74 (m, H-3), 3.99 (d, *J*=9.0 Hz, H-22), 4.36 (dt, *J*=4.4, 10.8 Hz, H-6), 4.49 (m, H-17), 4.51 (d, *J*=7.8 Hz, H-1 of GlcUA), 5.41 (br d, *J*=5.2 Hz, H-11).

CMR (CD$_3$OD, 125 MHz) : δ C-1) 36.9 (2) 30.1 (3) 78.9 (4) 30.2 (5) 49.0 (6) 78.4 (7) 40.7 (8) 35.5 (9) 146.4 (10) 39.9 (11) 118.0 (12) 42.7 (13) 42.3 (14) 54.1 (15) 37.3 (16) 83.6 (17) 67.5 (18) 14.4 (19) 19.7 (20) 81.5 (21) 26.2 (22) 83.2 (23) 75.0 (24) 44.6 (25) 28.4 (26) 22.7 (27) 18.9 (28) 11.2 **GlcUA** (1) 102.2 (2) 75.1 (3) 78.7 (4) 73.7 (4) 75.8 (6) 177.0.

Mass (FAB, Negative ion) : *m/z* 739 [M-Na]$^-$, 717 [M+H-2Na]$^-$.

Biological Activity: Cytotoxic against non-small-cell lung human carcinoma cell with IC$_{50}$ of 60 μg/ml.

Reference

1.   E. Palagiano, F. Zollo, L. Minale, L.G. Paloma, M. Iorizzi, P. Bryan, J. Mc Clintock, T. Hopkins, D. Riou and C. Roussakis, *Tetrahedron*, **51**, 12293 (1995).

DOWNEYOSIDE B

3β,20(R),23(S)-Trihydroxy-24(S)-methyl-6α-sulfooxy-5α-furost-9(11)-ene
3-O-β-D-glucuronopyranoside disodium salt

Source : *Henricia downeyae* (Echinasteridae, Starfish)
Mol. Formula : $C_{34}H_{52}O_{14}SNa_2$
Mol. Wt. : 762
$[\alpha]_D$ **:** -18.5° (MeOH)
Registry No. : [170894-37-8]

PMR (CD₃OD, 500 MHz) : δ 0.90 (d, *J*=6.5 Hz, 3xH-28), 0.94 (d, *J*=6.5 Hz, 3xH-26), 0.98 (d, *J*=6.8 Hz, 3xH-27), 1.07 (m, H-7), 1.08 (s, 3xH-19), 1.18 (s, 3xH-18), 1.23 (H-14), 1.26 (m, H-5), 1.32 (m, H-4A), 1.41 (H-16A), 1.42 (s, 3xH-21), 1.50 (H-1A), 1.51 (H-24), 1.65 (H-2A), 1.75 (m, H-1B), 2.01 (dd, *J*=5.2, 10.0 Hz, H-12A), 2.02 (m, H-25), 2.07 (m, H-2), 2.24 (m, H-8), 2.24 (H-16B), 2.28 (dd, *J*=5.2, 10.0 Hz, H-12B), 2.40 (H-4), 2.60 (m, H-7), 3.19 (t, *J*=7.8, 8.0 Hz, H-2 of GlcUA), 3.45 (H-3 and H-4 of GlcUA), 3.60 (H-5 of GlcUA), 3.74 (m, H-3), 3.80 (dd, *J*=9.3, 1.3 Hz, H-23), 3.93 (d, *J*=9.3 Hz, H-22), 4.36 (dt, *J*=4.4, 10.8 Hz, H-6), 4.49 (m, H-17), 4.51 (d, *J*=7.8 Hz, H-1 of GlcUA), 5.41 (br d, *J*=5.2 Hz, H-11).

CMR (CD₃OD, 125 MHz) : δ C-1) 36.9 (2) 30.1 (3) 78.9 (4) 30.2 (5) 49.0 (6) 78.4 (7) 40.7 (8) 35.6 (9) 146.4 (10) 40.0 (11) 118.0 (12) 42.5 (13) 42.3 (14) 54.2 (15) 37.2 (16) 83.5 (17) 68.1 (18) 14.4 (19) 19.7 (20) 81.2 (21) 26.1 (22) 82.9 (23) 72.8 (24) 42.6 (25) 31.3 (26) 21.1 (27) 21.6 (28) 10.0 **GlcUA** (1) 102.2 (2) 75.1 (3) 78.7 (4) 73.7 (5) 75.8 (6) 177.0.

Mass (FAB, Negative ion) : *m/z* 717 [M+H-2Na]⁻, 541 [M+H-2Na-GlcUA]⁻.

Biological Activity : Cytotoxic against non-small-cell lung human carcinoma cell with IC₅₀ of 36 μg/ml.

Reference

1. E. Palagiano, F. Zollo, L. Minale, L.G. Paloma, M. Iorizzi, P. Bryan, J. Mc Clintock, T. Hopkins, D. Riou and C. Roussakis, *Tetrahedron*, **51**, 12293 (1995).

COSTUS SPICATUS SAPONIN 1

3β,26-Dihydroxy-22α-methoxy-25R-furost-5-ene 3,26-*bis*-O-β-D-glucopyranoside

Source : *Costus spicatus* Swartz (Costaceae)
Mol. Formula : $C_{40}H_{66}O_{14}$
Mol. Wt. : 770
M.P. : 254-256°C
[α]$_D$: -115.0° (c=0.001, MeOH)
Registry No. : [223713-68-6]

IR (KBr) : 3430 (OH), 1053 (C-O), 913, 838 cm^{-1}.

PMR (C$_5$D$_5$N, 200 MHz) : δ 0.84 (s, 3xH-18), 0.98 (d, J=6.6 Hz, 3xH-27), 1.06 (s, 3xH-19), 1.20 (d, J=6.8 Hz, 1xH-21), 3.28 (s, 22-OCH$_3$), 4.82 (d, J=7.8 Hz, H-1 of Glc II), 4.91 (d, J=7.7 Hz, H-1 of Glc I), 5.30 (br s, H-6).

CMR (C$_5$D$_5$N, 50 MHz) : δ C-1) 37.5 (2) 29.9 (3) 77.8 (4) 38.7 (5) 140.5 (6) 121.5 (7) 32.2 (8) 31.7 (9) 50.3 (10) 36.9 (11) 20.9 (12) 39.8 (13) 40.5 (14) 56.3 (15) 32.1 (16) 80.8 (17) 63.7 (18) 16.3 (19) 18.9 (20) 40.5 (21) 16.8 (22) 112.7 (23) 30.5 (24) 27.9 (25) 33.8 (26) 74.9 (27) 17.1 (OCH$_3$) 47.1 **Glc I** (1) 102.6 (2) 75.4 (3) 78.7 (4) 71.8 (5) 78.4 (6) 62.9 **Glc II** (1) 105.1 (2) 75.2 (3) 78.6 (4) 71.8 (5) 78.5 (6) 62.9.

Mass (LSI-MS) : *m/z* 753 [M-OH]$^-$.

Note: The published assignment of peak at *m/z* 753 as [M-H]$^-$ peak is erroneous.

Reference

1.   B.P. Silva, R.R. Bernardo, J.P. Parente. *Fitoterapia*, **69**, 528 (1998).

DRACAENA SURCULOSA SAPONIN 5

(25S)-22α-Methoxy-1β,6β,26-trihydroxy-3α,5α-cyclofurostane 1-O-β-D-fucopyranoside-26-O-β-D-glucopyranoside

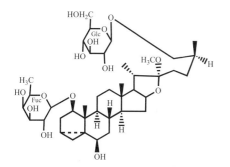

Source : *Dracaena surculosa* Lindle. (Agavaceae)
Mol. Formula : $C_{40}H_{66}O_{14}$
Mol. Wt. : 770
$[\alpha]_D^{26}$ **:** -56.0° (c=0.10, MeOH)
Registry No. : [463963-01-1]

IR (film) : 3388 (OH), 2930 (CH), 1074 cm^{-1}.

CMR (C_5D_5N, 500 MHz) : δ 0.57 (dd, *J*=7.9, 3.9 Hz, H-4A), 0.90 (s, 3xH-18), 1.06 (d, *J*=6.7 Hz, 3xH-27), 1.14 (m, H-3), 1.19 (d, *J*=6.9 Hz, 3xH-21), 1.51 (t-like, *J*=3.9 Hz, H-4B), 1.56 (d, *J*=6.4 Hz, 3xH-6 of Fuc), 1.64 (s, 3xH-19), 3.26 (s, OC*H₃*), 3.47 (br s, H-6), 4.37 (br d, *J*=5.0 Hz, H-1), 4.52 (q-like, *J*=7.2 Hz, H-16), 4.58 (d, *J*=7.8 Hz, H-1 of Fuc), 4.85 (d, *J*=7.8 Hz, H-1 of Glc).

CMR (C_5D_5N, 125 MHz) : δ C-1) 84.8 (2) 33.0 (3) 23.3 (4) 15.8 (5) 40.0 (6) 73.0 (7) 38.6 (8) 30.0 (9) 50.2 (10) 49.2 (11) 23.3 (12) 40.5 (13) 41.4 (14) 56.5 (15) 32.2 (16) 81.4 (17) 64.3 (18) 16.7 (19) 16.9 (20) 40.5 (21) 16.3 (22) 112.7 (23) 30.9 (24) 28.2 (25) 34.5 (26) 74.9 (27) 17.6 (OC*H₃*) 47.3 **Fuc** (1) 103.4 (2) 72.0 (3) 75.7 (4) 72.8 (5) 71.4 (6) 17.3 **Glc** (1) 105.1 (2) 75.2 (3) 78.6 (4) 71.8 (5) 78.5 (6) 62.9.

Mass (FAB, Negative ion) : *m/z* 769 [M-H]⁻.

Reference

1.   A. Yokosuka, Y. Mimaki and Y. Sashida, *Chem. Pharm. Bull.*, **50**, 992 (2002).

NOLINOFUROSIDE B

Nolinogenin 6β-Methoxy-3β,5-cyclo-(25S)-furost-an-1β,22α,26-triol
1-O-β-D-fucopyranoside-26-O-[β-D-glucopyranoside]

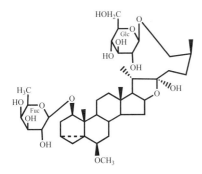

Source : *Nolina microcarpa* S. Wats. (Dracaenaceae)
Mol. Formula : $C_{40}H_{66}O_{14}$
Mol. Wt. : 770
$[\alpha]_D^{22}$: -44.2 ± 2° (c=1.50, Pyridine)
Registry No. : [402732-23-4]

IR (KBr) **:** 3200-3600 (OH), 910 cm^{-1}.

PMR (C_5D_5N, 250 MHz) **:** δ 0.50 (ddd, $J_{4,4}$=4.5 Hz, $J_{4,3}$=8.0 Hz, H-4), 0.91 (s, 3xH-18), 0.94 (m, H-3), 1.01 (d, $J_{27,25}$=6.5 Hz, 3xH-27), 1.31 (d, $J_{21,20}$=7.0 Hz, 3xH-21), 1.42 (s, 3xH-19), 1.54 (d, 3xH-6 of Fuc), 2.68 (m, H-6), 3.24 (s, 3xH-19), 3.46 (dd, $J_{26A,26B}$=9.5, $J_{26A,25}$=7.0 Hz, H-26eq), 3.75 (dq, $J_{5,6}$=6.5, H-5 of Fuc), 3.94 (m, H-5 of Glc), 4.01 (m, H-2 of Glc), 4.03 (m, H-4 of Fuc), 4.04 (dd, $J_{26B,25}$=5.5, H-26ax), 4.04 (m, H-3 of Fuc), 4.22 (m, H-4 of Glc), 4.24 (m, H-3 of Glc), 4.28 (m, H-2 of Fuc), 4.32 (m, H-1), 4.38 (dd, $J_{6A,6B}$=12.0, $J_{6,5}$=5.5 Hz, H-6A of Glc), 4.53 (d, $J_{1,2}$=7.5, H-1 of Fuc), 4.54 (dd, $J_{6,5}$=3.0 Hz, H-6B of Glc), 4.80 (d, $J_{1,2}$=7.5 Hz, H-1 of Glc), 4.97 (m, H-16).

CMR (C_5D_5N, 62.5 MHz) **:** δ C-1) 84.48 (2) 32.71 (3) 20.93 (4) 16.88 (5) 49.51 (6) 82.55 (7) 35.73 (8) 30.52 (9) 50.29 (10) 36.51 (11) 23.34 (12) 40.71 (13) 41.43 (14) 56.47 (15) 32.44 (16) 81.22 (17) 64.09 (18) 16.88 (19) 16.36 (20) 40.71 (21) 15.87 (22) 110.69 (23) 37.09 (24) 28.36 (25) 34.43 (26) 75.37 (27) 17.26 (6-OCH$_3$) 56.47 **Fuc** (1) 103.20 (2) 72.12 (3) 75.68 (4) 72.79 (5) 71.45 (6) 17.49 **Glc** (1) 105.13 (2) 75.24 (3) 78.38 (4) 71.86 (5) 78.62 (6) 62.96.

Mass (L.S.I.) **:** *m/z* 815 [M+2Na-H]$^+$.

Reference

1. G.V. Shevchuk, Yu. S. Vollerner, A.S. Shashkov, M.B. Gorovits and V. Ya. Chirva, *Khim. Prir. Soedin.*, 218 (1992); *Chem. Nat. Comp.*, **28**, 187 (1992).

TORVOSIDE E

(25*S*)-6α,26-Dihydroxy-22α-methoxy-5α-furostan-3-one 6-*O*-[β-D-quinovopyranoside]-26-*O*-β-D-glucopyranoside

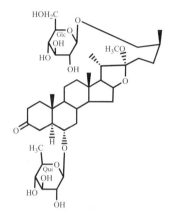

Source : *Solanum torvum* Swartz (Solanaceae)
Mol. Formula : $C_{40}H_{66}O_{14}$
Mol. Wt. : 770
$[α]_D^{27}$: -27.2° (c=2.70, MeOH)
Registry No. : [184777-16-0]

IR (KBr) : 3432 (OH), 1710 (C=O) cm^{-1}.

PMR (C_5D_5N, 400 MHz) **:** δ 0.83 (s, 3xH-18), 1.00 (s, 3xH-19), 1.07 (d, *J*=6.6 Hz, 3xH-27), 1.18 (d, *J*=6.6 Hz, 3xH-21), 1.66 (d, *J*=5.9 Hz, 3xH-6 of Qui), 3.23 (s, 3xOC*H*₃), 4.82 (d, *J*=8.1 Hz, H-1 of Glc), 4.87 (d, *J*=8.1 Hz, H-1 of Qui).

CMR (C_5D_5N, 100 MHz) **:** δ C-1) 38.7 (2) 38.1 (3) 210.7 (4) 39.9 (5) 52.4 (6) 79.8 (7) 41.1 (8) 34.5 (9) 53.2 (10) 36.8 (11) 21.3 (12) 39.7 (13) 41.4 (14) 56.0 (15) 32.1 (16) 81.2 (17) 64.2 (18) 16.5 (19) 12.6 (20) 40.5 (21) 16.3 (22) 112.6 (23) 30.9 (24) 28.2 (25) 34.0 (26) 75.0 (27) 17.6 (OCH₃) 47.3 **Qui** (1) 105.9 (2) 75.8 (3) 78.3 (4) 76.8 (5) 72.8 (6) 18.9 **Glc** (1) 105.1 (2) 75.2 (3) 78.7 (4) 71.8 (5) 78.5 (6) 62.9.

Mass (FAB, Positive ion) **:** *m/z* 739 [M-OMe]⁺, 577 [M-OMe-Glc]⁺.

Reference

1.   S. Yahara, T. Yamashita, N. Nozawa (nee Fujimura) and T. Nohara, *Phytochemistry*, **43**, 1069 (1996).

AGAVE AMERICANA SAPONIN 1

(25*R*)-3β,6α-Dihroxy-5α-spirostan-12-one 3,6-di-O-[β-D-glucopyranoside]

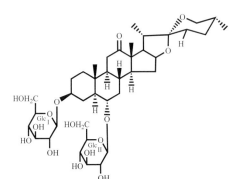

Source : *Agave americana* L. (Agavceae)
Mol. Formula : $C_{39}H_{62}O_{15}$
Mol. Wt. : 770
[α]$_D^{25}$: -57.1° (c=0.11, CHCl₃-MeOH)
Registry No. : [284677-34-5]

IR (film) : 3381 (OH), 2925 and 2865 (CH), 1686 (C=O), 1077, 1041 cm⁻¹.

PMR (C₅D₅N) : δ 0.72 (d, *J*=5.8 Hz, 3xH-27), 0.76 (s, 3xH-19), 0.78 (H-1ax), 0.95 (H-9), 1.05 (s, 3xH-18), 1.16 (H-7ax), 1.23 (H-5), 1.32 (H-1eq), 1.34 (d, *J*=6.9 Hz, 3xH-21), 1.37 (H-14), 1.45 (q-like, *J*=12.1 Hz, H-4ax), 1.54 (H-15β), 1.56 (2xH-24), 1.57 (H-25), 1.61 (H-23ax), 1.63 (H-2ax), 1.68 (H-23eq), 1.90 (H-20), 1.92 (H-8), 2.01 (H-2eq), 2.03 (H-15α), 2.22 (dd, *J*=13.9, 5.0 Hz, H-11eq), 2.37 (dd, *J*=13.9, 13.9 Hz, H-11ax), 2.65 (H-7eq), 2.70 (dd, *J*=8.5, 6.7 Hz, H-17), 3.40 (br d, *J*=12.1 Hz, H-4eq), 3.46 (dd, *J*=10.8, 3.3 Hz, H-26ax), 3.58 (dd, *J*=10.8, 10.8 Hz, H-26ax), 3.69 (ddd, *J*=10.8, 10.8, 4.7 Hz, H-6), 3.84 (ddd, *J*=8.8, 5.3, 2.4 Hz, H-5 of Glc I), 3.94 (H-3), 3.95 (ddd, *J*=8.8, 5.3, 2.5 Hz, H-5 of Glc II), 4.04 (dd, *J*=8.1, 7.6 Hz, H-2 of Glc II), 4.05 (dd, *J*=8.3, 7.6 Hz, H-2 of Glc I), 4.23 (dd, *J*=8.8, 8.3 Hz, H-3 of Glc I), 4.24 (dd, *J*=8.8, 8.8 Hz, H-4 of Glc II), 4.26 (dd, *J*=8.8, 8.8 Hz, H-4 of Glc I), 4.26 (dd, *J*=8.8, 8.8 Hz,, H-3 of Glc II), 4.32 (dd, *J*=11.8, 2.4 Hz, H-6A of Glc I), 4.36 (q-like, *J*=6.7 Hz, H-16), 4.41 (dd, *J*=11.6, 5.3 Hz, H-6A Glc II), 4.42 (dd, *J*=11.8, 2.4 Hz, H-6B of Glc I), 4.90 (d, *J*=7.8 Hz, H-1 of Glc II), 5.10 (d, *J*=7.6 Hz, H-1 of Glc I), 5.54 (dd, *J*=11.6, 2.5 Hz, H-6B of Glc II).

CMR (C₅D₅N) : δ C-1) 36.9 (2) 29.6 (3) 76.7 (4) 28.6 (5) 50.7 (6) 79.5 (7) 40.7 (8) 33.1 (9) 54.8 (10) 36.8 (11) 37.8 (12) 212.6 (13) 55.2 (14) 55.7 (15) 31.3 (16) 79.5 (17) 54.3 (18) 16.1 (19) 12.8 (20) 42.6 (21) 13.9 (22) 109.2 (23) 31.8 (24) 29.2 (25) 30.5 (26) 66.9 (27) 17.3 **Glc I** (1) 101.6 (2) 75.6 (3) 78.5 (4) 71.8 (5) 78.1 (6) 62.6 **Glc II** (1) 106.3 (2) 75.4 (3) 78.6 (4) 71.7 (5) 78.1 (6) 63.0.

Mass (FAB, Positive ion, H.R.) : *m/z* 771.4185 [(M+H)⁺, calcd. for 771.4167].

Reference

1.   A. Yokosuka, Y. Mimaki, M. Kuroda, Y.Sashida, *Planta Med.*, **66,** 393 (2000).

METANARTHECIUM SAPONIN 2

22ξ-Methoxy-furost-4-ene-2β,11α,26-triol-3-one 11-O-α-L-arabinofuranoside-26-O-β-D-glucopyranoside

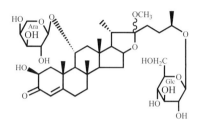

Source : *Metanarthecium lulteo-viride* Maxim. (Liliaceae)
Mol. Formula : C$_{39}$H$_{62}$O$_{15}$
Mol. Wt. : 770

Peracetate : [78536-10-4]

M.P. : 110-111.5°C; $[\alpha]_D^{21}$: -45.2° (c=0.42, MeOH)

UV (EtOH) : λ_{max} 240.5 (log ε, 13800) nm.

UV (MeOH) : λ_{max} 241 (log ε, 14100) nm.

IR (Nujol) : 1753, 1223 (OAc), 1697, 1618 cm^{-1}.

PMR (CDCl$_3$, 90.0 MHz) : δ 0.84 (s, 3xH-18), 0.88 (d, *J*=7.0 Hz, sec. C*H$_3$*), 1.02 (d, *J*=7.0 Hz, sec. C*H$_3$*), 1.30 (s, 3xH-19), 2.00 (s), 2.02, 2.04, 2.08, 2.17, 2.20 (each s) (8xOCOC*H$_3$*), 3.2-4.6 (m), 3.14 (s, 22-OC*H$_3$*), 4.9-5.3 (m, proton gemnial to acetate group), 5.32 (dd, *J*=4.0, 13.0, H-2α), 5.73 (s, H-4).

CD (c=0.1, dioxane) [θ] (nm) : O (367), +3900 (334) (sh), +4000 (324) (positive maximum); O (290), (c=0.067, MeOH) [θ] (nm) : +5300 (320) (positive maximum), -62700 (241) (negative maximum).

Reference

1.   I. Kitagawa and T. Nakanishi, *Chem. Pharm. Bull.*, **29**, 1299 (1981).

ANEMARRHENASAPONIN Ia

22-Methoxy-5β-furost-3β,15α-diol 3-O-β-D-glucopyranosyl-(1→2)-β-D-galactopyranoside

Source : *Anemarrhena asphodeloides* Bge. (Liliaceae)
Mol. Formula : $C_{40}H_{68}O_{14}$
Mol. Wt. : 772
Registry No. : [221317-02-8]

PMR (C_5D_5N) : δ 3.28 (OCH_3).

CMR (C_5D_5N) : δ C-1) 31.0 (2) 27.2 (3) 75.5 (4) 31.1 (5) 37.1 (6) 27.0 (7) 26.8 (8) 36.4 (9) 40.4 (10) 35.4 (11) 21.1 (12) 40.9 (13) 41.4 (14) 61.6 (15) 78.7 (16) 91.8 (17) 61.9 (18) 17.8 (19) 24.2 (20) 40.7 (21) 16.4 (22) 112.3 (23) 31.4 (24) 33.4 (25) 28.7 (26) 22.8 (27) 22.6 (OCH_3) 47.3 **Glc** (1) 106.2 (2) 77.0 (3) 78.0 (4) 71.7 (5) 78.4 (6) 62.8 **Gal** (1) 102.2 (2) 82.0 (3) 75.2 (4) 69.9 (5) 76.6 (6) 62.2.

Reference

1.   Z. Meng and S. Xu, *Zhongguo Yaowu Huaxue Zazhi* (*Chin. J. Med. Chem.*), **8**, 135 (1998).

CORDYLINE STRICTA SAPONIN 5

(25S)-22ξ-Methoxy-5α-furostane-3α,26-diol 3,26-bis-O-β-D-glucopyranoside

Source : *Cordyline stricta* (Agavaceae)
Mol. Formula : $C_{40}H_{68}O_{14}$
Mol. Wt. : 772
$[\alpha]_D^{27}$: -43.6° (c=0.11, MeOH)
Registry No. : [194218-99-0]

IR (KBr) : 3420 (OH), 2930 (CH), 1445, 1375, 1250, 1190, 1160, 1070, 1030, 1015, 895, 840 cm⁻¹.

PMR (C_5D_5N, 400 MHz) : δ 0.74 (s, 3xH-19), 0.81 (s, 3xH-18), 1.06 (d, *J*=6.7 Hz, 3xH-21), 1.17 (d, *J*=6.9 Hz, 3xH-27), 3.26 (s, OC*H₃*), 4.85 (d, *J*=7.7 Hz, H-1 of Glc), 4.91 (d, *J*=7.7 Hz, H-1 of Glc).

CMR (C_5D_5N, 100 MHz) : δ C-1) 32.7 (2) 25.6 (3) 73.1 (4) 35.0 (5) 39.5 (6) 28.8 (7) 32.4 (8) 35.2 (9) 54.0 (10) 36.1 (11) 20.8 (12) 40.1 (13) 41.1 (14) 56.5 (15) 32.0 (16) 81.4 (17) 64.4 (18) 16.6 (19) 11.6 (20) 40.5 (21) 16.3 (22) 112.7 (23) 30.9 (24) 28.2 (25) 34.5 (26) 75.0 (27) 17.6 (OC*H₃*) 47.3 **Glc I** (1) 102.6 (2) 75.4 (3) 78.7 (4) 71.8 (5) 78.4 (6) 62.9 **Glc II** (1) 105.1 (2) 75.2 (3) 78.6 (4) 71.8 (5) 78.5 (6) 62.9.

Mass (FAB, Negative ion) : *m/z* 772 [M]⁻.

Reference

1. Y. Mimaki, Y. Takaashi, M. Kuroda and Y. Sashida, *Phytochemistry*, **45**, 1229 (1997).

762

FILICINOSIDE A
(25*S*)-22α-Methoxy-5β-furostan-3β,26-diol 3-O-[β-D-glucopyranoside]-26-O-[β-D-glucopyranoside]

Source : *Asparagus filicinus* Buch.-Ham. (Liliaceae)
Mol. Formula : $C_{40}H_{68}O_{14}$
Mol. Wt. : 772
M.P. : 201-206°C
[α]$_D^{20}$: -47.3° (MeOH)
Registry No. : [156857-59-9]

Reference

1.   S.C. Sharma and N.K. Thakur, *Phytochemistry*, **36**, 469 (1994).

ALLIUM NUTANS SAPONIN 3
(25*R*)-Furost-5-en-1β,3β,22α,26-tetrol 1-O-β-D-galactopyranoside-26-O-β-D-glucopyranoside

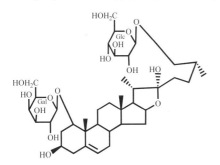

Source : *Allium nutans* L. (Liliaceae)
Mol. Formula : $C_{39}H_{64}O_{15}$
Mol. Wt. : 772
Registry No. : [241803-86-1]

PMR (CD$_3$OD, 62.5 MHz) : δ 0.87 (3xH-18, s), 0.98 (3xH-27, d, *J*=6.5 Hz), 1.02 (3xH-21, d, *J*=6.5 Hz), 1.13 (3xH-19, s), 3.20 (dd, *J*=7.5, 9.0 Hz, H-2 of Glc), 3.28 (ddd, *J*=2.5, 4.5, 9.0 Hz, H-5 of Glc), 3.30 (dd, *J*=9.0, 9.0 Hz, H-4 of Glc), 3.38 (dd, *J*=9.0, 9.0 Hz, H-3 of Glc), 3.43 (m, H-3), 3.47 (ddd, *J*=2.0, 3.5, 4.5 Hz, H-5 of Gal), 3.49 (dd, *J*=4.5, 9.0 Hz, H-3 of Gal), 3.50 (dd, *J*=7.8, 9.0 Hz, H-2 of Gal), 3.52 (dd, *J*=3.5, 11.5 Hz, H-1), 3.69 (dd, *J*=4.5, 9.0 Hz, H-6A of Glc), 3.75 (dd, *J*=4.5, 12.0 Hz, H-6A of Gal), 3.78 (dd, *J*=3.5, 12.0 Hz, H-6B of Gal), 3.86 (dd, *J*=2.0, 4.5 Hz, H-4 of Gal), 3.88 (dd, *J*=2.5, 9.0 Hz, H-6B of Glc), 4.27 (d, *J*=7.5 Hz, H-1 of Glc), 4.30 (d, *J*=7.8 Hz, H-1 of Gal), 4.39 (H-16, m), 5.57 (br d, *J*=5.4 Hz, H-6).

CMR (CD$_3$OD, 62.5 MHz) : δ C-1) 83.7 (2) 36.9 (3) 68.7 (4) 42.6 (5) 139.7 (6) 125.7 (7) 32.3 (8) 33.6 (9) 51.1 (10) 42.7 (11) 24.1 (12) 40.9 (13) 40.5 (14) 57.6 (15) 32.7 (16) 82.2 (17) 63.9 (18) 16.8 (19) 14.6 (20) 40.1 (21) 15.8 (22) 110.7 (23) 36.4 (24) 28.5 (25) 34.5 (26) 75.7 (27) 16.8 **Gal** (1) 101.9 (2) 72.1 (3) 74.8 (4) 63.6 (5) 76.0 (6) 62.1 **Glc** (1) 104.4 (2) 74.9 (3) 77.8 (4) 71.3 (5) 77.6 (6) 62.3.

Mass (L.S.I, Negative ion) : *m/z* 771 [M-H]⁻, 609 [M-H-Glc]⁻, 447 [M-H-2xGlc]⁻.

Reference

1.   L.S. Akhov, M.M. Musienko, S. Piacente, C. Pizza and W. Oleszek, *J. Agri. Food Chem.*, **47**, 3193 (1999).

NOLINOFUROSIDE D

25(*S*)-Furost-5-ene-1β,3β,22α,26-tetraol 1-O-[β-D-galactopyranoside]-26-O-[β-D-glucopyranoside]

Source : *Nolina microcarpa* S. Wats. (Dracaenaceae)
Mol. Formula : C$_{39}$H$_{64}$O$_{15}$
Mol. Wt. : 772
[α]$_D^{20}$: -60.0 ± 2° (c=1.0, Pyridine)
Registry No. : [144028-90-0]

IR (KBr) : 3200-3600 (OH), 910 (w, br) cm⁻¹.

PMR (C$_5$D$_5$N, 250/300 MHz) **:** δ 0.87 (s, 3xH-18), 0.97 (d, *J*=7.0 Hz, 3xH-27), 1.18 (d, *J*=7.0 Hz, 3xH-21), 1.20 (s, 3xH-19), 3.42 (dd, *J*$_{26A,26B}$=9.0 Hz, *J*$_{26A,25}$=6.5 Hz, H-26A), 3.85 (m, H-1ax), 3.88 (m, H-5 of Glc), 3.97 (dd, *J*$_{1,2}$=7.5 Hz, *J*$_{2,3}$=6.5 Hz, H-2 of Glc), 3.98 (m, H-3ax), 4.00 (m, H-5 of Gal), 4.02 (m, H-26B), 4.11 (dd, *J*$_{2,3}$=9.0 Hz, *J*$_{3,4}$=2.5 Hz, H-3 of Gal), 4.17 (m, H-3 of Glc), 4.19 (m, H-4 of Gal), 4.21 (m, H-3 of Glc), 4.32 (dd, *J*$_{6A,6B}$=11.5 Hz, *J*$_{6A,5}$=5.0 Hz, H-6A of Glc), 4.35 (m, 2xH-6 of Gal), 4.39 (dd, *J*$_{1,2}$=7.0 Hz, *J*$_{2,3}$=9.0 Hz, H-2 of Gal), 4.50 (m, H-16), 4.50 (dd, *J*$_{6A,6B}$=11.5, *J*$_{6B,5}$=11.5 Hz, H-6B of Glc), 4.75 (d, *J*=7.5 Hz, H-1 of Glc), 4.80 (d, *J*=4.0 Hz, H-1 of Gal).

CMR (C_5D_5N, 62.5/75 MHz) **:** δ C-1) 83.88 (2) 38.09 (3) 68.17 (4) 43.84 (5) 139.73 (6) 124.89 (7) 32.10 (8) 33.12 (9) 50.57 (10) 42.97 (11) 23.96 (12) 40.64 (13) 40.14 (14) 57.04 (15) 32.76 (16) 81.28 (17) 64.00 (18) 17.02 (19) 14.94 (20) 40.84 (21) 16.45 (22) 110.84 (23) 37.25 (24) 28.38 (25) 34.52 (26) 75.41 (27) 17.53 **Glc** (1) 105.24 (2) 75.31 (3) 78.55 (4) 71.80 (5) 78.68 (6) 62.91 **Gal** (1) 102.72 (2) 72.72 (3) 75.49 (4) 69.88 (5) 76.73 (6) 62.24.

Mass : m/z 754 $[M-H_2O]^+$.

Reference

1. G.V. Shevchuk, Yu. S. Vollerner, A.S. Shashkov and V.Ya Chirva, *Khim. Prir. Soedin.*, **27**, 678 (1991); *Chem. Nat. Comp.*, **27**, 597 (1991).

TUBEROSIDE R

(25*S*)-Furost-20(22)-en-2β,3β,5β,26-tetrol 3,26-*bis*-O-β-D-glucopyranoside

Source : *Allium tuberosum* L. (Liliaceae)
Mol. Formula : $C_{39}H_{64}O_{15}$
Mol. Wt. : 772
$[\alpha]_D^{24}$ **:** -18.2° (c=0.18, MeOH)
Registry No. : [651306-83-1]

IR (KBr) : 3400, 1448, 1381, 1078, 1039, 914, 856 cm^{-1}.

PMR (C_5D_5N, 400 MHz) : δ 0.76 (s, 3xH-18), 1.18 (d, *J*=7.6 Hz, 3xH-27), 1.22 (s, 3xH-19), 1.69 (s, 3xH-21), 2.53 (d, *J*=10.1 Hz, H-17), 3.57 (dd, *J*=7.0, 9.0 Hz, H-26A), 4.02 (m, H-5 of Glc I, H-5 of Glc II), 4.03 (m, H-2 of Glc I), H-2 of Glc II), 4.10 (m, H-2), 4.20 (m, H-26B), 4.25 (m, H-4 of Glc I, H-4 of Glc II), 4.30 (m, H-3 of Glc I, H-3 of Glc II), 4.44 (m, H-6A of Glc I, H-6A of Glc II), 4.59 (H-6B of Glc I, H-6B of Glc II), 4.78 (m, H-3), 4.88 (d, *J*=7.7 Hz, H-1 of Glc II), 4.90 (m, H-16), 5.13 (d, *J*=7.8 Hz, H-1 of Glc I).

CMR (C_5D_5N, 100 MHz) : δ C-1) 35.8 (2) 66.0 (3) 79.4 (4) 35.4 (5) 73.3 (6) 30.6 (7) 29.5 (8) 34.5 (9) 44.8 (10) 43.3 (11) 21.7 (12) 40.0 (13) 43.8 (14) 54.8 (15) 34.6 (16) 84.7 (17) 64.7 (18) 14.5 (19) 17.8 (20) 103.8 (21) 12.0 (22) 152.6 (23) 23.8 (24) 31.6 (25) 33.9 (26) 75.4 (27) 17.4 **Glc I** (1) 102.3 (2) 74.9 (3) 78.7 (4) 71.7 (5) 78.6 (6) 62.6 **Glc II** (1) 105.3 (2) 75.4 (3) 78.9 (4) 71.8 (5) 78.7 (6) 63.0.

Mass (FAB, Positive ion) : m/z 773 $[M+H]^+$. 611 $[M-H-Rha]^+$, 449 $[M+H-2xGlc]$.

Reference

1. S. Sang, S. Mao, A. Lao, Z. Chen and C.-T. Ho, *Food Chemistry*, **83**, 499 (2003).

METANARTHECIUM SAPONIN 1

(25*R*)-22ξ-Methoxy-furostane-2β,3β,11α,26-tetrol 11-O-α-L-arabinofuranoside-26-O-β-D-glucopyranoside

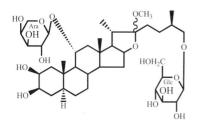

Source : *Metanarthecium lulteo-viride* Maxim. (Liliaceae)
Mol. Formula : $C_{39}H_{66}O_{15}$
Mol. Wt. : 774
M.P. : 167-169°C
$[\alpha]_D^{16}$: -38° (c=0.54, MeOH)

IR (KBr) : 3400 (OH) cm⁻¹.

Peracetate : [78536-10-4]

M.P. : 105-107°C; $[\alpha]_D^{25}$: -21° (c=0.5, MeOH)

IR (Nujol) : 1752, 1222 (OAc), 1045 (C-O-C) cm⁻¹.

PMR (CD₃OD) : δ 0.76 (s, 3xH-18), 0.87 (d, *J*=6.0 Hz, sec. C*H*₃), 0.99 (d, *J*=6.0 Hz, sec. C*H*₃), 1.12 (s, 3xH-19), 1.97 (s), 2.00 (s), 2.06 (s), 2.11 (s, 9xOCOC*H*₃), 3.10 (s, 22-OC*H*₃), 3.8-4.5 (unresolved m), 4.7-5.2 (m, proton geminal to acetate groups), 5.40 (m, H-3).

Mass : *m/z* 1121 [M⁺-MeOH].

Reference

1. I. Kitagawa, and T. Nakanishi, *Chem. Pharm. Bull.*, **29**, 1299 (1981).

DRACAENA SURCULOSA SAPONIN 1

(25S)-1β,3β,26-Trihydroxy-22α-methoxyfurost-5-ene-1,26-*bis*-O-[β-D-glucopyranoside]

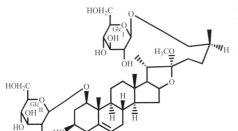

Source : *Dracaena surculosa* Lindle. (Agavaceae)
Mol. Formula : $C_{40}H_{66}O_{15}$
Mol. Wt. : 786
$[\alpha]_D^{25}$ **:** -104.0° (c=0.10, CHCl$_3$-MeOH, 1:1)
Registry No. : [244125-47-1]

IR (KBr) **:** 3400 (OH), 2920 (CH), 1450, 1370, 1150, 1090, 1060, 1020, 960, 940, 880, 830 cm^{-1}.

PMR (C_5D_5N, 400/500 MHz) **:** δ 0.88 (s, 3xH-18), 1.04 (d, *J*=6.7 Hz, 3xH-27), 1.13 (d, *J*=6.9 Hz, 3xH-21), 1.25 (s, 3xH-19), 3.25 (s, OC*H*$_3$), 3.84 (br m, *W*½=21.7 Hz, H-3), 3.94 (dd, *J*=11.7, 3.9 Hz, H-1), 4.84 (d, *J*=7.8 Hz, H-1 of Glc II), 4.97 (d, *J*=7.7 Hz, H-1 of Glc I), 5.58 (br d, *J*=5.6 Hz, H-6).

CMR (C_5D_5N, 100/125 MHz) **:** δ C-1) 83.3 (2) 37.8 (3) 68.0 (4) 43.7 (5) 139.5 (6) 124.7 (7) 31.9 (8) 32.9 (9) 50.3 (10) 42.8 (11) 23.8 (12) 40.3 (13) 40.5 (14) 56.8 (15) 32.3 (16) 81.3 (17) 64.3 (18) 16.7 (19) 14.8 (20) 40.5 (21) 16.2 (22) 112.7 (23) 31.0 (24) 28.1 (25) 34.5 (26) 74.9 (27) 17.5 (OCH$_3$) 47.3 **Glc I** (1) 101.6 (2) 75.3 (3) 78.6 (4) 72.4 (5) 78.1 (6) 63.6 **Glc II** (1) 105.1 (2) 75.2 (3) 78.6 (4) 71.7 (5) 78.5 (6) 62.9.

Mass (FAB, Negative ion) **:** *m/z* 785 [M-H]$^-$.

Reference

1.   A. Yokosuka, Y. Mimaki, Y. Sashida, *J. Nat. Prod.*, **63**, 1239 (2000).

DRACAENA SURCULOSA SAPONIN 4
(25S)-22α-Methoxy-1β,6β,26-trihydroxy-3α,5α-cyclofurostane 1,26-*bis*-O-β-D-glucopyranoside

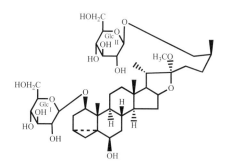

Source : *Dracaena surculosa* Lindle. (Agavaceae)
Mol. Formula : $C_{40}H_{66}O_{15}$
Mol. Wt. : 786
$[α]_D^{26}$ **:** -42.0° (c=0.10, MeOH)
Registry No. : [463963-00-0]

IR (film) : 3387 (OH), 2927 (CH), 1076 cm^{-1}.

PMR (C_5D_5N, 500 MHz) : δ 0.60 (dd, *J*=7.9, 4.0 Hz, H-4A), 0.82 (m, H-9), 0.85 (s, 3xH-18), 1.05 (d, *J*=6.7 Hz, 3xH-27), 1.13 (m, H-3), 1.17 (d, *J*=6.9 Hz, 3xH-21), 1.25 (ddd, *J*=13.0, 11.8, 2.5 Hz, H-7α), 1.58 (t-like, *J*=4.0 Hz, H-4B), 1.64 (s, 3xH-19), 1.80 (m, H-11α), 2.48 (m, H-8), 3.25 (s, OC*H₃*), 3.48 (br s, H-6), 4.41 (br d, *J*=4.9 Hz, H-1), 4.50 (q-like, *J*=7.0 Hz, H-16), 4.77 (d, *J*=7.8 Hz, H-1 of Glc I), 4.85 (d, *J*=7.8 Hz, H-1 of Glc II).

CMR (C_5D_5N, 125 MHz) : δ C-1) 85.0 (2) 33.0 (3) 23.2 (4) 15.8 (5) 40.0 (6) 73.0 (7) 38.5 (8) 30.0 (9) 50.0 (10) 49.1 (11) 23.2 (12) 40.4 (13) 41.3 (14) 56.5 (15) 32.2 (16) 81.4 (17) 64.3 (18) 16.7 (19) 16.9 (20) 40.4 (21) 16.2 (22) 112.7 (23) 30.9 (24) 28.2 (25) 34.4 (26) 74.9 (27) 17.6 (OCH₃) 47.3 **Glc I** (1) 103.1 (2) 74.9 (3) 78.9 (4) 72.0 (5) 78.2 (6) 63.1 **Glc II** (1) 105.1 (2) 75.2 (3) 78.6 (4) 71.7 (5) 78.5 (6) 62.9.

Mass (FAB, Negative ion) : *m/z* 785 [M-H]⁻.

Reference

1.   A. Yokosuka, Y. Mimaki and Y. Sashida, *Chem. Pharm. Bull.*, **50**, 992 (2002).

CORDYLINE STRICTA SAPONIN 6

22-O-Methyl-(25S)-5α-furostane-1β,3α,22ξ,26-tetrol 3,26-bis-O-β-D-glucopyranoside

Source : *Cordyline stricta* (Agavaceae)
Mol. Formula : $C_{40}H_{68}O_{15}$
Mol. Wt. : 788
$[\alpha]_D^{27}$: -36.4° (c=0.11, MeOH)
Registry No. : [194219-01-7]

IR (KBr) : 3420 (OH, 2910 (CH), 1445, 1370, 1250, 1155, 1065, 1015, 985, 935, 890, 835 cm^{-1}.

PMR (C_5D_5N, 400 MHz) : δ 0.87 (s, 3xH-18), 1.06 (d, J=7.0 Hz, 3xH-21), 1.07 (s, 3xH-19), 1.13 (d, J=6.9 Hz, 3xH-27), 3.26 (s, OCH$_3$), 4.84 (d, J=7.7 Hz, H-1 of Glc II), 4.96 (d, J=7.7 Hz, H-1 of Glc I).

CMR (C_5D_5N, 100 MHz) : δ C-1) 73.7 (2) 37.1 (3) 73.8 (4) 35.1 (5) 39.1 (6) 28.7 (7) 32.4 (8) 36.0 (9) 55.1 (10) 42.6 (11) 24.8 (12) 40.8 (13) 40.8 (14) 56.7 (15) 32.3 (16) 81.3 (17) 64.6 (18) 16.7 (19) 6.5 (20) 40.5 (21) 16.3 (22) 112.7 (23) 31.0 (24) 28.2 (25) 34.5 (26) 75.0 (27) 17.6 **Glc I** (1) 102.6 (2) 75.4 (3) 78.7 (4) 71.8 (5) 78.4 (6) 62.9 **Glc II** (1) 105.1 (2) 75.2 (3) 78.6 (4) 71.7 (5) 78.5 (6) 62.9.

Mass (FAB, Negative ion) : *m/z* 788 [M]$^-$.

Reference

1.   Y. Mimaki, Y. Takaashi, M. Kuroda and Y. Sashida, *Phytochemistry*, **45**, 1229 (1997).

ALLIUM MACROSTEMON SAPONIN 1

22α-Hydroxy-5β-furost-25(27)-ene-1β,3β,6β,26-tetraol 3-O-β-D-galactopyranoside-26-O-β-D-glucopyranoside

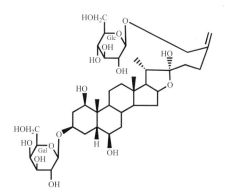

Source : *Allium macrostemon* Bunge (Liliaceae)
Mol. Formula : $C_{39}H_{64}O_{16}$
Mol. Wt. : 788
M.P. : 175-176°C
Registry No. : [500784-83-8]

IR (KBr) : 3414.8, 2931.1, 1631.9, 1452.3, 1076.6, 609.6 cm^{-1}.

PMR (C_5D_5N, 500 MHz) : δ 0.92 (s, 3xH-18)[a], 1.05 (s, 3xH-19)[a], 1.60 (d, *J*=6.9 Hz, 3xH-21), 3.98 (m, H-3), 4.34 (d, *J*=10.2 Hz, H-26ax), 4.38 (m), 4.50 (d, *J*=10.2 Hz, H-26eq), 4.56 (m, H-16), 4.60 (d, *J*=7.2 Hz, H-1), 4.76 (d, *J*=7.5 Hz, H-1 of Gal), 4.84 (d, *J*=7.8 Hz, H-1 of Glc), 4.96 (s, H-27A), 5.06 (s, H-27B).

CMR (C_5D_5N, 125 MHz) : δ C-1) 72.2 (2) 28.3 (3) 74.9 (4) 27.6 (5) 34.1 (6) 66.8 (7) 43.1 (8) 35.0 (9) 41.6 (10) 41.0 (11) 21.2 (12) 40.0 (13) 43.1 (14) 55.9 (15) 32.3 (16) 81.0 (17) 64.5 (18) 16.6 (19) 19.8 (20) 41.6 (21) 16.2 (22) 110.2 (23) 35.5 (24) 28.3 (25) 147.1 (26) 72.0 (27) 110.7 **Gal** (1) 104.7 (2) 71.5 (3) 74.9 (4) 71.5 (5) 75.2 (6) 63.7 **Glc** (1) 103.7 (2) 73.7 (3) 78.3 (4) 71.5 (5) 78.3 (6) 62.5.

Mass (FAB, Negative ion) : *m/z* 787 [M-H]$^-$, 626 [M-H-Glc]$^-$, 463 [Agl-H]$^-$.

Reference

1.   X. He, F. Qiu, Y. Shoyama, H. Tanaka and X. Yao, *Chem. Pharm. Bull.*, **50**, 653 (2002).

PARDARINOSIDE B

26-Acetoxy-22α-methoxy-(25R)-5α-furost-3β,17α-diol 3-O-[α-L-rhamnopyranosyl (1→2)-β-D-glucopyranoside]

Source : *Lilium pardarinum* (Liliaceae)
Mol. Formula : $C_{42}H_{70}O_{15}$
Mol. Wt. : 814
$[\alpha]_D^{27}$: -62.0° (c=0.50, MeOH)
Registry No. : [119152-53-3]

IR (KBr) : 3440 (OH), 2940 (CH), 1720 (C=O), 1455, 1370, 1240, 1125, 1050, 980, 905, 810 cm^{-1}.

PMR (C_5D_5N, 400 MHz) : δ 0.91 (s, 3xH-18)[a], 0.95 (s, 3xH-19)[a], 0.97 (d, J=6.7 Hz, 3xH-27), 1.28 (d, J=7.1 Hz, 3xH-21), 1.77 (d, J=6.2 Hz, 3xH-6 of Rha), 2.05 (3H, s, OCOCH_3), 2.55 (q, J=7.1 Hz, H-20), 3.21 (3H, s, OCH_3), 3.95 (2H, br m, H-3, H-5 of Glc), 4.05 (dd, J=10.8, 6.5 Hz, H-26A), 4.12 (dd, J=10.8, 6.1 Hz, H-26B), 4.61 (dd, J=9.4, 3.3 Hz, H-3 of Rha), 4.80 (br s, H-2 of Rha), 4.94 (H-5 of Rha overlapping with H_2O signal), 5.07 (d, J=7.2 Hz, H-1 of Glc), 6.37 (br s, H-1 of Rha).

CMR (C_5D_5N, 100.6 MHz) : δ C-1) 37.3 (2), 29.9 (3) 77.0 (4) 34.4 (5) 44.6 (6) 29.0 (7) 32.5[a] (8) 35.9 (9) 54.3 (10) 35.8 (11) 21.1 (12) 32.2[a] (13) 45.7 (14) 52.7 (15) 31.5[a] (16) 90.3 (17) 90.4 (18) 17.3 (19) 12.4 (20) 43.0 (21) 10.4 (22) 113.2 (23) 30.5 (24) 27.9 (25) 33.2 (26) 69.2 (27) 16.8 (OCH_3) 47.0 (OCOCH_3) 170.8 (OCOCH_3) 20.8 **Glc** (1) 99.8 (2) 79.6 (3) 78.1[b] (4) 71.9 (5) 78.2[b] (6) 62.8 **Rha** (1) 102.1 (2) 72.5 (3) 72.8 (4) 74.1 (5) 69.4 (6) 18.6.

Mass (SIMS) : m/z 783 [M-OMe]$^+$.

Reference

1.   H. Shimomura, Y. Sashida and Y. Mimaki, *Phytochemistry*, **28**, 3163 (1989).

ALLIUM KARATAVIENSE SAPONIN 7
(25R)-22ξ--Methoxy-5α-furostane-2α,3β,5α,6β,26-tetrol 2,26-di-O-β-D-glucopyaranoside

Source : *Allium karataviense* Regel (Liliaceae)
Mol. Formula : $C_{40}H_{68}O_{17}$
Mol. Wt. : 820
$[\alpha]_D^{27}$: -70° (c=0.10, MeOH)
Registry No. : [238398-02-2]

IR (KBr) : 3400 (OH), 2925 and 2880 (CH), 1455, 1370, 1240, 1155, 1060, 1020, 955, 885, 830, 815 cm^{-1}.

PMR (C_5D_5N, 400/500 MHz) : δ 0.89 (s, 3xH-18), 1.04 (d, J=6.7 Hz, 3xH-27), 1.21 (d, J=6.9 Hz, 3xH-21), 1.60 (s, 3xH-19), 3.27 (s, OCH_3), 4.90 (d, J=7.7 Hz, H-1 of Glc II), 5.21 (d, J=7.7 Hz, H-1 of Glc I).

CMR (C_5D_5N, 100/125 MHz) : δ C-1) 39.6 (2) 85.2 (3) 71.4 (4) 40.2 (5) 74.9 (6) 75.3 (7) 35.7 (8) 30.1 (9) 45.7 (10) 40.6 (11) 21.5 (12) 40.5 (13) 41.3 (14) 56.2 (15) 32.2 (16) 81.4 (17) 64.4 (18) 16.3 (19) 18.1 (20) 40.6 (21) 16.6 (22) 112.6 (23) 30.8 (24) 28.2 (25) 34.2 (26) 75.2 (27) 17.2 (OCH$_3$) 47.2 **Glc I** (1) 104.7 (2) 75.2 (3) 78.5 (4) 71.8 (5) 78.5 (6) 62.7 **Glc II** (1) 105.0 (2) 75.2 (3) 78.6 (4) 71.7 (5) 78.5 (6) 62.9.

Mass (FAB, Positive ion, H.R.) : *m/z* 843.4301 [M+Na]$^+$.

Reference

1.   Y. Mimaki, M. Kuroda, T. Fukasawa and Y. Sashida, *Chem. Pharm. Bull.*, **47**, 738 (1999).

PARDARINOSIDE A

26-Acetoxy-22α-methoxy-(25R)-5α-furost-3β,14α,17α-triol
3-O-[α-L-rhamnopyranosyl-(1→2)-β-D-glucopyanoside

Source : *Lilium pardarinum* (Liliaceae)
Mol. Formula : $C_{42}H_{70}O_{16}$
Mol. Wt. : 830
$[\alpha]_D^{27}$: -56.4° (c=0.5, MeOH)
Registry No. : [119152-52-2]

IR (KBr) : 3420 (OH), 2920, 2850 (CH), 1720 (C=O), 1450, 1365, 1235, 1120, 1040, 980 cm^{-1}.

PMR (C_5D_5N, 400 MHz) : δ 0.97 (s, 3xH-19), 0.99 (d, J=6.6 Hz, 3xH-27), 1.09 (s, 3xH-18), 1.34 (d, J=7.1 Hz, 3xH-21), 1.78 (d, J=6.2 Hz, 3xH-6 of Rha), 2.05 (s, OCOCH_3), 2.69 (q, J=7.1 Hz, H-20), 3.23 (3H, s, OCH_3), 3.96 (2H, br m, H-3 and H-5 of Glc), 4.07 (dd, J=10.8, 6.5 Hz, H-26A), 4.14 (dd, J=10.8, 6.0 Hz, H-26B), 4.61 (dd, J=9.3, 3.2 Hz, H-3 of Rha), 4.72 (dd, J=7.4, 6.0 Hz, H-16), 4.81 (br s, H-2 of Rha), 4.98 (dq, J=9.5, 6.2 Hz, H-5 of Rha), 5.07 (d, J=7.2 Hz, H-1 of Glc), 6.37 (br s, H-1 of Rha).

CMR: (C_5D_5N, 100.6 MHz) : δ C-1) 37.6 (2) 30.0 (3) 77.0 (4) 34.5 (5) 44.6 (6) 29.1 (7) 27.1^a (8) 39.8 (9) 46.6 (10) 36.1 (11) 20.4 (12) 27.0^a (13) 48.9 (14) 88.6 (15) 40.2 (16) 90.9 (17) 91.3 (18) 21.0 (19) 12.3 (20) 43.6 (21) 10.6 (22) 113.0 (23) 30.9 (24) 28.0 (25) 33.3 (26) 69.3 (27) 16.9 (OCH_3) 47.2 (OCOCH_3) 170.8 (OCOCH_3) 30.8 **Glc** (1) 99.9 (2) 79.7 (3) 78.2^b (4) 72.0 (5) 78.3^b (6) 62.9 **Rha** (1) 102.2 (2) 72.6 (3) 72.9 (4) 74.2 (5) 69.5 (6) 18.7.

Mass: (SIMS) : m/z 799 [M-OMe]$^+$, 780 [M-MeOH-H_2O]$^+$.

Reference

1.   H. Shimomura, Y. Sashida and Y. Mimaki, *Phytochemisry*, **28**, 3163 (1989).

REINECKEA CARNEA SAPONIN 5

1β,2β,3β,4β,5β,26-Hexahydroxy-22ξ-methoxyfurostane 3,26-*bis*-O-β-D-glucopyranoside

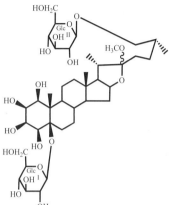

Source : *Reineckea carnea* Kunth (Liliaceae)
Mol. Formula : $C_{40}H_{68}O_{18}$
Mol. Wt. : 836
$[\alpha]_D^{25}$: -51.0° (c=0.10, MeOH)
Registry No. : [156665-31-5]

IR (KBr) : 3405 (OH), 2930 (CH), 1450, 1375, 1305, 1255, 1050, 885, 800, 700 cm^{-1}.

PMR (C_5D_5N, 400 MHz) : δ 0.82 (s, 3xH-18), 1.01 (d, J=6.6 Hz, 3xH-27), 1.18 (d, J=6.8 Hz, 3xH-21), 1.73 (s, 3xH-19), 3.27 (s, OCH_3), 4.50 (q-like, J=7.0 Hz, H-16), 4.85 (d, J=7.7 Hz, H-1 of Glc II), 5.32 (d, J=7.7 Hz, H-1 of Glc I).

CMR (C_5D_5N, 100 MHz) : δ C-1) 77.8 (2) 68.1 (3) 76.2 (4) 67.6 (5) 87.5 (6) 25.0 (7) 28.5 (8) 34.7 (9) 46.3 (10) 46.7 (11) 21.7 (12) 39.8 (13) 40.9 (14) 56.0 (15) 32.2 (16) 81.3 (17) 64.4 (18) 16.3 (19) 13.7 (20) 40.6 (21) 16.4 (22) 112.7 (23) 30.9 (24) 28.2 (25) 34.2 (26) 75.2 (27) 17.1 (OCH_3) 47.3 **Glc I** (1) 97.5 (2) 75.9 (3) 78.7^a (4) 71.9 (5) 78.8^a (6) 62.9 **Glc II** (1) 105.0 (2) 75.2 (3) 78.5 (4) 71.9 (5) 78.7 (6) 63.0.

Mass (FAB, Negative ion) : *m/z* 836 [M]$^-$, 673 [M-Glc]$^-$.

Biological Activity : The compound exhibited inhibitory activity on CAMP phosphodiesterase with IC$_{50}$=37.6×10^{-5} M.

Reference

1. T. Kanmoto, Y. Mimaki, Y. Sashida, T. Nikaido, K. Koike and T. Ohmoto, *Chem. Pharm. Bull.*, **42**, 926 (1994).

NOLINOFUROSIDE H

(25*S*)-Furost-5,20(22)-diene-1β,3β-ol 3-sulfate 1-O-β-D-fucopyranoside-26-O-β-D-glucopyranoside

Source : *Nolina microcarpa* S. Wats. (Dracaenaceae)
Mol. Formula : $C_{39}H_{61}O_{16}SNa$
Mol. Wt. : 840
[α]$_D^{20}$: -42.0 ± 2° (c=1.03, Pyridine)
Registry No. : [144721-25-5]

IR (KBr) **:** 3200-3600 (OH), 1200-1300 (ester bond) cm^{-1}.

PMR (C_5D_5N) **:** δ 0.64 (s, 3xH-18), 0.95 (d, $J_{27,25}$=7.0 Hz, 3xH-27), 1.03 (s, 3xH-19), 1.48 (s, 3xH-21), 1.48 (d, 3xH-6 of Fuc), 3.39 (dd, $J_{26,26}$=10.0 Hz, $J_{26,25}$=7.0 Hz, H-26A), 3.60 (q, $J_{5,6}$=6.5 Hz, H-5 of Fuc), 3.90 (m, H-5 of Glc), 3.95 (t, $J_{2,3}$=8.0 Hz, H-2 of Glc), 3.99 (br d, H-4 of Fuc), 4.04 (br d, H-26B), 4.13 (t, $J_{4,5}$=8.0 Hz, H-4 of Glc), 4.17 (t, $J_{3,4}$=8.0 Hz, H-3 of Fuc), 4.21 (t, $J_{3,4}$=8.0 Hz, H-3 of Glc), 4.22 (t, $J_{2,3}$=8.0 Hz, H-2 of Fuc), 4.23 (m, H-6A of Glc), 4.47 (dd, $J_{6A,6B}$=12.0 Hz, $J_{6B,5}$=2.5 Hz, H-6B of Glc), 4.58 (d, $J_{1,2}$=8.0 Hz, H-1 of Fuc), 4.75 (d, $J_{1,2}$=8.0 Hz, H-1 of Glc), 5.48 (br d, $J_{6,7}$=5.6 Hz, H-6).

CMR (C_5D_5N) **:** δ C-1) 83.78 (2) 32.52 (3) 75.28 (4) 40.52 (5) 138.25 (6) 126.25 (7) 32.81 (8) 33.88 (9) 50.55 (10) 43.26 (11) 23.78 (12) 40.52 (13) 42.84 (14) 55.40 (15) 32.19 (16) 84.62 (17) 64.81 (18) 14.65 (19) 14.65 (20) 103.91 (21) 11.90 (22) 152.40 (23) 23.78 (24) 31.53 (25) 34.85 (26) 75.34 (27) 17.28 **Glc** (1) 105.29 (2) 74.85 (3) 78.22 (4) 71.94 (5) 78.56 (6) 63.01 **Fuc** (1) 102.53 (2) 72.22 (3) 75.28 (4) 72.64 (5) 71.18 (6) 17.47.

Mass (L.S.I.) **:** *m/z* 863 [M+Na]$^+$, 743 [M+Na-NaHSO$_4$]$^+$, 841 [M+H]$^+$.

Reference

1.   G.V. Shevchuk, Yu.S. Vollerner, A.S. Shashkov, M.B. Gorovits and Y.Ya. Chirva, *Khim. Prir. Soedin.*, **27**, 801 (1991); *Chem. Nat. Comp.*, **27**, 706 (1991).

ALLIUM GIGANTEUM SAPONIN 8

(25*R*)-3-O-Acetyl-22ξ-methoxy-5α-furostane-2α,3β,5α,6β,26-pentol-2,26-*bis*-O-β-D-glucopyranoside

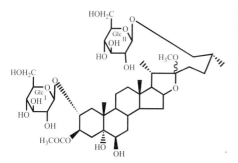

Source : *Allium giganteum* Regel (Liliaceae)
Mol. Formula : $C_{42}H_{70}O_{18}$
Mol. Wt. : 862
[α]$_D^{28}$: -72.2° (c=0.11, MeOH)
Registry No. : [156006-39-2]

IR (KBr) : 3400 (OH), 2925 (CH), 1710 (C=O), 1445, 1370, 1255, 1155, 1065, 1025, 950, 890 cm^{-1}.

PMR (C$_5$D$_5$N, 400 MHz) : δ 0.85 (s, 3xH-18), 1.01 (3H, d, J=6.6 Hz, 3xH-27), 1.17 (d, J=7.0 Hz, 3xH-21), 1.53 (s, 3xH-19), 2.35 (dd, J=12.0, 6.2 Hz, H-4), 2.40 (dd, J=2.8, 10.8 Hz, H-1), 2.81 (dd, J=12.0, 11.4 Hz, H-4), 3.24 (3H, s, OCH$_3$), 4.47 (q-like, J=7.4 Hz, H-16), 4.71 (ddd, J=10.8, 9.7, 5.9 Hz, H-2). 4.86 (d, J=7.8 Hz, H-1 of Glc II), 5.18 (d, J=7.7 Hz, H-1 of Glc I), 6.14 (ddd, J=11.4, 9.7, 6.2 Hz, H-3). Signals o the C-25 Isomer: δ 1.06 (d, J=6.6 Hz, 3xH-27), 1.15 (d, J=7.2 Hz, 3xH-21), 3.23 (s, OCH$_3$), 4.85 (d, J=7.8 Hz, H-1 of Glc II).

CMR (C$_5$D$_5$N, 100 MHz) : δ C-1) 38.8 (2) 77.6 (3) 75.0^a (4) 37.8 (5) 74.9 (6) 75.1^a (7) 35.7 (8) 30.0 (9) 45.6 (10) 40.4 (11) 21.4 (12) 40.2 (13) 41.3 (14) 56.1 (15) 32.3 (16) 81.4 (17) 64.4 (18) 16.3 (19) 17.9 (20) 40.4 (21) 16.6 (22) 112.6 (23) 30.8 (30.9) (24) 28.2 (25) 34.2 (34.5) (26) 75.2 (27) 17.2 (17.6) (OCOCH$_3$) 170.9 (OCOCH$_3$) 21.6 (OCH$_3$) 47.3 **Glc I** (1) 103.1 (2) 75.3 (3) 78.3^b (4) 71.8 (5) 78.5^b (6) 63.0 **Glc II** (1) 105.0 (105.1) (2) 75.2 (3) 78.6 (4) 71.8 (5) 78.5 (5) 62.9. Signals of C-25 isomer are given in paranthesis.

Mass (FAB, Negative ion) : m/z 862 [M]$^-$.

Biological Activity : The compound shows inhibitory activity on CAMP phosphor-diesterase with IC$_{50}$=0.5x10^{-5} M.

Reference

1.   Y. Mimaki, T. Nikaido, K. Matsumoto, Y. Sashida and T. Ohmoto, *Chem. Pharm. Bull.*, **42**, 710 (1994).

NOLINA RECURVATA SAPONIN 8

Furosta-5,20 (22),25(27)-triene-1β,3β,26-triol 1-O-[-α-L-rhamnopyranosyl-(1→2)-
α-L-arabinopyranoside]-26-O-β-D-glucopyranoside

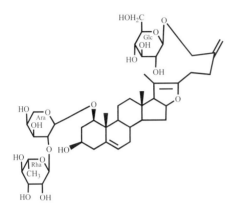

Source : *Nolina recurvata*[1] (Agavaceae), *Dracaena cochinchinesis* (Lour.) S.C. Chen.[2] (Agavaceae)
Mol. Formula : $C_{44}H_{68}O_{17}$
Mol. Wt. : 868
$[\alpha]_D^{26}$ **:** -24.0° (c=0.48, MeOH)
Registry No. : [180161-89-1]

IR (KBr)[1] : 3420 (OH), 2925 (CH), 1450, 1350, 1260, 1220, 1130, 1065, 1045, 980, 940, 905, 835, 815, 780, 700 cm^{-1}.

PMR (C_5D_5N, 400 MHz)[1] : δ 0.75 (s, 3xH-18), 1.45 (s, 3xH-19), 1.58 (s, 3xH-21), 1.75 (d, J=6.2 Hz, 3xH-6 of Rha), 4.74 (d, J=6.9 Hz, H-1 of Ara), 4.92 (d, J=7.9 Hz, H-1 of Glc), 5.06 and 5.36 (br s, 2xH-27), 5.60 (br d, J=5.5 Hz, H-6), 6.34 (br s, H-1 of Rha).

CMR (C_5D_5N, 100 MHz) : δ C-1) 83.5 (2) 37.4 (3) 68.2 (4) 43.9 (5) 139.6 (6) 124.7 (7) 32.1 (8) 32.9 (9) 50.4 (10) 43.1 (11) 24.2 (12) 40.2 (13) 42.9 (14) 55.1 (15) 34.7 (16) 84.5 (17) 64.6 (18) 14.5 (19) 15.1 (20) 104.2 (21) 11.8 (22) 151.5 (23) 31.1 (24) 24.7 (25) 146.3 (26) 71.7 (27) 111.6 **Ara** (1) 100.3 (2) 75.2 (3) 75.8 (4) 70.1 (5) 67.3 **Rha** (1) 101.7 (2) 72.5 (3) 72.7 (4) 74.2 (5) 69.4 (6) 19.0 **Glc** (1) 103.8 (2) 75.3 (3) 78.6 (4) 71.7 (5) 78.5 (6) 62.8.

Mass (FAB, Negative ion) : m/z 868 [M]⁻, 722 [M-Rha]⁻, 705 [M-Glc]⁻.

Biological Activity : The compound shows inhibitory activity on cyclic AMP phosphodiesterase with IC_{50}=127.0x10⁻⁵ M).[1]

References

1. Y. Mimaki, Y. Takaashi, M. Kuroda, Y. Sashida and T. Nikaido, *Phytochemistry*, **42**, 1609 (1996).

2. Z. Zhou, J. Wang and C. Yang, *Zhongkaoyao*, **30**, 801 (1999).

METANARTHECIUM GLYCOSIDE FG-3

(25*R*,5β)-3β,11α,22ξ,26-tetrahydroxyfurostane 11-O-[(2,3,4-tri-O-acetyl)-α-L-arabinopyranoside]-
26-O-β-D-glucopyranoside

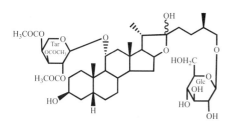

Source : *Metanarthecium luteo-viride* Maxim.
(Liliaceae)
Mol. Formula : $C_{44}H_{70}O_{17}$
Mol. Wt. : 870
Registry No. : [151802-01-6]

Mass (L.S.I.) : *m/z* 853 [M-OH]$^+$, 691 [M-Glc-OH]$^+$, 577 [M-OH-Tar-H$_2$O]$^+$, 415 [M-Tar-H$_2$O-Glc-OH]$^+$, 271, 259 (Tar), 217, 185, 159, 139, 97, 89.

Mass (L.S.I. with NaCl) : *m/z* 893 [M+Na]$^+$, 853, 731 [M-Glc+Na]$^+$, 691, 845, 617 [M-Tar-H$_2$O+Na]$^+$, 577, 531, 415, 271, 217, 171, 139, 97, 69, 43.

Reference

1.   Y. Ikenishi, S. Yoshimatsu, K. Takeda and Y. Nakagawa, *Tetrahedron*, **49**, 9321 (1993).

ASPAROSIDE B'

Furost-20(22)-en-3β,26-diol 3-O-[β-D-xylopyranosyl-(1→4)-β-D-glucopyranoside]-26-O-β-D-glucopyranoside

Source : *Asparagus meioclados* (Liliaceae)
Mol. Formula : $C_{44}H_{72}O_{17}$
Mol. Wt. : 872
M.P. : 278-280°C
[α]$_D$: -24.0° (c=2.34, MeOH-CHCl$_3$ 1:1)
Registry No. : [301643-62-9]

IR (KBr) **:** 3440, 2940, 1450, 1030 cm^{-1}.

PMR (C$_5$D$_5$N, 500 MHz) **:** δ 0.68 (s, CH_3), 0.85 (s, CH_3), 1.02 (d, *J*=6.4 Hz, 3xH-27), 1.61 (s, 3xH-21), 3.75 (m, 2xH-26), 3.90 (m, 1H), 4.82 (d, *J*=7.7 Hz, anomeric H), 4.90 (d, *J*=8.1 Hz, anomeric H), 5.14 (d, *J*=7.7 Hz).

CMR (C$_5$D$_5$N, 125 MHz) **:** δ C-1) 30.9 (2) 26.8 (3) 75.2 (4) 30.6 (5) 37.0 (6) 27.0 (7) 27.0 (8) 35.2 (9) 40.1 (10) 35.2 (11) 21.3 (12) 40.2 (13) 43.8 (14) 54.8 (15) 31.8 (16) 81.4 (17) 62.8 (18) 17.2 (19) 23.8 (20) 36.2 (21) 14.5 (22) 111.2 (23) 72.3 (24) 34.6 (25) 30.4 (26) 64.0 (27) 17.7 **Glc I** (1) 103.0 (2) 75.0 (3) 76.5 (4) 81.0 (5) 76.6 (6) 61.9 **Xyl** (1) 102.5 (2) 74.7 (3) 78.4 (4) 70.8 (5) 67.4 **Glc II** (1) 105.2 (2) 75.0 (3) 78.5 (4) 71.8 (5) 78.6 (6) 62.9.

Mass (E.I.) **:** *m/z* (rel.intens.) 578 (11), 415 (50), 397 (27), 181 (100), 163 (35).

Mass (E.S.I., Positive ion) **:** *m/z* 873 [M+H]$^+$, 741 [M+H-Xyl]$^+$, 579 [M+H-Xyl-Glc]$^+$.

Reference

1.   J. Feng, D.-F. Chen, Q.-Z. Sun, N. Nakamura, M. Hattori, *J. Asian Nat. Prod. Res.*, **4**, 221 (2002).

ASPARAGUS COCHINCHINENSIS SAPONIN 1

(25R)-Furosta-5,20-dien-3β,26-diol 3-O-[α-L-rhamnopyranosyl-(1→4)-β-D-glucopyranoside]-26-O-(β-D-glucopyranoside)

Source : *Asparagus cochinchinensis* (Lour.) Merr. (Liliaceae)
Mol. Formula : $C_{45}H_{74}O_{17}$
Mol. Wt. : 884
$[\alpha]_D^{25}$ **:** -76.4° (c=1.0, MeOH)
Registry No. : [117457-34-8]

PMR (CD$_3$OD, 250 MHz) : δ 0.74 (s, 3xH-18), 0.98 (d, J=6 Hz, 3xH-27), 1.08 (s, 3xH-19), 1.28 (d, J=6.0 Hz, 3xH-6 of Rha), 1.63 (s, 3xH-21), 4.27 (d, J=7.0 Hz, H-1 of Glc II), 4.42 (d, J=7.0 Hz, H-1 of Glc I), 4.87 (d, J=1.5 Hz, H-1 of Rha), 5.40 (m, H-6).

CMR (CD$_3$OD, 62.0 MHz) : δ C-1) 38.5 (2) 30.7 (3) 78.1 (4) 40.8 (5) 142.1 (6) 122.5 (7) 32.7 (8) 32.7 (9) 51.8 (10) 38.1 (11) 22.2 (12) 39.6 (13) 41.8 (14) 56.4 (15) 31.4 (16) 85.6 (17) 65.7 (18) 14.5 (19) 19.8 (20) 105.1 (21) 11.8 (22) 153.1 (23) 33.3 (24) 29.0 (25) 35.1 (26) 75.9 (27) 17.3 **Glc I** (1) 104.0 (2) 75.0 (3) 77.7 (4) 79.5 (5) 76.9 (6) 62.3 **Rha** (1) 103.0 (2) 72.3 (3) 72.5 (4) 73.8 (5) 70.8 (6) 17.9 **Glc II** (1) 104.6 (2) 75.2 (3) 78.2 (4) 71.8 (5) 77.8 (6) 62.9.

Mass (FAB, Negative ion) : m/z 883 [M-H]⁻, 721 [M-H-Glc]⁻, 575 [5-H-Glc-Rha]⁻, 413 [M-H-2xGlc-Rha]⁻, 395 [M-H-2xGlc-Rha-H$_2$O]⁻.

Reference

1.   Z.Z. Liang, R. Aquino, F.D. Simone, A. Dini, O. Schettino and C. Pizza, *Planta Med.*, **54,** 344 (1988).

SMILAX NIGRESCENS SAPONIN 1

Pseudodiosgenin 3-O-[α-L-rhamnopyranosyl-(1→2)-β-D-glucopyranoside]-26-O-β-D-glucopyranoside

Source : *Smilax nigrescens* Wang. et Tang. (Liliaceae)
Mol. Formula : $C_{45}H_{72}O_{17}$
Mol. Wt. : 884
Registry No. : [78229-03-5]

PMR : δ 1.65 (s, 3xH-21), 4.95 (d, *J*=7.1 Hz, H-1 of Glc), 5.00 (d, *J*=7.2 Hz, H-1 of Glc), 6.32 (br s, H-1 of Rha).

CMR (C_5D_5N) : δ C-1) 37.5 (2) 30.1 (3) 77.8 (4) 38.9 (5) 140.9 (6) 121.7 (7) 32.3 (8) 31.6 (9) 50.2 (10) 37.1 (11) 21.2 (12) 39.7 (13) 43.3 (14) 54.9 (15) 34.5 (16) 84.4 (17) 64.4 (18) 14.2 (19) 19.4 (20) 103.1 (21) 11.8 (22) 152.6 (23) 34.5 (24) 23.8 (25) 33.5 (26) 75.0 (27) 17.3 **Glc I** (1) 100.2 (2) 79.4 (3) 77.8 (4) 71.7 (5) 77.6 (6) 62.5 **Rha** (1) 101.9 (2) 72.0 (3) 72.4 (4) 73.7 (5) 69.3 (6) 18.6 **Glc II** (1) 104.8 (2) 75.1 (3) 78.5 (4) 71.7 (5) 78.3 (6) 62.8.

Reference

1.   Y. Ju and Z.J. Jia, *Chem. J. Chin. Univ.*, **12**, 1488 (1991).

TRITELEIA LACTEA SAPONIN 3

Nicotigenin 3-O-{α-L-rhamnopyranosyl-(1→2)-[α-L-rhamnopyranosyl-(1→4)]-β-D-glucopyranoside}

Source : *Triteleia lactea* (Liliaceae)
Mol. Formula : $C_{45}H_{72}O_{17}$
Mol. Wt. : 884

IR (KBr) : 3420 (OH), 2930 (CH), 1455, 1375, 1115, 1035 cm^{-1}.

PMR (C_5D_5N, 400/500 MHz) : δ 0.81 (s, 3xH-18), 1.05 (s, 3xH-19), 1.10 (d, *J*=6.9 Hz, 3xH-21), 1.35 (s, 3xH-27), 1.62 (d, *J*=6.2 Hz, 3xH-6 of Rha II), 1.76 (d, *J*=6.2 Hz, 3xH-6 of Rha I), 4.93 (d, *J*=7.9 Hz, H-1 of Glc), 5.31 (br d, *J*=5.2 Hz, H-6), 5.82 (br s, H-1 of Rha II), 6.36 (br s, H-1 of Rha I).

CMR (C_5D_5N, 100/125 MHz) : δ C-1) 37.5 (2) 30.2 (3) 78.0 (4) 39.0 (5) 140.9 (6) 121.8 (7) 32.3^a (8) 31.8 (9) 50.4 (10) 37.2 (11) 21.1 (12) 39.8 (13) 40.6 (14) 56.6 (15) 32.2^a (16) 81.1 (17) 62.7 (18) 16.1 (19) 19.4 (20) 38.5 (21) 15.2 (22) 120.3 (23) 33.8 (24) 32.6 (25) 85.7 (26) 70.1 (27) 24.1 **Glc** (1) 100.4 (2) 78.9 (3) 76.9 (4) 77.9 (5) 78.2 (6) 61.4 **Rha I** (1) 102.0 (2) 72.5 (3) 72.8 (4) 74.2 (5) 69.5 (6) 18.5^c **Rha II** (1) 103.0 (2) 72.5 (3) 72.9^b (4) 73.9 (5) 70.5 (6) 18.6^c.

Mass (FAB, Negative ion) : *m/z* 884 [M]$^-$.

Biological Activity : The compound shows inhibitory activity on cyclic AMP phosphodiesterase with IC$_{50}$ (10.4x10^{-5}M).

Reference

1.   Y. Mimaki, O. Nakamura, Y. Sashida, T. Nikaido and T. Ohmoto, *Phytochemistry*, **38**, 1279 (1995).

AFROMONTOSIDE

(25*R*)-Furost-5-ene-3β,22α-26-triol 3-O-[α-L-rhamnopyranosyl-(1→4)-β-D-glucopyranoside]-26-O-α-L-rhamnopyranoside

Source : *Dracaena afromontana* (Agavaceae)
Mol. Formula : $C_{45}H_{74}O_{17}$
Mol. Wt. : 886
M.P. : >300°C
$[\alpha]_D^{25}$ **:** -69.3° (c=2.2, C_5D_5N)
Registry No. : [91574-93-5]

UV : λ_{max} 231 nm.

IR (KBr) : 3500-3300, 1645, 1060 and 890-800 cm^{-1}.

PMR ([(CD$_3$)$_2$SO], 360 MHz) : δ 0.79 (s, 3H), 0.90 (d, 3H, *J*=7.0 Hz), 1.05 (d, 3H, *J*=7.0 Hz,), 1.10 (s, 3H), 1.14 (d, 3H, *J*=6.0 Hz), 1.18 (d, 3H, *J*=6.0 Hz,), 2.4 (dd, 2H, *J*=14.5, 5.5 Hz), 3.60 (m), 4.0 (dd, 2H, *J*=11.0, 4.0 Hz), 4.40 (d, *J*=9.0 Hz), 4.60 (d, *J*=4.0 Hz), 4.95 (d, *J*=4.0 Hz), 5.02 (s) and 5.32 (m, *J*=5.5 Hz).

Mass (C.I., Negative ion, CH$_4$/N$_2$O) : *m/z* 885 [M-H]⁻, 867 [M-H-H$_2$O]⁻, 721, 576 and 412.

Biological Activity : This compound showed significant inhibitory activity when tested in vitro at a conc. of 100 μg/ml against (KB) cells derived from human carcinoma of the nasopharynx.

Reference

1. K. Sambi-Reddy, M.S. Shekhani, D.E. Berry, D.G. Lynn and S.M. Hecht, *J. Chem. Soc. Perkin Trans. I*, 987 (1984).

ASPACOCHIOSIDE C

(5β,25*S*)-Furostan-20(32)-en-3β,26-diol 3-O-[α-L-rhamnopyranosyl-(1→4)-β-D-glucopyranoside]-
26-O-β-D-glucopyranoside

Source : *Asparagus cochinchinensis* (Lour.) Merr.
(Liliaceae)
Mol. Formula : $C_{45}H_{74}O_{17}$
Mol. Wt. : 886
M.P. : 140-141°C
[α]$_D^2$: -26.3° (c=0.09, MeOH)
Registry No. : [630055-42-4]

IR (KBr) : 3396, 2927, 1691, 1635, 1448, 1319, 1304, 1221, 1076, 1038, 1026, 910, 812 cm^{-1}.

PMR (C_5D_5N, 500 MHz) : δ 0.69 (s, 3xH-18), 0.84 (s, 3xH-19), 0.88 (m, H-14), 0.95 (m, H-11A), 1.03 (d, *J*=7.0 Hz, 3xH-27), 1.15 (m, H-9), 1.25 (m, H-7A), 1.32 (m, H-11B), 1.34 (m, H-24A), 1.42 (m, H-15A), 1.43 (m, H-1A), 1.50 (m, H-8), 1.51 (H-7B), 1.52 (m, H-2A), 1.62 (s, 3xH-21), 1.66 (m, H-6), 1.69 (m, 1B), 1.70 (m, H-2B), 1.73 (d, *J*=7.0 Hz, 3xH-6 of Rha), 1.74 (m, H12A), 1.76 (m, H-12B), 1.77 (m, H-4A), 1.78 (m, H-24B), 1.84 (m, H-6B), 1.86 (m, H-4B), 1.94 (m, H-25), 1.95 (m, H-5), 2.04 (m, H-15B), 2.21 (m, H-23A and H-23B), 2.48 (d, *J*=9.6 Hz, H-17), 3.47 (dd, *J*=10.0, 7.2 Hz, H-26A), 3.71 (br d, *J*=9.0 Hz, H-5 of Glc I), 3.96 (m, H-5 of Glc II), 3.98 (dd, *J*=7.5, 8.0 Hz, H-2 of Glc I), 4.00 (dd, *J*=8.0, 8.0 Hz, H-2 of Glc II), 4.10 (dd, *J*=10.0, 7.2 Hz, H-26B), 4.13 (br d, *J*=10.5 Hz, H-6A of Glc I), 4.22 (dd, *J*=8.0, 9.0 Hz, H-3 of Glc I), 4.23 (dd, *J*=8.0, 9.0 Hz, H-3 of Glc II), 4.25 (dd, *J*=9.0, 9.0 Hz, H-4 of Glc I), 4.27 (br d, *J*=10.5 Hz, H-6B fo Glc I), 4.28 (m, H-3), 4.34 (dd, *J*=9.0, 9.5 Hz, H-4 of Rha), 4.44 (br d, *J*=12.5 Hz, H-6A of Glc II), 4.49 (dd, *J*=9.0, 9.0 Hz, H-4 of Glc I), 4.56 (br d, *J*=9.5 Hz, H-3 of Rha), 4.58 (br d, *J*=12.5 Hz, H-6B of Glc II), 4.70 (br s, H-2 of Rha), 4.83 (m, H-16), 4.83 (d, *J*=8.0 Hz, H-1 of Glc II), 4.85 (d, *J*=7.5 Hz, H-1 of Glc I), 5.03 (dq, *J*=9.0, 7.0 Hz, H-5 of Rha), 5.92 (br s, H-1 of Rha).

CMR (C_5D_5N, 125 MHz) : δ C-1) 30.5 (2) 26.6 (3) 74.8 (4) 31.0 (5) 36.6 (6) 26.6 (7) 26.6 (8) 34.8 (9) 39.7 (01) 34.8 (11) 21.0 (12) 39.7 (13) 43.4 (14) 54.3 (15) 34.0 (16) 84.2 (17) 64.2 (18) 14.1 (19) 23.5 (20) 103.2 (21) 11.5 (22)

151.9 (23) 23.3 (24) 30.1 (25) 33.3 (26) 74.8 (27) 16.8 **Glc I** (1) 102.6 (2) 75.2 (3) 76.4 (4) 78.2 (5) 76.8 (6) 61.1 **Rha** (1) 102.2 (2) 72.2 (3) 72.4 (4) 73.6 (5) 69.9 (6) 18.2 **Glc II** (1) 104.8 (2) 74.8 (3) 78.2 (4) 71.3 (5) 78.2 (6) 62.4.

Mass (E.S.I., Positive ion) : m/z 887 [M+H]$^+$, 741 [M+H-Rha]$^+$, 725 [M+H-Glc], 579.

Mass (E.S.I., MS/MS) : m/z 887, 725, 579, 581, 435, 415, 399, 285, 273, 255, 163, 147, 129, 85, 71.

Mass (E.S.I., MS/MS) : m/z 725, 581, 435, 417, 399, 273, 225, 161, 147, 129, 85, 71.

Mass (E.S.I.,. Positive ion, H.R) : mz 887.5012 [(M+H)$^+$, calcd. for 887.5004].

Reference

1.   J.-G. Shi, G.-Q. Li, S.-Y. Huang, S.-Y. Mo, Y. Wang, Y.-C. Yang and W.-Y. Hu, *J. Asian Nat. Prod. Res.*, **6**, 99 (2004).

TRIBOL
(25R)-Furost-5-ene-3β,16,26-triol 3-O-{α-L-rhamnopyranosyl-(1→2)-[α-L-rhamnopyranosyl-(1→4)]-β-D-glucopyranoside}

Source : *Tribulus terrestris* L. (Zygophyllaceae)
Mol. Formula : $C_{45}H_{74}O_{17}$
Mol. Wt. : 886

IR (KBr) : 3411, 2926, 1678, 1042, 912, 879, 809 (intensity 912 < 837) cm^{-1}.

PMR (CD$_3$OD, 500 MHz) : δ 0.86 (s, 3xH-18), 0.93 (d, J=6.7 Hz, 3xH-27), 1.00 (H-9), 1.04 (d, J=7.6 Hz, 3xH-21), 1.06 (s, 3xH-19), 1.09 (H-1A), 1.10 (H-24A), 1.23 (H-12A), 1.25 (d, J=6.2 Hz, 3xH-6 of Rha II), 1.27 (d, J=6.2 Hz, 3xH-6 of Rha I), 1.33 (H-14 and H-15A), 1.50 (H-23A), 1.56 (H-7A and H-23B), 1.59 (H-8), 1.60 (H-2A, H-11, H-25), 1.67 (H-24), 1.77 (H-12A), 1.77 (br d, J=2.9 Hz, H-17), 1.88 (H-1A), 1.93 (H-2B), 1.97 (H-7), 2.02 (H-15), 2.30

(H-4A), 2.33 (ddd, J=2.6, 7.6, 8.1 Hz, H-20), 2.46 (H-4B), 3.32 (m, H-5 of Glc), 3.33 (dd, J=6.7, 10.5 Hz, H-26A), 3.40 (dd, J=9.1, 7.8 Hz, H-2 of Glc), 3.40 (dd, J=9.6, 9.6 Hz, H-4 of Rha II), 3.42 (dd, J=9.5, 9.5 Hz, H-4 of Rha I), 3.44 (dd, J=6.0, 10.7 Hz, H-26A), 3.52 (dd, J=9.1, 9.1 Hz, H-4 of Glc), 3.60 (H-3), 3.60 (dd, J=9.1, 9.1 Hz, H-3 of Glc), 3.63 (br dd, J=9.5, 3.3 Hz, H-3 of Rha I), 3.65 (dd, J=12.1, 3.1 Hz, H-6A of Glc), 3.67 (dd, J=9.6, 3.3 Hz, H-3 of Rha II), 3.80 (dd, J=12.1, 1.7 Hz, H-6B of Glc), 3.84 (dd, J=3.3, 1.6 Hz, H-2 of Rha I), 3.93 (dq, J=9.5, 6.2 Hz, H-5 of Rha I), 3.93 (dd, J=3.3, 1.5 Hz, H-2 of Rha II), 4.13 (dq, J=9.6, 6.2 Hz, H-5 of Rha II), 4.23 (ddd, J=4.4, 8.1, 9.7 Hz, H-22), 4.51 (d, J=7.8 Hz, H-1 of Glc), 4.85 (d, J=1.6 Hz, H-1 of Rha I), 5.21 (d, J=1.5 Hz, H-1 of Rha II), 5.39 (br d, J=5.2 Hz, H-6).

CMR (CD$_3$OD, 75 MHz) : δ C-1) 38.5 (2) 30.8 (3) 79.3 (4) 39.5 (5) 142.0 (6) 122.6 (7) 33.0 (8) 32.5 (9) 51.8 (10) 38.1 (11) 21.7 (12) 40.0 (13) 42.3 (14) 56.8 (15) 35.9 (16) 120.2 (17) 73.4 (18) 14.2 (19) 19.8 (20) 35.1 (21) 17.4 (22) 89.0 (23) 30.7 (24) 32.3 (25) 37.2 (26) 68.4 (27) 17.2 **Glc I** (1) 100.5 (2) 79.3 (3) 78.1 (4) 80.1 (5) 76.6 (6) 62.0 **Rha I** (1) 103.1 (2) 72.5 (3) 72.2 (4) 73.8 (5) 70.7 (6) 17.8 **Rha II** (1) 102.3 (2) 72.2 (3) 72.4 (4) 74.0 (5) 69.8 (6) 18.0.

Mass (E.S.I., Positive ion) : m/z 887 [M+H]$^+$, 869 [M+H-H$_2$O]$^+$, 433 [M+H-Glc-2xRha]$^+$, 415 [M+H-Glc-2xRha-H$_2$O]$^+$.

Reference

1. J. Conrad, D. Dinchev, I. Klaiber, S. Mika, I. Kostova and W. Kraus, *Fitoterapia*, **75**, 117 (2004).

CHINENOSIDE III

3β,26-Dihydroxy-(25*R*)-5α-furost-20(22)-en-6-one 3-O-[α-arabinopyranosyl-(1→6)-β-glucopyranoside]-
26-O-β-D-glucopyranoside

Source : *Allium chinense* (Liliaceae)
Mol. Formula : C$_{44}$H$_{70}$O$_{18}$
Mol. Wt. : 886
$[\alpha]_D$: -42.8° (c=0.59, Pyridine)
Registry No. : [172519-52-7]

IR (KBr) : 3400 (OH), 1700 (C=O), 1000-1100 (glycosyl C–O) cm^{-1}.

PMR (C₅D₅N, 400 MHz) : δ 0.65 (6H, s, 3xH-18 and 3xH-19), 1.03 (d, *J*=6.7 Hz, 3xH-27), 1.64 (s, 3xH-21), 2.45 (d, *J*=10.1 Hz, H-17), 3.61 (dd, *J*=9.5, 5.5 Hz, H-26A), 3.94 (m, H-26B), 4.02 (m, H-3), 4.75 (ddd, *J*=9.4, 7.9, 6.0 Hz, H-16), 4.82 (d, *J*=7.6 Hz, H-1 of Glc II), 4.95 (d, *J*=6.7 Hz, H-1 of Ara), 4.96 (d, *J*=7.6 Hz, H-1 of Glc I).

CMR (C₅D₅N, 100 MHz) : δ C-1) 36.8 (2) 29.5 (3) 76.6 (4) 26.9 (5) 56.4 (6) 209.5 (8) 37.1 (9) 53.6 (10) 40.9 (11) 21.6 (12) 39.3 (13) 43.9 (14) 54.7 (15) 34.0 (16) 84.1 (17) 64.3 (18) 14.3 (19) 13.0 (20) 103.4 (21) 11.7 (22) 152.5 (23) 23.6 (24) 31.4 (25) 33.4 (26) 74.9 (27) 17.4 **Glc I** (1) 102.0 (2) 75.1 (3) 78.5 (4) 72.3 (5) 77.0 (6) 69.7 **Ara** (1) 105.4 (2) 71.8 (3) 74.4 (4) 69.1 (5) 66.5 **Glc II** (1) 104.8 (2) 75.1 (3) 78.5 (4) 71.7 (5) 78.4 (6) 62.8.

Mass (F.D.) : *m/z* 909 [M+Na]⁺.

Reference

1.   J.-P. Peng, X.-S. Yao, Y. Tezuka and T. Kikuchi, *Phytochemistry*, **41**, 283 (1996).

DESGLUCORUSCOSIDE, RUSCOPONTICOSIDE E

1β,3β,22α,26-Tetrahydroxyfurosta-5,25(27)-diene 1-O-[α-L-rhamnopyranosyl-(1→2)-α-L-arabinopyranoside]-26-O-β-D-glucopyranoside

Source : *Ruscus aculeatus* L.[1] (Liliaceae),
R. ponticus Wor.[2], *Dracaena cochinensis* (Lour.)
S.C. Chen.[3] (Agavaceae)
Mol. Formula : C₄₄H₇₀O₁₈
Mol. Wt. : 886
[α]D²⁰ : -48.0° (c=1.0, H₂O)[1]
Registry No. : [50619-66-4]

PMR (CD₃OD, 500 MHz)[3] : δ 0.84 (s, 3xH-18), 1.20 (s, 3xH-19), 1.04 (3xH-21), 1.28 (d, *J*=5.8 Hz, 3xH-6 of Rha), 4.25 (d, *J*=7.2 Hz, H-1 of Ara), 4.25 (d, *J*=7.6 Hz, H-1 of Glc), 4.92 (s, H-27A), 5.09 (s, H-27B), 5.28 (s, H-1 of Rha), 5.54 (d, *J*=5.32 Hz, H-6).

CMR (C₅D₅N, 62.5 MHz)[2] : δ C-1) 83.71 (2) 37.88 (3) 68.19 (4) 43.69 (5) 139.50 (6) 124.71 (7) 33.11 (8) 31.96 (9) 50.50 (10) 42.91 (11) 24.00 (12) 40.42 (13) 40.66 (14) 56.73 (15) 32.56 (16) 81.22 (17) 63.75 (18) 16.74 (19) 14.97 (20) 40.66 (21) 16.17 (22) 110.82 (23) 37.27 (24) 28.28 (25) 147.17 (26) 72.39 (27) 110.42 **Ara** (1) 100.34 (2) 75.30 (3) 75.64 (4) 69.90 (5) 67.12 **Rha** (1) 101.50 (2) 72.39 (3) 72.07 (4) 74.17 (5) 69.34 (6) 18.82 **Glc** (1) 103.73 (2) 75.06 (3) 78.22 (4) 71.71 (5) 78.42 (6) 62.76.

Mass (E.S.I.)[4] : *m/z* 909 [M+Na]⁺, 869 [M+H-H₂O]⁺, 707 [M+H-H₂O-Glc]⁺, 591 [M+H-H₂O-Rha-Ara]⁺, 561 [M+H-H₂O-Glc-Rha]⁺, 429 [Agl+H]⁺.

Mass (E.S.I., Negative ion)[4] : *m/z* 885 [M-H]⁻.

References

1. E. Bombardelli, A. Bonati, B. Gabetta and G. Mustich, *Fitoterapia*, **43**, 3 (1972).

2. T. Sh. Korkashvili and V.S. Kikoladze, *Khim. Prir. Soedin.*, 435 (1991); *Chem. Nat. Comp.*, **27**, 379 (1991).

3. Z. Zhou, J. Wang and C. Yang, *Zhongkaoyao*, **30**, 801 (1999).

4. E. de Combarieu, M. Falzoni, N. Fuzzati, F. Gattesco, A. Giori, M. Lavati and R. Pace, *Fitoterapia*, **73**, 583 (2002).

METANARTHECIUM GLYCOSIDE FG-2
(25*R*,5β)-2β,3β,11α,22ξ,26-pentahydroxyfurostane 11-O-[(2,3,4-tri-O-acetyl)-α-L-arabinopyranoside]-26-O-β-D-glucopyranoside

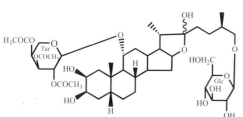

Source : *Metanarthecium luteo-viride* Maxim. (Liliaceae)
Mol. Formula : $C_{44}H_{70}O_{18}$
Mol. Wt. : 886
Registry No. : [151801-99-9]

Mass (L.S.I.) : *m/z* 869 [M-OH]⁺, 707 [M-Glc-OH]⁺, 593 [M-OH-Tar-H₂O]⁺, 431 [M-OH-Tar-Glc-H₂O]⁺, 413, 287, 259 (Tar), 199, 159, 139, 97, 69, 43.

Mass (L.S.I. + 1ng NaCl) : *m/z* 909 [M+Na]⁺, 869 [M-OH]⁺, 827, 707, 593, 431, 287, 259, 217, 195, 157, 121, 97, 69, 43.

Reference

1. Y. Ikenishi, S. Yoshimatsu, K. Takeda and Y. Nakagawa, *Tetrahedron*, **49**, 9321 (1993).

ALLIOFUROSIDE A

(25S)-Furost-5-ene-1β,3β,22α,26-tetraol 1-O-[α-L-rhamnopyranosyl-(1→2)-α-L-arabinopyranoside]-26-O-β-D-glucopyranoside

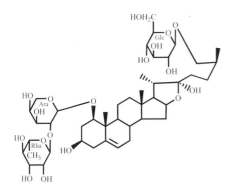

Source : *Allium cepa* L. (Liliceae)
Mol. Formula : $C_{44}H_{72}O_{18}$
Mol. Wt. : 888
M.P. : 164-166°C
$[\alpha]_D^{20}$ **:** -63.7±2° (c=1.11, Pyridine)
Registry No. : [105798-62-7]

22-O-Methyl ether :

M.P. : 160-171°C; $[\alpha]_D^{20}$: -75.3±2° (c=1.03, Pyridine)

IR (KBr) : 3300, 3500 (OH), 895 (weak, br) cm^{-1}.

PMR (C$_5$D$_5$N, 250 MHz) **:** δ 0.82 (s, 3xH-18), 1.01 (d, *J*=6.0 Hz, sec. CH$_3$), 1.10 (d, *J*=6.0 Hz, sec. CH$_3$), 1.43 (s, 3xH-19), 1.72 (d, *J*=6.0 Hz, 3xH-6 of Rha), 3.24 (s, OCH$_3$), 5.56 (d, *J*=5.0 Hz, H-6), 6.33 (br s, H-1 of Rha).

CMR (C$_5$D$_5$N, 62.5 MHz) **:** δ C-1) 83.81 (2) 37.51 (3) 68.37 (4) 43.98 (5) 139.73 (6) 124.85 (7) 33.28 (8) 32.16 (9) 50.58 (10) 43.07 (11) 24.16 (12) 40.42 (13) 40.68 (14) 56.90 (15) 32.56 (16) 81.51 (17) 64.40 (18) 16.78 (19) 15.20 (20) 40.68 (21) 17.69 (22) 112.88 (23) 31.18 (24) 28.32 (25) 34.62 (26) 75.10 (27) 17.69 (OCH$_3$) 47.48 **Ara** (1) 100.57 (2) 75.31 (3) 75.97 (4) 70.20 (5) 67.45 **Rha** (1) 101.73 (2) 72.66 (3) 72.66 (4) 74.35 (5) 69.54 (6) 19.10 **Glc** (1) 105.16 (2) 75.31 (3) 78.53 (4) 71.90 (5) 78.68 (6) 63.02.

Reference

1.   S.D. Kravets, Yu.S. Vollerner, M.B. Gorovits, A.S. Shashkov and N.K. Abuakirov, *Khim. Prir. Soedin.*, 188 (1986); *Chem. Nat. Comp.*, **22**, 174 (1986).

TRIGONEOSIDE VIII

(25R)-Furost-20(22)-en-2α,3β,26-triol 3-O-[β-D-xylopyranosyl-(1→6)-β-D-glucopyranoside]-26-O-β-D-glucopyranoside

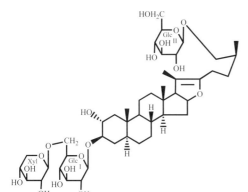

Source : *Trigonella foenum-graecum* L. (Leguminosae)
Mol. Formula : $C_{44}H_{72}O_{18}$
Mol. Wt. : 888
M.P. : 173-175°C
[α]$_D$: -28.6° (c=0.07, MeOH)
Registry No. : [425644-95-7]

IR (KBr) **:** 3390 (OH), 2923, 1627 (C=C), 1446, 1382, 1340, 1276, 1167, 1077, 1033, 921, 894, 816, 615 cm^{-1}.

PMR (C$_5$D$_5$N, 400 MHz) **:** δ 0.70 (s, 3xH-19), 0.73 (s, 3xH-18), 1.03 (d, J=6.8 Hz, 3xH-27), 1.62 (s, 3xH-21), 2.45 (d, J=10.0 Hz, H-17), 3.68 (m, H-3), 3.95 (m, H-2), 3.62 (m, H-26), 3.94 (m, H-26B), 4.79 (d, J=7.3 Hz, H-1 of Xyl), 4.83 (d, J=7.6 Hz, H-1 of Glc II), 4.96 (H-1 of Glc I).

CMR (C$_5$D$_5$N, 100 MHz) **:** δ C-1) 45.2 (2) 71.0 (3) 87.9 (4) 34.5 (5) 44.6 (6) 27.9 (7) 32.3 (8) 34.3 (9) 54.2 (10) 36.7 (11) 21.5 (12) 39.8 (13) 43.7 (14) 54.5 (15) 34.3 (16) 84.4 (17) 64.6 (18) 14.4 (19) 13.2 (20) 103.6 (21) 11.7 (22) 152.3 (23) 23.6 (24) 31.4 (25) 33.4 (26) 74.84 (27) 17.3 **Glc I** (1) 104.8 (2) 74.7 (3) 78.6 (4) 72.2 (5) 76.4 (6) 70.4 **Xyl** (1) 105.3 (2) 75.2 (3) 77.8 (4) 71.0 (5) 67.1 **Glc II** (1) 104.8 (2) 75.1 (3) 78.4 (4) 71.7 (5) 78.4 (6) 62.9.

Mass (FAB, Position ion) **:** *m/z* 889 [M+H]$^+$, 888 [M]$^+$, 757 [M-Xyl]$^+$, 727 [M-Glc]$^+$, 595 [M-Glc-Xyl]$^+$, 433 [M-Xyl-2xGlc]$^+$.

Reference

1.   M.-Y. Shang, S.-Q. Cai, S. Kadota and Y. Tezuka, *Yaoxue Xuebao (Acta Pharm. Sin.)*, **36**, 836 (2001).

TRIGONEOSIDE IIa

(25S)-5β-Furostane-3β,22ξ,26-triol 3-O-[β-D-xylopyranosyl-(1→6)]-β-D-glucopyranoside

Source : *Trigonella foenum-graecum* L.
(Legnuminosea)
Mol. Formula : $C_{44}H_{74}O_{18}$
Mol. Wt. : 890
[α]$_D$: -47.4° (c=1.77, Pyridine)
Registry No. : [187141-37-3]

IR (KBr) : 3400, 2928, 1078, 1040 cm^{-1}.

PMR (C$_5$D$_5$N, 270 MHz) : δ 0.85 (s, 3xH-18)^a, 0.87 (s, 3xH-19)^a, 1.03 (d, J=6.8 Hz, 3xH-27), 1.33 (d, J=6.7 Hz, 3xH-21), 1.93 (m, H-25), 1.98 (m, H-5), 2.24 (dq-like, H-20), 3.48 (dd, J=7.1, 9.5 Hz, H-26A), 4.08 (dd, J=5.6, 9.5 Hz, H-26B), 4.35, 4.82 (both m, 2xH-6 of Glc I), 4.41 (m, H-3), 4.80 (d, J=7.6 Hz, H-1 of Glc II), 4.87 (d, J=7.6 Hz, H-1 of Glc I), 4.98 (m, H-16), 5.04 (d, J=7.7 Hz, H-1 of Xyl).

CMR (C$_5$D$_5$N, 67.5 MHz) : δ C-1) 31.0 (2) 27.0 (3) 74.6 (4) 30.6 (5) 37.0 (6) 27.0 (7) 26.8 (8) 35.5 (9) 40.3 (10) 35.2 (11) 21.1 (12) 40.4 (13) 41.2 (14) 56.4 (15) 32.4 (16) 81.2 (17) 64.0 (18) 16.7 (19) 23.9 (20) 40.7 (21) 16.5 (22) 110.2 (23) 37.1 (24) 28.3 (25) 34.4 (26) 75.4 (27) 17.5 **Glc I** (1) 103.1 (2) 75.2 (3) 78.6 (4) 71.8 (5) 77.2 (6) 69.9 **Xyl** (1) 105.8 (2) 74.9 (3) 78.2 (4) 71.1 (5) 67.1 **Glc II** (1) 105.1 (2) 75.2 (3) 78.6 (4) 71.7 (5) 78.4 (6) 62.8.

Mass (FAB, Positive ion, H.R) : *m/z* 913.4789 [(M+Na)$^+$, calcd. for 913.4773].

Mass (FAB, Positive ion) : *m/z* 913 [M+Na]$^+$.

Mass (FAB, Negative ion) : *m/z* 889 [M-H]$^-$, 757 [M-C$_5$H$_9$O$_4$]$^-$, 727 [M-C$_6$H$_{11}$O$_5$]$^-$, 595 [M-C$_{11}$H$_{19}$O$_9$]$^-$.

Reference

1. M. Yoshikawa, T. Murakami, H. Komatsu, N. Murakami, J. Yamahara and H. Matsuda, *Chem. Pharm. Bull.*, **45**, 81 (1997).

TRIGONEOSIDE IIb

(25*R*)-5β-Furostane-3β,22ξ,26-triol 3-O-[β-D-xylopyranosyl-(1→6)-β-D-glucopyranoside]-26-O-β-D-glucopyranoside

Source : *Trigonella foenum-graecum* L. (Legnuminosea)
Mol. Formula : $C_{44}H_{74}O_{18}$
Mol. Wt. : 890
[α]$_D^{22}$: -45.3° (c=1.68, Pyridine)
Registry No. : [187141-38-4]

IR (KBr) : 3409, 2930, 1078, 1042 cm^{-1}.

PMR (C$_5$D$_5$N, 270 MHz) : δ 0.85 (s, 3xH-18)^a, 0.87 (s, 3xH-19)^a, 0.99 (d, *J*=6.7 Hz, 3xH-27), 1.35 (d, *J*=7.0 Hz, 3xH-21), 1.93 (m, H-25), 1.98 (m, H-5), 2.25 (m, H-20), 3.62 (dd, *J*=6.1, 9.5 Hz, H-26A), 3.94 (dd, *J*=7.0, 9.5 Hz, H-26B), 4.37 (dd, *J*=5.2, 11.3 Hz, H-6A of Glc I), 4.42 (m, H-3), 4.80 (d, *J*=7.7 Hz, H-1 of Glc II), 4.81 (dd, *J*=1.8, 11.3 Hz, H-6B of Glc I), 4.87 (dd, *J*=7.7 Hz, H-1 of Glc I), 5.00 (ddd,-like, H-16), 5.04 (d, *J*=7.3 Hz of Xyl).

CMR (C$_5$D$_5$N, 67.5 MHz) : δ C-1) 31.0 (2) 27.0 (3) 74.6 (4) 30.6 (5) 37.0 (6) 27.0 (7) 26.8 (8) 35.5 (9) 40.3 (10) 35.2 (11) 21.1 (12) 40.4 (13) 41.2 (14) 56.0 (15) 32.4 (16) 81.2 (17) 64.0 (18) 16.7 (19) 23.9 (20) 40.7 (21) 16.5 (22) 110.7 (23) 37.2 (24) 28.4 (25) 34.3 (26) 75.3 (27) 17.5 **Glc I** (1) 103.1 (2) 75.2 (3) 78.6 (4) 71.7 (5) 77.2 (6) 69.9 **Xyl** (1) 105.7 (2) 74.9 (3) 78.2 (4) 71.1 (5) 67.1 **Glc II** (1) 104.9 (2) 75.2 (3) 78.6 (4) 71.7 (5) 78.4 (6) 62.8.

Mass (FAB, Positive ion, H.R.) : *m/z* 913.4777 [(M+Na)$^+$, calcd. for 913.4773].

Mass (FAB, Positive ion) : *m/z* 913 [M+Na]$^+$.

Mass (FAB, Negative ion) : *m/z* 889 [M-H]$^-$, 757 [M-C$_5$H$_9$O$_4$]$^-$, 727 [M-C$_6$H$_{11}$O$_5$]$^-$, 595 [M-C$_{11}$H$_{19}$O$_9$]$^-$.

Reference

1.   M. Yoshikawa, T. Murakami, H. Komatsu, N. Murakami, J. Yamahara and H. Matsuda, *Chem. Pharm. Bull.*, **45**, 81 (1997).

TRILLIUM KAMTSCHATICUM SAPONIN Tf

17(20)-Dehydrokryptogenin 3-O-[α-L-rhamnopyranosyl-(1→2)-β-D-glucopyranoside]-26-O-[β-D-glucopyranoside]

Source : *Trillium kamtschaticum* Pall. (Liliaceae)
Mol. Formula : $C_{45}H_{70}O_{18}$
Mol. Wt. : 898
M.P. : 265-268°C (decomp.)
[α]$_D$: -80.1° (c=0.73, Pyridine)
Registry No. : [55972-79-7]

UV (EtOH) : λ_{max} 242 (ε, 8250) nm.

IR (KBr) : 3600-3300 (OH), 1720-1700 (C=O), 1630 (C=C) cm^{-1}.

PMR (C_5D_5N, 60 MHz) : δ 0.95 (s, 3xH-18), 1.02 (s, 3xH-19), 1.95 (s, 3xH-21).

Peracetate : colorless needle.

M.P. : 190-192°C; **[α]$_D^{30}$:** -67.1° (c=0.83, CHCl$_3$).

UV (EtOH) : λ_{max} 243 (ε, 10300) nm.

ORD (c=0.081, EtOH) [M] : +3500° (322, peak), -3240° (372) (trough) nm.

Mass (E.I.) : *m/z* 1318 [M]$^+$, 561 [$C_{24}H_{33}O_{15}$]$^+$, 392 [$C_{27}H_{36}O_2$]$^+$, 331 [$C_{14}H_{19}O_9$]$^+$, 273 [$C_{12}H_{17}O_7$]$^+$.

ORD (c=0.103, EtOH) [M] : +5140° (322, peak), -4760° (370) (trough) nm.

Reference

1.   T. Nohara, K. Miyahara and T. Kawasaki, *Chem. Pharm. Bull.*, **23**, 872 (1975).

ASCALONICOSIDE B

(25*R*)-Furost-5,20(22)-dien-1β,3β,26-triol 1-O-β-D-galactopyranoside-26-O-{α-L-rhamnopyranosyl-(1→2)-β-D-glucopyranoside}

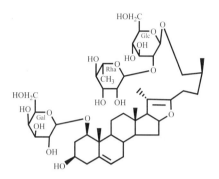

Source : *Allium ascalonicum* Hort. (Liliaceae)
Mol. Formula : $C_{45}H_{72}O_{18}$
Mol. Wt. : 900
$[\alpha]_D^{25}$: -8° (c=0.1, MeOH)
Registry No. : [473555-60-1]

PMR (CD₃OD, 500 MHz) : δ 0.70 (s, 3xH-18), 0.97 (d, *J*=6.6 Hz, 3xH-27), 1.09 (s, 3xH-19), 1.20 (m, H-14), 1.22* (H-12A), 1.28* (H-9), 1.30* (H-15A), 1.45 (m, H-11A), 1.50* (H-8), 1.55* (H-24A), 1.65 (m, H-24B), 1.70 (H-2A, H-12B), 1.78 (m, H-25), 1.89 (s, 3xH-21), 1.96* (H-7A), 1.98* (H-7B), 2.02* (H-15B), 2.12* (H-2B), 2.15* (H-23A), 2.20* (H-23B), 2.22* (H-4A), 2.30* (H-4B), 2.45 (d, *J*=5.5 Hz, H-17), 2.57 (dd, *J*=10.5, 2.5 Hz, H-11B), 3.30* (H-26A), 3.31* (H-3), 3.36 (H-1), 3.80 (dd, *J*=8.5, 6.9 Hz, H-26B), 4.70 (q, *J*=5.5 Hz, H-16), 5.55 (br d, *J*=3.2 Hz, H-6). * overlapped signals.

CMR (C₅D₅N, 125 MHz) : δ C-1) 84.2 (2) 37.1 (3) 69.2 (4) 43.1 (5) 141.1 (6) 126.2 (7) 32.5 (8) 33.5 (9) 51.9 (10) 43.0 (11) 25.0 (12) 41.0 (13) 43.2 (14) 56.3 (15) 31.5 (16) 84.6 (17) 65.3 (18) 15.0 (19) 15.5 (20) 104.2 (21) 12.3 (22) 143.0 (23) 34.2 (24) 25.4 (25) 34.5 (26) 76.0 (27) 17.5.

For PMR and CMR signals of the sugar moiety see Ascalonicoside A1

Mass (FAB, Negative ion, H.R.) : *m/z* 899.4652 [(M-H)⁻, requires 899.4639].

Reference

1.   E. Fattorusso, M. Iorizzi, V. Lanzotti and O. Taglialatela-scafati, *J. Agric. Food Chem.*, **50**, 5686 (2002).

LILIUM BROWNII SAPONIN 1

(22S,25S)-26-O-β-D-glucopyranosyl-22,25-epoxyfurost-5-en-3β,26-diol 3-O-[α-L-rhamnopyranosyl-(1→2)]-β-D-glucopyranoside]

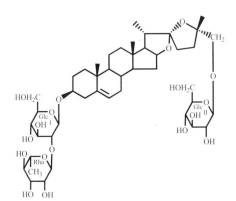

Source *: Lilium brownii* (Liliaceae)
Mol. Formula : $C_{45}H_{72}O_{18}$
Mol. Wt. : 900
$[\alpha]_D^{26}$ **:** -78.8° (c=0.16, MeOH)
Registry No. : [129777-35-1]

IR (KBr) : 3440 (OH), 2935 (CH), 1455, 1380, 1260, 1135, 1080, 1045, 915, 870, 840, 820, 705 cm^{-1}.

PMR (C_5D_5N 400 MHz) : δ 0.81 (s, 3xH-18), 1.05 (s, 3xH-19), 1.08 (d, J=6.9 Hz, 3xH-21), 1.40 (s, 3xH-27), 1.77 (d, J=6.2 Hz, 3xH-6 of Rha), 4.62 (dd, J=9.3, 3.2 Hz, H-3 of Rha), 4.71 (m, H-16), 4.80 (br d, J=3.2 Hz, H-2 of Rha), 4.96 (d, J=7.8 Hz, H-1 of Glc II), 4.99 (dq, J=9.5, 6.2 Hz, H-5 of Rha), 5.04 (d, J=7.1 Hz, H-1 of Glc I), 5.31 (br d, J=4.3 Hz, H-6), 6.37 (br s, H-1 of Rha).

CMR (C_5D_5N, 100.6 MHz) : δ C-1) 37.5 (2) 30.2 (3) 78.2 (4) 39.0 (5) 140.9 (6) 121.8 (7) 32.2^a (8) 31.7 (9) 50.3 (10) 37.2 (11) 21.1 (12) 39.9 (13) 40.5 (14) 56.5 (15) 32.3^a (16) 81.0 (17) 62.5 (18) 16.2 (19) 19.4 (20) 38.6 (21) 15.2 (22) 120.2 (23) 33.1^b (24) 34.0^b (25) 83.9 (26) 77.5 (27) 24.4 **Glc I** (1) 100.4 (2) 79.7 (3) 77.9^c (4) 71.9 (5) 78.0^c (6) 62.7 **Rha** (1) 102.1 (2) 72.6 (3) 72.9 (4) 74.2 (5) 69.5 (6) 18.7 **Glc II** (1) 105.5 (2) 75.4 (3) 78.4^d (4) 71.7 (5) 78.5^d (6) 62.7.

Mass (SI-MS) **:** m/z 939 [M+K]$^+$, 923 [M+Na]$^+$.

Mass (EIMS) : m/z (rel.intens.) 430 (3), 399 (40), 368 (5), 342 (19), 300 (15), 282 (31), 271 (32), 241 (7), 215 (8), 195 (16), 155 (100), 128 (79), 112 (60), 103 (79).

Reference

1.   Y. Mimaki and Y. Sashida, *Phytochemistry*, **29**, 2267 (1990).

RUSCUS ACULEATUS SAPONIN 3

22ξ-Methoxyfurosta-5,25(27)-diene-1β,3β,26-triol 1-O-{α-L-rhamnopyranosyl-(1→2)-
α-L-arabinopyranoside}-26-O-β-D-glucopyranoside

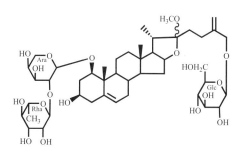

Source : *Ruscus aculeatus* L. (Liliaceae)
Mol. Formula : $C_{45}H_{72}O_{18}$
Mol. Wt. : 900
$[\alpha]_D^{26}$ **:** -54.0° (c=0.10, MeOH)
Registry No. : [139574-21-3]

PMR (C_5D_5N, 400 MHz) : δ 0.84 (s, 3xH-18), 1.13 (d, *J*=6.9 Hz, 3xH-21), 1.44 (s, 3xH-19), 1.74 (d, *J*=6.1 Hz, 3xH-6 of Rha), 3.25 (s, OC*H*₃), 4.16 (overlapping, H-3 and H-4 of Ara), 4.73 (d, *J*=7.1 Hz, H-1 of Ara), 4.92 (d, *J*=7.9 Hz, H-1 of Glc), 5.06 (br s, H-27A), 5.35 (br s, H-27B), 5.59 (br d, *J*=5.4 Hz, H-6), 6.34 (br s, H-1 of Rha).

CMR (C_5D_5N, 100 MHz) : δ C-1) 83.4 (2) 37.4 (3) 68.2 (4) 43.9 (5) 139.6 (6) 124.7 (7) 32.0 (8) 33.1 (9) 50.4 (10) 42.9 (11) 24.0 (12) 40.2 (13) 40.5 (14) 56.7 (15) 32.4 (16) 81.5 (17) 64.2 (18) 16.6 (19) 15.1 (20) 40.4 (21) 16.1 (22) 112.4 (23) 31.6 (24) 28.1 (25) 146.9 (26) 72.0 (27) 111.0 (OCH₃) 47.3 **Ara** (1) 100.3 (2) 75.2 (3) 75.9 (4) 70.1 (5) 67.3 **Rha** (1) 101.7 (2) 72.5 (3) 72.7 (4) 74.2 (5) 69.4 (6) 19.0 **Glc** (1) 103.9 (2) 75.2 (3) 78.5 (4) 71.7 (5) 78.6 (6) 62.9.

Mass (FAB, Negative ion) : *m/z* 899 [M-H]⁻.

Reference

1.   Y. Mimaki, M. Kuroda, A. Kameyama, A. Yokosuka and Y. Sashida, *Chem. Pharm. Bull.*, **46**, 298 (1998).

ANEMARRHENASAPONIN IV, TIMOSAPONIN B III
PSEUDOPROTOTIMOSAPONIN A-III

(5β,25S)-Furost-20(22)-en-3β,26-diol 3-O-[β-D-glucopyranosyl-(1→2)-β-D-galactopyranoside]-
26-O-β-D-glucopyranoside

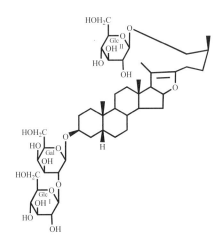

Source : *Anemarrhena asphodeloides* Bunge[1,2] (Liliaceae)
Mol. Formula : $C_{45}H_{74}O_{18}$
Mol. Wt. : 902
M.P. : 232-235°C[1]
[α]$_D^{26}$: -15.8° (c=0.75, Pyridine)[1]
Registry No. : [142759-74-8]

PMR (C₅D₅N, 270 MHz)[1] : δ 0.70 (s, C*H₃*), 1.00 (s, C*H₃*), 1.05 (d, *J*=6.2 Hz), 1.64 (s, 3xH-21), 4.85 (d, *J*=7.7 Hz), 4.94 (d, *J*=7.7 Hz), 5.31 (d, *J*=7.7 Hz).

CMR (C₅D₅N, 67.5 MHz)[1] : δ C-1) 31.0 (2) 26.9 (3) 75.2^a (4) 31.0 (5) 37.0 (6) 27.0 (7) 26.9 (8) 35.2 (9) 40.2 (10) 35.2 (11) 21.4 (12) 40.2 (13) 43.9 (14) 54.8 (15) 31.4 (16) 84.6 (17) 64.7 (18) 17.2 (19) 24.0 (20) 103.6 (21) 11.8 (22) 152.4 (23) 34.5 (24) 23.7 (25) 33.7 (26) 75.2 (27) 14.4 **Gal** (1) 102.6 (2) 81.9 (3) 76.9^b (4) 69.9 (5) 76.6^b (6) 62.2 **Glc I** (1) 106.1 (2) 75.6^a (3) 78.6 (4) 71.7 (5) 78.5^c (6) 62.9 **Glc II** (1) 105.2 (2) 75.6^a (3) 78.4^c (4) 71.7 (5) 78.1^c (6) 62.9.

Mass (F.D.)[1] : *m/z* 925 [M+Na]⁺, 902 [M]⁺, 740 [902-Hexosyl]⁺, 578 [740-Hexosyl]⁺.

Biological Activity : The compound (50 mg/kg, i.p.) showed significant hypoglycemic activity with 82.8 ± 3.4% decrease in blood glucose of alloxan-diabetic mice.[1]

References

1. N. Nakashima, I. Kimura, M. Kimura and H. Matsuura, *J. Nat. Prod.*, **56**, 345 (1993).

2. S. Saito, S. Nagase and K. Ichinose, *Chem. Pharm. Bull.*, **42**, 2342 (1994).

ANEMARSAPONIN B
Furost-20(22)-ene-3β,26-diol-3-O-[β-D-glucopyranosyl-(1→2)-β-D-galactopyranoside]-26-O-β-D-glucopyranoside

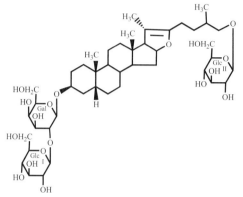

Source : *Anemarrhena asphodeloides* Bunge (Liliaceae)
Mol. Formula : $C_{45}H_{74}O_{18}$
Mol. Wt. : 902
M.P. : 226°C (decomp.)
Registry No. : [139051-27-7]

UV (MeOH) : λ_{max} 221 nm.

IR (KBr) : 3400 (OH), 2930, 2875 (CH), 1700 (double bond) 1070, 1030 (CO) cm^{-1}.

PMR (C_5D_5N, 300 MHz) : δ 0.66 (s, 3xH-18), 0.95 (s, 3xH-19), 1.00 (d, J=10.1 Hz, 3xH-27), 1.59 (s, 3xH-21), 2.31 (d, J=10.1 Hz, H-17), 4.80 (d, J=7.7 Hz, H-1 of Glc), 4.89 (d, J=7.6 Hz, H-1 of Gal), 5.25 (d, J=7.6 Hz, H-1 of Glc).

CMR (C_5D_5N, 75 MHz) : δ C-1) 31.0 (2) 27.1 (3) 75.2 (4) 31.0 (5) 37.0 (6) 26.9 (7) 26.9 (8) 35.2 (9) 40.3 (10) 35.2 (11) 1.3 (12) 40.1 (13) 43.8 (14) 54.8 (15) 31.4 (16) 84.6 (17) 64.7 (18) 14.4 (19) 24.0 (20) 103.5 (21) 11.8 (22) 152.4 (23) 34.4 (24) 23.6 (25) 33.7 (26) 75.2 (27) 17.2 **Gal** (1) 102.5 (2) 81.8 (3) 76.8 (4) 69.8 (5) 76.5 (6) 62.2 **Glc I** (1) 106.0 (2) 75.5 (3) 78.0 (4) 71.8 (5) 78.4 (6) 62.9 **Glc II** (1) 105.1 (2) 75.2 (4) 78.5 (4) 71.8 (5) 78.2 (6) 62.8.

Mass (FAB, Positive ion, NaCl added) : *m/z* 947 [M+2Na-H]⁺, 925 [M+Na]⁺, 907 [M+Na-H₂O]⁺.

Mass (FAB, Positive ion) : *m/z* 903 [M+H]⁺, 885 [M-H₂O+H]⁺, 741 [M-Glc+H]⁺, 579 [M-2Glc+H]⁺, 417 [Agl+H]⁺, 381 [Agl-H₂O+H]⁺.

Mass (E.I.) : *m/z* 578 [M-2Glc]⁺, 560 [M-2Glc-2H₂O]⁺, 416 [Agl]⁺, 398 [Agl-H₂O]⁺, 360 [Agl-2H₂O]⁺, 343, 329, 273, 255, 181, 159, 139.

Biological Activity : The compound inhibits PAF-induced rabbit platelet aggregation *in vitro* with IC_{50}=25 mm.

Reference

1.   D. Jun-xing and H.G. Yu, *Planta Med.*, **57**, 460 (1991).

ANEMARSAPONIN C

3β,26-Dihydroxy-5β-furost-29(22)-en 3-O-[β-D-glucopyranosyl-(1→2)-β-D-glucopyranoside]-26-O-β-D-glucopyranoside

Source : *Anemarrhena asphodeloides* Bge. (Liliaceae)
Mol. Formula : $C_{45}H_{74}O_{18}$
Mol. Wt. : 902
M.P. : 212°C (decomp.)
Registry No. : [185432-00-2]

IR (KBr) : 3354 (OH), 2929, 2850, 1691 ($\Delta^{20,22}$), 1075, 1037 (C-O) cm^{-1}.

PMR (C_5D_5N, D_2O, 400 MHz) : δ 0.71 (s, 3xH-18), 1.01 (s, 3xH-19), 1.08 (d, *J*=6.8 Hz, 3xH-27), 1.68 (s, 3xH-21), 2.54 (d, *J*=10.3 Hz, H-17), 3.96 (H-5 of Glc I), 4.01 (H-5 of Glc III), 4.02 (m, H-5 of Glc II), 4.06 (t, H-2 of Glc III), 4.08 (t, H-2 of Glc II), 4.18 (t, H-4 of Glc I and Glc II), 4.20 (t, H-4 of Glc II), 4.29 (t, H-3 of Glc II), 4.32 (t, H-3 of Glc III), 4.37 (t, H-2 of Glc I), 4.40 (t, H-3 of Glc I), 4.43 (H-6A of Glc II), 4.53 (H-6A of Glc I and Glc III), 4.64 (H-6B of Glc II), 4.86 (d, *J*=7.8 Hz, H-1 of Glc III), 4.99 (d, *J*=7.3 Hz, H-1 of Glc I), 5.49 (d, *J*=7.3 Hz, H-1 of Glc II).

CMR (C_5D_5N, 100 MHz) : δ C-1) 30.7 (2) 26.9 (3) 75.3 (4) 30.9 (5) 36.8 (6) 26.8 (7) 26.8 (8) 35.1 (9) 10.1 (10) 35.1 (11) 21.3 (12) 40.0 (13) 43.8 (14) 54.7 (15) 31.3 (16) 84.5 (17) 64.6 (18) 14.3 (19) 24.0 (20) 103.5 (21) 11.7 (22) 152.3 (23) 34.3 (24) 23.6 (25) 33.6 (26) 75.2 (27) 17.1 **Glc I** (1) 101.9 (2) 83.1 (3) 78.5 (4) 71.7 (5) 78.2 (6) 62.8 **Glc II** (1) 105.9 (2) 77.0 (3) 77.9 (4) 71.5 (5) 62.6 **Glc III** (1) 105.1 (2) 75.2 (3) 78.5 (4) 71.6 (5) 78.2 (6) 62.8.

Mass (FAB, Positive ion) : *m/z* 925 [M+Na]$^+$, 903 [M+H]$^+$, 741 [M+H-2xGlc]$^+$, 579 [M+H-2xGlc]$^+$, 417 [M+H-3xGlc]$^+$, 399 [M+H-3xGlc-H_2O]$^+$, 255, 185, 145.

Mass (E.I.) : *m/z* 416, 398 [416-H_2O]$^+$, 344, 343, 325, 287, 273, 255, 217, 201, 181, 163, 139, 109, 95.

Reference

1. B.P. Ma, J.X. Dong, B.J. Wang and X.Z. Yan, *Huaxue Xuebao* (*Acta Pharm. Sin.*), **31**, 271 (1996).

ASPARASAPONIN II

25S-Furost-5-ene-3β,22α,26-triol 3-O-[α-L-rhamnopyranosyl-(1→4)-β-D-glucopyranoside]-26-O-β-D-glucopyranoside

Source : *Asparagus officinalis* L. (Liliaceae)
Mol. Formula : $C_{45}H_{74}O_{18}$
Mol. Wt. : 902
M.P. : $188 \sim 192^{\circ}C$
$[\alpha]_D^{22}$: -74.6° (c=1.0, CH_3OH-$CHCl_3$)
Registry No. : [60433-66-1]

IR (KBr) **:** $3600 \sim 3200$ (br, OH) cm^{-1}.

Reference

1. K. Kawano, H. Sato and S. Sakamura, *Agric. Biol. Chem.*, 41, (1977).

DRACAENA CONCINNA SAPONIN 2

22ξ-Methoxy-5α-furost-25(27)-ene-1β-3α,26-triol 1-O-[-α-L-rhamnopyranosyl-(1→2)-α-L-arbinopyranoside]-26-O-β-D-glucopyranoside

Source : *Drcaena concinna* Kunth (Agavaceae)
Mol. Formula : $C_{45}H_{74}O_{18}$
Mol. Wt. : 902
$[\alpha]_D^{27}$: -37.5° (c=0.41, MeOH)
Registry No. : [209848-40-8]

IR (KBr) : 3430 (OH), 2925 (CH) cm^{-1}.

PMR (C$_5$D$_5$N, 400 MHz) : δ 0.83 (s, 3xH-18), 1.12 (d, J=6.8 Hz, 3xH-21), 1.25 (s, 3xH-19), 1.76 (d, J=6.1 Hz, 3xH-6 of Rha), 3.25 (s, OCH_3), 4.73 (d, J=7.7 Hz, H-1 of Ara), 4.91 (d, J=7.8 Hz, H-1 of Glc), 5.07 and 5.34 (each 1H, br s, H-27A and 27B), 6.39 (br s, H-1 of Rha).

CMR (C$_5$D$_5$N, 100 MHz) : δ C-1) 80.4 (2) 35.0 (3) 65.8 (4) 37.3 (5) 39.4 (6) 28.8 (7) 32.6 (8) 36.4 (9) 55.0 (10) 42.6 (11) 23.7 (12) 40.7 (13) 40.6 (14) 56.8 (15) 32.3 (16) 81.4 (17) 64.3 (18) 16.7 (19) 7.7 (20) 40.4 (21) 16.1 (22) 112.4 (23) 31.5 (24) 28.0 (25) 146.8 (26) 72.0 (27) 111.0 (OCH_3) 47.3 **Ara** (1) 100.4 (2) 74.9 (3) 76.3 (4) 70.3 (5) 67.5 **Rha** (1) 101.5 (2) 72.6 (3) 72.6 (4) 74.2 (5) 69.3 (6) 19.0 **Glc** (1) 103.8 (2) 75.1 (3)78.6 (4) 71.7 (5) 78.5 (6) 62.8.

Mass (FAB, Negative ion) : *m/z* 901 [M-H]$^-$.

Reference

1.   Y. Mimaki, M. Kuroda, Y. Takaashi and Y. Sashida, *Phytochemistry*, **47,** 1351 (1998).

DIOSCOREA GRACILLIMA SAPONIN A, PROTOBIOSIDE

(25*R*)-Furost-5-en-3β,22,26-triol 3-O-[α-L-rhamnopyranosyl-(1→2)-β-D-glucopyranoside]-26-O-[β-D-glucopyranoside]

Source : *Dioscorea gracillima* Miq.[1] (Dioscoraceae), *Ophiopogon planiscapus* Nakai[2] (Liliaceae), *Smilax china* L.[3] (Liliaceae)
Mol. Formula : $C_{45}H_{74}O_{18}$
Mol. Wt. : 902
M.P. : 194-197°C (decomp.)[2]
$[α]_D^{23}$: -70.4° (c=0.62, Pyridine)[2]
Registry No. : [55972-80-0]

IR (KBr)[3] : 3400, 1650, 1130, 1070, 1045, 905, 887, 833, 807 cm^{-1}.

PMR (C$_5$D$_5$N, 300 MHz)[3] : δ 0.86 (s, 3xH-18), 0.94 (d, *J*=6.0 Hz, 3xH-27), 1.01 (s, 3xH-19), 1.30 (d, *J*=6.8 Hz, 3xH-21), 1.73 (d, *J*=6.1 Hz, 3xH-6 of Rha), 4.98 (d, *J*=7.1 Hz, H-1 of Glc), 5.29 (br d, *J*=4.8 Hz, H-6), 6.31 (s, H-1 of Rha).

CMR (C$_5$D$_5$N, 75 MHz)[3] : δ C-1) 37.5 (2) 30.1 (3) 78.5^a (4) 40.0 (5) 140.8 (6) 121.7 (7) 32.1 (8) 31.6 (9) 50.3 (10) 37.1 (11) 21.0 (12) 38.9 (13) 40.7 (14) 56.5 (15) 32.3 (16) 81.1 (17) 63.7 (18) 16.4 (19) 19.4 (20) 40.6 (21) 16.4 (22) 110.6 (23) 37.1 (24) 28.1 (25) 34.1 (26) 75.1 (27) 17.4 **Glc I** (1) 100.3 (2) 79.5 (3) 77.8^b (4) 71.7 (5) 77.9^b (6) 62.6 **Rha** (1) 102.0 (2) 72.4 (3) 72.7 (4) 74.0 (5) 69.4 (6) 18.6 **Glc II** (1) 104.8 (2) 75.0 (3) 78.1 (4) 71.7 (5) 78.3^a (6) 62.8.

References

1. T. Kawasaki, T. Komori, K. Miyahara, T. Nohara, I. Hosokawa, K. Mihashi, *Chem. Pharm. Bull.*, **22**, 2164 (1974).

2. Y. Watanabe, S. Sanada, Y. Ida and J. Shoji, *Chem. Pharm. Bull.*, **31**, 3486 (1983).

3. S.W. Kim, K.C. Chung, K.H. Son and S.S. Kang, *Korean J. Pharmacog.*, **20**, 145 (1989).

MACROSTEMONOSIDE F

(25*R*)-5β-Furost-20(22)-ene-3β,26-diol 3-O-[β-D-glucopyranosyl-(1→2)-β-D-galactoside]-
26-O-β-D-glucopyranoside

Source : *Allium macrostemon* Bunge (Liliaceae)
Mol. Formula : $C_{45}H_{74}O_{18}$
Mol. Wt. : 902
M.P. : 196-198.5°C
Registry No. : [151215-11-1]

UV (MeOH) : λ_{max} 220 nm.

IR (KBr) : 3400 (OH), 2900, 2850, 1700 ($\Delta^{20(22)}$), 1070, 1030 (C-O) cm^{-1}.

PMR (C_5D_5N, 400 MHz) : δ 0.68 (s, 3xH-18), 0.95 (s, 3xH-19), 1.00 (d, *J*=6.3 Hz, 3xH-27), 1.61 (s, 3xH-21), 4.83 (d, *J*=7.6 Hz, H-1 of Glc), 4.92 (d, *J*=7.6 Hz, H-1 of Gal), 5.28 (d, *J*=7.6 Hz, H-1 of Glc).

CMR (C_5D_5N, 100 MHz) : δ C-1) 30.9 (2) 26.9 (3) 75.1 (4) 30.9 (5) 36.9 (6) 26.9 (7) 26.8 (8) 35.1 (9) 40.1 (10) 35.1 (11) 21.3 (12) 40.1 (13) 43.8 (14) 54.7 (15) 31.4 (16) 84.5 (17) 64.6 (18) 14.4 (19) 24.0 (20) 103.6 (21) 11.8 (22) 152.3 (23) 34.4 (24) 23.6 (25) 33.4 (26) 74.9 (27) 17.3 **Gal** (1) 102.5 (2) 81.8 (3) 76.8^a (4) 69.8 (5) 76.5^a (6) 61.8 **Glc I** (1) 106.0 (2) 75.5^a (3) 78.0^b (4) 71.6 (5) 78.4^b (6) 62.4 **Glc II** (1) 104.8 (2) 75.1 (3) 78.5^b (4) 71.6 (5) 78.3^b (6) 62.4.

Mass (FAB, Positive ion) : *m/z* 903 [M+H]$^+$, 741 [M-Glc+H]$^+$, 579 [M-Glcx2+H]$^+$, 417 [+H]$^+$, 399 [H$_2$O+H]$^+$.

Biological Activity: The compound strongly inhibits ADP-induced human platelet aggregation *in vitro* with $1C_{50}$=0.020 mm.

Reference

1.   J. Peng, X. Wang and X. Yao, *Yaoxue Xuebao* (*Acta Pharm. Sinica*), **28**, 526 (1993).

NOLINOFUROSIDE E

25(*S*)-Furost-5-ene-1β,3β,22α,26-tetraol 1-O-[α-L-rhamnopyranosyl-(1→2)-β-D-fucopyranoside]-26-O-[β-D-glucopyranoside]

Source : *Nolina microcarpa* S. Wats. (Dracaenaceae)
Mol. Formula : $C_{45}H_{74}O_{18}$
Mol. Wt. : 902
Registry No. : [144028-93-3]

Reference

1.   G.V. Shevchuk, Yu. S. Vollerner, A.S. Shashkov and V.Ya Chirva, *Khim. Prir. Soedin.*, **27**, 678 (1991); *Chem. Nat. Comp.*, **27**, 597 (1991).

NOLINOFUROSIDE F

(25S)-Furost-5-ene-1β,3β,22α,26-tetraol 1-O-[β-D-fucopuranoside]-3-O-[α-L-rhamnopyranoside]-26-O-[β-D-glucopyranoside]

Source : *Nolina microcarpa* S. Wats. (Dracaenaceae)
Mol. Formula : $C_{45}H_{74}O_{18}$
Mol. Wt. : 902
Registry No. : [144028-92-2]

22-O-Methyl ether : $C_{46}H_{76}O_{18}$

Isolated from *Nolina microcarpa* S. Wats (Dracaenaceae) Artefact?; $[\alpha]_D^{18}$: -54.0 ± 2° (c=1.01, Pyridine)

IR (KBr) : 3200-3600 (OH) cm^{-1}.

PMR (C_5D_5N, 250/300 MHz) : δ 0.81 (s, 3xH-18), 1.01 (d, J=7.0 Hz, 3xH-27), 1.08 (d, J=7.0 Hz, 3xH-21), 1.13 (s, 3xH-19), 1.54 (d, J=7.0 Hz, 3xH-6 of Fuc), 1.69 (d, J=6.0 Hz, 3xH-6 of Rha), 3.23 (s, OCH_3), 3.48 (dd, $J_{26A,26B}$=10.0 Hz, $J_{26,25}$=7.1 Hz, H-26A), 3.68 (dq, $J_{5,6}$=7.0 Hz, H-5 of Fuc), 3.74 (dd, $J_{1,2ax}$=12.0 Hz, $J_{1,2eq}$=4.2 Hz, H-1ax), 3.85 (m, H-3), 3.96 (m, H-5 of Glc), 4.04 (dd, $J_{1,2}$=8.0 Hz, $J_{2,3}$=9.0 Hz, H-2 of Glc), 4.05 (m, H-3 of Fuc), 4.05 (dd, $J_{26A,26B}$=10.0 Hz, $J_{26B,25}$=5.0 Hz, H-26B), 4.22 (t, $J_{3,4}$=$J_{4,5}$=9.0 Hz, H-4 of Glc), 4.25 (m, H-4 of Rha), 4.28 (t, $J_{2,3}$=$J_{3,4}$=9.0 Hz, H-3 of Glc), 4.31 (dd, $J_{1,2}$=7.8 Hz, $J_{2,3}$=8.5 Hz, H-2 of Fuc), 4.37 (dq, $J_{4,5}$=10.3 Hz, $J_{5,6}$=6.0 Hz, H-5 of Rha), 4.39 (dd, $J_{6A,6B}$=12.6 Hz, $J_{6A,5}$=4.7 Hz, H-6A of Glc), 4.42 (m, H-4 of Fuc), 4.50 (m, H-16), 4.51 (dd, $J_{2,3}$=4.0 Hz, $J_{3,4}$=9.5 Hz, H-3 of Rha), 4.55 (br s, H-2 of Rha), 4.57 (dd, $J_{6A,6B}$=12.6 Hz, $J_{6B,5}$=3.0 Hz, H-6B of Glc), 4.63 (d, J=8.0 Hz, H-1 of Glc), 4.68 (d, J=7.8 Hz, H-1 of Fuc), 5.51 (br d, J=5.5 Hz, H-6), 5.55 (br s, H-1 of Rha).

CMR (C_5D_5N, 62.5/75 MHz) : δ C-1) 83.77 (2) 35.89 (3) 73.64 (4) 39.78 (5) 138.51 (6) 125.73 (7) 32.07 (8) 33.10 (9) 50.72 (10) 43.04 (11) 23.88 (12) 40.62 (13) 40.42 (14) 57.05 (15) 32.48 (16) 81.47 (17) 64.49 (18) 16.78 (19) 14.65 (20) 40.64 (21) 16.20 (22) 112.84 (23) 31.06 (24) 28.27 (25) 34.56 (26) 75.02 (27) 17.54 (OCH_3) 47.39 **Fuc** (1) 102.53 (2) 72.26 (3) 75.28 (4) 72.53 (5) 71.22 (6) 17.36 **Rha** (1) 99.97 (2) 72.83 (3) 72.83 (4) 74.19 (5) 70.07 (6) 18.57 **Glc** (1) 105.12 (2) 75.28 (3) 78.43 (4) 71.93 (5) 78.66 (6) 63.03.

Mass (L.S.I.) : *m/z* 939 [M+Na]⁺.

Reference

1. G.V. Shevchuk, Yu. S. Vollerner, A.S. Shashkov and V.Ya Chirva, *Khim. Prir. Soedin.*, **27**, 678 (1991); *Chem. Nat. Comp.*, **27**, 597 (1991).

TORVOSIDE F
(25*S*)-6α,26-Dihydroxy-22α-methoxy-5α-furostan-3-one 6-O-[β-D-xylopyranosyl-(1→3)-β-D-quinovopyranoside]-26-O-β-D-glucopyranoside

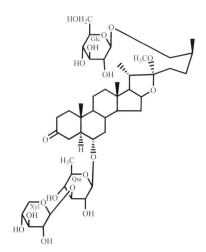

Source : *Solanum torvum* Swartz (Solanaceae)
Mol. Formula : $C_{45}H_{74}O_{18}$
Mol. Wt. : 902
$[\alpha]_D^{27}$: -18.8° (c=0.50, MeOH)
Registry No. : [184777-21-7]

PMR (C₅D₅N, 400 MHz) **:** δ 0.83 (s, 3xH-18), 1.00 (s, 3xH-19), 1.07 (d, *J*=6.6 Hz, 3xH-27), 1.17 (d, *J*=6.6 Hz, 3xH-21), 1.59 (d, *J*=5.9 Hz, 3xH-6 of Qui), 3.25 (s, OC*H₃*), 4.78 (d, *J*=7.3 Hz, H-1 of Glc), 4.87 (d, *J*=8.1 Hz, H-1 of Qui), 5.28 (d, *J*=8.1 Hz, H-1 of Xyl).

CMR (C₅D₅N, 100 MHz) **:** δ C-1) 38.7 (2) 38.1 (3) 210.6 (4) 39.8 (5) 52.5 (6) 79.9 (7) 40.9 (8) 34.4 (9) 53.2 (10) 36.8 (11) 21.3 (12) 39.7 (13) 41.4 (14) 56.0 (15) 32.0 (16) 81.2 (17) 64.2 (18) 16.4 (19) 12.6 (20) 40.5 (21) 16.3 (22) 112.6 (23) 30.9 (24) 28.2 (25) 33.9 (26) 74.9 (27) 17.6 (OCH₃) 47.3 **Qui** (1) 105.3 (2) 75.3 (3) 87.3 (4) 74.7 (5) 72.3 (6) 18.6 **Xyl** (1) 106.3 (2) 74.6 (3) 78.2 (4) 70.9 (5) 67.4 **Glc** (1) 105.1 (2) 75.3 (3) 78.6 (4) 71.8 (5) 78.2 (6) 62.9.

Mass (FAB, Positive ion) **:** *m/z* 871 [M-OMe]⁺, 739 [M-OMe-Xyl]⁺, 593 [M-OMe-Xyl-Qui]⁺.

Reference

1.   S. Yahara, T. Yamashita, N. Nozawa (nee Fujimura) and T. Nohara, *Phytochemistry*, **43**, 1069 (1996).

TRIGOFOENOSIDE A

(25*S*)-Furost-5-ene-3β,22ξ,26-triol 3-O-[α-L-rhamnopyranosyl-(1→2)-β-D-glucopyranoside]-26-O-β-D-glucopyranoside

Source : *Trigonella foenum-graecum* L. (Leguminosae)
Mol. Formula : $C_{45}H_{74}O_{18}$
Mol. Wt. : 902
M.P. : 219-221°C (decomp.)
[α]$_D$: -90.1° (c=1.0, Pyridine)
Registry No. : [99705-66-5]

Mass (FAB, Positive ion) : *m/z* 1036 [M+H+Cs]$^+$, 941 [M+K]$^+$, 925 [M+Na]$^+$.

22-Methyl ether : (Trigofoenoside A-1) [99664-40-1]

M.P. : 210-213°C (decomp.); **[α]$_D$:** -84.18° (c=1.0, Pyridine)

IR (KBr) : 3600-3200 (OH), 1150-1000 (C-O-C), cm^{-1}. No spiroketal bond.

PMR (DMSO-d_6, 400 MHz) : δ 1.76 (br s, 3xH-6 of Rha), 3.24 (s, OCH_3 H-22), 4.25 (d, *J*=7.1 Hz, H-1 of Glc), 4.57 (d, *J*=7.0 Hz, H-1 of Glc), 5.20 (br s, H-1 of Rha).

Reference

1.   R.K. Gupta, D.C. Jain and R.S. Thakur, *Phytochemistry*, **24**, 2399 (1985).

TRIGONEOSIDE XIIa

(25S)-Furostane-4-ene-3β,22ξ,26-triol 3-O-[α-L-rhamnopyranosyl-(1→2)-β-D-glucopyranoside]-26-O-β-D-glucopyranoside

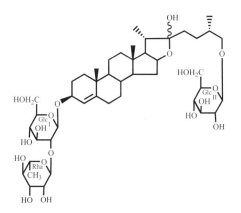

Source : *Trigonella foenum-graecum* L. (Fabaceae)
Mol. Formula : $C_{45}H_{74}O_{18}$
Mol. Wt. : 902
$[\alpha]_D^{22}$: -48.8° (c=0.6, MeOH)
Registry No. : [290348-00-4]

IR (KBr) : 3432, 2932, 1074, 1047 cm^{-1}.

PMR (C_5D_5N, 500 MHz) : δ 0.91, 1.07 (both s, 3xH-18, 3xH-19), 1.01 (d, J=7.2 Hz, 3xH-27), 1.29 (d, J=6.7 Hz, 3xH-21), 1.67 (d, J=5.8 Hz, 3xH-6 of Rha), 3.48, 4.05 (both dd-like, 2xH-26), 4.48 (dd-like, H-3), 4.77 (d, J=7.6 Hz, H-1 of Glc II), 4.91 (ddd-like, H-16), 4.99 (d, J=7.6 Hz, H-1 of Glc I), 5.81 (br s, H-4), 6.25 (br s, H-1 of Rha).

CMR (C_5D_5N, 125.0 MHz) : δ C-1) 35.8 (2) 27.8 (3) 75.5 (4) 121.5 (5) 147.2 (6) 33.5 (7) 32.6 (8) 36.0 (9) 54.7 (10) 37.7 (11) 21.1 (12) 40.2 (13) 41.1 (14) 56.2 (15) 32.4 (16) 81.1 (17) 63.9 (18) 16.7 (19) 18.9 (20) 40.8 (21) 16.4 (22) 110.7 (23) 37.1 (24) 28.3 (25) 34.4 (26) 75.3 (27) 17.5 **Glc I** (1) 101.7 (2) 78.2 (3) 78.6 (4) 72.1 (5) 79.7 (6) 62.9 **Rha** (1) 102.3 (2) 72.5 (3) 72.8 (4) 74.1 (5) 69.5 (6) 18.6 **Glc II** (1) 105.1 (2) 75.2 (3) 78.5 (4) 71.8 (5) 78.3 (6) 62.9.

Mass (FAB, Negative ion) : *m/z* 901 [M-H]$^-$, 755 [M-$C_6H_{11}O_4$]$^-$, 739 [M-$C_6H_{11}O_5$]$^-$, 593 [M-$C_{12}H_{21}O_9$]$^-$, 431 [M-$C_{18}H_{31}O_{14}$]$^-$.

Mass (FAB, Positive ion.) : *m/z* 925 [M+Na]$^+$.

Reference

1.   T. Murakami, A. Kishi, H. Matsuda and M. Yoshikawa, *Chem. Pharm. Bull.,* **48**, 994 (2000).

TRIGONEOSIDE XIIb

(25R)-Furost-4-ene-3β,22ξ,26-triol 3-O-[α-L-rhamnopyranosyl-(1→2)-β-D-glucopyranoside]-26-O-β-D-glucopyranoside

Source : *Trigonella foenum-graecum* L. (Fabaceae)
Mol. Formula : $C_{45}H_{74}O_{18}$
Mol. Wt. : 902
[α]$_D^{22}$: -48.2° (c=0.5, MeOH)
Registry No. : [290347-97-6]

IR (KBr) : 3432, 2932, 1071, 1048 cm^{-1}.

PMR (C$_5$D$_5$N, 500 MHz) : δ 0.92, 1.07 (both s, 3xH-18, 3xH-19), 0.99 (d, J=6.7 Hz, 3xH-27), 1.32 (d, J=6.7 Hz, 3xH-21), 1.68 (d, J=6.1 Hz, 3xH-6 of Rha), 3.68 (dd, J=6.1, 9.4 Hz, H-26A), 3.92 (m, H-26B), 4.49 (ddd-like, H-3), 4.78 (d, J=7.9 Hz, H-1 of Glc II), 4.92 (ddd-like, H-16), 4.99 (d, J=7.6 Hz, H-1 of Glc I), 5.81 (br s, H-4), 6.23 (br s, H-1 of Rha).

CMR (C$_5$D$_5$N, 125.0 MHz) : δ C-1) 35.8 (2) 27.8 (3) 75.5 (4) 121.5 (5) 147.3 (6) 32.5 (7) 33.5 (8) 35.8 (9) 54.7 (10) 37.7 (11) 21.1 (12) 40.2 (13) 41.1 (14) 56.2 (15) 32.4 (16) 81.1 (17) 63.9 (18) 16.7 (19) 18.9 (20) 40.7 (21) 16.4 (22) 110.7 (23) 37.2 (24) 28.4 (25) 34.3 (26) 75.3 (27) 17.5 **Glc I** (1) 101.6 (2) 78.3 (3) 78.6 (4) 72.0 (5) 79.6 (6) 62.9 **Rha** (1) 102.3 (2) 72.6 (3) 72.8 (4) 74.1 (5) 69.4 (6) 18.6 **Glc II** (1) 104.9 (2) 75.2 (3) 78.6 (4) 71.7 (5) 78.4 (6) 62.9.

Mass (FAB, Negative ion) : *m/z* 901 [M-H]$^-$, 755 [M-C$_6$H$_{11}$O$_4$]$^-$, 739 [M-C$_6$H$_{11}$O$_5$]$^-$, 593 [M-C$_{12}$H$_{21}$O$_9$]$^-$, 431 [M-C$_{18}$H$_{31}$O$_{14}$]$^-$.

Mass (FAB, Positive ion) : *m/z* 925 [M+Na]$^+$.

Reference

1.   T. Murakami, A. Kishi, H. Matsuda and M. Yoshikawa, *Chem. Pharm. Bull.,* **48**, 994 (2000).

TUBEROSIDE A (ALLIUM)

(25*S*)-5α-Furost-20(22)-ene-2α,3β,26-triol 3-O-{α-L-rhamnopyranosyl-(1→2)-β-D-glucopyranoside}-26-O-β-D-glucopyranoside

Source : *Allium tuberosum* Rottle ex spreng. (Liliaceae)
Mol. Formula : $C_{45}H_{74}O_{18}$
Mol. Wt. : 902
$[\alpha]_D^{22}$: -29.5° (c=0.27, MeOH)
Registry No. : [259810-65-6]

IR (KBr) **:** 3417, 1450, 1381, 1000-1100 cm^{-1}.

PMR (C$_5$D$_5$N, 400 MHz) **:** δ 0.76 (s, 3xH-18), 0.96 (s, 3xH-19), 1.10 (d, *J*=6.3 Hz, 3xH-27), 1.70 (s, 3xH-21), 1.73 (d, *J*=6.8 Hz, 3xH-6 of Rha), 2.54 (d, *J*=9.6 Hz, H-17), 3.52 (m, H-26A), 3.94 (m, H-3), 3.95 (m, H-5 of Glc II), 4.02 (m, H-2 of Glc II), 4.03 (m, H-5 of Glc I), 4.05 (m, H-4 of Glc I), 4.10 (m, H-26B), 4.12 (m, H-2), 4.20 (m, H-2 of Glc I), 4.21 (m, H-3 of Glc II), 4.22 (m, H-4 of Glc II), 4.25 (m, H-3 of Glc I), 4.32 (m, H-6A of Glc I), 4.33 (m, H-4 of Rha), 4.40 (m, H-6 of Glc I, H-6A of Glc II), 4.55 (m, H-6B of Glc II), 4.57 (m, H-3 of Rha), 4.81 (m, H-2 of Rha), 4.85 (d, *J*=7.7 Hz, H-1 of Glc II), 4.88 (m, H-5 of Rha), 4.89 (m, H-16), 5.06 (d, *J*=7.1 Hz, H-1 of Glc I), 6.30 (s, H-1 of Rha).

CMR (C$_5$D$_5$N, 100 MHz) **:** δ C-1) 46.0 (2) 70.9 (3) 85.5 (4) 33.8 (5) 44.9 (6) 28.4 (7) 32.7 (8) 34.8 (9) 54.6 (10) 37.1 (11) 21.8 (12) 40.0 (13) 44.0 (14) 54.9 (15) 34.6 (16) 84.7 (17) 64.8 (18) 14.6 (19) 13.7 (20) 103.8 (21) 12.0 (22) 152.6 (23) 31.6 (24) 23.9 (25) 33.9 (26) 75.5 (27) 17.4 **Glc I** (1) 101.4 (2) 78.3 (3) 79.7 (4) 72.1 (5) 78.5 (6) 62.8 **Glc II** (1) 105.3 (2) 75.3 (3) 78.7 (4) 71.9 (5) 78.6 (6) 63.0 **Rha** (1) 102.3 (2) 72.6 (3) 72.9 (4) 74.3 (5) 69.7 (6) 18.8.

Mass (FAB) **:** *m/z* 903 [M+H]$^+$, 741 [M+H-Glc]$^+$, 595 [M+H-Glc-Rha]$^+$, 433 [M+H-2xGlc-Rha]$^+$.

Reference

1.   S. Sang, A. Lao, H. Wang and Z. Chen, *Phytochemistry*, **52**, 1611 (1999).

TUBEROSIDE F' (SOLANUM)

Yamogenin 3-O-[α-L-rhamnopyranosyl-(1→2)-β-D-galactopyranoside]-26-O-β-D-glucopyranoside]

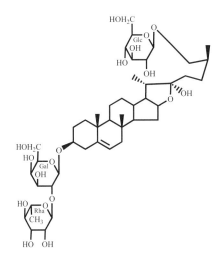

Source : *Solanum tuberosum* L. (Solanaceae)
Mol. Formula : $C_{45}H_{74}O_{18}$
Mol. Wt. : 902
M.P. : 152-153°C
$[α]_D^{20}$: -127.0° (c=2.5, MeOH)
Registry No. : [144071-72-7]

Reference

1.   P.K. Kintya, T.I. Prasol and N.E. Mashchenko, *Khim. Prir. Soedin.*, 730 (1991); *Chem. Nat. Comp.*, **27**, 646 (1991).

ASPARAGUS COCHINCHINENSIS SAPONIN ASP-IV

(5β,25S)-22ξ-Methoxy-furostane-3β,26-diol 3-O-[β-D-xylopyranosyl-(1→4)-β-D-glucopyranoside]-
26-O-β-D-glucopyranoside

Source : *Asparagus cochinchinensis* (Loureiro) Merrill (Liliaceae)

Mol. Formula : $C_{45}H_{76}O_{18}$

Mol. Wt. : 904

M.P. : 165-167°C (decomp.)

[α]$_D$: -22.9° (c=2.1, MeOH)

Registry No. : [72947-79-6]

PMR (C_5D_5N, 90 MHz) : δ 3.28 (s, OCH_3), 4.82 (d, J=7.0 Hz, anomeric H), 4.88 (d, J=6.0 Hz, anomeric H), 5.10 (d, J=6.5 Hz, anomeric H).

CMR (C_5D_5N, 22.15 MHz) : δ **Glc I** C-1) 102.8 **Glc II** (1) 104.7 **Xyl** (1) 105.3.

Reference

1.   T. Konishi and J. Shoji, *Chem. Pharm. Bull.*, **27**, 3086 (1979).

ASPACOCHIOSIDE A

**5β,25S-Furostan-3β,21α,26-triol 3-O-[α-L-rhamnopyranosyl-(1→4)-β-D-glucopyranoside
26-O-β-D-glucopyranoside]**

Source : *Asparagus cochinchinensis* (Lour.) Merr.
(Liliaceae)
Mol. Formula : $C_{45}H_{76}O_{18}$
Mol. Wt. : 904
M.P. : 212-213°C[1]
[α]$_D^{25}$: -48.5° (c=0.10, [(CH₃)₂CO : H₂O 1:1][1]
Registry No. : [557769-32-1]

IR (KBr)[1] : 3386 (br, OH) cm^{-1}.

PMR (C₅D₅N, 500 MHz)[2] : δ 0.82 (s, 3xH-18), 0.86 (s, 3xH-19), 0.95 (m, H-9), 1.02 (d, J=7.0 Hz, 3xH-27), 1.07 (m, H-14), 1.12 (m, H-12A), 1.24 (m, H-12B), 1.26 (m, H-11A), 1.28 (m, H-7A), 1.31 (d, J=7.0 Hz, 3xH-21), 1.38 (m, H-15A), 1.45 (m, H-5), 1.46 (m, H-1A), 1.48 (m, H-8), 1.53 (m, H-7B), 1.54 (m, H-2A), 1.67 (m, H-6A, 1.69 (d, J=7.0 Hz, 3xH-6 of Rha), 1.70 (m, H-1B and H-2B), 1.75 (m, H-11B), 1.77 (m, H-4), 1.88 (m, H-4B), 1.89 (m, H-6B), 1.90 (m, H-25), 1.92 (m, 2xH-24), 1.94 (m, H-23A), 1.95 (m, H-17), 2.03 (m, H-15B), 2.08 (m, H-23B), 2.22 (m, H-20), 3.46 (dd, J=10.0, 7.2 Hz, H-26A), 3.70 (br d, J=9.5 Hz, H-5 of Glc I), 3.93 (m, H-5 of Glc II), 3.99 (dd, J=8.0, 7.0 Hz, H-2 of Glc II), 4.02 (dd, J=7.5, 7.0 Hz, H-2 of Glc I), 4.07 (dd, J=10.0, 7.2 Hz, H-26B), 4.13 (br d, J=10.5 Hz, H-6A of Glc I), 4.21 (dd, J=7.5, 9.0 Hz, H-3 of Glc I), 4.25 (m, H-3, H-6B of Glc I), 4.25 (dd, J=9.0, 9.0 Hz, H-4 of Glc II), 4.35 (dd, J=9.0, 8.5 Hz, H-4 of Rha), 4.39 (br d, J=12.5 Hz, H-6A of Glc II), 4.48 (dd, J=9.0, 9.5 Hz, H-4 of Glc I), 4.54 (br d, J=12.5 Hz, H-6B of Glc II), 4.58 (br d, J=8.5 Hz, H-3 of Rha), 4.70 (br s, H-2 of Rha), 4.81 (d, J=7.0 Hz, H-1 of Glc II), 4.85 (d, J=7.0 Hz, H-1 of Glc I), 4.98 (m, H-16), 5.03 (dq, J=9.0, 7.0 Hz, H-5 of Rha), 5.92 (br s, H-1 of Rha).

CMR (C₅D₅N, 125 MHz)[2] : δ C-1) 30.5 (2) 27.0 (3) 74.6 (4) 30.9 (5) 37.0 (6) 27.0 (7) 26.8 (8) 35.5 (9) 40.3 (10) 35.2 (11) 21.2 (12) 40.4 (13) 41.2 (14) 56.4 (15) 32.4 (16) 81.2 (17) 64.2 (18) 16.7 (19) 23.9 (20) 40.7 (21) 16.5 (22) 10.6*

(23) 37.1 (24) 28.3 (25) 34.4 (26) 75.4 (27) 17.5 **Glc I** (1) 103.2 (2) 75.6 (3) 76.8 (4) 78.2 (5) 77.2 (6) 61.5 **Rha** (1) 102.7 (2) 72.7 (3) 72.8 (4) 74.0 (5) 70.3 (6) 18.3 **Glc II** (1) 105.2 (2) 75.2 (3) 78.6 (4) 71.7 (5) 78.5 (6) 62.8. * This value is erroneous and certainly a misprint.

Mass (E.S.I., Positive ion)^2 : m/z 927 [M+Na]$^+$., 905 [M+H]$^+$, 887 [M+H-H$_2$O]$^+$, 725.

Mass (E.S.I., Positive ion, H.R.)^2 : 927.4919 [(M+Na)$^+$, calcd. for 927.4929].

References

1.   Y.C. Yang, S.Y. Huang and J.G. Shi, *Chin. Chem. Lett.*, **13**, 1185 (2002).

2.   J.G. Shi, G.-Q. Li, S.-Y. Huang, S.-Y. Mo, Y. Wang, Y.-C. Yang and W.-Y. Hu, *J. Asian Natural Prod. Res.*, **6**, 99 (2004).

22-METHOXY - ASP-IV
(5β,25S)-22α-Methoxy-3β,26-Dihydroxyfurostane 3-O-{β-D-xylopyranosyl-(1→4)-β-D-glucopyranoside}-26-O-β-D-glucopyranoside

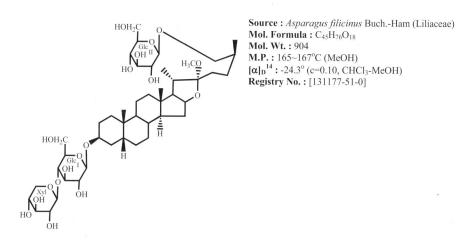

Source : *Asparagus filicinus* Buch.-Ham (Liliaceae)
Mol. Formula : C$_{45}$H$_{76}$O$_{18}$
Mol. Wt. : 904
M.P. : 165~167°C (MeOH)
[α]$_D^{14}$: -24.3° (c=0.10, CHCl$_3$-MeOH)
Registry No. : [131177-51-0]

IR (KBr) **:** 3400, 1050 cm^{-1}.

PMR (C$_5$D$_5$N, 400 MHz) **:** δ 0.81 (s, 3xH-19), 0.86 (s, 3xH-18), 1.06 (d, *J*=6.5 Hz, 3xH-27), 1.20 (d, *J*=5.6 Hz, 3xH-21), 3.28 (s, OC*H*$_3$ of H-22), 4.87 (d, *J*=7.8 Hz), 4.93 (d, *J*=7.6 Hz), 5.18 (d, *J*=7.6 Hz).

CMR (C$_5$D$_5$N, 100 MHz) : δ C-1) 30.6 (2) 26.9 (3) 74.7 (4) 30.9 (5) 36.9 (6) 26.9 (7) 26.7 (8) 35.5 (9) 40.5 (10) 35.2 (11) 21.1 (12) 40.2 (13) 41.2 (14) 56.4 (15) 32.1 (16) 81.1 (17) 64.4 (18) 16.3 (19) 23.8 (20) 40.2 (21) 16.5 (22) 112.7 (23) 30.9 (24) 28.2 (25) 34.4 (26) 75.1 (27) 17.5 (OCH$_3$) 47.4 **Glc I** (1) 102.9 (2) 75.0 (3) 76.5 (4) 81.5 (5) 76.4 (6) 61.9 **Xyl** (1) 105.5 (2) 74.9 (3) 78.3 (4) 70.7 (5) 67.3 **Glc II** (1) 105.0 (2) 75.0 (3) 78.5 (4) 71.8 (5) 78.4 (6) 62.9.

Reference

1.   Y. Ding and C.R. Yang, *Acta Pharm. Sinica*, **25**, 509 (1990).

NICOTIANOSIDE E
25(S)-5α-Furostan-3β,22α,26-triol 3-O-[α-L-rhamnopyranosyl-(1→2)-β-D-glucopyranoside]-26-O-[β-D-glucopyranoside]

Source : *Nicotiana tabacum* L. seeds (Solanaceae)[1]
Mol. Formula : C$_{45}$H$_{76}$O$_{18}$
Mol. Wt. : 904
M.P. : 178°C[1]
[α]$_D^{20}$: -78.0° (c=1.0, CH$_3$OH)[1]
Registry No. : [170678-02-1]

PMR (C$_5$D$_5$N, 300/500 MHz) : δ 0.70 (s, 3xH-19), 0.71 (s, 3xH-18), 1.09 (d, $J_{25,27}$=7.1 Hz, 3xH-27), 1.10 (d, $J_{20,21}$=6.8 Hz, 3xH-21), 1.67 (d, J=7.0 Hz, 3xH-6 of Rha), 3.59 (m, H-5 of Glc I), 3.78 (m, H-3), 3.82 (dd, $J_{3,4}$=9.8 Hz, H-3 of Glc I), 3.82 (t, H-4 of Glc II), 3.89 (dd, H-6A of Glc I), 3.89 (dd, $J_{2,3}$=8.0 Hz, H-2 of Glc II), 4.00 (dd, H-6A of Glc I), 4.05 (dd, $J_{1,2}$=7.5 Hz, $J_{2,3}$=7.5 Hz, H-2 of Glc I), 4.08 (dd, $J_{2,3}$=8.0 Hz, $J_{3,4}$=not reported, H-3 of Glc), 4.17 (dd, $J_{3,4}$=9.8 Hz, $J_{4,5}$=9.0 Hz, H-4 of Glc I), 4.20 (dd, $J_{3,4}$=9.5 Hz, $J_{4,5}$=9.0 Hz, H-4 of Rha), 4.23 (dd, H-6A of Glc II), 4.26 (dd, H-6B of Glc II), 4.38 (dd, $J_{2,3}$=2.5 Hz, $J_{3,4}$=9.5 Hz, H-3 of Rha), 4.43 (m, H-5 of Glc II), 4.54 (d, $J_{2,3}$=2.5 Hz, H-2 of Rha), 4.70 (d, $J_{1,2}$=7.5 Hz, H-1 of Glc II), 4.79 (m, H-5 of Rha), 4.85 (d, $J_{1,2}$=7.5 Hz, H-1 of Glc I), 5.62 (s, H-1 of Rha).

CMR (C$_5$D$_5$N, 75/125 MHz) : δ C-1) 37.3 (2) 30.0 (3) 77.5 (4) 35.9 (5) 44.8 (6) 29.1 (7) 32.3 (8) 35.3 (9) 53.9 (10) 36.0 (11) 21.4 (12) 39.1 (13) 40.8 (14) 56.8 (15) 32.3 (16) 79.0 (17) 63.5 (18) 17.0 (19) 12.6 (20) 44.8 (21) 13.9 (22)

112.9 (23) 34.6 (24) 28.1 (25) 37.2 (26) 75.4 (27) not reported **Glc I** (1) 100.0 (2) 78.9 (3) 78.0 (4) 71.4 (5) 76.6 (6) 61.1 **Rha** (1) 102.1 (2) 72.7 (3) 72.9 (4) 74.0 (5) 69.7 (6) 18.8 **Glc II** (1) 105.2 (2) 75.4 (3) 78.7 (4) 71.9 (5) 78.6 (6) 63.0.

Reference

1.   S.A. Shvets, P.K. Kintya and O.N. Gutsu, *Khim. Prir. Soedin.*, 737 (1994); *Chem. Nat. Comp.*, **30**, 684 (1994).

TORVOSIDE A

(25*S*)-5α-Furostan-3β,6α,22ξ,26-tetraol 6-O-[α-L-rhamnopyranosyl-(1→3)-β-D-quinovopyranoside]-26-O-β-D-glucopyranoside

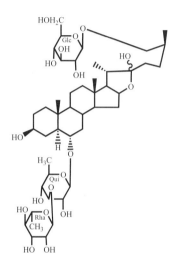

Source : *Solanum torvum* Swartz (Solanaceae)
Mol. Formula : $C_{45}H_{76}O_{18}$
Mol. Wt. : 904
$[\alpha]_D^{27}$: -38.3° (c=1.0, MeOH)
Registry No. : [184776-83-8]

PMR (C₅D₅N, 400 MHz) **:** δ 0.85 (6H, br s, 3xH-18, 3xH-19), 1.03 (d, *J*=5.2 Hz, 3xH-27), 1.30 (d, *J*=6.6 Hz, 3xH-21), 1.59 (d, *J*=6.9 Hz, 3xH-6 of Qui), 1.69 (d, *J*=5.5 Hz, 3xH-6 of Rha), 2.51 (br d, *J*=12.0 Hz, H-7), 4.74 (d, *J*=7.3 Hz, H-1 of Qui), 4.80 (2H, m, H-5 of Rha and H-1 of Glc), 6.26 (s, H-1 of Rha).

CMR (C₅D₅N, 100 MHz) : δ C-1) 37.6 (2) 32.1 (3) 71.5 (4) 32.9 (5) 51.1 (6) 79.1 (7) 41.2 (8) 34.2 (9) 53.7 (10) 36.5 (11) 21.1 (12) 40.0 (13) 40.9 (14) 56.1 (15) 31.9 (16) 80.9 (17) 63.6 (18) 16.2 (19) 13.4 (20) 40.4 (21) 16.6 (22) 110.4 (23) 36.8 (24) 28.1 (25) 34.0 (26) 75.2 (27) 17.3 **Qui** (1) 105.3 (2) 76.0 (3) 83.1 (4) 75.0 (5) 72.5 (6) 18.6 **Rha** (1) 102.7 (2) 72.5 (3) 72.3 (4) 73.9 (5) 69.7 (6) 18.4 **Glc** (1) 104.9 (2) 75.0 (3) 78.3 (4) 71.5 (5) 78.2 (6) 62.6.

816

Mass (FAB, Positive ion) : *m/z* 887 [M-OH]⁺, 595 [M-OH-Rha-Qui]⁺, 433 [M-OH-Rha-Qui-Glc]⁺.

Reference

1.   S. Yahara, T. Yamashita, N. Nozawa (nee Fujimura) and T. Nohara, *Phytochemistry*, **43**, 1069 (1996).

TORVOSIDE B

(25*S*)-22α-Methoxy-5α-furostan-3β,6α,26-triol 6-*O*-[β-D-xylopyranosyl-(1→3)-
β-D-quinovopyranoside]-26-*O*-β-D-glucopyranoside

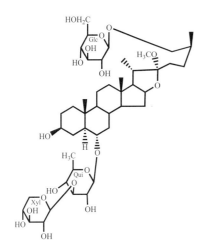

Source : *Solanum torvum* Swartz (Solanaceae)
Mol. Formula : C₄₅H₇₆O₁₈
Mol. Wt. : 904
[α]ᴅ²⁷ : -23.3° (c=2.80, MeOH)
Registry No. : [184686-00-8]

PMR (C₅D₅N, 400 MHz) : δ 0.79 (s, 3xH-18, 19), 0.89 (s, 3xH-19), 1.05 (d, *J*=6.6 Hz, 3xH-27), 1.15 (d, *J*=5.9 Hz, 3xH-21), 1.56 (d, *J*=5.9 Hz, 3xH-6 of Qui), 2.50 (br d, *J*=12.3 Hz, H-7), 3.23 (s, 22-OC*H₃*), 4.82 (2H, d, *J*=7.3 Hz, H-1 of Glc and H-1 of Qui), 5.21 (d, *J*=7.3 Hz, H-1 of Xyl).

CMR (C₅D₅N, 100 MHz) : δ C-1) 37.5 (2) 31.9 (3) 71.4 (4) 32.9 (5) 51.1 (6) 79.1 (7) 41.4 (8) 34.2 (9) 53.6 (10) 36.5 (11) 21.0 (12) 39.7 (13) 40.9 (14) 56.1 (15) 31.9 (16) 81.0 (17) 63.9 (18) 16.3 (19) 13.4 (20) 40.2 (21) 16.1 (22) 112.4 (23) 30.6 (24) 28.0 (25) 33.9 (26) 74.7 (27) 17.3 (OCH₃) 47.1 **Qui** (1) 104.9 (2) 74.5 (3) 87.4 (4) 74.6 (5) 72.1 (6) 18.4 **Xyl** (1) 106.1 (2) 74.9 (3) 77.7 (4) 70.6 (5) 67.1 **Glc** (1) 104.8 (2) 75.0 (3) 78.3 (4) 71.5 (5) 78.2 (6) 62.6.

Mass (FAB, Positive ion) : *m/z* 873 [M-OMe]⁺, 711 [M-OMe-Glc]⁺, 579 [M-OMe-Glc-Xyl]⁺, 433 [M-OMe-Glc-Xyl-Qui]⁺.

Reference

1.   S. Yahara, T. Yamashita, N. Nozawa (nee Fujimura) and T. Nohara, *Phytochemistry*, **43**, 1069 (1996).

TRIGONEOSIDE IIIa

(25*S*)-5α-Furostane-3β,22ξ,26-triol 3-O-[α-L-rhamnopyranosyl-(1→2)-β-D-glucopyranoside]-
26-O-β-D-glucopyranoside

Source : *Trigonella foenum-graecum* L. (Legnuminosea)
Mol. Formula : $C_{45}H_{76}O_{18}$
Mol. Wt. : 904
$[\alpha]_D^{27}$: -54.4° (c=0.66, Pyridine)
Registry No. : [187141-39-5]

IR (KBr) : 3413, 2932, 1076, 1048 cm^{-1}.

PMR (C_5D_5N, 270 MHz) : δ 0.88 (s, 3xH-18)^a, 0.89 (s, 3xH-19)^a, 1.03 (d, *J*=6.7 Hz, 3xH-27), 1.32 (d, *J*=7.1 Hz, 3xH-21), 1.77 (d, *J*=6.4 Hz, 3xH-6 of Rha), 1.93 (m, H-25), 2.23 (dq-like, H-20), 3.49 (dd, *J*=7.0, 9.5 Hz, H-26A), 3.99 (m, H-3), 4.09 (dd, *J*=5.8, 9.5 Hz, H-26B), 4.82 (d, *J*=7.7 Hz, H-1 of Glc II), 5.07 (d, *J*=7.4 Hz, H-1 of Glc I), 6.36 (br s, H-1 of Rha).

CMR (C_5D_5N, 67.5 MHz) : δ C-1) 37.3 (2) 30.0 (3) 77.0 (4) 34.5 (5) 44.6 (6) 29.0 (7) 32.5 (8) 35.3 (9) 54.5 (10) 36.0 (11) 21.3 (12) 40.3 (13) 41.1 (14) 56.4 (15) 32.4 (16) 81.2 (17) 64.0 (18) 16.7 (19) 12.5 (20) 40.7 (21) 16.5 (22) 110.6 (23) 37.2 (24) 28.3 (25) 34.3 (26) 75.4 (27) 17.5 **Glc I** (1) 99.9 (2) 78.1 (3) 78.3 (4) 72.0 (5) 79.7 (6) 62.9 **Rha** (1) 102.2 (2) 72.6 (3) 72.8 (4) 74.2 (5) 69.5 (6) 18.7 **Glc II** (1) 105.1 (2) 75.2 (3) 78.6 (4) 71.7 (5) 78.5 (6) 62.8.

Mass (FAB, Positive ion. H.R) : *m/z* 927.4967 [(M+Na)$^+$, calcd. for 927.4929].

Mass (FAB, Positive ion) : *m/z* 927 [M+Na]$^+$.

Mass (FAB, Negative ion) : *m/z* 903 [M-H]$^-$, 757 [M-$C_6H_{11}O_4$]$^-$, 741 [M-$C_6H_{11}O_5$]$^-$, 595 [M-$C_{12}H_{21}O_9$]$^-$.

Reference

1.   M. Yoshikawa, T. Murakami, H. Komatsu, N. Murakami, J. Yamahara and H. Matsuda, *Chem. Pharm. Bull.*, **45**, 81 (1997).

TRIGONEOSIDE IIIb

(25R)-5α-Furostane-3β,22ξ,26-triol 3-O-[α-L-rhamnopyransyl-(1→2)-β-D-glucopyranoside]-26-O-β-D-glucopyranoside

Source : *Trigonella foenum-graecum* L. (Legnuminosea)
Mol. Formula : $C_{45}H_{76}O_{18}$
Mol. Wt. : 904
$[\alpha]_D^{27}$: -21.7° (c=0.06, Pyridine)
Registry No. : [187141-40-8]

IR (KBr) : 3419, 2932, 1075, 1048 cm^{-1}.

PMR (C_5D_5N, 270 MHz) : δ 0.88 (s, 3xH-18)[a], 0.89 (s, 3xH-19)[a], 0.99 (d, J=6.7 Hz, 3xH-27), 1.34 (d, J=7.0 Hz, 3xH-21), 1.77 (d, J=6.4 Hz, 3xH-6 of Rha), 1.93 (m, H-25), 2.24 (dq-like, H-20), 3.63 (dd, J=5.8, 9.5 Hz, H-26A), 3.95 (dd, J=7.3, 9.5 Hz, H-26B), 4.01 (m, H-3), 4.83 (d, J=7.6 Hz, H-1 of Glc II), 5.08 (d, J=7.3 Hz, H-1 of Glc I), 6.37 (br s, H-1 of Rha).

CMR (C_5D_5N, 67.5 MHz) : δ C-1) 37.3 (2) 30.0 (3) 77.0 (4) 34.5 (5) 44.6 (6) 29.0 (7) 32.5 (8) 35.3 (9) 54.5 (10) 36.0 (11) 21.3 (12) 40.3 (13) 41.1 (14) 56.4 (15) 32.4 (16) 81.1 (17) 64.0 (18) 16.7 (19) 12.5 (20) 40.7 (21) 16.5 (22) 110.6 (23) 37.2 (24) 28.4 (25) 34.3 (26) 75.3 (27) 17.5 **Glc I** (1) 99.9 (2) 78.1 (3) 78.4 (4) 72.0 (5) 79.7 (6) 62.9 **Rha** (1) 102.2 (2) 72.6 (3) 72.9 (4) 74.2 (5) 69.5 (6) 18.7 **Glc II** (1) 105.0 (2) 75.2 (3) 78.6 (4) 71.8 (5) 78.5 (6) 62.8.

Mass (FAB, Positive ion, H.R) : m/z 927.4910 [(M+Na)$^+$, calcd. for 927.4929].

Mass (FAB, Positive ion) : m/z 927 [M+Na]$^+$.

Mass (FAB, Negative ion) : m/z 903 [M-H]$^-$, 757 [M-$C_6H_{11}O_4$]$^-$, 741 [M-$C_6H_{11}O_5$]$^-$, 595 [M-$C_{12}H_{21}O_9$]$^-$.

Reference

1. M. Yoshikawa, T. Murakami, H. Komatsu, N. Murakami, J. Yamahara and H. Matsuda, *Chem. Pharm. Bull.*, **45**, 81 (1997).

SMILAX SIEBOLDII SAPONIN 4
26-O-β-D-glucopyranosyl-3β,22ξ,26-trihydroxy-(25R)-5α-furostan-6-one
3-O-α-L-arabinopyranosyl-(1→6)-β-D-glucopyranoside

Source : *Smilax sieboldii* (Liliaceae)
Mol. Formula : $C_{44}H_{72}O_{19}$
Mol. Wt. : 904
$[\alpha]_D^{25}$: -44.8° (c=0.30, EtOH)
Registry No. : [143222-28-0]

UV (MeOH) : λ_{max} 285 (log ε, 75) nm.

CD (EtOH; c 9.62×10^{-4}) nm : θ 291 (-2911).

IR (KBr) : 3420 (OH), 2930 (CH), 1695 (C=O), 1445, 1375, 1255, 1170, 1155, 1070, 1040, 1005, 900, 775, 725, 695 cm^{-1}.

PMR (C_5D_5N, 400/500 MHz) : δ 0.65 (s, 3xH-19), 0.83 (s, 3xH-18), 0.99 (d, J=6.7 Hz, 3xH-27), 1.34 (d, J=6.8 Hz, 3xH-21), 4.39 (q-like, J=5.9 Hz, H-16), 4.81 (d, J=7.6 Hz, H-1 of Glc II), 4.98 (overlapping with H_2O signal, H-1 of Glc I and H-1 of Ara).

CMR (C_5D_5N, 100/125 MHz) : δ C-1) 36.7 (2) 29.5 (3) 76.8 (4) 27.0 (5) 56.4[a] (6) 209.7 (7) 46.8 (8) 37.3 (9) 53.7 (10) 40.8 (11) 21.5 (12) 39.6 (13) 41.1 (14) 56.3[a] (15) 32.0 (16) 80.8 (17) 63.8 (18) 16.5[b] (19) 13.1 (20) 40.5 (21) 16.4[b] (22) 110.6 (23) 37.1 (24) 28.3 (25) 34.2 (26) 75.2 (27) 17.4 **Glc I** (1) 102.1 (2) 75.2 (3) 78.5[c] (4) 71.9 (5) 77.0 (6) 69.7 **Ara** (1) 105.4 (2) 72.3 (3) 74.4 (4) 69.1 (5) 66.5 **Glc II** (1) 104.9 (2) 75.2 (3) 78.6[c] (4) 71.7 (5) 78.4[c] (6) 62.8.

Mass (SI, Positive ion) : m/z 927 $[M+Na]^+$, 887 $[M-OH]^+$.

Biological Activity : The compound shows inhibitory activity on Cyclic AMP phosphodiesterase with $IC_{50} >$ 500×10^{-5} M.

Reference

1.   S. Kubo, Y. Mimaki, Y. Sashida, T. Nikaido and T. Ohmoto, *Phytochemistry*, **31**, 2445 (1992).

TRIGONEOSIDE 1a

(25*S*)-5α-Furostane-2α,3β,22ξ,26-tetrol 3-O-[β-D-xylopyranosyl-(1→6)-β-D-glucopyranoside]-26-O-β-D-glucopyranoside

Source : *Trigonella foenum-graecum* L. (Fabaceae)
Mol. Formula : $C_{44}H_{74}O_{19}$
Mol. Wt. : 906
$[\alpha]_D^{26}$: -41.8° (c=0.34, Pyridine)
Registry No. : [187141-35-1]

IR (KBr) : 3405, 2930, 1078, 1040 cm^{-1}.

PMR (C$_5$D$_5$N, 270 MHz) : δ 0.72 (s, 3xH-18)[a], 0.86 (s, 3xH-19)[a], 1.03 (d, *J*=6.4 Hz, 3xH-27), 1.06 (m, H-5), 1.16 (m), 2.15 (dd, *J*=4.0, 10.6 Hz, 2xH-1), 1.30 (d, *J*=7.0 Hz, 3xH-21), 1.42 (m), 1.76 (dd, *J*=5.2, 11.6 Hz, 2xH-4), 1.93 (m, H-25), 2.22 (dq-like, H-20), 3.48 (dd, *J*=7.0, 9.5 Hz), 3.70 (m, H-3), 3.95 (m, H-2), 3.97 (m, H-6A of Glc I), 4.07 (m, 2xH-26), 4.38 (dd, *J*=5.4, 11.8 Hz, H-6A of Glc II), 4.54 (dd, *J*=2.6, 11.8 Hz, H-6B of Glc II), 4.80 (d, *J*=7.3 Hz, H-1 of Xyl), 4.81 (d, *J*=7.9 Hz, H-1 of Glc II), 4.95 (d, *J*=7.6 Hz, H-1 of Glc I), 4.98 (m, H-6B of Glc I).

CMR (C$_5$D$_5$N, 67.5 MHz) : δ C-1) 45.3 (2) 71.1 (3) 87.9 (4) 34.6 (5) 44.6 (6) 28.0 (7) 32.2 (8) 34.5 (9) 54.3 (10) 36.8 (11) 21.4 (12) 40.1 (13) 41.1 (14) 56.2 (15) 32.4 (16) 81.1 (17) 63.9 (18) 16.7 (19) 13.3 (20) 40.7 (21) 16.5 (22) 110.6 (23) 37.1 (24) 28.3 (25) 34.4 (26) 75.4 (27) 17.5 **Glc I** (1) 104.8 (2) 74.9 (3) 78.6 (4) 72.2 (5) 76.4 (6) 70.5 **Xyl** (1) 105.4 (2) 75.3 (3) 77.9 (4) 71.1 (5) 67.2 **Glc II** (1) 105.1 (2) 75.2 (3) 78.4 (4) 71.7 (5) 78.4 (6) 62.8.

Mass (FAB, Positive ion, H.R.) : *m/z* 424.4742 [(M+Na)$^+$, calcd. for 929.4722].

Mass (FAB, Positive ion) : *m/z* 929 [M+Na]$^+$.

Mass (FAB, Negative ion) : *m/z* 905 [M-H]$^-$, 773 [M-C$_5$H$_9$O$_4$]$^-$, 611 [M-C$_{11}$H$_{19}$O$_9$]$^-$.

Reference

1.   M. Yoshikawa, T. Murakami, H. Komatsu, N. Murakami, J. Yamahara and H. Matsuda, *Chem. Pharm. Bull.*, **45**, 81 (1997).

TRIGONEOSIDE 1b

(25*R*)-5α-Furostane-2α,3β,22ξ,26-tetral 3-O-[β-D-xylopyranosyl-(1→6)]-
β-D-glucopyranoside]-26-O-β-D-glucopyranoside

Source : *Trigonella foenum-graecum* L. (Legnuminosea)
Mol. Formula : C$_{44}$H$_{74}$O$_{19}$
Mol. Wt. : 906
[α]$_D^{24}$: -41.5° (c=0.37, Pyridine)
Registry No. : [187141-36-2]

IR (KBr) : 3405, 2930, 1082, 1049 cm^{-1}.

PMR (C$_5$D$_5$N, 270 MHz) : δ 0.73 (s, 3xH-18)^a, 0.87 (s, 3xH-19)^a, 0.99 (d, *J*=6.7 Hz, 3xH-27), 1.05 (m, H-5), 1.32 (d, *J*=6.7 Hz, 3xH-21), 1.93 (m, H-25), 2.23 (dq-like, H-20), 3.63 (dd, *J*=6.1, 9.2 Hz, H-26A), 3.70 (m, H-3), 3.95 (m, H-26B), 3.96 (m, H-2), 3.98 (m, H-6A of Glc I), 4.80 (d, *J*=6.1 Hz, H-1 of Xyl), 4.81 (d, *J*=7.6, Hz, H-1 of Glc II), 4.94 (d, *J*=7.6 Hz, H-1 of Glc I), 5.00 (m, H-6B of Glc I).

CMR (C$_5$D$_5$N, 67.5 MHz) : δ C-1) 45.3 (2) 71.1 (3) 87.9 (4) 34.6 (5) 44.7 (6) 28.0 (7) 32.2 (8) 34.5 (9) 54.3 (10) 36.8 (11) 21.4 (12) 40.1 (13) 41.1 (14) 56.2 (15) 32.4 (16) 81.1 (17) 64.0 (18) 16.7 (19) 13.3 (20) 40.7 (21) 16.4 (22) 110.6

(23) 37.2 (24) 28.4 (25) 34.3 (26) 75.3 (27) 17.5 **Glc I** (1) 104.8 (2) 74.8 (3) 78.6 (4) 72.2 (5) 76.5 (6) 70.4 **Xyl** (1) 105.4 (2) 75.3 (3) 77.9 (4) 71.1 (5) 67.1 **Glc II** (1) 104.9 (2) 75.2 (3) 78.4 (4) 71.7 (5) 78.4 (6) 62.8.

Mass (FAB, Positive ion) : m/z 929 [M+Na]$^+$.

Mass (FAB, Negative ion, H.R) : m/z 905.4722 [calcd. for (M-H)$^-$, 905.4722].

Mass (FAB, Negative ion) : m/z 905 [M-H]$^-$, 773 [M-C$_5$H$_9$O$_4$]$^-$, 611 [M-C$_{11}$H$_{19}$O$_9$]$^-$.

Reference

1. M. Yoshikawa, T. Murakami, H. Komatsu, N. Murakami, J. Yamahara and H. Matsuda, *Chem. Pharm. Bull.*, **45**, 81 (1997).

TRIGONEOSIDE XIb

(25R)-5α-Furostane-2α,3β,22ξ,26-tetraol 3-O-[β-D-xylopyranosyl-(1→4)-β-D-glucopyranoside]-26-O-β-D-glucopyranoside

Source : *Trigonella foenum-graecum* L. (Fabaceae)
Mol. Formula : C$_{44}$H$_{74}$O$_{19}$
Mol. Wt. : 906
[α]$_D^{23}$: -24.7° (c=0.2, MeOH)
Registry No. : [290347-58-9]

IR (KBr) : 3432, 2926, 1076, 1044 cm^{-1}.

PMR (C$_5$D$_5$N, 500 MHz) : δ 0.75, 0.87 (both s, 3xH-19, 3xH-18), 0.99 (d, J=6.7 Hz, 3xH-27), 1.31 (d, J=7.0 Hz, 3xH-21), 3.62, 3.93 (both m, 2xH-26), 3.84 (m, H-3), 3.95 (m, H-2), 4.79 (d, J=7.6 Hz, H-1 of Glc II), 4.92 (ddd-like, H-16), 5.01 (d, J=7.6 Hz, H-1 of Glc I), 5.10 (d, J=7.3 Hz, H-1 of Xyl).

CMR (C$_5$D$_5$N, 125.0 MHz) : δ C-1) 45.8 (2) 70.5 (3) 85.2 (4) 34.0 (5) 44.8 (6) 28.2 (7) 32.3 (8) 34.7 (9) 54.6 (10) 37.0 (11) 21.5 (12) 40.2 (13) 41.2 (14) 56.4 (15) 32.4 (16) 81.2 (17) 64.0 (18) 16.7 (19) 13.5 (20) 40.7 (21) 16.4 (22) 110.7 (23) 37.2 (24) 28.4 (25) 34.3 (26) 75.2 (27) 17.5 **Glc I** (1) 105.6 (2) 74.7 (3) 76.5 (4) 80.9 (5) 76.6 (6) 61.8 **Xyl** (1) 103.0 (2) 75.0 (3) 78.4 (4) 70.8 (5) 67.4 **Glc II** (1) 104.9 (2) 75.2 (3) 78.6 (4) 71.9 (5) 78.4 (6) 63.0.

Mass (FAB, Negative ion) : m/z 905 [M-H]$^-$, 773 [M-C$_5$H$_9$O$_4$]$^-$, 611 [M-C$_{11}$H$_{19}$O$_9$]$^-$.

Mass (FAB, Positive ion) : m/z 929 [M+Na]$^+$.

Reference

1.   T. Murakami, A. Kishi, H. Matsuda and M. Yoshikawa, *Chem. Pharm. Bull.,* **48**, 994 (2000).

COSTUS SPICATUS SAPONIN 2

3β,26-Dihydroxy-22α-methoxy-25*R*-furost-5-ene 3-O-[α-L-rhamnopyranosyl-(1→2)-β-D-glucopyranoside]-26-O-β-D-glucopyranoside

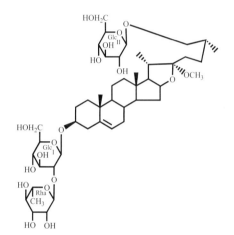

Source : *Costus speciosus* (Koen.) Sm.[1], *Costus spicatus* Swartz[2] (Costaceae)
Mol. Formula : C$_{46}$H$_{76}$O$_{18}$
Mol. Wt. : 916
M.P. : 245-249°C (decomp.)[1]; 184-186°C[2]
Registry No. : [84774-05-0]

IR (KBr)[2] : 3430 (OH), 1053 (C-O), 913, 838 cm^{-1}.

PMR (C$_5$D$_5$N, 200 MHz)[2] : δ 0.83 (s, 3xH-18), 0.98 (d, *J*=6.6 Hz, 3xH-27), 1.04 ((s, 3xH-19), 1.21 (d, *J*=6.8 Hz, 3xH-21), 1.79 (d, *J*=6.2 Hz, 3xH of Rha), 3.27 (s, 22-OC*H$_3$*), 4.83 (d, *J*=7.8 Hz, H-1 of Glc II), 4.98 (d, *J*=7.8 Hz, H-1 of Glc I), 5.31 (br s, H-6), 6.38 (br s, H-1 of Rha).

CMR (C$_5$D$_5$N, 50 MHz)[2] : δ C-1) 37.3 (2) 29.7 (3) 77.9 (4) 38.5 (5) 140.3 (6) 121.4 (7) 32.0 (8) 31.5 (9) 50.1 (10) 36.7 (11) 20.9 (12) 39.7 (13) 40.5 (14) 56.1 (15) 31.9 (16) 80.8 (17) 63.7 (18) 16.1 (19) 18.8 (20) 40.3 (21) 16.6 (22) 112.7 (23) 30.5 (24) 27.7 (25) 33.6 (26) 74.8 (27) 16.9 (OCH$_3$) 47.0 **Glc I** (1) 100.4 (2) 79.6 (3) 77.9 (4) 71.8 (5) 77.5 (6) 62.5 **Glc II** (1) 104.9 (2) 74.9 (3) 78.4 (4) 71.8 (5) 78.6 (6) 62.5 **Rha** (1) 102.0 (2) 72.5 (3) 72.8 (4) 74.2 (5) 69.4 (6) 18.6.

824

Mass (LSI-MS)² : *m/z* 915 [M-H]⁻.

References

1.	S.B. Singh and R.S. Thakur, *J. Nat. Prod.*, **45**, 667 (1982).

2.	B.P. Silva, R.R. Bernardo, J.P. Parente. *Fitoterapia*, **69,** 528 (1998).

COSTUS SPICATUS SAPONIN 3
3β,26-Dihydroxy-22-α-methoxy-25*R*-furost-5-ene 3-O-[α-L-rhamnopyranosyl-(1→4)-
β-D-glucopyranoside[-26-O-β-D-glucopyranoside

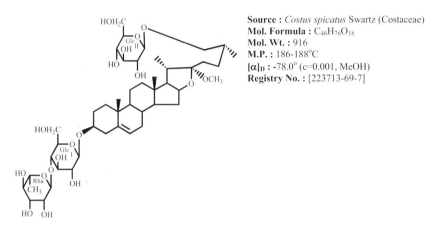

Source : *Costus spicatus* Swartz (Costaceae)
Mol. Formula : C₄₆H₇₆O₁₈
Mol. Wt. : 916
M.P. : 186-188°C
[α]_D : -78.0° (c=0.001, MeOH)
Registry No. : [223713-69-7]

IR (KBr) : 3430 (OH), 1053 (C-O), 913, 838 cm⁻¹.

PMR (C₅D₅N, 200 MHz) : δ 0.84 (s, 3xH-18), 0.99 (d, *J*=6.6 Hz, 3xH-27), 1.06 (s, 3xH-19), 1.20 (d, *J*=6.8 Hz, 3xH-21), 1.78 (d, *J*=6.3 Hz, 3xH-6 of Rha), 3.28 (s, 22-OCH₃), 4.84 (d, *J*=7.8 Hz, H-1 of Glc II), 4.93 (d, *J*=7.7 Hz, H-1 of Glc I), 5.33 (br s, H-6), 6.40 (br s, H-1 of Rha).

CMR (C₅D₅N, 50 MHz) : δ C-1) 37.2 (2) 29.7 (3) 77.9 (4) 38.5 (5) 140.3 (6) 121.4 (7) 32.2 (8) 31.5 (9) 49.9 (10) 36.9 (11) 20.7 (12) 39.8 (13) 40.5 (14) 56.1 (15) 31.9 (16) 80.8 (17) 63.7 (18) 16.1 (19) 18.8 (20) 40.3 (21) 16.6 (22) 112.7 (23) 30.3 (24) 27.7 (25) 33.6 (26) 74.8 (27) 16.9 (OCH₃) 46.9 **Glc I** (1) 100.3 (2) 75.2 (3) 76.8 (4) 79.6 (5) 76.9 (6) 61.6 **Rha** (1) 102.9 (2) 72.6 (3) 72.9 (4) 74.2 (5) 70.5 (6) 18.8 **Glc II** (1) 105.1 (2) 75.3 (3) 78.6 (4) 71.8 (5) 78.1 (6) 62.9.

Mass (LSI-MS) : *m/z* 915 [M-H]⁻.

Reference

1.   B.P. Silva, R.R. Bernardo and J.P. Parente, *Fitoterapia*, **69**, 528 (1998).

DRACAENA CONCINNA SAPONIN 1

22ξ-Methoxy-5α-furost-25 (27)-ene-1β-3α-triol 1-O-[α-L-rhmnopyranosyl-(1→2)-
β-D-fucopyrnoside]-26-O-β-D-glucpoyranoside

Source : *Dracaena concinna* Kunth. (Agavaceae)
Mol. Formula : $C_{46}H_{76}O_{18}$
Mol. Wt. : 916
$[\alpha]_D^{27}$: -45.0° (c=0.12, MeOH)
Registry No. : [209848-39-5]

IR (KBr) : 3420 (OH), 2920 (CH) cm^{-1}.

PMR (C_5D_5N, 400 MHz) : δ 0.85 (s, 3xH-18), 1.11 (d, *J*=6.8 Hz, 3xH-21), 1.25 (s, 3xH-19), 1.45 (d, *J*=6.3 Hz, 3xH-6 of Fuc), 1.75 (d, *J*=6.1 Hz, 3xH-6 of Rha), 3.25 (s, OC*H₃*), 4.73 (d, *J*=7.7 Hz, H-1 of Fuc), 4.91 (d, *J*=7.8 Hz, H-1 of Glc), 5.05 (br s, H-27A), 5.34 (br s, H-27B), 6.41 (br s, H-1 of Rha).

CMR (C_5D_5N, 100 MHz) : δ C-1) 81.0 (2) 35.5 (3) 65.9 (4) 37.4 (5) 39.4 (6) 28.8 (7) 32.6 (8) 36.6 (9) 55.3 (10) 42.6 (11) 23.7 (12) 40.6 (13) 40.7 (14) 57.0 (15) 32.3 (16) 81.5 (17) 64.3 (18) 16.9 (19) 7.7 (20) 40.4 (21) 16.0 (22) 112.4 (23) 31.5 (24) 28.1 (25) 146.8 (26) 72.0 (27) 111.0. (OC*H₃*) 47.3 **Fuc** (1) 100.2 (2) 74.3 (3) 77.0 (4) 73.3 (5) 71.0 (6) 17.1 **Rha** (1) 101.4 (2) 72.6 (3) 72.6 (4) 74.3 (5) 69.2 (6) 19.0 **Glc** (1) 103.8 (2) 75.1 (3) 78.6 (4) 71.7 (5) 78.5 (6) 62.8.

Mass (FAB, Negative ion) : *m/z* 915 [M-H]⁻.

Reference

1.   Y. Mimaki, M.Kuroda, Y. Takaashi and Y. Sashida, *Phytochemistry*, **47,** 1351 (1998).

ICOGENIN

(25S)-22-Methoxyfurost-5-en-3β,26-diol 3-O-[α-L-rhamnopyranosyl-(1→2)-[β-D-glucopyranosyl-(1→3)]-β-D-glucopyranoside}

Source : *Dracaena draco* (Agavaceae)
Mol. Formula : $C_{46}H_{76}O_{18}$
Mol. Wt. : 916
$[α]_D^{20}$ **:** -61.2 (c=0.4, EtOH)

IR (KBr) : 3420, 2920, 1040, 869, 837, 815 cm^{-1}.

PMR (C$_6$H$_5$N, 400 MHz) : δ 0.67 (br s, 3xH-27), 0.80 (s, 3xH-18), 0.89 (m, H-14), 0.96 (m, H-1A), 1.02 (s, 3xH-19), 1.11 (d, J=5.8 Hz, 3xH-21), 1.15 (m, H-9), 1.41 (m, H-11), 1.46 (m, H-7), 1.47 (m, H-15A), 1.55 (m, H-8, H-24), 1.55 (br s, H-25), 1.60 (m, H-23), 1.70 (m, H-1B), 1.70 (d, J=5.3 Hz, 3xH-6 of Rha), 1.79 (m, H-17), 1.85 (m, H-2ax, H-7B), 1.91 (m, H-20), 2.02 (m, H-15B), 2.06 (m, H-2B), 2.70 (m, H-4), 3.48 (m, H-26eq), 3.56 (s, OC*H*$_3$), 3.80 (H-5 of Glc I), 3.94 (H-2 and H-5 of Glc II), 4.00 (H-3, H-2 of Glc), 4.02 (H-4 of Glc I), 4.04 (H-4 of Glc II), 4.14 (br s, H-3 of Glc II), 4.15 (H-3 of Glc I), 4.20 (m, H-6A of Glc I), 4.28 (t, J=8.8 Hz, H-4 of Rha), 4.38 (m, H-6B of Glc I), 4.51 (H-16), 4.53 (d, J=9.3 Hz, H-3 of Rha), 4.53 (m, H-6B of Glc), 4.56 (m, H-26ax), 4.83 (br s, H-2 of Rha), 4.87 (H-1 of Glc I), 4.88 (d, J=4.3 Hz, H-5 of Rha), 5.04 (d, J=7.0 Hz, H-1 of Glc II), 5.33 (br d, H-6), 6.27 (br s, H-1 of Rha).

CMR (C$_5$D$_5$N, 75 MHz) : δ C-1) 38.70 (2) 31.28 (3) 79.56 (4) 39.90 (5) 142.05 (6) 123.15 (7) 33.53 (8) 32.91 (9) 51.49 (10) 38.35 (11) 22.31 (12) 41.09 (13) 41.68 (14) 57.86 (15) 33.41 (16) 82.35 (17) 64.06 (18) 17.55 (19) 20.61 (20) 43.19 (21) 16.23 (22) 110.50 (23) 33.02 (24) 30.45 (25) 31.78 (26) 68.10 (27) 18.53 (OC*H*$_3$) 50.85 **Glc I** (1) 101.20 (2) 79.80 (3) 90.35 (4) 72.70 (5) 79.00 (6) 63.60 **Rha** (1) 103.30 (2) 73.55 (3) 73.90 (4) 75.19 (5) 70.78 (6) 19.65 **Glc II** (1) 05.7 (2) 76.13 (3) 78.32 (4) 70.83 (5) 78.89 (6) 63.60.

Mass (FAB, Positive ion) : *m/z* 907 [M+Na-OCH$_3$]$^+$.

Biological Activity : Cytotoxic agaisnt myeloid leukemia cell line HL-60.

Reference

1.   J.C. Hernandez, F. Leon, J. Quintana, F. Estevez and J. Bermejo, *Bioorganic & Medicinal Chemistry*, **12**, 4423 (2004).

TRILLIUM KAMTSCHATICUM SAPONIN Td,
RHAPIS HUMILIS SAPONIN HSt₄

(25R)-Methoxyfurost-5-en-3β,26-diol 3-O-{α-L-rhamnopyranosyl-(1→2)-β-D-glucopyranoside-26-O-β-D-glucopyranoside}

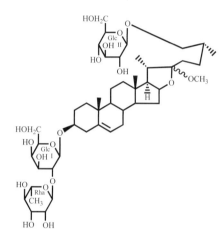

Source : *Trillium kamtschaticum* Pall.[1] (Liliaceae), *Rhapis humilis* Bl.[2] (Palmae), *Costus speciosus* (Koen.) Sm.[3] (Zingiberaceae), *Allium narcissiflorum*[4] (Liliaceae)
Mol. Formula : $C_{46}H_{76}O_{18}$
Mol. Wt. : 916
M.P. : 265-271°C (decomp.)[1], 178-180°C (decomp.)[2]
[α]$_D$: -83.1° (c=1.03, Pyridine)
Registry No. : [55916-50-2]

IR (KBr)[2] : 3460-3300 (OH) cm^{-1}.

IR[3] : 3650-3100, 1200-1000, 980, 920 (strong), 900 (weak), 845 cm^{-1}.

PMR (DMSO-d_6, 60 MHz)[3] : δ 1.5 (br s, 3xH-6 of Rha), 3.05 (s, OCH_3), 4.30 (d, *J*=7.0 Hz, 2xH-1 of Glc), 5.0 (br s, *W*½=5.0 Hz, H-1 of Rha).

CMR (C_5D_5N) : δ 18.5 (C-6 of Rha), 47.4 (OCH_3), 100.4 (C-1 of Glc II), 101.8 (C-1 of Rha), 104.7 (C-1 of Glc I).

References

1.   T. Nohara, K. Miyahara and T. Kawasaki, *Chem. Pharm. Bull.*, **23**, 872 (1975).

2.   Y. Hirai, S. Sanada, Y. Ida and J. Shoji, *Chem. Pharm. Bull.*, **32**, 4003 (1984).

3.   S.B. Singh and R.S. Thakur, *J. Nat. Prod.*, **45**, 667 (1982)

4.   Y. Mimaki, T. Saton, M. Ohmura and Y. Sashida, *Nat. Med.*, **50**, 308 (1996).

CHINENOSIDE V

3β,26-dihydroxy-23-hydroxymethyl-25(*R*)-5α-furost-20(22)-en-6-one 3-O-[α-L-arabinopyranosyl-(1→6)-β-D-glucopyranoside]-26-O-β-D-glucopyranoside

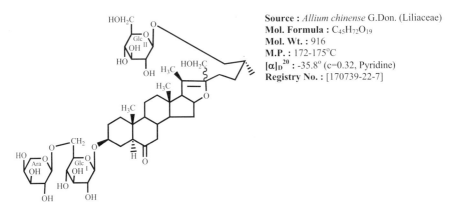

Source : *Allium chinense* G.Don. (Liliaceae)
Mol. Formula : $C_{45}H_{72}O_{19}$
Mol. Wt. : 916
M.P. : 172-175°C
$[\alpha]_D^{20}$: -35.8° (c=0.32, Pyridine)
Registry No. : [170739-22-7]

IR (KBr) : 3329.8, 2927.7, 2875.3, 2747.3, 1707.9, 1448.8, 1377.4, 1164.7, 1073.9, 907.1, 858.9, 777.6 cm^{-1}.

PMR (400 MHz) : δ 0.63 (s, 3xH-19), 0.66 (s, 3xH-18), 0.96* (H-14), 1.09* (H-9), 1.15 (H-1A), 1.17 (d, *J*=7.0 Hz, 3xH-27), 1.18 (H-11A), 1.33 (H-15A), 1.44* (H-11B), 1.57 (H-2A), 1.58* (H-1B), 1.61 (H-24), 1.67* (H-12), 1.69 (H-4A), 1.71* (H-8), 1.76 (s, 3xH-21), 1.88* (H-24), 1.94* (H-15B), 1.95* (H-8A), 2.11* (H-25), 2.12* (H-2B), 2.16 (dd, H-5), 2.26 (dd, H-7B), 2.32 (H-4B), 2.48 (d, *J*=10.4 Hz, H-17), 3.16 (m, H-23), 3.73 (d, *J*=11.1 Hz, H-5A of Ara), 3.78 (H-26A), 3.87 (H-28A), 3.90* (H-5 of Glc II), 3.92* (H-2 of Glc I), 3.97* (H-2 of Glc II), 3.99* (H-28B), 4.01* (H-3), 4.01* (H-5 of Glc I), 4.03* (H-26B), 4.07* (H-4 of Glc I), 4.11* (H-3 of Ara), 4.12* (H-4 of Glc II), 4.15* (H-3 of Glc I), 4.17* (H-3 of Glc II), 4.24* (H-5B of Ara), 4.28 (H-4 of Ara), 4.34 (dd, *J*=11.9, 5.3 Hz, H-6A of Glc II), 4.41 (dd, *J*=8.2, 6.7 Hz, H-2 of Ara), 4.53 (dd, *J*=11.9, 2.4 Hz, H-6B of Glc II), 4.73 (m, H-6B of Glc I), 4.92 (d, *J*=7.6 Hz, H-1 of Ara), 4.92 (d, *J*=7.6 Hz, H-1 of Glc I). *overlapped signals.

CMR (100 MHz) : δ C-1) 36.6 (2) 29.3 (3) 76.6 (4) 26.8 (5) 56.3 (6) 209.7 (7) 46.7 (8) 36.9 (9) 53.4 (10), 40.8 (11) 21.5 (12) 39.1 (13) 43.7 (14) 54.5 (15) 34.0 (16) 84.1 (17) 64.4 (18) 14.6 (19) 12.9 (20) 105.4 (21) 11.8 (22) 153.0 (23) 38.2 (24) 33.5 (5) 31.6 (26) 74.4 (27) 19.0 (28) 64.9 **Glc I** (1) 101.9 (2) 74.9 (3) 78.2 (4) 71.6 (5) 76.8 (6) 69.5 **Ara** (1) 105.2 (2) 72.1 (3) 74.2 (4) 68.9 (5) 66.3 **Glc II** (1) 104.7 (2) 74.9 (3) 78.3 (4) 71.5 (5) 78.2 (6) 62.8

Mass (FAB, Positive ion) : *m/z* 917 [M+H]$^+$.

Mass (FAB, Negative ion) : *m/z* 915 [M-H]⁻.

Reference

1.   J. Peng, X. Yao, Y. Tezuka, T. Kikuchi and T. Narui, *Planta Med.*, **62**, 465 (1996).

KINGIANOSIDE D
3β,22α,26-Trihydroxy-25(*R*)-furost-5-en-12-one 3-O-[β-D-glucopyranosyl-(1→4)-β-D-fucopyranoside]-26-O-β-D-glucopyranoside

Source : *Polygonatum kingianum* Coll. et Hemsl. (Liliaceae)

Mol. Formula : $C_{45}H_{72}O_{19}$

Mol. Wt. : 916

$[\alpha]_D^{20}$ **:** -13.4° (c=0.62, Pyridine)

Registry No. : [145854-05-3]

PMR (C_5D_5N, 270 MHz) : δ 0.95 (s, 3xH-19), 1.00 (d, *J*=6.2 Hz, 3xH-27), 1.17 (s, 3xH-18), 1.56 (d, *J*=6.6 Hz, 3xH-21), 1.63 (d, *J*=6.2 Hz, 3xH-6 of Fuc), 4.81 (d, *J*=7.7 Hz, H-1 of Fuc), 4.83 (d, *J*=7.7 Hz, H-1 of Glc II), 5.23 (d, *J*=7.3 Hz, H-1 of Glc I), 5.33 (br s, H-6).

CMR (C_5D_5N, 75 MHz) : δ C-1) 37.1 (2) 30.0 (3) 77.7 (4) 39.1 (5) 140.9 (6) 121.5 (7) 31.8 (8) 30.9 (9) 52.4 (10) 37.6 (11) 37.6 (12) 212.8 (13) 55.4 (14) 56.0 (15) 31.8 (16) 79.7 (17) 54.8 (18) 16.0 (19) 18.8 (20) 41.3 (21) 15.2 (22) 110.8 (23) 37.1 (24) 28.4 (25) 34.3 (26) 75.2 (27) 17.4 **Fuc** (1) 102.7 (2) 73.0 (3) 76.2ᵃ (4) 83.3 (5) 70.6 (6) 17.7 **Glc I** (1) 107.0 (2) 75.6ᵃ (3) 78.6 (4) 71.7 (5) 78.6 (6) 62.9 **Glc II** (1) 104.9 (2) 75.2 (3) 78.6 (4) 71.8 (5) 78.5 (6) 62.9.

Mass (FAB, Negative ion) : *m/z* 915 [M-H]⁻, 753 [M-Glc-H]⁻, 607 [M-Glc-Fuc-H]⁻.

Reference

1.   X.-C. Li, C.-R. Yang, M. Ichikawa, H. Matsuura, R. Kasai and K. Yamasaki, *Phytochemistry*, **31**, 3559 (1992).

TRIBULUS TERRESTRIS SAPONIN 5
(25*R*,*S*)-5α-Furostane-12-one-20(22)-en-3β,26-diol 3-O-[β-D-glucopyranosyl-(1→4)-β-D-galactopyranoside]-26-O-β-D-glucopyranoside

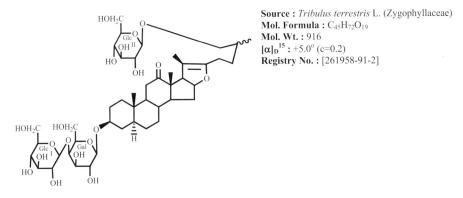

Source : *Tribulus terrestris* L. (Zygophyllaceae)
Mol. Formula : $C_{45}H_{72}O_{19}$
Mol. Wt. : 916
$[\alpha]_D^{15}$: +5.0° (c=0.2)
Registry No. : [261958-91-2]

IR (KBr) **:** 3426, 1701, 1620, 901 cm⁻¹.

PMR (CD₃OD-C₅D₅N, 400 MHz) **:** δ 4.32 (d, *J*=7.6 Hz), 4.44 (d, *J*=7.6 Hz), 4.60 (d, *J*=8.0 Hz).

CMR (CD₃OD-C₅D₅N, 100 MHz) **:** δ C-1) 38.1 (2) 30.6 (3) 78.2 (4) 35.6 (5) 46.1 (6) 29.9 (7) 32.3 (8) 35.8 (9) 57.5 (10) 37.7 (11) 39.3 (12) 216.5 (13) 58.9 (14) 56.0 (15) 34.8 (16) 84.3 (17) 57.5 (18) 14.9 (19) 12.5 (20) 104.7 (21) 11.7 (22) 154.2 (23) 24.4 (24) 33.2 (25) 34.6 (26) 76.3 (27) 17.6 **Gal** (1) 103.2 (2) 73.5 (3) 76.0 (4) 79.5 (5) 75.8 (6): 61.6 **Glc I** (1) 106.4 (2) 75.5 (3) 78.3 (4) 72.3 (5): 78.3 (6) 63.5 **Glc II** (1) 105.0 (2) 75.5 (3) 78.6 (4) 72.0 (5) 78.4 (6) 63.1.

Mass (E.S.I.) **:** *m/z* 939 [M+Na]⁺, 955 [M+K]⁺, 915 [M-H]⁻, 753 [M-Glc-H]⁻, 591 [M-2xGlc-H]⁻.

Reference

1.   L.F. Cai, J. Fengying, J.G. Zhang, F. Pei, Y. Xu, S. Liu and D. Xu, *Acta Pharm. Sin.*, **34**, 759 (1999).

TRIQUETROSIDE C1

Furost-5-en-1β,3β,22α,26-tetrol 3-O-[β-D-glucopyranoside]-26-O-[α-L-rhamnopyranosyl-(1→2)-β-D-glucopyranoside]

Source : *Allium triquetrum* L. (Alliaceae, Liliaceae)
Mol. Formula : $C_{45}H_{72}O_{19}$
Mol. Wt. : 916
[α]$_D^{25}$: -13.7° (c=0.1, MeOH)
Registry No. : [628728-01-8]

PMR (CD₃OD, 500 MHz) : δ 0.81 (s, 3xH-18), 0.95 (d, *J*=6.6 Hz, 3xH-27), 0.98 (d, *J*=6.6 Hz, 3xH-21), 1.02 (s, 3xH-19), 1.12* (H-9), 1.18 (m, H-14), 1.20* (H-12A), 1.23 (d, *J*=6.6 Hz, 3xH-6 of Rha), 1.25* (H-15A), 1.31* (H-24A), 1.32* (H-24B), 1.52 (m, H-8), 1.53 (m, H-11A), 1.62* (H-23A), 1.70* (H-2A), 1.72* (H-12B), 1.72 (m, H-25), 1.73* (H-23B), 1.74* (H-17), 1.95* (H-7A, H-15B), 1.96 (H-7B), 2.04* (H-20), 2.05* (H-2B), 2.24 (dd, *J*=11.5, 3.5 Hz, H-4A), 2.26 (dd, *J*=10.5, 2.5 Hz, H-11B), 2.36 (dd, *J*=11.5, 7.3 Hz, H-4B), 3.19 (t, *J*=7.5 Hz, H-2 of Glc II), 3.24 (dd, *J*=7.0, 6.4 Hz, H-4 of Glc II), 3.25* (H-1), 3.29 (m, H-5 of Glc I), 3.30 (m, H-5 of Glc II), 3.32* (H-26A), 3.36* (H-3 of Glc II), 3.38 (dd, *J*=6.5, 6.0 Hz, H-4 of Rha), 3.39* (H-3 of Glc I), 3.40 (t, *J*=7.5 Hz, H-2 of Glc I), 3.49* (H-3), 3.62 (d, *J*=6.5 Hz, H-3 of Rha), 3.64 (br d, *J*=11.5 Hz, H-6A of Glc I, H-6A of Glc II), 3.78 (dd, *J*=6.8, 6.5 Hz, H-4 of Glc I), 3.78 (br d, *J*=11.5 Hz, H-6B of Glc I), 3.82 (dd, *J*=9.5, 3.9 Hz, H-26B), 3.84 (br d, *J*=11.5 Hz, H-6B of Glc II), 3.91 (br s, H-2 of Rha), 4.09* (H-5 of Rha), 4.21 (d, *J*=7.5 Hz, H-1 of Glc II), 4.49 (d, *J*=7.5 Hz, H-1 of Glc I), 4.53 (q, *J*=5.5 Hz, H-16), 5.19 (br s, H-1 of Rha), 5.55 (br d, *J*=3.2 Hz, H-6). * overlapped signals.

CMR (CD₃OD, 125 MHz) : δ C-1) 74.1 (2) 43.2 (3) 82.3 (4) 42.5 (5) 143.7 (6) 128.5 (7) 35.1 (8) 36.0 (9) 54.6 (10) 44.5 (11) 27.3 (12) 43.4 (13) 46.7 (14) 60.2 (15) 35.3 (16) 84.5 (17) 66.7 (18) 19.3 (19) 16.0 (20) 43.2 (21) 18.3 (22) 114.0 (23) 39.2 (24) 32.9 (25) 37.7 (26) 74.8 (27) 19.5 **Glc I** (1) 103.5 (2) 76.2 (3) 77.7 (4) 78.4 (5) 74.4 (6) 64.5 **Glc II** (1) 107.1 (2) 77.5 (3) 74.6 (4) 74.2 (5) 78.5 (6) 65.2 **Rha** (1) 104.1 (2) 73.1 (3) 74.5 (4) 78.7 (5) 74.5 (6) 20.2.

Mass (FAB, Negative ion, H.R.) : *m/z* 915.4560 [(M-H)⁻, calcd. for 915.4569].

Reference

1.   G. Corea, E. Tattorusso and V. Lanzotti, *J. Nat. Prod.*, **66**, 1405 (2003).

TRIQUETROSIDE C2

Furost-5-en-1β,3β,22β,26-tetrol 3-O-[β-D-glucopyranoside]-26-O-α-L-rhamnopyranosyl-(1→2)-β-D-glucopyranoside]

Source : *Allium triquetrum* L. (Alliaceae, Liliaceae)
Mol. Formula : $C_{45}H_{72}O_{19}$
Mol. Wt. : 916
$[\alpha]_D^{25}$: -19.6° (c=0.1, MeOH)
Registry No. : [628728-02-9]

PMR (CD$_3$OD, 500 MHz) : δ 0.81 (s, 3xH-18), 0.95 (d, *J*=6.6 Hz, 3xH-27), 0.96 (d, *J*=6.6 Hz, 3xH-21), 1.02 (s, 3xH-19), 1.12* (H-9), 1.18 (m, H-14), 1.20* (H-2A), 1.25* (H-15A), 1.31* (H-24A), 1.32* (H-24B), 1.52 (m, H-8), 1.53 (m, H-11A), 1.62* (H-23A), 1.69* (H-17), 1.70 (H-2A), 1.72* (H-12B), 1.72 (m, H-25), 1.73* (H-23B), 1.95* (H-7A, H-15B), 1.96* (H-7B), 2.03* (H-2B), 2.04* (H-20), 2.24 (dd, *J*=11.5, 3.5 Hz, H-4A), 2.26 (dd, *J*=10.5, 2.5 Hz, H-11B), 2.36 (dd, *J*=11.5, 7.3 Hz, H-4B), 3.25* (H-1), 3.32* (H-26A), 3.48* (H-3), 3.82 (dd, *J*=9.5, 3.9 Hz, H-26B), 4.33 (q, *J*=5.5 Hz), 5.55 (br d, *J*=3.2 Hz, H-6). For signals of sugar protons see Triquetroside C1. * overlapped signals.

CMR (CD$_3$OD, 125 MHz) : δ C-1) 74.1 (2) 43.2 (3) 82.3 (4) 42.1 (5) 143.7 (6) 128.5 (7) 35.0 (8) 36.0 (9) 54.6 (10) 44.5 (11) 27.3 (12) 43.2 (13) 46.7 (14) 60.2 (15) 35.3 (16) 84.6 (17) 67.9 (18) 19.5 (19) 16.0 (20) 43.0 (21) 18.5 (22) 117.3 (23) 39.0 (24) 32.7 (25) 37.1 (26) 74.6 (27) 19.5. For signals of sugar carbons see Triquetroside C1.

Mass (FAB, Negative ion, H.R.) : *m/z* 915.4560 [(M-H)⁻, requires 915.4569].

Reference

1.   G. Corea, E. Tattorusso and V. Lanzotti, *J. Nat. Prod.*, **66**, 1405 (2003).

ASPARAGUS COCHINCHINENSIS SAPONIN ASP-V
(5β,25S)-22ξ-Methoxy-furostane-3β,26-diol 3-O-[α-L-rhamnopyranosyl-(1→6)-β-D-glucopyranoside]-26-O-β-D-glucopyranoside

Source : *Asparagus cochinchinensis* (Loureiro) Merill (Liliaceae)
Mol. Formula : $C_{46}H_{78}O_{18}$
Mol. Wt. : 918
M.P. : 150-156°C (decomp.)
[α]$_D$: -56.5° (c=1.0, MeOH)
Registry No. : [72947-80-9]

PMR (C_5D_5N, 90.0 MHz) : δ 3.28 (s, OCH_3), 4.83 (d, J=7.5 Hz, anomeric H), 4.96 (d, J=7.5 Hz, anomeric H), 5.52 (s, anomeric H).

CMR (C_5D_5N, 22.15 MHz) : δ **Glc I** C-1) 103.3 **Glc II** (1) 104.8 **Rha** (1) 102.2.

Reference

1.   T. Konishi and J. Shoji, *Chem., Pharm. Bull.*, **27**, 3086 (1979).

ASPACOCHIOSIDE B

(5β,25S)-22α-Methoxyfurostan-3β,26-diol 3-O-[α-L-rhamnopyranosyl-(1→4)-β-D-glucopyranoside]-26-O-β-D-glucopyranoside

Source : *Asparagus cochinchinensis* (Lour.) Merr. (Liliaceae)
Mol. Formula : $C_{46}H_{78}O_{18}$
Mol. Wt. : 918
M.P. : 199-200°C[1]
$[\alpha]_D^{25}$ **:** -64.7° (c=0.10, MeOH)[1]
Registry No. : [557769-33-2]

IR (KBr)[1] : 3404 (br, OH)cm^{-1}.

PMR (C$_5$D$_5$N, 500 MHz)[2] : δ 0.78 (s, 3xH-18), 0.80 (s, 3xH-19), 0.96 (m, H-6A), 1.00 (m, H-14), 1.02 (d, *J*=7.0 Hz, 3xH-27), 1.12 (m, H-11A), 1.13 (m, H-12A), 1.15 (d, *J*=7.0 Hz, 3xH-21), 1.25 (m, H-7A, H-12B), 1.34 (m, H-24A), 1.35 (m, H-11B), H-23A), 1.38 9m, H-15B), 1.44 (m, H-1A), 1,50 (m H-7B), 1.52 (m, H-8), 1.55 (m, H-2A), 1.69 (m, H-9), 1.69 (d, *J*=7.0 Hz, 3xH-6 of Rha), 1.72 (m, H-2B), 1.75 (m, H-4A, H-17), 1.78 (m, H-24B), 1.82 (m, H-6B), 1.85 (m, H-23B), 1.86 (m, H-25), 1.88 (m, H-4B), 1.92 (m, H-15B), 2.00 (m, H-5), 2.03 (m, H-1B), 2.20 (m, H-20), 3.24 (s, OC*H$_3$*), 3.50 (dd, *J*=10.0, 7.2 Hz, H-26A), 3.69 (br d, *J*=9.5 Hz, H-5 of Glc I), 3.91 (dd, *J*=7.0, 8.0 Hz, H-2 of Glc II), 3.95 (m, H-5 of Glc II), 3.98 (dd, *J*=8.0, 7.0 Hz, H-2 of Glc), 4.10 (br d, *J*=10.5 Hz, H-6A of Glc I), 4.20 (dd, *J*=8.0, 9.0 Hz, H-3 of Glc II), 4.21 (dd, *J*=8.0, 9.0 Hz, H-3 of Glc I), 4.25 (m, H-3), 4.27 (br d, *J*=10.5 Hz, H-6B of Glc I), 4.28 (dd, *J*=9.0, 9.0 Hz, H-4 of Glc II), 4.30 (dd, *J*=10.0, 7.2 Hz, H-26B), 4.32 (dd, *J*=9.0, 8.5 Hz, H-4 of Rha), 4.42 (br d, *J*=12.5 Hz, H-6A of Glc II), 4.47 (dd, *J*=9.0, 9.5 Hz, H-4 of Glc I), 4.50 (m, H-16), 4.57 (br d, *J*=8.5 Hz, H-3 of Rha), 4.57 (br d, *J*=12.5 Hz, H-6B of Glc II), 4.69 (br s, H-2 of Rha), 4.83 (d, *J*=7.0 Hz, H-1 of Glc II), 4.85 (d, *J*=7.0 Hz, H-1 of Glc I), 5.02 (dq, *J*=9.0, 7.0 Hz, H-5 of Rha), 5.91 (br s, H-1 of Rha).

CMR (C$_5$D$_5$N, 125 MHz)[1] : δ C-1) 30.5 (2) 27.0 (3) 74.4 (4) 30.9 (5) 37.0 (6) 26.9 (7) 26.8 (8) 35.5 (9) 39.5 (10) 35.2 (11) 21.1 (12) 41.1 (13) 41.3 (14) 56.3 (15) 32.2 (16) 81.4 (17) 64.4 (18) 16.6 (19) 23.8 (20) 40.5 (21) 16.5 (22) 112.5 (23) 31.1 (24) 28.2 (25) 34.5 (26) 74.9 (27) 17.6 (OCH$_3$) 47.4 **Glc I** (1) 102.9 (2) 75.5 (3) 76.7 (4) 78.1 (5) 77.1 (6)

61.5 **Rha** (1) 102.6 (2) 72.8 (3) 72.6 (4) 74.0 (5) 70.3 (6) 18.6 **Glc II** (1) 105.0 (2) 75.2 (3) 78.6 (4) 71.7 (5) 78.5 (6) 62.8.

Mass (E.S.I., Positive ion)^2 : *m/z* 941 [M+Na]$^+$, 919 [M+H]$^+$, 887 [M+H-MeOH]$^+$.

Mass (E.S.I., Positive ion, H.R.)^2 : *m/z* 919.5219 [(M+Na)$^+$, calcd. for 919.5266].

References

1.   Y.C. Yang, S.Y. Huang and J.G. Shi, *Chin. Chem. Lett.*, **13**, 1185 (2002).

2.   J.G. Shi, G.-Q. Li, S.-Y. Huang, S.-Y. Mo, Y. Wang, Y.-C. Yang and W.-Y. Hu, *J. Asian Nat. Prod. Res.*, **6**, 99 (2004).

ASCALONICOSIDE A1

(25*R*)-Furost-5(6)-en-1β,3β,22α,26-tetrol 1-O-β-D-galactopyranoside
26-O-[α-L-rhamnopyranosyl-(1→2)-β-D-glucopyranoside]

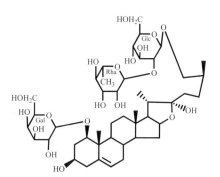

Source : *Allium ascalonicum* Hort. (Liliaceae)
Mol. Formula : C$_{45}$H$_{74}$O$_{19}$
Mol. Wt. : 918
[α]$_D^{25}$: -60° (c=0.1, MeOH)
Registry No. : [473555-58-7]

PMR (CD$_3$OD, 500 MHz) : δ 0.87 (s, 3xH-18), 0.99 (d, *J*=6.6 Hz, 3xH-27), 1.04 (d, *J*=6.6 Hz, 3xH-21), 1.13 (s, 3xH-19), 1.18 (m, H-14), 1.22* (H-12A), 1.28* (H-9), 1.28 (d, *J*=6.6 Hz, 3xH-6 of Rha), 1.30* (H-15A, H-24A), 1.38* (H-24B), 1.45 (m, H-11A), 1.50 (m, H-8), 1.63* (H-23A), 1.70* (H-2A, H-12B), 1.72* (H-23B), 1.78* (H-17), 1.78 (m, H-25), 1.96* (H-7A), 1.98* (H-7B), 1.99* (H-15B), 2.12* (H-2B), 2.13* (H-20), 2.22 (dd, *J*=11.5, 3.5 Hz, H-4A), 2.28 (dd, *J*=11.5, 7.3 Hz, H-4B), 2.57 (dd, *J*=10.5, 2.5 Hz, H-11B), 3.21 (t, *J*=7.5 Hz, H-2 of Glc), 3.27 (m, H-5 of Glc), 3.29* (H-26A), 3.30 (dd, *J*=7.0 Hz, 6.4 Hz, H-4 of Glc), 3.31* (H-3), 3.36* (H-1), 3.38* (H-3 of Glc), 3.40 (dd, *J*=6.5, 6.0 Hz, H- of Rha), 3.45* (H-5 of Gal), 3.47& (H-6A of Gal), 3.52 (br d, *J*=11.5 Hz, H-6A of Glc), 3.65* (H-6B of Glc), 3.68 (dd, *J*=6.8, 2.5 Hz, H-3 of Gal), 3.71* (H-2 of Gal), 3.72 (d, *J*=6.5 Hz, H-3 of Rha), 3.82 (dd, *J*=8.5, 6.9 Hz, H-26B), 3.85 (br d, *J*=11.5 Hz, H-6B of Glc), 3.87 (dd, *J*=3.2, 2.5 Hz,H-4 of Gal), 3.91 (br s, H-2 of Rha), 4.11 (dq, *J*=6.6, 6.0 Hz, H-5 of Rha), 4.26 (d, *J*=7.5 Hz, H-1 of Glc), 4.30 (d, *J*=7.5 Hz, H-1 of Gal), 4.59 (q, *J*=5.5 Hz, H-16), 5.32 (br s, H-1 of Rha), 5.59 (br d, *J*=3.2 Hz, H-6).

CMR (C₅D₅N, 125 MHz) : δ C-1) 84.2 (2) 37.1 (3) 68.9 (4) 43.2 (5) 141.1 (6) 126.2 (7) 32.5 (8) 33.5 (9) 51.5 (10) 43.0 (11) 24.9 (12) 40.8 (13) 41.0 (14) 57.7 (15) 33.0 (16) 82.2 (17) 65.3 (18) 17.1 (19) 15.5 (20) 40.5 (21) 16.0 (22) 112.0 (23) 36.5 (24) 31.7 (25) 35.1 (26) 75.8 (27) 17.5 **Gal** (1) 101.2 (2) 75.1 (3) 76.3 (4) 78.2 (5) 74.4 (6) 62.5 **Glc** (1) 104.3 (2) 76.1 (3) 74.1 (4) 71.9 (5) 77.0 (6) 66.0 **Rha** (1) 102.5 (2) 72.5 (3) 70.5 (4) 72.0 (5) 68.8 (6) 18.2.

Mass (FAB, Negative ion, H.R.) : *m/z* 917.4763 [(M-H)⁻, requires 916.4744].

Reference

1.   E. Fattorusso, M. Iorizzi, V. Lanzotti and O. Taglialatela-Scafati, *J. Agric. Food Chem.*, **50**, 5686 (2002).

ASCALONICOSIDE A2
(25*R*)-Furost-5-en-1β,3β,22β,26-tetrol 1-O-β-D-galactopyranoside
26-O-[α-L-rhamnopyranosyl-(1→2)-β-D-glucopyranoside]

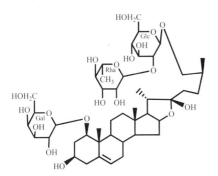

Source : *Allium ascalonicum* Hort. (Liliaceae)
Mol. Formula : C₄₅H₇₄O₁₉
Mol. Wt. : 918
[α]ᴅ²⁵ : -55° (c=0.1, MeOH)
Registry No. : [473555-62-3]

PMR (CD₃OD, 500 MHz) : δ 0.87 (s, 3xH-18), 0.99 (d, *J*=6.6 Hz, 3xH-27), 1.01 (d, *J*=6.6 Hz, 3xH-21), 1.13 (s, 3xH-19), 1.18 (m, H-14), 1.23* (H-12A), 1.28* (H-9), 1.30* (H-15A), 1.31* (H-24A), 1.34 (H-24B), 1.46 (m, H-11A), 1.50 (m, H-8), 1.61* (H-23A), 1.70 (H-2A), 1.73* (H-17), 1.75* (H-23B), 1.78 (m, H-25), 1.96* (H-7A), 1.98* (H-7B), 1.99* (H-15B), 2.10* (H-2B), 2.15* (H-20), 2.22 (dd, *J*=11.5, 3.5 Hz, H-4A), 2.27 (dd, *J*=11.5, 7.3 Hz, H-4B), 2.56 (dd, *J*=10.5, 2.5 Hz, H-11B), 3.29* (H-26A), 3.30* (H-3), 3.36* (H-1), 3.82 (dd, *J*=8.5, 6.9 Hz, H-26B), 4.39 (q, *J*=5.5 Hz, H-16), 5.59 (br d, *J*=3.2 Hz, H-6).

CMR (C₅D₅N, 125 MHz) : δ C-1) 84.2 (2) 37.1 (3) 68.9 (4) 43.2 (5) 141.1 (6) 126.2 (7) 32.3 (8) 33.5 (9) 51.5 (10) 43.0 (11) 24.9 (12) 40.7 (13) 41.0 (14) 52.7 (15) 33.0 (16) 82.2 (17) 65.2 (18) 17.0 (19) 15.3 (20) 40.2 (21) 16.3 (22) 115.5 (23) 36.2 (24) 31.6 (25) 35.0 (26) 76.0 (27) 17.5. * overlapped signals.

For PMR and CMR signals of the sugar moiety see Ascalonicoside A1.

Mass (FAB, Negative ion, H.R.) :: *m/z* 917.4722 [(M-H)⁻, requires 917.4744].

Reference

1.   E. Fattorusso, M. Iorizzi, V. Lanzotti and O. Taglialatela-Scafati, *J. Agric. Food Chem.*, **50**, 5686 (2002).

MACRANTHOSIDE I
(5β)-Furost-25(27)-ene-3β,22,26-triol 3-O-[β-D-glucopyranosyl-(1→6)-β-D-glucopyranoside]-26-O-β-D-glucopyranoside

Source : *Helleborus macranthus* Freyn (Ranunculaceae)
Mol. Formula : $C_{45}H_{74}O_{19}$
Mol. Wt. : 918
Registry No. : [90850-94-5]

IR (KBr) : 1640, 900, 875 cm^{-1}.

Reference

1.   R. Tschesche, R. Wagner and H.C. Jha, *Phytochemistry*, **23**, 695 (1984).

MACROSTEMONOSIDE L

2β,3β,26-Trihydroxy-25R-5β-furost-20(22)-ene 3-O-[β-D-glucopyranosyl-(1→2)-β-D-galactopyranoside]-
26-O-β-D-glucopyranoside

Source : *Allium macrostemon* Bunge (Liliaceae)
Mol. Formula : $C_{45}H_{74}O_{19}$
Mol. Wt. : 918
M.P. : 206-208°C
Registry No. : [159935-11-2]

PMR (C_5D_5N, 500 MHz) **:** δ 0.68 (s, CH_3), 0.95 (s, CH_3), 1.02 (d, *J*=6.7 Hz, 3xH-27), 1.61 (s, 3xH-21), 4.81 (d, *J*=7.8 Hz, anomeric H), 4.96 (d, *J*=7.7 Hz, anomeric H), 5.23 (d, *J*=7.7 Hz, anomeric H).

CMR (C_5D_5N, 125 MHz) **:** δ C-1) 40.1 (2) 67.3 (3) 82.1 (4) 31.9 (5) 36.6 (6) 26.3 (7) 26.9 (8) 35.3 (9) 41.4 (10) 37.1 (11) 21.5 (12) 40.6 (13) 43.9 (14) 54.7 (15) 31.5 (16) 84.6 (17) 64.7 (18) 14.5 (19) 23.9 (20) 103.7 (21) 11.9 (22) 152.4 (23) 34.4 (24) 23.7 (25) 33.5 (26) 75.0 (27) 17.4 **Gal** (1) 103.5 (2) 81.7 (3) 75.2 (4) 69.7 (5) 77.0 (6) 62.0^a **Glc I** (1) 106.3 (2) 77.0 (3) 78.0 (4) 71.8 (5) 78.5^b (6) 62.9^a **Glc II** (1) 104.9 (2) 75.2 (3) 78.6^b (4) 71.8 (5) 78.5^b (6) 62.9^a.

Mass (FAB, Positive ion) **:** *m/z* 941 [M+Na]$^+$.

Reference

1.   J. Peng, X. Yao, Y. Okada and T. Okuyama, *Chem. Pharm. Bull.*, **42**, 2180 (1994); *Acta Pharm. Sin.*, **29**, 526 (1994).

MELONGOSIDE O

(25*R*)-Furost-5-en-3β,22α,26-triol 3-O-[β-D-glucopyranosyl-(1→2)-β-D-glucopyranoside]-26-O-β-D-glucopyranoside

Source : *Solanum melongena* L. (Solanaceae)
Mol. Formula : $C_{45}H_{74}O_{19}$
Mol. Wt. : 918
M.P. : 183-184°C
[α]$_D^{20}$: -19° (c=1.0, CH_3OH)
Registry No. : [98464-54-1]

Reference

1.   P.K. Kintia and S.A Shvets, *Phytochemistry*, **24**, 1567 (1985).

OPHIOPOJAPONIN B

(25*R*)-3β,14α,22ξ ,26-Tetrahydroxyfurost-5-ene-3-O-[α-L-rhamnopyranosyl-(1→2)-β-D-glucopyranoside]-26-O-β-D-glucopyranoside

Source : *Ophiopogon japonicus*.Ker-Gawl.(Liliaceae)
Mol. Formula : $C_{45}H_{74}O_{19}$
Mol. Wt. : 918
M.P. : 198-201°C
[α]$_D^{21}$: -42.6° (c=0.16, C_5H_5N)
Registry No. : [313054-36-3]

IR (KBr) : 3400 (OH), 1620 (C=C), 1080 (C-O-C) cm^{-1}.

PMR (C$_5$D$_5$N, 400/500 MHz) : δ 0.98 (d, J=6.6 Hz, 3xH-27), 1.02 (s, 3xH-18), 1.13 (s, 3xH-19), 1.35 (d, J=6.8 Hz, 3xH-21), 1.77 (d, J=6.1 Hz, 3xH-6 of Rha), 3.64 (m, H-26), 3.94 (m, H-3), 4.38 (m, H-2 of Glc I), 4.81 (d, J=7.8 Hz, H-1 of Glc II), 5.01 (d, J=6.8 Hz, H-1 of Glc I), 5.36 (m, H-6), 5.43 (m, H-16), 6.38 (br s, H-1 of Rha).

CMR (C$_5$D$_5$N, 100 MHz) : δ C-1) 37.9 (2) 30.2 (3) 78.6 (4) 40.2 (5) 140.5 (6) 122.4 (7) 26.8 (8) 34.4 (9) 43.8 (10) 37.6 (11) 20.5 (12) 37.3 (13) 45.5 (14) 86.5 (15) 39.1 (16) 81.6 (17) 60.7 (18) 20.2 (19) 19.4 (20) 40.9 (21) 16.7 (22) 111.1 (23) 32.2 (24) 28.5 (25) 35.7 (26) 75.3 (27) 17.6 **Glc I** (1) 100.4 (2) 79.7 (3) 78.0 (4) 71.9 (5) 78.6 (6) 62.8 **Rha** (1) 102.1 (2) 72.6 (3) 72.9 (4) 74.2 (5) 69.5 (6) 18.7 **Glc II** (1) 104.9 (2) 75.3 (3) 78.4 (4) 71.9 (5) 78.2 (6) 62.8.

Mass (FAB, Negative ion) : m/z 917 [M-H]$^-$, 771 [M-H-Rha]$^-$, 755 [M-H-Glc]$^-$, 609 [M-H-Rha-Glc]$^-$.

Reference

1.   H.-F. Dai, J. Zhou, N.-H. Tan and Z.-T. Ding, *Zhiwu Xuebao* (*Acta Bot. Sin.*), **43**, 97 (2001).

POLYGONATUM SAPONIN 25-*EPI*-PO-8

3β,22α,26-Trihydroxyfurost-5-ene 3-O-[β-D-glucopyranosyl-(1→4)-β-D-galactopyranoside]-26-O-β-D-glucopyranoside]

Source : *Polygonatum officinale*[1] (Liliaceae),
P. kingianum Coll. et Hemsl.[2] (Liliaceae)
Mol. Formula : $C_{45}H_{74}O_{19}$
Mol. Wt. : 918
M.P. : 198-200°C[2]
$[\alpha]_D^{20}$: -59.4° (c=0.61, Pyridine)
Registry No. : [145867-19-2]

PMR (C_5D_5N, 270 MHz)[2] : δ 0.91 (s, 3xH-18 and 3xH-19), 0.99 (d, *J*=6.6 Hz, 3xH-27), 1.34 (*J*=7.0 Hz, 3xH-21), 4.82 (d, *J*=7.7 Hz, H-1 of Glc II), 4.90 (d, *J*=7.7 Hz, H-1 of Gal), 5.30 (d, *J*=7.7 Hz, H-1 of Glc I), 5.31 (br s, H-6).

CMR (C_5D_5N, 75 MHz)[2] : δ C-1) 37.5 (2) 30.3 (3) 78.1 (4) 39.3 (5) 141.0 (6) 121.7 (7) 32.3 (8) 31.7 (9) 50.4 (10) 37.1 (11) 21.1 (12) 39.9 (13) 40.8 (14) 56.6 (15) 32.5 (16) 81.3 (17) 63.9 (18) 16.4^a (19) 19.4 (20) 40.7 (21) 16.5^a (22) 110.7 (23) 37.2 (24) 28.4 (25) 34.3 (26) 75.2 (27) 17.5 **Gal** (1) 103.0 (2) 73.5 (3) 75.4^a (4) 80.0 (5) 76.0^a (6) 61.0 **Glc I** (1) 107.2 (2) 75.2^b (3) 78.8 (4) 72.4 (5) 78.5 (6) 63.2 **Glc II** (1) 105.0 (2) 75.3^b (3) 78.6^c (4) 71.8 (5) 78.5 (6) 62.9.

Mass (FAB, Negative ion): *m/z* 917 [M-H]⁻, 755 [M-Glc-H]⁻.

References

1.   M. Ono, K. Shoyama and T. Nohara, *Shoyakugaku Zasshi*, **42**, 135 (1988).

2.   X.-C. Li, C.-R. Yang, M. Ichikawa, H. Matsuura, R. Kasai and K. Yamasaki, *Phytochemistry*, **31**, 3559 (1992).

TIMOSAPONIN D

(25S)-5β-Furost-20(22)-en-2β,3β,26-triol 3-O-[β-D-glucopyranosyl-(1→2)-β-D-galactopyranoside]-26-O-β-D-glucopyranoside

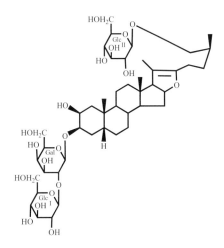

Source : *Anemarrhena asphodeloides* Bge. (Liliaceae)
Mol. Formula : $C_{45}H_{74}O_{19}$
Mol. Wt. : 918
M.P. : 186-190°C (decomp.)
Registry No. : [220095-97-6]

PMR (C_5D_5N, 300 MHz) : δ 0.66 (s, CH_3), 0.96 (s, CH_3), 1.03 (d, sec. CH_3), 1.59 (s, CH_3), 4.82 (d, anomeric H), 5.01 (d, anomeric H), 5.29 (d, anomeric H).

CMR (C_5D_5N, 75 MHz) : δ C-1) 40.1 (2) 67.3 (3) 82.1 (4) 31.9 (5) 36.6 (6) 26.3 (7) 26.9 (8) 35.3 (9) 41.4 (10) 37.1 (11) 21.5 (12) 40.6 (13) 43.8 (14) 54.7 (15) 31.4 (16) 84.6 (17) 64.7 (18) 14.4 (19) 23.9 (20) 103.6 (21) 11.8 (22) 152.4 (23) 34.4 (24) 23.7 (25) 33.7 (26) 75.3 (27) 17.2 **Glc I** (1) 106.3 (2) 77.0 (3) 78.1 (4) 71.7 (5) 78.6 (6) 62.9 **Gal** (1) 103.5 (2) 81.8 (3) 75.3 (4) 69.8 (5) 77.0 (6) 62.0 **Glc II** (1) 105.2 (2) 75.3 (3) 78.6 (4) 71.7 (5) 78.6 (6) 62.9.

Reference

1.   Z. Meng, X. Zhou and S. Xu, *Journal of Shenyang Pharmaceutical Univesity,* **15**, 254 (1998).

TRILLIUM KAMTSCHATICUM SAPONIN Te

Furost-5-en-3β,17α,22,26-tetrol 3-O-[α-L-rhamnopyranosyl-(1→2)-β-D-glucopyranoside]-26-O-[β-D-glucopyranoside]

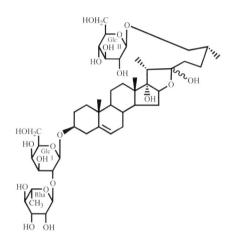

Source : *Trillium kamtschaticum* Pall.[1] (Liliaceae)
Mol. Formula : $C_{45}H_{74}O_{19}$
Mol. Wt. : 918
M.P. : 275-280°C (decomp.)[1]
[α]$_D$: -84.6° (c=1.02, Pyridine)[1]
Registry No. : [55972-78-6]

22-Methyl ether :

CMR (C_5D_5N, 22.5 MHz)[2] : δ C-1) 37.6 (2) 30.2 (3) 77.8 (4) 39.0 (5) 140.9 (6) 121.7 (7) 31.7 (8) 32.3 (9) 50.3 (10) 37.1 (11) 20.9 (12) 37.1 (13) 45.4 (14) 53.0 (15) 32.4 (16) 90.3 (17) 90.5 (18) 17.1 (19) 19.4 (20) 43.0 (21) 10.3 (22) 113.5 (23) 30.8 (24) 28.1 (25) 34.2 (26) 75.1 (27) 17.1 (OCH$_3$) 47.1 **Glc I** (1) 100.3 (2) 79.5 (3) 77.8 (4) 71.9^a (5) 78.1 (6) 62.7^b **Rha** (1) 101.9 (2) 72.8 (3) 72.5 (4) 74.1 (5) 69.4 (6) 18.5 **Glc II** (1) 104.9 (2) 75.1 (3) 78.5^c (4) 71.8^a (5) 78.3^c (6) 62.9^b.

Peracetate :

[α]$_D$: -38.7° (c=0.84, CHCl$_3$)[1]

References

1. T. Nohara, K. Miyahara and T. Kawasaki, *Chem. Pharm. Bull.*, **23**, 872 (1975).

2. K. Nakano, Y. Kashiwada, T. Nohara, T. Tomimatsu, H. Tsukatani and T. Kawasaki, *Yakugaku Zasshi*, **102**, 1031 (1982).

ASPARAGOSIDE E

(5β,25S)-Furostan-3β,22α,26-triol 3-O-[β-D-glucopyranosyl-(1→3)-β-D-glucopyranoside]-26-O-β-D-glucopyranoside

Source : *Asparagus officinalis* L. (Liliaceae)
Mol. Formula : $C_{45}H_{76}O_{19}$
Mol. Wt. : 920
M.P. : 254-260°C
[α]$_D^{20}$: -38.0° (c=1.05, H_2O)
Registry No. : [60267-25-6]

Reference

1.   G.M. Goryanu, V.V. Krokhmalyuk and P.K. Kintya, *Khim. Prir. Soedin.*, 823 (1976), *Chem. Nat. Comp.*, **12**, 743 (1976).

DISPOROSIDE C

(5β,25R)-Furostan-3β,22α,26-triol 3-O-[β-D-glucopyranosyl-(1→2)-β-D-glucopyranoside]-26-O-β-D-glucopyranoside

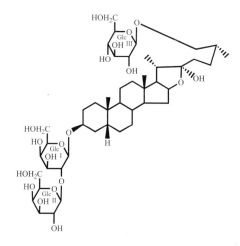

Source : *Disporopsis penyi* (Hua) Diels (Liliaceae)
Mol. Formula : $C_{45}H_{76}O_{19}$
Mol. Wt. : 920
$[\alpha]_D^{23}$: -40.9° (c=0.2, Pyridine)
Registry No. : [244779-39-3]

IR (KBr) : 3414, 2930, 1078, 1038 cm⁻¹.

PMR (C₅D₅N, 500 MHz) : δ 0.87 (s, 3xH-18), 0.98 (s, 3xH-19), 1.00 (d, J=5.0 Hz, 3xH-27), 1.33 (d, J=6.8 Hz, 3xH-21), 3.60 (dd, J=2.5, 12.1 Hz, H-26A), 4.00 (t, J=12.1 Hz, H-26B), 4.80 (d, J=7.8 Hz, H-1 of Glc III), 4.95 (d, J=7.5 Hz, H-1 of Glc I), 4.96 (m, H-16), 5.40 (d, J=7.7 Hz, H-1 of Glc II).

CMR (C₅D₅N, 125 MHz) : δ C-1) 30.8 (2) 27.1 (3) 75.3 (4) 32.5 (5) 36.9 (6) 26.9 (7) 27.1 (8) 35.3 (9) 40.4 (10) 35.6 (11) 21.3 (12) 40.4 (13) 41.4 (14) 56.5 (15) 31.0 (16) 81.3 (17) 64.1 (18) 16.8 (19) 24.1 (20) 40.7 (21) 16.5 (22) 110.7 (23) 31.0 (24) 28.4 (25) 34.5 (26) 75.5 (27) 17.5 **Glc I** (1) 101.9 (2) 83.2 (3) 78.0 (4) 71.7 (5) 78.2 (6) 62.9 **Glc II** (1) 105.9 (2) 77.0 (3) 78.5 (4) 71.85 (5) 78.5 (6) 62.9 **Glc III** (1) 104.9 (2) 75.3 (3) 78.6 (4) 71.7 (5) 78.5 (6) 62.9.

Mass (FAB, Negative ion) : *m/z* 919 [M-H]⁻, 757 [M-H-Glc]⁻.

Mass (FAB, Negative ion, H.R.) : *m/z* 919.4870 [(M-H)⁻, calcd. for 919.4903].

Reference

1.   Q.X. Yang, M. Xu, Y.-J. Zhang, H.-Z. Li and C.-R. Yang, *Helv. Chim. Acta*, **87**, 1248 (2004).

MELONGOSIDE N

(25R,5α)-Furostan-3β,22α,26-triol 3-O-[β-D-glucopyranosyl-(1→2)-β-D-glucopyranoside]-26-O-β-D-glucopyranoside

Source : *Solanum melongena* L. (Solanaceae)
Mol. Formula : $C_{45}H_{76}O_{19}$
Mol. Wt. : 920
M.P. : 187-189°C
[α]$_D^{20}$: -15° (c=1.0, MeOH)
Registry No. : [98524-46-0]

Reference

1.   P.K. Kintia and S.A Shvets, *Phytochemistry*, **24,** 1567 (1985).

OFFICINALISININ - I
(5β,25S)-Furostane-3β,22α,26-triol 3-O-[β-D-glucopyranosyl-(1→2)-β-D-glucopyranoside]-26-O-β-D-glucopyranoside

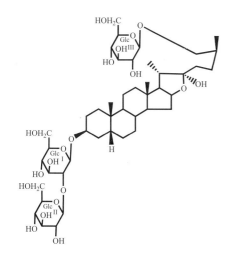

Source : *Asparagus officinalis* L. c.v. *altilis* (Liliaceae)
Mol. Formula : $C_{45}H_{76}O_{19}$
Mol. Wt. : 920
M.P. : 162-168°C (decomp.)
$[\alpha]_D^{25}$: -23.8° (c=0.92, MeOH)
Registry No. : [57944-18-0]

IR (KBr) **:** 3500-3300 (OH) cm^{-1}.

Reference

1.   K. Kawano, K. Sakai, H. Sato and S. Sakamura, *Agric. Biol. Chem.*, **39**, 1999 (1975).

PETUNIOSIDE L
(25R)-5α-Furostan-3β,22α,26-triol 3-O-[β-D-glucopyranosyl-(1→4)-β-D-galactopyranoside]-26-O-β-D-glucopyranoside

Source : *Petunia hybrida* L. (Solanaceae)
Mol. Formula : $C_{45}H_{76}O_{19}$
Mol. Wt. : 920
M.P. : 187-189°C
$[\alpha]_D^{20}$: -75° (c=1.0, H_2O)
Registry No. : [174630-02-5]

Reference

1. S.A. Shvets, A.M. Naibi and P.K. Kintya, *Khim. Prir. Soedin.*, **31**, 247 (1995), *Chem. Nat. Comp.*, **31**, 203 (1995).

TIMOSAPONIN B-II

(25*S*)-5β-Furostane-3β,22ξ,26-triol 3-O-[β-D-glucopyranosyl-(1→2)-β-D-galactopyrnoside]-26-O-β-D-glucopyranoside

Source : *Anemarrhena asphodeloides* Bunge[1,2] (Liliaceae)
Mol. Formula : $C_{45}H_{76}O_{19}$
Mol. Wt. : 920
M.P. : 243°C (decomp.)[2]
$[\alpha]_D^{25}$: -32.0° (c=1.95, C_5D_5N)
Registry No. : [136656-07-0]

IR (KBr)[2] : 3348 (OH), 2930, 2850, 1075, 1044 (C-O) cm^{-1}.

PMR (C_5D_5N, 400 MHz)[1] : δ 0.88 (s, CH_3), 0.99 (s, CH_3), 1.03 (d, J=6.6 Hz, sec. CH_3), 1.33 (d, J=6.8 Hz, sec. CH_3), 4.81 (d, J=7.8 Hz, H-1 of Glc II), 4.92 (d, J=7.60 Hz, H-1 of Glc I), 5.28 (d, J=7.8 Hz, H-1 of Gal).

CMR (C_5D_5N, 100 MHz)[1] : δ C-1) 30.8 (2) 27.0 (3) 75.2 (4) 30.9 (5) 36.9 (6) 27.0 (7) 26.8 (8) 35.5 (9) 40.3 (10) 35.3 (11) 21.2 (12) 40.5 (13) 41.3 (14) 56.5 (15) 32.4 (16) 81.2 (17) 64.0 (18) 16.7 (19) 24.0 (20) 40.7 (21) 16.5 (22) 110.6 (23) 37.1 (24) 28.3 (25) 34.4 (26) 75.4 (27) 17.5 **Gal** (1) 106.1 (2) 75.5 (3) 78.0 (4) 71.7 (5) 78.3 (6) 62.9 **Glc I** (1) 102.5 (2) 81.8 (3) 76.8 (4) 69.9 (5) 76.6 (6) 62.2 **Glc II** (1) 105.1 (2) 75.2 (3) 78.6 (4) 71.8 (5) 78.3 (6) 62.9.

Mass (FAB, Positive ion)[1,2] : m/z 943 [M+Na]$^+$, 903 [M+H-H_2O]$^+$, 741 [M+H-H_2O-Glc]$^+$, 579 [M+H-H_2O-2xGlc]$^+$, 417 [M+H-H_2O-2xGlc-Gal]$^+$, 399 [Agl+H_2OH-2xH_2O]$^+$, 255, 185, 145.

Mass (FAB, Negative ion)[1] : m/z 919 [M-H]$^-$, 757 [M-H-Glc]$^-$.

22-O-Methyl Ether : TIMOSAPONIN B-I [136565-73-6]; Isolated from *Anemarrhena asphodeloides* Bunge (Liliaceae).

References

1.   S. Nagumo, S.I. Kishi, T. Inoue, and M. Nagai, *Yakugaku Zasshi*, **111**, 306 (1991).

2.   B.P. Ma, J.X. Dong, B.J. Wang and X.Z. Yan, *Acta Pharm. Sinica*, **31**, 271 (1996).

TRIGOFOENOSIDE B
(25S)-22-O-methyl-5α-furostane-2α,3β,26-triol 3-O-[α-L-rhamnopyranosyl-(1→4)-β-D-glucopyranoside]-26-O-β-D-glucopyranoside

Source : *Trigonella foenum-graecum* L. (Leguminosae)
Mol. Formula : $C_{45}H_{76}O_{19}$
Mol. Wt. : 920
Registry No. : [99753-11-4]

22-Methyl Ether : [105344-38-5] **Trigofoenoside B-1**

M.P. : 198-200°C (decomp.); **[α]$_D$:** -62.10° (c=1.0, Pyridine).

IR (KBr) : 3500-3200 (OH), 2920, 2860, 1460, 1380, 1200-100 cm^{-1}. (C-O-C), no spiroketal bands.

PMR (CDSO-d_6, 400 MHz) : δ 1.67 (br s, 3xH-6 of Rha), 3.24 (s, OCH_3), 4.28 (d, J=7.1 Hz), 5.21 (br s).

Mass (FAB) : m/z 1067 [M+Cs]$^+$, 957 [M+Na]$^+$.

Reference

1.   R.K. Gupta, D.C Jain and R.S. Thakur, *Phytochemistry*, **25**, 2205 (1986)

TRIGONEOSIDE Xa

**(25S)-5α-Furostane-2α,3β,22ξ,26-tetraol 3-O-[α-L-rhamnopyranosyl-(1→2)-β-D-glucopyranoside]-
26-O-β-D-glucopyranoside**

Source : *Trigonella foenum-graecum* L. (Fabaceae)
Mol. Formula : $C_{45}H_{76}O_{19}$
Mol. Wt. : 920
[α]$_D^{22}$: -49.2° (c=0.6, MeOH)
Registry No. : [290347-38-5]

IR (KBr) : 3432, 2932, 1072, 1044 cm^{-1}.

PMR (C_5D_5N, 500 MHz) : δ 0.86, 0.88 (both s, 3xH-18, 3xH-19), 1.01 (d, J=6.2 Hz, 3xH-27), 1.28 (d, J=6.6 Hz, 3xH-21), 1.67 (d, J=5.6 Hz, 3xH-6 of Rha), 3.47 (dd-like, H-26A), 3.86 (m, H-3), 4.06 (m, H-26B), 4.08 (m, H-2), 4.76 (d, J=8.2 Hz, H-1 of Glc II), 4.90 (m, H-16), 5.01 (d, J=7.0 Hz, H-1 of Glc I), 6.27 (br s, H-1 of Rha).

CMR (C_5D_5N, 125.0 MHz) : δ C-1) 45.8 (2) 70.7 (3) 85.6 (4) 33.7 (5) 44.7 (6) 28.2 (7) 32.3 (8) 34.4 (9) 54.5 (10) 36.9 (11) 21.5 (12) 40.2 (13) 41.1 (14) 56.3 (15) 32.3 (16) 81.1 (17) 63.9 (18) 16.7 (19) 13.5 (20) 40.7 (21) 16.4 (22) 110.6 (23) 37.1 (24) 28.4 (25) 34.7 (26) 75.3 (27) 17.4 **Glc I** (1) 101.4 (2) 78.1 (3) 78.3 (4) 71.9 (5) 79.4 (6) 62.6 **Rha** (1) 102.1 (2) 72.4 (3) 72.7 (4) 74.1 (5) 69.4 (6) 18.5 **Glc II** (1) 105.0 (2) 75.1 (3) 78.5 (4) 71.8 (5) 78.2 (6) 62.9.

Mass (FAB, Negative ion) : m/z 919 [M-H]$^-$, 773 [M-C$_6$H$_{11}$O$_4$]$^-$, 757 [M-C$_6$H$_{11}$O$_5$]$^-$, 611 [M-C$_{12}$H$_{21}$O$_9$]$^-$, 449 [M-C$_{18}$H$_{31}$O$_{14}$]$^-$.

Mass (FAB, Positive ion) : m/z 943 [M+Na]$^+$.

Reference

1.   T. Murakami, A. Kishi, H. Matsuda and M. Yoshikawa, *Chem. Pharm. Bull.*, **48**, 994 (2000).

TRIGONEOSIDE Xb

(25R)-5α-Furostane-2α,3β,22ξ,26-tetraol 3-O-[α-L-rhamnopyranosyl-(1→2)-β-D-glucopyranoside]-26-O-β-D-glucopyranoside

Source : *Trigonella foenum-graecum* L. (Fabaceae)
Mol. Formula : C$_{45}$H$_{76}$O$_{19}$
Mol. Wt. : 920
$[\alpha]_D^{22}$: -51.5° (c=0.6, MeOH)
Registry No. : [290347-51-2]

IR (KBr) : 3432, 2932, 1075, 1044 cm^{-1}.

PMR (C$_5$D$_5$N, 500 MHz) : δ 0.86, 0.88 (both s, 3xH-18, 3xH-19), 0.98 (d, J=6.7 Hz, 3xH-27), 1.30 (d, J=7.0 Hz, 3xH-21), 1.68 (d, J=6.4 Hz, 3xH-6 of Rha), 3.61 (dd, J=6.4, 9.2 Hz, H-26A), 3.96 (m, H-26B), 3.86 (m, H-3), 4.06 (m, H-2), 4.77 (d, J=7.6 Hz, H-1 of Glc II), 4.91 (ddd-like, H-16), 5.01 (d, J=7.6 Hz, H-1 of Glc I), 6.27 (d, J=1.2 Hz, H-1 of Rha).

CMR (C$_5$D$_5$N, 125.0 MHz) : δ C-1) 45.8 (2) 70.7 (3) 85.6 (4) 33.7 (5) 44.8 (6) 28.2 (7) 32.3 (8) 34.3 (9) 54.5 (10) 36.9 (11) 21.5 (12) 40.2 (13) 41.2 (14) 56.4 (15) 32.4 (16) 81.1 (17) 64.0 (18) 16.7 (19) 13.5 (20) 40.7 (21) 16.3 (22) 110.6 (23) 37.1 (24) 28.4 (25) 34.7 (26) 75.2 (27) 17.4 **Glc I** (1) 101.4 (2) 78.1 (3) 78.2 (4) 72.0 (5) 79.5 (6) 62.6 **Rha** (1) 102.1 (2) 72.4 (3) 72.8 (4) 74.1 (5) 69.4 (6) 18.8 **Glc II** (1) 104.9 (2) 75.2 (3) 78.5 (4) 71.8 (5) 78.2 (6) 62.9.

Mass (FAB, Negative ion) : m/z 919 [M-H]$^-$, 773 [M-C$_6$H$_{11}$O$_4$]$^-$, 757 [M-C$_6$H$_{11}$O$_5$]$^-$, 611 [M-C$_{12}$H$_{21}$O$_9$]$^-$, 449 [M-C$_{18}$H$_{31}$O$_{14}$]$^-$.

Mass (FAB, Positive ion) : m/z 943 [M+Na]$^+$.

Reference

1.   T. Murakami, A. Kishi, H. Matsuda and M. Yoshikawa, *Chem. Pharm. Bull.*, **48**, 994 (2000).

ALLIUM GIGANTEUM SAPONIN 7
3-O-Benzoyl-22ξ-methoxy-(25*R*)-5α-furostane-2α,3β,5α,6β,26-pentol-2,26-*bis*-O-β-D-glucopyranoside

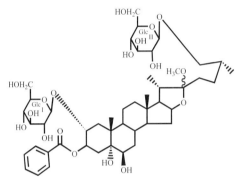

Source : *Allium giganteum* Regel (Liliaceae)
Mol. Formula : $C_{47}H_{72}O_{18}$
Mol. Wt. : 924
$[α]_D^{22}$: -84.9° (c=0.11, MeOH)
Registry No. : [156006-38-1]

IR (KBr) : 3390 (OH), 2940 (CH), 1720 (C=O), 1465, 1390, 1330, 1295, 1155, 1040, 895, 715 cm^{-1}.

PMR (C_5D_5N, 400 MHz) : δ 0.87 (s, 3xH-18), 1.01 (d, *J*=6.7 Hz, 3xH-27), 1.18 (d, *J*=7.1 Hz, 3xH-21), 1.58 (s, 3xH-19), 2.48 (dd, *J*=12.1, 12.1 Hz, H-1), 2.58 (dd, *J*=11.8, 6.3 Hz, H-4), 2.91 (dd, *J*=11.8, 11.0 Hz, H-4ax), 3.25 (s, OC*H*₃), 4.49 (q-like, *J*=7.2 Hz, H-16), 4.87 (d, *J*=7.7 Hz, H-1 of Glc II), 4.91 (ddd, *J*=12.1, 9.2, 5.8 Hz, H-2), 5.24 (d, *J*=7.7 Hz, H-1 of Glc I), 6.32 (ddd, *J*=11.0, 9.2, 6.3 Hz, H-3), 7.46 (H-3, H-4 and H-5 of benzoyl), 8.45 (2H, dd, *J*=7.8, 1.8 Hz, H-2 and H-6 of Benz). Signals of C-25-Epimer : δ 1.07 (d, *J*=6.7 Hz, 3xH-27), 1.16 (d, *J*=7.1 Hz, 3xH-21), 3.24 (s, OC*H*₃), 4.86 (d, *J*=7.8 Hz, H-1 of Glc II).

CMR (C_5D_5N, 100 MHz) : δ C-1) 38.9 (2) 77.7 (3) 76.4 (4) 37.9 (5) 75.0 (6) 75.0 (7) 35.7 (8) 30.0 (9) 45.6 (10) 40.5 (11) 21.5 (12) 40.5 (13) 41.4 (14) 56.1 (15) 32.3 (16) 81.4 (17) 64.4 (18) 16.3 (19) 18.0 (20) 40.5 (21) 16.6 (2) 112.6 (23) 30.8 (30.9) (24) 28.2 (25) 34.2 (34.5) (26) 75.3 (27) 17.2 (17.6) **Glc I** (1) 103.4 (2) 75.3 (3) 78.3^a (4) 71.8 (5) 78.5^a (6) 62.9 (OC*H*₃) 47.3 **Glc II** (1) 105.0 (105.1) (2) 75.2 (3) 78.6^b (4) 71.8 (5) 78.5^b (6) 62.9 **Benz** (1) 132.0 (2) 130.3 (3) 128.7 (4) 132.9 (5) 128.7 (6) 130.3 (7) 166.8. Values in paranthesis are of the C-25 isomer.

Mass (FAB, Negative ion) m/z : 924 [M]$^-$.

Biological Activity : The compound shows inhibitory activity on CAMP phosphodiesterase with IC_{50}=6.9x10^{-5} M.

Reference

1.   Y. Mimaki, T. Nikaido, K. Matsumoto, Y. Sashida and T. Ohmoto, *Chem. Pharm. Bull.*, **42**, 710 (1994).

METANARTHECIUM GLYCOSIDE FG-4
(25*R*)-2β-Acetoxy-11α,22ξ,26-trihydroxyfurost-4-en-3-one 11-O-[(2,3,4-tri-O-acetyl)-
α-L-arabinopyranoside]-26-O-β-D-glucopyranoside

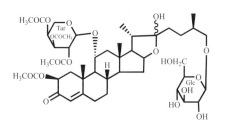

Source : *Metanarthecium luteo-viride* Maxim. (Liliaceae)
Mol. Formula : $C_{46}H_{68}O_{19}$
Mol. Wt. : 924
Registry No. : [151801-97-7]

Mass (L.S.I.) : *m/z* 907 [M-OH]$^+$, 745 [M-OH-Glc]$^+$, 685, 631 [M-OH-Tar-H$_2$O]$^+$, 573, 469 [M-OTar-Glc-OH]$^+$, 409, 325, 259 (Tar), 199, 159, 139, 97, 69, 43.

Mass (L.S.I. with NaCl) : *m/z* 947 [M+Na]$^+$, 907, 813, 785 [M-Glc+Na]$^+$, 745, 671 [M-Tar+Na]$^+$, 585, 469, 409, 325, 159, 139, 97, 69, 43.

Reference

1.   Y. Ikenishi, S. Yoshimatsu, K. Takeda and Y. Nakagawa, *Tetrahedron*, **49**, 9321 (1993).

854

METANARTHECIUM GLYCOSIDE FG-5

(25R)-2β-Acetoxy-11α,22ξ,26-trihydroxyfurostan-3-one 11-O-[(2,3,4-tri-O-acetyl)-
α-L-arabinopyranoside]-26-O-β-D-glucopyranoside

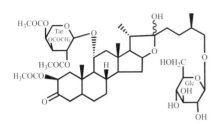

Source : *Metanarthecium luteo-viride* Maxim. (Liliaceae)
Mol. Formula : $C_{46}H_{70}O_{19}$
Mol. Wt. : 926
Registry No. : [151801-98-8]

Mass (L.S.I.) : *m/z* 909 [M-OH]⁺, 747 [M-OH-Glc]⁺, 687, 633 [M-OH-Tar-H₂O]⁺, 471 [M-OTar-Glc-OH]⁺, 411, 327, 259 (Tar), 185, 139, 97, 59, 43, 15.

Mass (L.S.I. with NaCl) : *m/z* 949 [M+Na]⁺, 909, 889, 815, 787 [M-Glc+Na]⁺, 673 [M-Tar-Na]⁺, 655, 587, 471, 411, 327, 259, 159, 139, 97, 69, 43.

Reference

1. Y. Ikenishi, S. Yoshimatsu, K. Takeda and Y. Nakagawa, *Tetrahedron*, **49**, 9321 (1993).

METANARTHECIUM GLYCOSIDE FG-1

(25R,5β)-2β-Acetoxy-3β,22ξ,11α,26-tetrahydroxyfurostane 11-O-[(2,3,4-tetra-O-acetyl)-
α-L-arabinopyranoside]-26-O-β-D-glucopyranoside

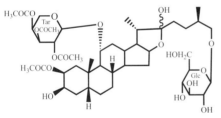

Source : *Metanarthecium luteo-viride* Maxim. (Liliaceae)
Mol. Formula : $C_{46}H_{72}O_{19}$
Mol. Wt. : 928
Registry No. : [151801-96-9]

Mass (L.S.I.) : *m/z* 911 [M-OH]⁺, 749 [M-Glc-OH]⁺, 689, 635 [M-OH-Tar-H₂O]⁺, 473 [M-Tar-Glc-H₂O-OH]⁺, 413, 328, 259 (Tar), 199, 159, 139, 97, 69, 45, 43.

Mass (L.S.I. with NaCl) : *m/z* 951 [M+Na]⁺, 911 [M-OH]⁺, 891, 817, 798, 749, 675, 635, 589, 473, 413, 329, 259, 217, 271, 171, 139, 117, 97, 69, 23, 43.

Reference

1.   Y. Ikenishi, S. Yoshimatsu, K. Takeda and Y. Nakagawa, *Tetrahedron*, **49**, 9321 (1993).

SOLANUM LYRATUM SAPONIN 2
(25*R*,*S*)-3β,26-Dihydroxy-22ξ-methoxyfurost-5-ene 3-O-[α-L-rhamnopyranosyl-(1→2)-β-D-glucuronopyranoside]-26-O-β-D-glucopyranoside

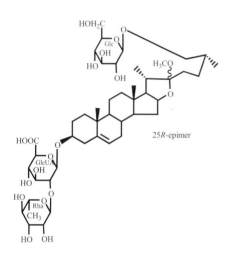

Source : *Solanum lyratum* Thunb. (Solanaceae)
Mol. Formula : C₄₆H₇₄O₁₉
Mol. Wt. : 930
[α]ᴅ¹⁸ : -89.0° (c=1.0, Pyridine)
Registry No. : [107783-53-9]

Isolated as a mixture of 25*R* and 25*S* epimers.

IR (KBr) : 3400, 1600 cm⁻¹.

Mass (F.D.) : *m/z* 969 [M+K]⁺, 414, 147.

Reference

1.   S. Yahara, M. Morooka, M. Ikeda, M. Yarrasaki and T. Nohara, *Planta Med.*, **52**, 496 (1986).

DRACAENA CONCINNA SAPONIN 3

22ξ-Methoxy-5α-furost-25(27)-ene-1β-3α,4α,26-tetrol 1-O-[α-L-rhmnopyranosyl-(1→2)-β-D-fucopyranoside]-26-O-β-D-glucopyranoside

Source : *Drcaena concinna* Kunth. (Agavaceae)
Mol. Formula : $C_{46}H_{76}O_{19}$
Mol. Wt. : 932
$[\alpha]_D^{29}$ **:** -64.0° (c=0.10, MeOH)
Registry No. : [209848-41-9]

IR (KBr) : 3420 (OH), 2915 (CH) cm^{-1}.

PMR (C_5D_5N, 400 MHz) : δ 0.85 (s, 3xH-18), 1.10 (d, *J*=6.9 Hz, 3xH-21), 1.33 (s, 3xH-19), 1.45 (d, *J*=6.3 Hz, 3xH-6 of Fuc), 1.74 (d, *J*=6.1 Hz, 3xH-6 of Rha), 2.00 (m, H-5), 2.29 (br dd, *J*=13.8, 11.7 Hz, H-2ax), 2.63 (ddd, *J*=13.8, 4.1, 3.4 Hz, H-2eq), 3.25 (s, OC*H*₃), 3.75 (dd, *J*=10.9, 2.5 Hz, H-4), 4.30 (br dd, *J*=3.4, 2.5 Hz, H-3), 4.49 (dd, *J*=11.7, 4.1 Hz, H-1), 4.78 (d, *J*=7.7 Hz, H-1 of Fuc), 4.91 (d, *J*=7.8 Hz, H-1 of Glc), 5.05 and 5.34 (each 1H, br s, H-27A and H-27B), 6.41 (br s, H-1 of Rha).

CMR (C_5D_5N, 100 MHz) : δ C-1) 80.2 (2) 33.5 (3) 71.2 (4) 69.9 (5) 45.6 (6) 23.1 (7) 32.4 (8) 36.4 (9) 55.5 (10) 42.6 (11) 23.3 (12) 40.6 (13) 40.8 (14) 57.0 (15) 32.3 (16) 81.5 (17) 64.3 (18) 16.9 (19) 9.0 (20) 40.4 (21) 16.0 (22) 112.4 (23) 31.5 (24) 28.0 (25) 146.8 (26) 72.0 (27) 111.0 (OC*H*₃) 47.3 **Fuc** (1) 100.2 (2) 74.3 (3) 77.0 (4) 73.3 (5) 71.1 (6) 17.1 **Rha** (1) 101.4 (2) 72.6 (3) 72.6 (4) 74.3 (5) 69.2 (6) 19.0 **Glc** (1) 103.8 (2) 75.1 (3)78.6 (4) 71.7 (5) 78.5 (6) 62.8.

Mass (FAB, Negative ion) : *m/z* 931 [M-H]$^-$.

Reference

1. Y. Mimaki, M. Kuroda, Y. Takaashi and Y. Sashida, *Phytochemistry*, **47**, 1351 (1998).

DRACAENA CONCINNA SAPONIN 4

22-O-Methyl-5α-furost-25(27)-ene-1β,3β,4α,22ξ,26-pentol 1-O-[α-L-rhamnopyranosyl-(1→2)-
β-D-fucopyranoside]-26-O-β-D-glucopyranoside

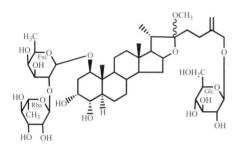

Source : *Dracaena concinna* Kunth (Agavaceae)
Mol. Formula : $C_{46}H_{76}O_{19}$
Mol. Wt. : 932
[α]$_D$: 64.8° (c=0.11, MeOH)
Registry No. : [207848-42-0]

IR (KBr) : 3425 (OH), 2915 (CH) cm^{-1}.

PMR (C$_5$D$_5$N, 400 MHz) : δ 0.84 (s, 3xH-18), 1.12 (d, J=6.8 Hz, 3xH-21), 1.28 (m, H-5), 1.33 (s, 3xH-19), 1.52 (d, J=6.3 Hz, 3xH-6 of Fuc), 1.73 (d, J=6.1 Hz, 3xH-6 of Rha), 2.34 (ddd, J=12.9, 12.9, 11.9 Hz, H-2ax), 2.79 (ddd, J=12.9, 5.0, 4.0 Hz, H-2eq), 3.25 (s, OCH$_3$), 3.73 (dd, J=12.9, 9.8 Hz, H-4), 3.87 (ddd, J=12.9, 9.8, 5.0 Hz, H-3), 4.05 (dd, J=11.9, 4.0 Hz, H-1), 4.78 (d, J=7.8 Hz, H-1 of Fuc), 4.91 (d, J=7.7 Hz, H-1 of Glc), 5.05 (br s, H-27A), 5.34 (br s, H-27B), 6.40 (br s, H-1 of Rha).

CMR (C$_5$D$_5$N, 100 MHz) : δ (C-1) 81.8 (2) 34.7 (3) 75.0 (4) 74.6 (5) 49.8 (6) 22.9 (7) 32.4 (8) 36.2 (9) 55.5 (10) 42.6 (11) 23.3 (12) 40.6 (13) 40.7 (14) 56.8 (15) 32.4 (16) 81.5 (17) 64.3 (18) 16.8 (19) 10.1 (20) 40.4 (21) 16.0 (22) 112.4 (23) 31.5 (24) 28.0 (25) 146.8 (26) 72.0 (27) 111.0 (OCH$_3$) 47.3 **Fuc** (1) 99.7 (2) 73.9 (3) 76.9 (4) 73.3 (5) 71.2 (6) 171.1 **Rha** (1) 101.4 (2) 72.5 (3) 72.5 (4) 74.2 (5) 69.2 (6) 19.0 **Glc** (1) 103.8 (2) 78.6 (4) 71.7 (5) 78.5 (6) 62.8.

Mass (FAB, Negative ion) : *m/z* : 931 [M-H]$^-$.

Reference

1. Y. Mamiki, M. Kuroda, Y. Takaashi, and Y. Sashida, *Phytochemistry*, **47**, 1351 (1998).

RUSCUS ACULEATUS SAPONIN 20
22-O-Methyl-(25R)-furost-5-ene-1β,3β,22ξ,26-tetrol 1-O-[α-L-rhamnopyranosyl-(1→2)-β-D-galactopyranoside]-26-O-β-D-glucopyranoside

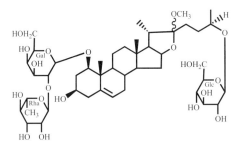

Source : *Ruscus aculeatus* L. (Liliaceae)
Mol. Formula : $C_{46}H_{76}O_{19}$
Mol. Wt. : 932
$[\alpha]_D^{25}$: -36.0° (c=0.10, MeOH)
Registry No. : [211379-08-7]

IR (KBr) **:** 3425 (OH), 2930 (CH), 1455, 1375, 1260, 1225, 1060, 980, 905, 835, 810 cm^{-1}.

PMR (C_5D_5N, 400 MHz) **:** δ 0.87 (s, 3xH-18), 0.99 (d, *J*=6.6 Hz, 3xH-27), 1.11 (d, *J*=6.8 Hz, 3xH-21), 1.44 (s, 3xH-19), 1.75 (d, *J*=6.1 Hz, 3xH-6 of Rha), 3.25 (s, OC*H₃*), 3.82 (m, H-3), 3.85 (dd, *J*=12.0, 3.9 Hz, H-1), 4.79 (d, *J*=7.7 Hz, H-1 of Gal), 4.84 (d, *J*=7.7 Hz, H-1 of Glc), 5.58 (br d, *J*=5.4 Hz, H-6), 6.39 (br s, H-1 of Rha).

CMR (C_5D_5N, 100 MHz) **:** δ C-1) 84.1 (2) 37.9 (3) 68.2 (4) 43.9 (5) 139.6 (6) 124.7 (7) 32.0 (8) 33.0 (9) 50.6 (10) 42.8 (11) 24.0 (12) 40.4 (13) 40.5 (14) 57.0 (15) 32.4 (16) 81.3 (17) 64.3 (18) 16.8 (19) 15.0 (20) 40.5 (21) 16.1 (22) 112.7 (23) 30.7 (24) 28.1 (25) 34.2 (26) 75.2 (27) 17.1 (OC*H₃*) 47.2 **Gal** (1) 100.6 (2) 75.0 (3) 76.8 (4) 70.4 (5) 76.3 (6) 61.9 **Rha** (1) 101.7 (2) 72.6 (3) 72.6 (4) 74.3 (5) 69.3 (6) 19.0 **Glc** (1) 105.0 (2) 75.2 (3) 78.6 (4) 71.8 (5) 78.6 (6) 62.9.

Mass (FAB, Negative ion) **:** *m/z* 931 [M-H]$^-$.

Reference

1.   Y. Mimaki, M. Kuroda, A. Kameyama, A. Yokosuka and Y. Sashida, *Phytochemistry*, **48**, 485 (1998).

KINGIANOSIDE C

3β,22ξ,26-Trihydroxy-25(*R*)-furost-5-en-12-one 3-*O*-[β-D-glucopyranosyl-(1→4)-β-D-galactopyranoside]-26-*O*-β-D-glucopyranoside

Source : *Polygonatum kingianum* Coll. et Hemsl. (Liliaceae)
Mol. Formula : $C_{45}H_{72}O_{20}$
Mol. Wt. : 932
[α]$_D^{20}$: -27.4° (c=0.37, Pyridine)
Registry No. : [145854-04-2]

PMR (C_5D_5N, 270 MHz) : δ 0.93 (s, 3xH-19), 0.99 (d, *J*=6.6 Hz, 3xH-27), 1.16 (s, 3xH-18), 1.55 (d, *J*=6.6 Hz, 3xH-21), 4.82 (d, *J*=7.7 Hz, H-1 of Glc II), 4.88 (d, *J*=7.7 Hz, H-1 of Gal), 5.28 (d, *J*=7.3 Hz, H-1 of Glc I), 5.31 (br s, H-6).

CMR (C_5D_5N, 75.5 MHz) : δ C-1) 37.0 (2) 30.0 (3) 77.8 (4) 39.1 (5) 140.8 (6) 121.5 (7) 31.8 (8) 30.9 (9) 52.4 (10) 37.6 (11) 37.6 (12) 212.8 (13) 55.3 (14) 55.9 (15) 31.8 (16) 79.7 (17) 54.8 (18) 16.0 (19) 18.8 (20) 41.3 (21) 15.2 (22) 110.8 (23) 37.1 (24) 28.4 (25) 34.3 (26) 75.2 (27) 17.4 **Gal** (1) 102.9 (2) 73.5 (3) 75.4^a (4) 80.0 (5) 75.9^a (6) 61.0 **Glc I** (1) 107.1 (2) 75.2^b (3) 78.7^c (4) 72.3 (5) 78.5 (6) 63.2 **Glc II** (1) 104.9 (2) 75.3^b (3) 78.6^c (4) 71.8 (5) 78.5 (6) 62.9.

Mass (FAB, Negative ion) : *m/z* 931 [M-H]$^-$, 769 [M-Glc-H]$^-$, 607 [M-Glc-Hexose-H]$^-$.

Reference

1.   X.-C. Li, C.-R. Yang, M. Ichikawa, H. Matsuura, R. Kasai and K. Yamasaki, *Phytochemistry* **31**, 3559 (1992).

MACROSTEMONOSIDE I

3β,22,26-Trihydroxy-5β-furost-25(27)-en-12-one 3-O-[β-D-glucopyranosyl-(1→2)-β-D-galactopyranoside]-26-O-[β-D-glucopyranoside]

Source : *Allium macrostemon* Bunge (Liliaceae)
Mol. Formula : $C_{45}H_{72}O_{20}$
Mol. Wt. : 932
M.P. : 224-227°C
[α]$_D^{20}$: -57.9° (c=0.61, Pyridine)
Registry No. : [162413-63-0]

PMR (C_5D_5N, 400 MHz) : δ 1.00 (s, CH_3), 1.13 (s, CH_3), 1.56 (d, *J*−6.8 Hz, 3xH-21), 4.90 (d, *J*=6.8 Hz, anomeric H), 4.92 (d, *J*=7.2 Hz, anomeric H), 5.07 (s, H-27A), 5.31 (d, *J*=7.8 Hz, anomeric H), 5.35 (s, H-27B).**CMR** (C_5D_5N, 100 MHz) : δ C-1) 30.7 (2) 26.6 (3) 75.3 (4) 30.8 (5) 36.6 (6) 26.8 (7) 26.5 (8) 34.7 (9) 42.0 (10) 35.8 (11) 37.8 (12) 213.2 (13) 56.0 (14) 56.0 (15) 31.7 (16) 79.9 (17) 55.1 (18) 16.2 (19) 23.2 (20) 41.3 (21) 15.2 (22) 110.5 (23) 37.9 (24) 28.4 (25) 147.2 (26) 72.1 (27) 110.8 **Gal** (1) 102.5 (2) 81.9 (3) 75.2 (4) 69.9 (5) 76.6 (6) 62.2 **Glc I** (1) 106.2 (2) 76.9 (3) 78.1 (4) 71.7 (5) 78.4 (6) 62.8 **Glc II** (1) 103.9 (2) 75.2 (3) 78.5 (4) 71.9 (5) 78.6 (6) 62.8.

Mass (FAB, Positive ion) : *m/z* 915 [M+H-H_2O]$^+$, 753 [M+H-H_2O-Glc]$^+$.591 [M+H-H_2O-2xGlc]$^+$, 429 [M+H-H_2O-2xGlc-Gal]$^+$, 391, 377, 309, 155, 135, 119, 103, 85, 77.

Mass (FAB, Positive ion + Na$^+$) : *m/z* 955 [M+Na]$^+$, 915, 309, 177, 155, 135, 119, 103, 85, 77, 73, 57.

Reference

1.   J. Peng, X. Yao, H. Kobayashi and C. Ma, *Planta Med.*, **61**, 58 (1995).

ANEMARSAPONIN E, TIMOSAPONIN B-I

(25S)-22-Methoxy-5β-furost-3β,26-diol 3-O-[β-D-glucopyranosyl-(1→2)-β-D-galactopyranoside]-26-O-β-D-glucopyranoside

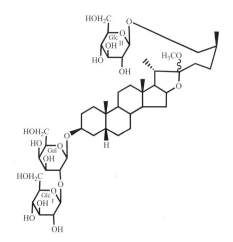

Source : *Anemarrhena asphodeloides* Bge.[1,2] (Liliaceae)
Mol. Formula : $C_{46}H_{78}O_{19}$
Mol. Wt. : 934
M.P. : 244°C (decomp.)[1]
Registry No. : [136565-73-6]

PMR (C_5D_5N, 400 MHz)[1] : δ 0.78 (s, 3xH-18), 0.95 (s, 3xH-19), 1.03 (d, J=6.0 Hz, 3xH-27), 1.16 (d, J=6.6 Hz, 3xH-21), 3.25 (s, OCH_3), 4.82 (d, J=7.7 Hz, H-1 of Glc), 4.90 (d, J=7.1 Hz, H-1 of Gal), 5.27 (d, J=7.7 Hz, H-1 of Glc).

CMR (C_5D_5N, 100 MHz)[1] : δ C-1) 30.9 (2) 27.0 (3) 75.2 (4) 31.0 (5) 36.9 (6) 26.7 (7) 26.7 (8) 35.5 (9) 40.2 (10) 35.2 (11) 21.0 (12) 40.5 (13) 41.2 (14) 56.4 (15) 32.1 (16) 81.4 (17) 64.4 (18) 16.5 (19) 24.0 (20) 41.2 (21) 16.4 (22) 112.6 (23) 30.9 (24) 28.2 (25) 34.4 (26) 75.2 (27) 17.5 (OCH_3) 47.3 **Gal** (1) 102.5 (2) 81.7 (3) 76.9 (4) 69.8 (5) 76.6 (6) 62.1 **Glc I** (1) 106.0 (2) 75.4 (3) 78.0 (4) 71.7 (5) 78.5 (6) 62.8 **Glc III** (1) 105.0 (2) 75.0 (3) 78.6 (4) 71.7 (5) 78.4 (6) 62.8.

Mass (FAB)[1] : *m/z* 957 [M+Na]$^+$, 933 [M-H]$^+$, 903 [M+H-MeOH]$^+$, 741 [M+H-MeOH-Glc]$^+$, 579 [M+H-MeOH-2xGlc]$^+$, 417 [M+H-MeOH-2xGlc-Gal]$^+$, 399 [M+H-MeOH-2xGlc-Gal-H_2O]$^+$.

References

1. B.P. Ma, J.X. Dong, B.J. Wang and X.Z. Yan, *Acta Pharm. Sinica*, **31**, 271 (1996).

2. J. Bian, S.S. Xu, S. Huang and Z.X. Wang, *J. Shenyang Pharm. Univ.*, **13**, 34 (1995).

ASPARAGUS AFRICANUS SAPONIN 3

(5β,25R)-22α-Methoxyfurostan-3β,26-diol 3-O-[β-D-glucopyranosyl-(1→2)-β-D-glucopyranoside]-26-O-β-D-glucopyranoside

Source : *Asparagus africanus* Lam. (Liliaceae)
Mol. Formula : $C_{46}H_{78}O_{19}$
Mol. Wt. : 934
M.P. : 160.2-161.4°C
[α]$_D^{20}$: -29° (c=0.14, MeOH)
Registry No. : [244779-38-2]

UV (MeOH) : λ_{max} 230 nm.

IR (KBr) : 3434 cm^{-1} (no spiroketal absorption).

PMR (C$_5$D$_5$N, 600 MHz) : δ 0.83 (s, 3xH-18), 0.99 (s, 3xH-19), 0.99 (H-7A), 1.05 (H-27), 1.06 (H-14), 1.08 (H-12A), 1.17 (3xH-21), 1.22 (H-6A and H-11A), 1.28 (H-7B), 1.31 (H-9), 1.34 (H-11B), 1.36 (H-15A), 1.46 (H-24A), 1.49 (H-4A), 1.51 (H-8), 1.51 (H-2A), 1.70 (H-12B), 1.72 (H-24B), 1.79 (H-17), 1.84 (2xH-1), 1.84 (H-4B), 1.85 (H-23A), 1.86 (H-6B), 1.88 (H-25), 1.93 (H-23B), 1.94 (H-2B), 1.98 (H-15B), 2.16 (H-5), 2.24 (H-20), 3.54 (H-26A), 3.58 (C$_{22}$-OCH$_3$), 3.84 (H-5 of Glc I), 3.91 (H-5 of Glc III), 3.93 (H-5 of Glc II), 3.99 (H-2 of Glc III), 4.03 (H-2 of Glc II), 4.07 (H-26B), 4.13 (H-4 of Glc I), 4.15 (H-4 of Glc III), 4.17 (H-3 of Glc III), 4.19 (H-2 of Glc I), 4.21 (H-3 of Glc II), 4.26 (H-4 of Glc II), 4.28 (H-3 of Glc I), 4.26 (H-4 of Glc II), 4.28 (H-3 of Glc I), 4.29 (H-6A of Glc I), 4.29 (H-3), 4.34 (H-6A of Glc III), 4.45 (H-6A of Glc II), 4.46 (H-6B of Glc I), 4.49 (H-6B of Glc II), 4.51 (H-6B of Glc III), 4.51 (H-16), 4.81 (d, *J*=7.4 Hz, H-1 of Glc III), 4.93 (d, *J*=7.9 Hz, H-1 of Glc I), 5.35 (d, *J*=7.4 Hz, H-1 of Glc II).

CMR (C$_5$D$_5$N, 100 MHz) : δ C-1) 30.9 (2) 26.9 (3) 75.3 (4), 31.0 (5) 37.0 (6) 27.1 (7) 26.9 (8) 35.6 (9) 40.4 (10) 35.3 (11) 21.2 (12) 40.3 (13) 41.3 (14) 56.5 (15) 32.2 (16) 81.5 (17) 64.5 (18) 16.5 (19) 24.0 (20) 40.6 (21) 16.3 (22) 112.7 (23) 31.0 (24) 28.3 (25) 34.5 (26) 75.0 (27) 17.6 (C$_{22}$-OCH$_3$) 49.7 **Glc I** (1) 101.79 (2) 83.3 (3) 78.4 (4) 71.8 (5) 78.0

(6) 62.9 **Glc II** (1) 106.0 (2) 77.0 (3) 78.0 (4) 71.9 (5) 78.5 (6) 63.0 **Glc III** (1) 102.0 (2) 75.2 (3) 78.5 (4) 71.9 (5) 78.5 (6) 63.0.

Mass (FAB, Positive ion) : *m/z* (rel.intens.) 936 [M+H]$^+$ (5), 904 [(M+H)-MeOH]$^+$ (100), 741 [(M+H)-MeOH-163]$^+$ (20), 579 [(M+H)-MeOH-325]$^+$ (37), 430 [(M+H)-MeOH-488]$^+$ (57).

Reference

1.   A. Debella, E. Haslinger, O. Kunert, G. Michl, D. Abebe, *Phytochemistry*, **51**, 1069 (1999).

HOSTA LONGIPES SAPONIN 11
(25*R*)-22ξ-Methoxy-5α-furostane-2α,3β,26-triol 3-O-[α-L-rhamnopyranosyl-(1→2)-β-D-galactopyranoside]-26-O-β-D-glucopyranoside

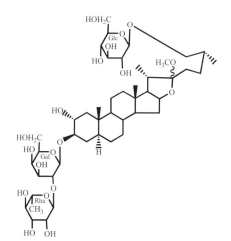

Source : *Hosta longipes* (Liliaceae)
Mol. Formula : C$_{46}$H$_{78}$O$_{19}$
Mol. Wt. : 934
[α]$_D^{25}$: -63.0° (c=0.10 CHCl$_3$-MeOH)
Registry No. : [178494-80-9]

IR (KBr) : λ$_{max}$ 3425 (OH), 2930 (CH), 1445, 1375, 1265, 1040, 950, 905, 890, 815, 780, 700 cm^{-1}.

PMR (C$_5$D$_5$N, 400 MHz) : δ 0.79 (s, 3xH-19), 0.88 (s, 3xH-19), 1.01 (d, *J*=6.6 Hz, 3xH-27), 1.18 (d, *J*=6.9 Hz, 3xH-21), 1.62 (d, *J*=6.2 Hz, 3xH-6 of Rha), 3.26 (s, OC*H$_3$*), 4.83 (d, *J*=7.8 Hz, H-1 of Glc), 4.98 (d, *J*=7.8 Hz, H-1 of Gal), 6.25 (d, *J*=1.0 Hz, H-1 of Rha).

CMR (C$_5$D$_5$N, 100 MHz) : δ C-1) 45.8 (2) 70.7 (3) 85.7 (4) 33.8 (5) 44.8 (6) 28.2 (7) 32.1 (8) 34.7 (9) 54.5 (10) 36.9 (11) 21.4 (12) 40.0 (13) 41.2 (14) 56.3 (15) 32.3 (16) 81.4 (17) 64.3 (18) 16.3 (19) 13.5 (20) 40.5 (21) 16.5 (22) 112.7 (23) 30.8 (24) 28.2 (25) 34.2 (26) 75.2 (27) 17.2 (OCH$_3$) 47.3 **Gal** (1) 101.9 (2) 76.9 (3) 76.4 (4) 70.7 (5) 76.3 (6) 62.2 **Rha** (1) 102.1 (2) 72.4 (3) 72.8 (4) 74.2 (5) 69.4 (6) 18.5 **Glc II** (1) 105.0 (2) 75.2 (3) 78.6$^+$ (4) 71.8 (5) 78.4$^+$ (6) 63.0.

Mass (FAB, Negative ion) *m/z* : 934 [M]⁻.

Reference

1.   Y. Mimaki, T. Kanmoto, M. Kuroda, Sashida, Y. Satomi, A. Nishino and Y. Nishino, *Phytochemistry*, **42**, 1065 (1996).

MACROSTEMONOSIDE G

3β,12β,22,26-Tetrahydroxy-5β-furost-25(27)-ene 3-O-[β-D-glucopyranosyl-(1→2)-β-D-galactopyranoside]-26-O-[β-D-glucopyranoside]

Source : *Allium macrostemon* Bunge (Liliaceae)
Mol. Formula : $C_{45}H_{74}O_{20}$
Mol. Wt. : 934
M.P. : 198-200°C
[α]_D²⁰ : -51.0° (c=0.41, Pyridine)
Registry No. : [162413-62-9]

PMR (C_5D_5N, 500 MHz) : δ 0.91 (H-7A)*, 0.97 (s, 3xH-19), 1.08 (H-14)*, 1.10 (s, 3xH-18), 1.21 (H-6)*, 1.24 (H-7B)*, 1.41 (H-9)*, 1.44 (H-1A, H-2A, H-4A)*, 1.49 (H-8)*, 1.51 (H-11A)*, 1.55 (H-15A)*, 1.59 (d, *J*=6.0, 7.0 Hz, 3xH-21), 1.72 (H-11B)*, 1.76 (H-6B)*, 1.79 (H-1B)*, 1.82 (H-4B)*, 1.90 (H-2B)*, 2.05 (H-15B)*, 2.16* (H-5), 2.27 (H-23)*, 2.33 (H-17), 2.49 (m, H-20), 2.70 (m, H-24), 3.53 (dd, *J*=9.2, 3.2 Hz, H-12), 3.82 (m, H-5 of Glc I), 3.91 (m, H-5 of Glc II), 4.00 (m, H-5 of Gal), 4.04 (H-2 of Glc II)*, 4.06 (H-2 of Glc I)*, 4.17 (H-3 of Glc I)*, 4.21 (H-3 of Glc II)*, 4.23 (H-4 of Glc II)*, 4.24 (H-3 of Gal)*, 4.28 (H-3)*, 4.29 (H-26A)*, 4.31 (H-4 of Glc I), 4.33 (H-6A of Glc II)*, 4.37 (H-6A of Glc I)*, 4.38 (H-6A of Gal)*, 4.40 (H-6B of Glc I)*, 4.42 (H-6B of Gal)*, 4.52 (H-6B of Glc II)*, 4.55 (H-4 of Gal)*, 4.59 (H-26B)*, 4.64 (t, *J*=8.0 Hz, H-2 of Gal), 4.87 (d, *J*=7.6 Hz, H-1 of Glc I), 4.89 (d, *J*=7.6 Hz, H-1 of Gal), 5.02 (s, H-27A), 5.04 (H-16)*, 5.26 (d, *J*=7.8 Hz, H-1 of Glc I), 5.32 (s, H-27B). * overlapped signals.**CMR** (C_5D_5N, 125 MHz) : δ C-1) 31.0 (2) 26.6 (3) 75.4 (4) 31.0 (5) 36.7 (6) 27.0 (7) 26.6 (8)

34.5 (9) 39.4 (10) 35.2 (11) 31.4 (12) 79.5 (13) 46.5 (14) 55.1 (15) 32.1 (16) 81.4 (17) 63.7 (18) 11.3 (19) 23.9 (20) 41.6 (21) 15.6 (22) 110.5 (23) 38.0 (24) 28.4 (25) 147.2 (26) 72.0 (27) 110.6 **Gal** (1) 102.5 (2) 81.8 (3) 75.2 (4) 69.8 (5) 76.5 (6) 62.1 **Glc I** (1) 106.1 (2) 76.9 (3) 78.0 (4) 71.7 (5) 78.4 (6) 62.7 **Glc II** (1) 103.9 (2) 75.2 (3) 78.5 (4) 71.6 (5) 78.5 (6) 62.7.

Mass (FAB, Positive ion) : m/z 917 $[M+H-H_2O]^+$, 575 $[M+H-2xH_2O-2xGlc]^+$.

Biological Activity : Inhibits ADP-induced human platelet aggregation *in vitro* (IC_{50}=0.871 μM).

22-O-Methyl ether : Macrostemonoside H

Artefact, **M.P. :** 187-189°C; $[\alpha]_D^{20}$: -45.8° (c=0.59, Pyridine)

Reference

1.   J. Peng, X. Yao, H. Kobayashi and C. Ma, *Planta Med.*, **61**, 58 (1995).

TRIBULUS TERRESTRIS SAPONIN 6

(23R,25R)-5α-Furostane-12-one-3β,22α,26-triol 3-O-[β-D-glucopyranosyl-(1→2)-β-D-galactopyranoside]-26-O-β-D-glucopyranoside

Source : *Tribulus terrestris* L. (Zygophyllaceae)
Mol. Formula : $C_{45}H_{74}O_{20}$
Mol. Wt. : 934
$[\alpha]_D^{15}$: -15.0° (c=0.2)
Registry No. : [261958-92-3]

IR (KBr) : 3420, 1706, 901 cm^{-1}.

PMR (CD₃OD-C₅D₅N, 400 MHz) : δ 4.60 (d, *J*=7.6 Hz), 4.65 (d, *J*=7.6 Hz), 5.08 (d, *J*=8.0 Hz).

CMR (CD₃OD-C₅D₅N, 100 MHz) : δ C-1) 36.6 (2) 29.7 (3) 78.5 (4) 34.5 (5) 44.4 (6) 28.5 (7) 31.7 (8) 34.2 (9) 55.7 (10) 36.2 (11) 37.9 (12) 212.9 (13) 55.5 (14) 55.7 (15) 31.7 (16) 79.6 (17) 55.0 (18) 16.2 (19) 11.6 (20) 41.2 (21) 15.5 (22) 110.7 (23) 37.1 (24) 28.33 (25) 34.23 (26) 75.17 (27) 17.37 **Gal** (1) 102.4 (2) 80.0 (3) 75.9 (4) 73.4 (5) 76.8 (6) 61.0 **Glc I** (1) 107.1 (2) 75.4 (3) 78.7 (4) 72.2 (5) 78.6 (6) 62.8 **Glc II** (1) 104.9 (2) 75.2 (3) 78.6 (4) 71.6 (5) 78.4 (6) 63.1.

Mass (E.S.I.) : *m/z* 957 [M+Na]⁺, 973 [M+K]⁺, 933 [M-H]⁻, 771 [M-Glc-H]⁻, 609 ₗM-2xGlc-H]⁻.

Reference

1. L.F. Cai, J. Fengying, J.G. Zhang, F. Pei, Y. Xu, S. Liu and D. Xu, *Acta Pharm. Sin.*, **34**, 759 (1999).

TRIBULUS TERRESTRIS SAPONIN 11
(25*S*)-5α-Furostane-12-one-3β,22α-26-triol 3-O-[β-D-glucopyranosyl-(1→2)-β-galactopyranoside]-26-O-β-D-glucopyranoside

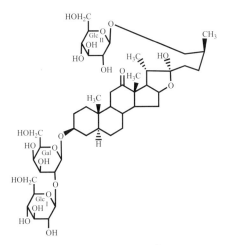

Source : *Tribulus terrestris* L. (Zygophylloceae)
Mol. Formula : C₄₅H₇₄O₂₀
Mol. Wt. : 934
[α]ᴅ¹⁵ : -19.2° (c=0.2, C₅D₅N)
Registry No. : [343265-99-6]

IR (KBr) : 3420, 2929, 1702, 901 cm⁻¹.

PMR (C₅D₅N, 400 MHz) : δ 4.60 (d, *J*=7.6 Hz), 4.65 (d, *J*=7.6 Hz,), 5.08 (d, *J*=8.0 Hz).

CMR (C₅D₅N, 100 MHz) : δ C-1) 36.6 (2) 29.7 (3) 78.5 (4) 34.5 (5) 44.4 (6) 28.6 (7) 31.7 (8) 34.3 (9) 55.7 (10) 36.2 (11) 37.9 (12) 212.9 (13) 55.5 (14) 55.8 (15) 31.7 (16) 79.6 (17) 55.0 (18) 16.2 (19) 11.6 (20) 41.2 (21) 15.2 (22)

110.7 (23) 37.0 (24) 28.26 (25) 34.34 (26) 75.30 (27) 17.41 **Gal** (1) 102.4 (2) 80.0 (3) 75.9 (4) 73.4 (5) 76.8 (6) 61.0 **Glc I** (1) 107.1 (2) 75.4 (3) 78.7 (4) 72.2 (5) 78.4 (6) 63.1 **Glc II** (1) 105.1 (2) 75.2 (3) 78.6 (4) 71.6 (5) 78.4 (6) 63.1.

Mass (E.S.I.) : m/z 957 [M+Na]$^+$, 973 [M+K]$^+$, 933 [M-H]$^-$, 771 [M-Glc-H]$^-$, 609 [M-2Glc-H]$^-$.

Reference

1.   L. Cai, Y. Wu, J. Zhang, F. Pei, Y, Xu, S. Xie and D. Xu, *Planta Med.*, **67**, 196 (2001).

ALLIUM TUBEROSUM SAPONIN 5
(25S)-3β,5β,6α,22ξ,26-Pentahydroxyfurostane 3-O-[α-L-rhamnopyranosyl-(1→4)-β-D-glucopyranoside]-26-O-β-D-glucopyranoside

Source : *Allium tuberosum* Rottler (Liliaceae)
Mol. Formula : $C_{45}H_{76}O_{20}$
Mol. Wt. : 936
$[\alpha]_D^{29}$ **:** -53.2° (c=0.2, Pyridine)

PMR (C_5D_5N, 500 MHz) : δ 0.88 (s, 3xH-18), 1.04 (d, J=6.7 Hz, 3xH-27), 1.15 (s, 3xH-19), 1.33 (d, J=6.7 Hz, 3xH-21), 1.71 (d, J=6,.1 Hz, 3xH-6 of Rha), 3.49 (dd, J=7.93, 7.94 Hz, H-26A), 4.07 (br d, J=10.8 Hz, H-6), 4.10 (m, H-26B), 4.65 (br s, $W\frac{1}{2}$=7.3 Hz, H-3), 4.82 (d, J=7.9 Hz, H-1 of Glc II), 4.97 (m, H-16), 5.03 (d, J=7.9 Hz, H-1 of Glc I), 5.87 (s, H-1 of Rha).

CMR (C_5D_5N, 125 MHz) : δ C-1) 35.8 (2) 29.1 (3) 79.2 (4) 35.0 (5) 73.0 (6) 66.1 (7) 35.5 (8) 34.6 (9) 44.6 (10) 43.0 (11) 21.8 (12) 40.1 (13) 40.9 (14) 56.3 (15) 32.4 (16) 81.2 (17) 63.7 (18) 16.6 (19) 17.5 (20) 40.6 (21) 16.4 (22) 110.6 (23) 37.1 (24) 28.3 (25) 34.4 (26) 75.3 (27) 17.5 **Glc I** (1) 101.9 (2) 74.8 (3) 76.7 (4) 78.5 (5) 77.5 (6) 61.1 **Rha** (1) 102.7 (2) 72.5 (3) 72. (4) 73.9 (5) 70.4 (6) 18.5 **Glc II** (1) 105.1 (2) 75.1 (3) 78.5 (4) 71.7 (5) 78.4 (6) 62.8.

Mass (FAB, Positive ion) : *m/z* 959 $[M+Na]^+$, 813 $[M+Na-Rha]^+$.

Mass (FAB, Positive ion, H.R.) : *m/z* 959.4794 $[(M+Na)^+$, requires 959.4828].

Reference

1. T. Ikeda, H. Tsumagari, M. Okawa and T. Nohara, *Chem. Pharm. Bull.*, **52**, 142 (2004).

MACROSTEMONOSIDE J

2β,3β,22,26-Tetrahydroxy-25(*R*),5β-furostane 3-O-[β-D-glucopyranoside]

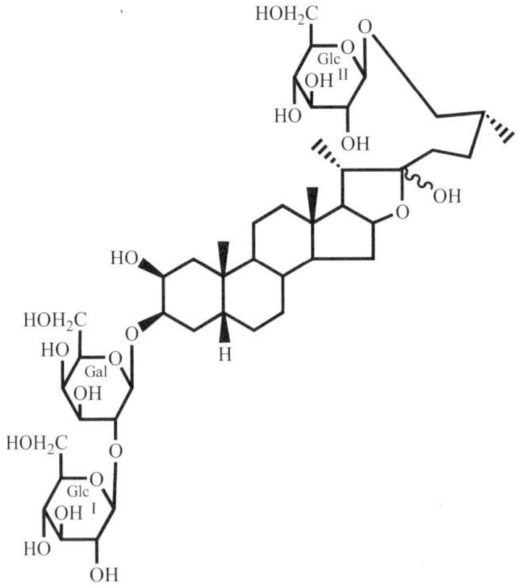

Source : *Allium macrostemon* Bunge (Liliaceae)
Mol. Formula : $C_{45}H_{76}O_{20}$
Mol. Wt. : 936
M.P. : 230-232°C
Registry No. : [159935-09-8]

PMR (C_5D_5N, 400 MHz) **:** δ 0.87 (s, CH_3), 0.97 (s, CH_3), 0.99 (d, J=6.6 Hz, sec. CH_3), 1.34 (d, J=6.8 Hz, sec. CH_3), 4.84 (d, J=7.8 Hz, anomeric H), 5.02 (d, J=7.8 Hz, anomeric H), 5.30 (d, J=7.6 Hz, anomeric H).**CMR** (C_5D_5N, 100.4 MHz) **:** δ C-1) 40.4 (2) 67.3 (3) 82.1 (4) 32.0 (5) 36.7 (6) 26.3 (7) 26.8 (8) 35.6 (9) 40.7 (10) 37.2 (11) 21.4 (12) 40.7 (13) 41.2 (14) 56.3 (15) 32.4 (16) 81.2 (17) 64.1 (18) 16.7 (19) 23.9 (20) 41.5 (21) 16.5 (22) 110.7 (23) 37.2 (24) 28.4 (25) 34.3 (26) 75.3 (27) 17.5 **Gal** (1) 103.5 (2) 81.9 (3) 75.2 (4) 69.8 (5) 77.0 (6) 62.0[a] **Glc I** (1) 106.3 (2) 77.0 (3) 78.1[b] (4) 71.8 (5) 78.5[b] (6) 62.9[a] **Glc II** (1) 105.0 (2) 75.2 (3) 78.7[b] (4) 71.8 (5) 78.5[b] (6) 62.9[a].

Mass (FAB, Positive ion) **:** *m/z* 919 $[M+H-H_2O]^+$, 595 $[M+H-H_2O-2xGlc]^+$, 433 $[Aglycone+H-H_2O]^+$, 415 $[Aglycone+H-2xH_2O]^+$, 287, 271, 253, 212, 185, 145, 115, 93, 73, 60, 53.

Mass (FAB, Positive ion + Na$^+$) : *m/z* 959 [M+Na]$^+$, 941 [M+Na-H$_2$O]$^+$, 919 [M+H-H$_2$O]$^+$, 595, 331, 309, 177, 155, 135, 119, 103, 85, 77, 73, 57.

22-O-Methyl ether : (Macrostemonoside K) [159935-10-1]

Artefact, **M.P.:** 219-221°C

Reference

1. J.P. Peng, X.S. Yao, Y. Okada and T. Okuyama, *Chem. Pharm. Bull.*, **42**, 2180 (1994); *Acta Pharm. Sin.*, **29**, 526 (1994).

TERRESTROSIN F
(25*R*)-5α-Furostane-2α,3β,22α,26-tetrol 3-O-[β-D-glucopyranosyl-(1→4)-
β-D-galactopyranoside]-26-O-β-D-glucopyranoside

Source : *Tribulus terrestris* L. (Zygophyllaceae)
Mol. Formula : C$_{45}$H$_{76}$O$_{20}$
Mol. Wt. : 936
[α]$_D^{20}$: -20.0° (c=0.60, C$_5$D$_5$N)
Registry No. : [193604-54-5]

PMR (C$_5$D$_5$N, 500 MHz) : δ 0.73 (s, 3xH-19), 0.86 (s, 3xH-18), 1.02 (d, *J*=6.7 Hz, 3xH-27), 1.30 (d, *J*=7.0 Hz, 3xH-21), 4.78 (d, *J*=7.9 Hz, H-1 of Glc II), 4.90 (d, *J*=7.6 Hz, H-1 of Gal), 5.24 (d, *J*=7.9 Hz, H-1 of Glc I).

CMR (C$_5$D$_5$N, 500 MHz) : δ C-1) 45.6 (2) 70.5 (3) 84.9 (4) 34.3 (5) 44.7 (6) 28.2 (7) 32.3 (8) 34.6 (9) 54.5 (10) 36.9 (11) 21.5 (12) 40.2 (13) 41.1 (14) 56.3 (15) 32.4 (16) 81.1 (17) 63.9 (18) 16.7 (19) 13.4 (20) 40.7 (21) 16.4 (22) 110.7

(23) 37.1 (24) 28.4 (25) 34.1 (26) 75.2 (27) 17.4 **Gal** (1) 103.6 (2) 73.1 (3) 75.8 (4) 80.1 (5) 75.9 (6) 60.9 **Glc I** (1) 107.2 (2) 75.4 (3) 78.8 (4) 72.3 (5) 78.5 (6) 63.2 **Glc II** (1) 104.9 (2) 75.2 (3) 78.4 (4) 71.8 (5) 78.6 (6) 62.9.

Mass (FAB, Negative ion, H.R) : *m/z* 935.4871 [(M-H)⁻, requires 935.4852].

Mass (FAB, Negative ion) : *m/z* 935 [M-H]⁻, 773 [M-Glc]⁻, 611 [M-Glc-Gal]⁻.

Reference

1.   Y. Wang, K. Ohtani, R. Kasai and K. Yamasaki, *Phytochemistry*, **45**, 811 (1997).

TIMOSAPONIN E 1

(25S,5β)-Furost-3β,15α,22,26-tetrol 3-O-[β-D-glucopyranosyl-(1→2)-β-D-galactopyranoside]-26-O-β-D-glucopyranoside

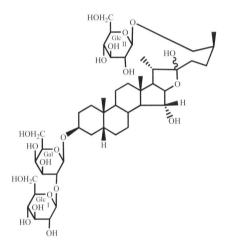

Source : *Anemarrhena asphodeloides* Bge. (Liliaceae)
Mol. Formula : C₄₅H₇₆O₂₀
Mol. Wt. : 936
M.P. : 164~167°C
Registry No. : [222018-46-4]

PMR (C₅D₅N, 300 MHz) : δ 0.95 (s, CH₃), 1.03 (s, CH₃), 1.32 (d, sec. CH₃), 1.54 (d, sec. CH₃), 4.83 (d, anomeric H), 4.91 (d, anomeric H), 5.26 (d, anomeric H).

CMR (C₅D₅N, 75 MHz) : δ C-1) 31.0 (2) 27.2 (3) 75.5 (4) 31.0 (5) 37.1 (6) 27.0 (7) 26.8 (8) 36.4 (9) 40.4 (10) 35.4 (11) 21.1 (12) 41.2 (13) 41.3 (14) 60.9 (15) 79.1 (16) 91.5 (17) 61.4 (18) 18.1 (19) 24.2 (20) 40.9 (21) 16.5 (22) 110.4 (23) 37.1 (24) 28.4 (25) 34.5 (26) 75.5 (27) 17.5 **Glc I** (1) 106.2 (2) 77.0 (3) 78.0 (4) 71.7 (5) 78.5 (6) 62.8 **Gal** (1) 102.5 (2) 81.9 (3) 75.2 (4) 69.9 (5) 76.6 (6) 62.2 **Glc II** (1) 105.2 (2) 75.2 (3) 78.6 (4) 71.7 (5) 78.4 (6) 62.8.

Mass (E.S.I.) : *m/z* 919.3 [M-OH]$^+$, 757.2 [M-OH-162]$^+$, 595.3 [M-OH-2x162]$^+$, 432.8 [M-OH-3x162]$^+$, 414.8 [M-3x162-H$_2$O]$^+$.

Reference

1.   Z.Y. Meng, S. Xu and L.H. Meng, *Yaoxue Xuebao* (*Acta Pharmaceutica Sinica*), **33**, 693 (1998).

RUSCUS ACULEATUS SAPONIN 12

22ξ-Methoxyfurosta-5,25(27)-diene-1β,3β,26-triol 1-O-{α-L-rhamnopyranosyl (1→2)-
4-O-acetyl-α-L-arabinopyranoside}-26-O-β-D-glucopyranoside

Source : *Ruscus aculeatus* L. (Liliaceae)
Mol. Formula : C$_{47}$H$_{74}$O$_{19}$
Mol. Wt. : 942
[α]$_D^{26}$: -38.0^0 (c=0.10, MeOH)
Registry No. : [205191-13-5]

IR (KBr) : 3425 (OH), 2930 (CH), 1735 (C=O), 1045 cm^{-1}.

PMR (C$_5$D$_5$N, 400 MHz) : δ 0.92 (s, 3xH-18), 1.16 (d, *J*=6.9 Hz, 3xH-21), 1.43 (s, 3xH-19), 1.79 (d, *J*=6.1 Hz, 3xH-6 of Rha), 2.00 (s, OCOC*H$_3$*), 3.26 (s, OC*H$_3$*), 4.69 (d, *J*=7.4 Hz, H-1 of Ara), 4.93 (d, *J*=7.6 Hz, H-1 of Glc), 5.07 and 5.35 (each 1H, br s, 2xH-27), 5.39 (br s, H-4 of Ara), 5.60 (br d, *J*=5.7 Hz, H-6), 6.32 (br s, H-1 of Rha).

CMR (C$_5$D$_5$N, 100 MHz) : δ C-1) 83.2 (2) 37.5 (3) 68.2 (4) 43.7 (5) 139.4 (6) 124.6 (7) 31.9 (8) 33.0 (9) 50.3 (10) 42.6 (11) 24.3 (12) 40.5 (13) 40.6 (14) 57.0 (15) 32.3 (16) 81.5 (17) 64.3 (18) 16.8 (19) 15.0 (20) 40.4 (21) 16.1 (22) 112.4 (23) 31.6 (24) 28.1 (25) 146.9 (26) 72.1 (27) 111.1 (OCH$_3$) 47.3 **Ara** (1) 100.0 (2) 75.3 (3) 73.2 (4) 73.7 (5) 64.3 **Rha** (1) 101.9 (2) 72.5 (3) 72.7 (4) 74.2 (5) 69.6 (6) 19.1 **Glc** (1) 103.9 (2) 75.9 (3) 78.6 (4) 71.7 (5) 78.6 (6) 62.9 (OCH$_3$) 170.8, 21.0.

Mass (FAB, Negative ion) : *m/z* 941 [M-H]$^-$.

Biological Activity : It possesses cytostatic activity on growth of Leukemia HL 60 cells, and shows 27.9% inhibition at 10 μg/ml sample concentration.

Reference

1.   Y. Mimaki, M. Kuroda, A. Kameyama, A. Yokosuka and Y. Sashida, *Chem. Pharm. Bull.*, **46**, 298 (1998).

PARDARINOSIDE G

26-Acetoxy-22α-methoxy-(25R)-5α-furost-3β,17α-diol 3-O-{α-L-rhamnopyranosyl-(1→2)]-[α-L-arabinopyranosyl-(1→3)]-β-D-glucopyranoside}

Source : *Lilium pardarinum* (Liliaceae)
Mol. Formula : $C_{47}H_{80}O_{19}$
Mol. Wt. : 948
$[\alpha]_D^{22}$: -40.8° (c=0.39, MeOH)
Registry No.: [125456-49-7]

IR (KBr) : 3425 (OH), 2932 (CH), 1736 (C=O), 1456, 1378, 1246, 1150, 1135, 1050, 1000, 912, 865, 838, 814, 782 cm^{-1}.

PMR (C_5D_5N, 400 MHz) : δ 0.90 (3xH-18)^a, 0.95 (s, 3xH-19)^a, 0.97 (d, *J*=6.7 Hz, 3xH-27), 1.28 (d, *J*=7.0 Hz, 3xH-21), 1.72 (d, *J*=6.1 Hz, 3xH-6 of Rha), 2.05 (3H, s, OCOC*H₃*), 2.54 (q , *J*=7.0 Hz, H-20), 3.21 (3H, s, OC*H₃*), 4.55 (dd, *J*=9.2, 3.1 Hz, H-3 of Rha), 4.87 (br s, H-2 of Rha), 4.44 (t, *J*=8.0 Hz, H-16), 4.90 (d, *J*=7.5 Hz, H-1 of Glc)^b, 4.98 (d, *J*=7.4 Hz, H-1 of Ara)^b, 6.22 (br s, H-1 of Rha).

CMR (C_5D_5N, 100.6 MHz) : δ C-1) 37.3 (2) 29.9 (3) 76.7 (4) 34.2 (5) 44.6 (6) 29.1 (7) 32.5^a (8) 36.0 (9) 54.3 (10) 35.8 (11) 21.1 (12) 32.3^a (13) 45.8 (14) 52.8 (15) 31.6^a (16) 90.4 (17) 90.4 (18) 17.3 (19) 12.5 (20) 43.0 (21) 10.4 (22) 113.2 (23) 30.6 (24) 28.0 (25) 33.3 (26) 69.2 (27) 16.8 (OC*H₃*) 47.0 (OCOC*H₃*) 170.8 (OCOC*H₃*) 20.8 **Glc** (1) 99.5 (2) 78.0^b (3) 88.0 (4) 69.6^c (5) 77.9^b (6) 62.6 **Rha** (1) 102.5 (2) 72.4^d (3) 72.8 (4) 74.1 (5) 69.4^c (6) 18.7 **Ara** (1) 105.5 (2) 72.3^d (3) 74.5 (4) 69.6^c (5) 67.7.

Mass (SIMS) : *m/z* 915 [M-MeOH-H]⁺, 503 [Aglycone-H]⁺.

Reference

1.   H. Shimomura, Y. Sashida and Y. Mimaki, *Phytochemistry*, **28**, 3163 (1989).

MACROSTEMONOSIDE H

22-Methoxy-5β-fursot-25(27)-ene-3β,12β,26-triol 3-O-[β-D-glucopyranosyl-(1→2)-
β-D-galactopyranoside]-26-O-β-D-glucopyranoside

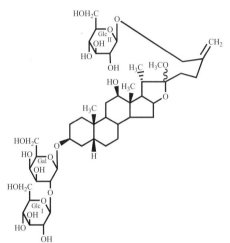

Source : *Allium macrostemon* Bunge (Liliaceae), artefact
Mol. Formula : C₄₆H₇₆O₂₀
Mol. Wt. : 948
M.P. : 187-190°C
[α]_D²⁰ : -45.8° (c=0.59, C₅D₅N)
Registry No. : [162413-64-1]

PMR (C₅D₅N, 500 MHz) **:** δ 0.90 (H-7A), 0.96 (s, 3xH-19), 1.04 (s, 3xH-18), 1.05* (H-14), 1.20 (H-6A), 1.25* (H-7B), 1.41* (H-9), 1.43 (H-1A), 1.43 (H-2A), 1.43* (H-4A), 1.44 (d, *J*=6.7 Hz, 3xH-21), 1.45 (H-11A), 1.47* (H-8), 1.50 (H-15), 1.72* (H-11B), 1.77* (H-6B), 1.79* (H-4B), 1.80* (H-1B), 1.90* (H-2B), 2.00* (H-15), 2.06 (H-23A), 2.13* (H-17), 2.15* (H-5), 2.16* (H-23B), 2.39 (m, H-24), 2.49 (m, H-20), 3.28 (s, OC*H*₃), 3.49 (dd, *J*=9.0, 3.1 Hz, H-12), 3.83 (m, H-5 of Glc I), 3.94 (m, H-5 of Glc II), 4.00 (m, H-5 of Gal), 4.05* (H-2 of Glc II), 4.07* (H-2 of Glc I), 4.17* (H-3 of Glc I), 4.19* (H-4 of Glc II), 4.22* (H-3 of Glc II), 4.23* (H-3 of Gal), 4.25* (H-3), 4.29* (H-4 of Glc I), 4.34 (H-26A), 4.34 (H-6A of Glc II), 4.37 (H-6A of Gal), 4.4 (H-6B of Glc I), 4.42* (H-6B of Gal), 4.47 (H-6B of Glc I), 4.40 (H-6 of Glc I), 4.55* (H-4 of Gal), 4.57* (H-6B of Glc II), 4.58* (H-16), 4.60* (H-26B), 4.63 (t, *J*=8.0 Hz, H-2 of Gal), 4.89 (d, *J*=7.6 Hz, H-1 of Gal), 4.91 (d, *J*=7.8 Hz, H-1 of Glc II), 5.03 (s, H-27A), 5.26 (d, *J*=7.6 Hz, H-1 of Glc I), 5.34 (s, H-27B). * overlapped signals.

874

CMR (C$_5$D$_5$N, 125 MHz) : δ C-1) 30.9 (2) 26.7 (3) 75.4 (4) 30.9 (5) 36.7 (6) 27.0 (7) 26.7 (8) 34.6 (9) 39.3 (10) 35.2 (11) 31.4 (12) 79.4 (13) 46.9 (14) 55.1 (15) 31.7 (16) 81.6 (17) 64.2 (18) 11.2 (19) 23.9 (20) 41.4 (21) 15.5 (22) 112.6 (23) 31.7 (24) 28.2 (25) 146.8 (26) 72.0 (27) 110.8 (OCH$_3$) 47.3 **Gal** (1) 102.6 (2) 81.9 (3) 75.2 (4) 69.8 (5) 76.6 (6) 62.1 **Glc I** (1) 106.1 (2) 76.9 (3) 78.0 (4) 71.6 (5) 78.4 (6) 62.7 **Glc II** (1) 103.8 (2) 75.2 (3) 78.5 (4) 71.6 (5) 78.5 (6) 62.7.

Mass (FAB, Positive ion) : m/z 947 [M-H]$^+$, 917 [M+H-MeOH]$^+$, 575 [M+H-MeOH-H$_2$O-2xGlc]$^+$.

Reference

1.   J. Peng, X. Yao, H. Kobayashi and C. Ma, *Planta Med.*, **61**, 58 (1995).

WATTOSIDE C
22ξ-Methoxy-25(S)-furost-5-en-1β,3β,26β-triol 3-O-[β-D-glucopyranosyl-(1→4)-β-D-galactopyranoside]-26-O-β-D-glucopyranoside

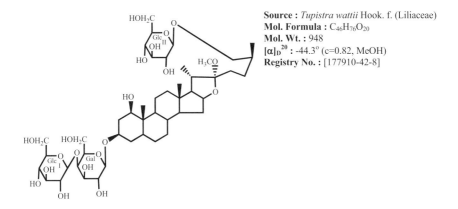

Source : *Tupistra wattii* Hook. f. (Liliaceae)
Mol. Formula : C$_{46}$H$_{76}$O$_{20}$
Mol. Wt. : 948
[α]$_D^{20}$: -44.3° (c=0.82, MeOH)
Registry No. : [177910-42-8]

PMR (C$_5$D$_5$N, 400 MHz) : δ 0.88 (s, 3xH-18), 1.04 (d, J=6.8 Hz, 3xH-27), 1.12 (d, J=6.8 Hz, 3xH-21), 1.20 (s, 3xH-19).

CMR (C$_5$D$_5$N, 100 MHz) : δ C-1) 78.5 (2) 41.2 (3) 75.6 (4) 39.8 (5) 139.4 (6) 125.1 (7) 32.8 (8) 33.0 (9) 51.4 (10) 43.7 (11) 24.3 (12) 40.5 (13) 40.6 (14) 56.9 (15) 32.4 (16) 81.4 (17) 64.6 (18) 16.8 (19) 13.8 (20) 40.8 (21) 16.5 (22) 112.8 (OCH$_3$) 47.6 (23) 30.2 (24) 28.3 (25) 34.5 (26) 75.5 (27) 17.5 **Gal** (1) 103.1 (2) 73.5 (3) 75.5 (4) 79.9 (5) 75.9 (6) 60.9 **Glc I** (1) 107.1 (2) 75.2 (3) 78.7 (4) 72.4 (5) 78.5 (6) 63.2 **Glc II** (1) 105.1 (2) 75.5 (3) 78.6 (4) 71.9 (5) 78.5 (6) 63.0.

Mass (FAB, Negative ion) : *m/z* 947 [M-H]⁻.

Reference

1. F. Yang, P. Shen, Y.-F. Wang, R.-Y. Zhang and C.-R. Yang, *Acta Botanica Yunnanica*, **23**, 373 (2001).

CAMASSIA CUSICKII SAPONIN 11

22-O-Methyl-(25R)-5α-furostan-3β,6α,22ξ-triol 6-O-[β-D-glucopyranosyl-(1→3)-β-D-glucopyranoside]-26-O-β-D-glucopyranoside

Source : *Camassia cusickii* S. Wats. (Liliaceae)
Mol. Formula : $C_{46}H_{78}O_{20}$
Mol. Wt. : 950
[α]$_D^{24}$: -21.8° (c=0.33, MeOH)
Registry No. : [141360-81-8]

IR (KBr) : 3420 (OH), 2925 (CH), 1450, 1375, 1260, 1155, 1070, 1025, 890 cm⁻¹.

PMR (C₅D₅N, 400 MHz) : δ 0.76 (s, 3xH-18), 0.85 (s, 3xH-19), 1.02 (d, *J*=6.6 Hz, 3xH-27), 1.17 (d, *J*=6.8 Hz, 3xH-21), 3.24 (s, OCH₃), 3.74 (ddd, *J*=10.7, 10.7, 4.5 Hz, H-6), 3.80 (m, H-3), 4.84 (d, *J*=7.7 Hz, H-1 of Glc I), 4.92 (d, *J*=7.6 Hz, H-1 of Glc III), 5.28 (d, *J*=7.8 Hz, H-1 of Glc II).

CMR (C₅D₅N, 100 MHz) : δ C-1) 37.8 (2) 32.2ᵃ (3) 70.7 (4) 33.3 (5) 51.4 (6) 79.6 (7) 41.3 (8) 34.1ᵇ (9) 54.0 (10) 36.7 (11) 21.2 (12) 40.0 (13) 41.1 (14) 56.3 (15) 32.0ᵃ (16) 81.3 (17) 64.2 (18) 16.2 (19) 13.6 (20) 40.5 (21) 16.6 (22) 112.6 (23) 30.7 (24) 28.3 (25) 34.2ᵇ (26) 75.2 (27) 17.2 (OCH₃) 47.3 **Glc I** (1) 105.4 (2) 74.6 (3) 89.1 (4) 69.9 (5) 77.7 (6)

62.5^c **Glc II** (1) 106.1 (2) 75.7 (3) 78.7^d (4) 71.7^e (5) 78.3^d (6) 62.6^c **Glc III** (1) 105.0 (2) 75.2 (3) 78.7^d (4) 71.8^e (5) 78.5^d (6) 63.0.

Mass (SIMS) : m/z 919 [M-OMe]$^+$.

Reference

1.   Y. Mimaki, Y. Sashida and K. Kawashima, *Chem. Pharm. Bull.*, **40**, 148 (1992).

TIMOSAPONIN E 2

(25S)-22-Methoxy-5β-furost-3β,15α,26-triol 3-O-[β-D-glucopyranosyl-(1→2)-β-D-galactopyranoside]-26-O-β-D-glucopyranoside

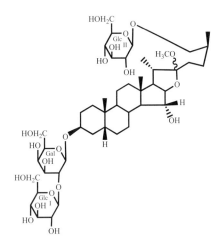

Source : *Anemarrhena asphodeloides* Bge. (Liliaceae)
Mol. Formula : $C_{46}H_{78}O_{20}$
Mol. Wt. : 950
M.P. : 180~185°C
Registry No. : [222018-47-5]

PMR (C_5D_5N, 300 MHz) : δ 0.86 (s, CH_3), 1.01 (s, CH_3), 1.15 (d, sec. CH_3), 1.48 (d), 3.20 (OCH_3).

CMR (C_5D_5N, 75 MHz) : δ C-1) 31.0 (2) 27.2 (3) 75.5 (4) 31.0 (5) 37.0 (6) 27.0 (7) 26.8 (8) 36.3 (9) 40.4 (10) 35.4 (11) 21.1 (12) 40.8 (13) 41.4 (14) 61.0 (15) 78.7 (16) 91.8 (17) 62.0 (18) 17.8 (19) 24.2 (20) 40.7 (21) 16.5 (22) 112.3 (23) 31.1 (24) 28.4 (25) 34.5 (26) 75.3 (27) 17.6 (OCH_3) 47.4 **Glc I** (1) 106.2 (2) 77.0 (3) 78.0 (4) 71.7 (5) 78.5 (6) 62.8 **Gal** (1) 102.5 (2) 81.9 (3) 75.0 (4) 69.9 (5) 76.6 (6) 62.2 **Glc II** (1) 105.1 (2) 75.0 (3) 78.6 (4) 71.7 (5) 78.5 (6) 62.8.

Reference

1   Z.Y. Meng, S. Xu and L.H. Meng, *Yaoxue Xuebao* (*Acta Pharmaceutica Sinica*), **33**, 693 (1998).

AMPELOSIDE Bf$_2$

(25R)-5α-Furostane-2α,3β,6β,22ξ,26-pentaol 3-O-[β-glucopyranosyl-(1$\rightarrow$4)-β-galactopyranoside]-26-O-β-glucopyranoside

Source : *Allium ampeloprasum* L. (Liliaceae)
Mol. Formula : $C_{45}H_{76}O_{21}$
Mol. Wt. : 952
[α]$_D^{21}$: -40.7° (c=1.03, C_5H_5N)
Registry No. : [118543-10-5]

PMR (C_5D_5N, 270 MHz) : δ 5.27 (d, J=7.69 Hz, anomeric H), 4.99 (d, J=7.7 Hz, anomeric H), 4.82 (d, J=7.70 Hz, anomeric H).

CMR (C_5D_5N, 67.80 MHz) : δ C-1) 47.1 (2) 70.4 (3) 85.0 (4) 31.7 (5) 47.8 (6) 69.9 (7) 40.7 (8) 29.8 (9) 54.5 (10) 36.9 (11) 21.3 (12) 40.1 (13) 41.1 (14) 56.0 (15) 32.4 (16) 81.0 (17) 63.8 (18) 17.1 (19) 16.3 (20) 40.5 (21) 16.6 (22) 110.6 (23) 36.9 (24) 28.3 (25) 34.1 (26) 75.2 (27) 17.4 **Gal** (1) 103.3 (2) 72.9 (3) 75.7^a (4) 79.9 (5) 75.8^a (6) 60.9 **Glc I** (1) 106.9 (2) 75.0 (3) 78.3^b (4) 72.1 (5) 78.3^b (6) 62.7^c **Glc II** (1) 104.7 (2) 75.0 (3) 78.4^b (4) 71.6 (5) 78.5^b (6) 63.0^c.

Mass (F.D.) : *m/z* 975 [M+Na]$^+$, 934 [M-H$_2$O]$^+$.

Reference

1. T. Morita, T. Ushiroguchi, N. Hayashi, H. Matsuura, Y. Itakura and T. Fuwa, *Chem. Pharm. Bull.*, **36**, 3480 (1988).

RUSCUS HYPOGLOSSUM SAPONIN 1

(25R)-22,26-Tetrahydroxyfurost-5-ene 1-O-[α-L-rhamnopyranosyl-(1→4)-(6-O-acetyl)-
β-D-galactopyranoside]-26-O-β-D-glucopyranoside

Source : *Ruscus hypoglossum* and *R. colchicus*
(Liliaceae)
Mol. Formula : $C_{47}H_{76}O_{20}$
Mol. Wt. : 960
M.P. : 186°C
Registry No. : [515835-89-9]

IR (KBr) : 3405, 2936, 1731, 1645, 1454, 1376, 1255, 1074, 983, 910, 837 cm^{-1}.

PMR (C$_5$D$_5$N, 300 MHz) **:** δ 0.90 (s, 3xH-18), 1.00 (d, *J*=6.6 Hz, 3xH-27), 1.27 (d, *J*=6.9 Hz, 3xH-21), 1.29 (m, H-14), 1.42 (s, 3xH-19), 1.46 (m, H-12A), 1.52 (m, H-15A), 1.61 (m, H-7A, H-8, H-9 and H-11A), 1.68 (m, H-24A), 1.69 (br d, H-12B), 1.73 (d, *J*=6.1 Hz, 3xH-6 of Rha), 1.91 (m, H-25), 1.96 (m, H-7B), 2.00 (s, OCOC*H*$_3$), 2.02 (d, H-23, H-24B), 2.05 (m, H-17), 2.07 (m, H-15B), 2.24 (m, H-20), 2.40 (dd, H-2A), 2.65 (dd, H-2B), 2.56 (dd, H-4A), 2.68 (dd, H-4B), 2.96 (m, H-11B), 3.62 (dd, H-26A), 3.78 (dd, *J*=11.8 Hz, H-1), 3.80 (m, H-3), 3.90 (m, H-5 of Glc), 3.92 (dd, H-26B), 3.94 (dd, H-2 of Glc), 3.95 (br dd, H-5 of Gal), 4.15 (m, H-3 of Gal), 4.18 (m, H-4 of Gal), 4.19 (m, H-3 and H-4 of Glc), 4.27 (dd, H-4 of Rha), 4.36 (dd, H-6A of Glc), 4.48 (dd, H-6B of Glc) 4.55 (dd, H-6A of Gal), 4.57 (H-2 of Gal),), 4.70 (dd, H-3 of Rha), 4.72 (d, *J*=8.1 Hz, H-1 of Gal), 4.80 (d, *J*=7.7 Hz, H-1 of Glc), 4.85 (dq, H-5 of Rha), 4.59 (br d, H-2 of Rha), 4.91 (dd, H-6B of Gal), 4.95 (m, H-16), 5.61 (br d, H-6), 6.36 (br, H-1 of Rha).

CMR (C$_5$D$_5$N, 75 MHz) : δ C-1) 84.8 (2) 38.0 (3) 68.3 (4) 43.8 (5) 139.6 (6) 124.8 (7) 32.2 (8) 33.2 (9) 50.7 (10) 42.9 (11) 24.2 (12) 40.4 (13) 40.7 (14) 57.2 (15) 32.8 (16) 81.2 (17) 64.0 (18) 17.0 (19) 15.0 (20) 40.8 (21) 16.3 (22) 110.7 (23) 37.2 (24) 28.4 (25) 34.3 (26) 75.3 (27) 17.4 **Gal** (1) 100.7 (2) 74.6 (3) 76.4 (4) 70.6 (5) 73.2 (6) 64.6 (OCOCH$_3$) 170.5 (OCOCH$_3$) 20.8 **Rha** (1) 101.7 (2) 72.7 (3) 72.5 (4) 74.3 (5) 69.3 (6) 18.9 **Glc** (1) 104.9 (2) 75.2 (3) 78.6 (4) 71.9 (5) 78.4 (6) 62.9.

Mass (E.S.I., CID, MS^2) : *m/z* (rel.intens.) 959 [(M-H)$^-$, 100], 917 [(M-H-COCH$_2$)$^-$, 40], 899 [M-H-AcOH]$^-$, 771 [M-H-COCH$_2$-Rha]$^-$, 755 [M-H-COCH$_2$-Glc]$^-$, 753 [M-H-AcOH-Rha]$^-$, 609 [771-Glc or 753-Rha]$^-$, 591 [753-Glc]$^-$, 447 [609-Glc].

Mass (E.S.I., Positive ion) : m/z 983 [M+Na]$^+$, 943 [M+H-H$_2$O], 781 [M+H-H$_2$O-Glc]$^+$, 431 [Agl+H]$^+$, 413 [Agl+H-H$_2$O]$^+$.

Reference

1.   E. de Combarieu, M. Falzoni, N. Fuzzati, F. Gattesco, A. Giori, M. Lovati and R. Pace, *Fitoterapia*, **73**, 583 (2002).

PARDARINOSIDE F

26-Acetoxy-22α-methoxy-(25R)-5α-furost-3β,14α,17α-triol 3-O-{α-L-rhamnopyranosyl-(1→2)-[α-L-arabinopyranosyl (1→3)]-β-D-glucopyanoside}

Source : *Lilium pardarinum* (Liliaceae)
Mol. Formula : C$_{47}$H$_{80}$O$_{20}$
Mol. Wt. : 964
$[\alpha]_D^{22}$: -32.8° (c=0.27, MeOH)
Registry No. : [125456-48-6]

IR (KBr) : 3435 (OH), 2936, 2874 (CH), 1724 (C=O), 1456, 1382, 1259, 1155, 1127, 1048, 1000, 940, 912, 868, 838, 812, 783 cm^{-1}.

PMR (C_5D_5N, 400 MHz) : δ 0.96 (s, 3xH-18)[a], 0.99 (d, *J*=6.5 Hz, 3xH-27), 1.08 (s, H-19)[a], 1.33 (d, *J*=7.0 Hz, 3xH-21), 1.73 (d, *J*=5.9 Hz, 3xH-6 of Rha), 2.05 (3H, s, OCOCH_3), 2.68 (q, *J*=7.0 Hz, H-20), 3.23 (3H, s, OCH_3), 4.55 (dd, *J*=9.2, 3.0 Hz, H-3 of Rha), 4.72 (t, *J*=6.5 Hz, H-16), 4.87 (br s, H-2 of Rha), 4.90 (d, *J*=7.7 Hz, H-1 of Glc)[b], 4.99 (d, *J*=6.9 Hz, H-1 of Ara)[b], 6.24 (br s, H-1 of Rha).

CMR (C_5D_5N, 100.6 MHz) : δ C-1) 37.5 (2) 29.9 (3) 76.6 (4) 34.2 (5) 44.5 (6) 29.1 (7) 27.0[a] (8) 39.7 (9) 46.8 (10) 36.1 (11) 20.3 (12) 26.9[a] (13) 48.8 (14) 88.6 (15) 40.1 (16) 90.9 (17) 91.3 (18) 21.0 (19) 12.3 (20) 43.5 (21) 10.6 (2) 112.9 (23) 30.9 (24) 28.0 (25) 33.3 (26) 69.2 (27) 16.9 (OCH_3) 47.2 (OCOCH_3) 170.8 (OCOCH_3) 20.8 **Glc** (1) 99.4 (2) 78.0[d] (3) 88.1 (4) 69.6[c] (5) 77.9[b] (6) 62.6 **Rha** (1) 102.5 (2) 72.4[d] (3) 72.9 (4) 74.1 (5) 69.4[c] (6) 18.7 **Ara** (1) 105.5 (2) 72.3[d] (3) 74.5 (4) 69.6[c] (5) 67.8.

Mass (SIMS) : *m/z* 931 [M-MeOH-H]$^+$, 519 [Agl-H]$^+$.

Reference

1. H. Shimomura, Y. Sashida and Y. Mimaki, *Phytochemisry*, **28**, 3163 (1989).

RUSCUS ACULEATUS SAPONIN 21
22-O-Methyl-(25*R*)-furost-5-ene-1β,3β,22ξ,26-tetrol 1-O-{α-L-rhamnopyranosyl-(1→2)-6-O-acetyl-β-D-galactopyranoside}-26-O-β-D-glucopyranoside

Source : *Ruscus aculeatus* L. (Liliaceae)
Mol. Formula : $C_{48}H_{78}O_{20}$
Mol. Wt. : 974
$[\alpha]_D^{25}$: -46.0° (c=0.10, MeOH)
Registry No. : [211379-11-2]

IR (KBr) : 3420 (OH), 2930 (CH), 1735 (C=O), 1445, 1370, 1245, 1130, 1065, 980, 905, 835, 810 cm^{-1}.

PMR (C$_5$D$_5$N, 400 MHz) : δ 0.89 (s, 3xH-18), 0.99 (d, *J*=6.6 Hz, 3xH-27), 1.14 (d, *J*=6.9 Hz, 3xH-21), 1.43 (s, 3xH-19), 1.73 (d, *J*=6.1 Hz, 3xH-6 of Rha), 2.03 (s, OCOC*H$_3$*), 3.24 (s, OC*H$_3$*), 3.78 (dd, *J*=11.9, 4.0 Hz, H-1), 3.83 (m, H-3), 4.74 (d, *J*=7.6 Hz, H-1 of Gal), 4.84 (d, *J*=7.7 Hz, H-1 of Glc), 5.63 (br d, *J*=5.5 Hz, H-6), 6.35 (br s, H-1 of Rha).

CMR (C$_5$D$_5$N, 100 MHz) : δ C-1) 84.8 (2) 38.0 (3) 68.2 (4) 43.8 (5) 139.6 (6) 124.8 (7) 32.1 (8) 33.2 (9) 50.6 (10) 42.8 (11) 24.1 (12) 40.2 (13) 40.6 (14) 57.1 (15) 32.5 (16) 81.4 (17) 64.4 (18) 16.8 (19) 15.0 (20) 40.6 (21) 16.2 (22) 112.7 (23) 30.9 (24) 28.2 (25) 34.2 (26) 75.2 (27) 17.1 (OC*H$_3$*) 47.2 **Gal** (1) 100.7 (2) 74.5 (3) 76.4 (4) 70.6 (5) 73.2 (6) 64.6 **Rha** (1) 101.7 (2) 72.5 (3) 72.7 (4) 74.2 (5) 69.3 (6) 19.0 **Glc** (1) 105.0 (2) 75.2 (3) 78.6 (4) 71.8 (5) 78.5 (6) 62.9 (OCOCH$_3$) 170.5 (OCOC*H$_3$*) 20.9.

Mass (FAB, Negative ion) : *m/z* 973 [M-H]⁻.

Reference

1.   Y. Mimaki, M. Kuroda, A. Kameyama, A. Yokosuka and Y. Sashida, *Phytochemistry*, **48**, 485 (1998).

PARDARINOSIDE D

26-Acetoxy-22α-methoxy-(25*R*)-5α-furost-3β,17α-diol 3-O-{α-L-rhamnopyranosyl-(1→2)-[β-D-glucopyanosyl-(1→4)]-β-D-glucopyanoside}

Source : *Lilium pardarinum* (Liliaceae)
Mol. Formula : C$_{48}$H$_{82}$O$_{20}$
Mol. Wt. : 978
[α]$_D^{22}$: -59.7° (c=0.79, MeOH)
Registry No. : [125477-04-5]

IR (KBr) : 3423 (OH), 2932, 2880 (CH), 1737 (C=O), 1456, 1377, 1310, 1260, 1246, 1061, 910, 813 cm^{-1}.

PMR (C_5D_5N, 400 MHz) : δ 0.90 (3H, s, H-18)^a, 0.95 (s, 3xH-19)^a, 0.97 (d, J=6.7 Hz, 3xH-27), 1.28 (d, J=7.0 Hz, 3xH-21), 1.75 (d, J=6.0 Hz, 3xH-6 of Rha), 2.05 (3H, s, OCOCH_3), 2.55 (q, J=7.0 Hz, H-20), 3.22 (3H, s, OCH_3), 4.74 (br s, H-2 of Rha), 4.95 [(H-1 of Glc I)^b, overlapped with H_2O signal], 5.13 (d, J=7.8 Hz, H-1 of Glc II)^b, 6.23 (br s, H-1 of Rha).

CMR (C_5D_5N, 100.6 MHz) : δ C-1) 37.3 (2) 29.9 (3) 77.3 (4) 34.5 (5) 44.7 (6) 29.0 (7) 32.5^a (8) 36.0 (9) 54.3 (10) 35.8 (11) 21.1 (12) 32.3^a (13) 45.8 (14) 52.8 (15) 31.6^a (16) 90.4 (17) 90.4 (18) 17.3 (19) 12.5 (20) 43.0 (21) 10.4 (22) 113.2 (23) 30.6 (24) 28.0 (25) 33.3 (26) 69.2 (27) 16.8 (OCH_3) 47.0 (COCH_3) 170.8 (COCH_3) 20.8 **Glc I** (1) 99.6 (2) 77.6^b (3) 76.2 (4) 82.2 (5) 77.7^b (6) 62.0^c **Rha** (1) 101.9 (2) 72.4 (3) 72.8 (4) 74.1 (5) 69.4 (6) 18.7 **Glc II** (1) 105.2 (2) 75.0 (3) 78.3^d (4) 71.3 (5) 78.5^d (6) 62.1^c.

Mass (SIMS) : m/z 945 [M-MeOH-H]$^+$, 503 [Agl-H]$^+$.

Reference

1.   H. Shimomura, Y. Sashida and Y. Mimaki, *Phytochemistry*, **28**, 3163 (1989).

OPHIOPOGON PLANISCAPUS SAPONIN G
25(R)-Furost-5-en-1β,3β,22,26-tetrol 1-O-[α-L-rhamnopyranosyl-(1→2)-α-L-(4-O-sulfo)-α-L-arabinopyranoside]-26-O-[β-D-glucopyranoside] sodium salt

Source : *Ophiopogon planiscapus* Nakai[1] (Liliaceae)
Mol. Formula : $C_{44}H_{71}O_{21}SNa$
Mol. Wt. : 990
M.P. : 212-215°C (decomp.)
$[α]_D^{23}$: -55.2° (c=0.80, Pyridine)
Registry No. : [88623-87-4]

IR (KBr) **:** 3600-3200 (OH), 1220 (S–O) cm^{-1}.

CMR (C$_5$D$_5$N, 100 MHz) **:** δ **Ara** C-1) 100.1 (2) 75.9 (3) 74.5 (4) 76.0 (5) 65.5 **Rha** (1) 101.2 (2) 72.1 (3) 72.1 (4) 73.9 (5) 69.4 (6) 18.7 **Glc** (1) 104.5 (2) 75.0 (3) 78.1 (4) 71.7 (5) 77.9 (6) 62.7. C$_1$-H *J* values (Ara) 159 Hz, (Rha) 172 Hz, (Glc) 156 Hz.

Reference

1.   Y. Watanabe, S. Sanada, Y. Ida and J. Shoji, *Chem. Pharm. Bull.*, **31**, 3486 (1983).

PARDARINOSIDE C

26-Acetoxy-22α-methoxy-(25R)-5α-furost-3β,14α,17α-triol 3-O-{α-L-rhamnopyranosyl-(1→2)-[β-D-glucopyranosyl-(1→4)]-β-D-glucopyranoside}

Source : *Lilium pardarinum* (Liliaceae)
Mol. Formula : C$_{48}$H$_{82}$O$_{21}$
Mol. Wt. : 994
[α]$_D^{22}$: -50.5° (c=1.17, MeOH)
Registry No. : [125477-03-4]

IR (KBr) : 3420 (OH), 2925 (CH), 1720 (C=O), 1450, 1365, 1240, 1050, 980, 900, 805 cm^{-1}.

PMR (C$_5$D$_5$N, 400 MHz) : δ 0.95 (s, 3xH-18)^a, 0.99 (3H, d, J=6.5 Hz, 3xH-27), 1.08 (s, 3xH-19)^a, 1.33 (d, J=7.0 Hz, 3xH-21), 1.75 (d, J=6.1 Hz, 3xH-6 of Rha), 2.06 (3H, s, OCOCH_3), 2.68 (q, J=7.0 Hz, H-20), 3.23 (3H, s, OCH_3), 4.55 (dd, J=9.3, 3.2 Hz, H-3 of Glc II), 4.72 (t, J=7.1 Hz, H-16), 4.73 (br s, H-2 of Rha), 4.91 (dq, J=9.4, 6.1 Hz, H-5 of Rha), 4.95 (d, J=7.2 Hz, H-1 of Glc I)^b, 5.12 (d, J=7.8 Hz, H-1 of Glc II)^b, 6.23 (br s, H-1 of Rha).

CMR (C$_5$D$_5$N, 100.6 MHz) : δ C-1) 37.5 (2) 29.9 (3) 77.2 (4) 34.4 (5) 44.5 (6) 29.0 (7) 27.0^a (8) 39.7 (9) 46.8 (10) 36.1 (11) 20.3 (12) 26.9^a (13) 48.8 (14) 88.6 (15) 40.1 (16) 90.8 (17) 91.3 (18) 21.0 (19) 12.3 (20) 43.5 (21) 10.6 (22) 112.9 (23) 30.8 (24) 28.0 (25) 33.3 (26) 69.2 (27) 16.8 (OCH_3) 47.1 (COCH_3) 170.8 (COCH_3) 20.8 **Glc I** (1) 99.5 (2) 77.6^b (3) 76.2 (4) 82.1 (5) 77.7^b (6) 62.0^c **Rha** (1) 101.9 (2) 72.4 (3) 72.7 (4) 74.1 (5) 69.4 (6) 18.6 **Glc II** (1) 105.2 (2) 74.9 (3) 78.3^d (4) 71.2 (5) 78.4^d (6) 62.1^c.

Mass (SIMS) : m/z 1016 [M+Na-H]$^+$, 961 [M-MeOH-H]$^+$, 519 [Agl-H]$^+$.

Reference

1.   H. Shimomura, Y. Sashida and Y. Mimaki, *Phytochemistry*, **28**, 3163 (1989).

RUSCUS ACULEATUS SAPONIN 8

22ξ-Methoxyfurosta-5,25(27)-dien-1β,3β,26-triol-1-O-{α-L-rhamnopyranosyl-(1→2)-(4-O-sulfo)-
α-L-arabinopyranoside}-26-O-β-D-glucopyranoside monosodium salt

Source : *Ruscus aculeatus* L. (Liliaceae)
Mol. Formula : $C_{45}H_{71}O_{21}SNa$
Mol. Wt. : 1002
$[\alpha]_D^{26}$: -52.0° (c=0.10, MeOH)
Registry No. : [205191-09-9]

IR (KBr) : 3425 (OH), 2920 (CH), 1255, 1225, 1070, 1045 cm^{-1}.

PMR (C$_5$D$_5$N + CD$_3$OD, 400 MHz) : δ 0.81 (s, 3xH-18), 1.10 (d, *J*=6.8 Hz, 3xH-21), 1.38 (s, 3xH-19), 1.67 (d, *J*=6.1 Hz, 3xH-6 of Rha), 3.23 (s, OC*H*$_3$), 4.19 (dd, *J*=9.2, 2.9 Hz, H-3 of Ara), 4.65 (d, *J*=7.2 Hz, H-1 of Ara), 4.83 (d, *J*=7.8 Hz, H-1 of Glc), 5.04 (br s, H-27A), 5.20 (br d, *J*=2.9 Hz, H-4 of Ara), 5.30 (br s, H-27B), 5.57 (ovrlaping with H$_2$O signal, H-6), 6.13 (br s, H-1 of Ara).

CMR (C$_5$D$_5$N+CD$_3$OD, 100 MHz) : δ C-1) 83.5 (2) 36.9 (3) 67.9 (4) 43.4 (5) 139.2 (6) 124.7 (7) 31.8 (8) 32.9 (9) 50.2 (10) 42.7 (11) 23.7 (12) 39.6 (13) 40.3 (14) 56.4 (15) 32.2 (16) 81.3 (17) 63.9 (18) 16.3 (19) 14.8 (20) 40.3 (21) 15.9 (22) 112.2 (23) 31.5 (24) 27.9 (25) 146.6 (26) 71.8 (27) 111.0 (OCH$_3$) 47.1 **Ara** (1) 100.0 (2) 75.7 (3) 74.6 (4) 75.8 (5) 65.5 **Rha** (1) 101.3 (2) 71.9 (3) 72.0 (4) 73.8 (5) 69.2 (6) 18.6 **Glc** (1) 103.4 (2) 74.7 (3) 78.0 (4) 71.3 (5) 78.0 (6) 62.4.

Mass (FAB, Negative ion) : *m/z* 947 [M-Na-OMe-H]$^-$.

Reference

1.   Y. Mimaki, M. Kuroda, A. Kameyama, A. Yokosuka and Y. Sashida, *Chem. Pharm. Bull.*, **46**, 298 (1998).

OPHIOPOGON OHWII SAPONIN O-6
25(*R*)-Furost-5-en-1β,3β,22,26-tetrol 1-O-[α-L-rhamnopyranosyl-(1→2)-
(4-O-sulfo)-β-D-fucopyranoside] sodium salt

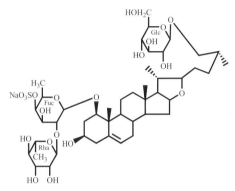

Source : *Ophiopogon ohwii* Okuyama (Liliaceae)
Mol. Formula : $C_{45}H_{73}O_{21}SNa$
Mol. Wt. : 1004
M.P. : 183-185°C (decomp.)
$[\alpha]_D^{23}$: -50.8° (c=0.31, Pyridine)
Registry No. : [94901-61-8]

IR (KBr) **:** 3600-3200 (OH), 1210 (S-O) cm^{-1}.

CMR (C_5D_5N, 25 MHz) **:** δ **Fuc** C-1) 100.1 (2) 76.0^a (3) 75.9^a (4) 79.0 (5) 70.5 (6) 17.3 **Rha** (1) 101.3 (2) 71.9^b (3) 72.2 (4) 74.2 (5) 69.2 (6) 18.8 **Glc** (1) 104.7 (2) 75.1 (3) 87.3 (4) 72.2^b (5) 78.1 (6) 62.9.

Reference

1.   Y. Watanabe, S. Sanada, Y. Ida and J. Shoji, *Chem. Pharm. Bull.*, **32**, 3994 (1984).

CHINENOSIDE II
3β,26-Dihydroxy-(25*R*)-5α-furost-20(22)-en-6-one 3-O-{β-xylopyranosyl-(1→4)-[α-arabinopyranosyl-(1→6)]-
β-glucopyranoside}-26-O-β-D-glucopyranoside

Source : *Allium chinense* G. Don (Liliaceae)
Mol. Formula : $C_{49}H_{78}O_{22}$
Mol. Wt. : 1018
$[\alpha]_D$: -63.1° (c=0.43, Pyridine)
Registry No. : [172589-64-9]

IR (KBr) **:** 3400 (OH), 1700 (C=O), 1000, 1100 (glycosyl C–O) cm^{-1}.

PMR (C$_5$D$_5$N, 400 MHz) **:** δ 0.64 (s, 3xH-19), 0.65 (s, 3xH-18), 1.01 (d, J=6.4 Hz, 3xH-27), 1.63 (s, 3xH-21), 2.45 (d, J=10.0 Hz, H-17), 3.59 (dd, J=8.8, 4.9 Hz, H-26), 3.90 (m, H-26), 3.90 (m, H-3), 4.75 (m, H-16), 4.77 (d, J=7.9 Hz, H-1 of Glc II), 4.87 (d, J=7.6 Hz, H-1 of Glc I), 4.98 (d, J=7.3 Hz, H-1 of Ara), 5.36 (d, J=7.9 Hz, H-1 of Xyl).

CMR (C$_5$D$_5$N, 100 MHz) **:** δ C-1) 36.5 (2) 29.2 (3) 76.8 (4) 26.7 (5) 56.3 (6) 209.4 (7) 46.7 (8) 37.0 (9) 53.4 (10) 40.7 (11) 21.5 (12) 39.2 (13) 43.8 (14) 54.6 (15) 33.9 (16) 84.0 (17) 64.2 (18) 14.1 (19) 12.9 (20) 103.2 (21) 11.6 (22) 152.4 (23) 23.4 (24) 31.2 (25) 33.2 (26) 74.7 (27) 17.2 **Glc I** (1) 101.8 (2) 74.5 (3) 75.9 (4) 79.6 (5) 78.1 (6) 67.8 **Xyl** (1) 104.8 (2) 74.8 (3) 78.1 (4) 70.7 (5) 66.9 **Ara** (1) 105.3 (2) 72.2 (3) 74.2 (4) 64.5 (5) 66.9 **Glc II** (1) 104.5 (2) 74.6 (3) 78.2 (4) 71.4 (5) 78.1 (6) 62.6.

Mass (F.D.) : m/z 1041 [M+Na]$^+$.

Reference

1.   J.-P. Peng, X.-S. Yao, Y. Tezuka and T. Kikuchi, *Phytochemistry*, **41**, 283 (1996).

ASPAFILIOSIDE C

(25S)-5β-Furost-3β,22,26-triol 3-O-{β-D-xylopyranosyl-(1→4)-[α-L-arabinopyranosyl-(1→6)]-β-D-glucopyranoside}-26-O-β-glucopyranoside

Source : *Asparagus filicinus* Buch.-Ham. (Liliaceae)
Mol. Formula : $C_{49}H_{82}O_{22}$
Mol. Wt. : 1022
M.P. : 178~180°C (MeOH)
[α]$_D^{14}$: -36.5° (c=0.09, CHCl₃-MeOH)
Registry No. : [131123-74-5]

IR (KBr) **:** 3400, 1040 cm⁻¹.

PMR (C₅D₅N, 400 MHz) **:** δ 0.71 (s, 3xH-19), 0.85 (s, 3xH-18), 1.01 (d, J=5.4 Hz, 3xH-27), 1.29 (d, J=5.6 Hz, 3xH-21), 4.72 (d, J=7.5 Hz), 4.78 (d, J=7.2 Hz), 4.96 (d, J=6.7 Hz), 5.29 (d, J=7.0 Hz).

CMR (C₅D₅N, 100 MHz) **:** δ C-1) 30.5 (2) 26.9 (3) 74.7 (4) 30.9 (5) 36.9 (6) 26.9 (7) 26.7 (8) 35.5 (9) 40.5 (10) 35.2 (11) 21.1 (12) 40.3 (13) 41.2 (14) 56.3 (15) 32.3 (16) 81.1 (17) 63.8 (18) 16.3 (19) 23.6 (20) 40.5 (21) 16.6 (22) 110.7 (23) 36.9 (24) 28.2 (25) 34.2 (26) 75.2 (27) 17.4 **Glc I** (1) 102.8 (2) 74.3 (3) 76.1 (4) 79.9 (5) 74.9 (6) 68.0 **Xyl** (1) 104.8 (2) 74.7 (3) 78.0 (4) 70.8 (5) 66.9 **Ara** (1) 105.2 (2) 72.3 (3) 74.7 (4) 69.5 (5) 66.9 **Glc II** (1) 104.7 (2) 74.9 (3) 78.3 (4) 71.5 (5) 78.1 (6) 62.7.

Mass (E.I.) **:** *m/z* 331, 259, 199, 157, 139, 115, 97, 69, 59, 55.

Reference

1.   Y. Ding and C.R. Yang, *Acta Pharm. Sin.*, **25**, 509 (1990).

ASPAFILIOSIDE D

(25*S*)-5β-Furost-3β,22,26-triol 3-O-{β-D-xylopyranosyl-(1→2)-[β-D-xylopyranosyl-(1→4)-
β-D-glucopyranoside}-26-O-β-D-glucopyranoside

Source : *Asparagus filicinus* Buch.-Ham. (Liliaceae)
Mol. Formula : $C_{49}H_{82}O_{22}$
Mol. Wt. : 1022
M.P. : 193-195°C
$[\alpha]_D^{20}$ **:** -13° (c=0.27, MeOH)
Registry No. : [618094-39-6]

UV (MeOH) : 223 (log ε, 4.06, 227 (log ε, 3.61) nm.

IR : 3415, 2927, 1635, 1452, 1378, 1152, 1041 cm^{-1}.

PMR (C$_5$D$_5$N, 500 MHz) : δ 0.99 (s, 3xH-18), 1.10 (s, 3xH-19), 1.14 (d, J=6.5 Hz, 3xH-27), 1.15 (m, H-14), 1.22 (m, H-11), 1.25 (m, H-6A, H-7), 1.35 (m, H-6), 1.38 (m, H-12), 1.42 (d, J=6.5 Hz, 3xH-21), 1.54 (m, H-15A), 1.58 (m, H-8), 1.60 (m, H-4A), 1.80 (m, H-24A), 1.85 (m, H-1, H-9), 1.95 (m, H-2, H-4B), 2.05 (m, H-25), 2.10 (m, H-17), 2.15 (m, H-15B), 2.18 (m, H-24B), 2.20 (m, H-5), 2.35 (m, H-20), 2.38 (m, H-23), 3.60 (t, J=8 Hz, H-26), 3.75 (m, H-5A of Xyl II), 3.82 (m, H-5B of Xyl II), 3.85 (m, H-5A of Xyl I), 3.90 (m, H-5 of Glc I), 4.05 (m, H-5 of Glc II), 4.08 (m, H-2 of Xyl II), 4.12 (m, H-2 of Glc II), 4.13 (m, H-2 of Xyl I), 4.23 (m, H-2 of Glc I), 4.25 (m, H-2 of Xyl II), 4.27 (m, H-3 of Glc II), 4.32 (m, H-3 of Xyl I, H-4 of Glc II), 4.35 (m, H-3 of Glc I, H-4 of Xyl I, H-4 of Xyl II), 4.37 (m, H-4 of Glc I), 4.40 (m, H-3), 4.45 (m, H-6A of Glc II), 4.55 (m, H-6A of Glc I, H-53 of Xyl I), 4.62 (m, H-6B of Glc II), 4.88 (d, J=7.6 Hz, H-1 of Glc II), 4.95 (d, J=7.5 Hz, H-1 of Glc I), 5.10 (d, J=7.3 Hz, H-16), 5.15 (d, J=7.7 Hz, H-1 of Xyl II), 5.38 (d, J=6.6 Hz, H-1 of Xyl I).

CMR (C$_5$D$_5$N, 125 MHz) : δ C-1) 30.5 (2) 27.1 (3) 74.9 (4) 31.0 (5) 37.4 (6) 27.3 (7) 27.3 (8) 35.9 (9) 40.6 (10) 35.6 (11) 21.5 (12) 40.7 (13) 41.6 (14) 56.7 (15) 32.7 (16) 81.6 (17) 64.2 (18) 17.0 (19) 24.2 (20) 41.0 (21) 17.1 (22) 111.1 (23) 36.5 (24) 28.6 (25) 34.7 (26) 75.7 (27) 17.8 **Glc I** (1) 101.0 (2) 82.3 (3) 76.8 (4) 80.7 (5) 76.7 (6) 61.8 **Xyl I** (1) 106.5 (2) 75.5 (3) 77.9 (4) 71.4 (5) 67.6 **Xyl II** (1) 105.7 (2) 75.2 (3) 71.1 (4) 76.8 (5) 67.5 **Glc II** (1) 105.3 (2) 75.5 (3) 78.5 (4) 72.0 (5) 78.7 (6) 63.1.

Mass (E.S.I., Negative ion) :: m/z 1021.9 [M-H]⁻.

Reference

1.   Y.F. Li, L.H. Hu, F.C. Lou and J.R. Hong, *Chin. Chem. Lett.*, **14**, 379 (2003).

PSEUDOPROTODIOSCIN
(25R)-Furosa-5,20-dien-3β,26-diol {3-O-α-L-rhamnopyranosyl-(1→2)-[α-L-rhamnopyranosyl-(1→4)]-β-D-glucopyranoside} 26-O-β-D-glucopyranoside

Source : *Trachycarpus wagnerianus* Becc.[1] (Palmae), *Asparagus cochinchinensis* (Loureiro Merill) Merr.[2] (Liliaceae), *Smilax china* L.[3] (Liliaceae), *Smilax menispermoidea*[4] (Liliaceae)
Mol. Formula : C$_{51}$H$_{82}$O$_{21}$
Mol. Wt. : 1030
M.P. : 174-176°C
[α]$_D$: -72.0° (MeOH)[3]
Registry No. : [102115-79-7]

IR (KBr)[1] : 3470-3280 (OH) cm^{-1}.

IR (KBr)[4] : 3500-3280 (OH), 1694 (C=C), 1638 (C=CH), 1044 (C–O), 837, 811 cm^{-1}.

PMR (C$_5$D$_5$N, 400 MHz)[4] : δ 1.61 (s, 3xH-21).

CMR (C$_5$D$_5$N, 100.16 MHz)[4] : δ C-1) 37.5 (2) 30.1 (3) 77.9 (4) 38.9 (5) 140.8 (6) 121.8 (7) 32.4 (8) 31.4 (9) 50.3 (10) 37.1 (11) 21.2 (12) 39.6 (13) 43.3 (14) 54.9 (15) 35.8 (16) 84.5 (17) 64.5 (18) 14.2 (19) 19.5 (20) 103.6 (21) 11.8 (22) 152.4 (23) 34.5 (24) 23.8 (25) 33.5 (26) 74.9 (27) 17.3 **Glc I** (1) 100.2 (2) 78.6 (3) 76.9 (4) 78.0 (5) 77.7 (6) 61.7 **Rha I** (1) 102.0 (2) 72.5 (3) 72.7 (4) 73.9 (5) 69.5 (6) 18.6 **Rha II** (1) 102.9 (2) 72.5 (3) 72.8 (4) 74.1 (5) 70.4 (6) 18.5 **Glc II** (1) 104.9 (2) 75.2 (3) 78.4 (4) 71.7 (5) 78.0 (6) 62.8.

Mass (FAB, Positive ion)[4] : *m/z* 1053 [M+Na]$^+$, 1037 [M+Li]$^+$.

Mass (FAB, Negative ion)[2] : *m/z* 1029 [M-H]$^-$, 883 [M-H-Rha]$^-$, 867 [M-H-Glc]$^-$, 721 [M-H-Glc-Rha]$^-$, 575 [M-H-2xRha]$^-$, 413 [M-H-2xGlc-2xRha]$^-$, 395 [Agl-H$_2$O]$^-$.

Biological Activity: The compound shows inhibitory activity on Cyclic AMP phosphodiesterase with IC$_{50}$=4.7x10^{-5} M.[3]

References

1.   Y. Hirai, S. Sanado, Y. Ida and J. Shoji, *Chem. Pharm. Bull.*, **34**, 82 (1986).

2.   Z.Z. Liang, R. Aquino, F.D. Simone, A. Dini, O. Schettino and C. Pizza, *Planta Med.*, **54**, 344 (1988).

3.   Y. Sashida, S. Kubo, Y. Mimaki, T. Nikaido and T. Ohmoto, *Phytochemistry*, **31**, 2439 (1992).

4.   Y. Ju and Z.-J. Jia, *Phytochemistry*, **31**, 1349 (1992).

TACCA CHANTRIERI SAPONIN 1

(25*S*)-Furosta-5,20(22)-dien-3β,26-diol 3-O-{α-L-rhamnopyranosyl-(1→2)-[α-L-rhamnopyranosyl-(1→3)]-β-D-glucopyranoside}-26-O-β-D-glucopyranoside

Source : *Tacca chantrieri* Andre. (Taccaceae)
Mol. Formula : $C_{51}H_{82}O_{21}$
Mol. Wt. : 1030
$[\alpha]_D^{26}$: -66.0° (c=0.10, CHCl₃-MeOH)

IR (film) : λ_{max} 3389 (OH), 2929 (CH), 1043 cm⁻¹.

PMR (C_5D_5N, 500 MHz) : δ 0.72 (s, 3xH-18), 1.04 (d, *J*=6.7 Hz, 3xH-27), 1.06 (s, 3xH-19), 1.64 (s, 3xH-21), 1.66 (d, *J*=6.2 Hz, 3xH-6 of Rha II), 1.76 (d, *J*=6.1 Hz, 3xH-6 of Rha I), 3.92 (m, *W*½=24.4 Hz, H-3), 4.84 (d, *J*=7.8 Hz, H-1 of Glc I), 4.92 (d, *J*=7.8 Hz, H-1 of Glc II), 5.33 (d, *J*=4.6 Hz, H-6), 5.76 (br s, H-1 of Rha II), 5.88 (br s, H-1 of Rha I).

CMR (C_5D_5N, 125 MHz) : δ C-1) 37.5 (2) 30.1 (3) 77.8 (4) 38.7 (5) 140.8 (6) 121.8 (7) 32.4 (8) 31.4 (9) 50.3 (10) 37.1 (11) 21.3 (12) 39.6 (13) 43.4 (14) 54.9 (15) 34.5 (16) 84.5 (17) 64.5 (18) 14.1 (19) 19.4 (20) 103.5 (21) 11.8 (22) 152.4 (23) 23.7 (24) 23.7 (25) 33.7 (26) 75.2 (27) 17.2 **Glc I** (1) 99.9 (2) 78.3 (3) 87.5 (4) 69.9 (5) 78.1 (6) 62.3 **Rha I** (1) 102.6 (2) 72.5 (3) 72.8 (4) 73.8 (5) 69.9 (6) 18.7 **Rha II** (1) 103.9 (2) 72.6 (3) 72.5 (4) 73.6 (5) 70.6 (6) 18.4 **Glc II** (1) 105.2 (2) 75.2 (3) 78.6 (4) 71.7 (5) 78.5 (6) 62.8.

Mass (FAB, Positive ion) : *m/z* 1053 [M+Na]⁺.

Reference

1.   A. Yokosuka, Y. Mimaki and Sashida, *Natural Medicines*, **56**, 208 (2002).

TACCAOSIDE A

(25*R*)-3,26-Dihydroxyfurost-5,20(22)-diene 3-O-{α-L-rhamnopyranosyl-(1→2)-
[α-L-rhamnopyranosyl-(1→3)]-β-D-glucopyranoside}-26-O-β-D-glucopyranoside

Source : *Tacca plantaginea* (Taccaceae)
Mol. Formula : $C_{51}H_{82}O_{21}$
Mol. Wt. : 1030
M.P. : 179~180°C
[α]$_D^{26}$: -58.5° (c=0.21, MeOH)
Registry No. : [475572-44-2]

PMR : δ 4.76 (d, *J*=6.2 Hz, H-1 of Glc II), 4.84 (d, *J*=7.72 Hz, H-1 of Glc I), 5.69 (s, H-1 of Rha I), 5.80 (s, H-1 of Rha I).

CMR (C₅D₅N, 125.77 MHz) **:** δ C-1) 37.7 (2) 30.6 (3) 78.5 (4) 39.8 (5) 140.9 (6) 121.9 (7) 32.6 (8) 31.6 (9) 50.4 (10) 37.2 (11) 21.4 (12) 38.8 (13) 43.6 (14) 55.1 (15) 34.6 (16) 84.6 (17) 64.6 (18) 14.3 (19) 19.5 (20) 103.6 (21) 11.9 (22) 152.5 (23) 33.8 (24) 23.8 (25) 31.6 (26) 75.3 (27) 17.3 **Glc I** (1) 99.9 (2) 78.0 (3) 87.6 (4) 70.7 (5) 77.9 (6) 62.4 **Rha I** (1) 102.6 (2) 72.6 (3) 72.5 (4) 73.6 (5) 69.9 (6) 18.5 **Rha II** (1) 103.9 (2) 72.8 (3) 72.5 (4) 73.9 (5) 70.7 (6) 18.7 **Glc II** (1) 105.2 (2) 75.3 (3) 78.6 (4) 71.8 (5) 78.0 (6) 62.9.

Mass (FAB, Negative ion) **:** *m/z* 883 [M-146-H]⁻, 867 [M-162-H]⁻, 721 [M-146-162-H]⁻, 575 [M-146-162-146-H]⁻.

Reference

1.   H.Y. Liu and C.X. Chen, *Chin. Chem. Lett.*, **13**, 633 (2002).

TRIBULOSAPONIN A

(25*S*)-5β-Furost-20(22)-en-3β,26-diol 3-*O*-α-L-rhamnopyranosyl-(1→2)-[α-L-rhamnopyranosyl-(1→4)]-β-D-glucopyranoside-26-*O*-β-D-glucopyranoside

Source : *Tribulus terrestris* Linn. (Zygophyllaceae)
Mol. Formula : $C_{51}H_{84}O_{21}$
Mol. Wt. : 1032
[α]$_D^{25}$: -73° (c=0.004, MeOH)
Registry No. : [311310-49-3]

IR (KBr) : 3404, 2948, 2395, 2340, 1649, 1455 cm^{-1}.

PMR (C$_5$D$_5$N, 500 MHz) : δ 0.70 (s, 3xH-18), 1.05 (d, *J*=6.5 Hz, 3xH-27), 1.08 (s, 3xH-19), 1.61 (d, *J*=6.1 Hz, 3xH-6 of Rha I), 1.64 (s, 3xH-21), 1.75 (d, *J*=6.1, 3xH-6 of Rha II), 2.49 (d, *J*=10.0 Hz, H-17), 3.49 (dd, *J*=7.1, 9.1 Hz, H-26A), 4.06 (dd, *J*=8.0, 11.5 Hz, H-26B), 4.24 (H-3), 4.83 (d, *J*=7.6 Hz, H-1 of Glc), 4.86 (H-16), 4.86 (d, *J*=7.3 Hz, H-1 of Glc II), 5.88 (br s, H-1 of Rha II), 6.55 (br s, H-1 of Rha I).

CMR (C$_5$D$_5$N, 125 MHz) : δ C-1) 31.1 (2) 27.0 (3) 76.3 (4) 31.1 (5) 35.4 (6) 27.0 (7) 31.1 (8) 37.4 (9) 40.5 (10) 35.4 (11) 21.5 (12) 40.3 (13) 44.1 (14) 55.0 (15) 34.6 (16) 84.8 (17) 64.9 (18) 14.6 (19) 24.0 (20) 103.0 (21) 12.0 (22) 152.5 (23) 23.8 (24) 31.6 (25) 33.8 (26) 75.4 (27) 17.3 **Glc I** (1) 102.0 (2) 78.7 (3) 78.6 (4) 77.1 (5) 77.1 (6) 61.9 **Rha I** (1) 101.6 (2) 72.6 (3) 72.9 (4) 74.0 (5) 69.6 (6) 18.5 **Rha II** (1) 102.4 (2) 72.8 (3) 72.9 (4) 74.1 (5) 70.5 (6) 18.8 **Glc II** (1) 105.3 (2) 75.3 (3) 78.7 (4) 71.8 (5) 78.6 (6) 63.0.

Mass (HRESIFT) : *m/z* 1056.5483 [M+Na]$^+$, 1033.5636 [(M)$^+$, calcd. for 1033.2153].

Reference

1.   E. Bedir and I.A. Khan, *J. Nat. Prod.*, **63**, 1699 (2000).

TUBEROSIDE T

(5α)-Furost-20(22)-en-3β,26-diol 3-O-{α-L-rhamnopyranosyl-(1→2)-[α-L-rhamnopyranosyl-(1→4)]-β-D-glucopyranoside}-β-D-glucopyranoside

Source : *Allium tuberosum* L. (Liliaceae)
Mol. Formula : C$_{51}$H$_{84}$O$_{21}$
Mol. Wt. : 1032
[α]$_D^{24}$: -29.4° (c=0.33, MeOH)
Registry No. : [651306-85-3]

PMR (C$_5$D$_5$N, 400 MHz) : δ 0.77 (s, 3xH-18), 0.90 (s, 3xH-19), 1.11 (d, *J*=6.5 Hz, 3xH-27), 1.69 (d, *J*=6.3 Hz, 3xH-6 of Rha I, 3xH-6 of Rha II), 1.70 (s, 3xH-21), 2.56 (d, *J*=10.0 Hz, H-17), 3.55 (dd, *J*=7.0, 9.0 Hz, H-26A), 3.80 (m, H-5 of Glc I), 4.00 (m, H-3), 4.01 (m, H-5 of Glc I), 4.09 (t, *J*=8.0 Hz, H-2 of Glc II), 4.11 (m, H-26B), 4.19 (m, H-6A of

Glc I), 4.28 (m, H-3 of Glc II, H-4 of Glc II), 4.30 (m, H-2 of Glc I, H-3 of Glc I), 4.34 (m, H-6B of Glc I), 4.39 (m, H-4 of Rha I, H-4 of Rha II), 4.42 (m, H-4 of Glc I), 4.44 (m, H-5A of Glc II), 4.59 (m, H-3 of Rha I, H-3 of Rha II), 4.60 (m, H-6B of Glc II), 4.76 (s, H-2 of Rha I, H-2 of Rha II), 4.88 (m, H-16), 4.90 (d, J=7.6 Hz, H-1 of Glc II), 4.96 (m, H-5 of Rha II), 5.04 (d, J=7.4 Hz, H-1 of Glc I), 5.90 (s, H-1 of Rha II), 6.42 (s, H-1 of Rha I).

CMR (C_5D_5N, 100 MHz) : δ C-1) 37.5 (2) 30.1 (3) 77.1 (4) 34.6 (5) 44.8 (6) 29.1 (7) 32.8 (8) 35.2 (9) 54.6 (10) 36.1 (11) 21.6 (12) 40.1 (13) 43.9 (14) 55.0 (15) 34.6 (16) 84.7 (17) 64.8 (18) 14.5 (19) 12.6 (20) 103.8 (21) 12.0 (22) 152.6 (23) 23.8 (24) 31.6 (25) 33.9 (26) 75.4 (27) 17.4 **Glc I** (1) 100.0 (2) 78.3 (3) 78.1 (4) 78.9 (5) 77.4 (6) 61.6 **Rha I** (1) 102.3 (2) 72.6 (3) 72.9 (4) 74.0 (5) 69.7 (6) 18.8 **Rha II** (1) 103.1 (2) 72.7 (3) 73.0 (4) 74.2 (5) 70.6 (6) 18.7 **Glc II** (1) 105.3 (2) 75.4 (3) 78.7 (4) 71.9 (5) 78.6 (6) 63.0.

Mass (FAB, Positive ion) : m/z 1033 [M+H]$^+$.

Reference

1.   S. Sang, S. Mao, A. Lao, Z. Chen and C.-T. Ho, *Food Chemistry*, **83**, 499 (2003).

NOLINA RECURVATA SAPONIN 7
22ξ-Methoxyforosta-5,25(27)-diene-1β,3β,26-triol 1-O-{α-L-rhamnopyranosyl-(1→2)-[β-D-xylopyranosyl-(1→3)]-α-L-arabinopyranoside}-26-O-β-D-glucopyranoside

Source : *Nolina recurvata*[1] (Agavaceae), *Ruscus aculeatus*[2] L. (Liliaceae)
Mol. Formula : $C_{50}H_{80}O_{22}$
Mol. Wt. : 1032
[α]$_D^{26}$: -46.6° (c=0.31, MeOH)[1]
Registry No. : [180161-88-0]

IR (KBr)[1] : 3420 (OH), 2900 (CH), 1440, 1365, 1035, 970, 900, 825, 800, 770, 690 cm^{-1}.

PMR (C$_5$D$_5$N, 400 MHz)[1] : δ 0.83 (s, 3xH-18), 1.13 (d, *J*=6.7 Hz, 3xH-21), 1.42 (s, 3xH-19), 1.73 (d, *J*=6.0 Hz, 3xH-6 of Rha), 3.25 (s, OC*H*$_3$), 4.74 (d, *J*=7.3 Hz, H-1 of Ara), 4.91 (overlapping with H$_2$O signal, H-1 of Glc), 4.98 (d, *J*=7.4 Hz, H-1 of Xyl), 5.05 and 5.34 (br s, 2xH-27), 5.60 (br d, *J*=5.4 Hz, H-6), 6.32 (br s, H-1 of Rha).

CMR (C$_5$D$_5$N, 100 MHz)[1] : δ C-1) 83.7 (2) 37.4 (3) 68.3 (4) 43.9 (5) 139.6 (6) 124.7 (7) 32.0 (8) 33.1 (9) 50.4 (10) 42.9 (11) 24.0 (12) 40.2 (13) 40.4 (14) 56.8 (15) 32.4 (16) 81.5 (17) 64.2 (18) 16.7 (19) 15.1 (20) 40.5 (21) 16.1 (22) 112.4 (23) 31.6 (24) 28.1 (25) 146.9 (26) 72.0 (27) 111.1 (OCH$_3$) 47.3 **Ara** (1) 100.4 (2) 74.2 (3) 84.5 (4) 69.6 (5) 67.1 **Rha** (1) 101.8 (2) 72.5 (3) 72.6 (4) 74.1 (5) 69.6 (6) 19.1 **Xyl** (1) 106.4 (2) 74.6 (3) 78.3 (4) 71.0 (5) 66.9 **Glc** (1) 103.9 (2) 75.2 (3) 78.6 (4) 71.8 (5) 78.5 (6) 62.9.

Mass (FAB, Negative ion)[1] : *m/z* 1032 [M]⁻, 901 [M-Xyl]⁻, 886 [M-Rha]⁻, 754 [M-Xyl-Rha]⁻.

Biological Activity : The compound shows inhibitory activity on cyclic AMP phosphodiesterase with IC$_{50}$ (45.6x10⁻⁵ M).[1]

References

1.   Y. Mimaki, Y. Takaashi, M. Kuroda, Y. Sashida and T. Nikaido, *Phytochemistry*, **42**, 1609 (1996).

2.   Y. Mimaki, M. Kuroda, A. Kameyama, A. Yokosuka and Y. Sashida, *Chem. Pharm. Bull.*, **46**, 298 (1998).

COSTUS SPIRALIS SAPONIN 1

(25*R*)-Furost-5-ene-3β,22ξ,26-triol 3-O-{β-D-apiofuranosyl-(1→2)-[α-L-rhamnopyranosyl-(1→4)]-β-D-glucopyranoside}-26-O-β-D-glucopyranoside}

Source : *Costus spiralis* Rosc. (Costaceae)
Mol. Formula : C$_{50}$H$_{82}$O$_{22}$
Mol. Wt. : 1034
M.P. : 216-218°C
[α]$_D^{25}$: -106° (c=0.1, MeOH)

IR (KBr) : 3430 (OH), 1050 (C-O), 913, 838, 813, 638 cm^{-1} (intensity 913 < 838, 25R-furostanol).

PMR (CDCl$_3$, 500 MHz) : δ 0.85 (s, 3xH-18), 0.98 (d, J=6.6 Hz, 3xH-27), 1.10 (s, 3xH-19), 1.22 (d, 3xH-21), 1.78 (d, 3xH-6 of Rha), 4.82 (d, J=7.8 Hz, H-1 of Glc II), 4.98 (d, J=7.7 Hz, H-1 of Glc I), 5.30 (br s, H-6), 5.95 (d, J=3.5 Hz, H-1 of Api), 6.30 (br s, H-1 of Rha).

CMR (C$_5$D$_5$N, 50 MHz) : δ C-1) 37.2 (2) 29.7 (3) 77.9 (4) 38.5 (5) 140.3 (6) 121.5 (7) 32.2 (8) 31.2 (9) 50.2 (10) 36.9 (11) 20.9 (12) 39.6 (13) 40.7 (14) 56.5 (15) 31.9 (16) 80.9 (17) 63.7 (18) 16.3 (19) 18.5 (20) 40.5 (21) 16.5 (22) 110.7 (23) 30.3 (24) 27.9 (25) 33.9 (26) 74.9 (27) 16.9 **Glc I** (1) 100.1 (2) 77.8 (3) 78.1 (4) 78.8 (5) 76.9 (6) 61.2 **Api** (1) 110.5 (2) 77.3 (3) 79.9 (4) 74.8 (5) 64.3 **Rha** (1) 102.7 (2) 72.3 (3) 72.2 (4) 73.7 (5) 69.3 (6) 18.2 **Glc II** (1) 104.5 (2) 74.8 (3) 78.1 (4) 71.8 (5) 78.8 (6) 62.7.

Mass (L.S.I., Negative ion) : m/z 1033 [M-H]$^-$.

Biological Activity: Anti-inflammatory.

Reference

1.   B.P. de Silva and J.P. Parente, *Z. Naturforsch.*, **59C**, 81 (2004).

COSTUS SPIRALIS SAPONIN 2
(25R)-Furost-5-ene-3β,22ξ,26-triol 3-O-{α-L-rhamnopyranosyl-(1→2)-[β-D-apiofuranosyl-(1→4)]-β-D-glucopyranoside}-26-O-β-D-glucopyranoside

Source : *Costus spiralis* Rosc. (Costaceae)
Mol. Formula : C$_{50}$H$_{82}$O$_{22}$
Mol. Wt. : 1034
M.P. : 218-220°C
$[\alpha]_D^{25}$: -110° (c=0.1, MeOH)

IR (KBr) : 3430 (OH), 1050 (C-O), 913, 838, 813, 638 cm^{-1} (intensity 913 < 838 cm^{-1} (25*R*-furastanol).

PMR (C$_5$D$_5$N, 200 MHz) : δ 0.85 (s, 3xH-18), 0.98 (d, *J*=6.6 Hz, 3xH-27), 1.10 (s, 3xH-19), 1.20 (d, *J*=6.8 Hz, 3xH-21), 1.76 (d, *J*=6.3 Hz, 3xH-6 of Rha), 4.84 (d, *J*=7.8 Hz, H-1 of Glc II), 4.96 (d, *J*=7.7 Hz, H-1 of Glc I), 5.30 (br s, H-6), 5.80 (br s, H-1 of Rha), 5.90 (d, *J*=3.5 Hz, H-1 of Api).

CMR (C$_5$D$_5$N, 50 MHz) : δ C-1) 37.1 (2) 29.5 (3) 77.9 (4) 38.5 (5) 140.3 (6) 121.4 (7) 32.2 (8) 31.3 (9) 49.9 (10) 36.7 (11) 20.7 (12) 39.5 (13) 40.5 (14) 56.3 (15) 31.7 (16) 80.7 (17) 63.5 (18) 16.2 (19) 18.3 (20) 40.3 (21) 16.3 (22) 110.5 (23) 30.1 (24) 27.7 (5) 33.7 (26) 74.7 (27) 16.7 **Glc I** (1) 100.0 (2) 77.8 (3) 78.0 (4) 78.8 (5) 76.7 (6) 61.2 **Rha** (1) 101.8 (2) 72.2 (3) 72.3 (4) 73.7 (5) 69.3 (6) 18.2 **Api** (1) 110.7 (2) 77.2; (3) 79.6 (4) 74.8 (5) 64.2 **Glc II** (1) 104.7 (2) 74.7 (3) 78.1 (4) 71.9 (5) 78.7 (6) 62.3.

Mass (L.S.I., Negative ion) : *m/z* 1033 [M-H]$^-$.

Bioloical Activity: Anti-inflammatory.

Reference

1.   B.P. de Silva and J.P. Parente, *Z. Naturforsch.*, **59C**, 81 (2004).

INDIOSIDE C

Furost-5-en-3β,22ξ ,26-triol 3-O-{α-L-rhamnopyranosyl-(1→2)-[β-D-xylopyranosyl-(1→3)]-β-D-galactopyranoside}-26-O-β-D-glucopyranoside

Source : *Solanum indicum* L. (Solanaceae)
Mol. Formula : C$_{50}$H$_{82}$O$_{22}$
Mol. Wt. : 1034
[α]$_D^{26}$: -37.0° (c=0.29, Pyridine)
Registry No. : [185332-90-5]

PMR (C$_5$D$_5$N, 400 MHz) : δ 0.85 (s, 3xH-18), 1.00 (d, J=6.6 Hz, 3xH-27), 1.06 (s, 3xH-19), 1.30 (d, J=7.0 Hz, 3xH-21), 1.70 (d, J=6.3 Hz, 3xH-6 of Rha), 4.85 (d, J=8.0 Hz, H-1 of Glc), 4.87 (d, J=8.1 Hz, H-1 of Gal), 5.06 (d, J=7.3 Hz, H-1 of Xyl), 5.31 (br s, H-6), 6.31 (s, H-1 of Rha).

CMR (CD$_3$OD, 100 MHz) : δ C-1) 37.5 (2) 29.9 (3) 78.3^a (4) 38.7 (5) 140.8 (6) 121.7 (7) 32.3 (8) 31.6 (9) 50.2 (10) 37.1 (11) 20.9 (12) 40.2 (13) 40.9 (14) 56.3 (15) 32.4 (16) 82.2 (17) 64.1 (18) 16.4 (19) 19.3 (20) 40.6 (21) 16.1 (22) 110.6 (23) 37.1 (24) 28.3 (25) 34.3 (26) 75.0 (27) 17.2.**Gal** (1) 100.4 (2) 76.4 (3) 85.0 (4) 70.3 (5) 75.2 (6) 62.3 **Xyl** (1) 102.2 (2) 72.5 (3) 72.8 (4) 74.1 (5) 69.4 (6) 18.6 **Rha** (1) (1) 106.8 (2) 74.7 (3) 77.4 (4) 71.0 (5) 67.1 **Glc** (1) 104.9 (2) 75.2 (3) 78.5^a (4) 71.6 (5) 78.5^a (6) 62.8.

Mass (FAB, Negative ion) : m/z 1033 [M-H]$^-$.

Reference

1. S. Yahara, T. Nakamura, Y. Someya, T. Matsumoto, T. Yamashita and T. Nohara, *Phytochemistry* **43**, 1319 (1996).

PARISAPONIN

Furost-5-en-3β,22ξ,26-triol 3-O-{α-L-rhamnopyranosyl-(1→2)-[α-L-arabinopyranosyl-(1→4)]-β-D-glucopyranoside}-26-O-β-D-glucopyranoside}

Source : *Paris polyphylla* var. *yunnonensis* (Liliaceae)
Mol. Formula : $C_{50}H_{82}O_{22}$
Mol. Wt. : 1034
$[\alpha]_D^{26}$ **:** -41.3° (c=1.1, MeOH)
Registry No. : [561007-63-4]

IR (KBr) : 3430 (OH), 2938 (CH_2), 1074 (–O–), 1055 (–O–), 756 (–CH_2–) cm^{-1}.

PMR (C_5D_5N, 500 MHz) : δ 0.89 (s, 3xH-18)[a], 1.05 (s, 3xH-19)[a], 0.98 (d, *J*=6.7 Hz, 3xH-27), 1.30 (d, *J*=6.7 Hz, 3xH-21), 1.71 (d, *J*=6.1 Hz, 3xH-6 of Rha), 3.59 (dd, *J*=5.0, 8.0, 9.2 Hz, H-26A), 3.92 (m, H-26B), 4.74 (d, H-1 of Glc II), 4.88 (d, *J*=7.3 Hz, H-1 of Glc I), 5.31 (br s, H-6), 5.81 (br s, H-1 of Ara(f), 6.15 (br s, H-1 of Rha).

CMR (C_5D_5N, 125 MHz) : δ C-1) 37.4 (2) 30.1 (3) 78.2 (4) 38.9 (5) 140.6 (6) 121.5 (7) 22.3 (8) 31.6 (9) 50.3 (10) 37.0 (11) 21.0 (12) 39.9 (13) 40.7 (14) 56.5 (15) 32.4 (16) 80.9 (17) 63.7 (18) 16.4 (19) 19.3 (20) 40.5 (21) 16.3 (22) 110.4 (23) 37.0 (24) 8.2 (25) 34.1 (26) 75.0 (27) 17.3 **Glc I** (1) 100.0 (2) 78.3 (3) 77.1 (4) 76.4 (5) 77.7 (6) 62.4 **Rha** (1) 101.5 (2) 72.5 (3) 72.1 (4) 73.9 (5) 69.2 (6) 18.5 **Ara(f)** 109.4 (2) 82.4 (3) 77.3 (4) 86.5 (5) 61.3 **Glc II** (1) 104.5 (2) 74.9 (3) 78.0 (4) 71.5 (5) 78.6 (6) 62.7.

Mass (FAB, Positive ion, H.R.) : 1057.5179 [(M+Na)$^+$, requires 1057.5195].

Mass (FAB, Negative ion) : *m/z* 1033 [M-H]$^-$.

Biological Activity : The compound inhibited EtOH-induced gastric mucosal injury by 15.6% at a dose of 5.0 mg/kg p.o.

Reference

1.　H. Matsuda, T. Morikawa, M. Yoshikawa, *Jpn. Kokai Tokkyo Koho* JY 2004 143125 (C) C07J 71/00 *Chem. Abstr.* **140**, 41291s (2004).

POLYPHYLLIN H

(25*R*)-Furost-5-en-3β,22α,26-triol 3-O-{α-L-rhamnopyranosyl-(1→3)-[α-L-arabinofuranosyl-(1→4)]-β-D-glucopyranoside}-26-O-β-D-glucopyranoside

Source : *Paris polyphylla* Sm. (Liliaceae)
Mol. Formula : C$_{50}$H$_{82}$O$_{22}$
Mol. Wt. : 1034
Registry No. : [76296-76-9]

Peracetate :

IR (KBr) : 3500 (OH), 1750, 1230 (COOMe) cm^{-1}.

PMR (CDCl$_3$, 60.0 MHz) : δ 1.9-2.2 (12xCOC*H*$_3$).

Reference

1.　S.B. Singh, R.S. Thakur and H.-R. Schulten, *Phytochemistry*, **21**, 2079 (1982).

CHINENOSIDE I

3β,22ξ,26-Trihydroxy-25(*R*)-5α-furostan-6-one 3-O-β-xylopyranosyl-(1→4)-[α-arabinopyranosyl-(1→6)]-
β-D-glucopyranoside-26-O-β-D-glucopyranoside

Source : *Allium chinense* G. Don (Liliaceae)
Mol. Formula : $C_{49}H_{80}O_{23}$
Mol. Wt. : 1036
$[\alpha]_D^{25}$: -42.2° (c=0.66, Pyridine)
Registry No. : [123961-66-2]

PMR (C_5D_5N, 300 MHz) : δ 0.65 (3H, s), 0.84 (s, CH_3), 0.99 (d, *J*=6.2 Hz, sec. CH_3), 1.35 (d, *J*=7.0 Hz, sec. CH_3), 4.83 (d, *J*=7.7 Hz), 4.95 (d, *J*=7.7 Hz), 5.01 (d, *J*=7.7 Hz), 5.50 (d, *J*=7.7 Hz, 4 x anomeric H).

CMR (C_5D_5N, 75 MHz) : δ C-1) 36.7 (2) 29.4 (3) 77.0 (4) 27.0 (5) 56.5ᵃ (6) 209.7 (7) 46.8 (8) 37.4 (9) 53.7 (10) 41.4 (11) 21.5 (12) 39.7 (13) 40.9 (14) 56.4ᵃ (15) 32.1 (16) 80.8 (17) 63.8 (18) 16.6 (19) 13.1 (20) 40.6 (21) 16.4 (22) 110.7 (23) 37.1 (24) 28.4 (25) 34.3 (26) 75.3 (27) 17.4 **Glc I** (1) 102.1 (2) 74.9ᵇ (3) 78.4ᶜ (4) 79.9 (5) 75.2ᵇ (6) 68.1 **Xyl** (1) 105.1ᵈ (2) 75.0ᵇ (3) 76.3 (4) 71.1 (5) 67.3 **Ara** (1) 104.9ᵈ (2) 72.5 (3) 74.9ᵇ (4) 69.8 (5) 67.3 **Glc II** (1) 105.7ᵈ (2) 74.5 (3) 78.6ᶜ (4) 71.7 (5) 78.4ᶜ (6) 62.8.

Mass (F.D.) : *m/z* 1057 [M-H_2O+K]⁺, 1041 [M-H_2O+Na]⁺, 925 [1057-pentose]⁺, 909 [M+Na-H_2O-pentose]⁺.

Reference

1.   H. Matsuura, T. Ushiroguchi, Y. Itakura and T. Fuwa, *Chem. Pharm. Bull.*, **37** 1390 (1989).

HELONIOPSIS ORIENTALIS SAPONIN 1

17-Dehydrokryptogenin 3-O-{α-L-rhamnopyranosyl-(1→2)-[α-L-rhamnopyranosyl-(1→4)]-β-D-glucopyranoside}-26-O-[β-D-glucopyranoside]

Source : *Heloniopsis orientalis* (Thunb.) C. Tanaka (Liliaceae)
Mol. Formula : $C_{51}H_{80}O_{22}$
Mol. Wt. : 1044
$[\alpha]_D^{22}$ **:** -94.0° (c=1.0, MeOH)
Registry No. : [121795-62-8]

UV (MeOH) : λ_{max} 247 (ε, 3580) nm.

IR (KBr) **:** 3400 (OH), 1700 (C=O), 1630 (C=C) cm^{-1}.

CMR (C_5D_5N, 200 MHz) **:** δ C-1) 37.1 (2) 30.0 (3) 77.8 (4) 38.8 (5) 140.9 (6) 121.3 (7) 31.7 (8) 30.8 (9) 49.9 (10) 37.0 (11) 20.9 (12) 38.7 (13) 43.4 (14) 50.5 (15) 36.0 (16) 210.4 (17) 142.5 (18) 15.7 (19) 19.3 (20) 145.6 (21) 16.7 (22) 205.6 (23) 37.9 (24) 27.9 (25) 33.3 (26) 75.0 (27) 17.4 **Glc I** (1) 100.2 (2) 78.8 (3) 76.8 (4) 78.0 (5) 77.8 (6) 61.3 **Rha I** (1) 101.8 (2) 72.6 (3) 72.4 (4) 73.7 (5) 69.4 (6) 18.5 **Rha II** (1) 102.8 (2) 72.7 (3) 72.4 (4) 74.0 (5) 70.3 (6) 18.4 **Glc II** (1) 104.7 (2) 75.0 (3) 78.3 (4) 71.4 (5) 78.5 (6) 62.8.

ORD (c=0.062, MeOH) : [M] + 4600 (322 nm) peak, -5200 (364 nm).

Mass (FAB, Positive ion) **:** *m/z* 1067 [M+Na]$^+$.

Reference

1.   K. Nakano, K. Murakami, Y. Takaishi, T. Tomimatsu and T. Nohara, *Chem. Pharm. Bull.*, **37**, 116 (1989).

NAMONIN E

(25*R*)-Furosta-5,20(22)-dien-1β,3β,26-triol 1-O-{α-L-rhamnopyranosyl-(1→2)-[β-D-xylopyranosyl-(1→3)]-4-O-acetyl-α-L-arabinopyranoside}-26-O-β-D-fucopyranoside

Source : *Dracaena angustifolia* Roxb. (Dracaenaceae)
Mol. Formula : $C_{51}H_{80}O_{22}$
Mol. Wt. : 1044
$[α]_D^{25}$: -49.4° (c=0.5, MeOH)
Registry No. : [352661-78-0]

PMR (C_5D_5N, 400 MHz) : δ 0.80 (s, 3xH-18), 1.02 (d, *J*=6.5 Hz, 3xH-27), 1.37 (s, 3xH-19), 1.68 (s, 3xH-21), 1.74 (d, *J*=6.3 Hz, 3xH-6 of Rha), 2.05 (s, OCOC*H*₃ of Ara), 3.62 (m, H-26A), 3.63 (m, H-5A of Xyl), 3.67 (m, H-5A of Ara), 3.75 (dd, *J*=11.0, 3.6 Hz, H-1), 3.83 (m, H-3), 3.93 (m, H-26B), 3.94 (m, H-5 of Glc), 3.97 (m, H-2 of Xyl), 4.03 (m, H-2 of Glc), 4.09 (t, *J*=8.6 Hz, H-3 of Xyl), 4.14 (m, H-4 of Xyl), 4.21 (m, H-2 of Ara), 4.22 (m, H-4 of Glc), 4.24 (m, H-3 of Glc), 4.28 (m, H-5B of Ara, H-4 of Rha), 4.30 (m, H-5B of Xyl), 4.32 (m, H-3 of Ara), 4.40 (dd, *J*=10.8, 2.0 Hz, H-6A of Glc), 4.54(dd, *J*=9.0, 3.5 Hz, H-3 of Rha), 4.57 (dd, *J*=10.8, 4.5 Hz, H-6B of Glc), 4.70 (m, H-5 of Rha), 4.78 (m, H-2 of Rha), 4.78 (m, H-16), 4.83 (d, *J*=8.0 Hz, H-1 of Glc), 4.86 (d, *J*=7.0 Hz, H-1 of Ara), 4.97 (d, *J*=7.2 Hz, H-1 of Xyl), 5.44 (dt, *J*=6.3, 9.0 Hz, H-4 of Ara), 5.59 (d, *J*=5.2 Hz, H-6), 6.31 (br s, H-1 of Rha).

CMR (C_5D_5N, 100 MHz) : δ C-1) 83.6 (2) 37.1 (3) 68.1 (4) 43.7 (5) 139.2 (6) 124.8 (7) 31.9 (8) 32.9 (9) 50.3 (10) 42.7 (11) 24.5 (12) 40.2 (13) 43.2 (14) 55.2 (15) 34.7 (16) 84.5 (17) 64.7 (18) 14.5 (19) 15.0 (20) 103.8 (21) 11.8 (22) 152.3 (23) 23.7 (24) 31.5 (25) 33.5 (26) 75.0 (27) 17.3 **Ara** (1) 99.8 (2) 76.4 (3) 82.5 (4) 71.2 (5) 63.3 (OCOC*H*₃) 21.0 (OCOCH₃) 170.1 **Rha** (1) 101.9 (2) 72.4 (3) 72.5 (4) 74.1 (5) 69.9 (6) 19.2 **Xyl** (1) 105.4 (2) 75.2 (3) 78.4 (4) 71.0 (5) 67.3 **Glc** (1) 104.9 (2) 75.1 (3) 78.6 (4) 71.8 (5) 78.5 (6) 62.9.

Mass (FAB, Negative ion) : *m/z* 1067.8 [M+Na]⁻.

Mass (FAB, Negative ion, H.R.) : *m/z* 1067.5052 [(M+Na)⁻, calcd. for 1067.5038].

Reference

1.   Q.L. Tran, Y. Tezuka, A.H. Banskata, Q.K. Tran, I. Saiki and S. Kadota, *J. Nat. Prod.*, **64**, 1127 (2001).

TRIQUETROSIDE B

Furost-5,20(22)-dien-1β,3β,26-triol 3-O-[α-L-rhamnopyranosyl-(1→2)-β-D-glucopyranoside[-26-O-[α-L-rhamnopyranosyl-(1→2)-β-D-glucopyranoside]

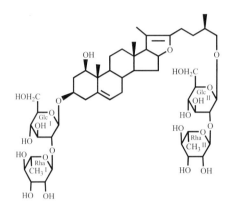

Source : *Allium triquetrum* L. (Alliaceae, Liliaceae)
Mol. Formula : $C_{51}H_{80}O_{22}$
Mol. Wt. : 1044
[α]$_D^{25}$: -20.2° (c=0.1, MeOH)
Registry No. : [628727-98-0]

PMR (CD₃OD, 500 MHz) : δ 0.83 (s, 3xH-18), 0.92 (d, J=6.6 Hz, 3xH-27), 1.02 (s, 3xH-19), 1.12* (H-9), 1.14 (m, H-14), 1.20* (H-12B), 1.25* (H-15A), 1.52 (m, H-11B), 1.53* (H-24A), 1.54* (H-8), 1.65 (m, H-24B), 1.71* (H-2A, H-12B), 1.72 (m, H-25), 1.88 (s, 3xH-21), 1.95* (H-7A), 1.96* (H-7B), 1.98* (H15B), 2.06* (H-2B), 2.14* (H-23A), 2.19* (H-23B), 2.25* (H-4A), 2.26 (dd, J=10.5, 2.5 Hz, H-11B), 2.40* (H-4B), 2.43 (d, J=5.5 Hz, H-17), 3.26* (H-1), 3.33* (H-26A), 3.49* (H-3), 3.81 (dd, J=9.5, 3.9 Hz, H-26B), 4.59 (q, J=5.9 Hz, H-16), 5.50 (br d, J=3.2 Hz, H-6). For signals of sugar protons see Triquetoroside A1. * overlapped signals.

CMR (CD₃OD, 125 MHz) : δ C-1) 74.0 (2) 43.4 (3) 82.4 (4) 41.9 (5) 143.5 (6) 128.4 (7) 35.1 (8) 36.1 (9) 54.8 (10) 44.4 (11) 27.4 (12) 43.6 (13) 48.8 (14) 58.4 (15) 33.9 (16) 86.8 (17) 67.7 (18) 17.6 (19) 16.5 (20) 105.4 (21) 14.3 (22) 154.8 (23) 37.0 (24) 26.5 (25) 36.5 (26) 74.5 (27) 19.7. For signals of sugar carbons see Triquetroside A1.

Mass (FAB, Negative ion, H.R.) : *m/z* 1043.5050 [(M-H)⁻, requires 1043.5040].

Reference

1.   G. Corea, E. Tattorusso and V. Lanzotti, *J. Nat. Prod.*, **66**, 1405 (2003).

RUSCUS ACULEATUS SAPONIN 9

22ξ-Methoxyfurosta-5,25(27)-diene-1β,3β,26-triol-1-O-{α-L-rhamnopyranosyl-(1→2)-(3-O-acetyl-4-O-sulfo)-
α-L-arabinopyranoside}-26-O-β-D-glucopyranoside monosodium salt

Source : *Ruscus aculeatus* (Liliaceae)
Mol. Formula : $C_{47}H_{73}O_{22}SNa$
Mol. Wt. : 1044
$[\alpha]_D^{26}$: -34.0° (c=0.10, MeOH)
Registry No. : [205191-10-2]

IR (KBr) : 3420 (OH), 2930 (CH), 1725 (C=O),1255, 1045 cm^{-1}.

PMR (C_5D_5N+CD$_3$OD, 400 MHz) : δ 0.82 (s, 3xH-18), 1.10 (d, J=6.9 Hz, 3xH-21), 1.36 (s, 3xH-19), 1.72 (d, J=6.1 Hz, 3xH-6 of Rha), 2.06 (s, OCOCH_3), 3.24 (s, OCH_3), 4.83 (d, J=6.7 Hz, H-1 of Ara), 4.87 (d, J=7.8 Hz, H-1 of Glc), 5.05 and 5.31 (each 1H, br s, 2xH-27), 5.43 (br d, J=3.7 Hz, H-4 of Ara), 5.46 (dd, J=8.3, 3.7 Hz, H-3 of Ara), 5.58 (br d, J=5.6 Hz, H-6), 5.72 (br s, H-1 of Rha).

CMR (C_5D_5N + CD$_3$OD, 100 MHz) : δ C-1) 83.8 (2) 37.1 (3) 67.9 (4) 43.6 (5) 139.2 (6) 124.9 (7) 31.9 (8) 33.0 (9) 50.2 (10) 42.8 (11) 23.9 (12) 39.7 (13) 40.4 (14) 56.5 (15) 32.3 (16) 81.4 (17) 63.9 (18) 16.4 (19) 14.8 (20) 40.4 (21) 16.0 (22) 112.3 (23) 31.6 (24) 28.0 (25) 146.8 (26) 72.0 (27) 111.0 (OCH_3) 47.2 **Ara** (1) 99.5 (2) 71.5 (3) 74.5 (4) 73.4 (5) 64.5 **Rha** (1) 101.8 (2) 72.0 (3) 72.0 (4) 73.7 (5) 70.0 (6) 18.7 **Glc** (1) 103.6 (2) 74.9 (3) 78.3 (4) 71.5 (5) 78.3 (6) 62.6 (OCH_3) 170.9, 21.0.

Mass (FAB, Negative ion) : *m/z* 1021 [M-Na-H]$^-$.

Reference

1.   Y. Mimaki, M. Kuroda, A. Kameyama, A. Yokosuka and Y. Sashida, *Chem. Pharm. Bull.*, **46**, 298 (1998).

ACULEATISIDE A

Nuatigenin 3-O-{α-L-rhamnopyranosyl-(1→2)-[α-L-rhamnopyranosyl-(1→4)]-β-D-glucopyranoside}-26-O-β-D-glucopyranoside

Source : *Solanum aculeatissimum* (Solanaceae)
Mol. Formula : $C_{51}H_{82}O_{22}$
Mol. Wt. : 1046
M.P. : 196-204° (decomp.)
$[\alpha]_D^{22}$: -96.7° (c=1.08, Pyridine)
Registry No. : [86848-73-9]

IR (KBr) : 3400 (OH), 919, 870, 840, 818 (spiroketal) cm^{-1}.

PMR (C_5D_5N, 100 MHz) : δ 0.81 (s, 3xH-18), 1.05 (s, 3xH-19), 1.08 (d, J=6.0 Hz, 3xH-21), 1.40 (s, 3xH-27), 1.62 (d, J=6.0 Hz, 3xH-6 of Rha), 1.76 (d, J=6.0 Hz, 3xH-6 of Rha).

CMR (C_5D_5N, 50.01 MHz) : δ C-1) 37.5 (2) 30.1 (3) 78.1 (4) 40.5 (5) 140.7 (6) 120.1 (7) 32.2 (8) 31.6 (9) 50.2 (10) 37.0 (11) 21.0 (12) 38.9 (13) 39.8 (14) 56.4 (15) 32.2 (16) 80.9 (17) 62.6 (18) 16.1 (19) 19.3 (20) 38.6 (21) 15.0 (22) 121.7 (23) 33.1[a] (24) 33.8[a] (25) 83.8 (26) 77.2 (27) 24.3 Glc I (1) 100.2 (2) 79.0 (3) 76.6 (4) 77.8 (5) 78.1 (6) 61.4 Rha I (1) 101.8 (2) 72.5[b] (3) 71.6[b] (4) 73.6[c] (5) 69.3[d] (6) 18.3[e] Rha II (1) 102.7 (2) 72.5[b] (3) 72.2[b] (4) 73.9[c] (5) 70.3[d] (6) 18.4[e] Glc II (1) 105.1 (2) 75.1 (3) 78.1 (4) 72.2 (5) 78.1 (6) 62.4.

Mass (F.D.) : m/z 1069 $[M+Na]^+$.

Reference

1.   R. Saijo, C. Fuke, K. Murakami, T. Nohara and T. Tomimatsu, *Phytochemistry*, 22, 733 (1983).

ANGUIVIOSIDE XV

Kryptogenin 3-O-{α-L-rhamnopyranosyl-(1→2)-[α-L-rhamnopyranosyl-(1→4)]-β-D-glucopyranoside}-26-O-β-D-glucopyranoside

Source : *Solanum anguivi* (Solanaceae)
Mol. Formula : $C_{51}H_{82}O_{22}$
Mol. Wt. : 1046
$[\alpha]_D^{26}$: -88.1° (c=0.6, MeOH).
Registry No. : [109617-40-5]

PMR (C_5D_5N, 500 MHz) : δ 0.69 (s, 3xH-18), 1.02 (d, *J*=6.7 Hz, 3xH-27), 1.05 (d, *J*=6.7 Hz, 3xH-21), 1.07 (s, 3xH-19), 1.64 (d, *J*=6.7 Hz, 3xH-6 of Rha), 1.78 (d, *J*=6.1 Hz, 3xH-6 of Rha), 4.98* (H-1 of Glc I, H-1 of Glc II), 5.31 (m, H-6), 6.41 (s, H-1 of Rha), 6.86 (s, H-1 of Rha).

CMR (C_5D_5N, 125 MHz) : δ C-1) 37.2 (2) 30.1 (3) 77.8 (4) 38.7 (5) 140.9 (6) 121.5 (7) 32.0 (8) 31.0 (9) 50.0 (10) 37.1 (11) 20.7 (12) 40.2 (13) 41.7 (14) 51.1 (15) 39.0 (16) 217.8 (17) 66.4 (18) 12.9 (19) 19.4 (20) 43.7 (21) 15.6 (22) 213.0 (23) 37.4 (24) 27.8 (25) 33.5 (26) 15.1 (27) 17.5 **Glc I** (1) 100.3 (2) 78.5 (3) 77.0 (4) 78.0 (5)77.8 (6) 61.3 **Rha I** (1) 102.1 (2) 72.5 (3) 72.8 (4) 73.9 (5) 69.5 (6) 18.5 **Rha II** (1) 102.9 (2) 72.6 (3) 72.8 (4) 73.9 (5) 70.4 (6) 18.7 **Glc II** (1) 100.3 (2) 75.2 (3) 78.1 (4) 71.5 (5) 78.9 (6) 62.3.

Mass (FAB, Positive ion) : *m/z* 1070 [M+Na+H]⁺.

Reference

1. T. Honbu, T. Ikeda, X.-H. Zhu, O. Yoshihara, M. Okawa, A.M. Nafady and T. Nohara, *J. Nat. Prod.*, **65**, 1918 (2002).

CORDYLINE STRICTA SAPONIN 13

22ξ-Methoxyfurosta-5,25(27)-diene-1β,3β,26-triol 1-O-{α-L-rhamnopyranosyl-(1→2)-[β-D-xylopyranosyl-(1→3)]-β-D-fucopyranoside}-26-O-β-D-glucopyranoside

Source : *Cordyline stricta* (Agavaceae)
Mol. Formula : $C_{51}H_{82}O_{22}$
Mol. Wt. : 1046
$[\alpha]_D^{27}$: -24.3° (c=0.12, MeOH)
Registry No. : [202347-03-3]

IR (KBr) : 3390 (OH), 2910 (CH), 1440, 1370,1300, 1150, 1055,1035, 975, 900, 825, 800 cm^{-1}.

PMR (C$_5$D$_5$N, 400/500 MHz) : δ 0.85 (s, 3xH-18), 1.11 (d, J=6.9 Hz, 3xH-21), 1.41 (s, 3xH-19), 1.52 (d, J=6.3 Hz, 3xH-6 of Fuc), 1.73 (d, J=6.1 Hz, 3xH-6 of Rha), 3.25 (s, OCH$_3$), 4.68 (d, J=7.8 Hz, H-1 of Fuc), 4.92 (overlapping with H$_2$O signal H-1 of Glc), 4.99 (d, J=7.5 Hz, H-1 of Xyl), 5.05 and 5.34 (each 1H, br s, 2xH-27), 5.60 (br d, J=5.6 Hz, H-6), 6.36 (br s, H-1 of Rha).

CMR (C$_5$D$_5$N, 100/125 MHz) : δ C-1) 84.3 (2) 38.1 (3) 68.3 (4) 43.9 (5) 139.7 (6) 124.7 (7) 32.0 (8) 33.0 (9) 50.6 (10) 42.8 (11) 24.0 (12) 40.4 (13) 40.4 (14) 57.1 (15) 32.4 (16) 81.5 (17) 64.2 (18) 16.8 (19) 15.0 (20) 40.5 (21) 16.1 (22) 112.4 (23) 31.5 (24) 28.1 (25) 146.9 (26) 72.0 (27) 111.0 (OCH$_3$) 47.3 **Fuc** (1) 100.5 (2) 73.5 (3) 85.6 (4) 72.7 (5) 70.8 (6) 17.1 **Rha** (1) 101.8 (2) 72.6 (3) 72.6 (4) 74.3 (5) 69.4 (6) 19.2 **Xyl** (1) 106.7 (2) 74.7 (3) 78.4 (4) 71.0 (5) 67.1 **Glc** (1) 103.9 (2) 75.2 (3) 78.6 (4) 71.7 (5) 78.5 (6) 62.9.

Mass (FAB, Negative ion) : *m/z* 1045 [M-H]$^-$, 899 [M-Rha]$^-$.

Reference

1.   Y. Mimaki, M. Kuroda, Y. Takaashi and Y. Sashida, *Phytochemistry*, **47,** 79 (1998).

PROTOGRACILLIN A

Furost-5,20(22)-dien-3β,26-diol 3-O-{α-L-rhamnopyranosyl-(1→2)-[β-D-glucopyranosyl-(1→3)]-β-D-glucopyranoside}-26-O-β-D-glucopyranoside

Source : *Dioscorea futschauensis* R. Kunth
(Dioscoreaceae)
Mol. Formula : $C_{51}H_{82}O_{22}$
Mol. Wt. : 1046
M.P. : 238-240°C
$[\alpha]_D^{24}$ **:** -45.6° (c=0.1, Pyridine]
Registry No. : [637349-03-2]

IR (KBr) : 3385 (OH), 2930, 1645, 1456, 1370, 1046 (glycosyl C-O) cm^{-1}.

PMR (C$_5$D$_5$N, 500 MHz) : δ 0.75 (s, 3xH-18), 0.86* (H-9), 0.94* (H-1A), 1.02 (d, *J*=6.0 Hz, 3xH-27), 1.06 (s, 3xH-19), 1.10* (H-14), 1.12 (H-12A), 1.46* (H-7A, H-11A), 1.48 (m, H-24A), 1.56* (H-8), 1.64 (s, 3xH-21), 1.71 (H-11B), 1.74* (H-1B), 1.74 (d, *J*=6.0 Hz, 3xH-6 of Rha), 1.76 (H-12B), 1.84* (H-2A), 1.85 (m, H-24B), 1.88* (H-7B), 1.94 (m, H-25), 2.10* (H-2B, H-15B), 2.23* (H-23), 2.44 (d, *J*=10.0 Hz, H-17), 2.72 (m, H-4A), 2.79 (m, H-4B), 3.62* (H-26A), 3.82 (m, H-5 of Glc I), 3.84 (m, H-3), 3.94* (H-5 of Glc III), 3.96* (H-26B), 4.00* (H-5 of Glc II), 4.02* (H-2 of Glc II), 4.03* (H-2 of Glc III), 4.05* (H-4 of Glc I), 4.10* (H-4 of Glc II), 4.16* (H-3 of Glc I), 4.18* (H-2 of Glc I, H-3 of Glc III), 4.19* (H-3 of Glc II, H-4 of Glc III), 4.26* (H-6A of Glc I), 4.28* (H-6A of Glc II),

4.30* (H-4 of Rha), 4.36 (H-6A of Glc III), 4.42* (H-6B of Glc I), 4.53* (H-6B of Glc II), 4.55* (H-3 of Rha), 4.57 (H-6B of Glc III), 4.83 (d, *J*=7.5 Hz, H-1 of Glc II), 4.83* (H-16), 4.86 (m, H-2 of Rha), 4.93* (H-5 of Rha), 4.94 (d, *J*=6.9 Hz, H-1 of Glc I), 5.09 (d, *J*=7.5 Hz, H-1 of Glc II), 5.32 (br s, H-6), 6.38 (br s, H-1 of Rha).

CMR (C_5D_5N, 125 MHz) : δ C-1) 37.5 (2) 30.1 (3) 78.5 (4) 39.0 (5) 141.2 (6) 122.3 (7) 32.5 (8) 31.5 (9) 50.3 (10) 37.2 (11) 21.3 (12) 39.7 (13) 43.5 (14) 55.0 (15) 34.5 (16) 84.5 (17) 64.6 (18) 14.1 (19) 19.4 (20) 103.6 (21) 11.8 (22) 152.6 (23) 23.7 (24) 31.5 (25) 33.6 (26) 75.2 (27) 17.4 **Glc I** (1) 100.0 (2) 77.1 (3) 89.5 (4) 69.6 (5) 77.9 (6) 62.5 **Rha** (1) 102.2 (2) 72.5 (3) 72.8 (4) 74.2 (5) 69.6 (6) 18.7 **Glc II** (1) 104.6 (2) 75.0 (3) 78.5 (4) 71.8 (5) 77.9 (6) 62.5 **Glc III** (1) 104.9 (2) 75.2 (3) 78.5 (4) 71.6 (5) 78.6 (6) 62.9.

Mass (FAB, Negative ion) : *m/z* 1045 [M-H]⁻, 833 [M-H-Glc]⁻, 737 [M-H-Rha-Glc]⁻, 575 [M-H-2xGlc]⁻, 413 [M-H-3xGlc-Rha].

Mass (FAB, Negative ion, H.R.) : 1045.4327 [(M-H)⁻, calcd. for 1045.4339].

Reference

1.   H.-W. Liu, S.-L. Wang, B. Cai, G.-X. Qu, X.-J. Yang, H. Kobayashi and X.-S. Yao, *J. Asian Nat. Prod. Res.*, **5**, 241 (2003).

ADSCENDOSIDE B
(25*S*)-Furost-5-en-3β,22α,26-triol 3-O-{α-L-rhamnopyranosyl-(1→4)-[α-L-rhamnopyranosyl-(1→6)]-β-D-glucopyranoside}-26-O-β-D-glucopyranoside

x

Source : *Asparagus adscendens* Roxb. (Liliaceae)
Mol. Formula : $C_{51}H_{84}O_{22}$
Mol. Wt. : 1048
$[\alpha]_D^{20}$: -92° (c=1.0, H_2O)
Registry No. : [91095-73-3]

IR (KBr) : 3400 (OH) cm^{-1}, no spiroketal abroptions.

21-O-Methyl Ether : Adscendoside A (artifact) [91095-72-6]

IR (KBr) : 3700 (OH) cm^{-1}, no spiroketal absorptions.

PMR (C$_5$D$_5$N) : δ 3.25 (s, OCH_3), 4.82 (d, J=6.0 Hz), 4.98 (br s), 5.10 (d, J=6.0 Hz), 5.25 (br s).

Reference

1.   S.C. Sharma and H.C. Sharma, *Phytochemistry*, **23**, 645 (1984).

ASPARAGUS PLUMOSUS SAPONIN 4
(25S)-Furost-5-en-3β,22α,26-triol 3-O-{α-L-rhamnopyranosyl-(1→2)-[α-L-rhamnopyranosyl-(1→3)]-β-D-glucopyranoside}-26-O-β-D-glucopyranoside

Source : *Asparagus plumosus* Baker (Liliaceae)
Mol. Formula : C$_{51}$H$_{84}$O$_{22}$
Mol. Wt. : 1048
M.P. : 193-195^{0}C
[α]$_D^{27}$: -81^0 (c=1.0, H$_2$O)
Registry No. : [98569-69-8]

IR (KBr) : λ_{max} 3400 (OH) cm^{-1}, no spiroketal absorptions.

Reference

1.   O.P. Sati and G. Pant, *J. Nat. Prod.*, **48**, 390 (1985).

ASPARASAPONIN I, TRIGONELLOSIDE C,
YAMOGENIN TRIOSIDE C, PROTONEODIOSCIN

Yamogenin 3-O-{α-L-rhamnopyranosyl-(1→2)-[α-rhamnopyranosyl-(1→4)-β-D-gluocopyranoside]-26-O-[β-D-glucopyranoside]}

Source : *Asparagus officinalis* L.[1] (Liliaceae), *Dioscorea collettii* Hook f. var. *hypoglauca* Pei et Ting[2] (Dioscoreaceae), *Asparagus dumosus* Baker[3] (Liliaceae), *Trigonella foenum-graecum* L.[4] (Fabaceae) etc.

Mol. Formula : $C_{51}H_{84}O_{22}$

Mol. Wt. : 1048

M.P. : 166-168°C (decomp.)[2]

[α]$_D^{13}$: -70.10° (c=0.01, Pyridine)[2]

Registry No. : [60478-69-5]

PMR (C$_5$D$_5$N, 500 MHz)[2] : δ 0.89 (s, 3xH-18), 1.02 (d, J=6.8 Hz, 3xH-27), 1.05 (s, 3xH-19), 1.30 (d, J=6.5 Hz, 3xH-21), 1.60 (d, J=6.0 Hz, 3xH-6 of Rha II), 1.74 (d, J=6.0 Hz, 3xH-6 of Rha I), 3.86 (m, H-3), 4.78 (d, J=7.6 Hz, H-1 of Glc II), 4.92 (d, J=7.2 Hz, H-1 of Glc I), 5.30 (br d, H-6), 5.80 (d, J=0.9 Hz, H-1 of Rha II), 6.34 (d, J=0.9 Hz, H-1 of Rha I).

CMR (C$_5$D$_5$N, 125 MHz)[2] : δ C-1) 37.5 (2) 30.2 (3) 78.2 (4) 39.0 (5) 140.9 (6) 121.8 (7) 32.4 (8) 31.7 (9) 50.4 (10) 37.2 (11) 21.0 (12) 40.0 (13) 40.8 (14) 56.6 (15) 32.5 (16) 81.1 (17) 63.9 (18) 16.4 (19) 19.4 (20) 40.7 (21) 16.4 (22) 110.7 (23) 37.2 (24) 28.3 (25) 34.4 (26) 75.4 (27) 17.4 **Glc I** (1) 100.3 (2) 77.8 (3) 78.0 (4) 78.8 (5) 76.9 (6) 61.4 **Rha I** (1) 102.0 (2) 72.5 (3) 72.8 (4) 74.1 (5) 69.5 (6) 18.5 **Rha II** (1) 102.9 (2) 72.5 (3) 72.7 (4) 73.9 (5) 70.4 (6) 18.6 **Glc II** (1) 105.1 (2) 75.2 (3) 78.4 (4) 71.8 (5) 78.6 (6) 62.9.

Mass (FAB, Positive ion)[2] : m/z 1071 [M+Na]$^+$, 1031 [M+H-H$_2$O]$^+$, 869 [M+H-H$_2$O-Glc]$^+$, 723 [M+H-H$_2$O-Glc-Rha]$^+$, 577 [M+H-H$_2$O-Glc-2xRha]$^+$, 415 [M+H-H$_2$O-2x Glc-2x Rha]$^+$.

Biological Activity : Bitter in taste.[1] Causes morphological abnormality of *Pyricularia oryzae* mycelia and shows cytotoxic activity against the cancer cell line of K 562 *in vitro*.[2]

22-O-Methyl Ether : Methyl Protoneodioscin, [60478-70-8]

artefact; **M.P. :** 224-266°C[4], [α]$_D$^{13} : -74.10° (c=0.01, Pyridine)[4]

References

1.   K. Kawano, H. Sato and S. Sakamura, *Agric. Biol. Chem.*, **41**, 1 (1977).

2.   K. Hu, A. Dong, X. Yao, H. Ksayashi, and S. Iwasaki, *Planta Med.*, **63**, 161 (1997).

3.   V.U. Ahmad, S.M. Khaliq-uz-Zaman, S. Shameel, S. Perveen and Z. Ali, *Phytochemistry*, **50**, 481 (1998).

4.   N.G. Bogacheva, V.I. Sheichenko and L.M. Kogan, *Khim. Farm. Zh.*, **11**, 65 (1977).

CORDYLINE STRICTA SAPONIN 12

(5α)-22ξ-Methoxyfurosta-25(27)-ene-1β,3β,26-triol 1-O-{α-L-rhamnopyranosyl-(1→2)-[β-D-xylopyranosyl-(1→3)]-β-D-fucopyranoside}-26-O-β-D-glucopyranoside

Source : *Cordyline stricta* (Agavaceae)
Mol. Formula : $C_{51}H_{84}O_{22}$
Mol. Wt. : 1048
$[\alpha]_D^{27}$: -20.0° (c=0.11, MeOH)
Registry No. : [202347-02-2]

IR (KBr) : 3420 (OH), 2925 (CH), 1445, 1370, 1160, 1065, 1040, 980, 910 cm^{-1}.

PMR (C_5D_5N, 400/500 MHz) : δ 0.81 (s, 3xH-18), 1.12 (d, *J*=6.7 Hz, 3xH-21), 1.23 (s, 3xH-19), 1.52 (d, *J*=6.3 Hz, 3xH-6 of Fuc), 1.74 (d, *J*=6.0 Hz, 3xH-6 of Rha), 3.25 (s, OCH₃), 4.60 (overlapped, H-1 of Fuc), 4.91 (d, *J*=7.8 Hz, H-1 of Glc), 4.99 (d, *J*=7.4 Hz, H-1 of Xyl), 5.05 and 5.34 (each 1H, br s, 2xH-27), 6.37 (br s, H-1 of Rha).

CMR (C_5D_5N, 100/125 MHz) : δ C-1) 82.6 (2) 37.7 (3) 67.9 (4) 39.7 (5) 43.2 (6) 28.9 (7) 32.5 (8) 36.5 (9) 55.2 (10) 41.5 (11) 23.6 (12) 40.6 (13) 40.7 (14) 56.8 (15) 32.4 (16) 81.5 (17) 64.3 (18) 16.8 (19) 8.7 (20) 40.4 (21) 16.0 (22) 112.4 (23) 31.5 (24) 28.1 (25) 146.9 (26) 72.0 (27) 111.1 (OCH₃) 47.3 **Fuc** (1) 99.7 (2) 73.6 (3) 85.8 (4) 72.7 (5) 70.8 (6) 17.1 **Rha** (1) 101.6 (2) 72.5 (3) 72.5 (4) 74.3 (5) 69.4 (6) 19.2 **Xyl** (1) 106.6 (2) 74.7 (3) 78.3 (4) 71.0 (5) 67.1 **Glc** (1) 103.8 (2) 75.1 (3) 78.6 (4) 71.7 (5) 78.5 (6) 62.8.

Mass (FAB, Negative ion) : *m/z* 1047 [M-H]⁻.

Reference

1.   Y. Mimaki, M. Kuroda, Y. Takaashi and Y. Sashida, *Phytochemistry*, **47**, 79 (1998).

COSTUS SPICATUS SAPONIN 4

3β,26-Dihydroxy-22α-methoxy-(25R)-furost-5-ene 3-O-{β-D-apiofuranosyl-(1→2)-
[α-L-rhamnopyranosyl-(1→4)]-β-D-glucopyranoside}-26-O-[β-D-glucopyranoside]

Source : *Costus spicatus* Swartz (Syn.
C. Cylindrecus Jacq.) (Costaceae)
Mol. Formula : $C_{51}H_{84}O_{22}$
Mol. Wt. : 1048
M.P. : 222-224°C
$[\alpha]_D^{20}$: -102° (c=0.001, MeOH)
Registry No. : [244095-89-4]

IR (KBr) : **3430** (OH), 1050 (C–O), 913, 838, 813, 638 cm^{-1}. The peak at 913 < the peak at 838 cm^{-1}.

PMR (C$_5$D$_5$N, 400 MHz) : δ 0.84 (s, 3xH-18), 0.98 (d, *J*=6.6 Hz, 3xH-27), 1.06 (s, 3xH-19), 1.20 (d, *J*=6.8 Hz, 3xH-21), 1.78 (d, *J*=6.3 Hz, 3xH-6 of Rha), 3.28 (s, OC*H*$_3$), 4.82 (d, *J*=7.8 Hz, H-1 of Glc II), 4.98 (d, *J*=7.8 Hz, H-1 of Glc I), 5.30 (br s, H-6), 5.95 (d, *J*=3.6 Hz, H-1 of Api), 6.28 (br s, H-1 of Rha).

CMR (C$_5$D$_5$N, 100 MHz) : δ C-1) 37.0 (2) 29.7 (3) 77.9 (4) 38.5 (5) 140.3 (6) 121.4 (7) 32.0 (8) 31.2 (9) 49.9 (10) 36.7 (11) 20.7 (12) 39.5 (13) 40.3 (14) 56.1 (15) 31.9 (16) 80.8 (17) 63.7 (18) 16.0 (19) 18.9 (20) 40.2 (21) 16.6 (22) 112.2 (23) 30.3 (24) 27.9 (25) 33.8 (26) 74.7 (27) 16.9 (OCH$_3$) 46.8 **Glc I** (1) 99.8 (2) 77.4 (3) 77.7 (4) 78.9 (5) 76.9 (6) 60.8 **Api** (1) 110.7 (2) 77.2 (3) 79.6 (4) 74.8 (5) 64.2 **Rha** (1) 102.4 (2) 72.2 (3) 72.3 (4) 73.6 (5) 69.9 (6) 18.2 **Glc II** (1) 104.5 (2) 74.7 (3) 78.1 (4) 71.9 (5) 78.9 (6) 62.3.

Mass (LSIMS, Negative ion) : *m/z* 1047 [M-H]⁻, 915 [M-H-Api]⁻, 901 [M-H-Rha]⁻, 885 [M-H-Glc]⁻.

Reference

1.   B.P. Da Silva, R.R. Bernardo and J.P. Parente, *Phytochemistry*, **51**, 931 (1999).

COSTUS SPICATUS SAPONIN 5

(25*R*)-3β,26-Dihydroxy-22α-methoxyfurost-5-ene 3-O-{β-D-apifuranosyl-(1→4)-[α-L-rhamnopyranosyl-(1→2)]-β-D-glucopyranoside}-26-O-β-D-glucopyranoside}-26-O-β-D-glucopyranoside

Source : *Costus spicatus* Swartz (Costaceae)
Mol. Formula : C₅₁H₈₄O₂₂
Mol. Wt. : 1048
M.P. : 220-222°C
[α]_D : -102.0° (c=0.001, MeOH)
Registry No. : [22369-35-8]

IR (KBr) : 3430, 1050, 913, 838, 813, 638 cm⁻¹.

PMR (C₅D₅N, 200 MHz) : δ 0.84 (s, 3xH-18), 0.98 (d, *J*=6.6 Hz, 3xH-27), 1.06 (s, 3xH-19), 1.20 (d, *J*=6.8 Hz, 3xH-21), 3.28 (s, OC*H₃*), 4.82 (d, *J*=7.8 Hz, H-1 of Glc II), 4.98 (d, *J*=7.7 Hz, H-1 of Glc I), 5.30 (br s, H-6), 5.95 (d, *J*=3.6 Hz, H-1 of Api), 6.28 (br s, H-1 of Rha)

CMR (C$_5$D$_5$N, 50 MHz) : δ C-1) 37.2 (2) 29.7 (3) 77.9 (4) 38.5 (5) 140.3 (6) 121.5 (7) 32.2 (8) 31.2 (9) 50.1 (10) 36.7 (11) 20.7 (12) 39.5 (13) 40.3 (14) 56.3 (15) 31.9 (16) 80.8 (17) 63.7 (18) 16.1 (19) 18.8 (20) 40.2 (21) 16.6 (22) 112.5 (23) 30.3 (24) 27.7 (25) 33.8 (26) 74.7 (27) 16.9 (OCH$_3$-22), 46.8 **Glc I** (1) 100.1 (2) 77.8 (3) 78.0 (4) 78.9 (5) 76.9 (6) 61.0 **Glc II** (1) 104.5 (2) 74.7 (3) 78.1 (4) 78.1 (5) 78.9 (6) 62.3 **Rha** (1) 101.5 (2) 72.2 (3) 72.3 (4) 73.4 (5) 69.1 (6) 18.1 **Api** (1) 110.2 (2) 77.2 (3) 79.6 (4) 74.8 (5) 64.2.

Mass (L.S.I., Negative ion) : *m/z* 1047 [M-H]⁻, 915 [(M-H)-Api]⁻, 901 [(M-H)-Rha]⁻, 885 [(M-H)-Glc]⁻.

Reference

1.   B.P. de Silva, R.R. Bernardo and J.P. Parente, *Planta Med.*, **65**, 285 (1999).

INDIOSIDE B
22α-Methoxyfurost-5-en-3β,26-diol 3-O-{α-L-rhamnopyranosyl-(1→2)-[β-D-xylopyranosyl-(1→3)]-β-D-glucopyranoside}-26-O-β-D-glucopyranoside

Source : *Solanum indicum* L. (Solanaceae)
Mol. Formula : C$_{51}$H$_{84}$O$_{22}$
Mol. Wt. : 1048
[α]$_D^{29}$: -55.2° (c=0.50, Pyridine)
Registry No. : [185332-78-9]

PMR (C_5D_5N, 400 MHz) : δ 0.78 (s, 3xH-18), 0.95 (d, *J*=6.6 Hz, 3xH-27), 1.01 (s, 3xH-19), 1.15 (d, *J*=7.0 Hz, 3xH-21), 1.70 (d, *J*=5.5 Hz, 3xH-6 of Rha), 3.22 (3H, s, OC*H*₃), 4.79 (d, *J*=7.7 Hz, H-1 of Glc II), 4.91 (2H, d, *J*=8.0 Hz, H-1 of Glc I and H-1 of Xyl), 5.29 (br s, H-6), 6.24 (s, H-1 of Rha).

CMR (CD₃OD, 100 MHz) : δ C-1) 37.3 (2) 29.9 (3) 78.3 (4) 40.0 (5) 140.6 (6) 121.7 (7) 32.1 (8) 31.5 (9) 50.1 (10) 37.0 (11) 20.9 (12) 38.5 (13) 40.6 (14) 56.4 (15) 32.1 (16) 81.2 (17) 64.0 (18) 16.1 (19) 19.3 (20) 40.3 (21) 16.2 (22) 112.5 (23) 30.6 (24) 28.0 (25) 34.0 (26) 75.0 (27) 17.2 (OC*H*₃) 47.1 **Glc I** (1) 99.7 (2) 77.5^a (3) 88.0 (4) 69.3 (5) 77.7 (6) 62.2 **Rha** (1) 102.2 (2) 72.3 (3) 72.6 (4) 73.8 (5) 69.4 (6) 18.6 **Xyl** (1) 105.2 (2) 74.5 (3) 77.2^a (4) 70.4 (5) 67.0 **Glc II** (1) (1) 104.7 (2) 75.0 (3) 78.3 (4) 71.5 (5) 78.3 (6) 62.7.

Mass (FAB, Positive ion) : *m/z* 1049 [M+H]$^+$.

Reference

1.   S. Yahara, T. Nakamura, Y. Someya, T. Matsumoto, T. Yamashita and T. Nohara, *Phytochemistry*, **43**, 1319 (1996).

LILIUM LONGIFLORUM SAPONIN 4
22-O-Methyl-26-O-β-D-glucopyranosyl-(25*R*)-furost-5-ene-3β,22ξ,26-triol
3-O-{α-L-rhamnopyranosyl-(1→2)-[α-L-arabinopyranosyl-(1→3)]-β-D-glucopyranoside}

Source : *Lilium longiflorum* (Liliaceae)
Mol. Formula : $C_{51}H_{84}O_{22}$
Mol. Wt. : 1048
[α]$_D^{26}$: -52.0° (c=0.20, MeOH)
Registry No. : [159690-20-7]

IR (KBr) : 3405 (OH), 2920 (CH), 1445, 1375, 1250, 1040, 905, 835, 805, 775 cm^{-1}.

PMR (C$_5$D$_5$N, 400 MHz) : δ 0.82 (s, 3xH-18), 1.01 (d, J=6.6 Hz, 3xH-27), 1.06 (s, 3xH-19), 1.20 (d, J=6.9 Hz, 3xH-21), 1.74 (d, J=6.2 Hz, 3xH-6 of Rha), 3.27 (s, OCH_3), 4.85 (d, J=7.8 Hz, H-1 of Glc II), 4.91 (d, J=7.8 Hz, H-1 of Glc I), 4.97 (d, J=7.2 Hz, H-1 of Ara), 5.35 (br d, J=4.3 Hz, H-6), 6.27 (br s, H-1 of Rha).

CMR (C$_5$D$_5$N, 100.6 MHz) : δ C-1) 37.5 (2) 30.1 (3) 77.8^a (4) 38.8 (5) 140.9 (6) 121.9 (7) 32.2^b (8) 31.7 (9) 50.4 (10) 37.2 (11) 21.1 (12) 39.8 (13) 40.8 (14) 56.6 (15) 32.4^b (16) 81.4 (17) 64.2 (18) 16.3 (19) 19.4 (20) 40.5 (21) 16.3 (22) 112.7 (23) 30.8 (24) 28.2 (25) 34.3 (26) 75.2 (27) 17.2 (OCH_3) 47.3 **Glc I** (1) 100.0 (2) 78.0 (3) 88.1 (4) 69.7^c (5) 77.7^a (6) 62.5 **Rha** (1) 102.4 (2) 72.5 (3) 72.9 (4) 74.1 (5) 69.4 (6) 18.7 **Ara** (1) 105.6 (2) 72.3 (3) 74.5 (4) 69.6^c (5) 67.8 **Glc II** (1) 105.0 (2) 75.2 (3) 78.7^d (4) 71.8 (5) 78.5^d (6) 63.0.

Mass (FAB, Negative ion) : *m/z* 1048 [M]$^-$.

Reference

1.   Y. Mimaki, O. Nakamura, Y. Sashida, Y. Satomi, A. Nishino and H. Nishino, *Phytochemistry*, **37**, 227 (1994).

LILIUM LONGIFLORUM SAPONIN 5
22-O-Methyl-26-O-β-D-glucopyranosyl-(25R)-furost-5-ene-3β,22,26-triol
3-O-{α-L-rhamnopyranosyl-(1→2)-[β-D-xylopyranosyl-(1→3)]-β-D-glucopyranoside}

Source : *Lilium longiflorum* (Liliaceae)
Mol. Formula : C$_{51}$H$_{84}$O$_{22}$
Mol. Wt. : 1048
[α]$_D^{28}$: -108.0° (c=0.10, MeOH)
Registry No. : [159812-01-8]

IR (KBr) : 3380 (OH), 2920 (CH), 1450, 1370, 1250, 1150, 1030, 890, 835, 810 cm^{-1}.

PMR (C$_5$D$_5$N, 400 MHz) : δ 0.82 (s, 3xH-18), 1.01 (d, J=6.6 Hz, 3xH-27), 1.07 (s, 3xH-19), 1.20 (d, J=6.9 Hz, 3xH-21), 1.77 (d, J=6.2 Hz, 3xH-6 of Rha), 3.28 (s, OCH_3), 4.86 (d, J=7.8 Hz, H-1 of Glc II), 4.99 (d, J=7.3 Hz, H-1 of Glc I), 5.00 (d, J=6.5 Hz, H-1 of Ara), 5.34 (br d, J=4.4 Hz, H-6), 6.33 (br s, H-1 of Rha).

CMR (C$_5$D$_5$N, 100.6 MHz) : δ C-1) 37.5 (2) 30.1 (3) 77.4 (4) 38.7 (5) 140.8 (6) 121.9 (7) 32.2^a (8) 31.7 (9) 50.3 (10) 37.2 (11) 21.1 (12) 39.8 (13) 40.8 (14) 56.6 (15) 32.3^a (16) 81.3 (17) 64.2 (18) 16.3 (19) 19.4 (20) 40.5 (21) 16.3 (22) 112.7 (23) 30.8 (24) 28.2 (25) 34.2 (26) 75.2 (27) 17.2 (OCH_3) 47.3 **Glc I** (1) 100.0 (2) 77.9 (3) 88.2 (4) 69.6 (5) 77.7 (6) 62.4 **Rha** (1) 102.4 (2) 72.4 (3) 72.8 (4) 74.1 (5) 69.5 (6) 18.7 **Xyl** (1) 105.4 (2) 74.7 (3) 78.4 (4) 70.7 (5) 67.3 **Glc II** (1) 105.0 (2) 75.2 (3) 78.6^c (4) 71.8 (5) 78.5^c (6) 62.9.

Mass (FAB, Negative ion) : *m/z* 1047 [M-H]$^-$.

Reference

1.   Y. Mimaki, O. Nakamura, Y. Sashida, Y. Satomi, A. Nishino and H. Nishino, *Phytochemistry*, **37**, 227 (1994).

PARIS POLYPHYLLA SAPONIN 1

(22ξ,25*R*)-22-Methoxyfurost-5-ene-3,26-diol 3-O-{α-L-arabinofuranosyl-(1→4)-[α-L-rhamnopyranosyl-(1→2)]-β-D-glucopyranoside}-26-O-β-D-glucopyranoside

Source : *Paris polyphylla* Sm. (Liliaceae)
Mol. Formula : C$_{51}$H$_{84}$O$_{22}$
Mol. Wt. : 1048
M.P. : 209-212°C
[α]$_D$: -135.6° (c=0.45, MeOH)
Registry No. : [78101-18-5]

IR (KBr) : 3600-3300 (OH) cm^{-1}.

PMR (C$_5$D$_5$N, 100 MHz) : δ 3.24 (OCH_3).

Reference

1.   M. Miyamura, K. Nakano, T. Nohara, T. Tomimatsu and T. Kawasaki, *Chem. Pharm. Bull.,* **30**, 712 (1982).

PROTODIOSCIN
(25*R*)-Furost-5-en-3β,22α,26-triol 3-O-{α-L-rhamnopyranosyl-(1→2)-
[α-L-rhamnopyranosyl-(1→4)]-β-D-glucopyranoside}-26-O-[β-D-glucopyranoside]

Source : *Dioscorea gracillima* Miq[1] (Dioscoraceae),
Melilotus tauricus Bieb. Ser.[2] (Fabaceae), *D. colletti*
Hook. f. var. *hypoglauca*[3], *Asparagus officinalis* L.[4]
(Liliaceae), *Smilax china* L.[5] (Liliaceae), *Costus speciosus*
(Koen), Sm.[6] (Zingiberaceae) etc.
Mol. Formula : C$_{51}$H$_{84}$O$_{22}$
Mol. Wt. : 1048
M.P. : 267-271°C (decomp.)[6]
Registry No. : [55056-80-9]

IR[6] : 3600-3100, 1150-1000, 980, 918 (strong), 898 (w), 840 cm^{-1}.

PMR (C_5D_5N, 250 MHz)[2] : δ 0.88 (s, 3xH-18), 1.03 (s, 3xH-19), 1.52 (d, 3xH-6 of Rha I), 1.67 (d, J=6.5 Hz, 3xH-6 of Rha II), 3.59 (m, H-5 of Glc I), 3.70 (m, H-3), 3.82 (dd, $J_{2,3}$=7.5 Hz, $J_{3,4}$=not given, H-3 of Glc I), 3.82 (m, $J_{5,6B}$=5.5 Hz, $J_{5,6A}$=2.5 Hz, H-5 of Glc II), 3.89 (dd, $J_{6A,5}$=2.5 Hz, $J_{6A,6B}$=11.5 Hz, H-6A of Glc I), 3.89 (dd, $J_{1,2}$=7.5, $J_{2,3}$=8.0 Hz, H-2 of Glc II), 4.01 (dd, $J_{6A,5}$=5.5 Hz, $J_{6A,6B}$=11.5 Hz, H-6B of Glc I), 4.05 (t, $J_{1,2}$=$J_{2,3}$=7.5 Hz, H-2 of Glc I), 4.08 (dd, H-4 of Glc II), 4.11 (dd, H-3 of Glc II), 4.17 (d, $J_{4,5}$=9.5 Hz, H-4 of Rha I), 4.17 (dd, $J_{4,5}$=9.0 Hz, H-4 of Glc I), 4.20 (dd, $J_{3,4}$=9.5 Hz, $J_{4,5}$=9.0 Hz, H-4 of Rha), 4.26 (dd, $J_{6A,6B}$=11.0 Hz, $J_{5,6A}$=2.5 Hz, H-6A of Glc II), 4.38 (dd, $J_{2,3}$=2.5 Hz, $J_{3,4}$=9.5 Hz, H-3 of Rha II), 4.43 (dd, $J_{6A,6B}$=11.0 Hz, $J_{5,6B}$=5.5 Hz, H-6B of Glc II), 4.47 (dd, $J_{2,3}$=2.0 Hz, $J_{3,4}$=9.5 Hz, H-3 of Rha I), 4.54 (d, $J_{2,3}$=2.5 Hz, H-2 of Rha II), 4.63 (dq, $J_{5,6}$=6.5 Hz, $J_{4,5}$=9.5 Hz, H-5 of Rha I), 4.65 (d, $J_{2,3}$=2.0 Hz, H-2 of Rha I), 4.71 (d, $J_{1,2}$=7.5 Hz, H-1 of Glc II), 4.79 (m, H-5 of Rha II), 4.85 (d, $J_{1,2}$=7.5 Hz, H-1 of Glc I), 5.62 (s, H-1 of Rha II), 6.14 (s, H-1 of Rha I).

CMR (C_5D_5N, 125 MHz)[3] : δ C-1) 37.5 (2) 30.2 (3) 78.2 (4) 39.0 (5) 140.9 (6) 121.8 (7) 32.4 (8) 31.7 (9) 50.4 (10) 37.2 (11) 21.1 (12) 40.0 (13) 40.8 (14) 56.6 (15) 32.5 (16) 81.1 (17) 63.9 (18) 16.5 (19) 19.4 (20) 40.7 (21) 16.5 (22) 110.7 (23) 37.2 (24) 28.3 (25) 34.3 (26) 75.2 (27) 17.5 **Glc I** (1) 100.3 (2) 77.8 (3) 78.0 (4) 78.8 (5) 76.9 (6) 61.4 **Rha I** (1) 102.0 (2) 72.5 (3) 72.8 (4) 74.2 (5) 69.5 (6) 18.5 **Rha II** (1) 102.9 (2) 72.5 (3) 72.7 (4) 73.9 (5) 70.4 (6) 18.6 **Glc II** (1) 104.9 (2) 75.2 (3) 78.4 (4) 71.8 (5) 78.6 (6) 62.9.

Mass (FAB, Positive ion)[3] : m/z 1071 [M+Na]$^+$, 1031 [M+H-H$_2$O]$^+$, 869 [M+H-H$_2$O-Glc]$^+$, 723 [M+H-H$_2$O-Glc-Rha]$^+$, 577 [M+H-H$_2$O-Glc-2x Rha]$^+$, 415 [M+H-H$_2$O-2x Glc-2x Rha]$^+$.

Mass (FAB, Negative ion)[4] : m/z 1047 [M-H]$^-$, 901 [M-Rha-H]$^-$, 885 [M-Glc-H]$^-$, 755 [M-2xRha-H]$^-$, 885 [M-Glc-H]$^-$, 755 [M-Rha-H]$^-$, 739 [M-Rha-Glc-H]$^-$, 593 [M-2xRha-Glc-H]$^-$, 413 [M-2xRha-2xGlc-H]$^-$.

Biological Activity : Causes morphological abnormality of *Pyricularia oryzae* mycelia, shows cytotoxic activity against the cancer cell line of K 562 in *vitro*.[3] It showed inhibitory effect on DNA, RNA and protein synthesis. It inhibited DNA synthesis at low concentration (IC_{50}=9 μM) in HL-60 cells. It cause irreversible DNA synthesis. It exhibited 50% inhibitory activity on RNA or protein at concentrations of 29 and 256 μM. It depressed the HL-60 cells growth in a dose dependent manner (IC_{50}=15 μM). Cytostatic concentration=32-64 μM, cytocidal concentration=64 μM. The relative inhibitory effects are DNA synthesis > RNA synthesis > protein synthesis.[4] Cytotoxic against sixty different human cancer lines.[7]

References

1. T. Kawasaki, T. Komori, K. Miyahara, T. Nohara, I. Hosokawa and K.M. Mihashi, *Chem. Pharm. Bull.*, **22**, 2164 (1974).

2. G.V. Khodakov, A.S. Shashkov, P.K. Kintya and Yu. A. Akimov, *Khim. Prir. Soedin*, **30**, 766 (1994); *Chem. Nat. Comp.*, 30, 713 (1994).

3. K. Hu, A. Dong, X. Yao, H. Kobayashi and S. Iwasaki, *Planta Med.*, **63**, 161 (1997).

4. Y. Shao, O. Poobrasert, E.J. Kenelly, C.-K. Chin, C.-T. Ho, M.-T. Huang, S.A. Garrison and G.A. Cordell, *Planta Med.*, **63**, 258 (1997).

5. S.W. Kim, K.C. Chung, K.H. Son and S.S. Kang, *Kor. J. Pharmacogn.*, **20**, 76 (1989).

6. S.B. Singh and R.S. Thakur, *J. Nat. Prod.*, **45**, 667 (1982).

7. K. Hu and X. Yao, *Planta Med.*, **68**, 297 (2002).

TRIBULOSAPONIN B

(25*S*)-5β-Furost-20(22)-en-3β,26-diol 3-O-[α-L-rhamnopyranosyl-(1→2)-[β-D-glucopyranosyl-(1→4)]-β-D-galactopyranoside]-26-O-β-D-glucoyranoside

Source : *Tribulus terrestris* Linn. (Zygophyllaceae)
Mol. Formula : $C_{51}H_{84}O_{22}$
Mol. Wt. : 1048
$[\alpha]_D^{25}$: -34° (c=0.004, MeOH)
Registry No. : [311310-52-8]

IR (KBr) : 3385, 2918, 2359, 1733, 1456, 1376 cm^{-1}.

PMR (C$_5$D$_5$N, 500 MHz) : δ 0.70 (s, 3xH-18), 0.98 (s, 3xH-19), 1.04 (d, J=6.3 Hz, 3xH-27), 1.61 (d, J=6.1 Hz, 3xH-6 of Rha), 1.65 (s, 3xH-21), 1.75 (d, J=6.1 Hz, 3xH-6 of Glc I), 2.49 (d, J=10.0 Hz, H-17), 3.49 (dd, J=7.1, 9.1 Hz, H-26A), 4.07 (dd, J=8.0, 11.5 Hz, H-26B), 4.23 (H-3), 4.84 (d, J=7.7, H-1 of Gal), 4.85 (H-16), 4.87 (d, J=7.0 Hz, H-1 of Glc II), 5.45 (d, J=7.6 Hz, H-1 of Glc I), 5.90 (br s, H-1 of Rha).

CMR (C$_5$D$_5$N, 125 MHz) : δ C-1) 30.8 (2) 26.8 (3) 75.3 (4) 30.7 (5) 35.2 (6) 26.8 (7) 26.8 (8) 36.7 (9) 40.1 (10) 35.2 (11) 21.3 (12) 40.1 (13) 43.8 (14) 54.7 (15) 33.6 (16) 84.5 (17) 64.6 (18) 14.4 (19) 24.0 (20) 103.5 (21) 11.8 (22) 152.3 (23) 23.6 (24) 31.4 (25) 33.6 (26) 75.2 (27) 17.1 **Gal** (1) 101.8 (2) 77.3 (3) 76.9 (4) 82.7 (5) 77.0 (6) 61 **Rha** (1) 102.3 (2) 72.4 (3) 72.7 (4) 73.9 (5) 70.1 (6) 18.4 **Glc I** (1) 105.6 (2) 75.2 (3) 78.4 (4) 71.6 (5) 78.5 (6) 62.8 **Glc II** (1) 105.1 (2) 76.3 (3) 78.4 (4) 71.8 (5) 77.8 (6) 62.8.

Mass (HRESIFT) : m/z 1072.5305 [M+Na]$^+$, 1049.5611 [M]$^+$ (calcd. 1049.2147).

Reference

1.   E. Bedir and I.A. Khan, *J. Nat. Prod.*, **63**, 1699 (2000).

TUBEROSIDE B (ALLIUM)

(25*S*)-5α-Furost-20(22)-ene-2α,3β,26-triol 3-O-{α-L-rhamnopyranosyl-(1→2)-[α-L-rhamnopyranosyl-(1→4)]-β-D-glucopyranoside}-26-O-β-D-glucopyranoside

Source : *Allium tuberosum* Rottl. Ex spreng. (Liliaceae)
Mol. Formula : $C_{51}H_{84}O_{22}$
Mol. Wt. : 1048
$[α]_D^{23}$: -29.5° (c=0.30, MeOH)
Registry No. : [259810-67-8]

IR (KBr) : 3417, 1452, 1383, 1000-1100 cm⁻¹.

PMR (C_5D_5N, 400 MHz) : δ 0.78 (s, 3xH-18), 0.96 (s, 3xH-19), 1.10 (d, *J*=6.5 Hz, 3xH-27), 1.66 (d, *J*=6.1 Hz, 3xH-6 of Rha II), 1.70 (s, 3xH-21), 1.72 (d, *J*=6.1 Hz, 3xH-6 of Rha I), 2.51 (d, *J*=9.9 Hz, H-17), 3.52 (dd, *J*=7.1, 9.1 Hz, H-26A), 3.83 (m, H-5 of Glc I), 3.92 (m, H-3), 3.99 (m, H-3 of Glc I, H-5 of Glc II), 4.05 (m, H-2 of Glc II), 4.10 (m, H-26B), 4.12 (m, H-2), 4.12 (m, H-6A of Glc I), 4.22 (m, H-2 of Glc I, H-3 and H-4 of Glc II), 4.32 (m, H-6B of Glc I), 4.35 (m, H-4 of Glc I), 4.36 (m, H-4 of Rha I, H-4 of Rha II), 4.41 (m, H-6A of Glc II), 4.52 (m, H-3 of Rha II), 4.56 (m, H-6B of Glc II), 4.60 (m, H-3 of Rha I), 4.69 (m, H-2 of Rha II), 4.85 (m, H-2 of Rha I), 4.87 (d, *J*=7.7 Hz, H-1 of Glc II), 4.87 (m, H-16), 4.89 (m, H-5 of Rha I), 4.90 (m, H-5 of Rha II), 5.05 (d, *J*=7.0 Hz, H-1 of Glc I), 5.82 (s, H-1 of Rha II), 6.36 (s, H-1 of Rha I).

CMR (C_5D_5N, 100 MHz) **:** δ C-1) 46.0 (2) 70.8 (3) 85.2 (4) 33.7 (5) 44.9 (6) 28.4 (7) 32.7 (8) 34.6 (9) 54.6 (10) 37.1 (11) 21.9 (12) 40.1 (13) 44.0 (14) 54.9 (15) 34.7 (16) 84.7 (17) 64.8 (18) 14.7 (19) 13.7 (20) 103.9 (21) 12.1 (22) 152.6 (23) 31.6 (24) 23.9 (25) 33.9 (26) 75.5 (27) 17.4 **Glc I** (1) 101.0 (2) 78.2 (3) 78.0 (4) 78.9 (5) 77.2 (6) 61.4 **Glc II** (1) 105.3 (2) 75.4 (3) 78.7 (4) 71.9 (5) 78.6 (6) 63.0 **Rha I** (1) 102.3 (2) 72.8 (30 72.6 (4) 74.0 (5) 69.7 (6) 18.7 **Rha II** (1) 103.1 (2) 72.9 (3) 72.7 (4) 74.2 (5) 70.6 (6) 18.7.

Mass (FAB) **:** *m/z* 1049 [M+H]⁺, 903 [M+H-Rha]⁺, 741 [M+H-Glc-Rha]⁺, 595 [M+H-Glc-2xRha]⁺, 433 [M+H-2xGlc-2xRha]⁺.

Reference

1.   S. Sang, A. Lao, H. Wang and Z. Chen, *Phytochemistry*, **52**, 1611 (1999).

TUBEROSIDE I

26-O-β-D-Glucopyranosyl-(5α,20S,25S)-furost-22-en-3β,20,26-triol 3-O-{α-L-rhamnopyranosyl-(1→2)-[α-L-rhamnopyranosyl-(1→4)]-β-D-glucopyranoside}

Source : *Allium tuberosum* Rottl. (Liliaceae)
Mol. Formula : $C_{51}H_{84}O_{22}$
Mol. Wt. : 1048
[α]$_D^{25}$ **:** -41.8° (c=0.28, MeOH)
Registry No. : [332838-01-4]

IR (KBr) **:** 3406 (OH), 1641 (C=C), 1041 (glycosidic linkage) cm⁻¹.

PMR (C$_5$D$_5$N, 400 MHz) : δ 0.95 (s, 3xH-18)[a], 0.96 (s, 3xH-19)[a], 1.15 (d, *J*=6.4 Hz, 3xH-27), 1.68 (d, *J*=6.2 Hz, 3xH-6 of Rha II), 1.79 (s, 3xH-21), 1.82 (d, *J*=6.1 Hz, 3xH-6 of Rha I), 3.51 (dd, *J*=7.0, 9.2 Hz), 3.60 (dd, *J*=7.0 Hz, H-26A), 3.79 (m, H-5 of Glc I), 4.00 (m, H-26B), 4.00 (m, H-5 of Glc II), 4.10 (m, H-2 of Glc II), 4.12 (m), 4.13 (m, H-6A of Glc I), 4.28 (m, H-2, H-3 of Glc I and H-3 of Glc II), 4.33 (m, H-4 of Glc II), 4.36 (m, H-6B of Glc I), 4.41 (m, H-4 of Rha I, H-4 of Rha II), 4.45 (m, H-4 of Glc I), 4.46 (m, H-6A of Glc II), 4.55 (m, H-23), 4.60 (m, H-6B of Glc II), 4.61 (m, H-3 of Rha II), 4.67 (m, H-3 of Rha I), 4.75 (s, H-2 of Rha II), 4.91 (d, *J*=7.9 Hz, H-1 of Glc II), 4.91 (s, H-2 of Rha I), 4.98 (m, H-5 of Rha I), 5.01 (m, H-5 of Rha II), 5.04 (d, *J*=7.2 Hz, H-1 of Glc I), 5.91 (s, H-1 of Rha II), 6.42 (s, H-1 of Rha I).

CMR (C$_5$D$_5$N, 100 MHz) : δ C-1) 37.4 (2) 30.1 (3) 77.1 (4) 34.6 (5) 44.8 (6) 29.1 (7) 32.4 (8) 34.8 (9) 54.4 (10) 36.0 (11) 21.0 (12) 39.7 (13) 40.8 (14) 57.0 (15) 33.6 (16) 84.4 (17) 68.1 (18) 13.9 (19) 12.5 (20) 76.9 (21) 22.0 (22) 163.7 (23) 91.5 (24) 29.8 (25) 34.8 (26) 75.5 (27) 17.6 **Glc I** (1) 100.0 (2) 78.1 (3) 78.3 (4) 78.9 (5) 77.1 (6) 61.6 **Rha I** (1) 102.3 (2) 72.7 (3) 73.0 (4) 74.2 (5) 70.6 (6) 18.8 **Rha II** (1) 103.1 (2) 72.6 (3) 72.9 (4) 74.0 (5) 69.6 (6) 18.7 **Glc II** (1) 105.3 (2) 75.4 (3) 78.8 (4) 71.8 (5) 78.6 (6) 62.9.

Reference

1.   S. Sang, S. Mao, A. Lao, Z. Chen and C.-T. Ho, *J. Agric. Food Chem.*, **49**, 1475 (2001).

TUBEROSIDE S
(25*S*)-Furost-20(22)-en-3β,26-diol 3-O-{β-D-glucopyranosyl-(1→2)-[α-L-rhamnopyranosyl-(1→4)]-β-D-glucopyranoside}-26-O-β-D-glucopyranoside

Source : *Allium tuberosum* L. (Liliaceae)
Mol. Formula : C$_{51}$H$_{84}$O$_{22}$
Mol. Wt. : 1048
[α]$_D^{24}$: -25.2° (c=0.32, MeOH)
Registry No. : [651306-84-2]

IR (KBr) : 3400, 1448, 1379, 1074 cm^{-1}.

PMR (C$_5$D$_5$N, 400 MHz) : δ 0.78 (s, 3xH-18), 1.05 (s, 3xH-19), 1.12 (d, *J*=6.6 Hz, 3xH-27), 1.70 (s, 3xH-21), 1.73 (d, *J*=6.2 Hz, 3xH-6 of Rha), 2.56 (d, *J*=10.1 Hz, H-17), 3.57 (dd, *J*=7.0, 7.9 Hz, H-26A), 3.70 (m, H-5 of Glc I), 4.05 (m, H-5 of Glc II, H-5 of Glc III), 4.11 (m, H-2 of Glc II), 4.12 (m, H-6A of Glc I), 4.15 (m, H-26B), 4.16 (m, H-2 of Glc III), 4.29 (m, H-6A of Glc I, H-6A of Glc III), 4.30 (m, H-2 of Glc II, H-4 of Glc III), 4.31 (m, H-3 of Glc II, H-4 of Glc II), 4.33 (m, H-3 of Glc I), 4.35 (m, H-3), 4.37 (m, H-4 of Rha), 4.46 (m, H-6A of Glc II), 4.51 (m, H-4 of Glc I), 4.55 (m, H-6A of Glc III), 4.58 (m, H-3 of Rha), 4.62 (m, H-6B of Glc II), 4.65 (m, H-6B of Glc III,), 4.74 (s, H-2 of Rha), 4.91 (m, H-16), 4.91 (d, *J*=7.7 Hz, H-1 of Glc II), 4.94 (d, *J*=7.1 Hz, H-1 of Glc I), 5.03 (m, H-5 of Rha), 5.51 (d, *J*=7.7 Hz, H-1 of Glc III), 5.96 (s, H-1 of Rha).

CMR (C$_5$D$_5$N, 100 MHz) : δ C-1) 30.9 (2) 27.0 (3) 75.6 (4) 31.0 (5) 36.9 (6) 27.1 (7) 26.9 (8) 35.4 (9) 40.4 (10) 35.4 (11) 21.5 (12) 40.3 (13) 44.0 (14) 54.9 (15) 34.6 (16) 84.8 (17) 64.9 (18) 14.6 (19) 24.2 (20) 103.8 (21) 12.0 (22) 152.6 (23) 23.8 (24) 31.6 (25) 33.9 (26) 75.4 (27) 17.4 **Glc I** (1) 102.0 (2) 83.0 (3) 77.2 (4) 77.5 (5) 76.6 (6) 61.5 **Glc II** (1) 105.3 (2) 75.4 (3) 78.7 (4) 71.9 (5) 78.7 (6) 63.0 **Rha** (1) 102.6 (2) 72.9 (3) 72.7 (4) 74.1 (5) 70.4 (6) 18.4 **Glc III** (1) 105.8 (2) 77.2 (3) 78.1 (4) 77.0 (5) 78.7 (6) 63.1.

Mass (E.S.I., Positive ion) : *m/z* 1072 [M+H+Na]$^+$.

Reference

1.   S. Sang, S. Mao, A. Lao, Z. Chen and C.-T. Ho, *Food Chemistry*, **83**, 499 (2003).

ANGUIVIOSIDE XI

(22*S*,23*S*,25*R*,26*S*)-3β,22α,26-Trihydroxy–23,26-epoxy-furost-5-ene 3-O-{α-L-rhamnopyranosyl-(1→2)-[β-D-xylopyranosyl-(1→3)]-β-D-glucopyranoside}-26-O-β-D-glucopyranoside

Source : *Solanum anguivi* (Solanaceae)
Mol. Formula : $C_{50}H_{80}O_{23}$
Mol. Wt. : 1048
[α]$_D^{26}$: -61.4° (c=0.7, MeOH)
Registry No. : [478035-45-9]

PMR (C$_5$D$_5$N, 500 MHz) : δ 0.84 (d, *J*=7.3 Hz, 3xH-27), 0.92 (s, 3xH-18), 1.04 (s, 3xH-19), 1.27 (d, *J*=6.7 Hz, 3xH-21),1.75 (d, *J*=6.1 Hz, 3xH-6 of Rha), 4.58* (H-23), 4.98 (d, *J*=6.7 Hz, H-1 of Glc I and H-1 of Xyl), 5.33 (d, *J*=8.6 Hz, H-1 of Glc II), 5.50 (s, H-26), 6.30 (s, H-1 of Rha). * overlapped signal.

CMR (C$_5$D$_5$N, 125 MHz) : δ C-1) 37.1 (2) 30.1 (3) 77.7 (4) 38.7 (5) 140.7 (6) 121.9 (7) 32.4 (8) 31.8 (9) 50.3 (10) 37.1 (11) 21.0 (12) 39.7 (13) 40.9 (14) 56.7 (15) 32.3 (16) 81.4 (17) 64.6 (18) 16.3 (19) 19.4 (20) 39.2 (21) 16.5 (22) 109.1 (23) 85.0 (24) 32.1 (25) 40.2 (26) 107.2 (27) 16.7 **Glc I** (1) 99.9 (2) 78.3 (3) 88.2 (4) 69.5 (5) 77.4 (6) 62.3 **Rha** (1) 102.4 (2) 72.4 (3) 72.8 (4) 74.1 (5) 69.5 (6) 18.7 **Xyl** (1) 105.4 (2) 74.6 (3) 78.7 (4) 70.7 (5) 67.2.

Mass (FAB, Positive ion) : *m/z* 1072 [M+Na+H]$^+$.

Reference

1. T. Honbu, T. Ikeda, X.-H. Zhu, O. Yoshihara, M. Okawa, A.M. Nafady and T. Nohara, *J. Nat. Prod.*, **65**, 1918 (2002).

CHINENOSIDE IV

3β,26-Dihydroxy-23-hydroxymethyl-25(*R*)-5α-furost-20(22)-en-6-one 3-O-{β-xylopyranosyl-(1→4)-
[α-L-arabinopyranosyl-(1→6)]-β-D-glucopyranoside}-26-O-β-D-glucopyranoside

Source : *Allium chinense* G. Don (Liliaceae)
Mol. Formula : $C_{50}H_{80}O_{23}$
Mol. Wt. : 1048
M.P. : 172-175°C
$[\alpha]_D^{20}$ **:** -35.8° (c=0.32, Pyridine)
Registry No. : [187144-80-5]

IR (KBr) : 3341, 2916.2 2875.2, 2875.3, 2747.8, 1708.1, 1448.2, 1371.3, 1254.6, 1161.8, 1040.6, 907.1, 859.8, 779.9 cm^{-1}.

PMR (400 MHz) : δ 0.63 (s, 3xH-19), 0.67 (s, 3xH-18), 0.97* (H-14), 1.09* (H-9), 1.09 (H-1A), 1.18 (d, *J*=6.7 Hz, 3xH-27), 1.34 (H-15A), 1.46 (H-11A), 1.54* (H-1B), 1.58 (H-2A), 1.62 (H-24A), 1.69 (H-4A), 1.69* (H-11B), 1.70* (H-12), 1.72* (H-8), 1.76 (s, 3xH-21), 1.88* (H-24B), 1.94* (H-15B), 1.95 (H-7A), 2.05* (H-2B), 2.13* (H-5), 2.13* (H-25), 2.28* (H-7B), 2.31* (H-4B), 2.47 (d, *J*=10.7, H-17), 2.31 (H-4B), 3.17 (m, H-23), 3.68 (d, *J*=11.6 Hz, H-5A of

Ara), 3.78 (H-26A), 3.83*(H-2 of Glc I), 3.87* (H-5 of Glc I), 3.87 (H-5 of Xyl), 3.87 (H-28A), 3.90* (H-3), 3.92* (H-5 of Glc II), 3.97* (H-2 of Xyl), 4.02, H-28B), 4.03* (H-26B), 4.14* (H-3 of Glc I), 4.15* (H-4 of Xyl), 4.16* (H-4 of Glc II), 4.18* (H-4 of Ara), 4.18* (H-3 of Glc II), 4.22* (H-5B of Ara), 4.24* (H-3 of Ara), 4.35* (H-6A of Glc II), 4.38* (H-4 of Glc), 4.39* (H-2 of Ara), 4.54 (dd, *J*=11.6, 2.1 Hz), H-6B of Glc II), 4.65 (dd, *J*=7.6, 2.4 Hz, H-6A of Glc I), 4.73* (H-16), 4.76 (m, H-6B of Glc I), 4.83 (d, *J*=7.6 Hz, H-1 of Glc II), 4.89 (d, *J*=7.6 Hz, H-1 of Glc), 5.01 (d, *J*=7.6 Hz, H-1 of Ara), 5.43 (d, *J*=7.6 Hz, H-1 of Xyl). * overlapped signal.

CMR (100 MHz) : δ C-1) 36.7 (2), 29.4 (3) 76.9 (4) 26.9 (5) 56.5 (6) 209.4 (7) 46.9 (8) 37.0 (9) 53.6 (10) 40.9 (11) 21.7 (12) 39.3 (13) 43.9 (14) 54.7 (15) 34.1 (16) 84.2 (17) 64.6 (18) 14.7 (19) 13.1 (20) 105.5 (21) 11.9 (22) 153.3 (23) 38.4 (24) 33.7 (25) 31.8 (26) 74.5 (27) 19.2 (28) 65.2 **Glc I** (1) 102.0 (2) 74.8 (3) 76.2 (4) 79.8 (5) 74.8 (6) 68.1 **Xyl** (1) 105.1 (2) 74.9 (3) 78.4 (4) 71.0 (5) 67.2 **Glc II** (1) 104.9 (2) 75.1 (3) 78.5 (4) 71.7 (5) 78.4 (6) 62.8.

Mass (FAB, Positive ion) : *m/z* 1049 [M+H]⁺.

Mass (FAB, Negative ion) : *m/z* 1047 [M-H]⁻.

Reference

1.   J. Peng, X. Yao, Y. Tezuka, T. Kikuchi and T. Narui, *Planta Med.*, **62**, 465 (1996).

RUSCOSIDE

1β,3β,22α,26-Tetrahydroxyfurosta-5,25(27)-diene 1-O-[β-D-glucopyranosyl-(1→3)-α-L-rhamnopyranosyl-(1→2)-α-L-arabinopyranoside]-26-O-β-D-glucopyranoside

Source : *Ruscus aculeatus* L. (Liliaceae)
Mol. Formula : C₅₀H₈₀O₂₃
Mol. Wt. : 1048
[α]ᴅ²⁰ : -42.0° (c=1.0, H₂O)¹
Registry No. : [51024-64-7]

Mass (E.S.I.)^2 : *m/z* 1071 [M+Na]$^+$, 1031 [M+H-H$_2$O]$^+$, 869 [M+H-H$_2$O-Glc]$^+$, 723 [M+H-H$_2$O-Glc-Rha]$^+$, 707 [M+H-H$_2$O-2xGlc]$^+$, 591 [M+H-H$_2$O-Glc-Rha-Ara]$^+$, 561 [M+H-H$_2$O-2xGlc-Rha]$^+$, 429 [Agl+H]$^+$.

Mass (E.S.I., Negative ion)^2 : *m/z* 1047 [M-H]$^-$.

Reference

1.   E. Bombardelli, A. Bonati, B. Gabetta and G. Mustich, *Fitoterapia*, **43**, 3 (1972).

2.   E.de Combarieu, M. Falzoni, N. Fuzzati, F. Gattesco, A. Giori, M. Lovati and R. Pace, *Fitoterapia*, **73**, 583 (2002).

YUCCA SCHIDIGERA SAPONIN 2

5β-(25*R*)-Furostan-20(22)-en-3β,26-diol-12-one 3-O-{β-D-glucopyranosyl-(1→2)-[β-D-xylopyranosyl-(1→3)]-β-D-glucopyranoside}-26-O-β-D-glucopyranoside

Source : *Yucca schidigera* Roezl ex ortgies (Agavaceae)
Mol. Formula : C$_{50}$H$_{80}$O$_{23}$
Mol. Wt. : 1048
M.P. : 193-195°C
$[\alpha]_D^{25}$: -3.6 (c=0.1, MeOH)

PMR (CD₃OD, 599.19 MHz) : δ 0.84 (s, 3xH-18)†, 0.96 (d, J=6.0 Hz, 3xH-27), 0.97 (s, 3xH-18)†, 1.09 (s, 3xH-19), 1.61 (s, 3xH-21), 3.27 (s, H-17), 3.43 (m, H-26A), 3.74 (m, H-26B), 4.14 (m, H-3), 4.72 (m, H-16) PMR of sugar superimpossible on those reported for Yacca Schidigera Saponin 1 (qv). † Two chemical shifts for 18-Me group are clearly erroneous.

CMR (CD₃OD, 150.86 MHz) : δ C-1) 30.2 (2) 27.1 (3) 75.6 (4) 30.8 (5) 36.6 (6) 27.0 (7) 26.8 (8) 35.4 (9) 43.0 (10) 36.3 (11) 38.4 (12) 216.3 (13) 58.6 (14) 55.6 (15) 34.1 (16) 83.8 (17) 56.8 (18) 14.2 (19) 23.4 (20) 104.3 (21) 11.2 (22) 153.5 (23) 23.8 (24) 33.7 (25) 31.6 (26) 75.4 (27) 16.9 **Glc I** (1) 100.5 (2) 78.7 (3) 87.6 (4) 69.9 (5) 77.5 (6) 62.5 **Glc II** (1) 103.3 (2) 75.8 (3) 77.7 (4) 72.3 (5) 78.1 (6) 63.4 **Xyl** (1) 104.8 (2) 74.9 (3) 78.2 (4) 70.8 (5) 66.8 **Glc III** (1) 104.6 (2) 75.4 (3) 77.4 (4) 71.5 (5) 77.8 (6) 62.4.

Mass (E.S.I., Negative ion, H.R.) : m/z 1047.5002 [(M-H)⁻, calcd. for 1047.5007], 915 [M-Xyl]⁻, 753 [M-Xyl-Glc]⁻, 591 [M-Xyl-2xGlc]⁻.

Reference

1.   W. Oleszek, M.Sitek, A. Stochmal, S. Piacente, C. Pizza and P. Cheeke, *J. Agric. Food Chem.*, **49**, 4392 (2001).

ASPARAGUS COCHINCHINENSIS SAPONIN ASP-VI
(5β,25*S*)-22ξ-Methoxy-furostane-3β,26-diol 3-O-{β-D-xylopyranosyl-(1→4)-[α-L-rhamnopyranosyl-(1→6)]-
β-D-glucopyranoside}-26-O-β-D-glucopyranoside

Source : *Asparagus cochinchinensis* (Lourerio) Merill (Liliaceae)
Mol. Formula : C₅₁H₈₆O₂₂
Mol. Wt. : 1050
M.P. : 165-168.5°C (decomp.)
[α]$_D$: -50° (c=1.0, MeOH)
Registry No. : [72947-81-0]

PMR (C_5D_5N, 90 MHz) : δ 3.29 (s, OCH_3), 4.84 (d, *J*=7.0 Hz, 3xanomeric H), 5.50 (s, anomeric H).

CMR (C_5D_5N, 22.15 MHz) : δ **Glc I** C-1) 104.6 **Glc II** (1) 104.6 **Rha** (1) 101.7 **Xyl** 104.9.

Reference

1.   T. Konishi and J. Shoji, *Chem., Pharm. Bull.*, **27**, 3086 (1979).

NEOPROTODIOSCIN, ALLIUM TUBEROSUM SAPONIN 4
(5α,25*R*)-Furostan-3β,22α,26-triol 3-O-{α-L-rhamnopyranosyl-(1→2)-[α-L-rhamnopyranosyl-(1→4)-
β-D-glucopyranoside]-26-O-β-D-glucopyranoside

Source : *Tribulus terresteris* L.[1] (Zygophyllaceae), *Allium tuberosum* Rottler[2] (Liliaceae)
Mol. Formula : $C_{51}H_{86}O_{22}$
Mol. Wt. : 1050
M.P. : 207-208°C(decomp.)[1]
Registry No. : [664366-25-0]

IR (KBr)[1] : 3420, 2933, 1653, 1457, 1382, 1130, 1041, 914, 841, 811 cm^{-1}.

PMR (C$_5$D$_5$N, 300 MHz)[1] : δ 0.88 (m, H-9), 0.90 (s, 3xH-18), 0.93 (br dd, H-5), 0.96 (m, H-1α), 1.01 (d, J=6.5 Hz, 3xH-27), 1.04 (s, 3xH-19), 1.08 (m, H-12α), 1.11 (m, H-14), 1.15 (m, 2xH-6), 1.34 (d, J=6.9 Hz, 3xH-21), 1.40 (m, H-4α), 1.48 (m, H-7α, 2xH-11), H-15α), 1.58 (br s, H-8), 1.62 (d, J=6.2 Hz, 3xH-6 of Rha II), 1.69 (m, H-24β), 1.74 (m, H-1β), 1.74 (d, J=6.2 Hz, 3xH-6 of Rha I), 1.82 (m, H-2α), 1.83 (m, H-12β), 1.87 (m, H-4b, H-7β), 1.92 (s, H-17), 1.94 (m, H-23β, H-25), 2.05 (m, H-15β, H-23α, H-24α), 2.07 (m, H-2β), 2.26 (m, H-20), 3.64 (dd, J=5.7, 9.4 Hz, H-26α), 3.71 (m, H-5 of Glc I), 3.92 (m, H-3), 3.95 (dd, J=5.7, 9.4 Hz, H-26β), 3.95 (m, H-5 of Glc II), 3.99 (m, H-2 of Glc II), 4.07 (m, H-6A of Glc I), 4.15 (m, H-6B of Glc I), 4.19 (m, H-2 of Glc I, H-3 of Glc II), 4.20 (m, H-3 of Glc I, H-4 of Glc II), 4.25 (m, H-4 of Rha II), 4.29 (m, H-4 of Glc I), 4.33 (m, H-6A of Glc II), 4.38 (m, H-6A of Glc II), 4.50 (m, H-3 of Rha II), 4.53 (m, H-6β of Glc II), 4.56 (dd, J=2.8, 9.0 Hz, H-3 of Rha I), 4.66 (m, H-2 of Rha II), 4.78 (m, H-2 of Rha I), 4.79 (d, J=7.5 Hz, H-1 of Glc II), 4.88 (m, H-5 of Rha II), 4.92 (m, H-5 of Rha I), 4.94 (m, H-16), 4.96 (d, J=7.5 Hz, H-1 of Glc I), 5.79 (d, J=1.0 Hz, H-1 of Rha II), 6.31 (d, J=0.9 Hz, H-1 of Rha I).

CMR (C$_5$D$_5$N, 75 MHz)[1] : δ C-1) 37.3 (2) 30.0 (3) 77.4 (4) 39.1 (5) 44.8 (6) 29.0 (7) 32.4 (8) 34.6 (9) 54.6 (10) 35.4 (11) 21.3 (12) 40.3 (13) 41.2 (14) 56.5 (15) 32.5 (16) 81.2 (17) 64.0 (18) 16.7 (19) 12.5 (20) 40.7 (21) 16.4 (22) 110.6 (23) 37.2 (24) 28.4 (25) 34.3 (26) 75.2 (27) 17.5 **Glc I** (1) 100.0 (2) 77.9 (3) 78.2 (4) 79.2 (5) 76.9 (6) 61.6 **Rha I** (1) 102.1 (2) 72.5 (3) 72.8 (4) 74.2 (5) 69.7 (6) 18.5 **Rha II** (1) 103.0 (2) 72.4 (3) 72.7 (4) 73.9 (5) 70.5 (6) 18.6 **Glc II** (1) 104.9 (2) 75.3 (3) 78.4 (4) 71.8 (5) 78.6 (6) 62.9.

Mass (E.S.I., Positive ion)[1] : m/z 1033 [M+H-H$_2$O]$^+$, 887 [M+H-H$_2$O-Rha]$^+$, 871 [M+H-H$_2$O-Glc]$^+$, 741 [M+H-H$_2$O-2xRha]$^+$, 725 [M+H-H$_2$O-Rha-Glc]$^+$, 579 [M+H-H$_2$O-2xRha-Glc]$^+$, 417 [M+H-H$_2$O-2xRha-2xGlc]$^+$.

Mass (E.S.I., Negative ion)[1] : m/z 1048 [M-H]$^-$, 1095 [M-H+HCOOH]$^-$.

References

1. E.De Combarieu, N. Fuzzati, M. Lovati and E. Mercalli, *Fitoterapia*, **74**, 583 (2003).

2. T. Ikeda, H. Tsumagari, M. Okawa and T. Nohara, *Chem. Pharm. Bull.*, **52**, 142 (2004).

NICOTIANOSIDE F

(25S)-5α-Furostan-3β,22α,26-triol 3-O-{α-L-rhamnopyranosyl-(1→2)-[α-L-rhamnopyranosyl-(1→4)]-β-D-glucopyranoside}-26-O-β-D-glucopyranoside

Source : *Nicotiana tabacum* L. (Solanceae)
Mol. Formula : $C_{51}H_{86}O_{22}$
Mol. Wt. : 1050
M.P. : 181-182°C
[α]$_D^{20}$: -75° (c=1.0, H$_2$O)
Registry No. : [174756-29-7]

CMR (C$_5$D$_5$N, 125 MHz) : δ C-1) 37.0 (2) 30.2 (3) 78.6 (4) 33.9 (5) 44.5 (6) 28.6 (7) 32.1 (8) 35.1 (9) 54.3 (10) 35.3 (11) 21.0 (12), 40.1 (13), 40.3 (14) 56.8 (15) 32.0 (16) 81.0 (17) 62.2 (18) 16.5 (19) 12.2 (20) 42.1 (21) 14.3 (22) 112.2 (23) 35.4 (24) 28.3 (25) 36.3 (26) 71.8 (27) 17.2 (OCH$_3$) 49.8* **Glc I** (1) 100.0 (2) 78.4 (3) 77.1 (4) 78.1 (5) 76.9 (6) 61.6 **Rha I** (1) 102.4 (2) 72.9 (3) 73.0 (4) 74.2 (5) 70.6 (6) 18.7 **Rha II** (1) 103.1 (2) 72.7 (3) 72.9 (4) 74.0 (5) 69.7 (6) 18.8 **Glc II** (1) 105.2 (2) 75.4 (3) 78.7 (4) 71.9 (5) 78.6 (6) 63.0. * This chmical shift is erroneous as the compound does not contain an OCH$_3$ group.

Reference

1. S.A. Shvets, P.K. Kintya, O.N. Gutsu, and V.I. Grishkovets, *Khim Prir. Soedin.*, **31**, 396 (1995); *Chem. Nat. Comp.*, **31**, 332 (1995).

TRIGOFOENOSIDE E-1

(25R)-22-Methoxy-5α-furostane-3β,26-diol 3-O-[α-L-rhamnopyranosyl-(1→2)-β-D-xylopyranosyl-(1→4)]-β-D-glucopyranoside-26-O-β-D-glucopyranoside

Source : *Trigonella foenum graecum* L. (Papilionaceae, Leguminosae)
Mol. Formula : $C_{51}H_{86}O_{22}$
Mol. Wt. : 1050
M.P. : 230-231°C
[α]$_D$: -57.9° (c=1.0, Pyridine)
Registry No. : [101910-70-7]

IR (KBr) : 3600-3240 cm^{-1} (OH), no spiroketal bands.

PMR (DMSO-d$_6$, 80 MHz) : δ 3.10 (s, OCH_3), 4.18 (d, *J*=6.5 Hz), 4.48 (d, *J*=8.0 Hz), 4.95 (d, *J*=7.0 Hz), 5.12 (br s), [4 x anomeric H].

Reference

1. R.K. Gupta, D.C. Jain and R.S. Thakur, *Ind. J. Chem.*, **24B**, 1215 (1985).

ASPARAGUS CURILLUS SAPONIN 5

(25*S*)-3β-Furostan-3β,22α,26-triol 3-O-{β-D-glucopyranosyl-(1→2)-[α-L-arabinopyranosyl-(1→4)]-
β-D-glucopyranoside}-26-O-β-D-glucopyranoside

Source : *Asparagus curillus* Buch.-Ham. (Liliaceae)
Mol. Formula : $C_{50}H_{84}O_{23}$
Mol. Wt. : 1052
$[\alpha]_D^{20}$: -51° (c=1.0, H_2O)
Registry No. : [84800-14-6]

IR (KBr) : 3400 (OH) cm^{-1}, no spiroketal absorptions.

Reference

1.   S.C. Sharma, O.P. Sati and R. Chand, *Phytochemistry*, **21**, 1711 (1982).

OFFICINALISININ - II

(5β,25S)-Furostane-3β,22α,26-triol 3-O-[β-D-glucopyranosyl-(1→2)-β-D-xylopyranosyl-(1→4)-β-D-glucopyranoside]-26-O-β-D-glucopyranoside

Source : *Asparagus officinalis* L. c.v. *altilis* (Liliaceae)
Mol. Formula : $C_{50}H_{84}O_{23}$
Mol. Wt. : 1052
M.P. : 175-182°C (decomp.)
[α]$_D^{25}$: -41.6° (c=0.625, MeOH)
Registry No. : [57944-19-1]

IR (KBr) **:** 3500-3300 (OH) cm^{-1}.

Reference

1.   K. Kawano, K. Sakai, H. Sato and S. Sakamura, *Agric. Biol. Chem.*, **39**, 1999 (1975).

YUCCA SCHIDIGERA SAPONIN 1

5β-(25*R*)-Furostan-3β,22α,26-triol 3-O-{β-D-glucopyranosyl-(1→2)-[β-D-xylopyranosyl-(1→3)]-
β-glucopyranoside}-26-O-β-D-glucopyranoside

Source : *Yucca schidigera* Roezl ex ortiges. (Agavaceae)
Mol. Formula : $C_{50}H_{84}O_{23}$
Mol. Wt. : 1052
M.P. : 235-236°C
$[\alpha]_D^{25}$: -43.25° (c=0.1, MeOH)

PMR (CD$_3$OD, 599.19 MHz) : δ 0.84 (s, 3xH-18), 0.98 (d, *J*=6.0 Hz, 3xH-27), 1.02 (s, 3xH-19), 1.03 (d, *J*=6.0 Hz, 3xH-21), 3.13 (dd, *J*=7.5, 9.0 Hz, H-2 of Glc II), 3.16 (dd, *J*=9.0, 9.0 Hz, H-4 of Glc II), 3.20 (dd, *J*=7.5, 9.0 Hz, H-2 of Glc III), 3.28 (t, *J*=11.0 Hz, H-5A of Xyl), 3.29 (ddd, *J*=2.5, 5.0, 9.0 Hz, H-5 of Glc II), 3.30 (dd, *J*=7.5, 9.0 Hz, H-2 of Xyl), 3.31 (dd, *J*=9.0, 9.0 Hz, H-4 of Glc III), 3.32 (ddd, *J*=2.5, 4.5, 9.0 Hz, H-5 of Glc II), 3.32 (dd, *J*=9.0, 9.0 Hz, H-4 of Glc I), 3.33 (dd, *J*=9.0, 9.0 Hz, H-3 of Xyl), 3.36 (m, H-26A), 3.36 (dd, *J*=9.0, 9.0 Hz, H-3 of Glc III), 3.39 (ddd, *J*=2.5, 5.0, 9.0 Hz, H-5 of Glc I), 3.39 (dd, *J*=9.0, 9.0 Hz, H-3 of Glc II), 3.54 (ddd, *J*=4.5, 9.0, 11.0 Hz, H-4 of Xyl), 3.65 (dd, *J*=4.5,11.5 Hz, H-6A of Glc II), 3.68 (dd, *J*=5.0, 12.0 Hz, H-6A of Glc I), 3.68 (dd, *J*=5.0, 11.5 Hz, H-6A of Glc III), 3.72 (dd, *J*=9.0, 9.0 Hz, H-3 of Glc I), 3.73 (dd, *J*=7.2, 9.0 Hz, H-2 of Glc I), 3.82 (m, H-26B), 3.85 (dd, *J*=4.5,11.5 Hz, H-6B of Glc II), 3.88 (dd, *J*=2.5, 11.5 Hz, H-6B of Glc I), 3.90 (dd, *J*=2.5, 12.0 Hz, H-6B of Glc I), 3.94 (dd, *J*=4.5, 11.0 Hz, H-5B of Xyl), 4.12 (m, H-3), 4.26 (d, *J*=7.5 Hz, H-1 of Glc III), 4.38 (m, H-16), 4.52 (d, *J*=7.2 Hz, H-1 of Glc I), 4.60 (d, *J*=7.5 Hz, H-1 of Xyl), 4.96 (d, *J*=7.5 Hz, H-1 of Glc II).

CMR (CD₃OD, 150.86 MHz) : δ C-1) 31.3 (2) 27.3 (3) 76.1 (4) 30.6 (5) 37.2 (6) 27.6 (7) 27.5 (8) 36.7 (9) 41.3 (10) 36.1 (11) 21.9 (12) 41.2 (13) 42.2 (14) 57.5 (15) 32.7 (16) 82.5 (17) 65.3 (18) 16.5 (19) 24.5 (20) 41.2 (21) 16.2 (22) 113.9 (23) 31.4 (24) 28.9 (25) 35.0 (26) 75.8 (27) 17.4 **Glc I** (1) 100.5 (2) 78.5 (3) 87.7 (4) 69.9 (5) 77.8 (6) 62.5 **Glc II** (1) 103.3 (2) 75.8 (3) 77.8 (4) 72.3 (5) 78.1 (6) 63.4 **Xyl** (1) 104.8 (2) 74.9 (3) 78.2 (4) 70.8 (5) 66.8 **Glc III** (1) 104.6 (2) 75.1 (3) 77.5 (4) 71.5 (5) 77.8 (6) 62.4.

Mass (E.S.I., Negative ion, H.R.) : *m/z* 1049.5166 [C₅₀H₈₁O₂₃ calcd. for 1049.5163].

Reference

1. W. Oleszek, M.Sitek, A. Stochmal, S. Piacente, C. Pizza and P. Cheeke, *J. Agric. Food Chem.*, **49**, 4392 (2001).

RUSCUS ACULEATUS SAPONIN 10

26ξ-Methoxyfurosta-5,25(27)-diene-1β,3β,26-triol 1-O-{α-L-rhamnopyranosyl-(1→2)-(3-O-acetyl)-4-O-[(2S,3S)-2-hydroxy-3-methylpentanoyl]-α-L-arabinopyranoside}-26-O-β-D-glucopyranoside

Source : *Ruscus aculeatus* L. (Liliaceae)
Mol. Formula : C₅₃H₈₄O₂₁
Mol. Wt. : 1056
[α]ᴅ²⁶ : -44.0° (c=0.10, MeOH)
Registry No. : [205191-11-3]

IR (KBr) : 3430 (OH), 2930 (CH), 1745 (C=O), 1045 cm⁻¹.

PMR (C₅D₅N, 400 MHz) : δ 0.94 (s, 3xH-18), 1.14 (t, *J*=7.4 Hz, 3xH-5 of HMP), 1.24 (d, *J*=6.9 Hz, 3xH-21), 1.31 (d, *J*=6.9 Hz, 3xH-6 of HMP), 1.40 (s, 3xH-19), 1.75 (d, *J*=6.2 Hz, 3xH-6 of Rha), 2.00 (s, OCOCH₃), 3.26 (s, OCH₃), 4.77 (d, *J*=7.3 Hz, H-1 of Ara), 4.93 (d, *J*=7.7 Hz, H-1 of Glc), 5.08 and 5.37 (each 1H, br s, 2xH-27), 5.45 (dd, *J*=9.7, 3.3 Hz, H-3 of Ara), 5.61 (br d, *J*=5.7 Hz, H-6), 5.66 (br s, H-1 of Rha), 5.68 (br dd, *J*=3.3, 1.9 Hz, H-4 of Ara).

CMR (C$_5$D$_5$N, 100 MHz) : δ C-1) 83.2 (2) 37.2 (3) 68.1 (4) 43.7 (5) 139.2 (6) 124.8 (7) 31.9 (8) 33.0 (9) 50.2 (10) 42.7 (11) 24.3 (12) 40.5 (13) 40.7 (14) 56.8 (15) 32.3 (16) 81.5 (17) 64.2 (18) 16.8 (19) 15.0 (20) 40.5 (21) 16.1 (22) 112.5 (23) 31.6 (24) 28.1 (25) 146.8 (26) 72.0 (27) 111.0 (OCH$_3$) 47.3 **Ara** (1) 99.5 (2) 73.1 (3) 74.9 (4) 70.1 (5) 64.1 **Rha** (1) 102.1 (2) 72.3 (3) 72.4 (4) 73.8 (5) 69.5 (6) 19.0 **HMP** (1) 174.7 (2) 75.6 (3) 39.7 (4) 24.7 (5) 12.2 (6) 16.1 **Glc** (1) 103.9 (2) 75.2 (3) 78.5 (4) 71.7 (5) 78.6 (6) 62.9 (OCH$_3$) 170.1, 20.7.

Mass (FAB, Negative ion) : m/z 1055 [M-H]$^-$.

Biological Activity : It showed potent cytostatic activity on growth of Leukemia HL 60 cells, and 92.4% inhibition at 10 μg/ml sample concentration. Its IC$_{50}$ value is 3.5 μg/ml.

Reference

1.   Y. Mimaki, M. Kuroda, A. Kameyama, A. Yokosuka and Y. Sashida, *Chem. Pharm. Bull.*, **46**, 298 (1998).

RUSCUS ACULEATUS SAPONIN 22
22-O-Methyl-(25R)-furost-5-ene-1β,3β,22ξ,26-tetrol 1-O-{α-L-rhamnopyranosyl-(1→2)-
3,4,6-tri-O-acetyl-β-D-galactopyranoside}-26-O-β-D-glucopyranoside

Source : *Ruscus aculeatus* (Liliaceae)
Mol. Formula : C$_{52}$H$_{82}$O$_{22}$
Mol. Wt. : 1058
$[\alpha]_D^{25}$: -160° (c=0.10, MeOH)
Registry No. : [211379-16-7]

IR (KBr) : 3430 (OH), 2930 (CH), 1745 (C=O), 1445, 1370, 1235, 1130, 1045, 980, 910, 835, 810 cm^{-1}.

PMR (C$_5$D$_5$N, 400 MHz) : δ 0.97 (s, 3xH-18), 1.00 (d, J=6.5 Hz, 3xH-27), 1.20 (d, J=6.8 Hz, 3xH-21), 1.40 (s, 3xH-19), 1.78 (d, J=6.1 Hz, 3xH-6 of Rha), 1.95, 1.98 and 2.08 (each s, 3xOCOCH$_3$), 3.26 (s, OCH$_3$), 3.77 (dd, J=12.0, 3.8 Hz, H-1), 3.83 (m, H-3), 4.85 (d, J=7.6 Hz, H-1 of Gal), 4.87 (d, J=7.5 Hz, H-1 of Glc), 5.50 (dd, J=9.9, 3.2 Hz, H-3 of Gal), 5.64 (br d, J=5.3 Hz, H-6), 5.69 (br s, H-1 of Rha), 5.76 (br d, J=3.2 Hz, H-4 of Gal).

CMR (C_5D_5N, 100 MHz) : δ C-1) 84.5 (2) 37.9 (3) 68.1 (4) 43.6 (5) 139.2 (6) 124.9 (7) 31.9 (8) 33.1 (9) 50.4 (10) 42.6 (11) 24.4 (12) 40.4 (13) 40.7 (14) 57.1 (15) 32.4 (16) 81.4 (17) 64.5 (18) 16.9 (19) 14.9 (20) 40.6 (21) 16.3 (22) 112.7 (23) 30.8 (24) 28.2 (25) 34.2 (26) 75.2 (27) 17.1 (OCH_3) 47.2 **Gal** (1) 99.8 (2) 72.8 (3) 75.2 (4) 68.5 (5) 70.7 (6) 62.2 **Rha** (1) 102.1 (2) 72.3 (3) 72.4 (4) 73.8 (5) 70.1 (6) 19.0 **Glc** (1) 105.0 (2) 75.2 (3) 78.6 (4) 71.8 (5) 78.5 (6) 62.9 (Ac) 170.6, 20.6 (O$COCH_3$) 170.1, 20.5 (O$COCH_3$) 170.1, 20.3.

Mass (FAB, Negative ion) : m/z 973 [M-H]⁻, 623 [M-Rha-Glc-3xAcetyl]⁻

Biological Activity : Potent cytostatic activity (82.5% inhibition at 10 μg/ml⁻¹) on leukemia HL-60 cells. Its I (C_{50} value=3.7 μg/ml⁻¹).

Reference

1.   Y. Mimaki, M. Kuroda, A. Kameyama, A. Yokosuka and Y. Sashida, *Phytochemistry*, **48**, 485 (1998).

DIOSCORESIDE C

(23S,25R)-3β,26-Dihydroxy-23-methoxy-furosta-5,20(22)-diene 3-O-{α-L-rhamnopyranosyl-(1→2)-[α-L-rhamnopyranosyl-(1→4)]-β-D-glucopyranoside}-26-O-β-D-glucopyranoside

Source : *Dioscorea panthaica* Prain et Burkill (Dioscoreaceae)
Mol. Formula : $C_{52}H_{84}O_{22}$
Mol. Wt. : 1060
M.P. : 180-182°C (decomp.)
[α]$_D^{25}$: -54° (c=0.005, C_5D_5N)
Registry No. : [344912-80-7]

IR (KBr) : 3420 (OH), 2930 (CH), 1642, 1450, 1375, 1340, 1225, 1110, 1070, 1045, 915, 890 cm^{-1}.

PMR (C_5D_5N, 300 MHz) : δ 0.69 (s, 3xH-18), 1.05 (s, 3xH-19), 1.09 (d, J=6.3 Hz, 3xH-27), 1.62 (d, J=6.0 Hz, 3xH-6 of Rha II), 1.76 (s, 3xH-21), 1.76 (d, J=5.7 Hz, 3xH-6 of Rha I), 3.33 (s, OCH_3 of H-23), 4.28 (m, H-23), 4.84 (d, J=7.8 Hz, H-26, H-1 of Glc II), 4.94 (d, J=7.2 Hz, H-1 of Glc I), 5.18 (br s, H-6), 5.86 (s, H-1 of Rha II), 6.40 (s, H-1 of Rha I).

CMR (C_5D_5N, 75 MHz) : δ C-1) 37.6 (2) 30.2 (3) 78.1 (4) 39.0 (5) 140.9 (6) 121.9 (7) 32.4 (8) 31.4 (9) 50.3 (10) 37.2 (11) 21.3 (12) 39.6 (13) 43.5 (14) 54.9 (15) 34.5 (16) 84.7 (17) 64.9 (18) 14.4 (19) 19.5 (20) 108.7 (21) 11.5 (22) 151.8 (23) 73.4 (24) 37.7 (25) 30.7 (26) 75.2 (27) 17.6 (OCH$_3$) 56.1 **Glc I** (1) 100.3 92) 78.7 (3) 77.0 (4) 78.6 (5) 77.9 (6) 61.3 **Rha I** (1) 102.1 (2) 72.6 (3) 72.8 (4) 74.2 (5) 69.6 (6) 18.7 **Rha II** (1) 102.9 (2) 72.6 (3) 72.9 (4) 73.9 (5) 70.5 (6) 18.5 **Glc II** (1) 105.0 (2) 75.2 (3) 78.6 (4) 71.8 (5) 78.0 (6) 62.7.

Mass (E.S.I., Positive ion) : m/z 1083 [M+Na]$^+$.

Mass (E.S.I., Negative ion) : m/z 1059 [M-H]$^-$, 913 [M-H-146]$^-$, 767 [M-H-146x2]$^-$.

Biological Activity : The compound showed cytotoxicity against cultured A375, L929 and HeLa cancer cell lines, *in vitro* with IC$_{50}$ 6.37±2.53 μM, 6.20±6.21 μM and 4.61±3.40 μM, respectively.

Reference

1.   M. Dong, X-Z. Feng, L-J. Wu, B-X. Wang and T. Ikejima, *Planta Med.*, **67**, 853 (2001).

DIOSCORESIDE E'

(20*R*,25*R*)-20-Methoxyfurost-22-en-3β,26-diol 3-O-{α-L-rhamnopyranosyl-(1→2)-[α-L-rhamnopyranosyl-(1→4)]-β-D-glucopyranoside}-26-O-β-D-glucopyranoside

Source : *Dioscorea panthaica* Prain et Burkill
(Dioscoreaceae)
Mol. Formula : $C_{52}H_{84}O_{22}$
Mol. Wt. : 1060
M.P. : 166-168°C
$[α]_D^{25}$: -61.2° (c=0.1, Pyridine)
Registry No. : [84953-46-4]

IR (KBr) : 3400 (OH), 1040 (C-O) cm^{-1}.

PMR (C$_5$D$_5$N, 300 MHz) : δ 0.82 (H-14), 0.84 (s, 3xH-18), 0.89 (H-9), 0.92 (H-1A), 1.03 (s, 3xH-19), 1.08 (d, 3xH-27), 1.09 (H-12A), 1.40 (s, 3xH-21), 1.44 (H-11), 1.50 (H-8), 1.53 (H-24), 1.62 (d, *J*=6.0 Hz, 3xH-6 of Rha II), 1.64 (H-24), 1.67 (H-1B), 1.76 (d, *J*=6.0 Hz, 3xH-6 of Rha I), 1.77 (H-25), 1.82 (H-2A), 1.86 (H-7), 1.91 (H-15A), 2.04 (H-15B), 2.05 (H-2B), 2.08 (H-17), 2.16 (H-12B), 2.72 (H-4A), 2.78 (H-4), 3.17 (s, OC*H*$_3$), 3.84 (H-3), 3.85 (H-5 of Glc), 4.00 (H-5 of Glc II), 4.03 (H-2 of Glc II), 4.06 (H-26), 4.10 (H-6A of Glc I), 4.22 (H-3 of Glc), 4.23 (H-6B of Glc I), 4.28 (H-3 of Glc II), 4.32 (H-4 of Rha II), 4.34 (H-4 of Glc II), 4.35 (H-6A of Glc I), 4.36 (H-4 of Rha I), 4.37 (H-23), 4.40 (H-4 of Glc I), 4.42 (H-2 of Glc I), 4.55 (H-3 of Rha II and H-6B of Glc II), 4.60 (H-3 of Rha I), 4.68 (H-2 of Rha II), 4.81 (H-2 of Rha I), 4.86 (d, *J*=7.2 Hz, H-1 of Glc II), 4.92 (H-5 of Rha II), 4.94 (H-5 of Rha I), 4.95 (d, *J*=7.5 Hz, H-1 of Glc I), 4.96 (H-16), 5.31 (H-6), 5.86 (br s, H-1 of Rha II), 6.39 (br s, H-1 of Rha I).

CMR (C$_5$D$_5$N, 75 MHz) : δ C-1) 37.6 (2) 30.3 (3) 78.2 (4) 39.0 (5) 140.9 (6) 121.8 (7) 32.1 (8) 31.1 (9) 50.1 (10) 37.1 (11) 20.6 (12) 39.3 (13) 40.4 (14) 56.8 (15) 33.6 (16) 84.1 (17) 66.8 (18) 13.7 (19) 19.5 (20) 82.5 (21) 15.4 (22) 157.3 (23) 96.4 (24) 29.7 (25) 35.0 (26) 75.3 (27) 17.6 (OCH$_3$) 48.9 **Glc I** (1) 100.3 (2) 78.6 (3) 77.0 (4) 78.7 (5) 77.9 (6) 61.4 **Rha I** (1) 102.2 (2) 72.6 (3) 72.8 (4) 74.2 (5) 69.7 (6) 18.8 **Rha II** (1) 103.0 (2) 72.6 (3) 72.9 (4) 74.0 (5) 70.5 (6) 18.6 **Glc II** (1) 105.0 (2) 75.3 (3) 78.5 (4) 71.8 (5) 78.0 (6) 62.9.

Mass (FAB, Negative ion) : *m/z* 1059 [M-H]$^-$, 897 [M-H-Glc]$^-$, 751 [M-H-Glc-Rha]$^-$, 605 [M-H-Glc-2xRha]$^-$, 443 [M-H-2xGlc-2xRha]$^-$.

Mass (FAB, Negative ion, H.R.) : *m/z* 1059.5364 [(M-H)$^-$, calcd. for 1059.5376].

Reference

1.   M. Dong, X.Z. Feng, B.X. Wang, T. Ikejima and L.J. Wu, *Pharmazie*, **59**, 294 (2004).

ASPARAGUS PLUMOSUS SAPONIN 3

(25*S*)-22α-Methoxyfurost-5-en-3β,26-diol 3-O-{α-L-rhamnopyranosyl-(1→2)-[α-L-rhamnopyranosyl-(1→3)]-β-D-glucopyranoside}-26-O-β-D-glucopyranoside

Source : *Asparagus plumosus* Bahar (Liliaceae)
Mol. Formula : $C_{52}H_{86}O_{22}$
Mol. Wt. : 1062
M.P. : 171-173°C
[α]$_D^{27}$: -78° (c=0.9, MeOH)
Registry No. : [98569-68-7]

IR (KBr) : λ_{max} 3400 (OH) cm^{-1}, no spiroketal absorptions.

Mass (F.D., Positive ion) : *m/z* (rel.intens.) 1085 [M+Na]$^+$ (0.55), 1063 [M+H]$^+$ (5.45), 1053 [(M+Na)-MeOH]$^+$ (1.65), 1031 [(M+H)-MeOH]$^+$ (37.7), 885 [(M+H)-MeOH-146]$^+$ (64.2), 869 [(M+H)-MeOH-162]$^+$ (40.7), 771 [(M+H)-292]$^+$ (6.99), 754 [(M+H)-308]$^+$ (6.17), 739 [(M+H)-MeOH-292]$^+$ (100), 722 [M-MeOH-308]$^+$ (56.6), 577 [(M+2H)-MeOH-454]$^+$ (50.9), 576 [(M+H)-MeOH-454]$^+$ (59), 414 [M-MeOH-616]$^+$ (6.71), 163 [Glc+H-H$_2$O]$^+$ (6.20), 147 [Rha+H-H$_2$O]$^+$ (22.3), 128 [Rha-2H$_2$O]$^+$ (23.1).

Mass (FAB, Positive ion) : *m/z* (rel.intens.) 1085 [M+Na]$^+$ (0.26), 1053 [(M+Na)-MeOH]$^+$ (0.30), 739 [(M+H)-MeOH-292]$^+$ (0.21), 631 [(M+Na)-454]$^+$ (0.19), 609 [(M+H)-454]$^+$ (0.24), 599 [(M+Na)-MeOH-454]$^+$ (0.31), 285 (1.0)^c, 271 (1.1)^c, 267 (1.3)^c 253 (5.0)^c.

Reference

1.   O.P. Sati and G. Pant, *J. Nat. Prod.*, **48**, 390 (1985).

METHYL (25*S*)-PROTODIOSCIN

(25*S*)-22-Methoxy-3β-hydroxyfurost-5-ene 3-O-{α-L-rhamnopyranosyl-(1→2)-[α-L-rhamnopyranosyl-(1→4)]-β-D-glucopyranoside}-26-O-β-D-glucopyranoside

Source : *Licuala spinosa* Wurmb. (Palmae)
Mol. Formula : $C_{52}H_{86}O_{22}$
Mol. Wt. : 1062
M.P. : 166-168°C (decomp.)
[α]$_D$: -73.0° (c=0.69, C_5D_5N)
Registry No. : [139490-48-5]

IR (KBr) : 3500-3300 (OH) cm^{-1}.

CMR (C_5D_5N, 100 MHz) : δ **Glc-I** C-1) 100.6 (2) 79.7 (3) 78.2 (4) 77.1 (5) 78.3 (6) 62.0 **Rha I** (1) 102.1 (2) 72.7 (3) 73.1 (4) 74.5 (5) 69.7 (6) 19.1 **Rha II** (1) 103.3 (2) 73.2 (3) 70.6 (4) 78.0 (5) 68.8 (6) 19.1 **Glc II** (1) 105.2 (2) 75.3 (3) 78.9 (4) 72.3 (5) 78.7 (6) 64.6. The signals of aglycone moiety are analogous to those of Methyl-25*S*-proto-Pb (q.v.).

Reference

1.   A. Asami, Y. Hirai and J. Shoji, *Chem. Pharm. Bull.*, **39**, 2053 (1991).

METHYL PROTODIOSCIN

(25*R*)-22α-Methoxyfurost-5-ene-3β,26-diol 3-*O*-{α-L-rhamnopyranosyl-(1→2)-[α-L-rhamnopyranosyl-(1→4)]-β-D-glucopyranoside}-26-*O*-β-D-glucopyranoside

Source : *Asparagus officinalis* L.[1] (Liliaceae), *Smilax menispermoidea*[2] (Liliaceae), *Chamaerops humalis* L.[3] (Palmae), *Trachycarpus wagnarianus* Becc.[3] (Palmae), *Smilax china* L.[4] (Liliaceae), *Costus speciosus* (Koen.) Sm.[5] (Zingiberaceae) etc.

Mol. Formula : $C_{52}H_{86}O_{22}$

Mol. Wt. : 1062

M.P. : 187-189°C[1]

[α]$_D$: -59.8° (c=0.56, MeOH)[1]

Registry No. : [54522-52-0]

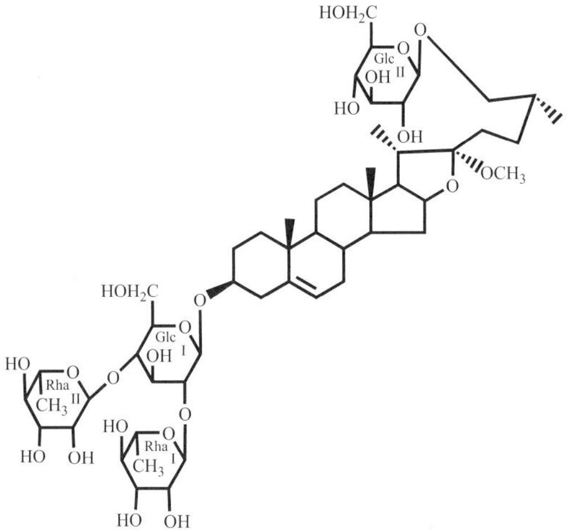

IR (KBr)[2] : 3600-3200 (OH) cm^{-1}.

IR : 3700-3050, 1130-1000, 980, 960, 920 (strong), 900 (weak), 840 cm^{-1}.

PMR (C$_5$D$_5$N, 500 MHz)[1] : δ 0.83 (s, 3xH-18), 0.91 (H-9), 1.00 (d, *J*=6.7 Hz, 3xH-27), 1.03 (H-1), 1.06 (s, 3xH-19), 1.12 (H-14), 1.21 (d, *J*=7.0 Hz, 3xH-21), 1.43 (H-11), 1.46 (H-24), 1.50 (H-7A), 1.66 (H-2), 1.66 (d, *J*=6.3 Hz, H-6 of Rha II), 1.68 (H 15Λ), 1.71 (H 12), 1.77 (H 8), 1.79 (H 17), 1.79 (d, *J*=6.2 Hz, 3xH-6 of Rha I), 1.88 (H-25), 2.01 (H-

7B), 2.09 (H-15B), 2.11 (H-23), 2.22 (H-20), 2.73 (H-4A), 2.82 (H-4B), 3.26 (OCH_3), 3.61 (H-26A), 3.68 (H-5 of Glc I), 3.89 (H-3), 4.00 (H-5 of Glc II), 4.05 (H-2 of Glc II), 4.09 (H-6A of Glc I), 4.23 (H-3 of Glc I), 4.24 (H-6B), 4.24 (H-4 of Glc II), 4.28 (H-2 of Glc I), 4.29 (H-3 of Glc II), 4.34 (H-4 of Rha II), 4.36 (H-4 of Rha I), 4.40 (H-4 of Glc I), 4.41 (H-6A of Glc II), 4.46 (H-16), 4.58 (H-6B of Glc II), 4.62 (H-3 of Rha), 4.68 (H-3 of Rha I), 4.70 (H-2 of Rha II), 4.84 (d, J=7.7 Hz, H-1 of Glc II), 4.86 (H-2 of Rha I), 4.93 (d, J=7.8 Hz, H-1 of Glc I), 4.95 (H-5 of Rha II), 4.97 (H-5 of Rha I), 5.30 (H-6), 5.82 (br s, H-1 of Rha II), 6.40 (br s, H-1 of Rha I).

CMR (C_5D_5N, 125 MHz)[1] : δ C-1) 37.6 (2) 30.3 (3) 78.6 (4) 39.1 (5) 140.9 (6) 121.9 (7) 32.4 (8) 31.7 (9) 50.4 (10) 37.2 (1) 21.2 (12) 39.8 (13) 40.8 (14) 56.6 (15) 32.3 (16) 81.4 (17) 64.3 (18) 16.4 (19) 19.5 (20) 40.6 (21) 16.3 (22) 112.7 (23) 30.9 (24) 28.2 (25) 34.3 (26) 75.3 (27) 17.2 (OCH_3) 47.4 **Glc I** (1) 100.3 (2) 78.7 (3) 78.0 (4) 77.9 (5) 77.0 (6) 61.3 **Glc II** (1) 105.1 (2) 75.3 (3) 78.6 (4) 71.8 (5) 78.1 (6) 62.9 **Rha I** (1) 102.1 (2) 72.6 (3) 72.8 (4) 74.0 (5) 69.6 (6) 18.6 **Rha II** (1) 102.9 (2) 72.6 (3) 72.9 (4) 74.2 (5) 70.5 (6) 18.8.

Mass (FAB, Positive ion)[7] : m/z 1062 [M]$^+$, 1031 [M-OCH$_3$]$^+$, 901 [M-Glc]$^+$, 871 [901-OCH$_3$+H]$^+$, 852 [870-H$_2$O]$^+$, 724 [870-Rha-H]$^+$, 560, 397.

Mass (FAB, Negative ion)[1] : m/z 1061 [M-H]$^-$.

Biological Activity : It showed weak inhibitory effect on DNA, RNA and protein synthesis in HL-60 cells in a dose dependent manner.[1] The compound shows inhibitory activity on cyclic AMP phosphodiesterate with IC$_{50}$=29.4x10^{-5} M. It causes morphological abnormality of *Pyricularia oryzae* mycelia and shows cytotoxic activity against the cancer cell line K$_{562}$ *in vitro*.[6]

References

1. Y. Shao, O. Poobrasert, E.J. Kennelly, C.K. Chin, C.T. Ho, M.T. Huang, S.A. Garrison and G.A. Cordell, *Planta Med.*, **63**, 258 (1997).

2. Y. Ju and Z.-J. Jia, *Phytochemistry*, **31**, 1349 (1992).

3. Y. Hirai, S. Sanada, Y. Ida and J. Shoji, *Chem. Pharm. Bull.*, **34**, 82 (1986).

4. Y. Sashida, S. Kubo, Y. Mimaki, T. Nikaido and T. Ohmoto, *Phytochemistry*, **31**, 2439 (1992).

5. S.B. Sing and R.S. Thakur, *J. Nat. Prod.*, **45**, 667 (1982).

6. K. Hu, A. Dong, X. Yao, H. Kobayashi and S. Iwasaki, *Planta Med.*, **63**, 161 (1997).

7. C. Yang and Z. Wang, *Yunnan Zhiwu Yanjiu* (*Acta Bot. Yunnanica*), **8**, 355 (1986).

METHYL PROTO-TACCAOSIDE

(25*R*)-22ξ-Methoxyfurost-5-en-3β,26-diol 3-O-{α-L-rhamnopyranosyl-(1→2)]-
[α-L-rhamnopyranosyl-(1→3)]-β-D-glucopyranoside}-26-O-β-D-glucopyranoside

Source : *Phoenix rupicola* T. Anderson (Palmae)
Mol. Formula : $C_{52}H_{86}O_{22}$
Mol. Wt. : 1062
M.P. : 180-183°C
[α]$_D$: 79.2° (c=1.05, C_5D_5N)
Registry No. : [138457-03-1]

IR (KBr) : 3600-3200 (OH) cm^{-1}.

CMR (C_5D_5N, 100 MHz) : δ C-1) 37.2 (2) 30.2 (3) 78.1 (4) 38.8 (5) 141.0 (6) 121.7 (7) 32.3 (8) 31.8 (9) 50.5 (10) 37.2 (11) 21.1 (12) 40.0 (13) 40.9 (14) 56.7 (15) 32.3 (16) 81.4 (17) 64.2 (18) 16.3 (19) 19.4 (20) 40.6 (21) 16.1 (22) 112.8 (23) 30.8 (24) 28.3 (25) 34.3 (26) 75.1 (27) 17.2 (OCH₃) 47.4 **Glc I** (1) 100.1 (2) 78.5 (3) 87.7 (4) 70.3 (5) 77.8 (6) 62.5 **Rha I** (1) 102.4 (2) 72.4 (3) 72.7 (4) 73.8^a (5) 69.8 (6) 18.5^b **Rha II** (1) 103.7 (2) 72.2 (3) 72.4 (4) 73.5^a (5) 70.6 (6) 18.3^b **Glc II** (1) 104.7 (2) 75.1 (3) 78.1 (4) 71.9 (5) 78.5 (6) 63.0.

Reference

1. K. Idaka, Y. Hirai and J. Shoji, *Chem. Pharm. Bull.*, **39**, 1455 (1991).

ABUTILOSIDE L

(22S,25S)-22,25-Epoxyfurost-5-en-3β,7β,26-triol 3-O-{α-L-rhamnopyranosyl-(1→2)-[α-L-rhamnopyranosyl-(1→4)]-β-D-glucopyranoside}-26-O-β-D-glucopyranoside

Source : *Solanum abutiloides* (Solanaceae)
Mol. Formula : $C_{51}H_{82}O_{23}$
Mol. Wt. : 1062
$[\alpha]_D^{25}$: -107.1° (c=1.15, MeOH)
Registry No. : [652144-57-5]

PMR (C₅D₅N, 400 MHz) : δ 0.90 (s, 3xH-18), 1.06 (s, 3xH-19), 1.08 (d, J=6.7 Hz, 3xH-21), 1.40 (s, 3xH-27), 1.51 (H-15A), 1.63 (d, J=6.1 Hz, 3xH-6 of Rha II), 1.76 (d, J=6.1 Hz, 3xH-6 of Rha), 1.83 (H-17), 2.01 (H-14), 2.51 (H-15B), 3.62 (br d, J=9.2 Hz, H-5 of Glc I), 3.78 (m, H-3), 3.88 (dd, J=9.8 Hz, H-26A), 3.85 (m, H-5 of Glc II), 3.96 (m, H-7), 4.04 (dd, J=7.8, 8.5 Hz, H-2 of Glc II), 4.12 (br d, J=12.2 Hz, H-6A of Glc I), 4.18 (d, J=9.8 Hz, H-26B), 4.23 (H-2 and H-3 of Glc I), 4.23 (dd, J=3.8, 11.6 Hz, H-6B of Glc I), 4.27 (dd, J=8.5, 8.5 Hz, H-4 of Glc II), 4.28 (dd, J=8.5, 8.5 Hz, H-3 of Glc II), 4.36 (dd, J=8.5, 8.5 Hz, H-4 of Rha II), 4.38 (dd, J=8.5, 8.5 Hz, H-4 of Rha I), 4.40 (dd, J=8.5, 8.5 Hz, H-4 of Glc I), 4.43 (dd, J=5.1, 11.6 Hz, H-6A of Glc II), 4.56 (dd, J=3.0, 8.5 Hz, H-3 of Rha II), 4.57 (dd, J=3.6, 11.6 Hz, H-6B of Glc II), 4.64 (dd, J=2.7, 8.5 Hz, H-3 of Rha), 4.71 (br s, H-2 of Rha II), 4.80 (m, H-16), 4.85 (br s, H-2 of Rha I), 4.90 (d, J=7.6 Hz, H-1 of Glc I), 4.95 (m, H-5 of Rha II), .4.95 (d, J=7.8 Hz, H-1 of Glc II), 4.98 (m, H-5), 5.79 (d, J=4.9 Hz, H-6), 5.85 (br s, H-1 of Rha II), 6.39 (br s, H-1 of Rha I).

CMR (C₅D₅N, 100 MHz) : δ C-1) 37.3 (2) 30.1 (3) 78.0 (4) 39.0 (5) 143.7 (6 125.9 (7) 64.5 (8) 37.9 (9) 42.5 (10) 38.0 (11) 21.0 (12) 39.8 (13) 40.4 (14) 49.7 (15) 32.4 (16) 81.2 (17) 62.6 (18) 16.3 (19) 18.4 (20) 38.7 (21) 15.3 (22) 120.1 (23) 33.2 (24) 34.1 (25) 83.8 (26) 77.5 (27) 24.5 **Glc I** (1) 100.3 (2) 77.3 (3) 77.8 (4) 78.6 (5) 77.0 (6) 61.4 **Rha I** (1) 102.0 (2) 72.5 (3) 72.8 (4) 74.1 (5) 69.6 (6) 18.7 **Rha II** (1) 102.9 (2) 72.6 (3) 72.8 (4) 73.9 (5) 70.5 (6) 18.6 **Glc II** (1) 105.4 (2) 75.4 (3) 78.4 (4) 71.6 (5) 78.6 (6) 62.7.

Mass (FAB, Positive ion) : m/z 1085 [M+Na]$^+$.

Mass (FAB, Positive ion) : 1085.5153 [(M+Na)$^+$, requires 1085.5145].

Mass (FAB, Negative ion) : m/z 1061 [M-H]$^-$.

Reference

1.   H. Yoshimitsu, M. Nishida and T. Nohara, *Phytochemistry*, **64**, 1361 (2003).

ACULEATISIDE B

Nuatigenin 3-O-{α-L-rhamnopyranosyl-(1→2)-[β-D-glucopyranosyl-(1→3)]-β-D-galactopyranoside}-26-O-β-D-glucopyranoside

Source : *Solanum aculeatissimum* (Solanaceae)
Mol. Formula : $C_{51}H_{82}O_{23}$
Mol. Wt. : 1062
$[α]_D^{21}$: -82.0° (c=1.02, Pyridine)
Registry No. : [86848-70-2]

IR (KBr) : 3350 (OH) cm^{-1}.

CMR (C$_5$D$_5$N, 50.01 MHz) : δ C-1) 37.5 (2) 30.1 (3) 78.1 (4) 40.5 (5) 140.9 (6) 120.2 (7) 32.2 (8) 31.7 (9) 50.3 (10) 37.1 (11) 21.1 (12) 38.6 (13) 39.8 (14) 56.5 (15) 32.2 (16) 80.9 (17) 62.7 (18) 16.2 (19) 19.4 (20) 38.6 (21) 15.1 (22) 121.6 (23) 33.1^a (24) 33.8^a (25) 83.8 (26) 77.3 (27) 24.3 **Gal** (1) 100.4 (2) 74.7^b (3) 85.0 (4) 69.9 (5) 74.9^b (6) 62.2^c **Rha** (1) 102.0 (2) 72.6^d (3) 72.2^d (4) 73.9 (5) 69.3 (6) 18.5 **Glc I** (1) 105.5 (2) 76.0 (3) 78.1 (4) 71.6 (5) 77.7 (6) 61.7^c **Glc II** (1) 105.1 (2) 75.2 (3) 78.1 (4) 71.6 (5) 78.1 (6) 62.4^c.

Mass (F.D.) : *m/z* 1085 [M+Na]$^+$.

Reference

1.   R. Saijo, C. Fuka, K. Murakami, T. Nohara and T. Tomimatsu, *Phytochemistry*, **22**, 733 (1983).

AVENACOSIDE A

Nuatigenin 3-O-{α-L-rhamnopyranosyl-(1→4)-[β-D-glucopyranosyl-(1→2)]-β-D-glucopyranoside}-26-O-β-D-glucopyranoside

Source : *Avena sativa* L.[1] (Gramineae)
Mol. Formula : $C_{51}H_{82}O_{23}$
Mol. Wt. : 1062
[α]$_D^{24}$: +52.9° (c=1.0, H$_2$O)[2]
Registry No. : [24915-65-9]

References

1.   R. Tschesche, M. Tauscher, H.-W. Fehlhaber and G. Wulff, *Chem. Ber.*, **102**, 2072 (1969).
2.   R. Tschesche and W. Schmidt, *Z. Naturforsch.*, **21B**, 896 (1966).

LILIUM BROWNII SAPONIN 2

Nuatigenin 3-O-{α-L-rhamnopyranosyl-(1→2)-[β-D-glucopyranosyl-(1→4)]-β-D-glucopyranoside}-26-O-β-D-glucopyranoside

Source : *Lilium brownii* var. *colchesteri*(Liliaceae)
Mol. Formula : $C_{51}H_{82}O_{23}$
Mol. Wt. : 1062
$[\alpha]_D^{28}$: -90.0° (c=0.10, MeOH).
Registry No. : [132922-46-4]

IR (KBr) : 3390 (OH), 2910, 2875 (CH), 1435, 1365, 1290, 1245, 1145, 1040, 895, 875, 850, 820, 895 cm^{-1}.

PMR (C_5D_5N-CD_3OD 4:1, 400/500 MHz) : δ 0.80 (s, 3xH-18), 1.03 (s, 3xH-19), 1.07 (d, J=6.9 Hz, 3xH-21), 1.36 (s, 3xH-27), 1.68 (d, J=6.3 Hz, 3xH-6 of Rha), 4.60 (dd, J=3.3, 1.6 Hz, H-2 of Rha), 4.67 (m, H-16), 4.80 (dq, J=9.5, 6.3 Hz, H-5 of Rha), 4.84 (d, J=7.7 Hz, H-1 of Glc III), 4.89 (d, J=7.5 Hz, H-1 of Glc I), 5.01 (d, J=7.8 Hz, H-1 of Glc II), 5.31 (br d, J=4.3 Hz, H-6), 6.07 (br s, H-1 of Rha).

CMR (C_5D_5N, 100/125 MHz) : δ C-1) 37.5 (2) 30.1 (3) 78.2 (4) 38.9 (5) 140.8 (6) 121.8 (7) 32.2^a (8) 31.6 (9) 50.3 (10) 37.1 (11) 21.1 (12) 39.8 (13) 40.5 (14) 56.5 (15) 32.3^a (16) 80.9 (17) 62.5 (18) 16.2 (19) 19.4 (20) 38.6 (21) 15.1 (22) 120.2 (23) 33.1 (24) 33.9 (25) 83.8 (26) 77.4 (27) 24.4 **Glc I** (1) 100.0 (2) 77.7 (3) 76.1 (4) 82.0 (5) 77.3 (6) 62.0^d **Glc II** (1) 101.7 (2) 72.4 (3) 72.7 (4) 74.1 (5) 69.4 (6) 18.6 **Rha** (1) 105.2 (2) 74.9 (3) 78.3^b (4) 71.2 (5) 78.5^b (6) 62.1^d **Glc III** (1) 105.4 (2) 75.3 (3) 78.4 (4) 71.6 (5) 78.4 (6) 62.7.

Mass (SI-MS) : *m/z* 1087 [M+Na]$^+$, 902 [M-Glucosyl+H]$^+$.

Reference

1.   Y. Mimaki and Y. Sashida, *Chem. Pharm. Bull.*, **38**, 3055 (1990).

RUSCUS ACULEATUS SAPONIN 4

22ξ-Methoxyfurosta-5,25(27)-diene-1β,3β,26-triol 1-O-{β-D-glucopyranosyl-(1→3)-
α-L-rhamnopyranosyl-(1→2)-α-L-arabinopyranoside}-26-O-β-D-glucopyranoside

Source : *Ruscus aculeatus* L. (Liliaceae)
Mol. Formula : $C_{51}H_{82}O_{23}$
Mol. Wt. : 1062
$[\alpha]_D^{26}$: -52.0° (c=0.10, MeOH)
Registry No. : [205190-07-7]

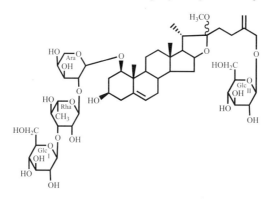

PMR (C_5D_5N, 400 MHz) : δ 0.84 (s, 3xH-18), 1.12 (d, *J*=6.9 Hz, 3xH-21), 1.46 (s, 3xH-19), 1.67 (d, *J*=6.1 Hz, 3xH-6 of Rha), 3.25 (s, OC*H₃*), 4.13 (overlapping, H-3 and H-4 of Ara), 4.63 (d, *J*=7.4 Hz, H-1 of Ara), 4.92 (d, *J*=7.8 Hz, H-1 of Glc II), 5.05 and 5.34 (each 1H, br s, 2xH-27), 5.58 (br d, *J*=5.5 Hz, H-6), 5.65 (d, *J*=7.8 Hz, H-1 of Glc I), 6.36 (br s, H-1 of Rha).

CMR (C_5D_5N, 100 MHz) : δ C-1) 84.7 (2) 38.1 (3) 68.2 (4) 43.9 (5) 139.6 (6) 124.8 (7) 32.0 (8) 33.0 (9) 50.5 (10) 42.9 (11) 23.9 (12) 40.4 (13) 40.5 (14) 56.7 (15) 32.3 (16) 81.4 (17) 64.2 (18) 16.6 (19) 15.1 (20) 40.5 (21) 16.1 (22) 112.4 (23) 31.6 (24) 28.1 (25) 146.8 (26) 72.0 (27) 111.0 (OCH₃) 47.3 **Ara** (1) 101.4 (2) 74.3 (3) 76.1 (4) 70.2 (5) 67.7 **Rha** (1) 101.1 (2) 72.0 (3) 82.7 (4) 73.3 (5) 69.3 (6) 18.7 **Glc I** (1) 106.5 (2) 76.1 (3) 78.3 (4) 71.7 (5) 78.4 (6) 62.6 **Glc II** (1) 103.8 (2) 75.1 (3) 78.5 (4) 71.7 (5) 78.6 (6) 62.9.

Mass (FAB, Negative ion) : *m/z* 1061 [M-H]⁻.

Reference

1.   Y. Mimaki, M. Kuroda, A. Kameyama, A. Yokosuka and Y. Sashida, *Chem. Pharm. Bull.*, **46**, 298 (1998).

TRIQUETROSIDE A1

Furost-5-en-1β,3β,22α,26-tetrol 3-O-[α-L-rhamnopyranosyl-(1→2)-β-D-glucopyranoside]-26-O-[α-L-rhamnopyranosyl-(1→2)-β-D-glucopyranoside]

Source : *Allium triquetrum* L. (Alliaceae, Liliaceae)
Mol. Formula : $C_{51}H_{82}O_{23}$
Mol. Wt. : 1062
$[\alpha]_D^{25}$: -33.3° (c=0.1, MeOH)
Registry No. : [628727-92-4]

PMR (CD₃OD, 500 MHz) : δ 0.81 (s, 3xH-18), 0.94 (d, *J*=6.6 Hz, 3xH-27), 0.98 (d, *J*=6.6 Hz, 3xH-21), 1.02 (s, 3xH-19), 1.12* (H-9), 1.18 (m, H-14), 1.20* (H-12A), 1.23 (d, *J*=6.6 Hz, 3xH-6 of Rha II), 1.24 (d, *J*=6.6 Hz, 3xH-6 of Rha I), 1.25* (H-15A), 1.30* (H-24A), 1.32* (H-24B), 1.53 (m, H-11B), 1.54 (m, H-8), 1.62* (H-23A), 1.71* (H-2A, H-12B), 1.72 (m, H-25), 1.73* (H-23B), 1.74* (H-17), 1.95* (H-7A, H-15B), 1.96* (H-7B), 2.06* (H-2B, H-20), 2.25 (dd, *J*=11.5, 3.5 Hz, H-4A), 2.26 (dd, *J*=10.5, 2.5 Hz, H-11B), 2.38 (dd, *J*=11.5, 7.3 Hz, H-4B), 3.19 (t, *J*=7.5 Hz, H-2 of Glc II), 3.24 (dd, *J*=7.0, 6.4 Hz, H-4 of Glc II), 3.26* (H-1), 3.28 (m, H-5 of Glc I), 3.30 (m, H-5 of Glc II), 3.32* (H-26A), 3.36 (dd, *J*=6.5, 6.0 Hz, H-4 of Rha I), 3.36* (H-3 of Glc II), 3.38* (H-3 of Glc I), 3.38 (dd, *J*=6.5, 6.0 Hz, H-4 of Rha II), 3.48 (t, *J*=7.5 Hz, H-2 of Glc I), 3.49* (H-3), 3.60 (d, *J*=6.5 Hz, H-3 of Rha I), 3.62 (d, *J*=6.5 Hz, H-3 of Rha II), 3.63 (br d, *J*=11.5 Hz, H-6A of Glc I), 3.64 (br d, *J*=11.5 Hz, H-6A of Glc II), 3.78 (dd, *J*=6.8, 6.5 Hz, H-4 of Glc I), 3.78 (br d, *J*=11.5 Hz, H-6B of Glc I), 3.82 (dd, *J*=9.5, 3.9 Hz, H-26B), 3.84 (br d, *J*=11.5 Hz, H-6B of Glc II), 3.89 (br s, H-2 of Rha I), 3.90* (H-5 of Rha I), 3.91 (br s, H-2 of Rha II), 4.09* (H-5 of Rha II), 4.21 (d, *J*=7.5 Hz, H-1 of Glc II), 4.47 (d, *J*=7.5 Hz, H-1 of Glc I), 4.53 (q, *J*=5.5 Hz, H-16), 4.82 (br s, H-1 of Rha I), 5.19 (br s, H-1 of Rha II), 5.55 (br d, *J*=3.2 Hz, H-6). * overlapped signals.

CMR (CD₃OD, 125 MHz) : δ C-1) 74.0 (2) 43.4 (3) 82.3 (4) 42.0 (5) 143.7 (6) 128.6 (7) 35.1 (8) 36.1 (9) 54.6 (10) 44.5 (11) 27.2 (12) 43.4 (13) 46.8 (14) 60.0 (15) 35.3 (16) 84.5 (17) 66.6 (18) 19.3 (19) 16.0 (20) 43.2 (21) 18.3 (22) 114.1 (23) 39.2 (24) 32.9 (25) 37.7 (26) 74.8 (27) 19.7 **Glc I** (1) 102.9 (2) 82.5 (3) 76.2 (4) 78.4 (5) 74.8 (6) 64.3 **Rha I** (1) 105.3 (2) 73.0 (3) 74.6 (4) 78.7 (5) 74.5 (6) 20.2 **Glc II** (1) 107.1 (2) 77.5 (3) 74.6 (4) 74.2 (5) 78.5 (6) 75.2 **Rha II** (1) 104.0 (2) 73.1 (3) 74.5 (4) 78.7 (5) 74.5 (6) 20.2.

Mass (FAB, Negative ion, H.R.) : *m/z* 1061.5136 [(M-H)⁻, requires 1061.5145].

Reference

1.   G. Corea, E. Tattorusso and V. Lanzotti, *J. Nat. Prod.*, **66**, 1405 (2003).

TRIQUETROSIDE A2

Furost-5-en-1β,3β,22β,26-tetrol 3-O-[α-L-rhamnopyranosyl-(1→2)-β-D-glucopyranoside]-26-O-[α-L-rhamnopyranosyl-(1→2)-β-D-glucopyranoside]

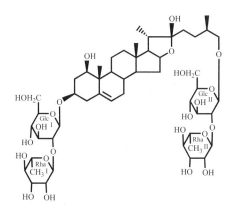

Source : *Allium triquetrum* L. (Alliaceae, Liliaceae)
Mol. Formula : $C_{51}H_{82}O_{23}$
Mol. Wt. : 1062
$[\alpha]_D^{25}$: -43.3° (c=0.1, MeOH)
Registry No. : [628727-95-7]

PMR (CD₃OD, 500 MHz) : δ 0.81 (s, 3xH-18), 0.94 (d, *J*=6.6 Hz, 3xH-27), 0.95 (d, *J*=6.6 Hz, 3xH-21), 1.02 (s, 3xH-19), 1.12* (H-9), 1.18 (m, H-14), 1.20* (H-12A), 1.25* (H-15A), 1.31* (H-24A), 1.32* (H-24B), 1.53 (m, H-11A), 1.54 (m, H-8), 1.62* (H-23A), 1.69* (H-17), 1.71* (H-3, H-12B), 1.72 (m, H-25), 1.73* (H-23B), 1.95* (H-7A), 1.96 (H-7B), 2.04* (H-2B), 2.06* (H-20), 2.26 (dd, *J*=11.5, 3.5 Hz, H-4A), 2.26 (dd, *J*=10.5, 2.5 Hz, H-11A), 2.37 (dd, *J*=11.5, 7.3 Hz, H-4B), 3.26* (H-1), 3.32* (H-26A), 3.48* (H-3), 3.82 (dd, *J*=9.5, 3.9 Hz, H-26B), 4.33 (q, *J*=5.5 Hz, H-16), 5.5 (br d, *J*=3.2 Hz, H-6). For signals of sugars protons see Triquetoroside A1. * overlapped signals.

CMR (CD₃OD, 125 MHz) : δ C-1) 74.0 (2) 43.4 (3) 82.3 (4) 42.0 (5) 143.7 (6) 128.5 (7) 35.0 (8) 36.1 (9) 54.6 (10) 44.5 (11) 27.2 (12) 43.2 (13) 46.8 (14) 60.0 (15) 35.3 (16) 84.6 (17) 67.7 (18) 19.5 (19) 16.0 (20) 43.0 (21) 18.5 (22) 117.2 (23) 39.0 (24) 32.7 (25) 37.1 (26) 74.6 (27) 19.7. For signals of sugar carbons see Triquetroside A1.

Mass (FAB, Negative ion, H.R.) : *m/z* 1061.5136 [(M-H)⁻, requires 1061.5145].

Reference

1.   G. Corea, E. Tattorusso and V. Lanzotti, *J. Nat. Prod.*, **66**, 1405 (2003).

BALANITOSIDE

3β,22,26-Trihydroxy-25*R*-furost-5-ene 3-O-[α-L-rhamnopyranosyl-(1→2)-β-D-glucopyranosyl-(1→4)-
β-D-glucopyranoside]-26-O-β-D-glucopyranoside

Source : *Balanites aegyptiaca* Del. (Zygophyllaceae)
Mol. Formula : $C_{51}H_{84}O_{23}$
Mol. Wt. : 1064
Registry No. : [145854-02-0]

IR (KBr) **:** 3400 (OH) cm^{-1}.

PMR (CD$_3$OD, 300 MHz) **:** δ 0.83 (s, 3xH-18), 0.95 (3xH-27, signal pattern unclear due to overlapping), 1.00 (d, *J*=6.9 Hz, 3xH-21), 1.05 (s, 3xH-19), 4.12 (m, H-16), 5.38 (br d, H-6), 1.23 (d, *J*=6.2 Hz, sec. 3xH-6 of Rha), 4.14 (m, H-16), 4.24 (d, *J*=7.8 Hz, anomeric H of Glc), 4.40 (d, *J*=7.8 Hz, H-1 of Glc), 4.51 (d, *J*=7.8 Hz, H-1 of Glc), 5.24 (br s, 3xH-6 of Rha), 5.38 (br d, H-6).

CMR (CD$_3$OD, 75.5 MHz) **:** δ C-1) 39.9 (2) 32.2 (3) 79.1 (4) 41.0 (5) 143.3 (6) 124.1 (7) 34.3 (8) 34.2 (9) 53.1 (10) 39.5 (11) 23.4 (12) 42.3 (13) 43.3 (14) 59.2 (15) 34.6 (16) 83.8 (17) 66.6 (18) 18.3 (19) 21.4 (20) 42.8 (21) 17.8 (22) 115.4 (23) 33.0 (24) 30.4 (25) 36.5 (26) 77.3 (27) 18.9 **Glc I** (1) 103.5 (2) 75.3 (3) 76.4 (4) 82.5 (5) 76.6 (6) 63.4

Glc II (1) 106.1 (2) 80.9 (3) 77.6 (4) 71.2 (5) 80.1 (6) 63.9 **Rha** (1) 101.9 (2) 73.1 (3) 73.6 (4) 73.8 (5) 70.6 (6) 19.5
Glc III (1) 106.0 (2) 76.6 (3) 79.3 (4) 72.8 (5) 79.5 (6) 64.2.

Mass (FAB-MS) **:** m/z 1103 $[M+K]^+$, 1088 $[M+Na+H]^+$.

Reference

1.   M. Hosny, T. Khalifa, I. Calis, A.D. Wright and O. Sticher, *Phytochemistry*, **31**, 3565 (1992).

CONVALLAMAROSIDE

1β,3β,22α,26-Tetrahydroxyfurost-25(27)-ene 3-O-[β-D-glucopyranosyl-(1→4)-α-L-rhamnopyranosyl-(1→4)-
α-L-rhamnopyranoside]-26-O-β-D-glucopyranoside

Source : *Convallaria majalis* L. (Liliaceae)
Mol. Formula : $C_{51}H_{84}O_{23}$
Mol. Wt. : 1064
$[α]_D^{20}$ **:** -53.1°
Registry No. : [52591-05-6]

Reference

1.   R. Tschesche, B.T. Tjoa, G. Wulff and R.V. Noronha, *Tetrahedron Lett.*, **49**, 5141 (1968).

DELTOSIDE

(25*R*)-3β,22α,26-Trihydroxyfurost-5-ene 3-O-{α-L-rhamnopyranosyl-(1→2)-[β-D-glucopyranosyl-(1→4)]-
β-D-galactopyranoside}-26-O-β-D-glucopyranoside

Source : *Allium nutans* L. (Liliaceae)
Mol. Formula : $C_{51}H_{84}O_{23}$
Mol. Wt. : 1064
Registry No. : [62751-68-2]

CMR (CD$_3$OD, 62.5 MHz) : δ C-1) 38.48 (2) 30.80 (3) 79.41 (4) 39.6 (5) 142.0 (6) 122.7 (7) 33.02 (8) 32.24 (9) 51.7 (10) 38.1 (11) 21.85 (12) 40.63 (13) 16.78 (14) 57.81 (15) 32.54 (16) 82.43 (17) 65.26 (18) 16.78 (19) 19.6 (20) 41.18 (21) 16.14 (22) 113.0 (23) 32.0 (24) 29.04 (25) 34.84 (26) 17.3 (27) 75.98 **Gal** (1) 100.5 (2) 78.7 (3) 77.9 (4) 81.1 (5) 76.3 (6) 62.3 **Rha** (1) 104.6 (2) 75.2 (3) 78.2 (4) 71.5 (5) 77.9 (6) 62.3 **Glc I** (1) 104.7 (2) 75.1 (3) 78.2 (4) 71.8 (5) 77.9 (6) 62.4 **Glc II** (1) 101.9 (2) 72.2 (3) 72.5 (4) 74.0 (5) 69.7 (6) 17.6.

Mass (LSI, Negative ion) : *m/z*: 1065 [M-H]⁻, 903 [M-H-Glc]⁻, 755 [M-H-Glc-Rha]⁻, 594 [M-H-2xGlc-Rha]⁻, 433 [M-H-2xGlc-Gal-Rha]⁻.

Reference

1.   L.S. Akhov, M.M. Musienko, S. Piacente, C. Pizza and W. Oleszek, *J. Agri. Food Chem.*, **47**, 3193 (1999).

HYPOGLAUCIN F

3β,22α-26,27-Tetrahydroxy-(25S)-furost-5-ene 3-O-{α-L-rhamnopyranosyl-(1→2)-[α-L-rhamnopyranosyl-(1→4)]-β-D-glucopyranoside}-26-O-[β-D-glucopyranoside]

Source : *Dioscorea collettii* Hook. f. var. *hypoglauca*
Polibin Pei et Ting
Mol. Formula : $C_{51}H_{84}O_{23}$
Mol. Wt. : 1064
M.P. : >300°C (decomp.)
[α]$_D$: -94.3° (c=0.01, Pyridine)
Registry No. : [189100-83-2]

IR (KBr) : 3400 (OH), 1000-1100 (glycosyl C–O) cm^{-1}.

PMR (C$_5$D$_5$N, 500 MHz) : δ 0.81 (s, 3xH-18), 0.89 (H-9)*, 0.98 (H-1A)*, 1.04 (s, 3xH-19), 1.06 (H-12A)*, 1.07 (H-14)*, 1.09 (d, *J*=6.9 Hz, 3xH-21), 1.41 (H-11)*, 1.43 (H-15A)*, 1.50 (H-7A)*, 1.55 (H-8)*, 1.60 (d, *J*=6.5 Hz, 3xH-6 of Rha II), 1.65 (2xH-23 and 2xH-24)*, 1.66 (H-12B)*, 1.74 (d, *J*=6.5 Hz, 3xH-6 of Rha I), 1.74 (H-1B)*, 1.76 (H-17)*, 1.86 (H-2A)*, 1.87 (H-7B)*, 1.92 (H-20)*, 2.02 (H-15B)*, 2.04 (H-25)*, 2.07 (H-2B)*, 2.70 (m, H-4A), 2.79 (m, H-4B), 3.45* (H-26A), 3.63 (m, H-5 of Glc I), 3.71 (H-27A)*, 3.86 (m, H-3), 3.92 (H-26B and H-5 of Glc II)*, 4.00 (H-2 of Glc II)*, 4.01 (H-27B)*, 4.07 (H-6A of Glc I)*, 4.19 (H-2 of Glc I, H-3 of Glc I, H-6B of Glc I, H-3 of Glc II)*, 4.20 (H-4 of Glc II)*, 4.29 (H-4 of Rha II)*, 4.31 (H-4 of Rha I)*, 4.34 (H-4 of Glc I)*, 4.37 (H-6A of Glc II), 4.49 (H-16)*, 4.50 (H-3 of Rha II)*, 4.54 (H-6B of Glc II)*, 4.59 (dd, *J*=3.5, 9.0 Hz, H-3 of Rha I), 4.64 (dd,

J=0.9, 3.5 Hz, H-2 of Rha II), 4.74 (d, *J*=7.8 Hz, H-1 of Glc II), 4.79 (dd, *J*=0.9, 3.5 Hz, H-2 of Rha II), 4.85 (H-5 of Rha II)*, 4.92 (d, *J*=7.2 Hz, H-1 of Glc I), 4.92 (H-5 of Rha I)*, 5.32 (br d, H-6), 5.80 (d, *J*=0.9 Hz, H-1 of Rha II), 6.34 (d, *J*=0.9 Hz, H-1 of Rha I).* overlapped signals.

CMR (C_5D_5N, 125 MHz) : δ C-1) 37.5 (2) 30.2 (3) 78.2 (4) 39.0 (5) 140.9 (6) 121.8 (7) 32.3 (8) 31.7 (9) 50.4 (10) 37.2 (11) 21.1 (12) 39.9 (13) 40.5 (14) 56.7 (15) 32.2 (16) 81.5 (17) 62.9 (18) 16.3 (19) 19.4 (20) 42.0 (21) 14.9 (22) 109.5 (23) 31.3 (24) 24.0 (25) 36.7 (26) 72.0 (27) 63.7 **Glc I** (1) 100.3 (2) 78.0 (3) 77.9 (4) 78.9 (5) 76.9 (6) 61.4 **Rha I** (1) 102.0 (2) 72.5 (3) 72.8 (4) 74.2 (5) 69.5 (6) 18.6 **Rha II** (1) 103.0 (2) 72.5 (3) 72.7 (4) 73.9 (5) 70.5 (6) 18.5 **Glc II** (1) 105.0 (2) 75.2 (3) 78.6 (4) 71.8 (5) 78.6 (6) 62.9.

Mass (FAB, Positive ion) : *m/z* 1047 [M+H-H_2O]$^+$, 885 [M+H-H_2O-Glc]$^+$, 739 [M+H-H_2O-Rha]$^+$, 721 [M+H-2xH_2O-Glc-Rha]$^+$, 575 [M+H-2xH_2O-Glc-2xRha]$^+$, 413 [M+H-2xH_2O-2xGlc-2xRha]$^+$.

Reference

1.   K. Hu, A.-J. Dong, X-Sheng Yao, H. Kobayashi and S. Iwasaki, *Phytochemistry*, **44**, 1339 (1997).

INDIOSIDE D

Furost-5-en-3β,22ξ ,26-triol 3-O-{α-L-rhamnopyranosyl-(1→2)-[β-D-glucopyranosyl-(1→3)]-β-D-galactopyranoside} 26-O-β-D-glucopyranoside

Source: *Solanum indicum* L. (Solanaceae)
Mol. Formula : $C_{51}H_{84}O_{23}$
Mol. Wt. : 1064
[α]$_D^{26}$: -75.2° (c=0.57, Pyridine)
Registry No. : [185332-91-6]

PMR (C$_5$D$_5$N, 400 MHz) : δ 0.82 (s, 3xH-18), 1.01 (d, J=7.0 Hz, 3xH-27), 1.06 (s, 3xH-19), 1.34 (d, J=7.0 Hz, 3xH-21), 1.69 (d, J=6.2 Hz, 3xH-6 of Rha), 4.65 (d, J=7.7 Hz, H-1 of Glc), 4.85 (d, J=8.1 Hz, H-1 of Glc), 4.98 (d, J=7.7 Hz, H-1 of Gal), 5.31 (br s, H-6), 6.30 (s, H-1 of Rha).

CMR (C$_5$D$_5$N, 100 MHz) : δ C-1) 37.3 (2) 29.9 (3) 78.4^a (4) 38.9 (5) 140.8 (6) 121.7 (7) 32.3 (8) 31.6 (9) 50.9 (10) 37.0 (11) 21.0 (12) 39.8 (13) 40.7 (14) 56.5 (15) 32.4 (16) 81.0 (17) 63.8 (18) 16.4 (19) 19.3 (20) 40.9 (21) 16.4 (22) 110.6 (23) 37.0 (24) 28.3 (25) 34.2 (26) 74.1 (27) 17.4 **Gal** (1) 100.6 (2) 76.6 (3) 84.8 (4) 70.8 (5) 75.1 (6) 62.8 **Glc I** (1) 101.9 (2) 72.5 (3) 72.8 (4) 74.1 (5) 69.4 (6) 18.5 **Rha** (1) 105.7 (2) 74.8 (3) 78.2^a (4) 71.5 (5) 78.3^a (6) 62.4 **Glc II** (1) 104.8 (2) 76.6 (3) 78.5^a (4) 71.6 (5) 78.7^a (6) 62.2.

Mass (FAB, Negative ion) : m/z 1063 [M-H]$^-$.

Reference

1.   S. Yahara, T. Nakamura, Y. Someya, T. Matsumoto, T. Yamashita and T. Nohara, *Phytochemistry* **43**, 1319 (1996).

OPHIOPOGON PLANISCAPUS SAPONIN F
25(*R*)-Furost-5-en-3β,22,26-triol 3-O-{α-L-rhamnopyranosyl-(1→2)-[β-D-glucopyranosyl-(1→4)-β-D-glucopyranoside]-26-O-β-D-glucopyranoside}

Source : *Ophiopogon planiscapus* Nakai[1] (Liliaceae),
Lilium hansonii[2] (Liliaceae)
Mol. Formula : C$_{51}$H$_{84}$O$_{23}$
Mol. Wt. : 1064
M.P. : 204-207°C (decomp.)[1]
[α]$_D^{23}$: -51.2° (c=0.72, Pyridine)[1]
Registry No. : [55639-71-9]

IR (KBr)[1] **:** 3600-3200 (OH) cm^{-1}.

CMR (C_5D_5N, 100 MHz)[2] **:** δ C-1) 37.5 (2) 30.2 (3) 78.2^a (4) 39.0 (5) 140.8 (6) 121.8 (7) 32.5^b (8) 31.7 (9) 50.4 (10) 37.2 (11) 21.1 (12) 40.0 (13) 40.8 (14) 56.6 (15) 32.4^b (16) 81.1 (17) 63.9 (18) 16.4 (19) 19.4 (20) 40.7 (21) 16.5 (22) 110.7 (23) 37.2 (24) 28.4 (25) 34.3 (26) 75.3 (27) 17.5 **Glc I** (1) 100.1 (2) 77.7 (3) 76.2 (4) 82.0 (5) 77.4 (6) 62.1^c **Rha** (1) 101.8 (2) 72.4 (3) 72.8 (4) 74.1 (5) 69.4 (6) 18.6 **Glc II** (1) 105.2 (2) 75.0 (3) 78.4^a (4) 71.3 (5) 78.3^a (6) 62.0^c **Glc III** (1) 104.9 (2) 75.2 (3) 78.6^a (4) 71.8 (5) 78.2^a (6) 62.9.

Biological Activity : The compound showed inhibitory activity on cyclic AMP-Phosphodiesterase with IC_{50} value of 48.5 x 10^{-5} M.[2]

References

1. Y. Watanabe, S. Sanada, Y. Ida and J. Shoji, *Chem. Pharm. Bull.*, **31**, 3486 (1983).

2. K. Ori, Y. Mimaki, K. Mito, Y. Sashida, T. Nikaido, T. Ohmoto and A. Masuto, *Phytochemistry*, **31**, 2767 (1992).

PROTOGRACILLIN

(25*R*)-Furost-5-ene-3β,22α,26-triol 3-O-{α-L-rhamnopyranosyl-(1→2)-[β-D-glucopyranosyl-(1→3)]-
β-D-glucopyranoside}-26-O-[β-D-glucopyranoside]

Source : *Dioscorea gracillima* Miq[1] (Dioscoreaceae),
Diocorea collettii Hook. f. var. *hypoglauca* Polibin Pei et
Ting[2] (Dioscoreaceae)
Mol. Formula : $C_{51}H_{84}O_{23}$
Mol. Wt. : 1064
M.P. : 254-256°C (decomp.)[2]
$[α]_D^{13}$: -72.9° (c=0.01, Pyridine)[2]
Registry No. : [54848-30-5]

PMR (C_5D_5N, 500 MHz)[2] : δ 0.89 (s, 3xH-18), 0.98 (d, *J*=6.7 Hz, 3xH-27), 1.05 (s, 3xH-19), 1.32 (d, *J*=6.5 Hz, 3xH-21), 1.74 (d, *J*=6.0 Hz, 3xH-6 of Rha), 3.94 (m, H-3), 4.79 (d, *J*=7.6 Hz, H-1 of Glc III), 4.93 (d, *J*=7.2 Hz, H-1 of Glc I), 5.09 (d, *J*=7.2 Hz, H-1 of Glc II), 5.32 (br d, H-6), 6.35 (d, *J*=0.9 Hz, H-1 of Rha).

CMR (C_5D_5N, 125 MHz)[2] : δ C-1) 37.5 (2) 30.1 (3) 77.9 (4) 38.7 (5) 140.8 (6) 121.9 (7) 32.4 (8) 31.7 (9) 50.3 (10) 37.2 (11) 21.1 (12) 40.0 (13) 40.7 (14) 56.6 (15) 32.5 (16) 81.1 (17) 63.9 (18) 16.5 (19) 19.4 (20) 40.8 (21) 16.5 (22) 110.7 (23) 37.0 (24) 28.4 (25) 34.3 (26) 75.2 (27) 17.5 **Glc I** (1) 100.0 (2) 77.0 (3) 89.5 (4) 69.6 (5) 77.9 (6) 62.4 **Rha** (1) 102.2 (2) 72.5 (3) 72.8 (4) 74.1 (5) 69.6 (6) 18.7 **Glc II** (1) 104.5 (2) 75.0 (3) 78.5 (4) 71.7 (5) 77.9 (6) 62.4 **Glc III** (1) 104.9 (2) 75.2 (3) 78.5 (4) 71.5 (5) 78.6 (6) 62.8.

968

Mass (FAB, Positive ion)[2] : *m/z* 1047 [M+H-H$_2$O]$^+$, 885 [M+H-H$_2$O-Glc]$^+$, 739 [M+H-H$_2$O-Glc-Rha]$^+$, 577 [M+H-H$_2$O-2xGlc-Rha]$^+$, 415 [M+H-H$_2$O-3xGlc-Rha]$^+$.

Biological Activity : Causes morphological abnormality of *Pyricularia oryzae* mycelia and shows cytotoxic activity against the cancer cell line of K 562 *in vitro*.[2]

References

1.   T. Kawasaki, T. Komori, K. Miyahara, T. Nohara, I. Hosokawa and K. Mihashi, *Chem. Pharm. Bull.*, **22,** 2164 (1974).

2.   K. Hu, A. Dong, X. Yao, H. Kobayashi and S. Iwasaki, *Planta Med.*, **63,** 161 (1997).

PROTONEOGRACILLIN

Yamogenin 3-O-{α-L-rhamnopyranosyl-(1→2)-[β-D-glucopyranosyl-(1→3)]-β-D-glucopyranoside}-26-O-[β-D-glucopyranoside]

Source : *Diocorea collettii* Hook. f. var. *hypoglauca* Polibin Pei et Ting (Dioscoreaceae)
Mol. Formula : C$_{51}$H$_{84}$O$_{23}$
Mol. Wt. : 1064
M.P. : 220-222°C (decomp.)[2]
[α]$_D$^{13} : -72.6° (c=0.01, Pyridine)[2]
Registry No. : [191334-50-6]

PMR (C_5D_5N, 500 MHz) : δ 0.89 (s, 3xH-18), 1.02 (s, d, J=6.7 Hz, 3xH-27), 1.05 (s, 3xH-19), 1.31 (d, J=6.5 Hz, 3xH-21), 1.74 (d, J=6.0 Hz, 3xH-6 of Rha), 3.94 (m, H-3), 4.79 (d, J=7.6 Hz, H-1 of Glc III), 4.93 (d, J=7.2 Hz, H-1 of Glc I), 5.09 (d, J=7.2 Hz, H-1 of Glc II), 5.32 (br d, H-6), 6.35 (d, J=0.9 Hz, H-1 of Rha).

CMR (C_5D_5N, 125 MHz) : δ (C-1) 37.5 (2) 30.1 (3) 77.9 (4) 38.8 (5) 140.8 (6) 121.9 (7) 32.4 (8) 31.8 (9) 50.4 (10) 37.2 (11) 21.1 (12) 40.0 (13) 40.7 (14) 56.7 (15) 32.5 (16) 81.2 (17) 63.9 (18) 16.5 (19) 19.4 (20) 40.7 (21) 16.5 (22) 110.7 (23) 37.2 (24) 28.3 (25) 34.5 (26) 75.2 (27) 17.5 **Glc I** (1) 100.1 (2) 77.1 (3) 89.5 (4) 69.6 (5) 77.9 (6) 62.5 **Rha** (1) 102.2 (2) 72.5 (3) 72.8 (4) 74.2 (5) 69.6 (6) 18.7 **Glc II** (1) 104.5 (2) 75.0 (3) 78.5 (4) 71.8 (5) 77.9 (6) 62.5 **Glc III** (1) 105.1 (2) 75.4 (3) 78.5 (4) 71.6 (5) 78.6 (6) 62.9.

Mass (FAB, Positive ion) : m/z 1047 [M+H-H_2O]$^+$, 885 [M+H-H_2O-Glc]$^+$, 739 [M+H-H_2O-Glc-Rha]$^+$, 577 [M+H-H_2O-2xGlc-Rha]$^+$, 415 [M+H-H_2O-3xGlc-Rha]$^+$.

Biological Activity : Causes morphological abnormality of *Pyricularia oryzae* mycelia and shows cytotoxic activity against the cancer cell line of K 562 *in vitro*.

References

1.   T. Kawasaki, *Chem. Pharm. Bull.*, **22**, 2164 (1974).

2.   K. Hu, A. Dong, X. Yao, H. Kobayashi, and S. Iwasaki, *Planta Med.*, **63**, 161 (1997).

PROTOZINGIBERENIN A, PROTOZINGIBERENSIS SAPONIN
(25*R*)-3β,22α,26-Trihydroxyfurost-5-ene 3-O-{α-L-rhamnopyranosyl-(1→3)-[β-D-glucopyranosyl-(1→2)]-β-D-glucopyranoside}-26-O-β-D-glucopyranoside

Source : *Dioscorea zingiberensis* Wright (Dioscoreaceae)
Mol. Formula : $C_{51}H_{84}O_{23}$
Mol. Wt. : 1064
M.P. : 246-248°C[2]
[α]$_D^{20}$: -80.9° (c=0.20)[2]
Registry No. : [89203-42-9]

IR (KBr)[2] : 3400 (OH), 1050 (OH) cm^{-1}.

CMR (DMSO-d$_6$, 22.5 MHz)[2] : δ C-1) 37.1 (2) 30.1 (3) 77.4 (4) 40.3 (5) 140.5 (6) 120.4 (7) 34.6 (8) 29.0 (9) 49.2 (10) 35.1 (11) 19.4 (12) 38.1 (13) 38.8 (14) 56.2 (15) 30.2 (16) 80.1 (17) 62.2 (18) 15.8 (19) 18.5 (20) 40.0 (21) 14.0 (22) 108.8 (23) 30.5 (24) 28.1 (25) 29.0 (26) 68.2 (27) 17.1 **Glc I** (1) 98.5 (2) 80.1 (3) 82.5 (4) 69.5 (5) 75.6 (6) 61.1 **Glc II** (1) 102.5 (2) 74.2 (3) 77.5 (4) 71.6 (5) 77.3 (6) 61.8 **Rha** (1) 101.6 (2) 71.0 (3) 71.3 (4) 74.0 (5) 69.0 (6) 17.5 **Glc III** (1) 104.0 (2) 73.2 (3) 77.4 (4) 71.6 (5) 77.3 (6) 61.8.

Reference

1. S. Tang, Y. Wu and Z. Pang, *Zhiwu Xuebao* (*Acta Botanica Sinica*), **25**, 556 (1983).

2. C.-L. Lau and Y.-Y. Chen, *Zhiwu Xuebao (Acta Batanica Sinica)*, **27**, 68 (1985).

SOLANUM NODIFLORUM SAPONIN SNF-11

3β,17α,22ξ,26-Tetrahydroxy-25R-furost-5-ene 3-O-{α-L-rhamnophyranosyl-(1→2)-
[α-L-rhamnopyranosyl-(1→4)]-β-D-glucopyranoside}-26-O-β-D-glucopyranoside

Source : *Solanum nodiflorum* (Solanaceae)
Mol. Formula : $C_{51}H_{84}O_{23}$
Mol. Wt. : 1064
[α]$_D$: -82.1° (c=0.5, Pyridine)
Registry No. : [260995-64-0]

PMR (C_5D_5N, 500 MHz) : δ 1.00 (s, 3xH-18), 1.02 (d, J=6.1 Hz, 3xH-27), 1.09 (s, 3xH-19), 1.38 (d, J=6.7 Hz, 3xH-21), 1.63 (d, J=6.1 Hz, 3xH-6 of Rha I), 1.76 (d, J=6.1 Hz, 3xH-6 of Rha II), 4.08 (d, J=8.6 Hz, H-6A of Glc I), 4.22 (d, m, H-6B of Glc II), 4.37 (m, H-6A of Glc II), 4.82 (d, J=7.9 Hz, H-1 of Glc II), 4.93 (overlapped H-1 of Glc I), 5.30 (br s, H-6), 5.85 (br s, H-1 of Rha I), 6.39 (br, H-1 of Rha II).

CMR (C_5D_5N, 125.0 MHz) : δ C-1) 37.6 (2) 30.2 (3) 78.1 (4) 39.0 (5) 140.8 (6) 121.0 (7) 32.0 (8) 34.3 (9) 50.2 (10) 37.2 (11) 21.0 (12) 36.9 (13) 45.1 (14) 53.1 (15) 32.5 (16) 90.5 (17) 90.8 (18) 17.3 (19) 19.4 (20) 43.6 (21) 10.5 (22) 111.4 (23) 32.3 (24) 28.0 (25) 32.1 (26) 75.3 (27) 17.5 **Glc** (1) 100.3 (2) 78.1 (3) 78.6 (4) 77.7 (5) 76.9 (6) 61.3 **Rha I** (1) 102.9 (2) 72.6 (3) 72.7 (4) 73.9 (5) 70.4 (6) 18.5 **Rha II** (1) 101.9 (2) 72.6 (3) 72.8 (4) 74.1 (5) 69.5 (6) 18.6 **Glc II** (1) 104.9 (2) 75.2 (3) 78.1 (4) 71.7 (5) 78.5 (6) 62.9.

Mass (FAB, Negative ion) : m/z 1063 [M-H]⁻.

Reference

1.   J. Ando, A. Miyazono, X.-H. Zhu, T. Ikeda and T. Nohara, *Chem. Pharm. Bull.*, **47**, 1794 (1999).

SPICATOSIDE B

(25*S*)-22-Methoxyfurost-5-en-1β,3β,26-triol 1-O-{β-D-glucopyranosyl-(1→2)-[β-D-xylopyranosyl-(1→3)]-
β-D-fucopyranosdie-26-O-β-D-glucopyranoside}

Source : *Liriope spicata* Lour. (Liliaceae)
Mol. Formula : $C_{51}H_{84}O_{23}$
Mol. Wt. : 1064
M.P. : 196-198°C
$[\alpha]_D^{20}$ **:** -88.7° (c=0.13, MeOH)
Registry No. : [128397-48-8]

IR (KBr) **:** 3420, 1070, 1050, 910, 840 cm^{-1}.

PMR (C$_5$D$_5$N, 200 MHz) **:** δ 0.72 (s, 3xH-18), 0.91 (d, *J*=6.2 Hz, 3xH-27), 1.01 (d, *J*=6.0 Hz, 3xH-21), 1.22 (s, 3xH-19), 1.40 (d, *J*=6.0 Hz, 3xH of Fuc), 3.12 (s, OC*H₃*), 4.73 (d, *J*=7.4 Hz, anomeric H), 5.15 (d, *J*=7.5 Hz, anomeric H), 5.31 (d, *J*=7.4 Hz, anomeric H), 5.45 (d, *J*=5.3 Hz, H-6).

CMR (C$_5$D$_5$N, 50.3 MHz) **:** δ C-1) 83.4 (2) 37.8 (3) 68.7 (4) 44.1 (5) 140.2 (6) 124.7 (7) 32.7 (8) 33.4 (9) 50.9 (10) 43.3 (11) 24.1 (12) 40.4 (13) 41.0 (14) 57.4 (15) 32.3 (16) 81.8 (17) 64.8 (18) 17.3 (19) 15.3 (20) 41.0 (21) 16.8 (22) 113.2 (23) 31.4 (24) 28.6 (25) 34.9 (26) 75.2 (27) 17.1 (OCH₃) 47.8 **Fuc** (1) 100.9 (2) 79.3 (3) 83.4 (4) 72.7^a (5) 71.5 (6) 17.6 **Xyl** (1) 106.7 (2) 75.6 (3) 78.7^b (4) 71.1 (5) 67.7 **Glc I** (1) 105.5 (2) 76.9 (3) 79.0 (4) 72.5^a (5) 78.9^b (6) 63.7 **Glc II** (1) 105.5 (2) 75.5 (3) 78.9 (4) 72.2 (5) 78.5 (6) 63.3.

Reference

1.   D.Y. Lee, K.H. Son, J.C. Do and S.S. Kang, *Arch. Pharm. Res.*, **12**, 295 (1989).

TRIGOFOENOSIDE F

(25R)-Furost-5-en-3β,22,26-triol 3-O-[α-L-rhamnopyranosyl-(1→2)-β-D-glucopyranosyl-(1→6)-β-D-glucopyranoside]-26-O-β-D-glucopyranoside

Source : *Trigonella foenum-graecum* L. (Fabaceae)
Mol. Formula : $C_{51}H_{84}O_{23}$
Mol. Wt. : 1064
M.P. : 233-236°C
Registry No. : [94714-56-4]

22-Methyl ether : (Trigofoenoside F-1)

M.P. : 256-258°C; **[α]$_D$:** -78.9° (Pyridine)

Reference

1.   R.K. Gupta, D.C. Jain and R.S. Thakur, *Phytochemistry*, **23**, 2605 (1984).

TRIGONEOSIDE IVa

(25*S*)-Furost-5-ene-3β,22ξ,26-triol 3-O-{α-L-rhamnopyranosyl-(1→2)-[β-D-glucopyranosyl-(1→4)]-β-D-glucopyranoside}-26-O-β-D-glucopyranoside

Source : *Trigonella foenum-graecum* L. (Leguminosae)
Mol. Formula : $C_{51}H_{84}O_{23}$
Mol. Wt. : 1064
$[\alpha]_D^{28}$: -56.5° (c=0.8, C_5D_5N)
Registry No. : [205760-54-9]

IR (KBr) : 3410, 2934, 1072, 1052 cm^{-1}.

PMR (C_5D_5N, 500 MHz) : δ 0.90 (s, 3xH-18), 1.03 (d, *J*=6.7 Hz, 3xH-27), 1.06 (s, 3xH-19), 1.33 (d, *J*=7.1 Hz, 3xH-21), 1.77 (d, *J*=6.4 Hz, 3xH-6 of Rha), 2.24 (dq, *J*=6.9, 6.9 Hz, H-20), 3.49 (dd, *J*=7.1, 9.5 Hz), 3.88 (m, H-3), 4.09 (dd, *J*=6.1, 9.5 Hz, 2xH-26), 4.82 (d, *J*=8.0 Hz, H-1 of Glc III), 4.95 (d, *J*=9.0 Hz, H-1 of Glc I), 5.14 (d, *J*=8.0 Hz, H-1 of Glc II), 5.29 (d-like, H-6), 6.26 (br, H-1 of Rha).

CMR (C_5D_5N, 125 MHz) : δ C-1) 37.5 (2) 30.1 (3) 78.1 (4) 38.9 (5) 140.7 (6) 121.8 (7) 32.3 (8) 31.7 (9) 50.3 (10) 37.1 (11) 21.1 (12) 39.9 (13) 40.8 (14) 56.6 (15) 32.5 (16) 81.1 (17) 63.8 (18) 16.5 (19) 19.4 (20) 40.7 (21) 16.5 (22) 110.7 (23) 37.1 (24) 28.3 (25) 34.4 (26) 75.4 (27) 17.4 **Glc I** (1) 100.0 (2) 77.3 (3) 76.2 (4) 81.9 (5) 77.7 (6) 62.1 **Rha** (1) 101.8 (2) 72.4 (3) 72.7 (4) 74.1 (5) 69.5 (6) 18.7 **Glc II** (1) 105.2 (2) 75.0 (3) 78.4 (4) 71.2 (5) 78.5 (6) 61.8 **Glc III** (1) 105.1 (2) 75.2 (3) 78.6 (4) 71.6 (5) 78.4 (6) 62.8.

Mass (FAB, Positive ion) : *m/z* 1087 [M+Na]$^+$.

Mass (FAB, Negative ion) : *m/z* 1063 [M-H]$^-$.

Reference

1.   M. Yoshikawa, T. Murakami, H. Komatsu, J. Yamahara and H. Matsuda, *Heterocycles*, **47**, 397 (1998).

TRILLIUM KAMTSCHATICUM SAPONIN Tj

25R-Furost-5-ene-3β,17α,22ξ,26-tetrol 3-O-{α-L-rhamnopyranosyl-(1→2)-[α-L-rhamnopyranosyl-(1→4)]-
β-D-glucopyranoside}-26-O-β-D-glucopyranoside

Source : *Trillium kamtschaticum* Pall. (Liliaceae)
Mol. Formula : $C_{51}H_{84}O_{23}$
Mol. Wt. : 1064
M.P. : 205-212°C (decomp.)[1]
[α]$_D$: -66.5° (c=1.05, C_5D_5N)[1]
Registry No. : [77663-50-4]

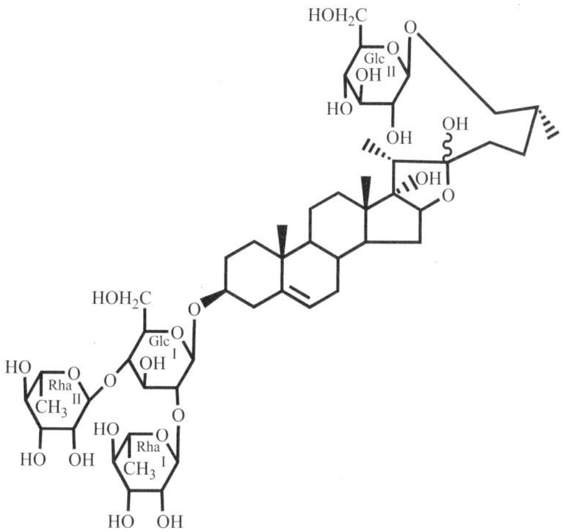

IR : no spiroketal absorptions.

22-Methyl Ether :

CMR (C_5D_5N, 22.5 MHz)[2] : δ C-1) 37.6 (2) 30.2 (3) 77.9 (4) 39.0 (5) 140.9 (6) 121.8 (7) 31.7 (8) 32.3 (9) 50.3 (10) 37.2 (11) 21.0 (12) 37.2 (13) 45.4 (14) 53.1 (15) 32.5 (16) 90.3 (17) 90.5 (18) 17.2 (19) 19.4 (20) 43.1 (21) 10.4 (22)

113.5 (23) 30.8 (24) 28.1 (25) 34.3 (26) 75.2 (27) 17.2 (OCH_3) 47.1 **Glc I** (1) 100.3 (2) 78.9 (3) 76.9 (4) 78.2 (5) 77.9 (6) 61.4 **Rha I** (1) 102.0 (2) 72.8 (3) 72.5 (4) 74.1[a] (5) 69.4[b] (6) 18.5[c] **Rha II** (10 102.9 (2) 72.8 (3) 72.5 (4) 73.9[a] (5) 70.4[b] (6) 18.6[c] **Glc II** (1) 104.9 (2) 75.2 (3) 78.4[d] (4) 71.8 (5) 78.2[d] (6) 63.0.

References

1. N. Fukuda, N. Imamura, E. Saito, T. Nohara and T. Kawasaki, *Chem. Phram. Bull.*, **29**, 325 (1981).

2. N. Nakano, Y. Kashiwada, T. Nohara, T. Tomimatsu, H. Tsukatani and T. Kawasaki, *Yakugaku Zasshi*, **102**, 1031 (1982).

TUBEROSIDE C (ALLIUM)

(25*S*)-5α-Furost-20(22)-ene-2α,3β,26-triol 3-O-{α-L-rhamnopyranosyl-(1→2)-
[β-D-glucopyranosyl-(1→3)]-β-D-glucopyranoside}-26-O-β-D-glucopyranoside

Source : *Allium tuberosum* Rottl. ex spreng. (Liliaceae)
Mol. Formula : $C_{51}H_{84}O_{23}$
Mol. Wt. : 1064
$[\alpha]_D^{23}$ **:** -19.8° (c=0.51, MeOH)
Registry No. : [259810-68-9]

IR (KBr) : 3400, 1450, 1381, 1000-1100 cm^{-1}.

PMR (C$_5$D$_5$N, 400 MHz) : δ 0.75 (s, 3xH-18), 0.98 (s, 3xH-19), 1.10 (d, J=6.6 Hz, 3xH-27), 1.69 (s, H-21), 1.80 (d, J=6.2 Hz, 3xH-6 of Rha), 2.51 (d, J=10.0 Hz, H-17), 3.55 (dd, J=6.8, 9.3 Hz, H-26A), 3.92 (m, H-3), 4.02 (m, H-5 of Glc I, H-5 of Glc II), 4.05 (m, H-5 of Glc III), 4.08 (m, H-2 of Glc II), 4.10 (m, H-26B), 4.11 (m, H-2 of Glc III), 4.15 (m, H-2), 4.21 (m, H-2 of Glc I), 4.22 (m, H-3 of Glc I), 4.24 (m, H-3 of Glc III), 4.30 (m, H-4 of Glc I, H-4 of Glc II, H-4 of Glc III), 4.31 (m, H-3 of Glc II), 4.38 (m, H-6A of Glc III, H-4 of Rha), 4.45 (m, H-6A of Glc II), 4.52 (m, H-6A of Glc I, H-6B of Glc III), 4.58 (m, H-6B of Glc II), 4.60 (m, H-3 of Rha), 4.81 (m, H-2 of Rha), 4.87 (m, H-16), 4.89 (d, J=7.7 Hz, H-1 of Glc II), 4.90 (m, H-5 of Rha), 5.05 (d, J=7.6 Hz, H-1 of Glc I), 5.17 (d, J=7.8 Hz, H-1 of Glc III), 5.58 (m, H-6B of Glc I), 6.25 (s, H-1 of Rha).

CMR (C$_5$D$_5$N, 100 MHz) : δ C-1) 46.0 (2) 70.8 (3) 85.0 (4) 33.6 (5) 44.8 (6) 28.3 (7) 32.6 (8) 34.5 (9) 54.6 (10) 37.1 (11) 21.8 (12) 40.0 (13) 43.9 (14) 54.8 (15) 34.6 (16) 84.8 (17) 64.9 (18) 14.6 (19) 13.7 (20) 103.8 (21) 12.0 (22) 152.6 (23) 31.6 (24) 23.8 (25) 33.9 (26) 75.4 (27) 17.4 **Glc I** (1) 100.6 (2) 77.8 (3) 82.0 (4) 71.4 (5) 76.6 (6) 61.9 **Glc II** (1) 105.3 (2) 75.4 (3) 78.6 (4) 71.9 (5) 78.7 (6) 63.0 **Glc III** (1) 105.3 (2) 75.1 (3) 78.4 (4) 71.4 (5) 77.8 (6) 62.3 **Rha** (1) 102.1 (2) 72.9 (3) 72.5 (4) 74.2 (5) 69.6 (6) 18.7.

Mass (E.S.I.) : m/z 1088 [M+Na]$^+$.

Reference

1.   S. Sang, A. Lao, H. Wang and Z. Chen, *Phytochemistry*, **52**, 1611 (1999).

TUBEROSIDE G

26-O-β-D-Glucopyranosyl-(5α,25S,20R)-furost-22-en-2α,3β,20,26-tetraol
3-O-{α-L-rhamnopyranosyl-(1→2)-[α-L-rhamnopyranosyl-(1→4)]-β-D-glucopyranoside}

Source : *Allium tuberosum* Rottl. (Liliaceae)
Mol. Formula : C$_{51}$H$_{84}$O$_{23}$
Mol. Wt. : 1064
[α]$_D^{17}$: -46.0° (c=0.30, MeOH)
Registry No. : [332837-99-7]

PMR (C_5D_5N, 400 MHz) : δ 0.95 (s, CH_3), 0.96 (s, CH_3), 1.43 (s, CH_3), 1.15 (d, J=6.4 Hz, 3xH-27), 1.69 (d, J=5.9 Hz, 3xH-6 of Rha II), 1.72 (d, J=6.0 Hz, 3xH-6 of Rha I), 3.75 (m, H-5 of Glc I), 3.96 (m, H-5 of Glc II), 4.02 (m, H-2 of Glc II), 4.08 (m, H-6A of Glc I), 4.23 (m, H-2 and H-3 of Glc I), 4.24 (m, H-3 and H-4 of Glc II), 4.25 (m, H-6B of Glc I), 4.35 (m, H-4 of Rha I, H-4 of Rha II), 4.37 (m, H-4 of Glc I), 4.41 (m, H-6A of Glc II), 4.55 (m, H-3 of Rha I), 4.57 (m, H-6B of Glc II), 4.61 (m, H-3 of Rha II), 4.69 (s, H-2 of Rha II), 4.84 (s, H-2 of Rha I), 4.85 (d, J=7.3 Hz, H-1 of Glc II), 4.91 (m, H-5 of Rha I), 4.92 (d, J=7.6 Hz, H-1 of Glc I), 5.01 (m, H-5 of Rha II), 5.85 (s, H-1 of Rha III), 6.38 (s, H-1 of Rha I).

CMR (C_5D_5N, 100 MHz) : δ C-1) 45.9 (2) 70.7 (3) 85.3 (4) 33.5 (5) 44.8 (6) 28.2 (7) 32.2 (8) 34.0 (9) 54.3 (10) 37.0 (11) 21.0 (12) 39.5 (13) 40.7 (14) 56.6 (15) 33.5 (16) 84.0 (17) 66.9 (18) 13.9 (19) 13.6 (20) 82.4 (21) 15.4 (22) 157.4 (23) 96.1 (24) 29.7 (25) 35.0 (26) 75.5 (27) 17.7 **Glc I** (1) 101.0 (2) 78.0 (3) 78.1 (4) 78.9 (5) 77.3 (6) 61.3 **Rha I** (1) 102.2 (2) 72.6 (3) 72.8 (4) 74.0 (5) 69.7 (6) 18.7 **Rha II** (1) 103.1 (2) 72.7 (3) 72.9 (4) 74.2 (5) 70.6 (6) 18.7 **Glc II** (1) 105.3 (2) 75.4 (3) 78.6 (4) 71.8 (5) 78.8 (6) 62.9.

Reference

1.   S. Sang, S. Mao, A. Lao, Z. Chen and C.-T. Ho, *J. Agric. Food Chem.*, **49**, 1475 (2001).

TUBEROSIDE H

26-O-β-D-Glucopyranosyl-(5α,25*S*,20*S*)-furost-22-en-2α,3β,20,26-tetraol 3-O-{α-L-rhamnopyranosyl-(1→2)-[α-L-rhamnopyranosyl-(1→4)]-β-D-glucopyranoside}

Source : *Allium tuberosum* Rottl. (Liliaceae)
Mol. Formula : $C_{51}H_{84}O_{23}$
Mol. Wt. : 1064
$[α]_D^{25}$: -41.8° (c=0.34, MeOH)
Registry No. : [332838-00-3]

IR (KBr) : 3421 (OH), 1637 (C=C), 1041 (glycosidic linkage) cm^{-1}.

PMR (C$_5$D$_5$N, 400 MHz) : δ 0.86 (s, CH_3), 0.90 (s, CH_3), 1.05 (d, J=6.0 Hz, 3xH-27), 1.62 (d, J=6.0 Hz, 3xH-6 of Rha II), 1.67 (d, J=6.0 Hz, 3xH-6 of Rha I), 1.67 (s, 3xH-21), 3.51 (dd, J=7.0, 9.2 Hz, H-26A), 3.75 (m, H-5 of Glc I), 3.92 (m, H-5 of Glc II), 4.02 (m, H-2 of Glc II), 4.07 (m, H-6A of Glc I), 4.12 (m, H-26B), 4.22 (m, H-3 and H-4 of Glc II), 4.23 (m, H-2 and H-3 of Glc I), 4.27 (m, H-6B of Glc I), 4.35 (m, H-4 of Rha I, H-4 of Rha II), 4.39 (m, H-4 of Glc I), 4.46 (m, H-6A of Glc II), 4.54 (m, H-3 of Rha II), 4.55 (m, O.P.), 4.58 (m, H-3 of Rha I), 4.61 (m, H-6B of Glc II), 4.67 (s, H-2 of Rha II), 4.84 (d, J=7.2 Hz, H-1 of Glc II), 4.84 (s, H-2 of Rha I), 4.91 (m, H-5 of Rha I), 4.96 (m, H-5 of Rha II), 5.01 (d, J=6.8 Hz, H-1 of Glc I), 5.83 (s, H-1 of Rha II), 6.37 (s, H-1 of Rha I).

CMR (C$_5$D$_5$N, 100 MHz) : δ C-1) 46.0 (2) 70.3 (3) 85.2 (4) 33.6 (5) 44.8 (6) 28.3 (7) 32.3 (8) 34.2 (9) 54.4 (10) 37.0 (11) 21.1 (12) 39.6 (13) 40.9 (14) 56.9 (15) 33.6 (16) 84.4 (17) 68.1 (18) 13.9 (19) 13.7 (20) 77.0 (21) 22.0 (22) 163.8 (23) 91.6 (24) 29.8 (25) 35.0 (26) 75.3 (27) 17.7 **Glc I** (1) 101.1 (2) 78.0 (3) 78.3 (4) 79.0 (5) 77.3 (6) 61.5 **Rha I** (1) 102.3 (2) 75.6 (3) 72.8 (4) 74.1 (5) 69.7 (6) 18.7 **Rha II** (1) 103.1 (2) 72.7 (3) 72.9 (4) 74.2 (5) 70.7 (6) 18.7 **Glc II** (1) 105.6 (2) 75.4 (3) 78.4 (4) 71.8 (5) 78.6 (6) 62.9.

Reference

1.   S. Sang, S. Mao, A. Lao, Z. Chen and C.-T. Ho, *J. Agric. Food Chem.*, **49**, 1475 (2001).

ZINGIBERENIN D, PROTOZINGIBERENIN B, TRIGOFOENOSIDE D

(25S)-3β,22,26-Trihydroxyfurost-5-ene 3-O-{α-L-rhamnopyranosyl-(1→2)-[β-D-glucopyranosyl-(1→3)]-β-D-glucopyranoside}-26-O-β-D-glucopyranoside

Source : *Dioscorea zingiberensis* Wright[1]
(Dioscoreaceae), *Trigonella foenum-graecum*[2] (Fabaceae)
Mol. Formula : $C_{51}H_{84}O_{23}$
Mol. Wt. : 1064
M.P. : 246-248°C[2]
$[\alpha]_D$: -73.2° (c=1.0, Pyridine)[2]
Registry No. : [89255-30-1], [99664-39-8]

22-Methyl ether : (Trigofoenoside D-1) [99664-41-2]

M.P. : 250-253°C (decomp.)[2]; $[\alpha]_D$: -77.6° (c=1.0, Pyridine)[2]

PMR (CD₃OD-CDCl₃, 400 MHz)[2] : δ 1.67 (br s, 3xH-1 of Rha), 3.12 (s, OCH₃), 4.21 (d, J=7.3 Hz), 4.56 (d, J=7.5 Hz), 4.92 (d, J=7.8 Hz), 5.10 (br s, 4xanomeric H).

References

1. S.R. Tang, Y.F. Wu and Z.-J. Pang, *Zhiwu Xuebao* (*Acta Bot. Sin.*), **25**, 556 (1983).

2. R.K. Gupta, D.C. Jain and R.S. Thakur, *Phytochemistry*, **24**, 2399 (1985).

ASPARAGUS CURILLUS SAPONIN 4

(25S)-22α-Methoxy-5β-furostan-3β,26-diol 3-O-{β-D-glucoyranosyl-(1→2)-[α-L-arabinopyranosyl-(1→4)]-β-D-glucopyranoside}-26-O-β-D-glucopyranoside

Source : *Asparagus curillus* Buch.-Ham. (Liliaceae)
Mol. Formula : $C_{51}H_{86}O_{23}$
Mol. Wt. : 1066
[α]$_D^{20}$: -47.0° (c=0.5, MeOH)
Registry No. : [84765-75-3]

IR (KBr) : 3400 (OH) cm^{-1}, no spiroketal absorptions.

PMR (C_5D_5N) : δ 3.25 (s, OCH_3), 4.75 (d, J=7.3 Hz), 5.00 (d, J=7.0 Hz), 6.65 (br s).

Reference

1.   S.C. Sharma, O.P. Sati and R. Chand, *Phytochemistry*, **21**, 1711 (1982).

ASPAROSIDE B, ASPARAGUS CURILLUS SAPONIN 6

(5β,25S)-3β,22α,26-Trihydroxyfurostane 3-O-{β-D-glucopyranosyl-(1→2)-[α-L-rhamnopyranosyl-(1→4)]-β-D-glucopyranoside]-26-O-β-D-glucopyranoside}

Source : *Asparagus adscendens* Roxb.[1] (Liliaceae),
Asparagus curillus Buch.-Ham.[2] (Liliaceae)
Mol. Formula : $C_{51}H_{86}O_{23}$
Mol. Wt. : 1066
M.P. : 170-174°C
[α]$_D^{20}$: -58°
Registry No. : [84633-36-3]

IR (KBr) : 3400 (OH) cm^{-1}. No spiroketal absorptions.

References

1. S.C. Sharma, R. Chand and O.P. Sati, *Phytochemistry*, **21**, 2075 (1982).

2. S.C. Sharma, O.M. Sati and R. Chand, *Planta Med.*, **47**, 117 (1983).

CURILLOSIDE H

(25S)-5β-Furostan-3β,22α,26-triol-3-O-{α-L-rhamnopyranosyl-(1→2)-[β-D-glucopyranosyl-(1→4)]-β-D-glucopyranoside}-26-O-β-D-glucopyranoside

Source : *Asparagus curillus* Buch.-Ham. (Liliaceae)
Mol. Formula : $C_{51}H_{86}O_{23}$
Mol. Wt. : 1066
M.P. : 181-183°C
[α]$_D^{20}$: -53.0° (c=1.0, MeOH)
Registry No. : [113834-22-3]

IR : 3400 (OH), cm^{-1}, no spiroketal absorptions.

Reference

1.   S.C. Sharma and H.C. Sharma, *Phytochemistry*, **33**, 683 (1993).

LILIUM HANSONII SAPONIN 1

(25*R*)-5α-Furostan-3β,22ξ,26-triol 3-O-{α-L-rhamnopyranosyl-(1→2)-[β-D-glucopyranosyl-(1→4)]-
β-D-glucopyranoside}-26-O-β-D-glucopyranoside

Source : *Lilium hansonii* (Liliaceae)
Mol. Formula : $C_{51}H_{86}O_{23}$
Mol. Wt. : 1066
$[\alpha]_D^{25}$: -48.8° (c=0.40, MeOH)
Registry No. : [145164-60-9]

IR (KBr) : 3400 (OH), 2920 (CH), 1445, 1375, 1300, 1255, 1155, 1035, 980, 900, 820, 800, 695 cm^{-1}.

PMR (C_5D_5N, 400 MHz) : δ 0.86 (s, 3xH-19), 0.88 (s, 3xH-18), 0.99 (d, *J*=6.5 Hz, H-27), 1.33 (d, *J*=7.0 Hz, 3xH-21), 1.73 (d, *J*=6.1 Hz, 3xH-6 of Rha), 4.80 (d, *J*=7.7 Hz, H-1 of Glc III), 4.98 (d, *J*=7.7 Hz, H-1 of Glc I), 5.11 (d, *J*=7.7 Hz, H-1 of Glc II), 6.20 (br s, H-1 of Rha).

CMR (C_5D_5N, 100 MHz) : δ C-1) 37.3 (2) 29.9 (3) 77.6^a (4) 34.5 (5) 44.7 (6) 29.0 (7) 32.4^b (8) 35.3 (9) 54.5 (10) 36.0 (11) 21.3 (12) 40.3 (13) 41.1 (14) 56.4 (15) 32.5^b (16) 81.1 (17) 64.0 (18) 16.7 (19) 12.4 (20) 40.7 (21) 16.4 (22) 110.6 (23) 37.2 (24) 28.4 (25) 34.3 (26) 75.3 (27) 17.5 **Glc I** (1) 99.6 (2) 77.7^a (3) 76.2 (4) 82.2 (5) 77.3 (6) 62.2^c **Rha** (1) 101.9 (2) 72.4 (3) 72.8 (4) 74.1 (5) 69.4 (6) 18.6 **Glc II** (1) 105.2 (2) 75.0 (3) 78.5^d (4) 71.3 (5) 78.3^d (6) 62.1^c **Glc III** (1) 104.9 (2) 75.2 (3) 78.6^d (4) 71.8 (5) 78.2^d (6) 62.9.

Mass (S.I.) : *m/z* 1049 [M-OH]$^+$, [M-Glc]$^+$.

Reference

1.   K. Ori, Y. Mimaki, K. Mito, Y. Sashida, T. Nikaido, T. Ohmoto and A. Masuko, *Phytochemistry*, **31**, 2767 (1992).

MELONGOSIDE P

(5α,25R)-Furostan-3β,22α,26-triol 3-O-β-D-glucopyranosyl-(1→2)-[α-L-rhamnopyranosyl-(1→3)]-β-D-glucopyranoside]-26-O-β-D-glucopyranoside

Source : *Solanum melongena* L. (Solanaceae)
Mol. Formula : $C_{51}H_{86}O_{23}$
Mol. Wt. : 1066
M.P. : 179-180°C
$[α]_D^{20}$: -75° (c=1.0, H_2O)
Registry No. : [98508-45-3]

Reference

1.   P.K. Kintia and S.A Shvets, *Phytochemistry*, **24**, 1567 (1985).

TRIGOFOENOSIDE C

(25R)-22-O-Methyl-5α-furostane-2α,3β,26-triol 3-O-{α-L-rhamnopyranosyl-(1→4)-[α-L-rhamnorapyranosyl-(1→2)]-β-D-glucopyranoside}-26-O-β-D-glucopyranoside

Source : *Trigonella-foenum-graecum* L. (Leguminosae)
Mol. Formula : $C_{51}H_{86}O_{23}$
Mol. Wt. : 1066
Registry No. : [99753-12-5]

22-Methyl Ether : [105317-71-3] **Trigofoenoside C-1**

M.P. : 210-212°C (decomp.); **[α]$_D$:** -64.1° (c=1.0, Pyridine)

IR (KBr) : 3600-3100 (OH), 2900, 1460, 1380, 1260-1150-1000 cm^{-1} (C-O-C), no sproketal bands.

PMR (CD$_3$OD-CDCl$_3$, 400 MHz) : δ 3.12 (s, OC*H*$_3$), 3.84 (d, *J*=7.0 Hz), 4.02 (d, *J*=8.0 Hz), 4.58 (br s), 4.94 (br s).

Reference

1.   R.K. Gupta, D.C Jain and R.S. Thakur, *Phytochemistry*, **25**, 2205 (1986).

TRIGONELLA FOENUM-GRAECUM SAPONIN 1
(25S,5α)-Furostan-3β,22,26-triol 3-O-{α-L-rhamnopyranosyl-(1→2)-[β-D-glucopyranosyl-(1→3)]-β-D-glucopyranoside}-26-O-β-D-glucopyranoside

Source : *Trigonella foenum-graecum* (Leguminosae)
Mol. Formula : $C_{51}H_{86}O_{23}$
Mol. Wt. : 1066
[α]$_D^{18}$: -81.8° (MeOH-EtOAc 1:1)
Registry No. : [74971-03-2]

22-Methyl ether :

M.P. : 242-246°C

IR (KBr) : 3600-3250 cm^{-1}.

PMR (C_5D_5N, 100 MHz) : δ 3.27 (s, OCH_3), 4.82 (d, J=7.3 Hz, anomeric H), 5.07 (d, J=7.0 Hz, anomeric H), 6.20 (s, H-1 of Rha).

Permethyl Derivative :

Mass (E.I.) : *m/z* 1230 [(M)$^+$ - MeOH], 1042, 1012, 822, 804, 617, 397, 361, 219, 189.

Reference

1.   R. Hardman, J. Kosugi and R.T. Parfitt, *Phytochemistry*, **19**, 698 (1980).

SMILAX SIEBOLDII SAPONIN 5

3β,22ξ,26-Trihydroxy-(25R)-5α-furostan-6-one 3-O-{β-D-glucopyranosyl-(1→4)-O-[α-L-arabinopyranosyl-(1→6)]-β-D-glucopyranoside}-26-O-β-D-glucopyranoside

Source : *Smilax sieboldii* Miq. (Liliaceae)
Mol. Formula : $C_{50}H_{82}O_{24}$
Mol. Wt. : 1066
$[\alpha]_D^{28}$: -46.3° (c=0.28, EtOH)
Registry No. : [143222-29-1]

UV (MeOH) : λ_{max} 280 (log ε, 102) nm.

CD (MeOH; c 1.26×10^{-3}) nm : θ: 293 (-2698).

IR (KBr) : 3425 (OH), 2940 (CH), 1700 (C=O), 1445, 1380, 1260, 1225, 1160, 1065, 1035, 900, 860, 775, 715, 695 cm^{-1}.

PMR (C_5D_5N, 400/500 MHz) : 0.65 (s, 3xH-19), 0.84 (s, 3xH-18), 0.99 (d, J=6.7 Hz, 3xH-27), 1.34 (d, J=6.9 Hz, 3xH-21), 4.81 (d, J=7.6 Hz, H-1 of Glc III), 4.92 (d, J=7.7 Hz, H-1 of Glc I), 5.09 (d, J=7.3 Hz, H-1 of Ara), 5.49 (d, J=7.9 Hz, H-1 of Glc II).

CMR (C_5D_5N, 100/125 MHz) : δ C-1) 36.7 (2) 29.4 (3) 77.0 (4) 27.0 (5) 56.5^a (6) 209.6 (7) 46.8 (8) 37.4 (9) 53.7 (10) 40.9 (11) 21.5 (12) 39.7 (13) 41.4 (14) 56.4^a (15) 32.0 (16) 80.8 (17) 63.8 (18) 16.5^b (19) 13.1 (20) 40.6 (21) 16.4^b (22) 110.7 (23) 37.1 (24) 28.4 (25) 34.2 (26) 75.2 (27) 17.4 **Glc I** (1) 102.0 (2) 74.7 (3) 76.5 (4) 81.1 (5) 74.9 (6) 68.4

Glc II (1) 104.9 (2) 75.2 (3) 78.5^c (4) 71.8^d (5) 78.2 (6) 62.6 **Ara** (1) 105.6 (2) 72.5 (3) 74.6 (4) 69.7 (5) 67.1 **Glc III** (1) 104.9 (2) 75.2 (3) 78.6^c (4) 71.7^d (5) 78.4^c (6) 62.8.

Mass (SI, Positive ion) : m/z 1106 [M+K+H]$^+$, 1089 [M+Na]$^+$.

Biological Activity : The compound shows inhibitory activity on cyclic AMP phosphodiesterase with IC$_{50}$ > 500x10^{-5} M.

Reference

1.   S. Kubo, Y. Mimaki, Y. Sashida, T. Nikaido and T. Ohmoto, *Phytochemistry*, **31**, 2445 (1992).

ASPAROSIDE A

(5β,25S)-3β,26-Dihydroxy-22α-methoxyfurostane 3-O-{β-D-glucopyranosyl-(1→2)-[α-L-rhamnopyranosyl-(1→4)]-β-D-glucopyranoside}-26-O-β-D-glucopyranoside

Source : *Aparagus adscendens* Roxb.[1], *Asparagus curillus* Buch.-Ham[2]. (Liliaceae)
Mol. Formula : C$_{51}$H$_{88}$O$_{23}$
Mol. Wt. : 1068
M.P. : 180-184°C (decomp.)
[α]$_D^{20}$: -53° (c=1.5, MeOH)
Registry No. : [84633-35-2]

IR (KBr)[2] : 3400 (OH) cm^{-1}. No spiroketal absorptions.

PMR (C$_5$D$_5$N)[1] : δ 3.25 (s, OCH_3), 4.82 (d, J=7.0 Hz, H-1 of Glc), 5.41 (d, J=7.1 Hz, H-1 of Glc), 5.90 (s, H-1 of Rha).

References

1.   S.C. Sharma, R. Chand and O.P. Sati, *Phytochemistry*, **21**, 2075 (1982).

2.   S.C. Sharma, O.M. Sati and R. Chand, *Planta Med.*, **47**, 117 (1983).

ABUTILOSIDE M
(22*S*,25*S*)-22,25-Epoxy-7β-methoxyfurost-5-ene-3β,26-diol 3-O-{α-L-rhamnopyranosyl-(1→2)-[α-L-rhamnopyranosyl-(1→4)]-β-D-glucopyranoside}-26-O-β-D-glucopyranoside

Source : *Solanum abutiloides* (Solanaceae)
Mol. Formula : C$_{52}$H$_{84}$O$_{23}$
Mol. Wt. : 1076
[α]$_D^{25}$: -110.9° (c=0.37, MeOH)
Registry No. : [652144-59-7]

PMR (C$_5$D$_5$N, 400 MHz) : δ 0.84 (s, 3xH-18), 1.02 (s, 3xH-19), 1.07 (d, J=6.7 Hz, 3xH-21), 1.42 (s, 3xH-27), 1.50 (H-15A), 1.63 (d, J=6.1 Hz, 3xH-6 of Rha II), 1.76 (H-17), 1.76 (d, J=6.1 Hz, 3xH-6 of Rha I), 1.82 (H-14), 2.15 (H-15B), 3.25 (m H-7), 3.26 (s, OCH_3), 3.62 (br d, J=9.2 , H-5 of Glc), 3.87 (m, H-3), 3.94 (d, J=9.8 Hz, H-26A), 3.96 (m, H-5 of Glc II), 4.06 (dd, J=7.6, 8.5 Hz, H-2 of Glc II), 4.10 (br d, J=12.5 Hz, H-6A of Glc I), 4.19 (br d, J=10.6 Hz, H-6B of Glc I), 4.21 (d, J=9.8 Hz, H-26B), 4.25 (H-2 and H-3 of Glc I), 4.29 (dd, J=8.5, 8.5 Hz, H-4 of Glc II),

4.32 (dd, J=8.5, 8.5 Hz, H-3 of Glc II), 4.37 (dd, J=8.5, 8.5 Hz, H-4 of Rha II), 4.40 (dd, J=8.5, 8.5 Hz, H-4 of Rha I), 4.42 (dd, J=8.5, 8.5 Hz, H-4 of Glc I), 4.44 (dd, J=5.1, 10.6 Hz, H-6A of Glc I), 4.55 (br d, J=10.6 Hz, H-6B of Glc II), 4.57 (dd, J=3.0, 8.5 Hz, H-3 of Rha II), 4.67 (dd, J=3.0, 8.5 Hz, H-3 of Rha I), 4.71 (br s, H-2 of Rha II), 4.75 (m, H-16), 4.88 (br s, H-2 of Rha), 4.96 (d, J=7.7 Hz, H-1 of Glc I), 4.98 (m, H-5 of Rha I and Rha II), 5.00 (d, J=7.6 Hz, H-1 of Glc II), 5.78 (d, J=3.7 Hz, H-6), 5.88 (br s, H-1 of Rha II), 6.40 (br s, H-1 of Rha I).

CMR (C_5D_5N, 100 MHz) : δ C-1) 37.2 (2) 30.1 (3) 78.1 (4) 39.2 (5) 146.0 (6) 121.3 (7) 73.2 (8) 37.1 (9) 43.1 (10) 38.1 (11) 21.0 (12) 39.5 (13) 40.4 (14) 49.0 (15) 32.3 (16) 81.1 (17) 62.6 (18) 16.0 (19) 18.3 (20) 38.7 (21) 15.3 (22) 120.2 (23) 33.2 (24) 34.1 (25) 83.9 (26) 77.6 (27) 24.5 (OCH₃) 56.4 **Glc I** (1) 100.6 (2) 77.8 (3) 78.0 (4) 78.5 (5) 76.9 (6) 61.2 **Rha I** (1) 102.0 (2) 72.6 (3) 72.9 (4) 74.2 (5) 69.6 (6) 18.7 **Rha II** (1) 102.9 (2) 72.6 (3) 72.8 (4) 74.0 (5) 70.4 (6) 18.6 **Glc II** (1) 105.3 (2) 75.4 (3) 78.5 (4) 71.6 (5) 78.5 (6) 62.7.

Mass (FAB, Negative ion) : m/z 1075 [M-H]⁻.

Reference

1. H. Yoshimitsu, M. Nishida and T. Nohara, *Phytochemistry*, **64**, 1361 (2003).

DIOSCORESIDE E

(23S,25R)-23-Methoxyfurost-5,20(22)-diene 3β,26-diol 3-O-{α-L-rhamnopyranosyl-(1→2)-[β-D-glucopyranosyl-(1→3)-β-D-glucopyranoside]-26-O-β-D-glucopyranoside

Source : *Dioscorea futschauensis* R. Kunth (Dioscoreaceae)
Mol. Formula : $C_{52}H_{84}O_{23}$
Mol. Wt. : 1076
M.P. : 256-258°C (decomp.)
[α]$_D^{24}$: -50.3° (c=0.05, Pyridine)
Registry No. : [465321-73-6]

IR (KBr) : 3420 (OH), 2950, 1640, 1380, 1040 (glycoside C-O) cm^{-1}.

PMR (C$_5$D$_5$N, 500 MHz) : δ 0.69 (s, 3xH-18), 0.88* (H-9, H-14), 0.96 (H-1A), 1.06 (s, 3xH-19), 1.09 (d, *J*=6 Hz, 3xH-27), 1.14 (H-12), 1.45* (H-11), 1.46 (H-7A), 1.48 (H-15A), 1.54* (H-8), 1.68 (H-24A), 1.70* (H-12B), 1.73* (H-1B), 1.75 (br s, 3xH-6 of Rha), 1.77 (s, 3xH-21), 1.86* (H-7B), 1.88 (H-2A), 2.07* (H-2B), 2.10* (H-15B), 2.24* (H-24B), 2.28 (m, H-25), 2.49 (d, *J*=10 Hz, H-17), 2.70 (m, H-4A), 2.78 (H-4B), 3.34 (s, OC*H$_3$*), 3.72* (H-26A), 3.82 (m, H-5 of Glc I), 3.83* (H-5 of Glc III), 3.86 (m, H-3), 4.00* (H-26B, H-5 of Glc II), 4.02* (H-2 of Rha, H-2 of Glc III), 4.03* (H-4 of Glc I), 4.10* (H-4 of Glc II), 4.18* (H-2 of Glc I, H-3 of Glc III), 4.19* (H-3 of Glc I, H-3 of Glc II, H-4 of Glc III), 4.20* (H-23), 4.24* (H-6A of Glc I), 4.25 (m, H-4 of Rha), 4.28* (H-6A of Glc II), 4.36* (H-6A of Glc III), 4.42* (H-6B of Glc I), 4.50* (H-6B of Glc III), 4.53* (H-6B of Glc II), 4.58* (H-3 of Rha), 4.84 (d, *J*=7.2 Hz, H-1 of Glc III), 4.86* (H-16, H-2 of Rha), 4.90* (H-5 of Rha), 4.92 (d, *J*=6.0 Hz, H-1 of Glc), 5.10 (d, *J*=7.5 Hz, H-1 of Glc II), 5.31 (br s, H-6), 6.38 (br s, H-1 of Rha). * overlapped signals.

CMR (C$_5$D$_5$N, 125 MHz) : δ C-1) 37.5 (2) 30.2 (3) 78.2 (4) 38.7 (5) 140.8 (6) 121.9 (7) 32.4 (8) 31.7 (9) 50.3 (10) 37.2 (11) 21.2 (12) 40.8 (13) 43.5 (14) 55.0 (15) 34.5 (16) 84.6 (17) 64.8 (18) 14.3 (19) 19.4 (20) 108.6 (21) 11.2 (22) 152.2 (23) 73.4 (24) 37.7 (25) 30.7 (26) 75.2 (27) 17.6 (OCH$_3$) 56.1 **Glc I** (1) 100.0 (2) 77.6 (3) 89.5 (4) 69.6 (5) 77.9 (6) 62.4 **Rha** (1) 102.2 (2) 72.5 (3) 72.8 (4) 74.1 (5) 69.6 (6) 18.7 **Glc II** (1) 104.5 (2) 75.0 (3) 78.5 (4) 71.7 (5) 77.9 (6) 62.4 **Glc III** (10) 104.9 (2) 75.2 (3) 78.5 (4) 71.5 (5) 78.6 (6) 62.8.

Mass (FAB, Negative ion) : *m/z* 1075 [M-H]$^-$, 913 [M-H-Glc]$^-$, 767 [M-H-Glc-Rha]$^-$, 605 [M-H-2xGlc-Rha]$^-$, 443 [M-H-3xGlc-Rha].

Mass (FAB, Negative ion, H.R.) : *m/z* 1075.5347 [(M-H)$^-$, calcd. for 1075.5354].

Reference

1. H.-W. Liu, S.-L. Wang, B. Cai, G.-X. Qu, X.-J. Yang, H. Kobayashi and X.-S. Yao, *J. Asian Nat. Prod. Res.*, **5**, 241 (2003).

LILIUM CANDIDUM SAPONIN 6

(25*R*)-3β,26-Dihydroxy-22ξ-methoxyfurost-5-ene 3-O-{α-L-rhamnopyranosyl-(1→2)-[β-D-glucopyranosyl-(1→6)]-β-D-glucopyranoside}-26-O-β-D-glucopyranoside

Source : *Lilium candidum* L. (Liliaceae)
Mol. Formula : $C_{52}H_{86}O_{23}$
Mol. Wt. : 1078
$[\alpha]_D^{29}$: -69.0° (c=0.29, MeOH)
Registry No. : [244160-64-3]

IR (KBr) : 3405 (OH), 2935 (CH), 1060, 1035 cm^{-1}.

PMR (C_5D_5N, 500 MHz) : δ 0.82 (s, 3xH-18), 1.01 (d, *J*=6.7 Hz, 3xH-27), 1.05 (s, 3xH-19), 1.20 (d, *J*=6.9 Hz, 3xH-21), 1.78 (d, *J*=6.1 Hz, 3xH-6 of Rha), 3.28 (s, OC*H*₃), 4.85 (d, *J*=7.7 Hz, H-1 of Glc III), 4.96 (d, *J*=7.7 Hz, H-1 of Glc I), 5.08 (d, *J*=7.9 Hz, H-1 of Glc II), 5.33 (br d, *J*=5.0 Hz, H-6), 6.32 (br s, H-1 of Rha).

CMR (C_5D_5N, 125 MHz) : δ C-1) 37.5 (2) 30.3 (3) 78.4 (4) 39.1 (5) 140.9 (6) 121.6 (7) 32.2 (8) 31.6 (9) 50.2 (10) 37.0 (11) 21.0 (12) 39.7 (13) 40.7 (14) 56.5 (15) 32.1 (16) 81.3 (17) 64.1 (18) 16.2 (19) 19.4 (20) 40.4 (21) 16.2 (22) 112.6 (23) 30.7 (24) 28.1 (25) 34.2 (26) 75.1 (27) 17.1 (OC*H*₃) 47.2 **Glc I** (1) 100.6 (2) 77.5 (3) 79.4 (4) 71.6 (5) 76.8

(6) 69.7 **Rha** (1) 102.0 (2) 72.4 (3) 72.8 (4) 74.1 (5) 69.4 (6) 18.6 **Glc II** (1) 105.3 (2) 75.1 (3) 78.4 (4) 71.6 (5) 78.5 (6) 62.7 **Glc III** (1) 104.9 (2) 75.1 (3) 78.3 (4) 71.7 (5) 78.4 (6) 62.8..

Mass (FAB, Negative ion) : m/z 1077 [M-H]⁻.

Reference

1.   Y. Mimaki, T. Satou, M. Kuroda, Y. Sashida and Y. Hatakeyama, *Phytochemistry*, **51**, 567 (1999).

METHYL PROTO-DELTONIN

3β,26-Dihydroxy-22ξ-methoxy-(25R)-furost-5-ene 3-O-{β-D-glucopyranosyl-(1→4)-[α-L-rhamnopyranosyl-(1→2)]-β-D-glucopyranoside}-26-O-β-D-glucopyranoside

Source : *Rhapis humilis* Bl.[1] (Palmae), *R. excelsa* Henry[1] (Palmae), *Lilium longiflorum*[2] (Liliaceae), *Sabal causiarum* Becc.[3] (Palmae), *Allium narcissiflorum*[4] (Liliaceae)

Mol. Formula : $C_{52}H_{86}O_{23}$

Mol. Wt. : 1078

M.P. : 196-199°C[1]

$[\alpha]_D^{20}$: -59.8° (c=1.00, Pyridine)[1]

Registry No. : [94992-07-1]

IR (KBr)[1] : 3480-3260 (OH) cm^{-1}.

CMR (C$_5$D$_5$N, 25 MHz)[1] : **Rha** δ C-1) 101.7 (6) 18.6 **Glc I** (1) 100.2 **Glc II** (1) 104.8 **Glc III** (1) 105.0.

References

1.   Y. Hirai, S. Sanada, Y. Ida and J. Shoji, *Chem. Pharm. Bull.*, **32**, 4003 (1984).

2.   Y. Mimaki, O. Nakamura, Y. Sashida, Y. Satomi, A. Nishino and H. Nishino, *Phytochemistry*, **37**, 227 (1994).

3.   K. Idaka, Y. Hirai and J. Shoji, *Chem. Pharm. Bull.*, **36**, 1783 (1988).

4.   Y. Mimaki, T. Satou, M. Ohmura and Y. Sashida, *Nat. Med.*, **50**, 308 (1996).

METHYLPROTOGRACILLIN
22α-Methoxy-(25*R*)-furost-5-en-3β,26-diol 3-O-{α-L-rhamnopyranosyl-(1→2)-[β-D-glucopyranosyl-(1→3)]-β-D-glucopyranoside}-26-O-β-D-glucopyranoside

Source : *Tamus communis* L.[1] (Dioscoreaceae), *Smilax china* L.[2] (Liliaceae), *Dracaena cambodiana*[3] etc (Agavaceae) etc.
Mol. Formula : C$_{52}$H$_{86}$O$_{23}$
Mol. Wt. : 1078
M.P. : 230-232°C[2]
[α]$_D^{25}$: -82.0° (c=0.03, C$_5$D$_5$N)[2]
Registry No. : [54522-53-1]

IR (KBr)[2] : 3400 (OH), 1640 (C=O), 1100-1000 (C-O), 915, 890, 838, 811 cm^{-1}.

PMR (C_5D_5N, 300 MHz)[2] : δ 0.81 (s, 3xH-18), 0.98 (d, J=6.6 Hz, 3xH-27), 1.03 (s, 3xH-19), 1.14 (d, J=6.8 Hz, 3xH-21), 1.65 (d, J=6.2 Hz, 3xH-6 of Rha), 3.23 (s, OCH_3), 4.82 (d, J=7.2 Hz, H-1 of Glc), 4.96 (d, J=7.7 Hz, H-1 of Glc), 5.32 (br d, J=4.1 Hz, H-6), 6.17 (s, H-1 of Rha).

CMR (C_5D_5N, 75.5 MHz)[2] : δ C-1) 37.4 (2) 29.9 (3) 78.3 (4) 39.7 (5) 140.7 (6) 121.6 (7) 32.0 (8) 31.6 (9) 50.3 (10) 37.0 (11) 20.9 (12) 38.6 (13) 40.7 (14) 56.5 (15) 32.2 (16) 81.2 (17) 64.0 (18) 16.2 (19) 19.3 (20) 40.3 (21) 16.1 (22) 112.5 (23) 30.6 (24) 28.1 (25) 34.1 (26) 75.1 (27) 17.1 (OCH_3) 47.2 **Glc** (1) 99.8 (2) 77.7 (3) 89.2 (4) 69.3 (5) 77.5 (6) 62.4 **Rha** (1) 101.9 (2) 72.1 (3) 72.5 (4) 73.9 (5) 69.4 (6) 18.4 **Glc II** (1) 104.2 (2) 74.7 (3) 76.9 (4) 71.3 (5) 78.2 (6) 62.4 **Glc II** (1) 104.7 (2) 74.9 (3) 78.0 (4) 71.7 (5) 78.3 (6) 62.8.

Mass (FAB, Positive ion) : m/z 1117 [M+K]$^+$, 1101 [M+Na]$^+$, 1079 [M+H]$^+$, 1047 [M+H-OCH$_3$]$^+$, 901 [(M+H-OCH$_3$)-146]$^+$, 885 [(M+H-OCH$_3$)-162]$^+$, 739 [(M+H-OCH$_3$)-308]$^+$, 723 [(M+H-OCH$_3$)-324]$^+$, 577 [(M+H-OCH$_3$)-470]$^+$, 415 [577-162], 397 [415-18].

References

1. R. Aquino, I. Behar, F.D. Simone, M.D. Agostino and C. Pizza, *J. Nat. Prod.*, **49**, 1096 (1986).

2. S.W. Kim, K.C. Chung, K.H. Son and S.S. King, *Kor. J. Pharmacogn.*, **20**, 76 (1989).

3. C. Yang and Z. Wang, *Yunnan Zhiwu Yanjiu*, **8**, 355 (1986).

TUBEROSIDE F (ALLIUM)
26-O-β-D-Glucopyranosyl-(5α,20R,25S)-20-O-methyl-5α-furost-22(23)en-2α,3β,20,26-tetraol 3-O-{α-L-rhamnopyranosyl-(1→2)-[α-L-rhamnopyranosyl-(1→4)]-β-D-glucopyranoside}

Source : *Allium tuberosum* Rottl. (Liliaceae)
Mol. Formula : $C_{52}H_{86}O_{23}$
Mol. Wt. : 1078
[α]$_D^{25}$: -27.8° (c=0.22, MeOH)
Registry No. : [332843-47-7]

IR (KBr) : 3415 (OH), 1641 (C=C), 1043 (glycosidic linkage) cm^{-1}.

PMR (C$_5$D$_5$N, 400 MHz) : δ 0.95 (s, CH_3), 0.96 (s, CH_3), 1.43 (s, CH_3), 1.15 (d, J=6.4 Hz, 3xH-27), 1.67 (d, J=6.2 Hz, 3xH-6 of Rha II), 1.75 (d, J=6.2 Hz, 3xH-6 of Rha I), 3.26 (s, OCH_3), 3.60 (dd, J=7.0, 9.2 Hz, H-26A), 3.84 (m, H-5 of Glc I), 4.02 (m, H-5 of Glc II), 4.12 (m, H-2 of Glc II), 4.14 (m, H-6A of Glc I), 4.21 (m, H-26B), 4.30 (m, H-2 and H-3 of Glc I), 4.31 (m, H-3 and H-4 of Glc II), 4.32 (m, H-6B of Glc I), 4.40 (m, H-4 of Rha I and H-4 of Rha II), 4.45 (m, H-4 of Glc I), 4.48 (m, H-6A of Glc II), 4.60 (m, H-6B of Glc II, H-23, H-3 of Rha II), 4.66 (m, H-3 of Rha I), 4.75 (s, H-2 of Rha II), 4.90 (s, H-2 of Rha I), 4.93 (d, J=7.2 Hz, H-1 of Glc I), 4.96 (m, H-5 of Rha I), 5.00 (m, H-5 of Rha II), 5.10 (d, J=7.6 Hz, H-1 of Glc I), 5.90 (s, H-1 of Rha II), 6.45 (s, H-1 of Rha I).

CMR (C$_5$D$_5$N, 100 MHz) : δ C-1) 45.9 (2) 70.7 (3) 85.2 (4) 33.6 (5) 44.8 (6) 28.2 (7) 32.2 (8) 34.0 (9) 54.3 (10) 37.0 (11) 21.0 (12) 39.5 (13) 40.7 (14) 56.6 (15) 33.5 (16) 84.0 (17) 66.9 (18) 14.0 (19) 13.6 (20) 82.5 (21) 15.4 (22) 157.4 (23) 96.2 (24) 29.8 (25) 35.0 (26) 75.5 (27) 17.7 **Glc I** (1) 101.0 (2) 78.0 (3) 78.1 94) 78.9 (5) 77.2 (6) 61.3 **Rha I** (1) 102.2 (2) 77.5 (3) 72.8 (4) 74.0 (5) 69.6 (6) 18.6 **Rha II** (1) 103.0 (2) 72.6 (3) 72.9 (4) 74.2 (5) 70.6 (6) 18.7 **Glc II** (1) 105.3 (2) 75.4 (3) 78.3 (4) 71.8 (5) 78.6 (6) 62.9.

Mass (E.S.I.) : m/z 1102 [M+Na]$^+$.

Reference

1.   S. Sang, S. Mao, A. Lao, Z. Chen and C.-T. Ho, *J. Agric. Food Chem.*, **49**, 1475 (2001).

ABUTILOSIDE N

(22*S*,25*S*)-22,25-Epoxyfurost-5-ene-3β,7β,26-triol 3-O-{α-L-rhamnopyranosyl-(1→2)-
[β-D-glucopyranosyl-(1→3)]-β-D-galactopyranoside}-26-O-β-D-glucopyranoside

Source : *Solanum abutiloides* (Solanaceae)
Mol. Formula : $C_{51}H_{82}O_{24}$
Mol. Wt. : 1078
$[\alpha]_D^{25}$ **:** -84.8° (c=0.24, MeOH)
Registry No. : [652144-60-0]

PMR (C_5D_5N, 400 MHz) : δ 0.89 (s, 3xH-18), 1.07 (s, 3xH-19), 1.08 (d, *J*=6.7 Hz, 3xH-21), 1.41 (s, 3xH-27), 1.52 (H-15A), 1.70 (d, *J*=6.1 Hz, 3xH-6 of Rha), 1.84 (H-17), 2.03 (H-14), 2.52 (H-15B), 3.89 (d, *J*=9.8 Hz, H-26A), 3.90 (m, H-3), 3.95 (H-5 of Rha), 3.97 (dd, *J*=7.8, 8.5 Hz, H-2 of Glc I), 3.97 (H-5 of Glc II), 3.99 (m, H-7), 4.02 (br t, *J*=6 Hz, H-5 of Gal), 4.04 (dd, *J*=7.6, 8.5 Hz, H-2 of Glc II), 4.19 (d, *J*=9.8 Hz, H-26B), 4.22 (dd, *J*=8.5, 8.5 Hz, H-3 and H-4 of Glc I), 4.26 (H-6A of Gal), 4.27 (dd, *J*=8.5, 8.5 Hz, H-4 of Glc II), 4.28 (dd, *J*=8.5, 8.5 Hz, H-3 of Glc II), 4.30 (dd, *J*=8.5, 8.5 Hz, H-4 of Rha), 4.33 (H-6A of Glc I), 4.34 (dd, *J*=2.9, 8.5 Hz, H-3 of Gal), 4.40 (dd, *J*=6.8, 11.5 Hz, H-6B of Gal), 4.43 (dd, *J*=5.1, 11.4 Hz, H-6A fo Glc II), 4.49 (br d, *J*=10.4 Hz, H-6B of Glc I), 4.56 (br d, *J*=10.3 Hz, H-6B of Glc II), 4.62 (dd, *J*=3.2, 8.5 Hz, H-3 of Rha), 4.71 (dd, *J*=7.8, 8.5 Hz, H-2 of Gal), 4.81 (m, H-16), 4.83 (d, *J*=2.9 Hz, H-4 of Gal), 4.88 (d, *J*=7.8 Hz, H-1 of Gal), 4.92 (br s, H-2 of Rha), 4.94 (m, H-5 of Rha), 4.96 (d, *J*=7.6 Hz, H-1 of Glc II), 5.22 (d, *J*=7.8 Hz, H-1 of Glc I), 5.82 (d, *J*=5.2 Hz, H-6), 6.32 (br s, H-1 of Rha).

CMR (C_5D_5N, 100 MHz) : δ C-1) 37.3 (2) 30.1 (3) 78.6 (4) 38.8 (5) 143.9 (6) 125.9 (7) 64.6 (8) 38.0 (9) 42.5 (10) 38.0 (11) 21.0 (12) 39.8 (13) 40.4 (14) 49.7 (15) 32.4 (16) 81.2 (17) 62.7 (18) 16.3 (19) 18.4 (20) 38.7 (21) 15.3 (22) 120.2 (23) 33.2 (24) 34.1 (25) 83.8 (26) 77.6 (27) 24.6 **Gal** (1) 100.4 (2) 75.0 (3) 84.8 (4) 70.6 (5) 76.6 (6) 61.4 **Rha** (1) 102.2 (2) 72.7 (3) 72.9 (4) 74.9 (5) 69.5 (6) 18.7 **Glc I** (1) 105.9 (2) 75.4 (3) 78.5 (4) 71.7 (5) 77.2 (6) 62.7 **Glc II** (1) 105.5 (2) 75.4 (3) 78 4 (4) 71 6 (5) 78.6 (6) 62.7.

Mass (FAB, Negative ion) : *m/z* 1077 [M-H]⁻.

Reference

1.   H. Yoshimitsu, M. Nishida and T. Nohara, *Phytochemistry*, **64**, 1361 (2003).

TERRESTROSIN K

(25*R*)-5α-Furost-20(22)-en-12-one-3β,26-diol 3-O-[β-D-galactopyranosyl-(1→2)-β-D-glucopyranosyl-(1→4)-β-D-galactopyranoside]-26-O-β-D-glucopyranoside

Source : *Tribulus terrestris* L. (Zygophyllaceae)
Mol. Formula : $C_{51}H_{82}O_{24}$
Mol. Wt. : 1078
[α]$_D^{25}$: +3.1° (c=1.31, C_5D_5N)
Registry No. : [193604-07-1]

PMR (C_5D_5N, 500 MHz) : δ 0.68 (s, 3xH-19), 0.92 (s, 3xH-18), 1.01 (d, *J*=6.6 Hz, 3xH-27), 1.72 (s, 3xH-21), 4.79 (d, *J*=7.6 Hz, H-1 of Glc II), 4.85 (d, *J*=7.3 Hz, H-1 of Gal I), 5.07 (d, *J*=7.6 Hz, H-1 of Gal II), 5.09 (d, *J*=7.6 Hz, H-1 of Glc I).

CMR (C$_5$D$_5$N, 125 MHz) : δ C-1) 36.7 (2) 29.8 (3) 77.9 (4) 34.8 (5) 44.6 (6) 28.6 (7) 31.1 (8) 34.8 (9) 55.6 (10) 36.3 (11) 38.2 (12) 212.9 (13) 57.6 (14) 54.2 (15) 33.8 (16) 83.0 (17) 56.2 (18) 14.1 (19) 11.8 (20) 103.2 (21) 11.5 (22) 153.2 (23) 23.7 (24) 31.4 (25) 33.5 (26) 74.9 (27) 17.3 **Gal I** (1) 102.4 (2) 73.1 (3) 75.7 (4) 80.3 (5) 75.2 (6) 60.5 **Glc I** (1) 105.2 (2) 85.0 (3) 77.4 (4) 72.1 (5) 78.0 (6) 63.2 **Gal II** (1) 107.2 (2) 74.3 (3) 74.2 (4) 70.8 (5) 77.4 (6) 63.0 **Glc II** (1) 104.8 (2) 75.1 (3) 78.6 (4) 71.8 (5) 78.4 (6) 62.9.

Mass (FAB, Negative ion, H.R.) : *m/z* 1077.5140 [(C$_{51}$H$_{82}$O$_{24}$-H)$^-$, requires 1077.5125].

Mass (FAB, Negative ion) : *m/z* 1077 [M-H]$^-$, 915 [M-Glc]$^-$, 753 [M-Glc-Gal]$^-$, 591 [M-Glc-Gal-Glc]$^-$.

Reference

1.   Y. Wang, K. Ohtani, R. Kasai and K. Yamasaki, *Phytochemistry*, **45**, 811 (1997).

CURILLOSIDE G
(25*S*)-22α-Methoxy-5β-furostane-3β,26-diol 3-O-{α-L-rhamnopyransyl-(1→2)-[β-D-glucopyranosyl-(1→4)]-β-D-glucopyranoside}-26-O-β-D-glucopyranoside

Source : *Asparagus curillus* Buch.-Ham. (Liliaceae)
Mol. Formula : C$_{52}$H$_{88}$O$_{23}$
Mol. Wt. : 1080
M.P. : 173-175°C
[α]$_D^{20}$: -58°
Registry No. : [150677-84-2]

IR : 3400 (OH) cm^{-1}. No spiroketal absorptions.

PMR (C$_5$D$_5$N) : δ 1.65 (d, *J*=6.0 Hz, 3xH-6 of Rha), 3.25 (s, OC*H*$_3$), 4.85 and 5.43 (both d, *J*=6.0 Hz, H-1 of 2 glucose units), 5.88 (s, H-1 of Rha).

Reference

1.   S.C. Sharma and H.C. Sharma, *Phytochemistry*, **33**, 683 (1993).

ALLIUMOSIDE B

25*R*-Furost-5-ene-3β,22α,26-triol 3-O-[β-D-glucopyranosyl-(1→3)-β-D-glucopyranosyl-(1→6)-β-D-glucopyranoside]-26-O-β-D-glucopyranoside

Source : *Allium narcissiflorum* (Liliaceae)
Mol. Formula : C$_{51}$H$_{84}$O$_{24}$
Mol. Wt. : 1080
M.P. : 270-272°C
[α]$_D$^{18} : -70° (c=0.7, CH$_3$OH)
Registry No. : [56126-14-8]

Reference

1.   V.V. Krokhmalyuk and P.K. Kintya, *Khim. Prir. Soedin.*, 55 (1976); *Chem. Nat. Comp.*, **12**, 46 (1976).

POLYFUROSIDE

25*R*-Furost-5-en-3,22,26-triol 3-O-[β-D-glucopyranosyl-(1→2)-[β-D-glucopyranosyl-(1→3)]-
β-D-glucopyranosyl-(1→4)-β-D-galactopyranoside]-26-O-β-D-glucopyranoside

Source : *Polygonatum officinale* All. Syn. *P. odoratum* Druce. (Liliaceae)
Mol. Formula : $C_{51}H_{84}O_{24}$
Mol. Wt. : 1080
M.P. : 252-258°C
[α]$_D^{20}$: -68.0° (c=1.0, Pyridine)
Registry No. : [108886-03-9]

IR (KBr) : 980, 915, 895, 840 cm^{-1}.

Reference

1. Z. Jaczko, P.E. Jansson and J. Sendra, *Planta Med.*, **53**, 52 (1987).

SOLAYAMOCINOSIDE C

**3β,22α,26-Trihydroxy-25S-furost-5-ene 3-O-[β-galactopyranosyl-(1→2)-
β-D-glucopyranosyl-(1→6)-β-D-glucopyranoside**

Source : *Solanunm dulcamara* L. (Solanaceae)
Mol. Formula : $C_{51}H_{84}O_{24}$
Mol. Wt. : 1080
Registry No. : [86933-73-5]

Reference

1   G. Willuhun and U. Köthe, *Arch. Pharm.* (Weinheim), **316**, 678 (1983).

TERRESTROSIN J

(25R,S)-Furost-5-ene-3β,22α,26-tetrol 3-O-[β-D-galactopyranosyl-(1→2)-β-D-glucopyranosyl-(1→4)-β-D-galactopyranoside]-26-O-β-D-glucopyranoside

Source : *Tribulus terrestris* L. (Zygophyllaceae)
Mol. Formula : $C_{51}H_{84}O_{24}$
Mol. Wt. : 1080
$[\alpha]_D^{25}$: -42.9° (c=0.79, C_5D_5N)
Registry No. : [193604-94-3]

Isolated as a mixture of 25*R* and 25*S* epimers.

PMR (C_5D_5N, 500 MHz) : **(25R)** δ 0.88 (s, 3xH-19), 0.90 (s, 3xH-18), 1.01 (d, *J*=6.6 Hz, 3xH-27), 1.31 (d, *J*=7.0 Hz, H-21), 4.77 (d, *J*=7.8 Hz, H-1 of Glc II), 4.87 (d, *J*=7.6 Hz, H-1 of Gal I), 5.08 (d, *J*=7.5 Hz, H-1 of Gal II), 5.10 (d, *J*=7.6 Hz, H-1 of Glc I), **(25S)** δ 0.88 (s, 3xH-19), 0.90 (s, 3xH-18), 0.98 (d, *J*=6.7 Hz, 3xH-27), 1.29 (d, *J*=7.0 Hz, 3xH-21).

CMR (C_5D_5N, 125 MHz) : δ C-1) 37.5 (2) 30.1 (3) 78.0 (4) 39.4 (5) 141.1 (6) 121.6 (7) 32.3 (8) 31.7 (9) 50.4 (10) 37.1 (11) 21.1 (12) 40.0 (13) 40.8 (14) 56.6 (15) 32.4 (16) 81.1 (17) 63.8 (18) 16.4 (19) 19.4 (20) 40.8 (21) 16.5 (22) 110.7 (23) 37.1 (24) 28.32 (28.26) (25) 34.36 (34.42) (26) 75.2 (75.3) (27) 17.4 **Gal I** (1) 102.8 (2) 73.2 (3) 75.8 (4) 80.3 (5) 75.2 (6) 60.5 **Glc I** (1) 105.2 (2) 85.1 (3) 77.4 (4) 72.2 (5) 78.0 (6) 63.3 **Gal II** (1) 107.3 (2) 74.4 (3) 74.2 (4) 70.9 (5) 77.4 (6) 63.1 **Glc II** (1) 104.9 (105.1) (2) 75.2 (3) 78.6 (4) 71.8 (5) 78.4 (6) 62.9. The values in parenthesis are of 25*S*-epimer

Mass (FAB, Negative ion, H.R) : *m/z* 1079.5280 [(C$_{51}$H$_{84}$O$_{24}$-H)$^-$, requires 1079.5273].

Mass (FAB, Negative ion) : *m/z* 1079 [M-H]$^-$, 917 [M-Glc]$^-$, 755 [M-Glc-Gal]$^-$, 593 [M-Glc-Gal-Glc]$^-$.

Reference

1.   Y. Wang, K. Ohtani, R. Kasai and K. Yamasaki, *Phytochemistry*, **45**, 811 (1997).

TRIBULUS TERRESTRIS SAPONIN 12
(25*S*)-5α-Furostane-12-one-3β,22α,26-triol 3-O-{β-D-glucopyranosyl-(1→4)-[α-rhamnopyranosyl-(1→2)]-β-D-galactopyranoside}-26-O-β-glucopyranoside

Source : *Tribulus terrestris* L. (Zygophylloceae)
Mol. Formula : C$_{51}$H$_{84}$O$_{24}$
Mol. Wt. : 1080
[α]$_D^{15}$: -23.0° (c=0.25, C$_5$D$_5$N)
Registry No. : [343266-01-3]

IR (KBr) : 3426, 2929, 1701, 901 cm^{-1}.

PMR (C$_5$D$_5$N, 400 MHz) : δ 1.47 (d, *J*=6.0 Hz), 4.57 (d, *J*=7.2 Hz,), 4.68 (d, *J*=7.2 Hz), 4.96 (d, *J*=7.6 Hz), 6.02 (s).

CMR (C_5D_5N, 100 MHz) : δ C-1) 36.6 (2) 29.7 (3) 75.6 (4) 34.4 (5) 44.3 (6) 28.6 (7) 31.7 (8) 34.3 (9) 55.7 (10) 36.3 (11) 38.0 (12) 212.7 (13) 55.5 (14) 55.8 (15) 31.7 (16) 79.6 (17) 55.0 (18) 16.2 (19) 11.8 (20) 41.2 (21) 15.2 (22) 110.7 (23) 37.0 (24) 28.3 (25) 34.3 (26) 75.3 (27) 17.4 **Gal** (1) 99.8 (2) 76.9 (3) 76.4 (4) 81.2 (5) 75.2 (6) 61.0 **Glc I** (1) 107.1 (2) 75.2 (3) 78.8 (4) 72.4 (5) 78.5 (6) 63.0 **Glc II** (1) 105.1 (2) 75.2 (3) 78.6 (4) 71.6 (5) 78.5 (6) 62.7 **Rha** (1) 102.3 (2) 72.1 (3) 72.7 (4) 74.0 (5) 69.4 (6) 18.6.

Mass (ES) m/z : 1103 [M+Na]$^+$, 1119 [M+K]$^+$, 1079 [M-H]$^-$, 917 [M-Glc-H]$^-$, 755 [M-2Glc-H]$^-$, 609 [M-2Glc-Rha]$^-$.

Reference

1.   L. Cai, Y. Wu, J. Zhang, F. Pei, Y, Xu, S. Xie and D. Xu, *Planta Med.*, **67**, 196 (2001).

ASPARAGOSIDE G

(5β,25S)-Furostan-3β,22α,26-triol 3-O-{β-D-glucopyranosyl-(1→3)-[β-D-glucopyranosyl-(1→4)-β-D-glucopyranoside}-26-O-β-D-glucopyranoside

Source : *Asparagus officinalis* L. (Liliaceae)
Mol. Formula : $C_{51}H_{86}O_{24}$
Mol. Wt. : 1082
M.P. : 170-174°C
[α]$_D^{20}$: -200° (c=0.6, H_2O)
Registry No. : [60267-27-8]

Reference

1.   G.M. Goryanu and P.K. Kintya, *Khim. Prir. Soedin.*, 762 (1976); *Chem. Nat. Comp.*, **12**, 684 (1976)

DISPOROSIDE D

(5β,25R)-Furostan-3β,5α,26-triol 3-O-{β-D-glucopyranosyl-(1→2)-[β-D-glucopyranosyl-(1→6)]-β-D-glucopyranoside}-26-O-β-D-glucopyranoside

Source : *Disporopsis penyi* (Hua) Diels (Liliaceae)
Mol. Formula : $C_{51}H_{86}O_{24}$
Mol. Wt. : 1082
[α]$_D^{23}$: -43.7° (c=0.4, Pyridine)
Registry No. : [770721-96-5]

IR (KBr) : 3415, 2930, 1453, 1076, 1040, 907 cm^{-1}.

PMR (C$_5$D$_5$N, 500 MHz) : δ 0.84 (s, 3xH-18), 0.97 (s, 3xH-19), 1.01 (d, *J*=6.7 Hz, 3xH-27), 1.33 (d, *J*=7.0 Hz, 3xH-21), 4.79 (d, *J*=7.7 Hz, H-1 of Glc IV), 4.89 (d, *J*=7.4 Hz, H-1 of Glc I), 5.00 (m, H-16), 5.11 (d, *J*=7.4 Hz, H-1 of Glc III), 5.35 (d, *J*=7.6 Hz, H-1 of Glc II).

CMR (C$_5$D$_5$N, 125 MHz) : δ C-1) 31.1 (2) 27.5 (3) 75.3 (4) 32.5 (5) 37.0 (6) 26.9 (7) 26.9 (8) 35.7 (9) 40.9 (10) 35.4 (11) 21.3 (12) 40.8 (13) 41.4 (14) 56.6 (15) 32.5 (16) 81.4 (17) 64.1 (18) 16.8 (19) 24.2 (20) 40.9 (21) 16.5 (22) 110.8 (23) 31.0 (24) 28.5 (25) 34.4 (26) 75.5 (27) 17.6 **Glc I** (1) 101.9 (2) 82.9 (3) 77.2 (4) 71.6 (5) 78.0 (6) 70.1 **Glc II** (1) 105.9 (2) 77.0 (3) 78.5 (4) 71.9 (5) 78.1 (6) 62.9 **Glc III** (1) 105.4 (2) 75.3 (3) 78.7 (4) 71.9 (5) 78.5 (6) 62.9 **Glc IV** (1) 104.9 (2) 75.3 (3) 78.7 (4) 71.6 (5) 78.1 (6) 62.9.

Mass (FAB, Negative ion) : *m/z* 1082 [M]$^-$, 919 [M-H-Glc]$^-$, 919 [M-H-Glc]$^-$, 757 [M-H-2xGlc]$^-$, 595 [M-H-3xGlc]$^-$.

Mass (FAB, Negative ion, H.R.) : m/z 1081.5364 [(M-H)$^-$, calcd. for 1081.5431].

Reference

1.   Q.X. Yang, M. Xu, Y.-J. Zhang, H.-Z. Li and C.-R. Yang, *Helv. Chim. Acta*, **87**, 1248 (2004).

LYCOPERISCUM ESCULENTUM SAPONIN TF-I, TOMATOSIDE A

(5α,25*S*)-Furostane-3β,22α,26-triol 3-O-[β-D-glucopyranosyl-(1→2)-β-D-glucopyranosyl-(1→4)-
β-D-galactopyranoside]-26-O-β-D-glucopyranoside

Source : *Lycoperiscum esculentum* Miller[1,2] (Solanaceae)
Mol. Formula : $C_{51}H_{86}O_{24}$
Mol. Wt. : 1082
M.P. : 247-250°C (decomp.)[2]
$[\alpha]_D^{20}$: -29.3 ± 2° (c=1.18, Pyridine)[2]
Registry No. : [75557-22-1]

IR (KBr)[1] : 3600 ~ 3200 (br, OH), 888 cm^{-1}.

References

1.   H. Sato and S. Sakamura, *Agr. Biol. Chem.*, **37**, 225 (1973).

2.   A.P. Schelochkoya, Yu. S. Vollerner and K.K. Koshoev, *Khim. Prir. Soedin.*, 533 (1980); *Chem. Nat. Comp.*, 386 (1980).

PETUNIOSIDE N
(25*R*)-5α-Furostan-3β,22α,26-triol 3-O-[β-D-glucopyranosyl-(1→2)-β-D-glucopyranosyl-(1→4)-β-D-galactopyranoside]-26-O-β-D-glucopyranoside

Source : *Petunia hybrida* L. (Solanaceae)
Mol. Formula : $C_{51}H_{86}O_{24}$
Mol. Wt. : 1082
M.P. : 192-193°C
[α]$_D^{20}$: -89° (c=1.0, H_2O)
Registry No. : [174693-29-9]

CMR (C_5D_5N, 75 MHz) : δ C-1) 37.4 (2) 29.9 (3) 83.8 (4) 34.8 (5) 45.1 (60 28.3 (7) 32.2 (8) 31.6 (9) 50.3 (10) 36.9 (11) 21.1 (12) 39.8 (13) 40.7 (14) 56.5 (15) 32.2 (16) 80.7 (17) 63.4 (18) 16.3 (19) 19.2 (20) 40.4 (21) 16.0 (22) 110.8

(23) 36.8 (24) 28.1 (25) 34.0 (26) 75.1 (27) 17.2 (OCH₃) 49.9 **Glc I** (1) 105.0 (2) 85.5 (3) 78.4 (4) 71.0 (5) 78.9 (6) 63.3 **Glc II** (1) 106.7 (2) 76.5 (3) 77.9 (4) 71.9 (5) 78.2 (6) 62.3 **Glc III** (1) 106.7 (2) 76.6 (3) 77.8 (4) 72.0 (5) 78.1 (6) 62.2 **Gal** (1) 103.65 (2) 72.9 (3) 75.7 (4) 80.5 (5) 75.75 (6) 60.8.

Reference

1. S.A. Shvets, A.M. Naibi and P.K. Kintya, *Khim. Prir. Soedin.*, **31**, 247 (1995), *Chem. Nat. Comp.*, **31**, 203 (1995).

PROTOYUCCOSIDE C

(25*S*)-5β-Furostan-3β,22α,26-triol 3-O-[α-D-galactopyranosyl-(1→2)-β-D-glucopyranosyl-(1→4)-β-D-glucopyranoside]-26-O-β-D-glucopyranoside

Source : *Yucca filamentosa* L.[1] (Liliaceae), *Solanum lyratum* Thunb.[2] (Solanaceae)
Mol. Formula : C₅₁H₈₆O₂₄
Mol. Wt. : 1082
M.P. : 182-184°C[1]
[α]ᴅ²⁰ : -30° (c=2.0, MeOH)[1]
Registry No. : [55826-89-6]

References

1.   I.P. Dragalin and P.K. Kintia, *Phytochemistry*, **14**, 1817 (1975).

2.   K. Murakami, R. Saijo, T. Nohara and T. Tomimatsu, *Yakugaku Zasshi*, **101**, 275 (1981).

TERRESTROSIN H

(25*R*,*S*)-5α-Furostane-3β,22α,26-tetrol 3-O-[β-D-galactopyranosyl-(1→2)-β-D-glucopyranosyl-(1→4)-β-D-galactopyranoside]-26-O-β-D-glucopyranoside

Source : *Tribulus terrestris* L. (Zygophyllaceae)
Mol. Formula : $C_{51}H_{86}O_{24}$
Mol. Wt. : 1082
[α]$_D^{24}$: -20.4° (c=0.54, C_5D_5N)
Registry No. : [193604-62-5]

Isolated as a mixture of 25*R* and 25*S* epimers.

PMR (C_5D_5N, 500 MHz) : **(25*R*)** δ 0.65 (s, 3xH-19), 0.85 (s, 3xH-18), 1.00 (d, *J*=6.6 Hz, 3xH-27), 1.29 (d, *J*=6.9 Hz, 3xH-21), **(25*RS*)** δ 4.75 (d, *J*=7.6 Hz, H-1 of Glc II), 4.85 (d, *J*=7.0 Hz, H-1 of Glc I), 5.05 (d, *J*=7.6 Hz, H-1 of Gal II), 5.08 (d, *J*=7.5 Hz, H-1 of Glc I), **(25*S*)** δ 0.65 (s, 3xH-19), 0.85 (s, 3xH-18), 0.96 (d, *J*=6.8 Hz, 3xH-27), 1.28 (d, *J*=6.9 Hz, 3xH-21).

CMR (C$_5$D$_5$N, 125 MHz) : δ C-1) 37.1 (2) 30.0 (3) 77.9 (4) 34.8 (5) 44.6 (6) 28.9 (7) 32.4 (8) 35.2 (9) 54.4 (10) 35.7 (11) 21.2 (12) 40.1 (13) 41.0 (14) 56.3 (15) 32.3 (16) 81.1 (17) 63.9 (18) 16.7 (19) 12.3 (20) 40.6 (21) 16.4 (22) 110.6 (23) 37.0 (24) 28.34 (28.30) (25) 34.2 (34.4) (26) 75.1 (75.2) (27) 17.4 **Gal I** (1) 102.3 (2) 73.1 (3) 75.7 (4) 80.4 (5) 75.2 (6) 60.4 **Glc I** (1) 105.2 (2) 84.9 (3) 77.4 (4) 72.2 (5) 77.9 (6) 63.2 **Gal II** (1) 107.2 (2) 74.4 (3) 74.1 (4) 70.8 (5) 77.4 (6) 62.9 **Glc II** (1) 104.9 (105.1) (2) 75.1 (3) 78.5 (4) 71.6 (5) 78.4 (6) 62.7. The values in parenthensis are of 25S-epimer.

Mass (FAB, Negative ion, H.R.) : m/z 1081.5400 [(C$_{51}$H$_{86}$O$_{24}$)$^-$, requires 1081.5428].

Mass (FAB, Negative ion) : m/z 1081 [M-H]$^-$, 919 [M-Glc]$^-$, 757 [M-Glc-Gal]$^-$, 595 [M-Glc-Gal-Glc]$^-$.

Reference

1.   Y. Wang, K. Ohtani, R. Kasai and K. Yamasaki, *Phytochemistry*, **45**, 811 (1997).

RUSCUS ACULEATUS SAPONIN 23
22-O-Methyl-(25R)-furost-5-ene-1β,3β,22ξ,26-tetrol 1-O-{β-D-glucopyranosyl-(1→3)-α-L-rhamnopyranosyl-(1→2)-β-D-galactopyranoside}-26-O-β-D-glucopyranoside

Source : *Ruscus aculeatus* L. (Liliaceae)
Mol. Formula : C$_{52}$H$_{86}$O$_{24}$
Mol. Wt. : 1094
[α]$_D^{25}$: -98.0° (c=0.10, MeOH)
Registry No. : [211379-19-0]

IR (KBr) : 3410 (OH), 2925 (CH), 1455, 1375, 1265, 1225, 1160, 1065, 980, 910, 835 cm^{-1}.

PMR (C$_5$D$_5$N, 400 MHz) : δ 0.86 (s, 3xH-18), 0.99 (d, *J*=6.6 Hz, 3xH-27), 1.10 (d, *J*=6.9 Hz, 3xH-21), 1.45 (s, 3xH-19), 1.67 (d, *J*=6.1 Hz, 3xH-6 of Rha), 3.25 (s, OC*H*$_3$), 3.75 (dd, *J*=11.9, 3.9 Hz, H-1), 3.82 (m, H-3), 4.71 (d, *J*=7.8 Hz, H-1 of Gal), 4.84 (d, *J*=7.7 Hz, H-1 of Glc II), 5.57 (br d, *J*=5.4 Hz, H-6), 5.69 (d, *J*=7.8 Hz, H-1 of Glc I), 6.40 (br s, H-1 of Rha).

CMR (C$_5$D$_5$N, 100 MHz) : δ C-1) 85.3 (2) 38.6 (3) 68.2 (4) 43.8 (5) 139.5 (6) 124.8 (7) 32.0 (8) 33.0 (9) 50.7 (10) 42.8 (11) 23.9 (12) 40.4 (13) 40.5 (14) 57.0 (15) 32.3 (16) 81.3 (17) 64.3 (18) 16.8 (19) 15.1 (20) 40.5 (21) 16.1 (22) 112.7 (23) 30.8 (24) 28.2 (25) 34.2 (26) 75.2 (27) 17.1 (OC*H*$_3$) 47.2 **Gal** (1) 101.2 (2) 74.0 (3) 76.8 (4) 70.4 (5) 76.4 (6) 62.6 **Rha** (1) 101.3 (2) 72.1 (3) 82.7 (4) 73.3 (5) 69.2 (6) 18.7 **Glc I** (1) 106.5 (2) 76.2 (3) 78.3 (4) 71.7 (5) 78.2 (6) 61.9 **Glc II** (1) 105.0 (2) 75.2 (3) 78.6 (4) 71.8 (5) 78.5 (6) 62.9.

Mass (FAB, Negative ion) : *m/z* 1093 [M-H]$^-$, 931 [M-Glc]$^-$.

Reference

1.   Y. Mimaki, M. Kuroda, A. Kameyama, A. Yokosuka and Y. Sashida, *Phytochemistry*, **48**, 485 (1998).

HOSTA LONGIPES SAPONIN 10
22-O-Methyl-(25*R*)-5α-furostane-2α,3β,22ξ,26-tetrol 3-O-{α-L-rhamnopyranosyl-(1→2)-[β-D-glucopyranosyl-(1→4)]-β-D-galactopyranoside}-26-O-β-D-glucopyranoside

Source : *Hosta longipes*[1], *H. sieboldii*[2] (Liliaceae)
Mol. Formula : C$_{52}$H$_{88}$O$_{24}$
Mol. Wt. : 1096
[α]$_D^{25}$: -53.0° (c=0.10, CHCl$_3$-MeOH)[1]
Registry No. : [178494-81-0]

IR (KBr)[1] : 3425 (OH), 2925 (CH), 1450, 1375, 1255, 1065, 1035, 900, 815, 695 cm^{-1}.

PMR (C_5D_5N, 400 MHz)[1] : δ 0.79 (s, 3xH-18), 0.91 (s, 3xH-19), 1.01 (d, J=6.6 Hz, 3H-27), 1.18 (d, J=6.9 Hz, 3xH-21), 1.62 (d, J=6.2 Hz, 3xH-6 of Rha), 3.26 (s, OCH_3), 4.85 (d, J=7.8 Hz, H-1 of Glc II), 4.94 (d, J=7.8 Hz, H-1 of Gal), 5.17 (d, J=7.8 Hz, H-1 of Glc I), 6.23 (br d, H-1 of Rha).

CMR (C_5D_5N, 100 MHz)[1] : δ C-1) 45.8 (2) 70.6 (3) 85.3 (4) 33.5 (5) 44.8 (6) 28.2 (7) 32.1 (8) 34.6 (9) 54.5 (10) 36.9 (11) 21.4 (12) 40.0 (13) 41.1 (14) 56.3 (15) 32.3 (16) 81.4 (17) 64.3 (18) 16.3 (19) 13.5 (20) 40.5 (21) 16.5 (22) 112.7 (23) 30.8 (24) 28.2 (25) 34.2 (26) 75.2 (27) 17.2 (OCH_3) 47.3 **Gal** (1) 101.2 (2) 76.9 (3) 76.4 (4) 81.3 (5) 75.5 (6) 60.9 **Rha** (1) 102.3 (2) 72.3 (3) 72.7 (4) 74.1 (5) 69.4 (6) 18.5 **Glc I** (1) 107.2 (2) 75.6 (3) 78.9^a (4) 72.2 (5) 78.6^a (6) 63.0 **Glc II** (1) 105.0 (2) 75.2 (3) 78.6^a (4) 71.8 (5) 78.5^b (6) 63.1.

Mass (FAB, Negative ion)[1] : m/z 1095 [M]$^-$.

Biological Activity : The compound exhibited cytostatic activity on leukemia HL-60 cells with the IC_{50}=3.0 μg/ml, it causes 98.6% cell growth inhibition at the sample concentration of 10 μg/ml.[2]

References

1. Y. Mimaki, T. Kanmoto, M. Kuroda, Y. Sashida, Y. Satomi, A. Nishino and H. Nishino, *Phytochemistry*, **42**, 1065 (1996).

2. Y. Mimaki, M. Kuroda, A. Kameyama, A. Yokosuka and Y. Sashida, *Phytochemistry*, **48**, 1361 (1998).

PRATIOSIDE B

3β,17α,22ξ,26-Tetrahydroxy-(25*R*)-furost-5-ene 3-O-[β-D-glucopyranosyl-(1→2)-β-D-glucopyranosyl-(1→4)-β-D-galactopyranoside]-26-O-β-D-glucopyranoside

Source : *Polygonatum prattii* Baker (Liliaceae)
Mol. Formula : $C_{51}H_{84}O_{25}$
Mol. Wt. : 1096
[α]$_D^{28}$: -52.5° (c=0.72, H_2O)
Registry No. : [150175-09-0]

PMR (C$_5$D$_5$N, 270/400 MHz) : δ 0.92 (s, 3xH-19), 1.01 (d, J=5.0 Hz, 3xH-27), 1.02 (s, 3xH-18), 1.40 (d, J=7.0 Hz, 3xH-21), 4.84 (d, J=7.7 Hz, H-1 of Glc III), 4.92 (d, J=7.7 Hz, H-1 of Gal), 5.17 (d, J=7.7 Hz, H-1 of Glc), 5.25 (d, J=7.7 Hz, H-1 of Glc), 5.27 (br s, H-6).

CMR (C$_5$D$_5$N, 67.5/100 MHz) : δ C-1) 37.5 (2) 30.3 (3) 78.2 (4) 39.3 (5) 141.0 (6) 121.7 (7) 32.4^a (8) 32.3 (9) 50.3 (10) 37.1 (11) 21.0 (12) 32.1^a (13) 45.1 (14) 53.0 (15) 31.9^a (16) 90.5 (17) 90.7 (18) 17.3 (19) 19.4 (20) 43.6 (21) 10.5 (22) 111.4 (23) 36.9 (24) 28.0 (25) 34.3 (26) 75.1 (27) 17.5 **Gal** (1) 102.7 (2) 73.3 (3) 75.6^b (4) 81.0 (5) 76.8^b (6) 60.5 **Glc I** (1) 105.2 (2) 86.1 (3) 78.2^c (4) 70.3 (5) 77.7^c (6) 61.6 **Glc II** (1) 107.0 (2) 75.2^d (3) 79.0^c (4) 71.8^e (5) 78.5^b (6) 63.2 **Glc III** (1) 105.0 (2) 75.3^d (3) 78.6^c (4) 71.7^e (5) 78.5^c (6) 62.9.

Mass (FAB, Negative ion, H.R.) : m/z 1095.5240 [(M-H)$^-$, calcd. 1095.5223].

Mass (FAB, Negative ion) : m/z 1095 [M-H]$^-$, 933 [M-Glc-H]$^-$.

Reference

1.   X.-C. Li, C.-R. Yang, H. Matsuura, R. Kasai and K. Yamasaki, *Phytochemistry*, **33**, 465 (1993).

TERRESTROSIN I

(25R,S)-5α-Furostane-12-one-3β, 22α, 26-tetrol 3-O-[β-D-galactopyranosyl-(1→2)-
β-D-glucopyranosyl-(1→4)-β-D-galactopyranoside]-26-O-β-D-glucopyranoside

Source : *Tribulus terrestris* L. (Zygophyllaceae)
Mol. Formula : $C_{51}H_{84}O_{25}$
Mol. Wt. : 1096
$[\alpha]_D^{24}$: -17.0° (c=0.53, C_5D_5N)
Registry No. : [193604-75-0]

Isolated as a mixture of 25R and 25S epimers.

PMR (C_5D_5N, 500 MHz) : **(25R)** δ 0.71 (s, 3xH-19), 1.00 (d, J=6.5 Hz, 3xH-27), 1.09 (s, 3xH-18), 1.52 (d, J=7.0 Hz, 3xH-21), **(25R,S)**, δ 4.78 (d, J=7.8 Hz, H-1 of Glc II), 4.86 (d, J=7.8 Hz, H-1 of Gal I), 5.08 (d, J=7.6 Hz, H-1 of Gal II), 5.11 (d, J=7.6 Hz, H-1 of Glc I), **(25S)** δ 0.71 (s, 3xH-19), 0.96 (d, J=6.5 Hz, 3xH-27), 1.09 (s, 3xH-18), 1.50 (d, J=7.0 Hz, 3xH-21).

CMR (C_5D_5N, 125 MHz) : δ C-1) 36.6 (2) 29.7 (3) 77.9 (4) 34.7 (5) 44.5 (6) 28.5 (7) 31.6 (8) 34.3 (9) 55.7 (10) 36.2 (11) 37.9 (12) 213.0 (13) 55.5 (14) 55.8 (15) 31.7 (16) 79.6 (17) 55.0 (18) 16.2 (19) 11.7 (20) 41.2 (21) 15.2 (22) 110.7 (23) 37.0 (24) 28.33 (28.26) (25) 34.2 (34.3) (26) 75.16 (75.22) (27) 17.41 (17.37) **Gal I** (1) 102.4 (2) 73.1 (3) 75.7 (4) 80.3 (5) 75.3 (6) 60.4 **Glc I** (1) 105.2 (2) 85.0 (3) 77.2 (4) 72.2 (5) 77.9 (6) 63.2 **Gal II** (1) 107.2 (2) 74.4 (3) 74.1 (4) 70.8 (5) 77.4 (6) 62.8 **Glc II** (1) 104.9 (105.0) (2) 75.2 (3) 78.5 (4) 71.7 (5) 78.4 (6) 62.9. The values in parenthesis are of 25S-epimer.

Mass (FAB, Negative ion, H.R) : m/z 1095.5220 [$(C_{51}H_{84}O_{25}$-H)$^-$, require 1095.5219].

Mass (FAB, Negative ion) : m/z 1095 [M-H]$^-$, 933 [M-Glc]$^-$, 771 [M-Glc-Gal]$^-$, 609 [M-Glc-Gal-Glc]$^-$.

Reference

1.   Y. Wang, K. Ohtani, R. Kasai and K. Yamasaki, *Phytochemistry*, **45**, 811 (1997).

TERRESTROSIN G

(25R,S)-5α-Furostane-2α,3β,22α,26-tetrol 3-O-[β-D-galactopyranosyl-(1→2)-β-D-glucopyranosyl-(1→4)-β-D-galactopyranoside]-26-O-β-D-glucopyranoside

Source : *Tribulus terrestris* L. (Zygophyllaceae)
Mol. Formula : $C_{51}H_{86}O_{25}$
Mol. Wt. : 1098
$[\alpha]_D^{24}$: -26.8° (c=0.75, C_5D_5N)
Registry No. : [193604-56-7]

Isolated as a mixture of 25R and 25S epimers.

PMR (C_5D_5N, 500 MHz) : (**25R**) δ 0.71 (s, 3xH-19), 0.84 (s, 3xH-18), 1.01 (d, *J*=6.7 Hz, 3xH-27), 1.29 (d, *J*=7.0 Hz, 3xH-21), 4.77 (d, *J*=7.8 Hz, H-1 of Glc II), 4.90 (d, *J*=7.8 Hz, H-1 of Gal I), 5.11 (d, *J*=7.5 Hz, H-1 of Gal II), 5.14 (d, *J*=7.6 Hz, H-1 of Glc I), (**25S**) δ 0.71 (s, 3xH-19), 0.84 (s, 3xH-18), 0.97 (d, *J*=6.7 Hz, 3xH-27), 1.28 (d, *J*=7.0 Hz, 3xH-21).

CMR (C$_5$D$_5$N, 125 MHz) : δ C-1) 45.5 (2) 70.4 (3) 85.0 (4) 34.2 (5) 44.7 (6) 28.1 (7) 32.3 (8) 34.5 (9) 54.4 (10) 36.8 (11) 21.4 (12) 40.1 (13) 41.0 (14) 56.2 (15) 32.3 (16) 81.1 (17) 63.9 (18) 16.7 (19) 13.4 (20) 40.6 (21) 16.4 (22) 110.6 (23) 37.1 (24) 28.32 (28.26) (25) 34.2 (34.4) (26) 75.1 (75.2) (27) 17.4 **Gal I** (1) 103.3 (2) 72.6 (3) 75.6 (4) 80.2 (5) 75.6 (6) 60.3 **Glc I** (1) 105.1 (2) 84.9 (3) 77.9 (4) 72.1 (5) 77.9 (6) 63.2 **Gal II** (1) 107.2 (2) 74.3 (3) 74.3 (4) 70.8 (5) 77.5 (6) 63.0 **Glc II** (1) 104.8 (105.1) (2) 75.2 (3) 78.5 (4) 71.6 (5) 78.4 (6) 62.8. The values in parenthesis are of 25*S*-epimer.

Mass (FAB, Negative ion, H.R.) : *m/z* 1097.5440 [(C$_{51}$H$_{86}$O$_{25}$-H)$^-$, requires 1097.5397].

Mass (FAB, Negative ion) : *m/z* 1097 [M-H]$^-$, 935 [M-Glc]$^-$, 773 [M-Glc-Gal]$^-$, 611 [M-Glc-Gal-Gal]$^-$.

Reference

1.   Y. Wang, K. Ohtani, R. Kasai and K. Yamaski, *Phytochemistry*, **45**, 811 (1997).

CAMASSIA CUSICKII SAPONIN 12

22ξ-Methoxy-(25*R*)-5α-furostan-3β,6α-diol 6-O-{β-D-glucopyranosyl-(1→2)-[β-D-glucopyranosyl-(1→3)]-β-D-glucopyranoside}-26-O-β-D-glucopyranoside

Source : *Camassia cusickii* S. Wats. (Liliaceae)
Mol. Formula : C$_{52}$H$_{88}$O$_{25}$
Mol. Wt. : 1112
[α]$_D^{24}$: -18.7° (c=0.31, MeOH)
Registry No. : [141360-82-9]

IR (KBr) : 3400 (OH), 2925 (CH), 1445, 1375, 1250, 1195, 1155, 1070, 1025, 890 cm^{-1}.

PMR (C_5D_5N, 400 MHz) : δ 0.78 (s, 3xH-18), 0.83 (s, 3xH-19), 1.01 (d, J=6.5 Hz, 3xH-27), 1.17 (d, J=6.9 Hz, 3xH-21), 3.24 (s, OCH_3), 3.67 (ddd, J=10.5, 10.5, 4.2 Hz, H-6), 3.90 (m, H-3), 4.84 (d, J=7.6 Hz, H-1 of Glc I), 4.92 (H-1 of Glc IV, overlapping with H_2O signal), 5.38 (d, J=7.8 Hz, H-1 of Glc II)^a, 5.74 (d, J=7.6 Hz, H-1 of Glc III)^a.

CMR (C_5D_5N, 100 MHz) : δ C-1) 37.9 (2) 32.2^a (3) 71.2 (4) 32.8 (5) 51.1 (6) 80.2 (7) 40.6 (8) 34.0^b (9) 54.0 (10) 36.6 (11) 21.2 (12) 40.0 (13) 41.1 (14) 56.3 (15) 32.0^a (16) 81.3 (17) 64.2 (18) 16.2 (19) 13.7 (20) 40.5 (21) 16.6 (22) 112.6 (23) 30.7 (24) 28.3 (25) 34.2^b (26) 75.2 (27) 17.2 (OCH_3) 47.3 **Glc I** (1) 103.5^c (2) 79.9 (3) 89.2 (4) 70.1 (5) 77.6 (6) 62.7^d **Glc II** (1) 104.0^c (2) 76.1^e (3) 78.6 (4) 71.7^g (5) 78.6 (6) 62.5^d **Glc III** (1) 104.7^c(2) 75.5^e (3) 78.6^f (4) 71.7^g (5) 77.6 (6) 62.4^d **Glc IV** (1) 104.9 (2) 75.2 (3) 78.6^f (4) 71.8^g (5) 78.5^f (6) 62.9.

Mass (SIMS) : m/z 1081 [M-OMe]$^+$.

Reference

1.   Y. Mimaki, Y. Sashida and K. Kawashima, *Chem. Pharm. Bull.*, **40**, 148 (1992).

CAPSICOSIDE C
22ξ-Methoxy-5α,25R-furostan-2α,3β,26-triol 3-O-[β-D-glucopyranosyl-(1→2)-β-D-glucopyranosyl-(1→4)-β-D-galactopyranoside]-26-O-β-D-glucopyranoside

Source : *Capsicum annuum* L. var. *conoides* and
Capsicum annuum L. var. *fasciculatum* (Solanaceae)
Mol. Formula : $C_{52}H_{88}O_{25}$
Mol. Wt. : 1112
[α]$_D$^{26} : -42.9° (c=0.53, Pyridine)
Registry No. : [160219-66-9]

PMR (C$_5$D$_5$N, 400 MHz) **:** δ 0.70 (s, CH_3), 0.79 (s, CH_3), 1.00 (d, J=6.6 Hz, sec. CH_3), 1.19 (d, J=7.0 Hz, sec. CH_3), 4.83 (d, J=7.7 Hz), 4.96 (d, J=7.7 Hz), 5.08 (d, J=7.7 Hz), 5.23 (d, J=7.3 Hz, 4 x anomeric H).

CMR (C$_5$D$_5$N, 100 MHz) **:** δ C-1) 45.0 (2) 69.9 (3) 84.0 (4) 33.5 (5) 44.1 (6) 27.6 (7) 31.6 (8) 34.0 (9) 53.8 (10) 36.4 (11) 20.8 (12) 39.3 (13) 40.6 (14) 55.7 (15) 31.7 (16) 80.9 (17) 63.7 (18) 16.0 (19) 12.9 (20) 40.0 (21) 15.8 (22) 112.2 (23) 30.3 (24) 27.7 (25) 33.6 (26) 74.8 (27) 16.6 (OCH$_3$) 46.8 **Gal** (1) 102.7 (2) 72.1 (3) 75.8 (4) 80.1 (5) 76.9 (6) 60.1 **Glc I** (1) 104.3 (2) 84.9 (3) 77.5 (4) 71.0 (5) 77.8 (6) 61.2 **Glc II** (1) 106.0 (2) 75.0 (3) 78.3 (4) 70.0 (5) 77.6 (6) 62.2 **Glc III** (1) 104.3 (2) 74.5 (3) 77.8 (4) 71.4 (5) 77.8 (6) 62.5.

Mass (FAB, Negative ion) **:** *m/z* 1111 [M-H]$^-$, 949 [M-H-Glc]$^-$, 787 [949-Glc]$^-$.

Reference

1. S. Yahara, T. Ura, C. Sakamoto and T. Nohara, *Phytochemistry*, **37**, 831 (1994).

AMPELOSIDE Bf₁

(25R)-5α-Furostane-2α,3β,6β,22ξ,26-pentaol 3-O-[β-glucopyranosyl-(1→3)-β-glucopyranosyl-(1→4)-β-galactopyranoside]-26-O-β-glucopyranoside

Source : *Allium ampeloprasum* L. (Liliaceae)
Mol. Formula : $C_{51}H_{86}O_{26}$
Mol. Wt. : 1114
$[\alpha]_D^{21}$: -25.1° (c=1.05, C_5H_5N)
Registry No. : [118524-13-3]

PMR (C_5D_5N, 270 MHz) : δ 5.26 (d, *J*=8.06 Hz, anomeric H), 5.23 (d, *J*=7.32 Hz, anomeric H), 4.98 (d, *J*=7.33 Hz, anomeric H), 4.81 (d, *J*=7.69 Hz, anomeric H).

CMR (C_5D_5N, 67.80 MHz) : δ C-1) 47.1 (2) 70.4 (3) 85.0 (4) 31.7 (5) 47.8 (6) 69.9 (7) 40.7 (8) 29.8 (9) 54.5 (10) 36.9 (11) 21.3 (12) 40.1 (13) 41.1 (14) 56.0 (15) 32.3 (16) 81.0 (17) 63.7 (18) 17.1 (19) 16.3 (20) 40.5 (21) 16.6 (22) 110.5 (23) 36.9 (24) 28.2 (25) 34.1 (26) 75.2 (27) 17.4 **Gal** (1) 103.2 (2) 72.8 (3) 75.4ᵃ (4) 79.8 (5) 75.6ᵃ (6) 60.9 **Glc I** (1) 106.2 (2) 74.5ᵃ (3) 88.1 (4) 70.2 (5) 78.4ᵇ (6) 62.4ᶜ **Glc II** (1) 105.4 (2) 75.0ᵃ (3) 78.2ᵇ (4) 71.5ᵈ (5) 78.4ᵇ (6) 62.7ᶜ **Glc III** (1) 104.7 (2) 74.9ᵃ (3) 78.1ᵇ (4) 71.6ᵈ (5) 77.8ᵇ (6) 62.7ᶜ.

Mass (F.D.) : *m/z* 1137 [M+Na]⁺, 1119 [M+Na-H₂O]⁺.

Reference

1.   T. Morita, T. Ushiroguchi, N. Hayashi, H. Matsuura, Y. Itakura and T. Fuwa, *Chem. Pharm. Bull.*, **36**, 3480 (1988).

AGAPANTHUS FUROSTANOL SAPONIN

26ξ-Methoxy-2α,3β,5α,9α,26-pentahydroxyfurostane 3-O-{α-L-rhamnopyranosyl-(1→2)-[β-D-galactopyranosyl-(1→3)]-β-D-glucopyranoside}-26-O-β-D-glucopyranoside

Source : *Agapanthus inapertus* (Liliaceae)
Mol. Formula : $C_{52}H_{88}O_{26}$
Mol. Wt. : 1128
$[\alpha]_D^{23}$: -54.0° (c=0.10, MeOH)
Registry No. : [155259-52-2]

IR (KBr) : 3380 (OH), 2910 (CH), 1440, 1365, 1295, 1250, 1035, 905, 885, 800 cm^{-1}.

PMR (C_5D_5N, 400/500 MHz) : δ 0.93 (s, 3xH-18), 1.01 (d, *J*=6.6 Hz, 3xH-27), 1.17 (d, *J*=6.9 Hz, 3xH-21), 1.26 (s, 3xH-19), 1.72 (d, *J*=6.2 Hz, H-6 of Rha), 3.26 (s, OC*H₃*), 4.86 (d, *J*=7.7 Hz, H-1 of Glc I and Glc II), 4.98 (overlapping with H_2O signal, H-1 of Gal), 6.32 (br s, H-1 of Rha).

CMR (C_5D_5N, 100/125 MHz) : δ C-1) 35.0 (2) 70.9 (3) 82.7 (4) 40.7 (5) 76.8 (6) 34.8 (7) 21.8 (8) 37.4 (9) 77.5 (10) 43.2 (11) 28.1 (12) 35.3 (13) 41.4 (14) 48.8 (15) 32.0 (16) 81.6 (17) 64.4 (18) 16.2 (19) 20.2 (20) 40.6 (21) 15.9 (22) 112.7 (23) 30.9 (24) 28.3 (25) 34.3 (26) 75.2 (26) 17.2 (OC*H₃*) 47.3 **Glc I** (1) 101.0 (2) 77.2 (3) 89.2 (4) 69.7 (5) 77.8 (6) 62.3 **Gal** (1) 102.2 (2) 72.4^a (3) 72.7 (4) 74.1 (5) 69.5 (6) 18.5 **Rha** (1) 105.1 (2) 72.5^a (3) 75.2 (4) 70.1 (5) 77.4 (6) 62.1 **Glc II** (1) 105.0 (2) 75.2 (3) 78.4^b (4) 71.9 (5) 78.6^b (6) 63.0.

Mass (FAB, Negative ion) : *m/z* 1127 [M-H]$^-$, 965 [M Glc]$^-$.

Reference

1. O. Nakamura, Y. Mimaki, Y. Sashida, T. Nikaido and T. Ohmoto, *Chem. Pharm. Bull.*, **41**, 1784 (1993).

OPHIOPOGON JABURAN SAPONIN J-6

Furost-5,25(27)-dien-1β,3β,22,26-tetrol 1-O-[α-L-rhamnopyranosyl-(1→2)-(4-O-sulfo)-
α-L-arabinopyranoside]-3,26-*bis*-O-[β-D-glucopyranoside] sodium salt

Source : *Ophiopogon jaburan* (Kunth) Lodd. (Liliaceae)
Mol. Formula : $C_{50}H_{79}O_{26}SNa$
Mol. Wt. : 1150
M.P. : 201-203°C (decomp.)
$[\alpha]_D^{24}$ **:** -18.6° (c=0.80, Pyridine)
Registry No. : [94901-60-7]

IR (KBr) **:** 3600-3200 (OH), 1220 (S-O) cm^{-1}.

CMR (C$_5$D$_5$N, 25 MHz) **:** δ **Ara** (C-1) 99.8 (2) 76.0 (3) 75.0 (4) 76.0 (5) 65.7 **Rha** (1) 101.3 (2) 72.1 (3) 72.1 (4) 74.6 (5) 69.5 (6) 18.8 **Glc I** (1) 102.3 (2) 75.0 (3) 78.1^a (4) 71.8 (5) 78.2^a (6) 62.7 **Glc II** (1) 103.6 (2) 75.0 (3) 78.3^a (4) 71.8 (5) 78.1^a (6) 62.9.

Reference

1. Y. Watanabe, S. Sanada, Y. Ida and J. Shoji, *Chem. Pharm. Bull.,* **32**, 3994 (1984).

PSEUDOPROTO-Pb

3β,26-Dihydroxyfurost-5,22-diene 3-O-{α-L-rhamnopyranosyl-(1→4)-α-L-rhamnopyranosyl-(1→4)-[α-L-rhamnopyranosyl-(1→2)]-β-D-glucopyranoside}-26-O-β-D-glucopyranoside

Source : *Trachycarpus wagnerianus* Becc. (Palmae)
Mol. Formula : $C_{57}H_{92}O_{25}$
Mol. Wt. : 1176
M.P. : 181-183°C
[α]$_D^{20}$: -84.4° (c=0.90, Pyridine)
Registry No. : [102100-46-9]

IR (KBr) : 3570-3210 (OH) cm^{-1}.

CMR (C_5D_5N, 25 MHz) : δ C-1) 37.7 (2) 30.3 (3) 78.4 (4) 39.2 (5) 141.1 (6) 121.8 (7) 32.6 (8) 31.7 (9) 50.6 (10) 37.3 (11) 21.5 (12) 39.9 (13) 43.6 (14) 55.2 (15) 34.7 (16) 84.6 (17) 64.8 (18) 14.3 (19) 19.5 (20) 103.6 (21) 11.8 (22) 152.6 (23) 33.7 (24) 23.9 (25) 31.7 (26) 75.3 (27) 17.4 **Rha I** (1) 102.1 (6) 18.5 **Rha II** (1) 102.4 (6) 18.7 **Rha III** (1) 103.2 (6) 18.9 **Glc I** (1) 100.6 **Glc II** (1) 104.9.

Reference

1.   Y. Hirai, S. Sanada, Y. Ida and J. Shoji, *Chem. Pharm. Bull.*, **34**, 82 (1986).

HELONIOPSIS ORIENTALIS SAPONIN Hc

3β,26-Dihydroxycholesta-5,17(20)-diene-16,22-dione 3-O-{α-L-rhamnopyranosyl-(1→4)-α-L-rhamnopyranosyl-(1→4)-[α-L-rhamnopyranosyl-(1→2)]-β-D-glucopyranoside}-26-O-β-D-glucopyranoside

Source : *Heloniopsis orientalis* (Thunb.)
C. Tanaka (Liliaceae)
Mol. Formula : $C_{57}H_{90}O_{26}$
Mol. Wt. : 1190
[α]$_D$: -109.4° (c=1.01, Pyridine)
Registry No. : [55916-47-7]

UV (EtOH) : λ_{max} 242 (ε, 9600) nm.

IR (KBr) : 3600-3200 (OH), 1730-1700 (C=O), 1630 (C=C) cm^{-1}.

PMR (C$_5$D$_5$N) : δ 0.95 (s, 3xH-19), 1.02 (s, 3xH-18), 1.95 (s, 3xH-21).

ORD (c=0.042, EtOH) : [M] + 3260° (322 nm, peak), -3620 (370 nm, trough).

Acetate :

[α]$_D$: -63.4° (c=0.73, CHCl$_3$)

Mass : *m/z* 503 $[C_{22}H_{31}O_{13}]^+$, 392 $[C_{27}H_{36}O_2]^+$, 331 $[C_{14}H_{19}O_9]^+$, 273 $[C_{12}H_{17}O_7]^+$.

Reference

1.   T. Nohara, Y. Ogata, M. Aritome, K. Miyahara and T. Kawasaki, *Chem. Pharm. Bull.*, **23**, 925 (1975).

TACCA CHANTRIERI SAPONIN 6

(25S)-3β,26-Dihydroxyfurosta-5,20(22)-diene 3-O-{α-L-rhamnopyranosyl-(1→2)-[β-D-glucopyranosyl-(1→4)-α-L-rhamnopyranosyl-(1→3)]-β-D-glucopyranoside}-26-O-β-D-glucopyranoside

Source : *Tacca chantrieri* André (Taccaceae)
Mol. Formula : $C_{57}H_{92}O_{26}$
Mol. Wt. : 1192
$[\alpha]_D^{25}$: -60° (c=0.10, CHCl$_3$-MeOH)
Registry No. : [469879-87-6]

IR (film) **:** 3389 (OH), 2932 (CH), 1041 cm^{-1}.

PMR (C$_5$D$_5$N, 500 MHz) **:** δ 0.71 (s, 3xH-18), 1.03 (d, J=6.6 Hz, 3xH-27), 1.06 (s, 3xH-19), 1.63 (s, 3xH-21), 1.69 (d, J=6.1 Hz, 3xH-6 of Rha II), 1.76 (d, J=6.2 Hz, 3xH-6 of Rha I), 3.91 (m, $W\frac{1}{2}$=27.0 Hz, H-3), 4.81 (m, H-16), 4.84 (d, J=7.8 Hz, H-1 of Glc III), 4.91 (d, J=7.8 Hz, H-1 of Glc I), 5.25 (d, J=7.8 Hz, H-1 of Glc II), 5.32 (br d, J=4.3 Hz, H-6), 5.77 (br s, H-1 of Rha II), 5.83 (br s, H-1 of Rha I).

CMR (C$_5$D$_5$N, 125 MHz) **:** δ C-1) 37.5 (2) 30.0 (3) 77.7 (4) 38.6 (5) 140.7 (6) 121.8 (7) 32.4 (8) 31.4 (9) 50.2 (10) 37.1 (11) 21.2 (12) 39.6 (13) 43.4 (14) 54.9 (15) 34.5 (16) 84.5 (17) 64.5 (18) 14.1 (19) 19.4 (20) 103.5 (21) 11.8 (22) 152.4 (23) 31.4 (24) 23.6 (25) 33.7 (26) 75.2 (27) 17.1 **Glc I** (1) 99.8 (2) 78.6 (3) 86.2 (4) 69.7 (5) 78.1 (6) 62.1 **Rha I** (1) 102.6 (2) 72.5 (3) 72.8 (4) 73.8 (5) 69.9 (6) 18.7 **Rha II** (1) 103.1 (2) 72.0 (3) 72.4 (4) 84.5 (5) 68.7 (6) 18.3 **Glc II** (1) 106.5 (2) 76.4 (3) 78.6 (4) 71.4 (5) 78.4 (6) 62.5 **Glc III** (1) 105.2 (2) 75.2 (3) 78.6 (4) 71.7 (5) 78.5 (6) 62.8.

Mass (FAB, Positive ion) **:** m/z 1215 [M+Na]$^+$.

Reference

1.   A. Yokosuka, Y. Mimaki and Y. Sashida, *J. Nat. Prod.*, **65**, 1293 (2002).

TACCAOSIDE B
(25R)-3,26-Dihydroxyfurost-5,20(22)-diene 3-O-{α-L-rhamnopyranosyl-(1→2)-[β-D-glucopyranosyl-(1→3)]-α-L-rhamnopyranosyl-(1→3)]-β-D-glucopyranoside}-26-O-β-D-glucopyranoside

Source : *Tacca plantaginea* (Taccaceae)
Mol. Formula : C$_{57}$H$_{92}$O$_{26}$
Mol. Wt. : 1192
M.P. : 193~194°C
[α]$_D$^{26} : -54.8° (c=0.17, MeOH)
Registry No. : [475572-46-4]

PMR : δ 4.79 (d, *J*=7.52 Hz, H-1 of Glc II), 4.86 (d, *J*=7.64 Hz, H-1 of Glc I), 5.26 (d, *J*=7.52 Hz, H-1 of Glc III), 5.69 (s, H-1 of Rha II), 5.74 (s, H-1 of Rha I).

CMR (C_5D_5N, 125.77 MHz) : δ C-1) 37.7 (2) 30.3 (3) 78.4 (4) 39.9 (5) 141.1 (6) 121.9 (7) 32.6 (8) 31.6 (9) 50.6 (10) 37.3 (11) 21.5 (12) 38.9 (13) 43.6 (14) 55.2 (15) 34.7 (16) 84.7 (17) 64.8 (18) 14.3 (19) 19.6 (20) 103.7 (21) 11.9 (22) 152.7 (23) 33.9 (24) 23.8 (25) 31.6 (26) 75.3 (27) 17.3 **Glc I** (1) 100.1 (2) 78.0 (3) 87.0 (4) 70.0 (5) 78.4 (6) 62.5 **Rha I** (1) 102.6 (2) 72.8 (3) 72.7 (4) 72.5 (5) 68.9 (6) 18.4 **Rha II** (1) 103.3 (2) 71.7 (3) 84.2 (4) 72.5 (5) 70.7 (6) 18.7 **Glc II** (1) 106.4 (2) 73.9 (3) 78.1 (4) 71.7 (5) 78.0 (6) 62.8 **Glc III** (1) 105.2 (2) 75.4 (3) 78.4 (5) 71.7 (5) 78.0 (6) 62.8.

Mass (FAB, Negative ion) : *m/z* 1046 [M-146]⁻, 1030 [M-162]⁻, 884 [M-162-146]⁻.

Reference

1.   H.Y. Liu and C.X. Chen, *Chin. Chem. Lett.*, **13**, 633 (2002).

DICHOTOMIN, PROTO-Pb
[(25*R*)-Furost-5-ene-3β,22α,26-triol 3-O-[α-L-rhamnopyranosyl-(1→4)-α-L-rhamnopyranosyl-(1→4)-[α-L-rhamnopyranosyl-(1→2)]-β-D-glucopyranoside]-26-O-β-D-glucopyranoside

Source : *Smilax aspera* L.[1] (Liliaceae), *Panicum dichotomiflorum*[2] (Graminae)
Mol. Formula : $C_{57}H_{94}O_{26}$
Mol. Wt. : 1194
Registry No. : [53093-47-3]

PMR (C$_5$D$_5$N-CD$_3$OD, 300.13 MHz)[2] : δ 0.71 (3xH-27), 0.90 (s, 3xH-18), 0.98 (3xH-21), 1.03 (s, 3xH-19), 1.45 (H-15A), 1.52 (3xH-6 of Rha II), 1.54 (3xH-6 of Rha III), 1.60 (H-25), 1.72 (3xH-6 of Rha I), 1.80 (H-17), 2.00 (H-5), 2.69 (H-4), 2.78 (H-4), 3.60 (H-26A), 3.60 (H-5 of Glc I), 3.86 (H-3), 3.92 (H-5 of Glc II), 3.93 (H-26B), 3.99 (H-2 of Glc II), 4.04 (H-6A of Glc I), 4.15 (H-2 of Glc I), 4.15 (H-3 of Glc I), 4.16 (H-6B of Glc I), 4.16 (H-4 of Glc II), 4.24 (H-3 of Glc II), 4.25 (H-5 of Rha III), 4.27 (H-4 of Rha III), 4.31 (H-6A of Glc II), 4.32 (H-4 of Glc I), 4.33 (H-4 of Rha I), 4.36 (H-4 of Rha II), 4.44 (H-3 of Rha III), 4.48 (H-3 of Rha II), 4.50 (H-6B of Glc I), 4.52 (H-2 of Rha II), 4.55 (H-16), 4.59 (H-3 of Rha I), 4.7 (H-1 of Glc I), 4.78 (H-5 of Rha II), 4.80 (H-2 of Rha I), 4.85 (H-2 of Rha III), 4.86 (H-5 of Rha I), 4.91 (H-1 of Glc I), 5.34 (H-6), 5.72 (H-1 of Rha II), 6.18 (H-1 of Rha III), 6.27 (H-1 of Rha I).

CMR (C$_5$D$_5$N, 75.47 MHz)[2] : δ C-1) 37.5 (2) 30.2 (3) 78.2 (4) 38.9 (5) 140.8 (6) 121.9 (7) 32.4^b (8) 31.6 (9) 50.3 (10) 37.3 (11) 21.1 (12) 39.9 (13) 40.8 (14) 56.6 (15) 32.1^b (16) 81.1 (17) 63.6 (18) 16.4 (19) 19.4 (20) 40.6 (21) 16.4 (22) 110.8 (23) 37.1 (24) 28.3 (25) 34.2 (26) 75.3 (27) 17.4 **Glc I** (1) 100.3 (2) 78.2 (3) 77.6 (4) 78.0 (5) 76.9 (6) 61.2 **Rha I** (1) 102.2 (2) 72.4 (3) 72.8 (4) 74.0 (5) 69.7 (6) 18.6 **Rha II** (1) 102.2 (2) 72.8 (3) 73.1 (4) 80.3 (5) 68.4 (6) 18.8 **Rha III** (1) 103.2 (2) 72.5 (4) 72.8 (4) 73.8 (5) 70.4 (6) 18.4 **Glc II** (1) 104.8 (2) 75.1 (3) 78.4 (4) 71.6 (5) 78.3 (6) 62.7.

Mass (LSI, Negative ion)[2] : *m/z* 1193 [M-H]⁻, 1047 [M-Rha]⁻, 901 [M-2xRha]⁻, 755 (M-3xRha]⁻, 593 [M-3xRha-Glc]⁻.

Biological Activity : Inhibits angiotensin-converting enzyme.[3]

References

1. R. Tschesche, A. Harz and J. Petricic, *Chem. Ber.,* **107**, 53 (1974).

2. S.C. Munday A.L. Wilkins, C.O. Miles and P.T. Holland, *J. Agric. Food Chem.*, **41**, 267 (1993).

3. Jpn Kokai Tokkyo Koho JP 04, 368, 336 [92, 368, 336], *Chem. Abstr.*, **118**, 154549c.

KALLSTROEMIN A

(25*R*)-Furost-5-ene-3β,22α,26-triol 3-O-[α-L-rhamnopyranosyl-(1→2)-α-L-rhamnopyranosyl-(1→2)-α-L-rhamnopyranosyl-(1→6)-β-D-glucopyranoside]-26-O-[β-D-glucopyranoside]

Source : *Kallstroemia pubescens* (G. Don) Dandy (Zygophyllaceae)
Mol. Formula : C$_{57}$H$_{94}$O$_{26}$
Mol. Wt. : 1194
M.P. : 235-238°C
[α]$_D$: 82.2° (Pyridine)
Registry No. : [64652-19-3]

Reference

1.   S.B. Mahato, N.P. Sahu, B.C. Pal and R.N. Chakravarti, *Indian J. Chem.*, **15B**, 445 (1977).

METHYLPROTO-RECLINATOSIDE

(25*R*)-22ξ-Methoxyfurost-5-en-3β,26-diol 3-O-{α-L-rhamnopyranosyl-(1→5)-α-L-arabinofuranosyl-(1→4)-[α-L-rhamnopyranosyl-(1→2)]-β-D-glucopyranoside}-26-O-β-D-glucopyranoside

Source : *Phoenix reclinata* N.J. Jacquin (Palmae)
Mol. Formula : $C_{57}H_{94}O_{26}$
Mol. Wt. : 1194
M.P. : 165-168°C (decomp.)
[α]$_D$: -44.2° (c=1.05, C_5D_5N)
Registry No. : [137295-54-6]

IR (KBr) : 3600-3200 (OH) cm^{-1}.

CMR of aglycone analogous to Methyl proto-taccaoside.

CMR (C_5D_5N, 100 MHz) : δ **Glc I** C-1) 100.5 (2) 79.0 (3) 76.7 (4) 78.6 (5) 77.7 (6) 61.8 **Rha I** (1) 102.1^a (2) 72.3 (3) 72.9^b (4) 74.3 (5) 69.9 (6) 18.6^c **Ara** (1) 110.0 (2) 83.3 (3) 78.7 (4) 83.6 (5) 67.9 **Rha II** (1) 101.8^a (2) 72.1 (3) 72.7^b (4) 74.0 (5) 69.5 (6) 18.5^c **Glc II** (1) 104.9 (2) 75.3 (3) 78.4 (4) 72.1 (5) 78.7 (6) 63.2.

Reference

1. K. Idaka, Y. Hirai and J. Shoji, *Chem. Pharm. Bull.*, **39**, 1455 (1991).

SOLAYAMOCINOSIDE A

3β,22α,26-Trihydroxy-25S-furost-5-ene 3-O-{α-L-arabinopyranosyl-(1→2)]-[β-D-galactopyranosyl-(1→3)]-[β-D-galactopyranosyl-(1→4)-rhamnopyranoside}-26-O-β-D-glucopyranoside

Source : *Solanum dulcamara* L. (Solanaceae)
Mol. Formula : $C_{56}H_{92}O_{27}$
Mol. Wt. : 1196
M.P. : 228-232°C
[α]$_D^{20}$: -27° (c=0.927, MeOH)
Registry No. : [86975-15-7]

Reference

1   G. Willuhn and U. Köthe, *Arch. Pharm.* (Weinheim), **316**, 678 (1983).

TRIGOFOENOSIDE G

(25*R*)-Furost-5-en-3β,22,26-triol 3-O-[α-L-rhamnopyranosyl-(1→2)-[β-D-xylopyranosyl-(1→4)]-β-D-glucpoyranosyl-(1→6)-β-D-glucopyranoside]-26-O-β-glucopyrnoside

Source : *Trigonella foenum-graecum* L.(Fabaceae)
Mol. Formula : $C_{56}H_{92}O_{27}$
Mol. Wt. : 1196
M.P. : 275-278°C
Registry No. : [94714-57-5]

22-Methyl ether : Trigofoenoside G-1

M.P. : 270-274°C; **[α]$_D$:** -79.2° (Pyridine)

Reference

1.   R.K. Gupta, D.C. Jain and R.S. Thakur, *Phytochemistry*, **23**, 2605 (1984).

ASPARAGUS CURILLUS SAPONIN 9

(5β,25S)-Furostan-3β,22α,26-triol 3-O-{β-D-glucopyranosyl-(1→2)-[α-L-rhamnopyranosyl-(1→4)]-
[α-L-arabinopyranosyl-(1→6)]-β-D-glucopyranoside}-26-O-β-D-glucopyranoside

Source : *Asparagus curillus* Buch.-Ham. (Liliaceae)
Mol. Formula : $C_{56}H_{94}O_{27}$
Mol. Wt. : 1198
[α]$_D^{25}$: -66.0°C
Registry No. : [98973-43-4]

IR (KBr) : 3400 (OH) cm^{-1}. No spiroketal absorption.

Reference

1.   O.P. Sati and S.C. Sharma, *Pharmazie*, **40**, 417 (1985).

ASPAROSIDE D
(5β,25S)-3β,22α,26-Trihydroxyfurostane 3-O-{β-D-glucopyranosyl-(1→2)-[α-L-arabinopyranosyl-(1→4)]-α-L-rhamnopyranoside}

Source : *Asparagus adscendens* Roxb. (Liliaceae)
Mol. Formula : $C_{56}H_{94}O_{27}$
Mol. Wt. : 1198
M.P. : 161-166°C (decomp.)
[α]$_D^{20}$: -66° (c=1.5, Pyridine)
Registry No. : [83946-26-3]

IR (KBr) : 3400 (OH) cm^{-1}.

Reference

1.   S.C. Sharma, R. Chand, B.S. Bhatti and O.P. Sati, *Planta Med.*, **46**, 48 (1982).

KALLSTROEMIN B

(25*R*),22α-Methoxyfurost-5-en-3β,26-diol 3-O-[α-L-rhamnopyranosyl-(1→2)-α-L-rhamnopyranosyl-(1→2)-
α-L-rhamnopyranosyl-(1→6)-β-D-glucopyranoside]-26-O-[β-D-glucopyranoside]

Source : *Kallstroemia pubescens* (G. Don) Dandy
(Zygophyllaceae)
Mol. Formula : $C_{58}H_{96}O_{26}$
Mol. Wt. : 1208
M.P. : 232-235°
[α]$_D$: 79.5° (Pyridine)
Registry No. : [64652-20-6]

Reference

1.   S.B. Mahato, N.P. Sahu, B.C. Pal and R.N. Chakravarti, *Indian J. Chem.*, **15B**, 445 (1977).

METHYL (25*S*)-PROTO-Pb

(25*S*)-22ξ–Methoxyfurost-5-en-3β,26-diol 3-O-{α-L-rhamnopyranosyl-(1→4)-α-L-rhamnopyranosyl-(1→4)-
[α-L-rhamnopyranosyl-(1→2)]-β-D-glucopyranoside}-26-O-β-D-glucopyranoside

Source : *Phoenix canariensis* Hort. ex Chabaud[1], *P.
humilis* Royle var. *hanceana* Becc.[1] (Palmae)
Mol. Formula : $C_{58}H_{96}O_{26}$
Mol. Wt. : 1208
M.P. : 177-179°C (decomp.)
[α]$_D^{22}$: -87.2° (c=0.35, C_5D_5N)
Registry No. : [139406-75-0]

IR (KBr) : λ_{max} 3500-3300 (OH) cm^{-1}.

CMR (C$_5$D$_5$N, 100 MHz) : δ C-1) 38.1 (2) 30.7 (3) 78.6 (4) 39.5 (5) 141.0 (6) 121.8 (7) 32.9 (8) 32.3 (9) 50.9 (10) 37.7 (11) 21.7 (12) 40.3 (13) 41.0 (14) 57.1 (15) 32.8 (16) 81.7 (17) 63.4 (18) 16.7 (19) 20.0 (20) 41.3 (21) 18.1 (22) 112.9 (23) 35.0 (24) 32.7 (25) 31.5 (26) 64.6 (27) 16.8 (OCH$_3$) 47.8 **Glc I** (1) 100.6 (2) 80.6 (3) 78.4 (4) 77.2 (5) 78.9 (6) 61.9 **Rha I** (1) 102.2 (2) 72.7 (3) 73.1 (4) 74.5 (5) 69.8 (6) 19.3 **Rha II** (1) 103.2 (2) 73.1 (3) 70.6 (4) 78.0 (5) 68.8 (6) 19.1 **Rha III** (1) 102.5 (2) 72.9 (3) 73.0 (4) 74.3 (5) 73.4 (6) 18.9 **Glc II** (1) 105.1 (2) 75.5 (3) 78.9 (4) 72.3 (5) 78.6 (6) 64.6.

Reference

1. A. Asami, Y. Hirai and J. Shoji, *Chem. Pharm. Bull.*, **39**, 2053 (1991).

METHYL PROTO-Pb

3β,26-Dihydroxy-22ξ-methoxyfurost-5-ene 3-O-{α-L-rhamnopyranosyl-(1→2)-[α-L-rhamnopyranosyl-(1→4)-α-L-rhamnopyranosyl-(1→4)]-β-D-glucopyranoside-26-O-β-D-glucopyranoside

Source : *Paris polyphylla* Sm.[1] (Liliaceae), *Trachycarpus fortunei* (Hook) H. Wendl.[2] (Palmae), *Chamaerops humilis* L.[3] (Palmae), *Trachycarpus wagnerianus* Becc[3] (Palmae), *Smilax krausiana*[4] (Liliaceae) etc.

Mol. Formula : $C_{58}H_{96}O_{26}$

Mol. Wt. : 1208

M.P. : 189-190°C[2]

$[\alpha]_D^{16}$: -86.4° (c=1.03, C_5D_5N)[2]

Registry No. : [78101-21-0]

IR (KBr)[2] : 3600-3250 (OH) cm⁻¹.

PMR (CD₃OD, 300 MHz)[4] : δ 0.83 (s, 3xH-18), 0.95 (d, *J*=7.0 Hz, 3xH-27), 1.00 (d, *J*=7.0 Hz, 3xH-21), 1.05 (s, 3xH-19), 1.24 (d, *J*=7.0 Hz, 3xH-6 of Rha I), 1.25 (d, *J*=7.0 Hz, 3xH-6 of Rha III), 1.28 (d, *J*=7.0 Hz, 3xH-6 of Rha II), 3.14 (s, OC*H₃*), 3.17 (dd, *J*=9.0, 8.0 Hz, H-2 of Glc II), 3.60 (H-3), 3.92 (br d, *J*=1.5 Hz, H-2 of Rha I and Rha III), 4.03 (dq, *J*=9.5, 7.0 Hz, H-5 of Rha III), 4.12 (dq, *J*=9.5, 7.0 Hz, H-5 of Rha I), 4.23 (d, *J*=8.0 Hz, H-1 of Glc I), 4.37 (dt, *J*=8.0, 7.0 Hz, H-16), 4.49 (d, *J*=8.0 Hz, H-1 of Glc I), 4.83 (H-1 of Rha III), 5.18 (d, *J*=1.5 Hz, H-1 of Rha III), 5.19 (d, *J*=1.5 Hz, H-1 of Rha I), 5.38 (br d, *J*=4.0 Hz, H-6).

CMR (C$_5$D$_5$N, 25 MHz)[3] : δ 18.3 (C-6 of Rha), 18.5 (C-6 of Rha), 18.8 (C-6 of Rha), 47.4 (OCH$_3$), 100.4 (C-1 of Glc), 102.0 (C-1 of Rha), 102.3 (C-1 of Rha), 103.0 (C-1 of Rha), 104.7 (C-1 of Rha).

References

1.   M. Miyamura, K. Nakano, T. Nohara, T. Tomimatsu and T. Kawasaki, *Chem. Pharm. Bull.*, **30**, 712 (1982).

2.   Y. Hirai, S. Sanada, Y. Id and J. Shoji, *Chem. Pharm. Bull.*, **32**, 295 (1984).

3.   Y. Hirai, S. Sanada, Y. Ida and J. Shoji, *Chem. Pharm. Bull.*, **34**, 82 (1986).

4.   C. Lavaud, G. Massiot and L. Le Men-Olivier, *Planta Med. Phytother.*, **26**, 64 (1993).

COSTUSOSIDE J

(25*R*)-Furostan-5-en-3β,22α,26-triol 3-O-{β-D-glucopyranosyl-(1→2)-α-L-rhamnopyranosyl-(1→2)-[α-L-rhamnopyranosyl-(1→4)]-β-D-glucopyranoside}-26-O-β-D-glucopyranoside

Source : *Costus speciosus* (Koen.) Sm. (Costaceae)
Mol. Formula : C$_{57}$H$_{94}$O$_{27}$
Mol. Wt. : 1210
M.P. : 248-250°C (decomp.)
[α]$_D^{25}$: -79.1° (c=1.32, C$_5$D$_5$N)
Registry No. : [82844-07-3]

IR (KBr) : 3800-3100, 1180-990, 980, 910 (str.), 895 (w), 840 cm^{-1}.

PMR (DMSO-d_6, 60.0 MHz) : δ 4.12 (br s, $W\frac{1}{2}$=7.0 Hz), 4.25 (br s, $W\frac{1}{2}$=7.0 Hz), 4.32 (br s, $W\frac{1}{2}$=7.0 Hz), 4.75 (br s, $W\frac{1}{2}$=5.0 Hz), 5.0 (br s, $W\frac{1}{2}$=5.0 Hz), 5.25 (m, H-6).

Reference

1.   S.B. Singh and R.S. Thakur, *Phytochemistry*, **21**, 911 (1982).

HELONIOPSIS ORIENTALIS SAPONIN Hd,
TRILLIUM KAMTSCHATICUM SAPONIN Th

3β,17α,22,26-Tetrahydroxyfurost-5-ene 3-O-{α-L-rhamnopyranosyl-(1→4)-α-L-rhamnopyranosyl-(1→4)-[α-L-rhamnopyranosyl-(1→2)]-β-D-glucopyranoside}-26-O-β-D-glucopyranoside

Source : *Heloniopsis orientalis* (Thunb.) C. Tanaka[1] (Liliaceae), *Trillium kamtschaticum* Pall.[2] (Liliaceae), *Paris quadrifolia* L.[3] (Liliaceae)
Mol. Formula : $C_{57}H_{94}O_{27}$
Mol. Wt. : 1210
M.P. : 194-200°C (decomp.)[2]
[α]$_D$: -92.1° (c=0.83, Pyridine)[1]
Registry No. : [55916-46-6]

IR (KBr)[3] : 3400 (OH) cm^{-1}.

Acetate :

M.P. : 135-139°C[1]; **[α]$_D$:** -61.9° (c=0.42, CHCl$_3$)[1]

Mass (E.I.)[1] : m/z 503 (C$_{22}$H$_{31}$O$_{13}$)$^+$, 412 (C$_{27}$H$_{40}$O$_3$)$^+$, 394 (C$_{27}$H$_{38}$O$_2$)$^+$, 331 (C$_{14}$H$_{19}$O$_9$)$^+$, 273 (C$_{12}$H$_{17}$O$_2$)$^+$.

22-Methyl ether :

CMR (C$_5$D$_5$N, 22.5 MHz)[4] : δ C-1) 37.6 (2) 29.9 (3) 77.7 (4) 39.0 (5) 140.9 (6) 121.8 (7) 31.7 (8) 32.3 (9) 50.3 (10) 37.1 (11) 20.9 (12) 37.1 (13) 45.4 (14) 53.0 (15) 32.4 (16) 90.3 (17) 90.5 (18) 17.1 (19) 19.4 (20) 43.0 (21) 10.3 (22) 113.5 (23) 30.8 (24) 28.1 (25) 34.2 (26) 75.1 (27) 17.4 (OCH$_3$) 47.1 **Glc I** (1) 100.3 (2) 80.3 (3) 76.9 (4) 78.0 (5) 77.7 (6) 61.3 **Rha I** (1) 102.1^a (2) 72.8 (3) 72.5^b (4) 74.0^c (5) 69.4^d (6) 18.6^e **Rha II** (1) 102.2^a (2) 73.1 (3) 72.8 (4) 78.3 (5) 68.4 (6) 18.6^e **Rha III** (1) 103.1 (2) 72.8 (3) 72.4^b (4) 73.9^c (5) 70.3^d (6) 18.8^e **Glc II** (1) 104.9 (2) 75.1 (3) 78.5 (4) 71.8 (5) 78.3 (6) 62.9.

References

1. T. Nohara, Y. Ogata, M. Aritome, K. Miyahara and T. Kawasaki, *Chem. Pharm. Bull.*, **23**, 925 (1975).

2. N. Fukuda, N. Imamura, E. Saito, T. Nohara and T., Kawasaki, *Chem. Pharm. Bull.*, **29**, 325 (1981).

3. T. Nohara, Y. Ito, H. Seike, T. Komori, M. Moriyama, Y. Gomita and T. Kawasaki, *Chem. Pharm. Bull.*, **30**, 1851 (1982).

4. K. Nakano, Y. Kashiwada, T. Nohara, T. Tomimatsu, H. Tsukolani and T. Kawasaki, *Yakugaku Zasshi*, **102**, 1031 (1982)

METHYL (25*S*)-PROTO-LOUREIROSIDE

(25*S*)-22-Methoxyfurost-5-en-3β,26-diol 3-O-{β-D-glucopyranosyl-(1→5)-α-L-arabinofuranosyl-(1→4)-[α-L-rhamnopyranosyl-(1→2)]-β-D-glucopyranoside}-26-O-β-D-glucopyranoside

Source : *Phoenix humilis* Royle var. *hanceana* Becc. (Palmae)
Mol. Formula : C$_{57}$H$_{94}$O$_{27}$
Mol. Wt. : 1210
M.P. : 185-187°C (decomp.)
[α]$_D$: -78.3° (c=0.37, C$_5$D$_5$N)
Registry No. : [139491-20-6]

IR (KBr) : λ_{max} 3500-3300 (OH) cm^{-1}.

CMR (C$_5$D$_5$N, 100 MHz) : d C-1) 38.1 (2) 30.7 (3) 78.6 (4) 39.5 (5) 141.0 (6) 121.8 (7) 32.9 (8) 32.3 (9) 50.9 (10) 37.7 (11) 21.7 (12) 40.3 (13) 41.0 (14) 57.1 (15) 32.8 (16) 81.7 (17) 63.4 (18) 16.7 (19) 20.0 (20) 41.3 (21) 18.1 (22) 112.9 (23) 35.0 (24) 32.7 (25) 31.5 (26) 64.6 (27) 16.8 (OCH3) 47.8 **Glc I** (1) 100.5 (2) 78.9 (3) 78.1 (4) 76.9 (5) 78.5 (6) 62.0 **Rha** (1) 102.2 (2) 73.1 (3) 72.6 (4) 74.5 (5) 79.8 (6) 19.1 **Ara** (1) 109.9 (2) 83.4 (3) 78.6 (4) 84.6 (5) 70.6 **Glc II** (1) 105.4 (2) 75.3 (3) 78.3 (4) 72.1 (5) 77.9 (6) 63.4 **Glc III** (1) 105.1 (2) 75.5 (3) 78.9 (4) 72.3 (5) 78.6 (6) 64.6.

Reference

1.   A. Asami, Y. Hirai and J. Shoji, *Chem. Pharm. Bull.*, **39**, 2053 (1991).

METHYL PROTO-LOUREIROSIDE

(25R)-22ξ-Methoxyfurost-5-en-3β,26-diol 3-O-{β-D-glucopyranosyl-(1→5)-α-L-arabinofuranosyl-(1→4)]-
[α-L-rhamnopyranosyl-(1→2)]-β-D-glucopyranoside}-26-O-β-D-glucopyranoside

Source : *Phoenix loureirii* Kunth. (Palmae)
Mol. Formula : $C_{57}H_{94}O_{27}$
Mol. Wt. : 1210
M.P. : 183-186°C (decomp.)
[α]$_D$: -84.1° (c=1.06, C_5D_5N)
Registry No. : [137297-53-5]

IR (KBr) : 3600-3200 (OH) cm^{-1}.

CMR of the aglycone analogous to that of Methyl Proto-taccaoside

CMR (C_5D_5N, 100 MHz) : δ **Glc I** C-1) 100.5 (2) 78.9 (3) 76.9 (4) 78.5^a (5) 77.9 (6) 61.9 **Rha** (1) 102.2 (2) 72.6 (3) 73.1 (4) 74.5 (5) 69.7 (6) 19.1 **Ara** (1) 109.9 (2) 83.4 (3) 78.6^a (4) 84.3 (5) 70.6 **Glc II** (1) 105.4^b (2) 75.5 (3) 78.1 (4) 72.1^e (5) 78.5^a (6) 63.2^d **Glc III** (1) 105.0^b (2) 75.5 (3) 78.2 (4) 72.3^c (5) 78.9 (6) 63.9^d.

Reference

1.   K. Idaka, Y. Hirai and J. Shoji, *Chem. Pharm. Bull.*, **39**, 1455 (1991).

TRIBULUS TERRESTRIS SAPONIN 2

(25ξ,5α)-3β,26-Dihydroxyfurost-20(22)-en-12-one 3-O-{β-D-xylopyranosyl-(1→3)-[β-D-galactopyranosyl-(1→2)]-β-D-glucopyranosyl-(1→4)-β-D-glucopyranoside}-26-O-β-D-glucopyranoside

Source : *Tribulus terrestris* L. (Zygophyllceae)
Mol. Formula : $C_{56}H_{90}O_{28}$
Mol. Wt. : 1210
M.P. : 218-220°C
[α]$_D$: 3.82° (c=0.3, Pyridine)
Registry No. : [180050-94-6]

UV : λ_{max} 202, 240, 255 nm.

IR (KBr) : 3400 (br) 1705 cm^{-1} (CO).

PMR (C_5D_5N, 400/600 MHz) : δ 0.63 (s, 3xH-19), 0.75 (H-1A), 0.77 (H-7A), 0.85 (H-5), 0.87 (H-9), 0.90 (s, 3xH-18), 1.00 (d, *J*=6.6 Hz, 3xH-27), 1.12 (H-14), 1.15 (H-6), 1.30 (H-1B), 1.32 (H-4A), 1.42 (H-24A), 1.50 (H-7B), 1.55 (H-2A), 1.55 (H-15A), 1.42 (H-24a), 1.63 (H-8), 1.72 (s, 3xH-21), 1.79 (H-24B), 1.80 (H-4B), 1.90 (H-25), 1.95 (H-2B), 2.12 (H-15B), 2.20 (H-11A), 2.20 (H-23), 2.30 (H-11B), 3.38 (d, *J*=10.2 Hz, H-17), 3.55 (H-5A of Xyl), 3.60 (H-26A), 3.74 (H-4 of Glc II), 3.85 (H-3), 3.85 (H-5 of Glc II), 3.87 (H-26B), 3.88 (H-3 of Gal), 3.91 (H-2 of Xyl), 3.92 (H-5 of Glc III), 3.96 (H-3 of Xyl), 4.00 (H-6A of Glc II), 4.01 (H-5 of Glc I and H-2 of Glc III and H-5 of Gal), 4.04

(H-4 of Xyl), 4.08 (H-3 of Glc I) 4.11 (H-3 of Glc II), 4.15 (H-6A of Gal), 4.17 (H-5B of Xyl), 4.18 (H-4 of Glc III), 4.20 (H-3 of Glc III), 4.25 (H-4 of Gal), 4.32 (H-2 of Glc I), 4.35 (H-6A of Glc I), 4.37 (H-6B of Glc III), 4.49 (H-6B of Glc II), 4.51 (H-2 of Glc II), 4.53 (H-6B of Glc I), 4.54 (H-4 of Glc I), 4.57 (H-2 of Gal), 4.59 (H-6A of Glc III), 4.63 (H-6A of Gal), 4.69 (H-16), 4.81 (d, J=8.0 Hz, H-1 of Glc III), 4.87 (d, J=7.2 Hz, H-1 of Glc I), 5.04 (d, J=7.2 Hz, H-1 of Xyl), 5.15 (d, J=8.0 Hz, H-1 of Glc II), 5.45 (d, J=8.0 Hz, H-1 of Gal).

CMR (C_5D_5N, 100/150 MHz) : δ C-1) 36.35 (2) 29.38 (3) 76.99 (4) 34.38 (5) 43.61 (6) 28.25 (7) 31.50 (8) 33.76 (9) 55.21 (10) 35.93 (11) 37.83 (12) 212.47 (13) 57.22 (14) 53.84 (15) 33.45 (16) 82.63 (17) 55.89 (18) 13.81 (19) 11.96 (20) 102.83 (21) 11.96 (22) 152.80 (23) 23.36 (24) 31.07 (25) 33.15 (26) 74.78 (27) 16.94 **Glc I** (1) 101.97 (2) 72.69 (3) 75.43 (4) 75.06 (5) 75.06 (6) 62.54 **Glc II** (1) 105.17 (2) 80.70 (3) 85.31 (4) 70.36 (5) 77.26 (6) 62.60 **Glc III** (1) 104.53 (2) 74.83 (3) 78.25 (4) 71.41 (5) 78.11 (6) 62.24 **Gal** (1) 104.95 (2) 73.41 (3) 73.66 (4) 70.05 (5) 75.05 (6) 60.19 **Xyl** (1) 104.47 (2) 74.83 (3) 78.11 (4) 70.05 (5) 66.80.

Mass (FAB, Positive ion) : m/z (rel.intens.) 1249 [(M+K)$^+$, 33], 1233 [(M+Na)$^+$, 42], 1210 [(M)$^+$, 18].

Reference

1.   G. Wu, S. Jiang, F. Jiang, D.Zhu, H.Wu and S. Jiang, *Phytochemistry*, **42** ,1677 (1996).

ASPARAGUS COCHINCHINENSIS SAPONIN ASP-VII
(5β,25S)-22ξ-Methoxyfurostane-3β,26-diol 3-O-{β-D-glucopyranosyl-(1→2)-[β-D-xylopyranosyl-(1→4)]-[α-L-rhamnopyranosyl-(1→6)]-β-D-glucopyranoside}-26-O-β-D-glucopyranoside

Source : *Asparagus cochinchinensis* (Loureiro) Merill (Liliaceae)
Mol. Formula : $C_{57}H_{96}O_{27}$
Mol. Wt. : 1212
M.P. : 179-181°C (decomp.)
[α]$_D$: -26.8° (c=1.2, MeOH)
Registry No. : [72947-82-1]

PMR (C₅D₅N, 90 MHz) : δ 3.28 (s, OCH₃), 4.83 (d, J=7.5 Hz, anomeric H), 5.39 (d, J=7.0 Hz, anomeric H), 5.53 (s, anomeric H).

CMR (C₅D₅N, 22.15 MHz) : δ **Glc I** C-1) 101.7 **Glc II** (1) 104.7 **Rha** (1) 101.7 **Xyl** (1) 104.9.

Reference

1.   T. Konishi and J. Shoji, *Chem., Pharm. Bull.*, **27**, 3086 (1979).

ASPARAGUS CURILLUS SAPONIN 8
(5β,25S)-22α-Methoxyfurostan-3β,26-diol 3-O-{β-D-glucopyranosyl-(1→2)-[α-L-rhamnopyranosyl-(1→4)]-[α-L-arabinopyranosyl-(1→6)]-β-D-glucopyranoside}-26-O-β-D-glucopyranoside]

Source : *Asparagus curillus* Buch.-Ham. (Liliaceae)
Mol. Formula : C₅₇H₉₆O₂₇
Mol. Wt. : 1212
[α]$_D^{25}$: -78.0°
Registry No. : [98973-42-3]

IR (KBr) : 3400 (OH) cm^{-1}, no spiroketal absorption.

Reference

1.   O.P. Sati and S.C. Sharma, *Pharmazie*, **40,** 417 (1985).

ASPAROSIDE C

(5β,25S)-3β,26-Dihydroxy-22α-methoxyfurostane 3-O-{β-D-glucopyranosyl-(1→2)-[α-L-arabinopyranosyl-(1→4)]-α-L-rhamnopyranosyl-(1→6)-β-D-glucopyranoside}-26-O-β-D-glucopyranoside

Source : *Asparagus adscendens* Roxb. (Liliaceae)
Mol. Formula : C$_{57}$H$_{96}$O$_{27}$
Mol. Wt. : 1212
M.P. : 167-172°C
[α]$_D$^{20} : -75° (c=1.0, MeOH)
Registry No. : [83946-27-4]

IR (KBr) : 3400 (OH) cm^{-1}.

Reference

1.   S.C. Sharma, R. Chand, B.S. Bhatti and O.P. Sati, *Planta Med.*, **46**, 48 (1982).

PROTOASPIDISTRIN, TIMOSAPONIN C₁

(25R)-3β,22ξ,26-Trihydroxyfurost-5-ene 3-O-{β-D-glucopyranosyl-(1→2)-[β-D-xylopyranosyl-(1→3)]-β-D-glucopyranosyl-(1→4)-β-D-galactopyranoside}-26-O-β-D-glucopyranoside

Source : *Aspidistra elatior* Blume (Liliaceae)
Mol. Formula : $C_{56}H_{92}O_{28}$
Mol. Wt. : 1212
M.P. : 210-214°C (decomp.)
[α]$_D^{15.4}$: -64.0° (c=1.00, C_5D_5N)
Registry No. : [84272-82-2]

IR (KBr) : 3600-3300 (OH) cm^{-1}.

PMR (C$_5$D$_5$N, 100 MHz) : δ 0.91 (s, 2xCH_3).

CMR (C$_5$D$_5$N, 25.0 MHz) : δ C-1) 37.4 (2) 29.9 (3) 78.5 (4) 39.1 (5) 141.0 (6) 121.4 (7) 32.2 (8) 31.6 (9) 50.3 (10) 36.9 (11) 21.1 (12) 39.8 (13) 40.7 (14) 56.5 (15) 32.2 (16) 80.7 (17) 63.4 (18) 16.3 (19) 19.2 (20) 40.4 (21) 16.0 (22) 110.7 (23) 36.8 (24) 28.1 (25) 34.0 (26) 75.1^a (27) 17.2 **Gal** (1) 102.5 (2) 72.8 (3) 75.1^a (4) 79.3 (5) 75.7 (6) 60.6 **Glc I** (1) 104.4 (2) 80.7 (3) 87.0 (4) 70.5 (5) 77.6 (6) 62.3 **Glc II** (1) 104.4 (2) 74.8^a (3) 78.1 (4) 70.1 (5) 77.1 (6) 62.6 **Xyl** (1) 104.4 (2) 75.2^a (3) 78.1 (4) 71.1 (5) 66.9 **Glc III** (1) 104.4 (2) 74.8^a (3) 78.1 (4) 71.6 (5) 77.8 (6) 62.6.

Reference

1.   Y. Hirai, T. Konishi, S. Sanada, Y. Ida and J. Shoji, *Chem. Pharm. Bull.*, **30**, 3476 (1982).

SAPONOSIDE B

(25*R*)-Furost-5-en-3β,22α,26-triol 3-O-{β-D-glucopyranosyl-(1→4)-[β-D-xylopyranosyl-(1→2)]-β-D-glucopyranosyl-(1→4)-β-D-galactopyranoside]-26-O-β-D-glucopyranoside

Source : *Polygonatum multiflorum* L. (Liliaceae)
Mol. Formula : $C_{56}H_{92}O_{28}$
Mol. Wt. : 1212
Registry No. : [78886-82-5]

Reference

1.   Z. Janeczko, *Acta Polon. Pharm.*, **XXXVII**, 559 (1980).

TIMOSAPONIN H₁

(25S)-Furost-5-ene3β,22,26-triol 3-O-{β-D-glucopyranosyl-(1→2)-[β-D-xylopyranosyl-(1→3)]-β-D-glucopyranosyl-(1→4)-β-D-galactopyranoside}-26-O-β-D-glucopyranoside

Source : *Anemarrhena asphodeloides* Bunge[1] (Liliaceae), *Aspidistra elatior* Blume[2] (Liliaceae)
Mol. Formula : $C_{56}H_{92}O_{28}$
Mol. Wt. : 1212
Registry No. : [274692-73-8]

PMR (C_5D_5N, 300 MHz)[1] : δ 0.69 (s, 3xH-18)ᵃ, 0.82 (s, 3xH-19)ᵃ, 0.88 (d), 1.23 (d, 3xH-27), 4.79 (d, *J*=7.7 Hz, H-1 of Glc III), 4.88 (d, *J*=7.5 Hz, H-1 of Gal), 5.18 (d, *J*=7.9 Hz, H-1 of Glc I), 5.23 (d, *J*=7.6 Hz, H-1 of Xyl), 5.57 (d, *J*=7.0~8.0 Hz, H-1 of Glc II).

CMR (C_5D_5N, 75 MHz)[1] : δ C-1) 37.4 (2) 30.1 (3) 78.7 (4) 39.2 (5) 141.0 (6) 121.7 (7) 32.4 (8) 31.6 (9) 50.3 (10) 37.1 (11) 21.1 (12) 40.0 (13) 40.8 (14) 56.6 (15) 32.5 (16) 81.4 (17) 63.8 (18) 16.5 (19) 19.4 (20) 40.8 (21) 16.5 (22) 110.7 (23) 37.0 (24) 28.3 (25) 34.5 (26) 75.1 (27) 17.5 **Gal** (1) 102.8 (2) 73.2 (3) 75.3 (4) 79.9 (5) 76.3 (6) 60.6 **Glc I** (1) 104.8 (2) 81.1 (3) 86.7 (4) 70.8 (5) 77.7 (6) 62.5 **Glc II** (1) 105.0 (2) 75.3 (3) 78.5 (4) 70.5 (5) 77.7 (6) 62.8 **Xyl** (1) 105.2 (2) 75.6 (3) 78.6 (4) 71.0 (5) 67.4 **Glc III** (1) 104.9 (2) 75.4 (3) 78.6 (4) 71.7 (5) 78.2 (6) 63.03.

Mass (ESI)[1] : *m/z* 1233.4 [M-H₂O+K]⁺, 1101.2 [M-H₂O+K-132]⁺, 939.3 [M-H₂O+K-132-162]⁺, 777.3 [M-H₂O+K-132-2x162]⁺, 576.7 [M+H₂O-132-3x162]⁺, 414.9 [M-H₂O-132-4x162]⁺.

References

1.   Z. Meng, S. Xu, W. Li and Y. Sha, *Zhongguo Yaowu Huaxue Zazhi* (*Chin. J. Med. Chem.*), **9**, 294 (1999).

2.   Q.X. Yang and C.R. Yang, *Yunnan Zhiwu Yanjiu* (*Acta Bot. Yunnanica*), **22**, 109 (2000).

ASPARAGOSIDE H

(25*S*)-3β,22α,26-Trihydroxyfurostane 3-O-{β-D-xylopyranosyl-(1→4)-β-D-glucopyranosyl-(1→4)-β-D-glucopyranosyl-(1→3)]-β-D-glucopyranoside}

Source : *Asparagus officinalis* L. (Liliaceae)
Mol. Formula : $C_{56}H_{94}O_{28}$
Mol. Wt. : 1214
M.P. : 146-150°C
[α]$_D^{20}$: -200° (c=0.6, H_2O)
Registry No. : [60267-28-9]

Reference

1.   G.M. Goryanu and P.K. Kintya, *Khim. Prir. Soedin*, 810 (1977); *Chem. Nat. Comp.* **13**, 682 (1977).

FILICINOSIDE D

(5β,25S)-Furostan-3β,22α,26-triol 3-O-{β-D-galactopyranosyl-(1→4)-β-D-xylopyranosyl-(1→6)]-
β-D-glucopyranosyl-(1→4)-β-D-glucopyranoside}-26-O-β-D-glucopyranoside

Source : *Asparagus filicinus* Buch.-Ham. (Liliaceae)
Mol. Formula : $C_{56}H_{94}O_{28}$
Mol. Wt. : 1214
M.P. : 168-172°C
[α]$_D^{20}$: -48.0° (Pyridine)
Registry No. : [173395-17-0]

Reference

1.   S.C. Sharma and N.K. Thakur, *Phytochemistry*, **41**, 599 (1996).

PROTO-DEGALACTOTIGONIN

$5\alpha,25R$-Furostan-3β,22ξ,26-triol 3-O-{β-D-glucopyranosyl-(1→2)-[β-D-xylopyranosyl-(1→3)]-β-D-glucopyranosyl-(1→4)-β-D-galactopyranoside}-26-O-β-D-glucopyranoside

Source : *Capsicum annuum* L.[1] var. *conoides* and *Capsicum annuum* L. var. *fasciculatum* (Solanaceae), *Solanum nigrum* L.[2] (Solanaceae)
Mol. Formula : $C_{56}H_{94}O_{28}$
Mol. Wt. : 1214
$[\alpha]_D^{30}$: -14.8° (c=0.10, Pyridine)
Registry No. : [126643-25-2]

PMR (C_5D_5N, 400 MHz) : δ 0.64 (s, CH_3), 0.88 (s, CH_3), 1.98 (d, *J*=6.6 Hz, sec. CH_3), 1.34 (d, *J*=6.6 Hz, sec. CH_3), 4.82, 4.94, 5.23, 5.57 (d, *J*=7.7 Hz, 4 x anomeric H), 5.13 (d, *J*=7.3 Hz, anomeric H).

CMR (C_5D_5N, 100 MHz) : δ C-1) 36.7 (2) 29.4 (3) 78.5 (4) 34.0 (5) 44.2 (6) 28.4 (7) 31.9 (8) 34.3 (9) 54.0 (10) 34.7 (11) 20.8 (12) 39.7 (13) 40.6 (14) 55.9 (15) 31.9 (16) 80.6 (17) 63.4 (18) 16.2 (19) 11.8 (20) 40.2 (21) 15.9 (22) 110.1 (23) 35.3 (24) 27.9 (25) 33.8 (26) 74.7 (27) 16.9 **Gal** (1) 101.8 (2) 72.7 (3) 75.1 (4) 79.3 (5) 75.7 (6) 60.7 **Glc I** (1)

104.3 (2) 80.8 (3) 86.3 (4) 69.9 (5) 78.2 (6) 61.9 **Glc II** (1) 104.3 (2) 75.0 (3) 78.1 (4) 71.3 (5) 77.7 (6) 62.3 **Xyl** (1) 104.6 (2) 74.7 (3) 78.1 (4) 70.5 (5) 66.8 **Glc III** (1) 104.3 (2) 74.5 (3) 77.6 (4) 70.7 (5) 77.1 (6) 62.3.

References

1.   S. Yahara, T. Ura, C. Sakamoto and T. Nohara, *Phytochemistry*, **37**, 831 (1994).

2.   R. Saijo, K. Murakami, T. Nohara, T. Tomimatsu, A. Sato and K. Matsuoka, *Yakugaku Zasshi*, **102**, 300 (1982).

TIMOSAPONIN I$_1$
(5α,25S)-Furostane-3β,22,26-triol 3-O-{β-D-glucopyranosyl-(1→2)-[β-D-xylopyranosyl-(1→3)]-β-D-glucopyranosyl-(1→4)-β-D-galactopyranoside}-26-O-β-D-glucopyranoside

Source : *Anemarrhena asphodeloides* Bge. (Liliaceae)
Mol. Formula : $C_{56}H_{94}O_{28}$
Mol. Wt. : 1214
Registry No. : [274693-78-6]

PMR (C_5D_5N, 300 MHz) : δ 0.61 (s, 3xH-18)[a], 0.85 (s, 3xH-19)[a], 1.01 (d, sec. CH_3), 1.31 (d, sec. CH_3), 4.80 (d), 4.89 (d), 5.21 (d), 5.25 (d), 5.56 (d, J=7.0 Hz, 5 x anomeric H).

CMR (C$_5$D$_5$N, 75 MHz) : δ C-1) 37.2 (2) 29.9 (3) 77.4 (4) 34.8 (5) 44.7 (6) 29.0 (7) 32.4 (8) 35.3 (9) 54.4 (10) 35.8 (11) 21.3 (12) 40.2 (13) 41.1 (14) 56.4 (15) 32.4 (16) 81.2 (17) 64.0 (18) 16.8 (19) 12.3 (20) 40.7 (21) 16.5 (22) 110.6 (23) 37.0 (24) 28.4 (25) 34.5 (26) 75.3 (27) 17.5 **Gal** (1) 102.5 (20 73.2 (3) 75.1 (4) 80.0 (5) 75.4 (6) 60.7 **Glc I** (1) 104.9 (2) 81.4 (3) 86.7 (4) 70.5 (5) 78.7 (6) 62.5 **Glc II** (1) 105.2 (2) 75.6 (3) 78.8 (4) 70.8 (5) 77.6 (6) 63.0 **Xyl** (1) 105.2 (2) 76.3 (3) 77.7 (4) 71.0 (5) 67.4 **Glc III** (1) 105.0 (2) 75.2 (3) 78.6 (4) 71.7 (5) 78.4 (6) 62.8.

Reference

1.   Z. Meng, S. Xu, W. Li and Y. Sha, *Zhongguo Yaowu Huaxue Zazhi* (*Chin. J. Med. Chem.*), **9**, 294 (1999).

TIMOSAPONIN I$_2$

(5α,25S)-22-Methoxyfurostane-3β,26-diol 3-O-{β-D-glucopyranosyl-(1→2)-[β-D-xylopyranosyl-(1→3)]-β-D-glucopyranosyl-(1→4)-β-D-galactopyranoside}-26-O-β-D-glucopyranoside

Source : *Anemarrhena asphodeloides* Bge.[1] (Liliaceae), *Capsicum annuum* L. var. *acuminatum*[2] (Solanaceae)
Mol. Formula : C$_{57}$H$_{96}$O$_{28}$
Mol. Wt. : 1228
[α]$_D$: -52.5° (c=0.1, Pyridine)[2]
Registry No. : [274693-79-7]

PMR (C$_5$D$_5$N, 500.13 MHz)[2] : δ 0.50 (m, H-9), 0.64 (s, 3xH-19), 0.79 (s, 3xH-18), 1.01 (d, *J*=6.8 Hz, 3xH-27), 1.20 (d, *J*=6.8 Hz, 3xH-21), 3.32 (s, OC*H*$_3$), 3.62 (m, 2xH-26).

CMR (C$_5$D$_5$N, 75 MHz)[1] : δ C-1) 37.2 (2) 29.6 (3) 77.3 (4) 34.8 (5) 44.6 (6) 28.9 (7) 32.1 (8) 35.2 (9) 54.4 (10) 35.8 (11) 21.2 (12) 40.0 (13) 41.1 (14) 56.3 (15) 32.1 (16) 81.4 (17) 64.4 (18) 16.5 (19) 12.3 (20) 40.5 (21) 16.4 (22) 112.6 (23) 31.0 (24) 28.2 (25) 34.2 (26) 75.3 (27) 17.6 (OCH$_3$) 47.3 **Gal** (1) 102.4 (20 73.2 (3) 75.1 (4) 79.9 (5) 75.4 (6) 60.6 **Glc I** (1) 105.0 (2) 81.4 (3) 86.7 (4) 70.5 (5) 78.7 (6) 62.4 **Glc II** (1) 105.2 (2) 75.6 (3) 78.7 (4) 70.8 (5) 77.7 (6) 63.0 **Xyl** (1) 105.2 (2) 76.3 (3) 77.7 (4) 71.0 (5) 67.4 **Glc III** (1) 105.0 (2) 75.2 (3) 78.6 (4) 71.7 (5) 78.5 (6) 62.9.

References

1. Z. Meng, S. Xu, W. Li and Y. Sha, *Zhongguo Yaowu Huaxue Zazhi* (*Chin. J. Med. Chem.*), **9**, 294 (1999).

2. M. Iorizzi, V. Lanzotti, G. Ranalli, S. de Marino and F. Zollo, *J. Agric. Food Chem.*, **50**, 4310 (2002).

UTTROSIDE B

(25R)-5α-Furostane-3β,22α,26-triol 3-O-{β-D-glucopyranosyl-(1→2)-[β-D-xylopyranosyl-(1→3)]-β-D-glucopyranosyl-(1→4)-β-D-galactopyranoside}-26-O-β-D-glucopyranoside

Source : *Solanum nigrum* L. (Solanaceae)
Mol. Formula : C$_{56}$H$_{94}$O$_{28}$
Mol. Wt. : 1214
M.P. : 210-215°C (decomp.)
[α]$_D^{20}$: -46.0° (c=1.0, C$_5$D$_5$N)
Registry No. : [88048-09-3]

IR (KBr) : 3400-3450 (OH) cm^{-1}, no spiroketal absorption.

Reference

1.   S.C. Sharma, R. Chand, O.P. Sati and A.K. Sharma, *Phytochemistry*, **22**, 1241 (1983).

RUSCUS ACULEATUS SAPONIN 24

22-O-Methyl-(25R)-furost-5-ene-1β,3β,22ξ,26-tetrol 1-O-{β-D-glucopyranosyl-(1→3)-α-L-rhamnopyranosyl-(1→2)-3,4,6-tri-O-acetyl-β-D-galactopyranoside}-26-O-β-D-glucopyranoside

Source : *Ruscus aculeatus* L. (Liliaceae)
Mol. Formula : C$_{58}$H$_{92}$O$_{27}$
Mol. Wt. : 1220
[α]$_D^{25}$: -40.0° (c=0.10, MeOH)
Registry No. : [211379-22-5]

IR (KBr) : 3430 (OH), 2930 (CH), 1745 (C=O), 1450, 1370, 1245, 1070, 1045, 980, 910, 835, 805 cm^{-1}.

PMR (C$_5$D$_5$N, 400 MHz) : δ 0.96 (s, 3xH-18), 1.00 (d, J=6.6 Hz, 3xH-27), 1.19 (d, J=6.9 Hz, 3xH-21), 1.43 (s, 3xH-19), 1.72 (d, J=6.1 Hz, 3xH-6 of Rha), 1.94 (s, OCOCH$_3$), 1.98 (s, OCOCH$_3$) and 2.04 (s, OCOCH$_3$), 3.26 (s, OCH$_3$), 3.72 (dd, J=12.1, 3.8 Hz, H-1), 3.82 (m, H-3), 4.77 (d, J=7.7 Hz, H-1 of Gal), 4.85 (d, J=7.7 Hz, H-1 of Glc II), 5.43 (br d, J=5.3 Hz, H-6), 5.62 (d, J=7.6 Hz, H-1 of Glc I), 5.67 (br s, H-1 of Rha), 5.78 (br d, J=3.4 Hz, H-4 of Gal).

CMR (C$_5$D$_5$N, 100 MHz) : δ C-1) 85.4 (2) 38.3 (3) 68.1 (4) 43.6 (5) 139.1 (6) 124.9 (7) 31.9 (8) 33.1 (9) 50.6 (10) 42.7 (11) 24.2 (12) 40.4 (13) 40.7 (14) 57.2 (15) 32.4 (16) 81.4 (17) 64.5 (18) 16.9 (19) 15.0 (20) 40.6 (21) 16.3 (22) 112.7 (23) 30.8 (24) 28.2 (25) 34.2 (26) 75.2 (27) 17.1 (OCH$_3$) 47.2 **Gal** (1) 100.1 (2) 71.8 (3) 75.2 (4) 68.4 (5) 70.8 (6) 62.2 **Rha** (1) 101.7 (2) 71.6 (3) 82.3 (4) 72.9 (5) 69.8 (6) 18.7 **Glc I** (1) 106.5 (2) 76.1 (3) 78.4 (4) 71.7 (5) 78.4 (6)

62.5 **Glc II** (1) 105.0 (2) 75.2 (3) 78.6 (4) 71.8 (5) 78.5 (6) 62.9 (CH_3CO) 170.6, 170.2, 170.2 (CH_3CO) 20.6, 20.5, 20.3.

Mass (FAB, Negative ion) **:** m/z 1219 [M-H]$^-$.

Reference

1. Y. Mimaki, M. Kuroda, A. Kameyama, A. Yokosuka and Y. Sashida, *Phytochemistry*, **48**, 485 (1998).

COSTUSOSIDE I

(25*R*)-22α-Methoxyfurosta 5-en-3β,26-diol 3-O-{β-D-glucopyranosyl-(1→2)-α-L-rhamnopyranosyl-(1→2)-[α-L-rhamnopyranosyl-(1→4)]-β-D-glucopyranoside}-26-O-β-D-glucopyranoside

Source : *Costus speciosus* (Koen.) Sm. (Costaceae)
Mol. Formula : $C_{58}H_{96}O_{27}$
Mol. Wt. : 1224
M.P. : 220-224°C (decomp.)
[α]$_D^{25}$: -76.5° (c=1.52, C_5D_5N)
Registry No. : [82844-06-2]

IR (KBr) : 3600-3100, 1120-1000, 980, 915 (S), 895 (W), 840 cm^{-1}.

PMR (DMSO-d_6, 60.0 MHz) : δ 3.10 (s, OCH_3), 4.10 (br s, $W\frac{1}{2}$=7.0Hz), 4.20 (br s, $W\frac{1}{2}$=7.5 Hz), 4.3 (br s, $W\frac{1}{2}$=6.5 Hz), 4.8 (br s, $W\frac{1}{2}$=5.0 Hz), 5.0 (br s, $W\frac{1}{2}$=4.5 Hz), 5.25 (br s, H-6).

Reference

1.   S.B. Singh and R.S. Thakur, *Phytochemistry*, **21**, 911 (1982).

METHYL PROTO-CAUSIAROSIDE I

22ξ-Methoxy-25(*R*)-furost-5-en-3β,26-diol 3-O-{α-L-rhamnopyranosyl-(1→4)-β-D-glucopyranosyl-(1→4)-[α-L-rhamnopyranosyl-(1→2)]-β-D-glucopyranoside}-26-O-β-D-glucopyranoside

Source : *Sabal causiarum* Becc. (Palmae)
Mol. Formula : $C_{58}H_{96}O_{27}$
Mol. Wt. : 1224
M.P. : 192-194°C (decomp.)
[α]$_D^{20}$: -72.6° (c=1.02, C_5D_5N)
Registry No. : [116424-72-7]

IR (KBr) : 3600-3200 (OH) cm^{-1}.

CMR (C$_5$D$_5$N, 25/100 MHz) : δ C-1) 37.9 (2) 30.6 (3) 78.8 (4) 39.4 (5) 140.7 (6) 121.8 (7) 32.6 (8) 32.1 (9) 50.7 (10) 37.5 (11) 21.5 (12) 40.2 (13) 41.2 (14) 56.9 (15) 32.7 (16) 81.5 (17) 64.5 (18) 16.8 (19) 19.9 (20) 40.9 (21) 16.8 (22) 112.7 (23) 31.2 (24) 28.7 (25) 34.7 (26) 75.4 (27) 17.7 (22-OCH$_3$) 47.7 **Glc I** (1) 100.1 (2) 78.4 (3) 76.3 (4) 82.0 (5) 77.8 (6) 61.3 **Rha I** (1) 101.7 (2) 72.6 (3) 72.9 (4) 74.1 (5) 77.8 (6) 19.1 **Glc II** (1) 105.0 (2) 75.4 (3) 77.3 (4) 77.4 (5) 76.6 (6) 62.1 **Rha II** (1) 102.7 (2) 72.7 (3) 73.0 (4) 74.3 (5) 70.6 (6) 19.0 **Glc III** (1) 105.0 (2) 75.3 (3) 78.1 (4) 72.0 (5) 78.6 (6) 63.2.

Reference

1.   K. Idaka, Y. Hirai and J. Shoji, *Chem. Pharm. Bull.*, **36**, 1783 (1988).

TACCA CHANTRIERI SAPONIN 3

(25S)-3β,26-Dihydroxy-22α-methoxyfurost-5-ene 3-O-{α-L-rhamnopyranosyl-(1→2)-[β-D-glucopyranosyl-(1→4)-α-L-rhamnopyranosyl-(1→3)]-β-D-glucopyranoside}-26-O-β-D-glucopyranoside

Source : *Tacca chantrieri* André (Taccaceae)
Mol. Formula : C$_{58}$H$_{96}$O$_{27}$
Mol. Wt. : 1224
[α]$_D^{25}$: -82.0° (c=0.10, CHCl$_3$-MeOH)
Registry No. : [290809-76-6]

IR (KBr) : 3400 (OH), 2930 (CH)< 1040 cm^{-1}.

PMR (C₅D₅N, 500 MHz) **:** δ 0.82 (s, 3xH-18), 1.04 (s, 3xH-19), 1.05 (d, J=6.7 Hz, 3xH-27), 1.17 (d, J=6.9 Hz, 3xH-21), 1.69 (d, J=6.2 Hz, 3xH-6 of Rha II), 1.76 (d, J=6.2 Hz, 3xH-6 of Rha I), 3.26 (s, OCH_3), 3.78 (ddd, J=9.1, 4.5, 2.3 Hz, H-5 of Glc I), 3.78 (ddd, J=9.4, 4.9, 2.3 Hz, H-5 of Glc II), 3.91 (m, $W\frac{1}{2}$=26.3 Hz, H-3), 3.97 (ddd, J=9.1, 5.3, 2.5 Hz, H-5 of Glc III), 4.05 (dd, J=8.7, 7.8 Hz, H-2 of Glc I), 4.05 (dd, J=8.7, 7.7 Hz, H-2 of Glc III), 4.08 (dd, J=9.1, 9.1 Hz, H-4 of Glc I), 4.09 (dd, J=8.8, 7.8 Hz, H-2 of Glc II), 4.17 (dd, J=9.1, 8.7 Hz, H-3 of Glc I), 4.21 (dd, J=9.0, 8.8 Hz, H-3 of Glc II), 4.22 (dd, J=9.1, 9.1 Hz, H-4 of Glc III), 4.25 (dd, J=9.1, 8.7 Hz, H-3 of Glc III), 4.27 (dd, J=9.4, 9.0 Hz, H-4 of Glc II), 4.31 (dd, J=9.7, 9.3 Hz, H-4 of Rha I), 4.34 (dd, J=11.5, 4.9 Hz, H-6A of Glc II), 4.36 (dd, J=11.8, 4.5 Hz, H-6A of Glc I), 4.39 (dd, J=11.8, 5.3 Hz, H-6A of Glc III), 4.41 (dd, J=11.5, 2.3 Hz, H-6B of Glc II), 4.43 (dd, J=9.1, 9.1 Hz, H-4 of Rha II), 4.44 (dd, J=11.8, 2.3 Hz, H-6B of Glc I), 4.45 (m, H-16), 4.51 (dd, J=9.3, 3.2 Hz, H-3 of Rha I), 4.57 (dd, J=9.1, 3.3 Hz, H-3 of Rha II), 4.57 (dd, J=11.8, 2.5 Hz, H-6B of Glc III), 4.73 (br d., J=3.2 Hz, H-2 of Rha I), 4.80 (br d, J=3.3 Hz, H-2 of Rha II), 4.83 (dq, J=9.1, 6.2 Hz, H-5 of Rha II), 4.85 (d, J=7.7 Hz, H-1 of Glc III), 4.87 (dq, J=9.7, 6.2 Hz, H-5 of Rha I), 4.89 (d, J=7.8 Hz, H-1 of Glc I), 5.24 (d, J=7.8 Hz, H-1 of Glc II), 5.33 (br d, J=4.6 Hz, H-6), 5.75 (br s, H-1 of Rha II), 5.81 (br s, H-1 of Rha I).

CMR (C₅D₅N, 125 MHz) **:** δ C-1) 37.5 (2) 30.0 (3) 77.8 (4) 38.6 (5) 140.8 (6) 121.8 (7) 32.2 (8) 31.6 (9) 50.2 (10) 37.1 (11) 21.0 (12) 39.7 (13) 40.8 (14) 56.5 (15) 32.3 (16) 81.3 (17) 64.1 (18) 16.3 (19) 19.3 (20) 40.4 (21) 16.2 (22) 112.7 (23) 30.9 (24) 28.1 (25) 34.4 (26) 74.9 (27) 17.5 (OCH_3) 47.3 **Glc I** (1) 99.8 (2) 78.6 (3) 86.2 (4) 69.7 (5) 78.0 (6) 62.2 **Rha I** (1) 102.6 (2) 72.5 (3) 72.8 (4) 73.8 (5) 69.9 (6) 18.6 **Rha II** (1) 103.1 (2) 72.0 (3) 72.3 (4) 84.4 (5) 68.7 (6) 18.3 **Glc II** (1) 106.5 (2) 76.4 (3) 78.5 (4) 71.4 (5) 78.4 (6) 62.5 **Glc III** (1) 105.0 (2) 75.2 (3) 78.6 (4) 71.7 (5) 78.5 (6) 62.9.

Mass (FAB, Negative ion) **:** m/z 1223 [M-H]⁻.

Reference

1.   A. Yokosuka, Y. Mimaki and Y. Sashida, *J. Nat. Prod.*, **65**, 1293 (2002).

AVENACOSIDE B

Nuatigenin 3-O-{β-D-glucopyranosyl-(1→3)-β-D-glucopyranosyl-(1→2)-[α-L-rhamnopyranosyl-(1→4)]-β-D-glucopyranoside}-26-O-β-D-glucopyranoside

Source : *Avena sativa* L.[1] (Gramineae)
Mol. Formula : C₅₇H₉₂O₂₈
Mol. Wt. : 1224
$[\alpha]_D^{24}$ **:** +52.0° (c=1.0, H₂O)[2]
Registry No. : [35920-91-3]

References

1.   R. Tschesche, M. Tauscher, H.-W. Fehlhaber and G. Wulff, *Chem. Ber.*, **102**, 2072 (1969).

2.   R. Tschesche and W. Schmidt, *Z. Naturforsch.*, **21B**, 896 (1966).

3.   R. Tschesche and P. Lauven, *Chem. Ber.*, **104**, 3549 (1971).

MACROSTEMONOSIDE E

(25*R*)-5α-Furost-20-(22)-ene-3β,26-diol 3-O-{β-D-glucopyranosyl-(1→2)-[β-D-glucopyranosyl-(1→3)]-β-D-glucopyranosyl-(1→4)-β-D-galactopyranoside}-26-O-β-D-glucopyranoside

Source : *Allium macrostemon* Bunge (Liliaceae)
Mol. Formula : $C_{57}H_{94}O_{28}$
Mol. Wt. : 1226
M.P. : 227.5-230°C
Registry No. : [151140-39-5]

IR (KBr) : 3400 (OH), 2900, 2850, 1725, ($\Delta^{20(22)}$), 1090, 1070, 1020(C-O) cm^{-1}.

PMR (C$_5$D$_5$N, 400 MHz) : δ 0.67 (s, 3xH-18), 0.72 (s, 3xH-19), 1.02 (d, J=6.6 Hz, 3xH-27), 1.64 (s, 3xH-21), 4.81 (d, J=7.8 Hz, H-1 of Glc), 4.85 (d, J=7.6 Hz, H-1 of Gal), 5.12 (d, J=8.0 Hz, H-1 of Glc), 5.26 (d, J=7.8 Hz, H-1 of Glc), 5.54 (d, J=7.6 Hz, H-1 of Glc).

CMR (C$_5$D$_5$N, 100 MHz) : δ C-1) 37.3 (2) 29.9 (3) 77.6 (4) 34.9 (5) 44.8 (6) 28.9 (7) 32.6 (8) 35.1 (9) 54.8 (10) 35.8 (11) 21.5 (12) 40.0 (13) 43.8 (14) 54.5 (15) 31.5 (16) 84.5 (17) 64.7 (18) 14.4 (19) 12.3 (20) 103.6 (21) 11.8 (22) 152.5 (23) 34.4 (24) 23.7 (25) 33.5 (26) 74.9 (27) 17.3 **Gal** (1) 102.5 (2) 73.2 (3) 75.6^a (4) 80.1 (5) 76.0 (6) 60.6 **Glc I** (1) 104.8^b (2) 81.4 (3) 88.5 (4) 70.8^c (5) 77.9^d (6) 62.4^e **Glc II** (1) 104.8^b (2) 75.3^a (3) 78.6 (4) 71.8^e (5) 77.5^d (6) 62.4^e **Glc III** (1) 104.5^b (2) 75.3^a (3) 78.5 (4) 71.6^c (5) 78.3 (6) 62.9^e **Glc IV** (1) 104.9^b (2) 75.2^a (3) 78.6 (4) 71.1^c (5) 78.5 (6) 63.0^e.

Mass (FAB, Negative ion) : m/z 1225 [M-H]$^-$, 1063 [M-Glc-H]$^-$, 901 [M-Glc x2H]$^-$, 577 [M-Glcx4H]$^-$.

Biological Activity: The compound strongly inhibits ADP-induced human platelet aggregation *in vitro* with 1C$_{50}$=0.417 mmol.

Reference

1.   J. Peng, X. Wang and X. Yao, *Xaoxue Xuebao* (*Acta Pharm. Sinica*), **28**, 526 (1993).

METHYL PROTOASPIDISTRIN

(25*R*)-22ξ-Methoxyfurost-5-en-3β,26-diol 3-O-{β-D-glucopyranosyl-(1→2)-[β-D-xylopyranosyl-(1→3)]-
β-D-glucopyranosyl-(1→4)-β-D-glucopyranoside}-26-O-β-D-glucopyranoside

Source : *Aspidistra elatior* Blume[1] (Liliaceae), *Reineckea carnea* Kunth[2] (Liliaceae)
Mol. Formula : $C_{57}H_{94}O_{28}$
Mol. Wt. : 1226
M.P. : 202-207°C[1]
[α]$_D^{51.5}$: -63.4° (c=1.01, C_5D_5N)[1]
Registry No. : [90352-17-3]

IR (KBr)[1] : 3600-3300 (OH) cm^{-1}.

PMR (C_5D_5N, 100 MHz)[1] : δ 0.85 (s, CH_3), 0.92 (s, CH_3), 3.30 (s, OCH_3).

CMR (C_5D_5N, 25.0 MHz)[1] : δ C-1) 37.6 (2) 30.2 (3) 78.6 (4) 39.4 (5) 141.2 (6) 121.6 (7) 32.4 (8) 31.8 (9) 50.5 (10) 37.2 (11) 21.2 (12) 39.9 (13) 40.9 (14) 56.8 (15) 32.4 (16) 81.2 (17) 64.2 (18) 16.3 (19) 19.5 (20) 40.6 (21) 16.2 (21) 112.8 (22) 30.9 (23) 28.3 (24) 34.3 (25) 75.3^a (26) 17.2 (27) 47.4 **Gal** (1) 102.8 (2) 73.1 (3) 75.3^a (4) 79.7 (5) 76.0 (6) 60.7 **Glc I** (1) 104.9^b (2) 81.2 (3) 87.2 (4) 70.4 (5) 77.9 (6) 62.7 **Glc II** (1) 104.9^b (2) 75.1^a (3) 78.6 (4) 70.5 (5) 77.5 (6) 63.0^c **Xyl** (1) 104.9^b (2) 75.6 (3) 78.6 (4) 71.3 (5) 67.2 **Glc III** (1) 104.9^b (2) 75.1^a (3) 78.6 (4) 71.9 (5) 78.2 (6) 63.1^c.

Biological Activity : The compound shows inhibitory activity on CAMP phosphodiesterase with $IC_{50}=32.2 \times 10^{-5}$ M.

References

1. Y. Hirai, T. Konishi, S. Sanada, Y. Ida and J. Shoji, *Chem. Pharm. Bull.,* **30,** 3476 (1982).

2. T. Kanmoto, Y. Mimaki, Y. Sashida, T. Nikaido, K. Koike and T. Ohmoto, *Chem. Pharm. Bull.,* **42**, 926 (1994).

SIBIRICOSIDE A
22-Methoxy-25(S)-furost-5-en-3β,26-diol 3-O-{β-D-glucopyranosyl-(1→2)-[β-D-xylopyranosyl-(1→3)]-β-D-glucopyranosyl-(1→4)-β-D-galactopyranoside}-26-O-β-D-glucopyranoside

Source : *Polygonatum sibiricum* Redoute (Liliaceae)
Mol. Formula : $C_{57}H_{94}O_{28}$
Mol. Wt. : 1226
M.P. : 218-221°C
[α]$_D^{20}$: -51.0° (c=0.15, MeOH)
Registry No. : [128820-35-9]

IR (KBr) : 3400, 1630, 1150, 1030, 1010, 885, 860 cm^{-1}.

PMR (C$_5$D$_5$N, 200 MHz) : δ 0.69 (s, 3xH-18), 0.75 (s, 3xH-19), 0.94 (d, *J*=6.3 Hz, 3xH-27), 1.06 (d, *J*=6.7 Hz, 3xH-21), 3.14 (s, OC*H*$_3$), 4.75 (d, *J*=7.8 Hz, anomeric H), 5.09 (d, *J*=7.8 Hz, anomeric H), 5.14 (d, *J*=7.8 Hz, anomeric H), 5.34(m, H-6), 5.47 (d, *J*=6.8 Hz, anomeric H).

CMR (C$_5$D$_5$N, 50.3 MHz) : δ C-1) 37.6 (2) 30.3 (3) 78.8 (4) 39.4 (5) 141.2 (6) 121.8 (7) 32.4 (8) 31.8 (9) 50.4 (10) 37.1 (11) 21.2 (12) 39.9 (13) 40.9 (14) 56.7 (15) 32.4 (16) 81.5 (17) 64.3 (18) 16.4 (19) 19.5 (20) 40.6 (21) 16.5 (22) 112.9 (23) 31.2 (24) 28.4 (25) 34.7 (26) 75.1 (27) 17.7 (OCH$_3$) 47.5 **Gal** (1) 102.9 (2) 73.4 (3) 75.4 (4) 80.1 (5) 76.4 (6) 60.7 **Glc I** (1) 105.1 (2) 81.5 (3) 86.9 (4) 70.9 (5) 78.8 (6) 62.6 **Xyl** (1) 105.3 (2) 75.7 (3) 77.8 (4) 70.7 (5) 67.5 **Glc II** (1) 105.3 (2) 75.4 (3) 78.3 (4) 71.2 (5) 77.8 (6) 63.1 **Glc III** (1) 105.3 (2) 75.4 (3) 78.8 (4) 71.9 (5) 78.3 (6) 63.1.

Reference

1.   K.H. Son, J.C. Do and S.S. Kang, *J. Nat. Prod.*, **53**, 333 (1990).

TIMOSAPONIN B-IV
(25*S*)-5β-Furost-20(22)-ene-3β,26-diol 3-O-{β-D-glucopyranosyl-(1→2)-[β-D-glucopyranosyl-(1→3)]-β-D-glucopyranosyl-(1→4)-β-D-galactopyranoside}-26-O-β-D-glucopyranoside

Source : *Anemarrhena asphodeloides* Bge. (Liliaceae)
Mol. Formula : C$_{57}$H$_{94}$O$_{28}$
Mol. Wt. : 1226
M.P. : 243°C (decomp.)
Registry No. : [215050-64-9]

CMR : δ 153.5 (C-20), 103.5 (C-22).

Reference

1.   Z. Meng and S. Xu, *Shenyang Yaoke Daxue Xuebao* (*J. Shenyang Pharm. Univ.*) **15**, 130 (1998).

TIMOSAPONIN H₂

(25S)-22-Methoxyfurost-5-ene-3β,26-diol 3-O-{β-D-glucopyranosyl-(1→2)-[β-D-xylopyranosyl-(1→3)]-β-D-glucopyranosyl-(1→4)-β-D-galactopyranoside}-26-O-β-D-glucopyranoside

Source : *Anemarrhena asphodeloides* Bge. (Liliaceae)
Mol. Formula : $C_{57}H_{94}O_{28}$
Mol. Wt. : 1226
Registry No. : [274693-15-1]

CMR (C_5D_5N, 75 MHz) **:** δ C-1) 37.5 (2) 30.0 (3) 78.6 (4) 39.2 (5) 141.0 (6) 121.6 (7) 32.1 (8) 31.6 (9) 50.2 (10) 37.0 (11) 21.0 (12) 39.7 (13) 40.8 (14) 56.5 (15) 32.2 (16) 81.4 (17) 64.2 (18) 16.4 (19) 19.4 (20) 40.5 (21) 16.3 (22) 112.7 (23) 31.0 (24) 28.2 (25) 34.5 (26) 75.2 (27) 17.6 (OCH₃) 47.4 **Gal** (1) 102.7 (2) 73.2 (3) 75.3 (4) 79.8 (5) 76.2 (6) 60.6 **Glc I** (1) 104.9 (2) 81.4 (3) 86.7 (4) 70.7 (5) 77.6 (6) 62.5 **Glc II** (1) 105.1 (2) 75.0 (3) 78.6 (4) 70.5 (5) 77.6 (6) 62.9 **Xyl** (1) 105.1 (2) 75.5 (3) 78.7 (4) 71.0 (5) 67.4 **Glc III** (1) 104.9 (2) 75.1 (3) 78.7 (4) 71.8 (5) 78.2 (6) 62.9.

Reference

1.   Z. Meng, S. Xu, W. Li and Y. Sha, *Zhongguo Yaowu Huaxue Zazhi* (*Chin. J. Med. Chem.*), **9**, 294 (1994).

TRIGONEOSIDE XIIIa

(25S)-Furost-5-ene-3β,22ξ,26-triol 3-O-[α-L-rhamnopyranosyl-(1→2)-[β-D-glucopyranosyl-(1→3)-
β-D-glucopyranosyl-(1→4)]-β-D-glucopyranoside]-26-O-β-D-glucopyranoside

Source : *Trigonella foenum-graecum* L. (Fabaceae)
Mol. Formula : $C_{57}H_{94}O_{28}$
Mol. Wt. : 1226
$[\alpha]_D^{26}$: -31.4° (c=0.5, MeOH)
Registry No. : [290348-13-9]

IR (KBr) : 3410, 2934, 1072, 1006 cm^{-1}.

PMR (C$_5$D$_5$N, 500 MHz) : δ 0.90, 1.06 (both s, 3xH-18, 3xH-19), 1.03 (d, J=6.7 Hz, 3xH-27), 1.31 (d, J=6.7 Hz, 3xH-21), 1.74 (d, J=6.1 Hz, 3xH-6 of Rha), 3.49, 4.06 (both dd-like, 2xH-26), 3.87 (dd-like, H-3), 4.78 (d, J=7.9 Hz, H-1 of Glc IV), 4.91 (d, J=5.9 Hz, H-1 of Glc I), 4.93 (dd-like, H-16), 5.05 (d, J=8.0 Hz, H-1 of Glc II), 5.22 (d, J=6.7 Hz, H-1 of Glc III), 5.30 (br. s, H-6), 6.17 (br s, H-1 of Rha).

CMR (C$_5$D$_5$N, 125.0 MHz) : δ C-1) 37.6 (2) 30.2 (3) 78.4 (4) 39.0 (5) 140.9 (6) 121.8 (7) 32.4 (8) 31.8 (9) 50.5 (10) 37.2 (11) 21.2 (12) 40.0 (13) 40.8 (14) 56.7 (15) 32.5 (16) 81.2 (17) 63.9 (18) 16.5 (19) 19.4 (20) 40.7 (21) 16.4 (22) 110.7 (23) 37.2 (24) 28.3 (25) 34.4 (26) 75.5 (27) 17.4 **Glc I** (1) 100.1 (2) 77.3 (3) 76.2 (4) 81.5 (5) 77.6 (6) 61.7 **Rha** (1) 101.7 (2) 72.4 (3) 72.8 (4) 74.2 (5) 69.4 (6) 18.6 **Glc II** (1) 104.5 (2) 73.7 (3) 88.3 (4) 69.4 (5) 78.4 (6) 61.8 **Glc III** (1) 105.8 (2) 75.2 (3) 78.3 (4) 71.8 (5) 78.6 (6) 62.6 **Glc IV** (1) 105.1 (2) 75.2 (3) 78.6 (4) 71.7 (5) 78.4 (6) 62.9.

Mass (FAB, Positive ion) : m/z 1249 [M+Na]$^+$.

Mass (FAB, Negative ion) : m/z 1225 [M-H]$^-$.

Reference

1. T. Murakami, A. Kishi, H. Matsuda and M. Yoshikawa, *Chem. Pharm. Bull.,* **48**, 994 (2000).

FILICINOSIDE C

(25*S*)-22α-Methoxy-5β-furostan-3β,26-diol 3-O-{β-D-galactopyranosyl-(1→4)-β-D-xylopyranosyl-(1→6)]-
β-D-glucopyranosyl-(1→4)-β-D-glucopyranoside}-26-O-β-D-glucopyranoside

Source : *Asparagus filicinus* Buch.-Ham. (Liliaceae)
Mol. Formula : $C_{57}H_{96}O_{28}$
Mol. Wt. : 1228
M.P. : 179-183°C
[α]$_D^{20}$: -46.0° (MeOH)
Registry No. : [173356-81-5]

Reference

1. S.C. Sharma and N.K. Thakur, *Phytochemistry*, **41**, 599 (1996).

SARSAPARILLOSIDE
(25*S*)-Furostan-3β,22α,26-triol 3-*O*-{β-D-glucopyranosyl-(1→2)-[α-D-rhamnopyranosyl-(1→4)]-
β-D-glucopyranosyl-(1→6)-β-D-glucopyranoside}-26-*O*-β-D-glucopyranoside

Source : *Smilax aristolochiaefolia* Mill. (Liliaceae)
Mol. Formula : $C_{57}H_{96}O_{28}$
Mol. Wt. : 1228
$[\alpha]_D^{20}$: -44° (c=0.86, H_2O)
Registry No. : [24333-07-1]

Reference

1.   R. Tschesche, G. Lüdke and G. Wulff., *Tetrahedron Lett.*, 2785 (1967); *Chem. Ber.*, **102**, 1253 (1969).

UTTROSIDE A

(25*R*)-22α-Methoxy-5α-furostane-3β,26-diol 3-O-{β-D-glucopyranosyl-(1→2)-[β-D-xylopyranosyl-(1→3)]-β-D-glucopyranosyl-(1→4)-β-D-galactopyranoside}-26-O-β-D-glucopyranoside

Source : *Solanum nigrum* L.[1,2] (Solanaceae), *Digitalis cariensis* Boiss ex Jaub & Spach[3] (Scrophulariaceae)
Mol. Formula : $C_{57}H_{96}O_{28}$
Mol. Wt. : 1228
M.P. : 220-225°C (decomp.)[1]
[α]$_D^{20}$: -49.0° (c=1.0, MeOH)[1]
Registry No. : [82003-86-9]

IR (KBr)[1] : 3400 (OH) cm⁻¹, no spiroketal absorption's.

PMR (CD₃OD, 400 MHz)[3] : δ 0.68 (m, H-9), 0.81 (s, 3xH-18), 0.85 (s, 3xH-19), 0.93* (H-4A), 0.94 (d, *J*=6.6 Hz, 3xH-26), 0.98* (H-1A), 0.99 (d, *J*=7.0 Hz, 3xH-21), 1.09* (H-5), 1.12* (H-14), 1.13* (H-12, H-24), 1.23* (H-15), 1.29* (H-23A), 1.30* (H-11), 1.32 (m, H-6), 1.51* (H-2A), 1.53* (H-11B), 1.57 (H-8), 1.59* (H-24B), 1.62* (H-7), 1.65* (H-23B), 1.70* (H-4 and H-17), 1.72* (H-1B, H-12B), 1.73* (H-25), 1.81* (H-7B), 1.86 (H-2B), 1.95* (H-15B), 2.16 (m, H-20), 3.13 (s, OC*H₃*), 3.18 (dd, *J*=7.8, 8.5 Hz, H-2 of Glc III), 3.23* (H-2 of Xyl), 3.24 (H-5 of Glc II), 3.25* (H-4 of Xyl, H-5A of Xyl, H-4 of Glc III), 3.30* (H-3 of Glc II), 3.32* (H-3 of Glc III), 3.33 (H-5 of Glc III), 3.37* (H-26A), 3.46* (H-5 of Glc I), 3.48 (H-3 of Gal), 3.49* (H-5 of Gal), 3.50* (H-4 of Glc II), 3.55* (H-2 of

Glc II and H-6A), 3.56* (H-3 of Xyl), 3.58 (H-6A of Gal), 3.65* (H-6A of Glc III), 3.66* (H-2 of Gal), 3.67 (H-3), 3.70* (H-26B, H-6A of Glc I and H-3 of Glc I), 3.79 (dd, J=8.2, 8.6 Hz, H-2 of Glc I), 3.85* (H-6B of Glc II), 3.86* (H-4 of Glc I), 3.87* (H-6B of Glc II), 3.80* (H-6B of Gal), 3.90 (dd, J=11.9, 5.4 Hz, H-5B of Xyl), 3.93* (HB-6B of Glc I), 4.00 (br d, J=3.1 Hz, H-4 of Gal), 4.23 (d, J=7.8 Hz, H-1 of Glc III), 4.35 (dd, J=14.5, 7.4 Hz, H-16), 4.37 (d, J=7.8 Hz, H-1 of Gal), 4.58 (d, J=8.2 Hz, H-1 of Glc I), 4.60 (d, J=8.2 Hz, H-1 of Xyl), 4.87 (d, J=7.8 Hz, H-1 of Glc II). * Multiplicity unclear due to overlapping.

CMR (CD$_3$OD, 100 MHz)[3] : δ C-1) 38.2 (2) 30.4 (3) 79.4 (4) 33.5 (5) 46.0 (6) 29.9 (7) 31.4 (8) 36.5 (9) 55.8 (10) 36.8 (11) 22.1 (12) 41.1 (13) 42.1 (14) 57.5 (15) 32.7 (16) 82.4 (17) 65.2 (18) 17.0 (19) 12.8 (20) 41.2 (21) 165.1 (22) 113.9 (23) 35.4 (24) 28.9 (25) 35.0 (26) 76.0 (27) 17.3 (OCH$_3$) 47.6 **Gal** (1) 102.6 (2) 72.9 (3) 75.3 (4) 80.1 (5) 75.7 (6) 61.0 **Glc I** (1) 104.9 (2) 80.9 (3) 87.5 (4) 70.5 (5) 74.8 (6) 62.4 **Glc II** (1) 104.8 (2) 73.4 (3) 77.5 (4) 71.0 (5) 78.1 (6) 63.2 **Xyl** (1) 104.9 (2) 75.2 (3) 77.2 (4) 70.5 (5) 67.2 **Glc III** (1) 104.6 (2) 75.2 (3) 77.9 (4) 71.7 (5) 78.3 (6) 62.8.

Mass (F.D.)[2] : m/z 1214 [M-CH$_2$]$^+$.

Mass (E.S.I.)[3] : m/z 1267 [M+K]$^+$, 1251 [M+Na]$^+$.

References

1. S.C. Sharma, R. Chand, O.P. Sati and A.K. Sharma, *Phytochemistry*, **22**, 1241 (1983).

2. R. Saijo, K. Murakami, T. Nohara, T. Tomimatsu, A. Sato and K. Matsuoka, *Yakugaku Zasshi*, **102**, 300 (1982).

3. H. Kirmizibekmez, D. Tasdemir, T. Ersöz, C.M. Ireland and I. Calis, *Pharmazie*, **57**, 716 (2002).

KARATAVIOSIDE C

(25R)-Furost-5-ene-2α,3β,22α,26-tetraol 3-O-{β-D-glucopyranosyl-(1→2)-[β-D-xylopyranoside-(1→3)]-β-D-glucopyranosyl-(1→4)-β-D-galactopyranoside-26-O-β-D-glucopyranoside

Source : *Allium karataviense* Rgl. (Liliaceae)
Mol. Formula : C$_{56}$H$_{92}$O$_{29}$
Mol. Wt. : 1228
M.P. : 241-245°C (decomp.)
[α]$_D^{20}$: -54.4±2° (c=1.14, Pyridine)
Registry No. : [75080-54-7]

IR (KBr) : 3300-3500 (OH), 895 (weak, br.) cm⁻¹.

Reference

1.   Y.S. Vollerner, M.B. Gorovits, T.T. Gorovits and N.K. Abubakirov, *Khim. Prir. Soedin.,* 355 (1980); *Chem. Nat. Comp.,* **16**, 264 (1980).

POLIANTHOSIDE D

(5α,25*R*)-Furost-3β,22α,26-triol-12-one 3-O-{β-D-glucopyranosyl-(1→2)-[β-D-xylopyranosyl-(1→3)]-
β-D-glucopyranosyl-(1→4)-β-D-galactopyranoside}-26-O-β-D-glucopyranoside

Source : *Polianthes tuberosa* L. (Agavaceae)
Mol. Formula : $C_{56}H_{92}O_{29}$
Mol. Wt. : 1228
[α]$_D^{18.1}$: -23.21° (c=0.0474, Pyridine)
Registry No. : [655233-63-9]

IR (KBr) : 3407, 2927, 1704, 1373, 1160, 1071, 1040, 894 cm^{-1}.

PMR (C$_5$D$_5$N, 500 MHz) : δ 0.65 (s, 3xH-18), 0.96 (d, J=6.3 Hz, 3xH-27), 1.11 (s, 3xH-19), 1.51 (d, J=6.0 Hz, 3xH-21), 3.77 (H-3), 3.82 (m, H-26A), 4.21 (H-26B), 4.47 (m, H-16), 4.77 (d, J=7.7 Hz, H-1 of Glc III), 4.84 (d, J=6.6 Hz, H-1 of Gal), 5.14 (d, J=6.9 Hz, H-1 of Glc I), 5.18 (d, J=7.4 Hz, H-1 of Xyl), 5.51 (d, J=6.3 Hz, H-1 of Glc II).

CMR (C$_5$D$_5$N, 125 MHz) : δ C-1) 36.8 (2) 29.8 (3) 77.8 (4) 34.8 (5) 44.6 (6) 28.7 (7) 31.9 (8) 34.5 (9) 55.8 (10) 36.4 (11) 38.2 (12) 213.3 (13) 55.9 (14) 56.0 (15) 31.9 (16) 79.8 (17) 55.2 (18) 16.4 (19) 11.9 (20) 41.3 (21) 15.4 (22) 111.0 (23) 37.2 (24) 28.5 (25) 34.4 (26) 75.3 (27) 17.6 **Gal** (1) 102.5 (2) 73.3 (3) 75.6 (4) 80.0 (5) 75.4 (6) 60.8 **Glc I** (1) 105.2 (2) 81.4 (3) 87.0 (4) 70.6 (5) 77.6 (6) 63.1 **Glc II** (1) 104.9 (2) 76.2 (3) 77.8 (4) 71.1 (5) 78.6 (6) 62.5 **Xyl** (1) 105.0 (2) 75.2 (3) 78.6 (4) 70.8 (5) 67.4 **Glc III** (1) 105.0 (2) 75.5 (3) 78.5 (4) 71.8 (5) 78.6 (6) 62.9.

Mass (FAB, Negative ion) : m/z 1228 [M]$^-$, 1066 [M-Glc]$^-$, 771 [M-H-Xyl-2xGlc]$^-$.

Mass (FAB, Negative ion) : m/z 1227.5735 [(M-H)$^-$, calcd. for 1227.5646].

Biological Activity: Cytotoxic against HeLa cells with IC$_{50}$ 7.86 μg/mL.

Reference

1.   J.-M. Jin, Y.-J. Zhang and C.-R. Yang, *J. Nat. Prod.*, **67**, 5 (2004).

TRIBULUS TERRESTRIS SAPONIN 3

(5α,25ξ)-Furost-3β,22ξ,26-triol-22-one 3-O-{β-D-xylopyranosyl-(1→3)}-[β-D-galactopyranosyl-(1→2)]-
β-D-glucopyranosyl-(1→4)-β-D-glucopyranoside}-26-O-β-D-glucopyranoside

Source : *Tribulus terrestris* L. (Zygophyllceae)
Mol. Formula : $C_{56}H_{92}O_{29}$
Mol. Wt. : 1228
M.P. : 211-213°C
[α]$_D$: -46.7° (c=0.01, Pyridine)
Registry No. : [180050-92-4]

UV : λ_{max} 199, 240, 255 nm.

IR (KBr) : 3400 (br) 1705 cm^{-1} (CO).

PMR (C_5D_5N, 400/500 MHz) : δ 0.62 (s, 3xH-19), 0.70 (H-1A), 0.71 (H-7A), 0.82 (H-5), 0.92 (d, *J*=6.5 Hz, 3xH-27), 1.10 (H-6), 1.10 (s, 3xH-18), 1.21 (H-1B), 1.25 (H-14), 1.40 (H-2A),1.45 (H-4A), 1.51 (H-15A), 1.52 (H-7B), 1.52 (d, *J*=6.4 Hz, 3xH-21), 1.65 (H-4B), 1.79 (H-8), 1.82 (H-23), 1.89 (H-9), 1.90 (H-25), 1.98 (H-15B), 2.00 (H-24) 2.01 (H-2B), 2.17 (H-20), 2.20 (H-11A), 2.35 (H-11B), 2.88 (H-17), 3.57 (H-26A), 3.58 (H-5A of Xyl), 3.75 (H-4 of Glc II), 3.88 (H-5 of Glc II), 3.90 (H-3), 3.90 (H-26B), 3.90 (H-3 of Gal), 3.92 (H-2 of Xyl), 3.97 (H-4 of Glc III), 4.01 (H-5 of Glc III and H-4 of Gal), 4.02 (H-6A of Glc II), 4.04 (H-2 of Glc III), 4.07 (H-4 of Xyl) 4.09 (H-3 of Glc I), 4.11 (H-

3 of Glc II), 4.17 (H-6A of Gal), 4.19 (H-5B of Xyl), 4.20 (H-3 of Xyl), 4.22 (H-3 of Glc III and H-4 of Gal), 4.28 (H-2 of Glc I), 4.39 (H-6A of Glc III), 4.49 (H-6A of Glc I), 4.49 (H-6B of Glc II), 4.51 (H-2 of Glc II), 4.55 (H-4 of Glc I), 4.56 (H-6B of Glc I), 4.60 (H-2 of Gal), 4.66 (H-6B of Glc III), 4.68 (H-6B of Gal), 4.79 (d, J=7.7 Hz, H-1 of Glc III), 4.85 (d, J=7.2 Hz, H-1 of Glc I), 4.90 (H-16), 5.10 (d, J=7.2 Hz, H-1 of Xyl), 5.15 (d, J=7.6 Hz, H-1 of Glc II), 5.45 (d, J=7.7 Hz, H-1 of Gal).

CMR (C_5D_5N, 100/150 MHz) : δ C-1) 36.49 (2) 29.62 (3) 77.28 (4) 34.57 (5) 44.22 (6) 28.46 (7) 31.55 (8) 34.13 (9) 55.32 (10) 36.13 (11) 37.88 (12) 213.06 (13) 55.70 (14) 55.48 (15) 31.56 (16) 79.56 (17) 55.68 (18) 16.14 (19) 11.57 (20) 41.44 (21) 15.18 (22) 110.5 (23) 34.45 (24) 28.27 (25) 24.13 (26) 74.90 (27) 17.32 **Gal** (1) 105.25 (2) 73.67 (3) 73.88 (4) 70.19 (5) 77.29 (6) 60.43 **Glc I** (1) 102.22 (2) 72.96 (3) 75.01 (4) 79.48 (5) 74.85 (6) 62.28 **Glc II** (1) 105.43 (2) 85.01 (3) 85.36 (4) 70.63 (5) 77.41 (6) 62.96 **Xyl** (1) 104.67 (2) 74.16 (3) 78.38 (4) 70.48 (5) 67.11 **Glc III** (1) 104.80 (2) 75.19 (3) 78.48 (4) 71.56 (5) 78.49 (6) 62.68.

Mass (FAB, Positive ion) : m/z (rel.intens.) 1267 [$(M+K)^+$, 26], 1251 [$(M+Na)^+$, 28].

Reference

1.   G. Wu, S. Jiang, F. Jiang, D.Zhu, H.Wu and S. Jiang, *Phytochemistry*, **42**, 1677 (1996).

PURPUREAGITOSIDE
(5α,25*R*)-Furostane-2α,3β,22α,26-tetrol 3-O-{β-D-glucopyranosyl-(1→2)-[β-D-xylopyranosyl-(1→3)-β-D-glucopyranosyl-(1→4)-β-D-galactopyranoside]-26-O-β-D-glucopyranoside}

Source : *Digitalis purpurea* (Scrophulariaceae), isolated admixed with 22-O-methyl derivative
Mol. Formula : $C_{56}H_{94}O_{29}$
Mol. Wt. : 1230
Registry No. : [54191-26-3]

Biological Activity : Hypocholestermic activity,[2] inhibits growth of tobacco virus at higher concentration.[3]

References

1.   R. Tschesche, M. Javellana and G. Wulff, *Chem. Ber.*, **107**, 2828 (1974).

2.   G.M. Goryani, V.A. Bobeiko, I.V. Suetina and N.E. Mashchenko, *Khim. Pharm. Zh.*, **15**, 55 (1981).

3.   I.T. Balashova, T.D. Verdervskaya and P.K. Kintya, *S.-kh. Biol.,* 83 (1984); *Chem. Abstr.*, **101**, 2211 x (1984).

TACCA CHANTRIERI SAPONIN 7

(25*S*)-3β,26-Dihydroxyfurosta-5,20(22)-diene 3-O-{α-L-rhamnopyranosyl-(1→2)-|β-D-glucopyranosyl-(1→4)-α-L-rhamnopyranosyl-(1→3)]-6-O-acetyl-β-D-glucopyranoside}-26-O-β-D-glucopyranoside

Source : *Tacca chantrieri* André (Taccaceae)
Mol. Formula : $C_{59}H_{94}O_{27}$
Mol. Wt. : 1234
[α]$_D^{25}$: -42° (c=0.10, CHCl$_3$-MeOH)
Registry No. : [469879-88-7]

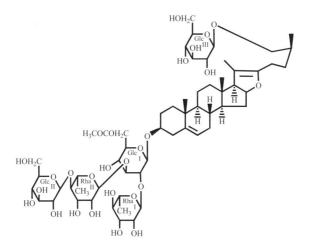

IR (film) : 3389 (OH), 2933 (CH), 1739 (C=O), 1044 cm^{-1}.

PMR (C$_5$D$_5$N, 500 MHz) : δ 0.72 (s, 3xH-18), 1.04 (d, J=6.8 Hz, 3xH-27), 1.05 (s, 3xH-19), 1.63 (s, 3xH-21), 1.67 (d, J=6.2 Hz, 3xH-6 of Rha II), 1.75 (d, J=6.1 Hz, 3xH-6 of Rha I), 2.01 (s, OCOCH_3), 3.90 (m, W½=20.7 Hz, H-3), 4.82 (m, H-16), 4.83 (d, J=7.7 Hz, H-1 of Glc III), 4.85 and 4.72 (m, 2xH-6 of Glc I), 4.87 (d, J=7.8 Hz, H-1 of Glc I), 5.23 (d, J=7.8 Hz, H-1 of Glc II), 5.32 (br d, J=4.5 Hz, H-6), 5.76 (br s, H-1 of Rha II), 5.82 (br s, H-1 of Rha I).

CMR (C$_5$D$_5$N, 125 MHz) : δ C-1) 37.6 (2) 30.1 (3) 78.4 (4) 38.7 (5) 140.8 (6) 121.8 (7) 32.4 (8) 31.4 (9) 50.3 (10) 37.1 (11) 21.2 (12) 39.6 (13) 43.4 (14) 54.9 (15) 34.5 (16) 84.5 (17) 64.5 (18) 14.1 (19) 19.4 (20) 103.5 (21) 11.8 (22) 152.4 (23) 31.4 (24) 23.6 (25) 33.7 (26) 75.2 (27) 17.1 **Glc I** (1) 100.1 (2) 78.6 (3) 85.6 (4) 69.9 (5) 74.7 (6) 64.2 **Rha I** (1) 102.7 (2) 72.4 (3) 72.7 (4) 73.8 (5) 70.0 (6) 18.2 **Rha II** (1) 103.2 (2) 72.0 (3) 72.4 (4) 84.5 (5) 68.7 (6) 18.2 **Glc II** (1) 106.5 (2) 76.4 (3) 78.6 (4) 71.4 (5) 78.4 (6) 62.5 **Glc III** (1) 105.1 (2) 75.2 (3) 78.6 (4) 71.7 (5) 78.4 (6) 62.8 (OCOCH$_3$) 170.8.

Mass (FAB, Positive ion) : m/z 1257 [M+Na]$^+$.

Mass (MALDI-TOF, H.R.) : m/z 1257.5891 [M+Na]$^+$.

Reference

1.   A. Yokosuka, Y. Mimaki and Y. Sashida, *J. Nat. Prod.*, **65**, 1293 (2002).

CHLOROMALOSIDE B

22ξ-Methoxy-25(S)-5α-furostane-3β,26-diol 3-O-{β-D-glucopyranosyl-(1→2)-[β-D-xylopyranosyl-(1→3)-β-D-glucopyranosyl-(1→4)]-β-D-galactopyranoside}-26-O-β-D-glucopyranoside

Source : *Chlorophytum malayense* Ridley[1]
(Anthericaceae)
Mol. Formula : $C_{57}H_{94}O_{29}$
Mol. Wt. : 1242
M.P. : 219-222°C
[α]$_D^{21}$: -31.9° (c=0.517, Pyridine)
Registry No. : [132998-90-4]

PMR (C_5D_5N, 400 MHz)[1] : δ 0.666 (s, 3xH-19), 1.048 (d, J=6.9 Hz, 3xH-27), 1.056 (s, 3xH-18), 1.390 (d, J=6.4 Hz, 3xH-21), 3.265 (s, OCH_3), 4.835 (d, J=8.0 Hz, H-1 of Gal), 4.857 (H-1 of Glc III), 5.156 (d, J=7.6 Hz, H-1 of Glc I), 5.218 (d, J=8.1 Hz, H-1 of Glc II), 5.558 (d, J=7.0 Hz, H-1 of Xyl).

CMR (C_5D_5N, 100 MHz)[1] : δ C-1) 36.7 (2) 29.7 (3) 77.3 (4) 34.7 (5) 44.5 (6) 28.6 (7) 31.4 (8) 34.4 (9) 55.7 (10) 36.3 (11) 37.9 (12) 212.8 (13) 55.5 (14) 55.9 (15) 31.7 (16) 79.8 (17) 55.3 (18) 16.0 (19) 11.7 (20) 41.1 (21) 15.0 (22) 112.8 (23) 30.9 (24) 28.2 (25) 34.4 (26) 74.6 (27) 17.5 (OCH_3) 47.4 **Gal** (1) 102.5 (2) 73.1 (3) 75.6[a] (4) 79.9 (5) 76.1[a] (6) 60.7 **Glc I** (1) 105.0 (2) 81.2 (3) 87.2 (4) 70.4[b] (5) 77.5[c] (6) 62.5[d] **Glc II** (1) 105.0 (2) 75.2[e] (3) 78.6 (4) 71.1 (5) 77.8[c] (6) 62.9[d] **Xyl** (1) 104.7 (2) 75.3[e] (3) 78.6 (4) 70.7[b] (5) 67.3.

Mass (FAB)[1] : *m/z* 1265 [M+Na]⁺, 1249 [M+Li]⁺.

Biological Activity : Shows inhibitory effect on human spermatozoa *in vitro* at a concentration of 1 mg/ml.[2]

References

1.   X.-C. Li, D.-Z, Wang and C.-R. Yang, *Phytochemistry*, **29**, 3893 (1990).

2.   Y.-F. Wang, X.-C. Li, H.-Y. Yang, J.-J. Wang and C.-R. Yang, *Planta Med.*, **62**, 130 (1996).

POLYGONATUM SIBIRICUM SAPONIN PS-II

22-Methoxy-25(*S*)-furost-5-en-3β,14α,26-triol 3-O-{β-D-glucopyranosyl-(1→2)-[β-D-xylopyranosyl-(1→3)]-β-D-glucopyranosyl-(1→4)-β-D-galactopyranoside}-26-O-β-D-glucopyranoside

Source : *Polygonatum sibiricum* Redoute (Liliaceae)
Mol. Formula : $C_{57}H_{94}O_{29}$
Mol. Wt. : 1242
M.P. : 202-204°C
[α]$_D^{20}$: -54.2° (c=0.08, MeOH)
Registry No. : [128820-37-1]

IR (KBr) : 3400, 1620, 1150, 1075, 1040, 1010, 885, 870 cm⁻¹.

PMR (C$_5$D$_5$N, 200 MHz) : δ 0.84 (s, 3xH-18), 0.92 (s, 3xH-19), 0.96 (d, J=6.4 Hz, 3xH-27), 1.10 (d, J=6.7 Hz, 3xH-21), 3.12 (s, OCH$_3$), 4.76 (d, J=7.7 Hz, anomeric H), 5.08 (d, J=8.2 Hz, anomeric H), 5.13 (d, J=7.6 Hz, anomeric H), 5.33 (m, H-6), 5.47 (d, J=6.6 Hz, anomeric H).

CMR (C$_5$D$_5$N, 50.3 MHz) : δ C-1) 37.9 (2) 30.4 (3) 78.8 (4) 40.2 (5) 140.5 (6) 122.4 (7) 26.8 (8) 35.7 (9) 43.7 (10) 37.6 (11) 20.2 (12) 32.0 (13) 45.6 (14) 86.9 (15) 39.5 (16) 82.3 (17) 61.3 (18) 20.1 (19) 19.5 (20) 40.6 (21) 16.8 (22) 113.5 (23) 31.2 (24) 28.4 (25) 34.7 (26) 75.2 (27) 17.7 (OCH$_3$) 47.5 **Gal** (1) 102.9 (2) 73.4 (3) 75.4 (4) 80.1 (5) 76.4 (6) 60.7 **Glc I** (1) 105.1 (2) 81.5 (3) 86.9 (4) 70.9 (5) 78.8 (6) 62.6 **Xyl** (1) 105.4 (2) 75.7 (3) 77.8 (4) 70.7 (5) 67.5 **Glc II** (1) 105.4 (2) 75.4 (3) 78.4 (4) 71.2 (5) 77.8 (6) 63.1 **Glc III** (1) 105.4 (2) 75.3 (3) 78.8 (4) 71.9 (5) 78.4 (6) 63.1.

Reference

1.   K.H. Son, J.C. Do and S.S. Kang, *J. Nat. Prod.*, **53**, 333 (1990).

PROTOYUCCOSIDE E

(5β,25S)-3β,22α,26-Trihydroxyfurostane 3-O-{β-D-galactopyranosyl-(1→2)-[β-D-galactopyranosyl-(1→4)]-β-D-glucopyranosyl-(1→6)-β-D-glucopyranoside}-26-O-β-D-glucopyranoside

Source : *Yucca filamentosa* L. (Agavaceae)
Mol. Formula : C$_{57}$H$_{96}$O$_{29}$
Mol. Wt. : 1244
M.P. : 150-152°C[1]
[α]$_D^{20}$: -29° (c=2.75, MeOH)[1]
Registry No. : [55750-40-8]

Peracetate :

Mass : *m/z* 331, 243, 242, 200, 169, 157, 145, 141, 140, 115, 109, 103 and 98 (tetraacetylglucose), 129 ($C_7H_{13}O_2$) and 97.

References

1.   I.P. Dragalin and P.K. Kintya, *Khim. Prir. Soedin.*, 806 (1975); *Chem. Nat. Comp.*, **11**, 821 (1975).

2.   G.V. Lazur'evskii, P.K. Kintya and I.P. Dragalin, *Dokl. Akad. Nauk SSSR*, **221,** 481 (1975).

TIMOSAPONIN B-V

(25*S*)-22-Hydroxy-5β-furostane-3β,26-diol 3-O-{β-D-glucopyranosyl-(1→2)-[β-D-glucopyranosyl-(1→3)]-β-D-glucopyranosyl-(1→4)-β-D-galactopyranoside}-26-O-β-D-glucopyranoside

Source : *Anemarrhena asphodeloides* Bge. (Liliaceae)
Mol. Formula : $C_{57}H_{96}O_{29}$
Mol. Wt. : 1244
Registry No. : [215051-32-4]

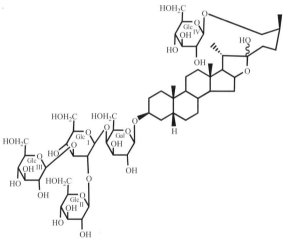

CMR : δ 110.7 (C-20).

Reference

1.   Z. Meng and S. Xu, *Shenyang Yaoke Daxue Xuebao* (*J. Shenyang Pharm. Univ.*) **15**, 130 (1998).

ALLIUM SCHUBERTII SAPONIN 5

(25R&S)-5α-Furostan-2α,3β,6β,22ξ,26-pentol 3-O-{β-D-glucopyranosyl-(1→2)-[β-D-xylopyranosyl-(1→3)]-β-D-glucopyranosyl-(1→4)-β-D-galactopyranoside}-26-O-β-D-glucopyranoside

Source : *Allium schubertii* (Liliaceae)
Mol. Formula : $C_{56}H_{94}O_{30}$
Mol. Wt. : 1246
[α]$_D^{21}$: -34.5° (c=0.11, Pyridine)
Registry No. : [147802-39-9] (25R)
[147852-68-4] (25S)

IR (KBr) **:** 3430 (OH), 2930 (CH), 1440, 1260, 1160, 1070, 1030 cm^{-1}.

PMR (C$_5$D$_5$N, 400 MHz) **:** δ 0.89 (s, 3xH-18), 0.98 [d, *J*=6.6 Hz, H-27 of (25R)-isomer], 1.03 [d, *J*=6.6 Hz, 3xH-27 of (25S)-isomer], 1.30 (s, 3xH-19), 1.32 (d, *J*=6.8 Hz, 3xH-21), 4.83 (d, *J*=7.8 Hz, H-1 of Glc III), 4.98 (d, *J*=7.7 Hz, H-1 of Gal), 5.23 (d, *J*=7.9 Hz, H-1 of Glc I), 5.26 (d, *J*=7.7 Hz, H-1 of Xyl), 5.59 (d, *J*=7.7 Hz, H-1 of Glc II).

CMR (C$_5$D$_5$N, 100.6 MHz) **:** δ C-1) 47.2 (2) 70.8^a (3) 84.7 (4) 32.0^b (5) 47.9 (6) 70.0 (7) 40.8^c (8) 30.0 (9) 54.6 (10) 37.1 (11) 21.4 (12) 40.2^c (13) 41.2 (14) 56.2 (15) 32.5^b (16) 81.1^d (17) 63.9 (18) 16.4 (19) 17.2 (20) 40.7 (21) 16.7 (22) 110.6 (23) 37.1 (24) 28.3 (25) 34.4 (26) 75.4/75.3 (27) 17.5 **Gal** (1) 103.2 (2) 72.5 (3) 75.6 (4) 79.4 (5) 76.1 (6) 60.6 **Glc I** (1) 104.7 (2) 81.3^d (3) 87.0 (4) 70.5^a (5) 77.6 (6) 62.8 **Glc II** (1) 104.8 (2) 75.1 (3) 78.2^e (4) 71.4 (5) 78.6^e

(6) 62.8 **Xyl** (1) 104.9 (2) 75.7 (3) 78.5^e (4) 70.4^a (5) 67.3 **Glc III** (1) 105.2 (2) 75.2 (3) 78.5^e (4) 71.7 (5) 78.7^e (6) 62.9.

Mass (FAB, Negative ion) **:** m/z 1245 [M-H]⁻, 1113 [M-Xyl]⁻, 1083 [M-Glc]⁻, 933 [M-OH-Xyl-Glc]⁻, 789 [M-Xyl-Glcx2]⁻.

Reference

1.   K. Kawashima, Y. Mimaki and Y. Sashida, *Phytochemistry*, **32**, 1267 (1993).

HOSTA SIEBOLDII SAPONIN 5
(25R)-2α,3β,26-Trihydroxy-22ξ-methoxy-5α-furost-9(11)-en-12-one 3-O-{β-D-glucopyranosyl-(1→2)-[β-D-xylopyranosyl-(1→3)]-β-D-galactopyranosyl-(1→4)-β-D-galactopyranoside}

Source : *Hosta sieboldii* (Liliaceae)
Mol. Formula : C$_{57}$H$_{92}$O$_{30}$
Mol. Wt. : 1256
$[\alpha]_D^{25}$: -24.0° (c=0.10, MeOH)
Registry No. : [213771-48-3]

UV (MeOH) : λ$_{max}$ 237 (log ε, 4.05) nm.

IR (KBr) : 3420 (OH), 2925 (CH), 1660 (C=O), 1455, 1370, 1300, 1260, 1160, 1070, 1040, 915, 890 cm⁻¹.

PMR (C$_5$D$_5$N, 400/500 MHz) : δ 0.90 (s, 3xH-19), 0.99 (s, 3xH-18), 1.00 (d, *J*=6.3 Hz, 3xH-27), 1.46 (d, *J*=6.8 Hz, 3xH-21), 3.28 (s, OC*H$_3$*), 4.85 (d, *J*=7.7 Hz, H-1 of Glc III), 4.92 (d, *J*=7.9 Hz, H-1 of Gal), 5.21 (d, *J*=7.8 Hz, H-1 of Glc I), 5.25 (d, *J*=7.8 Hz, H-1 of XyI), 5.60 (d, *J*=7.8 Hz, H-1 of Glc II), 5.93 (br s, H-11).

CMR (C$_5$D$_5$N, 100/125 MHz) : δ C-1) 43.4 (2) 70.2 (3) 83.4 (4) 33.7 (5) 42.4 (6) 27.1 (7) 32.4 (8) 36.1 (9) 170.5(10) 40.5 (11) 120.0 (12) 204.2 (13) 51.7 (14) 52.5 (15) 31.6 (16) 80.4 (17) 55.6 (18) 15.2 (19) 19.3 (20) 41.2 (21) 14.7 (22) 112.9 (23) 30.5 (24) 28.2 (25) 34.2 (26) 75.2 (27) 17.1 (OCH$_3$) 47.4 **Gal** (1) 103.2 (2) 72.5 (3) 75.5 (4) 79.5 (5) 75.7 (6) 60.6 **Glc I** (1) 104.7 (2) 81.2 (3) 87.0 (4) 70.4 (5) 77.6 (6) 62.9 **Glc II** (1) 104.7 (2) 76.0 (3) 78.1 (4) 71.3 (5) 78.4 (6) 62.7 **Xyl** (1) 104.9 (2) 75.1 (3) 78.7 (4) 70.7 (5) 67.3 **Glc III** (1) 104.9 (2) 75.2 (3) 78.4 (4) 71.7 (5) 78.6 (6) 62.9.

Mass (FAB, Positive ion) : *m/z* 1279 [M+Na]$^+$.

Mass (FAB, Negative ion) : *m/z* 1255 [M-H]$^-$.

Reference

1.   Y. Mimaki, M. Kuroda, A. Kameyama, A. Yokosuka and Y. Sashida, *Phytochemistry*, **48**, 1361 (1998).

HOSTA SIEBOLDII SAPONIN 4
(25*R*)-2α,3β,26-Trihydroxy-22ξ-methoxy-5α-furostan-12-one 3-O-[β-D-glucopyranosoyl-(1→2)-
[β-D-xylopyranosyl-(1→3)]-β-D-glucopyranosyl-(1→4)-β-D-galactopyranoside}-26-O-β-D-glucopyranoside

Source : *Hosta sieboldii* (Liliaceae)
Mol. Formula : C$_{57}$H$_{94}$O$_{30}$
Mol. Wt. : 1258
[α]$_D^{25}$: -60.0° (c=0.10, MeOH)
Registry No. : [213771-43-8]

IR (KBr) : 3400 (OH), 2925 (CH), 1695 (C=O), 1445, 1415, 1370, 1255, 1155, 1065, 1030, 885, 795 cm^{-1}.

PMR (C_5D_5N, 400/500 MHz) : δ 0.73 (s, 3xH-19), 1.00 (d, J=6.6 Hz, 3xH-27), 1.05 (s, 3xH-18), 1.41 (d, J=6.8 Hz, 3xH-21), 3.26 (s, OCH_3), 4.85 (d, J=7.7 Hz, H-1 of Glc III), 4.91 (d, J=7.9 Hz, H-1 of Gal), 5.21 (d, J=7.9 Hz, H-1 of Glc I), 5.25 (d, J=7.8 Hz, H-1 of XyI), 5.59 (d, J=7.8 Hz, H-1 of Glc II).

CMR (C_5D_5N, 100/125 MHz) : δ C-1) 45.0 (2) 70.1 (3) 83.8 (4) 33.9 (5) 44.4 (6) 27.8 (7) 31.3 (8) 33.6 (9) 55.3 (10) 37.2 (11) 38.0 (12) 212.5 (13) 55.7 (14) 55.5 (15) 31.5 (16) 79.9 (17) 55.6 (18) 16.0 (19) 12.8 (20) 41.1 (21) 15.0 (22) 112.7 (23) 30.7 (24) 28.2 (25) 34.2 (26) 75.2 (27) 17.1 (OCH$_3$) 47.3 **Gal** (1) 103.2 (2) 72.5 (3) 75.5 (4) 79.4 (5) 75.7 (6) 60.6 **Glc I** (1) 104.7 (2) 81.2 (3) 87.0 (4) 70.4 (5) 77.6 (6) 62.9 **Glc II** (1) 104.7 (2) 76.0 (3) 78.1 (4) 71.3 (5) 78.5 (6) 62.7 **Glc III** (1) 104.9 (2) 75.2 (3) 78.5 (4) 71.7 (5) 78.6 (6) 62.9 **Xyl** (1) 104.9 (2) 75.1 (3) 78.7 (4) 70.7 (5) 67.3.

Mass (FAB, Positive ion) : m/z 1281 [M+Na]$^+$.

Mass (FAB, Negative ion) : m/z 1257 [M-H]$^-$.

Reference

1.   Y. Mimaki, M. Kuroda, A. Kameyama, A. Yokosuka and Y. Sashida, *Phytochemistry*, **48,** 1361 (1998).

TIMOSAPONIN B-VI
(25S)-22-Methoxy-5β-furostane-3β,26-diol 3-O-{β-D-glucopyranosyl-(1→2)-[β-D-glucopyranosyl-(1→3)]-β-D-glucopyranosyl-(1→4)-β-D-galactopyranoside}-26-O-β-D-glucopyranoside

Source : *Anemarrhena asphodeloides* Bge. (Liliaceae)
Mol. Formula : $C_{58}H_{98}O_{29}$
Mol. Wt. : 1258
Registry No. : [215051-42-6]

CMR : δ 112.7 (C-22), 31.0 (C-23).

Reference

1.   Z. Meng and S. Xu, *Shenyang Yaoke Daxue Xuebao* (*J. Shenyang Pharm. Univ.*), **15**, 130 (1998).

PROTOERUBOSIDE B

(25*R*,5α)-Furostane-3β,6β,22ξ,26-tetrol 3-O-{β-D-glucopyranosyl-(1→2)-[β-D-glucopyranosyl-(1→3)]-
β-D-glucopyranosyl-(1→4)-galactopyranoside}-26-O-β-D-glucopyranoside

Source : *Allium sativum* L. (Liliaceae)
Mol. Formula : $C_{57}H_{96}O_{30}$
Mol. Wt. : 1260
[α]$_D^{27}$: -36.5° (c=0.7, C_5D_5N)
Registry No. : [118543-11-6]

PMR (C$_5$D$_5$N, 270 MHz)[1] : δ 4.82 (d, J=7.9 Hz), 4.95 (d, J=7.3 Hz), 5.16 (d, J=7.6 Hz), 5.31 (d, J=7.6 Hz), 5.58 (d, J=7.0 Hz, 5 x anomeric H).

CMR (C$_5$D$_5$N, 67.5 MHz)[1] : δ C-1) 38.9 (2) 30.0 (3) 77.9 (4) 32.8 (5) 48.0 (6) 70.8 (7) 40.3 (8) 30.6 (9) 54.7 (10) 36.1 (11) 21.3 (12) 40.8 (13) 41.2 (14) 56.3 (15) 32.5 (16) 81.1 (17) 64.0 (18) 16.7 (19) 16.4 (20) 40.6 (21) 16.0 (22) 110.6 (23) 37.2 (24) 28.4 (25) 34.2 (26) 75.3 (27) 17.5 **Gal** (1) 102.3 (2) 73.2 (3) 75.6^a (4) 80.3 (5) 76.1 (6) 60.6 **Glc I** (1) 105.0^b (2) 81.5 (3) 88.5 (4) 70.8^c (5) 78.0^d (6) 63.0^e **Glc II** (1) 104.9^b (2) 75.3^a (3) 78.6^f (4) 71.7^c (5) 77.5^d (6) 62.3^e **Glc III** (1) 104.5^b (2) 75.2^a (3) 78.6^f (4) 71.6^c (5) 78.6^f (6) 62.3^e **Glc IV** (1) 104.9^b (2) 75.2^a (3) 78.4^f (4) 71.7^c (5) 78.4^f (6) 62.8^e.

Biological Activity : Antifungal.[2]

Reference

1.   H. Matsuura, T. Ushiroguchi, Y. Itakura, N. Hayashi and T. Fuwa, *Chem. Phram. Bull.*, **36**, 3659 (1988).

TACCA CHANTRIERI SAPONIN 4

(25*S*)-3β,26-Dihydroxy-22α-methoxyfurost-5-ene 3-O-{α-L-rhamnopyranosyl-(1→2)-[β-D-glucopyranosyl-(1→4)-α-L-rhamnopyranosyl-(1→3)]-6-O-acetyl-β-D-glucopyranoside}-26-O-β-D-glucopyranoside

Source : *Tacca chantrieri* André (Taccaceae)
Mol. Formula : $C_{60}H_{98}O_{28}$
Mol. Wt. : 1266
$[\alpha]_D^{25}$: -106.0° (c=0.10, CHCl₃-MeOH)
Registry No. : [290809-77-7]

IR (KBr) : 3400 (OH), 2930 (CH), 1730 (C=O), 1040 cm⁻¹.

PMR (C₅D₅N, 500 MHz) : δ 0.81 (s, 3xH-18), 1.02 (s, 3xH-19), 1.05 (d, *J*=6.7 Hz, 3xH-27), 1.17 (d, *J*=6.9 Hz, 3xH-21), 1.67 (d, *J*=6.2 Hz, 3xH-6 of Rha II), 1.75 (d, *J*=6.2 Hz, 3xH-6 of Rha I), 2.01 (s, OCOC*H₃*), 3.26 (s, OC*H₃*), 3.89 (m, *W*½=24.6 Hz, H-3), 4.46 (m, H-16), 4.85 (d, *J*=7.8 Hz, H-1 of Glc III), 4.86 and 4.72 (m, 2xH-6 of Glc I), 4.87 (d, *J*=7.6 Hz, H-1 of Glc I), 5.28 (d, *J*=7.8 Hz, H-1 of Glc II), 5.32 (br d, *J*=4.7 Hz, H-6), 5.76 (br s, H-1 of Rha II), 5.82 (br s, H-1 of Rha I).

CMR (C₅D₅N, 125 MHz) : δ C-1) 37.5 (2) 30.1 (3) 78.4 (4) 38.7 (5) 140.8 (6) 121.8 (7) 32.2 (8) 31.6 (9) 50.3 (10) 37.1 (11) 21.0 (12) 39.7 (13) 40.8 (14) 56.5 (15) 32.3 (16) 81.3 (17) 64.1 (18) 16.3 (19) 19.3 (20) 40.4 (21) 16.2 (22) 112.7 (23) 30.9 (24) 28.1 (25) 34.4 (26) 74.9 (27) 17.5 (OCH₃) 47.3 **Glc I** (1) 100.1 (2) 78.6 (3) 85.6 (4) 69.9 (5) 74.7 (6) 64.1 **Rha I** (1) 102.7 (2) 72.4 (3) 72.7 (4) 73.7 (5) 70.0 (6) 18.6 **Rha II** (1) 103.2 (2) 72.0 (3) 72.4 (4) 84.4 (5) 68.7

(6) 18.2 **Glc II** (1) 106.5 (2) 76.4 (3) 78.6 (4) 71.4 (5) 78.4 (6) 62.5 **Glc III** (1) 105.1 (2) 75.2 (3) 78.6 (4) 71.7 (5) 78.5 (6) 62.9 (OCOCH₃) 170.8.

Mass (FAB, Negative ion) **:** m/z 1265 [M-H]⁻.

Reference

1.   A. Yokosuka, Y. Mimaki and Y. Sashida, *J. Nat. Prod.*, **65**, 1293 (2002).

CAPSICOSIDE B

22ξ-Methoxy-5α,25*R*-furostan-2α,3β,26-triol 3-O-{β-D-glucopyranosyl-(1→2)-[β-D-glucopyranosyl-(1→3)]-β-D-glucopyranosyl-(1→4)-β-D-galactopyranoside}-26-O-β-D-glucopyranoside

Source : *Capsicum annuum* L. var. *conoides* and *Capsicum annuum* L. var. *fasciculatum* (Solanaceae)
Mol. Formula : $C_{58}H_{98}O_{30}$
Mol. Wt. : 1274
$[\alpha]_D^{26}$ **:** -49.3° (c=0.73, Pyridine)
Registry No. : [160219-65-8]

PMR (C$_5$D$_5$N, 400 MHz) : δ 0.68 (s, C*H$_3$*), 0.78 (s, C*H$_3$*), 1.00 (d, *J*=6.6 Hz, sec. C*H$_3$*), 1.18 (d, *J*=6.6 Hz, sec. C*H$_3$*), 3.30 (s, OC*H$_3$*), 4.81 (d, *J*=7.7 Hz), 4.93 (d, *J*=7.3 Hz), 5.08 (d, *J*=7.3 Hz), 5.25 (d, *J*=7.7 Hz), 5.53 (d, *J*=8.3 Hz, 5 x anomeric H).

CMR (C$_5$D$_5$N, 100 MHz) : δ C-1) 44.6 (2) 69.7 (3) 83.1 (4) 33.0 (5) 43.7 (6) 27.3 (7) 31.3 (8) 33.8 (9) 53.5 (10) 36.1 (11) 20.6 (12) 39.1 (13) 40.4 (14) 55.7 (15) 31.4 (16) 80.2 (17) 63.4 (18) 15.8 (19) 12.6 (20) 39.7 (21) 15.6 (22) 112.1 (23) 30.3 (24) 27.4 (25) 33.3 (26) 74.5 (27) 16.4 (OCH$_3$) 46.7 **Gal** (1) 102.6 (2) 72.3 (3) 75.5 (4) 79.5 (5) 77.0 (6) 60.9 **Glc I** (1) 103.8 (2) 81.2 (3) 87.8 (4) 70.1 (5) 78.0 (6) 62.0 **Glc II** (1) 104.1 (2) 74.7 (3) 77.6 (4) 71.0 (5) 79.0 (6) 62.2 **Glc III** (1) 104.2 (2) 75.4 (3) 78.0 (4) 71.2 (5) 78.0 (6) 62.6 **Glc IV** (1) 104.5 (2) 74.9 (3) 78.1 (4) 71.3 (5) 78.0 (6) 62.2.

Mass (FAB, Negative ion) : *m/z* 1273 [M-H]⁻, 1259 [M-Me]⁻, 1111 [M-H-Glc]⁻.

Reference

1. S. Yahara, T. Ura, C. Sakamoto and T. Nohara, *Phytochemistry*, **37**, 831 (1994).

BALANITSIN

(25*R*)-Furost-5-ene-3β,22ξ,26-triol 3-O-{α-L-rhamnopyranosyl-(1→2)-α-L-rhamnopyranosyl-(1→4)-[β-D-xylopyranosyl-(1→2)]-β-D-xylopyranosyl-(1→2)-β-D-xylopyranoside}-26-O-β-D-glucopyranoside

Source : *Balanites aegyptiaca* (Balanitaceae)
Mol. Formula : C$_{60}$H$_{98}$O$_{29}$
Mol. Wt. : 1282
[α]$_D^{23}$: +25.2° (c=0.31, MeOH)
Registry No. : [212369-13-6]

PMR (C₅D₅N, 400 MHz) : δ 0.80 (s, 3xH-18), 0.88 (d, J=7.0 Hz, 3xH-27), 1.02 (s, 3xH-19), 1.4 (d, J=6.8 Hz, 3xH-21), 1.75 (6H, d, J=6.1 Hz, 2x3xH-6 of Rha), 4.81 (d, J=6.5 Hz, H-1 of Glc), 5.10 (d, J=7.0 Hz, H-1 of Xyl), 5.13 (d, J=7.2 Hz, H-1 of Xyl), 5.22 (d, J=7.0 Hz, H-1 of Xyl), 6.22 (br s, H-1 of Rha), 6.24 (br s, H-1 of Rha).

CMR (C₅D₅N, 100 MHz) : δ C-1) 37.5ᵃ (2) 30.1 (3) 78.4 (4) 38.9 (5) 140.8 (6) 121.8 (7) 32.3 (8) 31.7 (9) 50.1 (10) 37.1ᵃ (11) 21.1 (12) 40.0ᵇ (13) 40.3ᵇ (14) 57.0 (15) 32.8 (16) 81.2 (17) 62.1 (18) 16.4 (19) 19.3 (20) 40.7 (21) 16.4 (22) 110.7 (23) 37.1 (24) 29.0 (25) 35.1 (26) 75.2 (27) 17.4 **Glc** (1) 104.4 (2) 75.3 (3) 78.6 (4) 71.7 (5) 78.2 (6) 62.8 **Rha I** (1) 100.0 (2) 78.8 (3) 71.2 (4) 73.9 (5) 69.4 (6) 18.6 **Rha II** (1) 101.7 (2) 72.4 (3) 72.7 (4) 74.9 (5) 69.4 (6) 18.6 **Xyl I** (1) 104.6 (2) 81.6 (3) 77.3 (4) 70.9 (5) 68.5 **Xyl II** (1) 104.6 (2) 81.6 (3) 76.1 (4) 78.0 (5) 67.3 **Xyl III** (1) 106.9 (2) 75.4 (3) 78.0 (4) 70.9 (5) 68.5.

Mass (FAB, Positive ion) : *m/z* 1283 [M+H]⁺, 1137, 975, 829, 433.

Reference

1.   M.S. Kamel, *Phytochemistry*, **48**, 755 (1998).

TERRESTRININ B

(5α)-Furostan-3β,22α,26-triol 3-O-{β-D-xylopyranosyl-(1→2)-[β-D-xylopyranosyl-(1→3)]-
β-D-glucopyranosyl-(1→4)-[α-L-rhamnopyranosyl-(1→2)]-β-D-galactopyranoside}

Source : *Tribulus terrestris*.Linn. (Zygophyllaceae)
Mol. Formula : $C_{61}H_{102}O_{31}$
Mol. Wt. : 1330
M.P. : 268-270°C
[α]$_D^{20}$: -17.1° (c=0.51, Pyridine)
Registry No. : [637004-31-0]

IR (KBr) : 3406, 1641, 1452, 1382, 1161, 1074, 980 cm^{-1}.

PMR (CDCl$_3$, 400 MHz) : δ 0.94 (s, 3xH-19), 0.96 (s, 3xH-18), 1.05 (d, *J*=6.5 Hz, 3xH-27), 1.40 (d, *J*=6.5 Hz, 3xH-21), 1.78 (d, *J*=5.8 Hz, H-6 of Rha), 3.98 (m, H-3), 4.81 (d, *J*=7.8 Hz, anomeric H), 4.85 (d, *J*=8.0 Hz, anomeric H), 5.01 (br s, H-16), 5.25 (d, *J*=7.6 Hz, anomeric H), 5.42 (d, *J*=7.7 Hz, anomeric H), 6.20 (s, H-1 of Rha).

CMR (C$_5$D$_5$N, 100 MHz) : δ C-1) 37.2 (2) 29.9 (3) 76.9 (4) 34.2 (5) 44.6 (6) 28.9 (7) 32.4 (8) 35.2 (9) 54.4 (10) 35.9 (11) 21.2 (12) 39.9 (13) 41.1 (14) 56.3 (15) 32.0 (16) 81.3 (17) 64.3 (18) 16.4 (19) 12.3 (20) 40.5 (21) 16.3 (22) 112.5 (23) 30.8 (24) 28.1 (25) 34.2 (26) 75.1 (27) 17.1 **Gal** (1) 100.1 (2) 76.6 (3) 76.6 (4) 81.5 (5) 75.8 (6) 60.4 **Rha** (1) 102.0 (2) 172.5 (3) 72.7 (4) 74.0 (5) 69.4 (6) 18.5 **Glc I** (1) 105.0 (2) 81.3 (3) 87.6 (4) 70.4 (5) 77.8 (6) 62.8 **Xyl** (1) 105.3 (2) 75.1 (3) 78.6 (4) 70.7 (5) 67.3 **Xyl II** (1) 105.9 (2) 75.1 (3) 79.1 (4) 70.9 (5) 67.6 **Glc II** (1) 105.0 (2) 75.2 (3) 78.5 (4) 71.7 (5) 78.8 (6) 62.9.

Mass (E.S.I.) : *m/z* 1313 [M+H-H$_2$O]$^+$.

Reference

1.   J.W. Huang, C.-H. Tan, S.-H. Jiang and D.-Y. Zhu, *J. Asian Nat. Prod. Res.*, **5**, 285 (2003).

FLORIBUNDASAPONIN F

(25R)-Furost-5-ene-3β,22α-26-triol 3-O-[α-L-rhamnopyranosyl-(1→3)-α-L-rhamnopyranosyl-(1→3)-α-L-rhamnopyranosyl-(1→3)-α-L-rhamnopyranosyl-(1→4)-β-D-glucopyranoside]-26-O-[β-D-glucopyranoside]

Source : *Dioscorea floribunda* Mart. et Gal.
(Dioscoreaceae)
Mol. Formula : C$_{63}$H$_{104}$O$_{30}$
Mol. Wt. : 1340
M.P. : 226-229°C (decomp.)
[α]$_D$: -66° (C$_5$H$_5$N)
Registry No. : [68406-06-4]

IR (Nujol) : No spiroketal absorption.

Reference

1.   S.B. Mahato, N.P. Sahu and B.C. Pal, *Indian Journal of Chemistry*, **16**(B), 350 (1978).

TRIBULUS CISTOIDES SAPONIN 7
22-Methoxy-(3β,5α,25S)-fu rostan-3,26-diol-3-O-{β-D-xylopyranosyl-(1→2)-[β-D-xylopyranosyl-(1→3)]-β-D-glucopyranosyl-(1→4)-[α-L-rhamnopyranosyl-(1→2)]-β-D-galactopyranoside}-26-O-β-D-glucopyranoside

Source : *Tribulus cistoides* (Zygophyllaceae)
Mol. Formula : $C_{62}H_{104}O_{31}$
Mol. Wt. : 1344
M.P. : 208-211°C
[α]$_D^{20}$: -58° (c=0.8, Pyridine)
Registry No. : [155408-09-6]

IR (KBr) : 3400, 2930 cm^{-1}.

PMR (C$_5$D$_5$N, 400 MHz) : δ 0.46-0.54 (1H, m), 0.74-2.07 (40H, m) within 0.79 (s, 3xH-18)[a], 0.86 (s, 3xH-19)[a], 1.06 and 1.18 (3H each; d, J=7.0 Hz) (3xH-21 and 3xH-27), 1.73 (d, J=6.0 Hz, 3xH-6 of Rha), 2.18-2.26 (1H, m), 3.26 (s, OCH_3), 3.47-3.56 (2H, m), 3.69 (1H, dd, $J_1 \sim J_2 \sim 10.0$ Hz), 3.82-4.33 (20H, m), 4.39-4.62 (8H, m), 4.66-4.98 (7H, m), within 4.87 and 4.88 (1H each, d, J=8.0 Hz), 5.00-5.04 (m) overlapped with HOD–peak at δ 5.01, 5.26 and 5.45 (1H, d, J=8.0 Hz) and 6.21 (br s) [6 x anomeric H].

CMR (C_5D_5N, 100 MHz) : δ C-1) 37.1 (2) 29.9 (3) 78.7 (4) 34.3^a (5) 44.6 (6) 28.9 (7) 32.1^b (8) 35.2 (9) 54.4 (10) 35.9 (11) 21.1 (12) 39.9 (13) 41.1 (14) 56.3 (15) 32.4^b (16) 81.3^c (17) 64.3 (18) 16.3^d (19) 12.3 (20) 40.4 (21) 16.4^d (22) 112.6 (23) 30.9 (24) 28.1 (25) 34.4^a (26) 74.9 (27) 17.5 (OCH$_3$) 47.3 **Gal** (1) 100.0 (2) 81.3^c (3) 75.8 (4) 79.0 (5) 77.7 (6) 60.4 **Glc I** (1) 105.8 (2) 81.4^c (3) 87.5 (4) 70.4^e (5) 76.9 (6) 62.8 **Xyl I** (1) 105.0 (2) 75.0 (3) 76.5^f (4) 70.7^e (5) 67.3^g **Xyl II** (1) 105.0 (2) 75.0 (3) 76.6^f (4) 70.8^e (5) 67.6^g **Rha** (1) 101.9 (2) 72.4^h (3) 72.6^h (4) 73.9 (5) 69.3 (6) 18.4 **Glc II** (1) 105.3 (2) 75.2 (3) 78.5^i (4) 71.7 (5) 78.6^i (6) 62.8.

Mass (FAB, Negative ion) : *m/z* (rel.intens.) 1344 [M]$^-$ (54), 1343 [M-H]$^-$ (100), 1212 [M-132]$^-$ (37), 1211 [M-132-H]$^-$ (69), 1197 [M-146-H]$^-$ (13), 1181 [M-162-H]$^-$ (11), 1079 (16), 917 [M-426-H]$^-$ (32), 609 [M-734-H]$^-$ (21).

Reference

1.   H. Achenbach, H. Hübner, W. Brandt and M. Reiter, *Phytochemistry*, **35**, 1527 (1994).

TRIBULUS TERRESTRIS SAPONIN 9
(5α,25R)-22α-Methoxy-furostan-3β,26-diol 3-O-{β-D-xylopyranosyl-(1→2)-[β-D-xylopyranosyl-(1→3)]-β-D-glucopyranosyl-(1→4)-[α-L-rhamnopyranosyl-(1→2)]-β-D-galactopyranoside}

Source : *Tribulus terrestris* L. (Zygopyllaceae)
Mol. Formula : $C_{62}H_{104}O_{31}$
Mol. Wt. : 1344
M.P. : 217°C
[α]$_D^{20}$: -62.87° (c=0.25, MeOH)
Registry No. : [300817-18-9]

IR (KBr) : 3450 (br, OH), 1050 (br, CO) cm^{-1}.

PMR (C$_5$D$_5$N, 400 MHz) : δ 0.75 (s, 3xH-18), 0.81 (s, 3xH-19), 0.96 (d, J=6.4 Hz, 3xH-27), 1.16 (d, J=6.7 Hz, 3xH-21), 1.69 (d, J=5.7 Hz, 3xH-6 of Rha), 3.23 (s, OCH_3), 3.90 (H-3), 4.83 (d, J=7.6 Hz, H-1 of Glc II), 4.93 (d, J=7.6 Hz, H-1 of Glc II), 4.97 (H-1 of Glc I), 5.23 (d, J=7.7 Hz, H-1 of Xyl), 5.40 (d, J=7.2 Hz, H-1 of Xyl II), 6.17 (s, H-1 of Rha).

CMR (C$_5$D$_5$N, 100 MHz) : δ C-1) 37.1 (2) 29.8 (3) 76.9 (4) 34.1 (5) 44.5 (6) 28.9 (7) 32.3 (8) 35.1 (9) 54.3 (10) 35.8 (11) 21.1 (12) 39.9 (13) 41.0 (14) 56.2 (15) 32.0 (16) 81.2 (17) 64.2 (18) 16.3 (19) 12.3 (20) 40.4 (21) 16.2 (22) 112.5 (23) 30.7 (24) 28.0 (25) 34.1 (26) 75.0 (27) 17.0 (OCH_3) 47.1 **Gal** (1) 100.0 (2) 76.5 (3) 76.5 (4) 81.2 (5) 75.6 (6) 60.3 **Glc I** (1) 104.8 (2) 81.2 (3) 87.5 (4) 70.2 (5) 77.6 (6) 62.7 **Xyl I** (1) 105.2 (2) 74.9 (3) 78.5 (4) 70.6 (5) 67.6 **Xyl II** (1) 105.6 (2) 74.9 (3) 78.8 (4) 70.7 (5) 67.2 **Rha** (1) 101.8 (2) 72.3 (3) 72.5 (4) 73.8 (5) 69.2 (6) 18.3 **Glc II** (1) 104.8 (2) 75.0 (3) 78.3^a (4) 71.6 (5) 78.5^a (6) 62.7.

Mass (E.S.I.) : m/z 1369 [M+Na]$^+$, 1392 [M+2Na]$^+$, 696 [M+2Na]$^{+2}$.

Reference

1.   Y.-X. Xu, H.-S. Chen, H.-Q. Liang, Z.-B. Gu, W.-Y Liu, W.-N. Leung, T.-J. Li, *Planta Med.*, **66,** 545 (2000).

CHLOROMALOSIDE E

(5α,25S),3β,22α,26-Trihydroxyfurostan-12-one 3-O-{α-L-arabinopyranosyl-(1→2)-[β-D-xylopyranosyl-(1→3)]-
β-D-glucopyranosyl-(1→4)-α-L-rhamnopyranosyl-(1→2)-β-D-galactopyranoside}-26-O-β-D-glucopyranoside

Source : *Chlorophytum malayense* Ridley (Liliaceae)
Mol. Formula : C$_{61}$H$_{100}$O$_{32}$
Mol. Wt. : 1344
[α]$_D^{20}$: -27.4° (c=0.40, Pyridine)
Registry No. : [306759-87-5]

IR (KBr) : 3410 (br), 1700 (C=O) cm^{-1}.

PMR (C$_5$D$_5$N, 400 MHz) : δ 0.85 (s, 3xH-18), 0.99 (d, J=6.4 Hz, 3xH-27), 1.09 (s, 3xH-19), 1.50 (d, J=6.3 Hz, 3xH-21), 1.67 (d, J=5.3 Hz, 3xH-6 of Rha), 4.80 (d, J=7.2 Hz, H-1 of Gal), 4.85 (d, J=7.0 Hz, H-1 of Glc II), 5.00 (d, J=7.3 Hz, H-1 of Glc I), 5.19 (d, J=6.4 Hz, H-1 of Xyl), 5.30 (d, J=7.2 Hz, H-1 of Ara), 6.25 (br s, H-1 of Rha).

CMR (C$_5$D$_5$N, 100 MHz) : δ C-1) 37.12 (2) 29.82 (3) 77.45 (4) 34.77 (5) 44.63 (6) 28.77 (7) 31.81 (8) 34.47 (9) 55.87 (10) 36.51 (11) 38.12 (12) 213.15 (13) 55.87 (14) 55.99 (15) 31.81 (16) 79.81 (17) 55.13 (18) 16.31 (19) 11.97 (20) 41.35 (21) 15.28 (22) 110.89 (23) 30.66 (24) 28.40 (25) 34.47 (26) 75.25 (27) 17.56 **Gal** (1) 100.30 (2) 77.45 (3) 76.26 (4) 79.81 (5) 75.14 (6) 60.55 **Glc I** (1) 105.08 (2) 81.49 (3) 88.04 (4) 70.46 (5) 77.67 (6) 62.94 **Rha** (1) 101.68 (2) 72.41 (3) 72.76 (4) 74.05 (5) 69.74 (6) 18.52 **Xyl** (1) 105.08 (2) 75.14 (3) 78.61 (4) 70.78 (5) 67.34 **Ara** (1) 105.77 (2) 73.28 (3) 74.70 (4) 69.34 (5) 67.34 **Glc II** (1) 105.0 (2) 75.14 (30 78.61 (4) 71.85 (5) 78.39 (6) 62.94.

Mass (FAB, Positive ion, H.R.) : m/z 1343.6173 [(M-H)$^-$, requires 1343.6119].

Mass (FAB, Negative ion) : m/z 1343 [M-H]$^-$, 1212 [M-Xyl or M-Ara]$^-$, 1066 [(M-Xyl-Rha or M-Ara-Rha]$^-$, 918 [M-Xyl-Glc-Ara]$^-$, 772 [M-Xyl-Glc-Ara-Rha]$^-$.

Reference

1.	Q.-X. Yang and C.-R. Yang, *Yunnan Zhiwu Yanjiu (Acta Bot. Yunnanica)*, **22**, 191 (2000).

POLIANTHOSIDE F

(5α,25*R*)-Furost-3β,22α,26-triol 3-O-{β-D-xylopyranosyl-(1→3)-[β-D-xylopyranosyl-(1→2)-β-D-xylopyranosyl-(1→3)]-β-D-glucopyranosyl-(1→4)-β-D-galactopyranoside}-26-O-β-D-glucopyranoside

Source : *Polianthes tuberosa* L. (Agavaceae)
Mol. Formula : $C_{61}H_{102}O_{32}$
Mol. Wt. : 1346
$[\alpha]_D^{19.8}$: -37.18° (c=0.0390, Pyridine)
Registry No. : [655233-76-4]

IR (KBr) : 3413, 2927, 1699, 1373, 1161, 1074, 1040, 894 cm^{-1}.

PMR (C$_5$D$_5$N, 500 MHz) : δ 0.61 (s, 3xH-19), 0.86 (s, 3xH-18), 0.96 (d, *J*=6.8 Hz, 3xH-27), 1.32 (d, *J*=6.9 Hz, 3xH-21), 3.77 (H-3), 3.93 (m, H-26A), 4.38 (m, H-26B), 4.87 (d, *J*=7.7 Hz, H-1 of Glc III), 4.91 (d, *J*=6.9 Hz, H-1 of Gal), 4.94 (q-like, *J*=7.3 Hz, H-16), 5.08 (d, *J*=7.3 Hz, H-1 of Xyl II), 5.14 (d, *J*=7.7 Hz, H-1 of Xyl I), 5.17 (d, *J*=7.7 Hz, H-1 of Glc I), 5.56 (d, *J*=7.7 Hz, H-1 of Glc II).

CMR (C$_5$D$_5$N, 125 MHz) : δ C-1) 37.3 (2) 30.0 (3) 77.8 (4) 34.9 (5) 44.8 (6) 29.1 (7) 32.6 (8) 35.4 (9) 54.6 (10) 35.9 (11) 21.4 (12) 40.8 (13) 40.3 (14) 56.5 (15) 32.6 (16) 81.3 (17) 64.0 (18) 16.9 (19) 12.4 (20) 41.2 (21) 16.6 (22) 110.8 (23) 37.3 (24) 28.5 (25) 34.4 (26) 75.3 (27) 17.6 **Gal** (1) 102.6 (2) 73.2 (3) 75.2 (4) 79.8 (5) 75.3 (6) 60.8 **Glc I** (1) 104.9 (2) 80.8 (3) 86.9 (4) 70.5 (5) 77.6 (6) 63.0 **Glc II** (1) 104.0 (2) 75.2 (3) 86.9 (4) 69.2 (5) 78.5 (6) 62.2 **Xyl I** (1) 104.9 (2) 75.3 (3) 78.6 (4) 70.8 (5) 67.3 **Xyl II** (1) 106.2 (2) 75.5 (3) 77.6 (4) 70.8 (5) 67.2 **Glc III** (1) 104.9 (2) 75.5 (3) 78.6 (4) 78.6 (5) 71.8 (6) 62.9.

Mass (FAB, Negative ion) : *m/z* 1346 [M]⁻, 1214 [M-Xyl]⁻, 1051 [M-H-Xyl-Glc]⁻, 919 [M-H-Xyl-Glc-Xyl]⁻, 757 [M-H-Xyl-Glc-Xyl-Glc]⁻.

Mass (FAB, Negative ion, H.R.) : *m/z* 1345.619 [(M-H)⁻, requires 1345.6276].

Reference

1.   J.-M. Jin, Y.-J. Zhang and C.-R. Yang, *J. Nat. Prod.*, **67**, 5 (2004).

FLORIBUNDASAPONIN E
(25*R*)-22α-Methoxyfurost-5-ene-3β,26-diol 3-O-[α-L-rhamnopyranosyl-(1→3)-α-L-rhamnopyranosyl-(1→3)-α-L-rhamnopyranosyl-(1→3)-α-L-rhamnopyranosyl-(1→4)-β-D-glucopyranoside]-26-O-[β-D-glucopyranoside]

Source : *Dioscorea floribunda* Mart. et Gal. (Dioscoreaceae)
Mol. Formula : C$_{64}$H$_{106}$O$_{30}$
Mol. Wt. : 1354
M.P. : 226-229°C (decomp.)
[α]$_D$: -66° (C$_5$H$_5$N)
Registry No. : [68406-05-3]

IR (Nujol) : No spiroketal absorption.

PMR (CDCl$_3$, 60 MHz) : δ 3.18 (s, OCH_3).

Reference

1.   S.B. Mahato, N.P. Sahu and B.C. Pal, *Ind. J. Chem.*, **16**(B), 350 (1978).

ALLIUMOSIDE C

(25R)-3β,22α,26-Trihydroxy-furostane 3-O-{α-L-rhamnopyranosyl-(1→4)-α-L-rhamnopyranosyl-(1→4)-α-L-rhamnopyranosyl-(1→6)-β-D-galactopyranosyl-(1→6)-β-D-glucopyranoside}-26-O-β-D-glucopyranoside

Source : *Allium narcissiflorum* (Liliaceae)
Mol. Formula : $C_{63}H_{104}O_{31}$
Mol. Wt. : 1356
Registry No. : [56126-13-7]

Reference

1.   G.V. Lazurevski, V.V. Krokhmalyuk and P.K. Kintya, *Dokl. Akad. Nauk. SSSR*, **221**, 744 (1975).

IPHEION UNIFLORUM SAPONIN 1

3β,26-Dihydroxy-22α-methoxy-5α-furost-25(27)-en-2-one 3-O-{α-L-arabinopyranosyl-(1→2)-[β-D-xylopyranosyl-(1→3)]-β-D-glucopyranosyl-(1→4)-[α-L-rhamnopyranosyl-(1→2)]-β-D-galactopyranoside}-26-O-β-D-glucopyranoside

Source : *Ipheion uniflorum* (Liliaceae)
Mol. Formula : $C_{62}H_{100}O_{32}$
Mol. Wt. : 1356
$[\alpha]_D^{28}$: -72.0° (c=0.10, CHCl$_3$-MeOH)
Registry No. : [156788-86-2]

IR (KBr) **:** 3425 (OH), 2930 (CH), 1720 (C=O), 1450, 1370, 1165, 1070, 1040, 900, 775 cm^{-1}.

PMR (C$_5$D$_5$N-CD$_3$OD 10:1, 500 MHz) **:** δ 1.71 (d, J=6.1 Hz, 3xH-6 of Rha), 3.54 (br d, J=12.5 Hz, H-5A of Ara), 3.60 (dd, J=10.8, 10.8 Hz, H-5A of Xyl), 3.75 (H-5 of Glc I), 3.79 (dd, J=9.0, 9.0 Hz, H-4 of Glc I), 3.86 (dd, J=8.3, 7.8 Hz, H-2 of Xyl), 3.86 (H-5 of Glc II), 3.94 (dd, J=8.7, 3.7 Hz, H-3 of Ara), 3.96 (dd, J=8.3, 8.3 Hz, H-3 of Xyl), 3.97 (H-2 of Glc II), 3.97 (H-6A of Glc I), 3.97 (H-5 of Gal), 4.03 (ddd, J=10.8, 8.3, 3.1 Hz, H-4 of Xyl), 4.03 (dd, J=9.0, 9.0 Hz, H-3 of Glc I), 4.10 (H-3 of Gal), 4.10 (H-4 of Ara), 4.10 (H-4 of Glc), 4.13 (H-6A of Gal), 4.13 (H-3 of Glc II), 4.17 (dd, J=9.4, 9.4 Hz, H-4 of Rha), 4.17 (dd, J=9.0, 8.1 Hz, H-2 o f Glc I), 4.17 (dd, J=10.8, 3.1 Hz, H-5B of Xyl), 4.28 (dd, J=10.8, 5.8 Hz, H-6A of Glc II), 4.29 (dd, J=8.7, 7.7 Hz, H-2 of Ara), 4.38 (dd, J=9.4, 2.9 Hz, H-3 of Rha), 4.42 (H-2 of Gal), 4.42 (H-4 of Gal), 4.43 (H-6B of Glc I), 4.47 (br d, J=10.8 Hz, H-6B of Glc II), 4.57 (dd, J=12.5, 3.5 Hz, H-5B of Ara), 4.58 (H-6B of Gal), 4.67 (br d, J=2.9 Hz, H-2 of Rha), 4.72 (dq, J=9.4, 6.1 Hz, H-5 of Rha), 4.84 (d, J=7.9 Hz, H-1 of Glc II), 4.88 (d, J=8.1 Hz, H-1 of Gal), 4.90 (d, J=8.1 Hz, H-1 of Glc I), 5.15 (d, J=7.8 Hz, H-1 of Xyl), 5.24 (d, J=7.7 Hz, H-1 of Ara), 6.17 (br s, H-1 of Rha).

PMR (C$_5$D$_5$N, 500 MHz) **:** δ 0.76 (s, 3xH-18), 0.83 (s, 3xH-19), 1.16 (d, J=6.9 Hz, 3xH-21), 1.77 (d, J=6.1 Hz, 3xH-6 of Rha), 3.27 (s, OCH_3), 4.90 (d, J=8.1 Hz, H-1 of Glc II), 4.91 (d, J=7.7 Hz, H-1 of Gal), 4.96 (d, J=7.8 Hz, H-1 of Glc I), 5.21 (d, J=7.7 Hz, H-1 of Xyl), 5.31 (d, J=7.8 Hz, H-1 of Ara), 6.28 (br s, H-1 of Rha).

CMR (C$_5$D$_5$N, 100 MHz) **:** δ C-1) 53.2 (2) 205.1 (3) 80.0 (4) 35.8 (5) 44.0 (6) 27.8 (7) 32.1 (8) 34.5 (9) 53.7 (10) 41.1 (11) 21.2 (12) 39.7 (13) 41.1 (14) 56.1 (15) 32.0 (16) 81.4 (17) 64.2 (18) 16.4 (19) 12.7 (20) 40.5 (21) 16.2 (22) 112.4 (23) 31.7 (24) 28.2 (25)146.9 (26) 72.1 (27) 111.2 (OCH_3) 47.4 **Gal** (1) 99.8 (2) 77.1 (3) 76.2 (4) 81.4 (5) 75.5 (6) 60.4 **Rha** (1) 101.7 (2) 72.3 (3) 72.7 (4) 74.0 (5) 69.9 (6) 18.5 **Glc I** (1) 105.5 (2) 81.6 (3) 88.0 (4) 70.4 (5) 77.7 (6) 62.9 **Ara** (1) 105.9 (2) 73.3 (3) 74.8 (4) 69.8 (5) 67.3 **Xyl** (1) 105.0 (2) 75.1 (3) 78.7 (4) 70.7 (5) 67.3 **Glc II** (1) 103.9 (2) 75.2 (3) 78.6 (4) 71.8 (5) 78.5 (6) 62.9.

Mass (FAB, Negative ion) **:** m/z 1356 [M]$^-$, 1223 [M Ara (or Xyl)]$^-$, 1210 [M Rha]$^-$, 929 [M Ara Xyl Glc]$^-$.

Biological Activity : The compound shows inhibitory activity on CAMP phosphodiesterae with IC_{50}=89.3×10^{-5} M.

Reference

1.   O. Nakamura, Y. Mimaki, Y. Sashida, T. Nikaido and T. Ohmoto, *Chem. Pharm. Bull.*, **42**, 1116 (1994).

IPHEION UNIFLORUM SAPONIN 2

3β,26-Dihydroxy-22α-methoxy-(25S)-5α-furostan-2-one 3-O-{α-L-arabinopyranosyl-(1→2)-[β-D-
xylopyranosyl-(1→3)]-β-D-glucopyranosyl-(1→4)-[α-L-rhamnopyranosyl-(1→2)]-β-D-galactopyranoside}-
26-O-β-D-glucopyranoside

Source : *Ipheion uniflorum* (Liliaceae)
Mol. Formula : $C_{62}H_{102}O_{32}$
Mol. Wt. : 1358
[α]$_D^{28}$: -62.0° (c=0.10, CHCl$_3$-MeOH)
Registry No. : [156788-87-3]

IR (KBr) : 3430 (OH), 2925 (CH), 1720 (C=O), 1440, 1370, 1155, 1070, 1040 cm^{-1}.

PMR (C$_5$D$_5$N, 400/500 MHz) : δ 0.77 (s, 3xH-18), 0.83 (s, 3xH-19), 1.05 (d, J=6.6 Hz, 3xH-27), 1.16 (d, J=6.8 Hz, 3xH-21), 1.77 (d, J=6.1 Hz, 3xH-6 of Rha), 3.26 (s, OCH$_3$), 4.84 (d, J=7.8 Hz, H-1 of Glc II), 4.90 (d, J=7.6 Hz, H-1 of Gal), 4.96 (d, J=7.8 Hz, H-1 of Glc I), 5.21 (d, J=7.7 Hz, H-1 of Xyl), 5.31 (d, J=7.7 Hz, H-1 of Ara), 6.29 (br s, H-1 of Rha).

CMR (C$_5$D$_5$N, 100 MHz) **:** δ C-1) 53.3 (2) 205.1 (3) 80.1 (4) 35.8 (5) 44.0 (6) 27.8 (7) 32.1 (8) 34.5 (9) 53.7 (10) 41.1 (11) 21.3 (12) 39.7 (13) 41.1 (14) 56.1 (15) 32.0 (16) 81.3 (17) 64.3 (18) 16.4 (19) 12.7 (20) 40.5 (21) 16.3 (22) 112.7 (23) 31.0 (24) 28.2 (25) 34.5 (26) 74.9 (27) 17.5 (OCH$_3$) 47.4 **Gal** (1) 99.8 (2) 77.1 (3) 76.2 (4) 81.4 (5) 75.6 (6) 60.5 **Rha** (1) 101.7 (2) 72.3 (3) 72.7 (4) 74.1 (5) 69.9 (6) 18.5 **Glc I** (1) 105.5 (2) 81.6 (3) 88.0 (4) 70.4 (5) 77.7 (6) 63.0 **Ara** (1) 105.9 (2) 73.4 (3) 74.9 (4) 69.8 (5) 67.4 **Xyl** (1) 105.1 (2) 75.1 (3) 78.8 (4) 70.7 (5) 67.4 **Glc II** (1) 105.0 (2) 75.3 (3) 78.7 (4) 71.9 (5) 78.5 (6) 63.0.

Mass (FAB, Negative ion) **:** *m/z* 1358 [M]⁻, 1225 [M-Ara (or Xyl)]⁻, 1094 [M-Ara-Xyl]⁻, 934 [M-Ara-Xyl-Glc]⁻.

Biological Activity : The compound exhibited inhibitory activity on CAMP phosphodiesterae with IC$_{50}$=41.2×10^{-5} M.

Reference

1.   O. Nakamura, Y. Mimaki, Y. Sashida, T. Nikaido and T. Ohmoto, *Chem. Pharm. Bull.*, **42**, 1116 (1994).

IPHEION UNIFLORUM SAPONIN 3

3β,26-Dihydroxy-22α-methoxy-(25R)-5α-furostan-2-one 3-O-{α-L-arabinopyranosyl-(1→2)-[β-D-xylopyranosyl-(1→3)]-β-D-glucopyranosyl-(1→4)-[α-L-rhamnopyranosyl-(1→2)]-β-D-galactopyranoside}-26-O-β-D-glucopyranoside

Source : *Ipheion uniflorum* (Liliaceae)
Mol. Formula : C$_{62}$H$_{102}$O$_{32}$
Mol. Wt. : 1358
[α]$_D^{28}$: -40.0° (c=0.10, CHCl$_3$-MeOH)
Registry No. : [156857-31-7]

IR (KBr) : 3410 (OH), 2925 (CH), 1720 (C=O), 1445, 1370, 1255, 1155, 1065, 1045, 910, 890, 780 cm^{-1}.

PMR (C$_5$D$_5$N, 400/500 MHz) : δ 0.77 (s, 3xH-18), 0.84 (s, 3xH-19), 1.01 (d, J=6.7 Hz, 3xH-27), 1.19 (d, J=6.8 Hz, 3xH-21), 1.77 (d, J=6.1 Hz, 3xH-6 of Rha), 3.27 (s, OCH_3), 4.84 (d, J=7.8 Hz, H-1 of Glc II), 4.91 (d, J=7.7 Hz, H-1 of Gal), 4.96 (d, J=7.8 Hz, H-1 of Glc I), 5.22 (d, J=7.7 Hz, H-1 of Xyl), 5.32 (d, J=7.7 Hz, H-1 of Ara), 6.29 (br s, H-1 of Rha).

CMR (C$_5$D$_5$N, 100 MHz) : δ C-1) 53.3 (2) 205.1 (3) 80.1 (4) 35.8 (5) 44.0 (6) 27.8 (7) 32.1 (8) 34.5 (9) 53.7 (10) 41.1 (11) 21.3 (12) 39.7 (13) 41.1 (14) 56.1 (15) 32.1 (16) 81.3 (17) 64.3 (18) 16.4 (19) 12.7 (20) 40.6 (21) 16.3 (22) 112.7 (23) 30.9 (24) 28.3 (25) 34.3 (26) 75.2 (27) 17.2 (OCH$_3$) 47.3 **Gal** (1) 99.8 (2) 77.1 (3) 76.2 (4) 81.4 (5) 75.6 (6) 60.5 **Rha** (1) 101.7 (2) 72.3 (3) 72.7 (4) 74.1 (5) 69.9 (6) 18.5 **Glc I** (1) 105.5 (2) 81.6 (3) 88.0 (4) 70.4 (5) 77.7 (6) 62.9 **Ara** (1) 105.9 (2) 73.4 (3) 74.9 (4) 69.8 (5) 67.4 **Xyl** (1) 105.0 (2) 75.1 (3) 78.8 (4) 70.7 (5) 67.4 **Glc II** (1) 105.0 (2) 75.2 (3) 78.7 (4) 71.9 (5) 78.5 (6) 63.0.

Mass (FAB, Negative ion) : m/z 1358 [M]$^-$, 1225 [M-Ara (or Xyl)]$^-$, 1094 [M-Ara-Xyl]$^-$, 1080 [M-Ara-(or Xyl)-Rha]$^-$, 932 [M-Ara-Xyl-Glc]$^-$, 785 [M-Ara-Xyl-Glc-Rha]$^-$.

Biological Activity : The compound exhibited inhibitory activity on CAMP phosphodiesterae with IC$_{50}$=14.5×10^{-5} M.

Reference

1.   O. Nakamura, Y. Mimaki, Y. Sashida, T. Nikaido and T. Ohmoto, *Chem. Pharm. Bull.*, **42**, 1116 (1994).

POLIANTHOSIDE E

(5α,25*R*)-Furost-3β,22α,26-triol-12-one 3-O-{β-D-xylopyranosyl-(1→3)-[β-D-glucopyranosyl-(1→2)-
β-D-xylopyranosyl-(1→3)]-β-D-glucopyranosyl-(1→4)-β-D-galactopyranoside-26-O-β-D-glucopyranoside

Source : *Polianthes tuberosa* L. (Agavaceae)
Mol. Formula : $C_{61}H_{100}O_{33}$
Mol. Wt. : 1360
[α]$_D^{18.1}$: -23.53° (c=0.340, Pyridine)
Registry No. : [655233-69-5]

IR (KBr) : 3435, 2928, 1707, 1424, 1375, 1161, 1075, 1040, 894 cm^{-1}.

PMR (C$_5$D$_5$N, 500 MHz) : δ 0.63 (s, 3xH-18), 0.96 (d, *J*=6.5 Hz, 3xH-27), 1.11 (s, 3xH-19), 1.52 (d, *J*=6.7 Hz, 3xH-21), 3.77 (H-3), 3.84 (m, H-26A), 4.21 (m, H-26B), 4.52 (m, H-16), 4.77 (d, *J*=7.7 Hz, H-1 of Glc III), 4.83 (d, *J*=7.0 Hz, H-1 of Gal), 5.05 (d, *J*=7.0 Hz, H-1 of Xyl II), 5.11 (d, *J*=7.7 Hz, H-1 of Xyl I), 5.14 (d, *J*=7.0 Hz, H-1 of Glc I), 5.53 (d, *J*=6.2 Hz, H-1 of Glc II).

CMR (C_5D_5N, 125 MHz) : δ C-1) 36.7 (2) 29.7 (3) 77.8 (4) 34.7 (5) 44.6 (6) 28.7 (7) 31.8 (8) 34.3 (9) 55.7 (10) 36.4 (11) 38.1 (12) 213.2 (13) 55.8 (14) 56.0 (15) 31.8 (16) 79.8 (17) 55.1 (18) 16.4 (19) 11.8 (20) 41.3 (21) 15.4 (22) 110.8 (23) 37.2 (24) 28.5 (25) 34.3 (26) 75.2 (27) 17.5 **Gal** (1) 102.5 (2) 73.2 (3) 75.2 (4) 79.8 (5) 75.2 (6) 60.8 **Glc I** (1) 105.0 (2) 80.8 (3) 86.8 (4) 70.5 (5) 77.6 (6) 63.0 **Glc II** (1) 104.0 (2) 75.2 (3) 86.8 (4) 69.2 (5) 78.5 (6) 62.2 **Xyl I** (1) 105.0 (2) 75.2 (3) 78.6 (4) 70.8 (5) 67.2 **Xyl II** (1) 106.2 (2) 75.4 (3) 77.3 (4) 70.8 (5) 67.2 **Glc III** (1) 105.0 (2) 75.4 (3) 78.6 (4) 71.8 (5) 78.6 (6) 62.9.

Mass (FAB, Negative ion) : m/z 1360 [M]⁻, 1228 [M-Xyl]⁻, 1066 [M-Xyl-Glc]⁻, 934 [M-Xyl-Glc-Xyl]⁻, 771 [M-H-2xXyl-2xGlc]⁻.

Mass (FAB, Negative ion, H.R.) : m/z 1459.6039 [(M-H)⁻, calcd. for 1359.6069].

Biological Activity: Cytotoxic against HeLa cells with IC_{50}=5.21 μg/mL.

Reference

1.   J.-M. Jin, Y.-J. Zhang and C.-R. Yang, *J. Nat. Prod.*, **67**, 5 (2004).

YUCCA GLORIOSA SAPONIN YG - 4

(22ξ,25R)-Furost-5α-spirostan-2α,3β,22,26-tetraol 3-O-{β-D-xylopyranosyl-(1→3)-β-D-glucopyranosyl-(1→2)-[β-D-xylopyranosyl-(1→3)]-β-D-glucopyranosyl-(1→4)-β-D-galactopyranoside}-26-O-β-D-glucopyranoside

Source : *Yucca gloriosa* L. (Liliaceae)
Mol. Formula : $C_{61}H_{102}O_{33}$
Mol. Wt. : 1362
$[α]_D^{19}$: +21.0° (c=1.0, Pyridine)
Registry No. : [119459-82-4]

IR (KBr) **:** 3400 (OH) cm^{-1}.

Mass (FAB, Positive ion) **:** m/z 1401 [M+K]$^{+}$.

Reference

1. K. Nakano, E. Matsuda, K. Tsurumi, T. Yamasaki, K. Murakami, Y. Takaishi and T. Tomimatsu, *Phytochemistry*, **27**, 3235 (1988).

METHYL PROTO-RHAPSISSAPONIN

3β,26-Dihydroxy-22ξ-methoxyfurost-5-ene 3-O-{β-D-glucopyranosyl-(1→4)-α-L-rhamnopyranosyl-(1→4)-
α-L-rhamnopyranosyl-(1→4)-[α-L-rhamnopyranosyl-(1→2)]-β-D-glucopyranoside}-26-O-β-D-glucopyranoside

Source : *Chamaerops humilis* L. (Palmae)
Mol. Formula : $C_{64}H_{106}O_{31}$
Mol. Wt. : 1370
M.P. : 196-199°C
$[\alpha]_D^{20}$ **:** -92.4° (c=1.02, C_5D_5N)
Registry No. : [102100-45-8]

IR (KBr) : 3460-3320 (OH) cm^{-1}.

CMR (C_5D_5N, 25 MHz) : **Rha I** δ C-1) 102.0 (6) 18.2 **Rha II** (1) 102.3 (6) 18.5 **Rha III** (1) 102.6 (6) 18.8 **Glc I** (1)
100.4 **Glc II** (1) 104.8 **Glc III** (1) 106.5 (OCH₃) 47.3.

Reference

1.   Y. Hirai, S. Sanada, Y. Ida and J. Shoji, *Chem. Pharm. Bull.*, **34**, 82 (1986).

METHYL PROTO-RUPICOLASIDE

(25*R*)-22ξ-Methoxyfurost-5-en-3β,26-diol 3-O-{β-D-glucopyranosyl-(1→2)-α-L-rhamnopyranosyl-(1→4)-α-L-rhamnopyranosyl-(1→4)-[α-L-rhamnopyranosyl-(1→2)]-β-D-glucopyranoside}-26-O-β-D-glucopyranoside

Source : *Phoenix rupicola* T. Anderson (Palmae)
Mol. Formula : $C_{64}H_{106}O_{31}$
Mol. Wt. : 1370
M.P. : 195-199°C (decomp.)
[α]$_D$: -82.6° (c=0.71, C_5D_5N)
Registry No. : [137232-01-0]

IR (KBr) : 3600-3200 (OH) cm^{-1}.

CMR each signal of the aglycone moeity was analogous to that of Methyl proto-taccaoside.

CMR (C_5D_5N, 100 MHz) : δ **Glc I** C-1) 100.4 (2) 81.0 (3) 76.7 (4) 78.4 (5) 77.6 (6) 61.5 **Rha I** (1) 102.0^a (2) 72.3 (3) 72.6 (4) 74.1 (5) 69.4 (6) 18.5^b **Rha II** (1) 102.3^a (2) 69.9 (3) 72.4 (4) 78.0 (5) 68.3 (6) 18.5^b **Rha III** (1) 102.1^a (2) 82.4 (3) 72.7 (4) 74.1 (5) 73.1 (6) 18.2^b **Glc II** (1) 107.0 (2) 75.6 (3) 78.0 (4) 71.8 (5) 78.4 (6) 62.9 **Glc III** (1) 104.7 (2) 75.0 (3) 78.0 (4) 71.8 (5) 78.4 (6) 62.9.

Reference

1.   K. Idaka, Y. Hirai and J. Shoji, *Chem. Pharm. Bull.*, **39**, 1455 (1991).

ALLIUMOSIDE D

**(25S)-Furost-5-en-3β,22α,26-triol 3-O-{α-L-rhamnopyranosyl-(1→4)-α-L-rhamnopyranosyl-(1→6)-
β-D-glucopyranosyl-(1→2)-[β-D-glucopyranosyl-(1→3)]-β-D-glucopyranoside}-26-O-β-D-glucopyranoside**

Source : *Allium narcissiflorum* (Liliaceae)
Mol. Formula : $C_{63}H_{104}O_{32}$
Mol. Wt. : 1372
M.P. : 268-269°C
$[\alpha]_D^{18}$: -88.0° (c=0.85, MeOH)
Registry No. : [56190-05-7]

Reference

1.   V.V. Krokhmalyuk and P.K. Kintya, *Khim. Prir. Soedin.*, 184 (1976); *Chem. Nat. Comp.*, **12**, 165 (1976).

CESTRUM NOCTURNUM SAPONIN 7

(25R)-2α,3β,26-Trihydroxyfurost-5,20(22)-diene 3-O-{β-D-glucopyranosyl-(1→3)-β-D-glucopyranosyl-(1→2)-[β-D-xylopyranosyl-(1→3)]-β-D-glucopyranosyl-(1→4)-β-D-galactopyranoside}-26-O-β-D-glucopyranoside

Source : *Cestrum nocturnum* L. (Solanaceae)
Mol. Formula : $C_{62}H_{100}O_{33}$
Mol. Wt. : 1372
[α]$_D^{28}$: -46.0° (c=0.10, MeOH)
Registry No. : [479078-21-2]

IR (film) : 3442 (OH), 2925 (CH), 1650, 1453, 1434, 1376, 1159, 1077, 1039, 894 cm^{-1}.

PMR (C_5D_5N, 500 MHz) : δ 0.69 (s, 3xH-18), 0.94 (s, 3xH-19), 1.01 (d, *J*=6.6 Hz, 3xH-27), 1.61 (s, 3xH-21), 4.83 (d, *J*=7.8 Hz, H-1 of Glc IV), 4.92 (d, *J*=7.7 Hz, H-1 of Gal), 5.12 (d, *J*=7.8 Hz, H-1 of Xyl), 5.16 (d, *J*=7.7 Hz, H-1 of Glc III), 5.17 (d, *J*=7.7 Hz, H-1 of Glc I), 5.30 (br d, *J*=4.1 Hz, H-6), 5.57 (d, *J*=7.7 Hz, H-1 of Glc II).

CMR (C_5D_5N, 125 MHz) : δ C-1) 45.7 (2) 70.0 (3) 84.4 (4) 37.5 (5) 140.0 (6) 121.9 (7) 32.2 (8) 30.8 (9) 50.1 (10) 37.8 (11) 21.3 (12) 39.5 (13) 43.3 (14) 54.7 (15) 34.4 (16) 84.4 (17) 64.4 (18) 14.1 (19) 20.4 (20) 103.5 (21) 11.7 (22) 152.3 (23) 23.6 (24) 31.4 (25) 33.4 (26) 74.9 (27) 17.3 **Gal** (1) 103.2 (2) 72.6 (3) 75.3 (4) 79.1 (5) 75.6 (6) 60.6 **Glc I** (1) 104.3 (2) 80.6 (3) 86.9 (4) 70.3 (5) 77.4 (6) 62.8 **Glc II** (1) 103.8 (2) 74.6 (3) 87.7 (4) 69.6 (5) 77.7 (6) 62.1 **Glc**

III (1) 105.3 (2) 75.5 (3) 77.9 (4) 71.5 (5) 78.4 (6) 62.4 **Xyl** (1) 104.8 (2) 75.1 (3) 78.4 (4) 70.6 (5) 67.1 **Glc IV** (1) 104.8 (2) 75.1 (3) 78.5 (4) 71.7 (5) 78.3 (6) 62.8.

Mass (TOF, Positive ion, H.R.) **:** m/z 1395.6025 [(M+Na)$^+$ requires 1395.6045].

Mass (TOF, Negative ion, H.R.) **:** m/z 1371.6121 [(M-H)$^-$ requires 1371.6069].

Reference

1. Y. Mimaki, K. Watanabe, H. Sakagami and Y. Sashida, *J. Nat. Prod.*, **65**, 1863 (2002).

AGAVOSIDE G

(25R)-3β,22α,26-Trihydroxy-5α-furostan-12-one 3-O-{β-D-xylopyranosyl-(1→2)-[α-L-rhamnopyranosyl-(1→3)-β-D-glucopyranosyl-(1→4)]-β-D-glucopyranosyl-(1→4)-β-D-galactopyranoside}-26-O-β-D-glucopyranoside

Source : *Agave americana* L.[1,2,3] (Agavaceae)
Mol. Formula : $C_{62}H_{102}O_{33}$
Mol. Wt. : 1374
M.P. : 128-130°C
[α]$_D^{20}$: -90° (c=1.0, CHCl$_3$)
Registry No. : [58546-20-6]

IR : 1700 (C=O), 900 (br) cm^{-1}.

Reference

1   P.K. Kintya, V.A. Bobeiko and A.P. Gulya, *Khim. Prir. Soedin.*, 486 (1976); *Chem. Nat. Comp.*, **12**, 427 (1976).

CAPSICOSIDE D

5α,25*R*-Furost-3β,22ξ,26-triol 3-O-{β-D-glucopyranosyl-(1→4)-β-D-glucopyranosyl-(1→2)-[β-D-xylopyranosyl-(1→3)]-β-D-glucopyranosyl-(1→4)-β-D-galactopyranoside}-26-O-β-D-glucopyranoside

Source : *Capsicum annuum* L. var. *conoides*[1] and *Capsicum annuum* L. var. *fasciculatum*[1] , *C. annuum* L. var. *acuminatum*[2] (Solanaceae)
Mol. Formula : C$_{62}$H$_{104}$O$_{33}$
Mol. Wt. : 1376
[α]$_D^{28}$: -53.9^0 (c=0.50, Pyridine)
Registry No. : [160260-25-3]

PMR (C_5D_5N, 400 MHz)[1] : δ 0.75 (s, CH_3), 0.78 (s, CH_3), 1.04 (d, J=6.2 Hz, sec. CH_3), 1.22 (d, J=6.6 Hz, sec. CH_3), 4.64 (d, J=8.0 Hz), 4.77 (d, J=7.3 Hz), 4.82 (d, J=8.1 Hz), 5.38 (d, J=8.0 Hz, 4 x anomeric H), other anomeric H signals were overlapped with H_2O.

PMR (C_5D_5N+H_2O, 500 MHz)[2] : δ 0.54 (m, H-9), 0.66 (s, 3xH-19), 0.81 (s, 3xH-18), 0.99 (d, J=6.8 Hz, 3xH-27), 1.32 (d, J=6.8 Hz, 3xH-21), 3.63 (m, 2xH-26).

CMR (C_5D_5N, 100 MHz)[1] : δ C-1) 36.6 (2) 29.1 (3) 78.3 (4) 33.9 (5) 44.2 (6) 28.1 (7) 31.4 (8) 34.5 (9) 53.8 (10) 35.2 (11) 20.7 (12) 40.2 (13) 40.5 (14) 55.6 (15) 31.7 (16) 80.9 (17) 62.3 (18) 16.1 (19) 11.8 (20) 39.6 (21) 15.4 (22) 110.7 (23) 35.6 (24) 27.3 (25) 33.3 (26) 75.0 (27) 16.6 **Gal** (1) 101.3 (2) 71.9 (3) 75.7 (4) 79.7 (5) 75.9 (6) 60.3 **Glc I** (1) 102.4 (2) 79.4 (3) 86.1 (4) 70.2 (5) 76.5 (6) 61.3 **Glc II** (1) 103.2 (2) 73.5 (3) 74.7 (4) 79.3 (5) 74.1 (6) 60.6 **Glc III** (1) 103.2 (2) 73.7 (3) 76.5 (4) 69.4 (5) 76.5 (6) 61.0 **Glc IV** (1) 103.3 (2) 73.9 (3) 76.5 (4) 68.8 (5) 76.5 (6) 61.3 **Xyl** (1) 103.4 (2) 73.6 (3) 76.5 (4) 70.0 (5) 65.7.

Mass (FAB, Positive ion)[2] : m/z 1399 [M+Na]$^+$, 1237 [M+Na-Glc]$^+$, 1075 [M+Na-2xGlc]$^+$.

Mass (FAB, Negative ion)[1] : m/z 1375 [M-H]$^-$.

References

1. S. Yahara, T. Ura, C. Sakamoto and T. Nohara, *Phytochemistry*, **37**, 831 (1994).

2. M. Iorizzi, V. Lanzotti, G. Ranalli, S. de Marino and F. Zollo, *J. Agric. Food Chem.*, **50**, 4310 (2002).

LANATIGOSIDE

(25*R*)-5α-Furostan-3β,22α-26-triol 3-O-{β-D-glucopyranosyl-(1→3)-β-D-galactopyranosyl-(1→2)-β-D-xylopyranosyl-(1→3)-β-D-glucopyranosyl-(1→4)-β-D-galactopyranoside}-26-O-β-D-glucopyranoside

Source : *Digitalis lanata* Ehrh. (Scrophulariaceae)
Mol. Formula : $C_{62}H_{104}O_{33}$
Mol. Wt.: 1376
M.P. : 260-283°C
[α]$_D^{21}$: -18.6° (c=0.74, $CHCl_3$/MeOH)
Registry No. : [40526-01-0]

Reference

1. R. Tschesche, L. Seidel, S.C. Sharma and G. Wulff, *Chem. Ber.*, **105**, 3397 (1972).

POLIANTHOSIDE G

(5α,25R)-Furostane-3β,22α,26-triol 3-O-{β-D-xylopyranosyl-(1→3)-[β-D-glucopyranosyl-(1→2)-
[β-D-glucopyranosyl-(1→3)]-β-D-glucopyranosyl-(1→4)-β-D-galactopyranoside}-
26-O-β-D-glucopyranoside

Source : *Polianthes tuberosa* L. (Agavaceae)
Mol. Formula : $C_{62}H_{104}O_{33}$
Mol. Wt. : 1376
$[\alpha]_D^{19.7}$: -35.26° (c=0.0390, Pyridine)
Registry No. : [655233-84-4]

IR (KBr) : 3412, 2927, 1704, 169, 1453, 1425, 1376, 1160, 1075, 1040, 894 cm^{-1}.

PMR (C$_5$D$_5$N, 500 MHz) : δ 0.63 (s, 3xH-19), 0.86 (s, 3xH-18), 0.96 (d, J=6.6 Hz, 3xH-27), 1.31 (d, J=6.3 Hz, 3xH-21), 3.78 (H-3), 3.81 (m, H-26A), 4.34 (m, H-26B), 4.78 (d, J=7.7 Hz, H-1 of Glc IV), 4.86 (d, J=4.4 Hz, H-1 of Gal), 4.92 (m, H-16), 5.07 (H-1 of Xyl I), 5.11 (d, J=5.5 Hz, H-1 of Glc I), 5.18 (d, J=7.7 Hz, H-1 of Glc II).

CMR (C$_5$D$_5$N, 125 MHz) : δ C-1) 37.3 (2) 30.0 (3) 77.8 (4) 34.9 (5) 44.8 (6) 29.1 (7) 32.5 (8) 35.4 (9) 54.6 (10) 35.9 (11) 21.4 (12) 40.8 (13) 40.3 (14) 56.5 (15) 32.5 (16) 81.2 (17) 64.0 (18) 16.9 (19) 12.4 (20) 41.2 (21) 16.6 (22) 110.8 (23) 37.3 (24) 28.5 (25) 34.3 (26) 75.2 (27) 17.6 **Gal** (1) 102.5 (2) 73.2 (3) 75.2 (4) 80.1 (5) 75.5 (6) 60.8 **Glc I** (1) 104.9 (2) 81.0 (3) 88.5 (4) 70.8 (5) 77.6 (6) 62.9 **Glc II** (1) 104.0 (2) 75.2 (3) 86.9 (4) 69.2 (5) 78.5 (6) 62.2 **Glc III** (1) 104.6 (2) 75.5 (3) 78.6 (4) 71.8 (5) 78.5 (6) 62.5 **Xyl** (1) 106.1 (2) 75.2 (3) 77.6 (4) 70.8 (5) 67.2 **Glc IV** (1) 104.9 (2) 75.5 (3) 78.6 (4) 71.7 (5) 78.5 (6) 62.9.

Mass (FAB, Negative ion) : m/z 1376 [M]$^-$, 1214 [M-Glc]$^-$, 1082 [M-Xyl-Glc]$^-$, 757 [M-H-Xyl-3xGlc]$^-$.

Mass (FAB, Negative ion, H.R.) : m/z 1375.6420 [(M-H)$^-$, requires 1375.6381].

Reference

1.   J.-M. Jin, Y.-J. Zhang and C.-R. Yang, *J. Nat. Prod.*, **67**, 5 (2004).

SATIVOSIDE-R$_1$

3β,22ξ,26-Trihydroxy-5α-furostane 3-O-{β-D-glucopyranosyl-(1→3)-β-D-glucopyranosyl-(1→2)-[β-D-xylopyranosyl-(1→3)]-β-D-glucopyranosyl-(1→4)-β-D-galactopyranoside}-26-O-β-D-glucopyranoside

Source : *Allium sativum* L. (Liliaceae)
Mol. Formula : $C_{62}H_{104}O_{33}$
Mol. Wt. : 1376
[α]$_D^{26}$: -45.0° (c=0.59, Pyridine)
Registry No. : [126594-43-2]

PMR (C$_5$D$_5$N, 270 MHz) : δ 0.65 (s, CH$_3$), 0.88 (s, CH$_3$), 0.99 (d, J=6.6 Hz, sec. CH$_3$), 1.34 (d, J=6.6 Hz, sec. CH$_3$), 4.82 (d, J=7.7 Hz), 4.89 (d, J=7.3 Hz), 5.12 (d, J=8.7 Hz), 5.15 (d, J=8.7 Hz), 5.18 (d, J=8.3 Hz), 5.54 (d, J=7.4 Hz, 6 x anomeric H).

CMR (C$_5$D$_5$N, 67.8 MHz) : δ C-1) 37.2 (2) 29.9 (3) 77.6^a (4) 34.8 (5) 44.7 (6) 29.0 (7) 32.4 (8) 35.3 (9) 54.5 (10) 35.8 (11) 21.3 (12) 40.2 (13) 41.1 (14) 56.4 (15) 32.4 (16) 81.1 (17) 64.0 (18) 16.7 (19) 12.3 (20) 40.7 (21) 16.4 (22) 110.6 (23) 37.2 (24) 28.4 (25) 34.3 (26) 75.3 (27) 17.5 **Gal** (1) 102.5 (2) 73.1 (3) 75.4^b (4) 79.6 (5) 75.6^b (6) 60.7 **Glc I** (1) 104.9^c (2) 80.8 (3) 86.8 (4) 70.6^d (5) 78.0^a (6) 62.2^e **Glc II** (1) 104.9^c (2) 74.7 (3) 87.4 (4) 69.4 (5) 78.2^a (6) 62.5^e **Glc III** (1) 104.0^c (2) 75.2^b (3) 78.5^a (4) 70.4^d (5) 78.5^d (6) 62.9^e **Xyl** (1) 105.4 (2) 75.4^b (3) 78.6^a (4) 71.5^a (5) 67.3 **Glc IV** (1) 104.9^c (2) 75.2^h (3) 78.6^a (4) 71.7^d (5) 78.5^a (6) 62.9^c.

Reference

1.   H. Matsuura, T. Ushiroguchi, Y. Itakura and T. Fuwa, *Chem. Pharm. Bull.*, **37**, 2741 (1989).

TACCA CHANTRIERI SAPONIN 5

(25*S*)-3β,26-Dihydroxy-22α-methoxyfurost-5-ene 3-O-{α-L-rhamnopyranosyl-(1→2)-[β-D-glucopyranosyl-(1→4)-α-L-rhamnopyranosyl-(1→3)]-β-D-glucopyranoside}-26-O-[β-D-glucopyranosyl-(1→6)-β-D-glucopyranoside]

Source : *Tacca chantrieri* André (Taccaceae)
Mol. Formula : $C_{64}H_{106}O_{32}$
Mol. Wt. : 1386
[α]$_D^{25}$: -54° (c=0.10, CHCl$_3$-MeOH)
Registry No. : [469879-86-5]

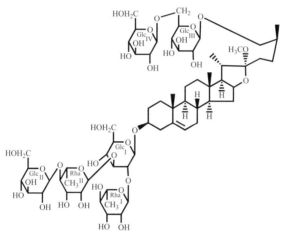

IR (film) : 3388 (OH), 2934 (CH), 1047 cm^{-1}.

PMR (C$_5$D$_5$N, 500 MHz) : δ 0.83 (s, 3xH-18), 1.03 (s, 3xH-19), 1.04 (d, *J*=6.7 Hz, 3xH-27), 1.18 (d, *J*=6.8 Hz, 3xH-21), 1.69 (d, *J*=6.2 Hz, 3xH-6 of Rha II), 1.75 (d, *J*=6.2 Hz, 3xH-6 of Rha I), 3.28 (s, OC*H$_3$*), 3.90 (m, *W½*=24.3 Hz, H-3), 4.45 (m, H-16), 4.77 (d, *J*=7.8 Hz, H-1 of Glc III), 4.88 (d, *J*=7.9 Hz, H-1 of Glc I), 5.11 (d, *J*=7.9 Hz, H-1 of Glc IV), 5.23 (d, *J*=7.8 Hz, H-1 of Glc II), 5.32 (br d, *J*=4.0 Hz, H-6), 5.74 (br s, H-1 of Rha II), 5.80 (br s, H-1 of Rha I).

CMR (C₅D₅N, 125 MHz) : δ C-1) 37.4 (2) 30.0 (3) 77.8 (4) 38.6 (5) 140.7 (6) 121.8 (7) 32.2 (8) 31.6 (9) 50.2 (10) 37.1 (11) 21.0 (12) 39.7 (13) 40.8 (14) 56.5 (15) 32.3 (16) 81.3 (17) 64.2 (18) 16.4 (19) 19.3 (20) 40.5 (21) 16.2 (22) 112.7 (23) 31.0 (24) 28.1 (25) 34.3 (26) 74.8 (27) 17.5 (OCH₃) 47.4 **Glc I** (1) 99.8 (2) 78.6 (3) 86.2 (4) 69.7 (5) 78.0 (6) 62.1 **Rha I** (1) 102.6 (2) 72.3 (3) 72.7 (4) 73.7 (5) 69.9 (6) 18.6 **Rha II** (1) 103.1 (2) 72.0 (3) 72.4 (4) 84.3 (5) 68.7 (6) 18.2 **Glc II** (1) 106.4 (2) 76.4 (3) 78.5 (4) 71.4 (5) 78.4 (6) 62.5 **Glc III** (1) 104.8 (2) 75.0 (3) 78.5 (4) 71.6 (5) 77.2 (6) 70.1 **Glc IV** (1) 105.4 (2) 75.1 (3) 78.4 (4) 71.5 (5) 78.4 (6) 62.7.

Mass (FAB, Negative ion) : *m/z* 1385 [M-H]⁻.

Reference

1.   A. Yokosuka, Y. Mimaki and Y. Sashida, *J. Nat. Prod.*, **65**, 1293 (2002).

LANAGITOSIDE

(25*R*)-5α-Furostan-2α,3β,22α,26-tetraol 3-O-{β-D-glucopyranosyl-(1→3)-β-D-galactopyranosyl-(1→2)-[β-D-xylopyranosyl-(1→3)]-β-D-glucopyranosyl-(1→4)-β-D-galactopyranoside}-26-O-β-D-glucopyranoside

Source : *Digitalis lanata* Ehrh. (Scrophulariaceae)
Mol. Formula : C₆₂H₁₀₄O₃₄
Mol. Wt. : 1392
M.P. : 268-292°C
[α]ᴅ²¹ : -28.4° (c=0.73, CHCl₃/MeOH)
Registry No. : [40526-02-1]

Reference

1.   R. Tschesche, L. Seidel, S.C. Sharma and G. Wulff, *Chem. Ber.*, **105**, 3397 (1972).

CESTRUM NOCTURNUM SAPONIN 6

(25*R*)-2α,3β,26-Trihydroxy-22α-methoxyfurost-5-ene 3-O-{β-D-glucopyranosyl-(1→3)-β-D-glucopyranosyl-
(1→2)-[β-D-xylopyranosyl-(1→3)]-β-D-glucopyranosyl-(1→4)-β-D-galactopyranoside}-
26-O-β-D-glucopyranoside

Source : *Cestrum nocturnum* L. (Solanaceae)
Mol. Formula : $C_{63}H_{104}O_{34}$
Mol. Wt. : 1404
$[\alpha]_D^{27}$: -60° (c=0.10, MeOH)
Registry No. : [479078-20-1]

IR (film) : 3417 (OH), 2935 (CH), 2903 (CH), 1453, 1433, 1376, 1259, 1159, 1077, 1040, 892 cm^{-1}.

PMR (C$_5$D$_5$N, 500 MHz) : δ 0.79 (s, 3xH-18), 0.94 (s, 3xH-19), 1.00 (d, *J*=6.6 Hz, 3xH-27), 1.18 (d, *J*=6.9 Hz, 3xH-21), 3.26 (s, OC*H*$_3$), 4.85 (d, *J*=7.8 Hz, H-1 of Glc IV), 4.92 (d, *J*=7.7 Hz, H-1 of Gal), 5.10 (d, *J*=7.8 Hz, H-1 of Glc III), 5.17 (d, *J*=7.7 Hz, H-1 of Xyl), 5.20 (d, *J*=7.9 Hz, H-1 of Glc I), 5.31 (br d, *J*=4.3 Hz, H-6), 5.57 (d, *J*=7.3 Hz, H-1 of Glc II).

CMR (C$_5$D$_5$N, 125 MHz) : δ C-1) 45.7 (2) 70.0 (3) 84.6 (4) 37.6 (5) 140.1 (6) 121.9 (7) 32.1 (8) 31.0 (9) 50.2 (10) 37.9 (11) 21.1 (12) 39.6 (13) 40.8 (14) 56.4 (15) 32.2 (16) 81.3 (17) 64.1 (18) 16.2 (19) 20.4 (20) 40.5 (21) 16.3 (22) 112.6 (23) 30.8 (24) 28.2 (25) 34.2 (26) 75.2 (27) 17.1 (OCH$_3$) 47.3 **Gal** (1) 103.3 (2) 72.6 (3) 75.4 (4) 79.0 (5) 75.7 (6) 60.6 **Glc I** (1) 104.4 (2) 80.6 (3) 87.0 (4) 70.4 (5) 77.5 (6) 62.9 **Glc II** (1) 103.9 (2) 74.7 (3) 87.7 (4) 69.7 (5) 77.7 (6) 62.3 **Glc III** (1) 105.4 (2) 75.6 (3) 78.0 (4) 71.5 (5) 78.4 (6) 62.5 **Xyl** (1) 104.9 (2) 75.2 (3) 78.5 (4) 70.6 (5) 67.2 **Glc IV** (1) 105.0 (2) 75.2 (3) 78.4 (4) 71.8 (5) 78.6 (6) 62.9.

Mass (TOF, Positive ion, H.R.) : *m/z* 1427.6340 [(M+Na)$^+$ requires 1427.6307].

Mass (TOF, Negative ion, H.R.) : *m/z* 1403.6366 [(M-H)$^-$ requires 1403.6331].

Biological Activity : Cytotoxic against human oral squamous cell carcinoma HSC-2 cells.

Reference

1.   Y. Mimaki, K. Watanabe, H. Sakagami and Y. Sashida, *J. Nat. Prod.*, **65**, 1863 (2002).

TUROSIDE C

(25*S*)-5α-Furostan-2α,3β,6β,22α,26-pentaol 3-O-{β-D-xylopyranosyl-(1→3)-[β-D-glucopyranosyl-(1→2)-β-D-glucopyranosyl-(1→2)]-β-D-glucopyranosyl-(1→4)-β-D-galactopyranoside}-26-O-β-D-glucopyranoside

Source : *Allium turcomanicum* Rgl. (Liliaceae)
Mol. Formula : C$_{63}$H$_{104}$O$_{35}$
Mol. Wt. : 1420
M.P. : 192-196°C (decomp.)
[α]$_D^{22}$: 69.5±3° (c=0.46, CHCl$_3$-MeOH)
Registry No. : [72947-92-3]

IR (KBr) : 3300-3500 (OH), 900 cm^{-1}.

Reference

1.   G.V. Pirtskhalava, M.B. Gorovits, T.T. Gorovits and N.K. Abubakirov, *Khim. Prir. Soedin.*, 514 (1979); *Chem. Nat. Comp.*, **15**, 446 (1979).

CAPSICOSIDE A

5α,25*R*-Furostan-2α,3β,22ξ,26-tetraol 3-O-{β-D-glucopyranosyl-(1→3)-β-D-glucopyranosyl-(1→2)-[β-D-glucopyranosyl-(1→3)]-β-D-glucopyranosyl-(1→4)-β-D-galactopyranoside}-26-O-β-D-glucopyranoside

Source : *Capsicum annuum* L. var. *conoides*[1], *Capsicum annuum* L. var. *fasciculatum*[1] (Solanaceae)
Mol. Formula : $C_{63}H_{106}O_{35}$
Mol. Wt. : 1422
[α]$_D^{28}$: -37.1° (c=0.80, Pyridine)[1]
Registry No. : [54999-56-3]

PMR (C$_5$D$_5$N, 400 MHz)[1] : δ 0.70 (s, CH$_3$), 0.86 (s, CH$_3$), 1.02 (d, *J*=6.9 Hz, sec. CH$_3$), 1.30 (d, *J*=6.6 Hz, sec. CH$_3$), 4.84 (d, *J*=7.0 Hz, anomeric H), 4.93 (d, *J*=7.7 Hz, anomeric H), 5.30 (d, *J*=8.0 Hz, anomeric H), 5.53 (d, *J*=7.7 Hz, anomeric H), other anomeric H overlapped with H$_2$O.

CMR (C$_5$D$_5$N, 100 MHz)[1] : δ C-1) 45.4 (2) 69.7 (3) 83.9 (4) 33.9 (5) 44.4 (6) 27.9 (7) 32.1 (8) 34.4 (9) 54.2 (10) 36.7 (11) 21.3 (12) 40.0 (13) 40.9 (14) 56.1 (15) 32.1 (16) 81.0 (17) 63.0 (18) 16.6 (19) 13.2 (20) 40.5 (21) 16.3 (22) 110.5

(23) 36.7 (24) 28.2 (25) 34.1 (26) 75.0 (27) 17.3 **Gal** (1) 103.0 (2) 72.4 (3) 75.8 (4) 79.5 (5) 77.3 (6) 60.5 **Glc I** (1) 104.3 (2) 81.0 (3) 87.6 (4) 70.3 (5) 77.8 (6) 61.9 **Glc II** (1) 104.5 (2) 73.9 (3) 88.2 (4) 70.6 (5) 77.8 (6) 62.4 **Glc III** (1) 105.0 (2) 75.5 (3) 78.4 (4) 71.1 (5) 78.2 (6) 62.7 **Glc IV** (1) 103.7 (2) 75.0 (3) 78.4 (4) 71.5 (5) 78.4 (6) 62.8 **Glc V** (1) 104.7 (2) 75.0 (3) 78.4 (4) 71.3 (5) 78.4 (6) 62.6.

Mass (FAB, Negative ion)[1] : m/z 1421 [M-H]⁻, 1259 [M-H-Glc]⁻, 1097 [1259-Glc]⁻, 935 [1097-Glc]⁻, 773 [935-Glc]⁻.

22-O-Methyl ether : 22-O-Methyl Capsicoside A [335316-95-5]

Isolated from *Capsicum annuum* L. var. *acuminatum*[2] (Solanaceae);
Mol. Formula : $C_{64}H_{108}O_{35}$; **Mol. Wt. :** 1436; $[\alpha]_D^{25}$: -49.7° (c=1.0, Pyridine)

PMR (C_5D_5N, 500.13 MHz)[2] : δ 0.57 (br t, H-9), 0.69 (s, 3xH-19), 0.79 (s, 3xH-18), 0.98 (m, H-5, H-12A and H-14), 1.02 (d, J=6.6 Hz, 3xH-27), 1.05 (H-6A), 1.17 (H-6B), 1.18 (H-1A), 1.19 (H-11A), 1.20 (d, J=6.6 Hz, 3xH-21), 1.32 (H-24A), 1.34 (H-8), 1.36 (H-15A), 1.47 (H-11B), 1.49 (H-4A), 1.50 (H-7), 1.63 (H-12B), 1.76 (H-17), 1.78 (H-23A), 1.80 (H-24B), 1.87 (H-4B), 1.92 (H-25), 1.98 (H-15B), 2.03 (H-23B), 2.18 (H-1B), 2.22 (m, H-20), 3.32 (s, OCH₃), 3.63 (H-26A), 3.96 (m, H-3), 3.98 (m, H-2), 4.00 (H-26B), 4.47 (m, H-16). The signals of the sugar are virtually identical to those of capsicoside F (qv).

CMR (C_5D_5N, 125.73 MHz)[2] : δ C-1) 44.5 (2) 69.7 (3) 83.0 (4) 32.8 (5) 43.6 (6) 27.3 (7) 31.2 (8) 33.7 (9) 53.7 (10) 36.0 (11) 20.5 (12) 39.0 (13) 40.3 (14) 55.3 (15) 31.3 (16) 80.6 (17) 63.4 (18) 15.7 (19) 12.5 (20) 39.6 (21) 15.5 (22) 112.0 (23) 29.4 (24) 27.2 (25) 33.2 (26) 74.5 (27) 16.3 (OCH₃) 46.6. The signals of the sugars are virtually identical to those of Capsicoside F (qv).

References

1.   S. Yahara, T. Ura, C. Sakamoto and T. Nohara, *Phytochemistry*, **37**, 831 (1994).

2.   M. Iorizzi, V. Lanzotti, G. Ranalli, S.D. Marino and F. Zollo, *J. Agric. Food Chem.*, **50**, 4310 (2002).

SATIVOSIDE-B₁
(25*R*)-3β,6β,22ξ,26-Tetrahydroxy-5α-furostane 3-O-{β-D-glucopyranosyl-(1→3)-β-D-glucopyranosyl-(1→2)-[β-D-glucopyranosyl-(1→3)]-β-D-glucopyranosyl-(1→4)-β-D-galactopyranoside}-26-O-β-D-glucopyranoside

Source : *Allium sativum* L. (Liliaceae)
Mol. Formula : $C_{63}H_{106}O_{35}$
Mol. Wt. : 1422
$[\alpha]_D^{26}$: -40.0° (c=0.39, H₂O)
Registry No. : [126594-42-1]

PMR (C₅D₅N, 270 MHz) : δ 0.91 (s, CH_3), 0.99 (d, J=6.3 Hz, sec. CH_3), 1.24 (s, CH_3), 1.35 (d, J=6.6 Hz, sec. CH_3), 4.85 (d, J=7.7 Hz), 4.95 (d, J=7.3 Hz), 5.17 (2xH-1 d, J=7.3 Hz), 5.25 (J=7.3 Hz), 5.57 (d, J=6.4 Hz, 6 x anomeric H).

CMR (C₅D₅N, 67.8 MHz) : δ C-1) 38.8 (2) 30.0 (3) 78.0^a (4) 32.8 (5) 48.0 (6) 70.8 (7) 40.3 (8) 30.6 (9) 54.7 (10) 36.2 (11) 21.3 (12) 40.9 (13) 41.2 (14) 56.3 (15) 32.5 (16) 81.1 (17) 64.0 (18) 16.7 (19) 16.4 (20) 40.7 (21) 16.0 (22) 110.7 (23) 37.2 (24) 28.4 (25) 34.3 (26) 75.3 (27) 17.5 **Gal** (1) 102.3 (2) 73.2 (3) 75.5^b (4) 80.1 (5) 75.5^b (6) 60.7 **Glc I** (1) 104.1^c (2) 80.9 (3) 88.6 (4) 70.9 (5) 78.0^a (6) 62.0^d **Glc II** (1) 104.5 (2) 74.7 (3) 87.5 (4) 69.2 (5) 77.5^a (6) 62.3^d **Glc III** (1) 104.8^c (2) 75.2^b (3) 78.6^d (4) 71.5^e (5) 78.6^a (6) 62.5^d **Glc IV** (1) 105.0^c (2) 75.2^b (3) 78.5^a (4) 71.5^e (5) 78.5^a (6) 62.8^d **Glc V** (1) 105.3^c (2) 75.2^b (3) 78.6^a (4) 71.7^e (5) 78.5^a (6) 62.8^d.

Reference

1.   H. Matsuura, T. Ushiroguchi, Y. Itakura and T. Fuwa, *Chem. Pharm. Bull.*, **37,** 2741 (1989).

TRIGONEOSIDE VIIb

(25R)-Furost-5-ene-3β,22ξ,26-triol 3-O-{β-D-xylopyranosyl-(1→4)-[β-D-xylopyranosyl-(1→6)-β-D-glucopyranosyl-(1→3)-β-D-glucopyranosyl-(1→4)-[α-L-rhamnopyranoside}-26-O-β-D-glucopyranoside

Source : *Trigonella foenum-graecum* L. (Leguminosae)
Mol. Formula : $C_{67}H_{110}O_{36}$
Mol. Wt. : 1490
[α]$_D^{27}$: -55.9° (c=1.7, C_5D_5N)
Registry No. : [205761-69-9]

IR (KBr) : 3409, 2930, 1161, 1071, 1044 cm^{-1}.

PMR (C_5D_5N, 500 MHz) : δ 0.90 (s, 3xH-18), 0.99 (d, *J*=6.8 Hz, 3xH-27), 1.05 (s, 3xH-19), 1.32 (d, *J*=7.0 Hz, 3xH-21), 1.74 (d, *J*=6.5 Hz, 3xH-6 of Rha), 2.24 (dq, *J*=6.4, 6.4 Hz, H-20), 3.62 (dd, *J*=5.8, 9.5 Hz), 3.84 (m, H-3), 3.92 (m, 2xH-26), 4.79 (d, *J*=8.0 Hz, H-1 of Glc IV), 4.91 (d-like, H-1 of Glc I), 5.02 (d, *J*=6.4 Hz, H-1 of Glc II), 5.04 (d, *J*=7.3 Hz, H-1 of Glc III), 5.15 (d, *J*=8.6 Hz, H-1 of Xyl I), 5.25 (d, *J*=7.7 Hz, H-1 of Xyl II), 5.30 (br s, H-6), 6.17 (br s, H-1 of Rha).

CMR (C_5D_5N, 125 MHz) : δ C-1) 37.6 (2) 30.2 (3) 78.3 (4) 39.0 (5) 140.9 (6) 121.8 (7) 32.4 (8) 31.8 (9) 50.5 (10) 37.2 (11) 21.2 (12) 40.0 (13) 40.8 (14) 56.7 (15) 32.5 (16) 81.1 (17) 63.9 (18) 16.5 (19) 19.4 (20) 40.7 (21) 16.4 (22) 110.7 (23) 37.2 (24) 28.4 (25) 34.3 (26) 75.3 (27) 17.5 **Glc I** (1) 100.1 (2) 77.3 (3) 76.2 (4) 81.6 (5) 77.6 (6) 61.7 **Rha**

(1) 101.8 (2) 72.4 (3) 72.8 (4) 74.2 (5) 69.4 (6) 18.6 **Glc II** (1) 104.5 (2) 73.7 (3) 88.5 (4) 69.1 (5) 77.9 (6) 61.6 **Glc III** (1) 105.5 (2) 75.2 (3) 76.1 (4) 80.1 (5) 75.4 (6) 68.5 **Xyl I** (1) 105.2 (2) 74.8 (3) 78.2 (4) 70.9 (5) 67.3 **Xyl II** (1) 105.5 (2) 75.1 (3) 78.4 (4) 71.7 (5) 67.1 **Glc IV** (1) 104.9 (2) 75.2 (3) 78.6 (4) 71.8 (5) 78.4 (6) 62.9.

Mass (FAB, Positive ion) : m/z 1513 [M+Na]$^{+}$.

Mass (FAB, Negative ion) : m/z 1489 [M-H]$^{-}$, 1357 [M-C$_5$H$_9$O$_4$]$^{-}$, 1327 [M-C$_6$H$_{11}$O$_5$]$^{-}$, 1063 [M-C$_{16}$H$_{27}$O$_{13}$]$^{-}$.

Reference

1.   M. Yoshikawa, T. Murakami, H. Komatsu, J. Yamahara and H. Matsuda, *Heterocycles*, **47**, 397 (1998).

FUNKIOSIDE I

(25R)-Furost-5-en-3β,22α,26-triol 3-O-{α-L-rhamnopyranosyl-(1→4)-α-L-rhamnopyranosyl-(1→4)-[β-D-xylopyranosyl-(1→3)]-β-D-glucopyranosyl-(1→2)-β-D-glucopyranosyl-(1→4)-β-D-galactopyranoside}-26-O-β-D-glucopyranoside

Source : *Funkia ovata* Spr. (Liliaceae)
Mol. Formula : C$_{68}$H$_{112}$O$_{36}$
Mol. Wt. : 1504
Registry No. : [60454-83-3]

Reference

1.   G.V. Lazur'evskii, P.K. Kintya and N.E. Mashchenko, *Dokl. Akad. Nauk SSSR, Biochem.*, **230**, 476 (1976).

TRIGONEOSIDE VIb

**Furost-5,25(27)-diene-3β,22ξ,26-triol 3-O-{β-D-xylopyranosyl-(1→4)-[β-D-glucopyranosyl-(1→6)]-
β-D-glucopyranosyl-(1→3)-β-D-glucopyranosyl-(1→4)-[α-L-rhamnopyranosyl-(1→2)]-β-D-glucopyranoside}-
26-O-β-D-glucopyranoside**

Source : *Trigonella foenum-graecum* L. (Leguminosae)
Mol. Formula : $C_{68}H_{110}O_{37}$
Mol. Wt. : 1518
$[\alpha]_D^{26}$: -54.7° (c=0.3, C_5D_5N)
Registry No. : [205761-18-8]

IR (KBr) : 3404, 2932, 1655, 1159, 1073, 1044 cm^{-1}.

PMR (C_5D_5N, 500 MHz) : δ 0.89 (s, 3xH-18), 1.06 (s, 3xH-19), 1.33 (d, *J*=6.7 Hz, 3xH-21), 1.77 (d, *J*=6.1 Hz, 3xH-6 of Rha), 2.26 (m, H-20), 3.87 (m, H-3), 4.37, 4.62 (both m, 2xH-26), 4.89 (d, *J*=7.9 Hz, H-1 of Glc V), 4.94 (d-like,

H-1 of Glc I), 5.06 (m), 5.07 (d, J=8.6 Hz, H-1 of Glc II), 5.21 (d, J=7.0 Hz, H-1 of Glc III), 5.22 (d, J=7.0 Hz, H-1 of Glc IV), 5.27 (d, J=7.7 Hz, H-1 of Xyl), 5.30 (br s, H-6), 5.34 (br s, 2xH-27), 6.22 (br s, H-1 of Rha).

CMR (C_5D_5N, 125 MHz) : δ C-1) 37.4 (2) 30.1 (3) 78.2 (4) 38.9 (5) 140.7 (6) 121.8 (7) 32.3 (8) 31.6 (9) 50.3 (10) 37.1 (11) 21.1 (12) 39.9 (13) 40.7 (14) 56.6 (15) 32.4 (16) 81.1 (17) 63.7 (18) 16.4 (19) 19.3 (20) 40.6 (21) 16.3 (22) 110.3 (23) 37.9 (24) 28.3 (25) 147.2 (26) 72.0 (27) 110.7 **Glc I** (1) 99.9 (2) 77.2 (3) 76.1 (4) 81.3 (5) 77.5 (6) 62.5 **Rha** (1) 101.7 (2) 72.3 (3) 72.7 (4) 74.0 (5) 69.4 (6) 18.6 **Glc II** (1) 104.4 (2) 73.7 (3) 88.0 (4) 69.0 (5) 77.8 (6) 61.4 **Glc III** (1) 105.3 (2) 75.0 (3) 76.0 (4) 79.9 (5) 75.1 (6) 68.3 **Xyl** (1) 105.2 (2) 74.8 (3) 78.2 (4) 70.8 (5) 67.2 **Glc IV** (1) 104.8 (2) 75.3 (3) 78.1 (4) 71.5 (5) 78.1 (6) 61.6 **Glc V** (1) 103.8 (2) 75.2 (3) 78.5 (4) 71.6 (5) 78.4 (6) 62.7.

Mass (FAB, Positive ion) : m/z 1541 $[M+Na]^+$.

Mass (FAB, Negative ion) : m/z 1517 $[M-H]^-$, 1385 $[M-C_5H_9O_4]^-$, 1355 $[M-C_6H_{11}O_5]^-$, 1061 $[M-C_{17}H_{29}O_{14}]^-$, 899 $[M-C_{23}H_{39}O_{19}]^-$.

Reference

1.	M. Yoshikawa, T. Murakami, H. Komatsu, J. Yamahara and H. Matsuda, *Heterocycles*, **47**, 397 (1998).

AGAVASAPONIN H, AGAVOSIDE H
(25R)-3β,22α,26-Trihydoxy-5α-furostan-12-one 3-O-{α-L-rhamnopyranosyl-(1→4)-[α-L-rhamnopyranosyl-(1→3)-[β-D-xlopyranosyl-(1→2)]-β-D-glucopyranosyl-(1→4)-β-D-glucopyranosyl-(1→4)-β-D-galactopyranoside}-26-O-β-D-glucopyranoside

Source : *Agave americana* L.[1,2,3] (Agavaceae)
Mol. Formula : $C_{68}H_{112}O_{37}$
Mol. Wt. : 1520
M.P. : 228-230°C
$[α]_D^{20}$: -113.0° (c=0.62, MeOH)
Registry No. : [58546-21-7]

Reference

1. B. Wilkomirski, V.A. Bobeyko and P.K. Kintia, *Phytochemistry*, **14**, 2657 (1975).

TRIGONEOSIDE Va

(25S)-Furost-5-ene-3β,22ξ,26-triol 3-O-{β-D-xylopyranosyl-(1→4)-[β-D-glucopyranosyl-(1→6)]-β-D-glucopyranosyl-(1→3)-β-D-glucopyranosyl-(1→4)-[α-L-rhamnopyranosyl-(1→2)]-β-D-glucopyranoside}-26-O-β-D-glucopyranoside

Source : *Trigonella foenum-graecum* L. (Leguminosae)
Mol. Formula : $C_{68}H_{112}O_{37}$
Mol. Wt. : 1520
$[\alpha]_D^{26}$: -53.8° (c=0.3, C_5D_5N)
Registry No. : [205760-61-8]

IR (KBr) : 3400, 2931, 1159, 1074, 1044 cm^{-1}.

PMR (C$_5$D$_5$N, 500 MHz) : δ 0.90 (s, 3xH-18), 1.04 (d, J=6.7 Hz, 3xH-27), 1.07 (s, 3xH-19), 1.33 (d, J=6.7 Hz, 3xH-21), 1.78 (d, J=6.1 Hz, 3xH-6 of Rha), 2.24 (dq, J=7.0, 7.0 Hz, H-20), 3.50 (dd, J=7.1, 9.2 Hz), 3.88 (m, H-3), 4.09 (m, 2xH-26), 4.83 (d, J=7.7 Hz, H-1 of Glc V), 4.95 (d-like, H-1 of Glc I), 5.08 (d-like, H-1 of Glc II), 5.24 (d, J=8.0 Hz, H-1 of Glc III), 5.28 (d, J=8.0 Hz, H-1 of Glc IV), 5.29 (br s, H-6), 5.33 (d, J=7.6 Hz, H-1 of Xyl), 6.25 (br s, H-1 of Rha).

CMR (C$_5$D$_5$N, 125 MHz) : δ C-1) 37.5 (2) 30.2 (3) 78.2 (4) 38.9 (5) 140.8 (6) 121.8 (7) 32.4 (8) 31.7 (9) 50.4 (10) 37.1 (11) 21.1 (12) 39.9 (13) 40.8 (14) 56.6 (15) 32.5 (16) 81.1 (17) 63.8 (18) 16.5 (19) 19.4 (20) 40.7 (21) 16.5 (22) 110.7 (23) 37.2 (24) 28.3 (25) 34.4 (26) 75.4 (27) 17.5 **Glc I** (1) 99.9 (2) 77.2 (3) 76.3 (4) 81.4 (5) 77.6 (6) 62.6 **Rha** (1) 101.8 (2) 72.4 (3) 72.8 (4) 74.1 (5) 69.5 (6) 18.7 **Glc II** (1) 104.5 (2) 73.8 (3) 88.1 (4) 69.1 (5) 77.9 (6) 61.4 **Glc III** (1) 105.4 (2) 75.1 (3) 76.1 (4) 80.0 (5) 75.3 (6) 68.4 **Xyl** (1) 105.3 (2) 74.9 (3) 78.4 (4) 70.9 (5) 67.3 **Glc IV** (1) 105.0 (2) 75.4 (3) 78.3 (4) 71.5 (5) 78.3 (6) 61.6 **Glc V** (1) 105.1 (2) 75.2 (3) 78.6 (4) 71.7 (5) 78.5 (6) 62.8.

Mass (FAB, Positive ion) : m/z 1543 [M+Na]$^+$.

Mass (FAB, Negative ion) : m/z 1519 [M-H]$^-$, 1387 [M-C$_5$H$_9$O$_4$]$^-$, 1357 [M-C$_6$H$_{11}$O$_5$]$^-$, 1063 [M-C$_{17}$H$_{29}$O$_{14}$]$^-$, 901 [M-C$_{23}$H$_{39}$O$_{19}$]$^-$.

Reference

1.   M. Yoshikawa, T. Murakami, H. Komatsu, J. Yamahara and H. Matsuda, *Heterocycles*, **47**, 397 (1998).

TRIGONEOSIDE Vb

(25R)-Furost-5-ene-3β,22ξ,26-triol 3-O-{β-D-xylopyranosyl-(1→4)-[β-D-glucopyranosyl-(1→6)]-β-D-glucopyranosyl-(1→3)-β-D-glucopyranosyl-(1→4)-[α-L-rhamnopyranosyl-(1→2)]-β-D-glucopyranoside}-26-O-β-D-glucopyranoside

Source : *Trigonella foenum-graecum* L. (Leguminosae)
Mol. Formula : $C_{68}H_{112}O_{37}$
Mol. Wt. : 1520
$[\alpha]_D^{24}$ **:** -55.6° (c=0.5, C_5D_5N)
Registry No. : [205760-65-2]

IR (KBr) : 3400, 2934, 1159, 1074, 1050 cm^{-1}.

PMR (C_5D_5N, 500 MHz) : δ 0.90 (s, 3xH-18), 0.99 (d, *J*=6.4 Hz, 3xH-27), 1.06 (s, 3xH-19), 1.34 (d, *J*=6.8 Hz, 3xH-21), 1.77 (d, *J*=6.4 Hz, 3xH-6 of Rha), 2.25 (dq, *J*=7.0, 7.0 Hz, H-20), 3.62 (dd, *J*=6.1, 9.8 Hz), 3.88 (m, H-3), 3.92 (m, 2xH-26), 4.81 (d, *J*=7.9 Hz, H-1 of Glc V), 4.95 (d, *J*=6.7 Hz, H-1 of Glc I), 5.08 (d, *J*=8.6 Hz, H-1 of Glc II), 5.22 (d, *J*=9.2 Hz, H-1 of Glc III), 5.23 (d, *J*=8.0 Hz, H-1 of Xyl), 5.28 (d, *J*=8.0 Hz, H-1 of Xyl), 5.30 (br s, H-6), 6.22 (br s, H-1 of Rha).

CMR (C_5D_5N, 125 MHz) : δ C-1) 37.5 (2) 30.1 (3) 78.2 (4) 38.9 (5) 140.8 (6) 121.8 (7) 32.3 (8) 31.7 (9) 50.3 (10) 37.1 (11) 21.1 (12) 39.9 (13) 40.8 (14) 56.6 (15) 32.5 (16) 81.1 (17) 63.8 (18) 16.4 (19) 19.4 (20) 40.7 (21) 16.5 (22) 110.7 (23) 37.1 (24) 28.3 (25) 34.3 (26) 75.3 (27) 17.5 **Glc I** (1) 99.9 (2) 77.2 (3) 76.2 (4) 81.4 (5) 77.6 (6) 62.6 **Rha**

(1) 101.8 (2) 72.4 (3) 72.7 (4) 74.1 (5) 69.4 (6) 18.6 **Glc II** (1) 104.5 (2) 73.7 (3) 88.1 (4) 69.1 (5) 77.8 (6) 61.4 **Glc III** (1) 105.3 (2) 75.0 (3) 76.1 (4) 80.0 (5) 75.2 (6) 68.3 **Xyl** (1) 105.2 (2) 74.8 (3) 78.3 (4) 70.8 (5) 67.3 **Glc IV** (1) 104.9 (2) 75.3 (3) 78.2 (4) 71.5 (5) 78.2 (6) 61.6 **Glc V** (1) 104.9 (2) 75.1 (3) 78.5 (4) 71.7 (5) 78.4 (6) 62.8.

Mass (FAB, Positive ion) : m/z 1543 [M+Na]$^+$.

Mass (FAB, Negative ion) : m/z 1519 [M-H]$^-$, 1387 [M-C$_5$H$_9$O$_4$]$^-$, 1357 [M-C$_6$H$_{11}$O$_5$]$^-$, 1063 [M-C$_{17}$H$_{29}$O$_{14}$]$^-$, 901 [M-C$_{23}$H$_{39}$O$_{19}$]$^-$.

Reference

1.   M. Yoshikawa, T. Murakami, H. Komatsu, J. Yamahara and H. Matsuda, *Heterocycles*, **47**, 397 (1998).

TRIGONEOSIDE VIIIb

(25R)-5α-Furostone-3β,22ξ,26-triol 3-O-{β-D-xylopyranosyl-(1→4)-[β-D-glucopyranosyl-(1→6)]-β-D-glucopyranosyl-(1→3)-glucopyranosyl-(1→4)-[α-L-rhamnopyranosyl-(1→2)]-β-D-glucopyranoside}-26-O-β-D-glucopyranoside

Source : *Trigonella foenum-graecum* L. (Leguminosae)
Mol. Formula : C$_{68}$H$_{114}$O$_{37}$
Mol. Wt. : 1522
[α]$_D^{23}$: -44.0° (c=0.3, C$_5$D$_5$N)
Registry No. : [205761-70-2]

IR (KBr) : 3416, 2932, 1073, 1044 cm^{-1}.

PMR (C$_5$D$_5$N, 500 MHz) : δ 0.87 (s, 3xH-18), 0.89 (s, 3xH-19), 0.99 (d, J=6.4 Hz, 3xH-27), 1.34 (d, J=6.8 Hz, 3xH-21), 1.76 (d, J=6.1 Hz, 3xH-6 of Rha), 2.24 (dq, J=6.8, 6.8 Hz, H-20), 3.63 (dd, J=5.8, 9.2 Hz), 3.92 (m, H-3), 3.94 (m, 2xH-26), 4.83 (d, J=8.0 Hz, H-1 of Glc V), 4.97 (d, J=7.0 Hz, H-1 of Glc I), 5.08 (d, J=7.7 Hz, H-1 of Glc II), 5.24 (d, J=7.9 Hz, H-1 of Glc III), 5.28 (d, J=7.9 Hz, H-1 of Glc IV), 5.32 (d, J=7.9 Hz, H-1 of Xyl), 6.22 (br s, H-1 of Rha).

CMR (C$_5$D$_5$N, 125 MHz) : δ C-1) 37.2 (2) 29.9 (3) 77.2 (4) 34.4 (5) 44.6 (6) 29.0 (7) 32.5 (8) 35.3 (9) 54.5 (10) 35.9 (11) 21.3 (12) 40.2 (13) 41.1 (14) 56.4 (15) 32.4 (16) 81.1 (17) 64.0 (18) 16.7 (19) 12.4 (20) 40.7 (21) 16.5 (22) 110.6 (23) 37.2 (24) 28.4 (25) 34.3 (26) 75.3 (27) 17.5 Glc I (1) 99.5 (2) 77.5 (3) 76.3 (4) 81.6 (5) 77.6 (6) 62.6 Rha (1) 101.9 (2) 72.4 (3) 72.8 (4) 74.1 (5) 69.4 (6) 18.7 Glc II (1) 104.6 (2) 73.8 (3) 88.1 (4) 69.1 (5) 77.9 (6) 61.6 Glc III (1) 105.4 (2) 75.1 (3) 76.1 (4) 80.0 (5) 75.3 (6) 68.4 Xyl (1) 105.3 (2) 74.9 (3) 78.4 (4) 70.9 (5) 67.3 Glc IV (1) 105.0 (2) 75.4 (3) 78.3 (4) 71.5 (5) 78.3 (6) 61.6 Glc V (1) 104.9 (2) 75.2 (3) 78.6 (4) 71.7 (5) 78.5 (6) 62.8.

Mass (FAB, Positive ion) : m/z 1545 [M+Na]$^+$.

Mass (FAB, Negative ion) : m/z 1521 [M-H]$^-$, 1389 [M-C$_5$H$_9$O$_4$]$^-$, 1359 [M-C$_6$H$_{11}$O$_5$]$^-$, 1065 [M-C$_{17}$H$_{29}$O$_{14}$]$^-$, 903 [M-C$_{23}$H$_{39}$O$_{19}$]$^-$.

Reference

1.   M. Yoshikawa, T. Murakami, H. Komatsu, J. Yamahara and H. Matsuda, *Heterocycles*, **47**, 397 (1998).

ALLIUMOSIDE E

(25S)-Furost-5-en-3β,22α,26-triol 3-O-{β-D-glucopyranosyl-(1→4)-α-L-rhamnopyranosyl-(1→4)-α-L-rhamnopyranosyl-(1→6)-β-D-glucopyranosyl-(1→2)-[β-D-glucopyranosyl-(1→3)]-β-D-glucopyranoside}-26-O-β-D-glucopyranoside

Source : *Allium narcissiflorum* (Liliaceae)
Mol. Formula : C$_{69}$H$_{114}$O$_{37}$
Mol. Wt. : 1534
M.P. : 286-288°C
[α]$_D$^{18} : -83.0° (c=0.6, MeOH)
Registry No. : [56625-83-3]

Reference

1.   V.V. Krokhmalyuk and P.K. Kintya, *Khim. Prir. Soedin.*, 184 (1976); *Chem. Nat. Comp.*, **12**, 165 (1976).

BESHORNOSIDE

(25*R*)-5α-Furostan-3β,22α,26-triol 3-O-{α-L-rhamnopyranosyl-(1→4)-β-D-glucopyranosyl-(1→2)-[α-L-rhamnopyranosyl-(1→4)-β-D-glucopyranosyl-(1→3)]-β-D-glucopyranosyl-(1→4)-β-D-galactopyranoside}-26-O-β-D-glucopyranoside

Source : *Beshorneria yuccoides* (Agavaceae)
Mol. Formula : $C_{69}H_{116}O_{37}$
Mol. Wt. : 1536
M.P. : 219-221°C
[α]$_D^{20}$: -25° (c=1.0, H_2O)
Registry No. : [82659-24-3]

Reference

1.   P.K. Kintia, V.A. Bobeyko, I.P. Dragalin and S.A. Shvets, *Phytochemistry*, **21**, 1447 (1982).

FURCREAFUROSTATIN

(5α,25R)-Furostan-3β,22ξ,26-triol 3-O-{β-D-glucopyranosyl-(1→3)-β-D-glucopyranosyl-(1→2)-[α-L-rhamnopyranosyl-(1→4)-β-D-glucopyranosyl-(1→3)]-β-D-glucopyranosyl-(1→4)-β-D-galactopyranoside}-26-O-β-D-glucopyranoside

Source : *Furcraea selloa* var. *marginata* (Agavaceae)
Mol. Formula : $C_{69}H_{116}O_{38}$
Mol. Wt. : 1552
$[\alpha]_D^{20}$ **:** +96.6° (c=0.10, H_2O)

IR (KBr) : 3385 (OH) cm^{-1}.

PMR (C$_5$D$_5$N, 500 MHz) : δ 0.49 (H-9), 0.66 (s, 3xH-19), 0.78 (H-7A), 0.88 (H-5), 0.89 (s, 3xH-18), 1.01 (H-14), 1.02 (d, J=6.8 Hz, 3xH-27), 1.04 (H-12A), 1.13 (H-6, H-24), 1.19 (H-11A), 1.34 (d, J=6.8 Hz, 3xH-27), 1.04 (H-12A), 1.13 (H-6, H-24), 1.19 (H-11A), 1.34 (d, J=6.8 Hz, 3xH-21), 1.36 (H-4A, H-23A), 1.40 (H-11B), 1.52 (H-1B), 1.54 (H-7B), 1.61 (H-2A), 1.66 (H-12B), 1.69 (d, J=5.4 Hz, 3xH-6 of Rha), 1.76 (H-15), 1.80 (H-4B), 1.92 (H-8, H-25), 1.94 (H-17), 1.97 (H-23B), 2.01 (H-15B), 2.03 (H-1B), 2.05 (H-2B), 2.23 (H-20), 3.63 (H-26A), 3.74 (H-4 of Glc I), 3.79 (H-3), 3.79 (H-5 of Glc II), 3.80 (H-5 of Glc III), 3.92 (H-5 of Glc I, H-5 of Glc IV, H-5 of Glc V), 3.96 (H-5 of Gal, H-26B), 3.99 (H-6A of Glc I), 4.01 (H-2 of Glc IV), 4.03 (H-3 of Gal), 4.05 (H-2 of Glc II, H-6A of Glc III), 4.06 (H-3 of Glc III), 4.08 (H-3 of Glc II, H-2 of Glc III), 4.09 (H-4 of Glc III), 4.12 (H-2 of Glc V), 4.13 (H-3 of Glc I, H-4 of Glc IV), 4.16 (H-3 of Glc IV), 4.22 (H-6A of Gal), 4.23 (H-6B of Glc III), 4.24 (H-3 of Glc V and H-4 of Glc V), 4.26 (H-6A of Glc IV), 4.27 (H-6A of Glc II), 4.32 (H-2 of Glc I), 4.33 (H-4 of Glc II, H-4 of Rha), 4.40 (H-2 of Gal, H-6A of Glc V), 4.44 (H-6B of Glc II), 4.45 (H-6B of Glc I), 4.49 (H-6B of Glc IV), 4.53 (H-3 of Rha), 4.57 (H-6B of Glc V), 4.58 (H-16, H-4 of Gal), 4.63 (H-2 of Rha), 4.67 (H-6B of Gal), 4.85 (d, J=7.6 Hz, H-1 of Glc V), 4.88 (d, J=7.6 Hz, H-1 of Glc V), 4.88 (d, J=7.6 Hz, H-1 of Gal), 4.94 (H-5 of Rha), 5.10 (d, J=7.6 Hz, H-1 of Glc III), 5.12 (d, J=8.1 Hz, H-1 of Glc I), 5.18 (d, J=7.6 Hz, H-1 of Glc II), 5.54 (d, J=7.1 Hz, H-1 of Glc III), 5.76 (s, H-1 of Rha).

CMR (C$_5$D$_5$N, 125 MHz) : δ C-1) 37.2 (2) 29.9 (3) 77.3 (4) 34.9 (5) 44.7 (6) 29.0 (7) 32.5 (8) 34.3 (9) 54.5 (10) 35.9 (11) 21.3 (12) 40.2 (13) 41.1 (14) 56.5 (15) 30.9 (16) 79.9 (17) 64.0 (18) 16.9 (19) 12.3 (20) 40.7 (21) 16.6 (22) 110.7 (23) 32.1 (24) 28.2 (25) 34.3 (26) 75.4 (27) 17.2 **Gal** (1) 102.5 (2) 73.2 (3) 75.3 (4) 79.9 (5) 78.6 (6) 60.7 **Glc I** (1) 104.9 (2) 80.9 (3) 88.1 (4) 70.7 (5) 77.5 (6) 63.0 **Glc II** (1) 104.3 (2) 75.3 (3) 76.5 (4) 78.0 (5) 77.4 (6) 62.2 **Rha** (1) 102.7 (2) 72.5 (3) 72.7 (4) 74.0 (5) 70.4 (6) 18.6 **Glc III** (1) 104.1 (2) 74.7 (3) 88.1 (4) 69.3 (5) 78.1 (6) 61.2 **Glc IV** (1) 105.7 (2) 75.7 (3) 78.1 (4) 71.6 (5) 78.5 (6) 62.5 **Glc V** (1) 105.0 (2) 75.6 (3) 78.7 (4) 71.8 (5) 78.5 (6) 63.0.

Mass (E.S.I., Positive ion) : *m/z* 1553 [M+H]$^+$, 1407 [M-Rha+H]$^+$, 1391 [M-Glc+H]$^+$, 1245 [M-Rha-Glc+H]$^+$, 1229 [M-Glc-Glc+H]$^+$, 1083 [M-Rha-2xGlc+H]$^+$.

Reference

1.   J.L. Simmons-Boyce, W.F. Tinto, S. McLean and W.F. Reynolds, *Fitoterapia*, **75**, 634 (2004).

CAMASSIA LEICHTLINII SAPONIN 6

(25*R*)22ξ-Methoxy-5α-furostan-3β,26-diol 3-O-{β-D-glucopyranosyl-(1→3)-β-D-glucopyranosyl-(1→2)-[α-L-rhamnopyranosyl-(1→4)-β-D-glucopyranosyl-(1→3)]-β-D-glucopyranosyl-(1→4)-β-D-galactopyranoside}-26-O-β-D-glucopyranoside

Source : *Camassia leichtlinii* (Liliaceae)
Mol. Formula : C$_{70}$H$_{118}$O$_{38}$
Mol. Wt. : 1566
[α]$_D^{25}$: -44.0° (c=0.10, MeOH)
Registry No. : [354584-07-9]

IR (film) : 3483 and 3311 (OH), 2931 (CH), 1449, 1419, 1376, 1261, 1155, 1072, 1039, 894 cm^{-1}.

PMR (C_5D_5N, 500 MHz) : δ 0.65 (s, 3xH-18), 0.80 (s, 3xH-19), 1.00 (d, J=6.6 Hz, 3xH-27), 1.19 (d, J=6.8 Hz, 3xH-21), 1.68 (d, J=6.2 Hz, 3xH-6 of Rha), 3.27 (s, OCH_3), 4.85 (d, J=7.7 Hz, H-1 of Gal and Glc V), 5.08 (d, J=7.6 Hz, H-1 of Glc IV), 5.13 (d, J=7.8 Hz, H-1 of Glc I), 5.19 (d, J=7.7 Hz, H-1 of Glc III), 5.54 (d, J=7.7 Hz, H-1 of Glc II), 5.77 (br s, H-1 of Rha).

CMR (C_5D_5N, 125 MHz) : δ C-1) 37.2 (2) 29.9 (3) 77.4 (4) 34.8 (5) 44.6 (6) 28.9 (7) 32.4 (8) 35.2 (9) 54.4 (10) 35.8 (11) 21.2 (12) 40.0 (13) 41.1 (14) 56.3 (15) 32.1 (16) 81.3 (17) 64.3 (18) 16.3 (19) 12.3 (20) 40.5 (21) 16.5 (22) 112.6 (23) 30.8 (24) 28.2 (25) 34.2 (26) 75.2 (27) 17.1 (OCH$_3$) 47.2 **Gal** (1) 102.4 (2) 73.1 (3) 75.6 (4) 80.0 (5) 75.3 (6) 60.7 **Glc I** (1) 104.7 (2) 80.8 (3) 88.1 (4) 70.5 (5) 77.4 (6) 62.9 **Glc II** (1) 104.0 (2) 74.6 (3) 88.1 (4) 69.3 (5) 78.0 (6) 62.0 **Glc III** (1) 104.2 (2) 75.5 (3) 76.5 (4) 77.9 (5) 77.2 (6) 61.1 **Rha** (1) 102.6 (2) 72.5 (3) 72.7 (4) 73.9 (5) 70.3 (6) 18.5 **Glc IV** (1) 105.6 (2) 75.6 (3) 77.9 (4) 71.5 (5) 78.4 (6) 62.5 **Glc V** (1) 105.0 (2) 75.2 (3) 78.6 (4) 71.7 (5) 78.5 (6) 63.0.

Mass (FAB, Positive ion) : m/z 1558 [M+Na-OCH$_3$]$^+$.

Mass (FAB, Negative ion) : m/z 1565 [M-H]$^-$, 1419 [M-H-Rha]$^-$, 1403 [M-H-Glc]$^-$, 1257 [M-H-Rha-Glc]$^-$, 1241 [M-H-2xGlc]$^-$, 1095 [M-H-Rha-2xGlc]$^-$, 1079 [M-H-3xGlc]$^-$, 933 [M-H-4xGlc]$^-$.

Biological Activity : The compound showed cytotoxic activity against human oral squamous cell carcinoma (HSC-2) with LD_{50} value 4.7 µg/ml and normal human gingival fibroblasts (HGF) with LD_{50} value of 34 µg/ml.

Reference

1.   M. Kuroda, Y. Mimaki, F. Hasegawa, A. Yoksuka, Y. Sashida and H. Sakagami, *Chem. Pharm. Bull.,* **49**, 726 (2001).

DIOSCOREA PSEUDOJAPONICA SAPONIN 1

22α-Methoy-3β,26-dihydroxyfurost-5-ene 3-O-{α-L-rhamnopyranosyl-(1→4)-α-L-rhamnopyranosyl-(1→4)-
[α-L-rhamnopyranosyl-(1→2)-β-D-glucopyranoside}-26-O-β-D-glucopyranoside

Source : *Dioscorea pseudojaponica* Yamamoto[1]
(Dioscoreaceae), *D. cayenensis* Lam.-Holl.[2]
Mol. Formula : $C_{58}H_{96}O_{26}$
Mol. Wt. : 1208
M.P. : 189-190°C (decomp.)[1]
$[\alpha]_D^{16}$ **:** -86.4° (c=0.05, MeOH)[1]
Registry No. : [89826-97-1]

PMR (C$_5$D$_5$N, 500 MHz)[1] : δ 0.83 (s, 3xH-18), 1.06 (s, 3xH-19), 1.20 (d, *J*=6.8 Hz, 3xH-21), 1.60 (d, *J*=5.0 Hz, 3xH-6 of Rha II), 1.61 (d, *J*=5.5 Hz, 3xH-6 of Rha III), 1.78 (d, *J*=6.1 Hz, 3xH-6 of Rha I), 3.29 (s, 21-OC*H*$_3$), 3.89 (m, H-3), 4.86 (d, *J*=7.7 Hz, H-1 of Glc II), 4.96 (d, *J*=7.2 Hz, H-1 of Glc I), 5.34 (br d, *J*=5.3 Hz, H-6), 5.85 (br s, H-1 of Rha II), 6.31 (br s, H-1 of Rha III), 6.42 (br s, H-1 of Rha I).

CMR (C$_5$D$_5$N, 125 MHz)[1] : δ C-1) 38.0 (2) 30.5 (3) 78.6 (4) 39.4 (5) 141.3 (6) 122.3 (7) 32.8 (8) 32.1 (9) 50.8 (10) 37.6 (11) 21.5 (12) 40.2 (13) 41.3 (14) 57.1 (15) 32.6 (16) 81.8 (17) 64.6 (18) 16.7 (19) 19.9 (20) 41.0 (21) 16.7 (22) 113.2 (23) 31.3 (24) 28.6 (25) 34.7 (26) 75.7 (27) 17.6 (OC*H*$_3$) 47.8 **Glc I** (1) 100.8 (2) 78.2 (3) 78.5 (4) 79.1 (5) 77.5 (6) 61.7 **Rha I** (1) 102.7 (2) 73.0 (3) 73.3 (4) 74.6 (5) 70.1 (6) 18.9 **Rha II** (1) 102.7 (2) 73.3 (3) 73.3 (4) 80.9 (5) 70.8 (6) 19.1 **Rha III** (1) 103.8 (2) 73.0 (3) 73.3 (4) 74.5 (5) 70.1 (6) 19.3 **Glc II** (1) 105.4 (2) 75.7 (3) 78.6 (4) 72.2 (5) 79.0 (6) 63.3.

Mass (FAB, Positive ion)[1] : *m/z* 1231 [M+Na]$^+$, 1215 [M-CH$_3$-OH+K]$^+$, 1177 [M+H-CH$_3$OH]$^+$, 1031 [M+H-CH$_3$OH-Rha]$^+$, 885 [M+H-CH$_3$OH-2xRha]$^+$, 869 [M+H-CH$_3$OH-Glc-Rha]$^+$, 739 [M+H-CH$_3$OH-3xRha]$^+$, 723 [M+H-CH$_3$OH-Glc-2xRha]$^+$, 577 [M+H-CH$_3$OH-Glc-3xRha]$^+$, 415 [M-H-CH$_3$OH-2xGlc-3xRha]$^+$.

References

1. D.-J. Yang, T.-J. Lu and L.-S. Hwang, *J. Agric. Food Chem.*, **51**, 6438 (2003).

2. M. Sautour, A.-C. Mitane-Offer, T. Miyamoto, A. Dongmo and M.A.-Lacaille-Dubois, *Planta Med.*, **70**, 90 (2004).

MULTIFIDOSIDE
Nuatigenin 3-O{β-D-glucopyranosyl-(1→4)-α-L-rhamnopyranosyl-(1→4)-[α-L-glucopyranosyl-(1→2)]-β-D-glucopyranosyl}-26-O-β-D-glucopyranoside

Source : *Veronica fuhsii* and *V. multifida*
(Scrophulariaceae)
Mol. Formula : C$_{57}$H$_{92}$O$_{27}$
Mol. Wt. : 1208
[α]$_D^{20}$: -78.0° (c=0.1, MeOH)
Registry No. : [474646-73-6]

IR (KBr) : 3418 (OH), 2927 (CH), 1456, 1045, 910, 870, 820 cm^{-1}.

PMR (CD$_3$OD, 300/500 MHz) : δ 0.81 (s, 3xH-18), 0.99 (d, J=7.0 Hz, 3xH-21), 1.05 (s, 3xH-19), 1.23 (s, 3xH-27), 1.05 (s, 3xH-19), 1.23 (s, 3xH-27), 1.24 (d, J=6.1 Hz, 3xH-6 of Rha II), 1.33 (d, J=6.1 Hz, 3xH-6 of Rha I), 1.76 (t, J=7.0 Hz, H-17), 3.19 (dd, J=7.9, 9.0 Hz, H-2 of Glc II), 3.22 (dd, J=7.9, 9.0 Hz, H-2 of Glc III), 3.26* (H-5 of Glc II and Glc III), 3.28* (H-4 of Glc II and Glc III), 3.32* (H-5 of Glc I), 3.37* (H-3 of Glc II and Glc III), 3.38* (H-2 of Glc I and H-4 of Rha I), 3.48 (d, J=10.0 Hz, H-26A), 3.54 (t, J=7.5 Hz, H-4 of Glc I), 3.57* (H-3 of Glc I), 3.60 (t, J=9.2 Hz, H-3), 3.65 (dd, J=12.0, 6.0 Hz, H-6A of Glc I), 3.65* (H-4 of Rha II), 3.67* (H-3 of Rha I and Rha II), 3.70 (dd, J=12.0, 6.0 Hz, H-6A of Glc II and Glc III), 3.79 (dd, J=12.0, 2.0 Hz, H-6B of Glc I), 3.85 (dd, J=3.4, 1.7 Hz, H-2 of Rha II), 3.86 (d, J=10.0 Hz, H-26B), 3.87 (dd, J=12.0, 2.0 Hz, H-6B of Glc II and Glc III), 3.92 (dd, J=3.4, 1.7 Hz, H-2 of Rha I), 4.02 (dq, J=10.0, 6.1 Hz, H-5 of Rha II), 4.12 (dd, J=10.0, 6.1 Hz, H-5 of Rha I), 4.30 (d, J=7.9 Hz, H-1 of Glc III), 4.45 (t, J=7.3 Hz, H-16), 4.49 (d, J=7.9 Hz, H-1 of Glc I), 4.58 (d, J=7.9 Hz, H-1 of Glc II), 4.85 (br s, H-1 of Rha II), 5.19 (d, J=1.5 Hz, H-1 of Rha I), 5.37 (d, J=4.9 Hz, H-6). * overlapped signals.

CMR (CD$_3$OD, 75/125 MHz) : δ C-1) 38.5 (2) 30.7 (3) 79.2 (4) 39.4 (5) 141.8 (6) 122.5 (7) 32.8 (8) 32.8 (9) 51.7 (10) 38.0 (11) 21.9 (12) 40.9 (13) 41.5 (14) 57.7 (15) 33.1 (16) 82.1 (17) 63.2 (18) 16.5 (19) 19.8 (20) 39.3 (21) 15.0 (22) 121.7 (23) 33.5 (24) 33.7 (25) 85.2 (26) 77.5 (27) 24.2 **Glc I** (1) 100.4 (2) 79.3 (3) 78.0 (4) 79.6 (5) 76.6 (6) 61.9 **Rha I** (1) 102.2 (2) 72.1 (3) 72.3 (4) 73.9 (5) 69.7 (6) 17.9 **Rha II** (1) 102.6 (2) 72.2 (3) 72.3 (4) 83.1 (5) 69.2 (6) 17.9 **Glc II** (1) 105.5 (2) 76.1 (3) 78.2 (4) 71.6 (5) 78.0 (6) 62.7 **Glc III** (1) 104.9 (2) 75.2 (3) 77.7 (4) 71.4 (5) 78.1 (6) 62.7.

Mass (FAB, H.R.) : m/z 1231.5 [M+Na]$^+$.

Reference

1.   M. Ozipek, I. Saracoglu, Y. Ogihara and I. Calis, *Z. Naturforsch.*, **57C**, 603 (2002).

PROTOPOLYGONATOSIDE E'

(25R)-Furost-5-en-3β,22α,26-triol 3-O-[β-D-glucopyranosyl-(1→3)-β-D-glucopyranosyl-(1→4)-β-D-galactopyranosyl-(1→3)-β-D-glucopyranoside]-26-O-β-D-glucopyranoside

Source : *Polygonatum latifolium* Jasq. Desf. (Liliaceae)
Mol. Formula : $C_{57}H_{94}O_{29}$
Mol. Wt. : 1242
M.P. : 186-190°C
[α]$_D^{20}$: -120.0° (c=1.0, H_2O)
Registry No. : [68127-10-6]

IR : 3400, 900 (br) cm^{-1}.

Reference

1.   P.K. Kintya, A.I. Stamova, L.B. Bakinovskii and V.V. Krokhmalyuk, *Khim. Prir. Soedin.*, 350 (1978); *Chem. Nat. Comp.*, **14**, 290 (1978).

PROTOISOERUBOSIDE B

3β,6β,22ξ,26-Tetrahydroxy-5α,25(S)-furostane 3-O-{β-D-glucopyranosyl-(1→2)-[β-D-glucopyranosyl-(1→3)]-β-D-glucopyranosyl-(1→4)-β-D-galactopyranoside}-26-O-β-D-glucopyranoside

Source : *Allium sativum* L. (Liliaceae)
Mol. Formula : $C_{57}H_{96}O_{30}$
Mol. Wt. : 1260
M.P. : 218-220°C
$[\alpha]_D^{20}$: -26.0° (c=0.1, Pyridine)
Registry No. : [186545-34-6]

IR (KBr) : 3412.6, 2931.8, 1641.2, 1452.5, 1370.3, 1159.6, 1073.5, 1037.9, 893.2, 635.8, 533.6 cm⁻¹.

PMR (C_5D_5N, 500 MHz) : δ 0.90 (s, CH_3), 1.03 (d, J=6.5 Hz, sec. CH_3), 1.23 (s, CH_3), 4.82 (d, J=7.8 Hz, anomeric H), 4.95 (d, J=7.8 Hz, anomeric H), 5.16 (d, J=7.8 Hz, anomeric H), 5.31 (d, J=7.8 Hz, anomeric H), 5.58 (d, J=7.9 Hz, anomeric H).

CMR (C_5D_5N, 125 MHz) : δ C-1) 38.9 (2) 30.1 (3) 77.9 (4) 32.8 (5): 48.0 (6) 70.8 (7) 40.3 (8) 30.6 (9) 54.7 (10) 36.2 (11) 21.2 (12) 40.9 (13) 41.2 (14) 56.3 (15) 32.5 (16) 81.1 (17) 64.0 (18) 16.7 (19) 16.5 (20) 40.7 (21): 16.0 (22) 110.6 (23) 37.1 (24) 28.3 (25) 34.4 (26) 75.3 (27) 17.5 **Gal** (1) 102.3 (2) 73.2 (3) 75.6ᵃ (4) 80.3 (5) 76.1 (6) 60.6 **Glc I** (1)

105.0^b (2) 81.5 (3) 88.5 (4) 70.8^c (5) 78.0^d (6) 63.0^e **Glc II** (1) 104.9^b (2) 75.3^a (3) 78.6^f (4) 71.7^c (5) 77.5^d (6) 62.3^e **Glc III** (1) 104.5^b (2) 75.3^a (3) 78.6^f (4) 71.6^c (5) 78.6^f (6) 62.3^e **Glc IV** (1) 104.9^b (2) 75.2^a (3) 78.4^f (4) 71.7^c (5) 78.4^f (6) 62.8^e.

Mass (FAB, Positive ion) : *m/z* 1283 [M+Na]$^+$, 1243 [M+H-H$_2$O]$^+$, 919 [M+H-H$_2$O-2xGlc]$^+$, 757 [M+H-H$_2$O-3xGlc]$^+$, 595 [M+H-H$_2$O-4xGlc]$^+$, 577 [M+H-2xH$_2$O-4xGlc]$^+$.

Biological Activity : Inhibits fibrinolysis.

Reference

1. J.P. Peng, H. Chen, Y.Q. Qiao, L.A. Ma, T. Narui, H. Suzuki, T. Okuyama and H. Kobayashi, *Yaoxue Xuebao, Acta Pharm. Sinica*, **31**, 607 (1996).

CAPSICOSIDE F

(5α,25R)-2α,3β,26-Trihydroxyfurost-20(22)-ene 3-O-{β-D-glucopyranosyl-(1→3)-β-D-glucopyranosyl-(1→2)-[β-D-glucopyranosyl-(1→2)-β-D-glucopyranosyl-(1→3)]-β-D-glucopyranosyl-(1→4)-β-D-galactopyranoside}-26-O-β-D-glucopyranoside

Source : *Capsicum annuum* L. var. *acuminatum* (Solanaceae)

Mol. Formula : C$_{63}$H$_{104}$O$_{34}$

Mol. Wt. : 1404

[α]$_D^{25}$: -42.0° (c=0.07, Pyridine)

Registry No. : [117742-23-1]

PMR (C$_5$D$_5$N+D$_2$O, 500.13 MHz) : δ 0.57 (br t, H-9), 0.68 (s, 3xH-19), 0.70 (s, 3xH-18), 1.00 (m, H-5, H-14), 1.01 (d, J=6.8 Hz, 3xH-27), 1.02 (H-12A), 1.05 (H-6A), 1.17 (H-6B), 1.18 (H-1A), 1.22 (H-11A), 1.34 (H-8), 1.40 (H-15A, H-23A), 1.45 (H-11B), 1.49 (H-4A), 1.50 (H-7, H-24A), 1.62 (s, 3xH-21), 1.67 (H-12B), 1.73 (H-24B), 1.78 (H-25), 1.87 (H-4B), 2.18 (H-1B), 2.41 (H-15B), 2.45 (d, J=10.0 Hz, H-17), 2.45 (H-23B), 3.60 (m, H-26A), 3.80 (H-5 of Gal), 3.84 (H-5 of Glc I), 3.93 (H-5 of Glc IV), 3.95 (H-4 of Glc I, H-5 of Glc II), 3.96 (H-3), 3.97 (H-5 of Glc III), 3.98 (m, H-2, H-26B, H-4 of Glc II, H-6A of Glc II), 3.99 (H-5 and H-6A of Glc V), 4.02 (H-2 of Glc III), 4.05 (H-2 of Glc II), 4.06 (H-2 of Glc IV), 4.07 (H-2 of Glc V), 4.17 (H-3 of Gal), 4.20 (H-6B of Glc II, H-6A of Glc III, H-6A of Glc IV), 4.21 (H-4 of Glc III, H-4 of Glc V), 4.23 (H-3 of Glc II, H-4 of Glc IV), 4.26 (H-2 of Glc I), 4.28 (H-3 of Glc III), 4.30 (H-3 of Glc IV), 4.31 (H-3 of Glc V), 4.32 (H-6A of Gal, H-3 of Glc I), 4.36 (H-6 of Glc I, H-6B of Glc II), 4.42 (H-6B of Glc IV), 4.44 (H-6B of Glc I, H-6B of Glc III), 4.54 (H-6B of Gal), 4.55 (H-4 of Gal), 4.56 (H-2 of Gal), 4.73 (H-16), 4.86 (d, J=7.8 Hz, H-1 of Glc V), 4.96 (d, J=7.8 Hz, H-1 of Gal I), 5.12 (d, J=8.0 Hz, H-1 of Glc I), 5.16 (d, J=7.8 Hz, H-1 of Glc III), 5.35 (d, J=7.8 Hz, H-1 of Glc II), 5.59 (d, J=7.8 Hz, H-1 of Glc IV).

CMR (C$_5$D$_5$N, 125.76 MHz) : δ C-1) 44.3 (2) 69.7 (3) 83.0 (4) 33.2 (5) 43.5 (6) 27.3 (7) 31.1 (8) 33.6 (9) 53.5 (10) 36.0 (11) 20.7 (12) 39.2 (13) 43.2 (14) 55.0 (15) 34.2 (16) 84.3 (17) 64.5 (18) 14.2 (19) 12.4 (20) 103.3 (21) 11.5 (22) 152.0 (23) 33.4 (24) 23.7 (25) 31.5 (26) 74.8 (27) 17.3 **Gal** (1) 102.0 (2) 71.8 (3) 74.2 (4) 78.9 (5) 76.3 (6) 60.3 **Glc I** (1) 103.3 (2) 87.0 (3) 79.9 (4) 69.5 (5) 76.9 (6) 61.8 **Glc II** (1) 102.3 (2) 73.3 (3) 86.9 (4) 69.0 (5) 76.9 (6) 61.2 **Glc III** (1) 104.1 (2) 74.9 (3) 77.5 (4) 70.3 (5) 77.5 (6) 61.6 **Glc IV** (1) 103.5 (2) 74.8 (3) 77.5 (4) 70.8 (5) 77.5 (6) 61.6 **Glc V** (1) 103.8 (2) 74.5 (3) 77.5 (4) 70.7 (5) 77.3 (6) 61.7.

Mass (FAB, Positive ion, H.R.) : m/z 1427.6326 [(M+Na)$^+$, requires 1427.6307], 1265 [M+Na-Glc]$^+$, 1103 [M+Na-2xGlc]$^+$.

Reference

1.   M. Iorizzi, V. Lanzotti, G. Ranalli, S. de Marino and F. Zollo, *J. Agric. Food Chem.*, **50**, 4310 (2002).

CAPSICOSIDE G

(5α,25*R*)-3β,22ξ,26-Trihydroxyfurostane 3-O-{β-D-glucopyranosyl-(1→3)-β-D-glucopyranosyl-(1→2)-β-D-glucopyranosyl-(1→3)-[β-D-glucopyranosyl-(1→4)]-β-D-galactopyranoside}-26-O-β-D-glucopyranoside

Source : *Capsicum annuum* L. var. *acuminatum* (Solanaceae)
Mol. Formula : $C_{63}H_{106}O_{34}$
Mol. Wt. : 1406
$[\alpha]_D^{25}$: -55.2° (c=0.5, Pyridine)
Registry No. : [446290-91-1]

PMR (C$_5$D$_5$N, 500.13 MHz) : δ 0.50 (br t, H-9), 0.68 (s, 3xH-19), 0.78 (H-7), 0.80 (H-1A), 0.83 (s, H-18), 0.91 (m, H-5), 0.99 (d, J=6.8 Hz, 3xH-27), 1.00 (m, H-14), 1.04 (H-12A), 1.06 (H-6), 1.20 (H-11B), 1.33 (d, J=6.8 Hz, 3xH-21), 1.38 (H-24), 1.39 (H-15), 1.40 (H-4A), 1.41 (H-11B), 1.51 (H-7B), 1.53 (H-1B), 1.70 (H-12B), 1.74 (H-17), 1.80 (H-23), 1.85 (H-4B), 1.94 (H-25), 1.95 (H-8), 2.06 (H-20), 2.14 (H-2), 3.64 (m, H-26A), 4.00 (H-26B), 4.49 (m, H-16), 4.62 (H-3). Chemical shifts of sugar moiety virtually identical to those of Capsicoside F (q.v.).

CMR (C$_5$D$_5$N, 125.76 MHz) : δ C-1) 35.8 (2) 28.9 (3) 79.6 (4) 33.6 (5) 43.7 (6) 27.0 (7) 31.2 (8) 33.0 (9) 53.3 (10) 34.8 (11) 20.1 (12) 38.9 (13) 40.0 (14) 55.3 (15) 30.9 (16) 80.2 (17) 62.1 (18) 15.1 (19) 12.2 (20) 40.0 (21) 15.5 (22) 109.8 (23) 34.7 (24) 26.5 (25) 33.1 (26) 74.3 (27) 16.2 **Gal** (1) 102.0 (2) 71.8 (3) 74.2 (4) 78.9 (5) 76.3 (6) 60.3 **Glc I** (1) 103.3 (2) 87.0 (3) 79.9 (4) 69.5 (5) 76.9 (6) 61.8 **Glc II** (1) 102.3 (2) 73.3 (3) 86.9 (4) 69.0 (5) 76.9 (6) 61.2 **Glc III** (1) 104.1 (2) 74.9 (3) 77.5 (4) 70.7 (5) 77.5 (6) 61.6 **Glc IV** (1) 103.5 (2) 74.8 (3) 77.5 (4) 70.8 (5) 77.5 (6) 61.6 **Glc V** (1) 103.8 (2) 74.5 (3) 77.5 (4) 70.7 (5) 77.3 (6) 61.7.

Mass (FAB, Positive ion, H.R.) : *m/z* 1429.6485 [(M+Na)$^+$, requires 1429.6463], 1267 [M+Na-Glc]$^+$, 1105 [M+Na-2xGlc]$^+$, 943 [M+Na-3xGlc]$^+$.

Reference

1.   M. Iorizzi, V. Lanzotti, G. Ranalli, S. de Marino and F. Zollo, *J. Agric. Food Chem.*, **50**, 4310 (2002).

AGAVE ATTENUATA SAPONIN 1
(5β)-22α-Methoxyfurostan-3β,26-diol 3-O-{β-D-glucopyranosyl-(1→2)-β-D-glucopyranosyl-(1→2)-β-D-glucopyranosyl-(1→2)-[β-D-glucopyranosyl-(1→3)]-β-D-glucopyranosyl-(1→4)-β-D-galactopyranoside}-26-O-β-glucopyranoside

Source : *Agave attenuate* Salm-Dyck (Agavaceae)
Mol. Formula : C$_{64}$H$_{108}$O$_{34}$
Mol. Wt. : 1420
M.P. : 225-235°C (decomp.)
Registry No. : [475204-18-3]

IR (KBr) : 3435, 3936, 1641, 1592, 1513, 1461, 1403, 1381, 1313, 1241, 1184, 1055, 985, 915, 845, 815 (Intens. 915>814, 25S-spirostanol) cm^{-1}.

PMR (C$_5$D$_5$N, 200 MHz) : δ 0.82 (s, 3xH-18), 0.88 (s, 3xH-19), 1.04 (d, J=5.9 Hz, 3xH-27), 1.07 (d, J=5.3 Hz, 3xH-21), 1.40 (m, H-5), 3.28 (s, OCH_3), 3.63 (m, H-3), 4.70 (d, J=7.8 Hz, H-1 of Glc II), 4.75 (d, J=7.5 Hz, H-1 of Glc I), 4.80 (d, J=7.7 Hz, H-1 of Glc V), 4.82 (d, J=7.3 Hz, H-1 of Glc III), 4.90 (d, J=6.7 Hz, H-1 of Gal), 5.28 (d, J=7.3 Hz, H-1 of Glc IV).

CMR (C$_5$D$_5$N, 50 MHz) : δ C-1) 36.40 (2) 30.45 (3) 77.88 (4) 34.77 (5) 35.05 (6) 26.45 (7) 26.32 (8) 34.75 (9) 40.18 (10) 35.05 (11) 20.73 (12) 39.76 (13) 39.76 (14) 55.92 (15) 40.78 (16) 80.74 (17) 62.34 (18) 15.96 (19) 23.51 (20) 40.78 (21) 15.86 (22) 112.19 (23) 31.68 (24) 27.64 (25) 27.82 (26) 74.67 (27) 15.96 (OCH_3) 46.88 **Gal** (1) 101.95 (2) 73.28 (3) 74.93 (4) 80.11 (5) 74.93 (6) 61.62 **Glc I** (1) 104.84 (2) 80.81 (3) 85.55 (4) 71.20 (5) 77.82 (6) 62.23 **Glc II** (1) 104.53 (2) 74.67 (3) 85.55 (4) 69.31 (5) 77.46 (6) 62.27 **Glc III** (1) 107.31 (2) 75.90 (3) 77.84 (4) 71.31 (5) 78.20 (6) 62.42 **Glc IV** (1) 105.49 (2) 74.93 (3): 78.43 (4) 71.30 (5) 78.43 (6) 62.51 **Glc V** (1) 104.85 (2) 74.87 (3) 78.04 (4) 71.82 (5) 78.42 (6) 62.33.

Mass (L.S.I., Negative ion) : m/z 1419 [M-H]$^-$.

Biological Activity : Anti-inflammatory, no hemolytic effect.

Reference

1.   B.P. de Silva, A.C. de Souza, G.M. Silva, T.P. Mendes and J.P. Parente, *Z. Naturforsch.*, **57C**, 423 (2002).

22-*O*-METHYLCAPSICOSIDE G

Source : *Capsicum annuum* L. var. *acuminatum*
(Solanaceae)
Mol. Formula : $C_{64}H_{108}O_{34}$
Mol. Wt. : 1420
$[\alpha]_D^{25}$: -59.3° (c=0.4, Pyridine)
Registry No. : [446290-92-2]

PMR ($C_5D_5N+D_2O$, 500.13 MHz) : δ 0.51 (m, H-9), 0.66 (s, 3xH-19), 0.81 (s, 3xH-18), 1.03 (d, *J*=6.8 Hz, 3xH-27), 1.21 (d, *J*=6.8 Hz, 3xH-21), 3.32 (s, OC*H₃*), 3.64 (m, 2xH-26).

CMR ($C_5D_5N+D_2O$, 125.76 MHz) : δ C-1) 36.0 (2) 28.7 (3) 79.6 (4) 33.6 (5) 43.6 (6) 27.1 (7) 31.2 (8) 33.1 (9) 53.3 (10) 42.9 (11) 20.2 (12) 38.9 (13) 40.0 (14) 55.2 (15) 31.0 (16) 80.5 (17) 63.1 (18) 15.2 (19) 11.2 (20) 39.3 (21) 15.5 (22) 112.0 (23) 29.6 (24) 27.7 (25) 34.1 (26) 74.2 (27) 16.0 (OCH_3) 46.5.

Biological Activity : The compound has antifungal and anti-yeast activities.

Reference

1.   M. Iorizzi, V. Lanzotti, G. Ranalli, S. de Marino and F. Zollo, *J. Agric. Food Chem.*, **50**, 4310 (2002).

CAPSICOSIDE E

22ξ-Methoxy-2α,3β,26-trihydroxy-5α-furost-25(27)-ene 3-O-{β-D-glucopyranosyl-(1→2)-[β-D-glucopyranosyl-(1→3)]-β-D-glucopyranosyl-(1→4)-β-D-galactopyranoside}-26-O-β-D-glucopyranoside

Source : *Capsicum annuum* L. var. *acuminatum* (Solanaceae)
Mol. Formula : $C_{64}H_{106}O_{35}$
Mol. Wt. : 1434
$[α]_D^{25}$: -57.4° (c=0.2, Pyridine)
Registry No. : [446290-90-0]

PMR (C$_5$D$_5$N+D$_2$O, 500.13 MHz) : δ 0.51 (br t, H-9), 0.66 (s, 3xH-19), 0.76 (s, 3xH-18), 0.92 (m, H-14), 0.98 (m, H-5), 1.00 (H-12A), 1.03 (H-6A), 1.15 (H-1A), 1.17 (H-6B), 1.17 (d, J=6.6 Hz, 3xH-21), 1.20 (H-11A), 1.31 (H-15A), 1.33 (H-8), 1.43 (H-11B), 1.49 (H-4A, H-7), 1.63 (H-12B), 1.78 (H-17), 1.88 (H-4B), 1.96 (H-15B), 2.05 (H-23A), 2.16 (H-1B), 2.17 (H-23B), 2.22 (m, H-20), 2.40 (m, H-24), 3.32 (s, OCH_3), 3.80 (H-5 of Gal), 3.84 (H-5 of Glc I), 3.92 (m, H-3), 3.93 (H-5 of Glc IV), 3.95 (H-4 of Glc I), 3.97 (H-5 of Glc III), 3.98 (m, H-2, H-4, H-5 and H-6A of Glc II, H-5 of Glc V), 4.00 (H-6A of Glc V), 4.02 (H-2 of Glc III), 4.05 (H-2 of Glc II), 4.06 (H-2 of Glc IV and H-2 of Glc V), 4.17 (H-3 of Gal), 4.18 (H-6A of Glc IV), 4.20 (H-6B of Glc II and H-6A of Glc III), 4.21 (H-4 of Glc III), 4.21 (H-4 of Glc III), 4.23 (H-3 of Glc II, H-4 of Glc IV and H-4 of Glc V), 4.26 (H-2 of Glc I), 4.28 (H-3 of Glc III and H-3 of Glc V), 4.30 (H-3 of Glc IV), 4.32 (H-6A of Gal, H-3 of Glc I), 4.36 (H-6A of Glc I, H-6B of Glc V), 4.38 (H-26A), 4.40 (H-6B of Glc IV), 4.44 (H-6B of Glc I, H-6B of Glc III), 4.46 (m, H-16), 4.54 (H-6B of Gal), 4.55 (H-4 of Gal), 4.56 (H-2 of Gal), 4.65 (dd, J=11.0, 2.1 Hz, H-26B), 4.90 (d, J=7.8 Hz, H-1 of Glc V), 4.96 (d, J=7.8 Hz, H-1 of Gal), 5.06 (br s, H-27A), 5.10 (d, J=8.0 Hz, H-1 of Glc I), 5.14 (d, J=7.8 Hz, H-1 of Glc III), 5.32 (d, J=7.8 Hz, H-1 of Glc II), 5.36 (br s, H-27B), 5.59 (d, J=7.8 Hz, H-1 of Glc IV).

CMR (C$_5$D$_5$N+D$_2$O, 125.76 MHz) : δ C-1) 45.5 (2) 69.6 (3) 82.8 (4) 33.0 (5) 43.3 (6) 27.0 (7) 31.0 (8) 33.4 (9) 53.1 (10) 35.7 (11) 20.2 (12) 38.7 (13) 40.0 (14) 55.0 (15) 31.6 (16) 81.8 (17) 63.4 (18) 15.7 (19) 12.2 (20) 39.2 (21) 15.5 (22) 111.7 (23) 31.5 (24) 28.8 (25) 145.9 (26) 71.6 (27) 109.8 (OCH_3) 46.5 **Gal** (1) 101.6 (2) 71.6 (3) 74.1 (4) 78.8 (5) 76.4 (6) 60.2 **Glc I** (1) 102.9 (2) 86.7 (3) 80.4 (4) 69.1 (5) 76.5 (6) 61.6 **Glc II** (1) 102.3 (2) 73.3 (3) 86.8 (4) 68.9 (5) 76.8 (6) 61.2 **Glc III** (1) 103.7 (2) 74.7 (3) 77.3 (4) 70.3 (5) 77.5 (6) 61.5 **Glc IV** (1) 103.7 (2) 74.8 (3) 77.5 (4) 70.8 (5) 77.6 (6) 61.6 **Glc V** (1) 103.6 (2) 74.6 (3) 77.0 (4) 70.5 (5) 77.0 (6) 61.4.

Mass (FAB, Positive ion, H.R.) : m/z 1457.6423 [(M+Na)$^+$, calcd. for 1457.6412], 1295 [M+Na-Glc]$^+$, 1133 [M+Na-2xGlc]$^+$.

Biological Activity : Anti-yeast activity.

Reference

1.   M. Iorizzi, V. Lanzotti, G. Ranalli, S. de Marino and F. Zollo, *J. Agric. Food Chem.*, **50**, 4310 (2002).

AETHIOSIDE B

Source : *Solanum aethiopicum* (Solanaceae)
Mol. Formula : $C_{52}H_{80}O_{20}$
Mol. Wt. : 1024
$[\alpha]_D$: -36.8° (MeOH)
Registry No. : [603111-80-4]

PMR (C_5D_5N) : δ 0.95 (s, 3xH-18), 1.06 (d, *J*=6.7 Hz, 3xH-27), 1.14 (s, 3xH-19), 1.77 (d, *J*=6.1 Hz, 3xH-6 of Rha), 2.34 (s, 3xH-21), 4.87 (d, *J*=7.9 Hz, H-1 of Glc II), 5.39 (br s, H-6), 6.33 (s, H-1 of Rha), 7.04 (d, *J*=7.3 Hz, H-16), 7.52 (d, *J*=7.3 Hz, H-22).

CMR (C_5D_5N) : δ C-1) 37.4 (2) 30.2 (3) 78.5 (4) 38.8 (5) 141.1 (6) 121.9 (7) 32.5 (8) 31.1 (9) 50.6 (10) 37.2 (11) 21.3 (12) 37.1 (13) 47.2 (14) 57.8 (15) 31.3 (16) 140.8 (17) 151.9 (18) 17.4 (19) 19.4 (20) 131.3 (21) 14.7 (22) 139.8 (23) 32.0 (24) 35.5 (25) 34.2 (26) 75.0 (27) 16.6 (16') 122.9 (22') 127.5 **Glc** (1) 100.1 (2) 77.8 (3) 88.3 (4) 77.5 (5) 74.7 (6) 62.5 **Rha** (1) 102.4 (2) 72.4 (3) 72.9 (4) 74.2 (5) 69.5 (6) 18.5 **Xyl** (1) 105.4 (2) 75.0 (3) 78.0 (4) 70.7 (5) 67.3.

Mass (FAB, Negative ion) : *m/z* 1023 [M-H]⁻.

Reference

1. C. Tagawa, M. Okawa, T. Ikeda, T. Yoshida and T. Nohara, *Tetrahedron Lett.*, **44**, 4839 (2003).

AETHIOSIDE A

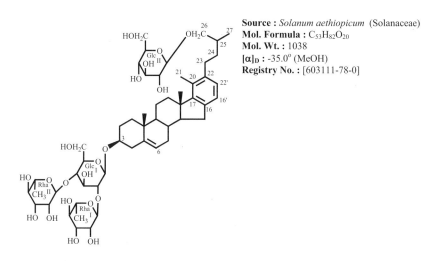

Source : *Solanum aethiopicum* (Solanaceae)
Mol. Formula : $C_{53}H_{82}O_{20}$
Mol. Wt. : 1038
[α]$_D$: -35.0° (MeOH)
Registry No. : [603111-78-0]

PMR (C_5D_5N) : δ 0.96 (s, CH_3), 1.12 (s, CH_3), 2.34 (s, CH_3), 0.99 (d, J=6.7 Hz, sec. CH_3), 5.38 (br s, H-6), 4.87 (d, J=7.9 Hz, anomeric H), 5.85 (s, H-1 of Rha), 6.39 (s, H-1 of Rha), 7.04 (d, J=7.3 Hz, aromatic H), 7.11 (d, J=7.3 Hz, aromatic H).

CMR (C_5D_5N) : δ C-1) 37.4 (2) 30.2 (3) 78.6 (4) 39.1 (5) 141.1 (6) 121.8 (7) 32.5 (8) 31.0 (9) 50.6 (10) 37.2 (11) 21.3 (12) 37.0 (13) 47.2 (14) 57.7 (15) 32.0 (16) 140.8 (17) 51.9 (18) 17.4 (19) 19.4 (20) 131.3 (21) 14.7 (22) 139.8 (23) 31.3 (24) 35.5 (25) 34.2 (26) 75.0 (27) 16.6 (16') 122.9 (22') 127.5 **Glc I** (1) 100.3 (2) 78.2 (3) 76.9 (4) 78.0 (5) 77.8 (6) 61.4 **Rha I*** (1) 102.1 (2) 72.5 (3) 72.8 (4) 74.2 (5) 69.5 (6) 18.5 **Rha II*** (1) 103.0 (2) 72.6 (3) 72.9 (4) 73.9 (5) 70.5 (6) 18.6 **Glc II** (1) 105.0 (2) 75.3 (3) 78.7 (4) 71.8 (5) 78.8 (6) 62.9.

* The signals of Rha I and Rha II are interchangeable with each other.

Mass (FAB, Negative ion) : *m/z* 1037 [M-H]⁻.

Reference

1. C. Tagawa, M. Okawa, T. Ikeda, T. Yoshida and T. Nohara, *Tetrahedron Lett.*, **44**, 4839 (2003).

AETHIOSIDE C

Source : *Solanum aethiopicum* (Solanaceae)
Mol. Formula : $C_{54}H_{82}O_{22}$
Mol. Wt. : 1082
$[\alpha]_D$: -63.8° (MeOH)
Registry No. : [603111-81-5]

PMR (C_5D_5N) : δ 0.99 (s, 3xH-18), 1.07 (d, *J*=6.7 Hz, 3xH-27), 1.11 (s, 3xH-19), 1.62 (d, *J*=6.1 Hz, 3xH-6 of Rha I)^b, 1.78 (d, *J*=6.1 Hz, 3xH-6 of Rha II)^b, 2.40 (s, 3xH-21), 4.87 (d, *J*=7.9 Hz, H-1 of Glc I), 4.96 (d, *J*=6.7 Hz, H-1 of Glc II), 5.39 (br s, H-6), 5.79 (s, H-1 of Rha I)^b, 6.32 (s, H-1 of Rha II)^b.

CMR (C_5D_5N) : δ C-1) 37.4 (2) 30.2 (3) 78.3 (4) 38.9 (5) 141.0 (6) 121.9 (7) 33.3 (8) 31.0 (9) 50.5 (10) 37.1 (11) 21.3 (12) 36.9 (13) 46.9 (14) 57.2 (15) 32.0 (16) 143.4 (17) 153.3 (18) 16.5 (19) 19.4 (20) 135.1 (21) 15.1 (22) 140.2 (23) 31.3 (24) 35.3 (25) 34.1 (26) 75.1 (27) 17.3 (16') 123.0 (22') 129.5 (COOH) 170.7 **Glc I** (1) 100.2 (2) 78.1 (3) 76.7 (4) 78.0 (5) 77.7 (6) 61.3 **Rha** (1) 102.0 (2) 71.7 (3) 72.2 (4) 73.9 (5) 69.5 (6) 18.4 **Rha II** (1) 102.8 (2) 72.4 (3) 72.6 (4) 73.6 (5) 70.4 (6) 18.6 **Glc II** (1) 104.7 (2) 75.0 (3) 78.9 (4) 71.7 (5) 78.9 (6) 62.7.

Reference

1.   C. Tagawa, M. Okawa, T. Ikeda, T. Yoshida and T. Nohara, *Tetrahedron Lett.*, **44**, 4839 (2003).

SPIROSTANE

Basic skeleton

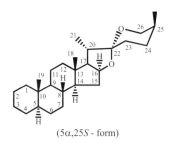

(5α,25S - form)

DESGLUCODESRHAMNORUSCIN
Ruscogenin 1-O-α-L-arabinopyranoside

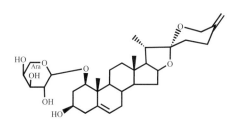

Source : *Ruscus aculeatus* L.[1,2], *R. hypoglossum* L.[2]
(Liliaceae)
Mol. Formula : $C_{32}H_{48}O_8$
Mol. Wt. : 560
Registry No. : [39491-39-9]

Mass (E.S.I., Positive ion) **:** *m/z* 583 [M+Na]⁺, 561 [M+H]⁺, 429 [Agl+H]⁺, 411 [Agl+H-H₂O]⁺.

Mass (E.S.I., Negative ion) **:** *m/z* 559 [M-H]⁻.

Biological Activity : Pharmacologically active,[3] Anti-inflammatory.[4]

References

1. E. Bombardelli, A. Bonati, B. Gabetta and G. Mustich, *Fitoterapia*, **42**, 127 (1971).

2. E. de Combarieu, M. Falzoni, N. Fuzzati, F. Gattesco, A. Giori, M. Lovati and R. Pace, *Fitoterapia*, **73**, 583 (2002).

3. A. Di Lazzaro, A. Morana, C. Schiraldi, A. Martino, C. Ponzone and M. De Rosa, *J. Mol. Cat. B ; Enzym.*, **11**, 307 (2001).

4. E. Bombardelli, A. Bonati, B. Gabetta, *Ger. Offen.*, 2, 202, 393 (Cl. 07c, A 61k) (German Patent) 28, Sept. 1972, *Chem. Abstr.*, **78**, 30161f (1973).

SOLANUM CHRYSOTRICHUM SAPONIN SC-3

(25*R*)-5α-Spirostan-3β-ol 6α-O-β-D-xylopyranoside

Source : *Solanum chrysotrichum* Schldh. (Solanaceae)
Mol. Formula : $C_{32}H_{52}O_8$
Mol. Wt. : 564
M.P. : 167-168°C
Registry No. : [478298-13-4]

IR : 3500-3300 (OH), 2925 (CH) cm^{-1}.

Tetra-acetate :

PMR (CDCl$_3$, 400 MHz) **:** δ 0.67 (H-9), 0.74 (s, 3xH-18), 0.78 (d, *J*=6.4 Hz, 3xH-27), 0.84 (s, 3xH-19), 0.95 (d, *J*=6.4 Hz, 3xH-21), 1.06 (H-1A), 1.07 (H-7A, H-14A), 1.2 (H-11A), 1.21 (H-2A), 1.21 (H-14B), 1.25 (H-15A), 1.4 (H-24A), 1.45 (H-11B), 1.46 (H-24B), 1.47 (H-8), 1.47 (H-23A), 1.6 (H-25), 1.61 (H-4), 1.61 (H-12A), 1.65 (H-1B, H-2B), 1.65 (dd, *J*=8.4, 6.8 Hz, H-17), 1.75 (H-5), 1.76 (H-20), 1.79 (H-23B), 1.99 (H-12B), 2.01 (H-7B, H-15B), 3.28 (ddd, *J*=10.2, 10.0, 4.2 Hz, H-6), 3.28 (dd, *J*=11.0, 10.0 Hz, H-26α), 3.28 (dd, *J*=11.0, 9.5 Hz, H-5α of Xyl), 3.46 (dd, *J*=11.0, 2.4 Hz, H-26β), 4.07 (dd, *J*=11.0, 5.0 Hz, H-5β of Xyl), 4.39 (ddd, *J*=7.5, 7.4, 7.0 Hz, H-16), 4.45 (d, *J*=7.2 Hz, H-1 of Xyl), 4.45 (ddd, *J*=9.5, 9.2 Hz, H-4 of Xyl), 4.62 (dddd, *J*=11.0, 10.0, 6.0, 5.5 Hz, H-3), 4.89 (dd, *J*=9.2, 7.2 Hz, H-2 of Xyl), 5.14 (t, *J*=9.2 Hz, H-3 of Xyl).

CMR (CDCl$_3$, 100 MHz) **:** δ C-1) 36.70 (2) 27.1 (3) 71.5 (4) 28.45 (5) 49.96 (6) 80.60 (7) 39.90 (8) 33.80 (9) 53.43 (10) 36.38 (11) 20.72 (12) 39.60 (13) 40.75 (14) 56.06 (15) 31.30 (16) 80.60 (17) 62.20 (18) 17.10 (19) 13.20 (20) 41.81 (21) 14.49 (22) 109.15 (23) 30.20 (24) 28.70 (25) 31.6 (26) 66.8 (27) 16.40 **Xyl** (1) 102.39 (2) 71.74 (3) 72.26 (4) 69.15 (5) 62.24.

Mass (FAB, Positive ion) **:** *m/z* 733 [M+H]$^+$, 457 [M-Xyl]$^+$, 397 [M-Xyl-CH$_3$COO]$^+$.

Mass (FAB, Positive ion, H.R.) : *m/z* 733.4141 [(M+H)$^+$, calcd. for 733.4163].

Reference

1.   A. Zamilpa, J. Tortoriello, V. Navarro, G. Delgado and L. Alvarez, *J. Nat. Prod.*, **65**, 1815 (2002).

YANONIN
Yanogenin 3-O-α-L-arabinopyranoside

Source : *Dioscorea tokoro* Makino (Dioscoraceae)
Mol. Formula : $C_{32}H_{52}O_8$
Mol. Wt. : 564
M.P. : 238-240°C (decomp.)
[α]$_D^{15}$: -14.5°
Registry No. : [5491-96-3]

Reference

1.   T. Kawasaki and T. Yamauchi, *Yakugaku Zasshi*, **83**, 757 (1963).

3-*EPI*-SCEPTRUMGENIN 3-O-β-D-GLUCOPYRANOSIDE

Source : Crude drug *Senshoku shichikon, Rhodea japonica* (Thunb.) Roth (Liliaceae)
Mol. Formula : $C_{33}H_{50}O_8$
Mol. Wt. : 574
M.P. : 236-238°C
[α]$_D^{20}$: -109° (c=0.11, CHCl$_3$)
Registry No. : [66251-03-4]

IR (KBr) : 3350 (OH), 1655 (ϵ=CH$_2$), 980, 960, 918, 900 (spriostane ring), 880 (c=CH$_2$), 840 (Δ^5) cm^{-1}.

Mass (E.I) : m/z (rel.intens.) 574 [M$^+$, 3.2], 412 [M$^+$-Glc-H$_2$O, 8], 342, 300 (12), 282 (80), 271 (12), 137 [C$_9$H$_{13}$O, 100].

Tetraacetate :

M.P. : 174-177°C, [α]$_D^{20}$: -168.0° (c=0.38, CHCl$_3$).

IR (KBr) : 1740 (OCOCH$_3$), 1650 (C=CH$_2$), 1235 (OCOCH$_3$), 980, 960, 918, 897 (spiroketal) 875 (C=CH$_2$), 840 (Δ^5) cm^{-1}.

PMR (CDCl$_3$, 100 MHz) : δ 0.81 (s, 3xH-19), 1.03 (s, 3xH-18), 2.04 (s, 2xOCOCH_3), 2.06 (s, OCOCH_3), 2.12 (s, OCOCH_3), 4.57 (d, J=7.0 Hz, H-1 of Glc), 4.80 (br s, $W\frac{1}{2}$=6.0 Hz, 2xH-27), 4.92-5.40 (m, 4xCH-OCOCH$_3$ and H-6).

Mass (E.I.) : m/z 742 [M]$^+$.

Reference

1. M. Takahira, Y. Kondo, G. Kusano and S. Nozoe, *Tetrahedron Lett.*, 3647 (1977); *Yakugaku Zasshi*, **99**, 264 (1979).

CAPSICOSIDE A$_3$, ATROPOSIDE B
Diosgenin 3-O-β-D-galactopyranoside

Source : *Capsicum annuum* L.[1], *Solanum unguiculatum* (A.) Rich.[2] (Solanaceae)
Mol. Formula : C$_{33}$H$_{52}$O$_8$
Mol. Wt. : 576
M.P. : 251-254°C[3]
[α]$_D$: -108° (c=0.5, MeOH)[3]
Registry No. : [14270-72-5]

PMR (C$_5$D$_5$N, 300 MHz)[2] : δ 0.70 (d, J=6.5 Hz, 3xH-27), 0.90 (s, 3xH-19), 1.37 (d, J=7.0 Hz, 3xH-21), 4.98 (d, J=8.0 Hz, H-1 of Gal), 5.29 (br s, H-1 of H-6).

CMR (C$_5$D$_5$N, 75.0 MHz)[2] : δ C-1) 37.0 (2) 30.0 (3) 77.0 (4) 39.2 (5) 140.8 (6) 121.4 (7) 31.8 (8) 30.9 (9) 52.3 (10) 37.6 (11) 21.1 (12) 39.9 (13) 40.8 (14) 56.0 (15) 31.6 (16) 81.1 (17) 62.9 (18) 16.4 (19) 19.4 (20) 42.0 (21) 15.0 (22) 109.3 (23) 31.8 (24) 29.3 (25) 30.6 (26) 66.9 (27) 17.3 **Gal** (1) 103.2 (2) 72.7 (3) 75.5 (4) 70.4 (5) 77.0 (6) 62.6.

Mass (FAB, Negative ion)[2] : m/z 575 [M-H]$^-$.

References

1. E.V. Gutsu, P.K. Kintya and G.V. Lazur'evskii, *Khim. Prir. Soedin.*, 242 (1987); *Chem. Nat. Comp.*, **23**, 202 (1987).

2. F.A. Abbas, *Sci. Pharm.*, **69**, 219 (2001).

3. P. Bite and T. Rettegi, *Acta Chim. Sci. Hung.*, **52**, 79 (1967).

COLLETTINSIDE I
Yamogenin 3-O-β-D-glucopyranoside

Source : *Dioscorea collettii* Hook. f. (Dioscoreaceae)
Mol. Formula : $C_{33}H_{52}O_8$
Mol. Wt. : 576
M.P. : 258-261°C[1]
$[\alpha]_D^{20}$: -96.0° (c=0.1, MeOH)[1]
Registry No. : [60134-72-7]

IR (KBr)[1] : 3380 (br, bonded OH), 1010-1060 (br, OH), 980, 913 > 890 (25S-spirostane) cm^{-1}.

Mass (F.D.)[1] : *m/z* 576 [M]$^+$, 396.

Mass (E.I.)[1] : *m/z* 396 (23.6), 282 (80.5), 267 (12.8), 253 (22.4), 139 (100), 115 (27.1).

Peracetate :

Yellowish granular crystals; **M.P.:** 117-120°C[1].

Reference

1. Y. Minghe and C. Yanyong, *Planta Med.*, **49**, 38 (1983).

3-*EPI*-DIOSGENIN 3-O-β-D-GLUCOPYRANOSIDE

Source : Crude drug *Senshoku shichikon, Rhodea japonica* Roth (Liliaceae)
Mol. Formula : $C_{33}H_{52}O_8$
Mol. Wt. : 576
M.P. : 218-221°C
[α]$_D^{20}$: -122.0° (c=0.26, CHCl$_3$)
Registry No. : [66289-51-8]

IR (KBr) : 3380 (OH), 980, 960, 920, 895 (spirostane ring) 840 (Δ^5) cm^{-1}.

Mass (E.I.) : m/z 576[M$^+$, 4.3], 414 [M$^+$-Glc-H$_2$O, 12], 342 (4), 300 (12), 282 (80), 231 (12), 253 (16) 139 [C$_9$H$_{15}$O, 100].

Tetraacetate :

M.P. : 212-215°; **[α]$_D^{26}$:** -180° (c=0.42, CHCl$_3$).

IR (KBr) : 1740, 1235, (OCOCH$_3$), 988, 960, 911, 890 (spiroketal), 840 (Δ^5) cm^{-1}.

PMR (CDCl$_3$, 100 MHz) : δ 0.79 (s, 3xH-19), 1.01 (s, 3xH-18), 2.03 (s, 2xOCOC*H*$_3$), 2.06 (s, OCOC*H*$_3$), 2.11 (s, OCOC*H*$_3$), 4.58 (d, *J*=7.0 Hz, aromatic H), 4.92-5.40 (m, 4xC*H*-OCOC*H*$_3$ and H-6), 5.01 (t, *J*=8.0 Hz, H-3 of Glc), 5.16 (t, *J*=8.0 Hz, H-2 of Glc), 4.30 (q, *J*=15.0, 4.0 Hz, H-6A of Glc), 3.92 (q, *J*=15.0, 2.0 Hz, H-6B of Glc), 5.20 (m, H-4 of Glc), 3.72 (m, H-5 of Glc).

Mass (E.I.) : m/z 744 [M]$^+$.

Reference

1. M. Takahira, Y. Kondo, G. Kusano and S. Nozoe, *Tetrahedron Lett.*, 3647 (1977); *Yakugaku Zasshi*, **99**, 264 (1979).

HISPININ A
(25S,5α)-6α-Hydroxyspirostan-3-one 6-O-α-L-rhamnopyranoside

Source : *Solanum hispidum* Pers. (Solanaceae)
Mol. Formula : $C_{33}H_{52}O_8$
Mol. Wt. : 576
M.P. : 224-226°C
[α]$_D$: -44° (C_5D_5N)
Registry No. : [72165-07-2]

Mass : *m/z* 576 [M$^+$].

Reference

1.   A.K. Chokravanty, C.R. Saha and S.C. Pakrashi, *Phytochemistry*, **18**, 902 (1979).

LIRIOPE PLATYPHYLLA SAPONIN A
Ruscogenin 3-O-α-L-rhamnopyranoside

Source : *Liriope platyphylla* Wang. et Tang. (Liliaceae)
Mol. Formula : $C_{33}H_{52}O_8$
Mol. Wt. : 576
M.P. : 226-228°C (decomp.)
[α]$_D^{19}$: -107° (c=1.00, C_5D_5N)
Registry No. : [87436-70-2]

IR (KBr) : 3600-3200 (OH), 982, 920, 902, 865 cm^{-1} (intensity 920>902, 25*R*-spiroketal).

PMR (C$_5$D$_5$N, 100 MHz) : δ 0.72 (br d, sec. C*H*$_3$), 0.91 (s, tert. C*H*$_3$), 1.11 (d, *J*=6.0 Hz, sec. C*H*$_3$), 1.27 (s, tert. C*H*$_3$), 1.55 (d, *J*=6.0 Hz, sec. C*H*$_3$).

CMR (C$_5$D$_5$N, 100 MHz) : δ C-1) 78.0 (2) 41.1 (3) 73.9 (4) 39.8 (5) 139.4 (6) 125.1 (7) 33.1 (8) 32.5 (9) 51.5 (10) 43.9 (11) 24.3 (12) 40.7 (13) 40.4 (14) 57.1 (15) 32.6 (16) 81.2 (17) 63.4 (18) 16.7 (19) 13.8 (20) 42.2 (21) 15.1 (22) 109.4 (23) 32.1 (24) 29.5 (25) 30.8 (26) 67.1 (27) 17.4 **Rha** (1) 100.1 (2) 72.9 (3) 72.9 (4) 74.3 (5) 69.9 (6) 18.6.

Reference

1.   Y. Watanabe, S. Sanada, Y. Ida and J. Shoji, *Chem. Pharm. Bull.*, **31**, 1980 (1983).

NOLINOSPIROSIDE C

(25*S*)-Ruscogenin 1-O-β-D-fucopyranoside

Source : *Nolina microcarpa* S. Wats. (Dracaenacea)
Mol. Formula : C$_{33}$H$_{52}$O$_8$
Mol. Wt. : 576
M.P. : 209-210°C
$[\alpha]_D^{20}$: -98.0 ± 2° (c=0.96, Pyridine)
Registry No. : [128397-27-3]

IR (KBr) : 3200-2600 (OH), 910 < 930, 990 (25*S*-spiroketal), 860 cm^{-1}.

PMR (C$_5$D$_5$N, 250/300 MHz) : δ 0.80 (s, 3xH-18), 1.01 (d, $J_{21,20}$=7.0, 3xH-21), 1.01 (d, $J_{27,25}$=7.0 Hz, 3xH-27), 1.16 (s, 3xH-19), 1.50 (d, $J_{6,5}$=6.5 Hz, 3xH-6 of Fuc), 3.29 (br d, H-26A), 3.69 (m, H-5 of Fuc), 3.70 (dd, $J_{1ax,2ax}$=11.5 Hz, $J_{1ax,2ex}$=4.5 Hz, H-1ax), 3.81 (m, H-3ax), 3.95 (m, H-4 of Fuc), 3.99 (dd, $J_{3,4}$=3.5 Hz, H-3 of Fuc), 3.99 (dd, $J_{26B,25}$=4.0 Hz, $J_{26A,26B}$=11.5 Hz, H-26B), 4.21 (dd, $J_{2,3}$=9.0 Hz, H-2 of Fuc), 4.44 (m, H-16), 4.64 (d, $J_{1,2}$=8.0 Hz, H-1 of Fuc), 5.53 (br d, $J_{6,7}$=5.5 Hz, H-6).

CMR (C$_5$D$_5$N, 62.5/75 MHz) : δ C-1) 84.00 (2) 38.16 (3) 68.07 (4) 43.81 (5) 139.68 (6) 124.73 (7) 32.44 (8) 33.09 (9) 50.69 (10) 42.92 (11) 23.84 (12) 40.46 (13) 40.26 (14) 57.12 (15) 32.03 (16) 81.25 (17) 62.99 (18) 16.87 (19) 14.88 (20) 42.51 (21) 14.88 (22) 109.76 (23) 26.43 (24) 26.25 (25) 27.58 (26) 65.08 (27) 16.35 **Fuc** (1) 102.60 (2) 72.13 (3) 75.32 (4) 72.46 (5) 71.25 (6) 17.46.

Mass : *m/z* 576 [M]$^+$.

Reference

1.   G.V. Shevchuk, Yu.S. Vollerner, A.S. Shashkov and V. Ya. Chirva, *Khim. Prir. Soedin.*, **27**, 672 (1991); *Chem. Nat. Comp.*, **27**, 592 (1991).

TRILLIN, MELONGOSIDE B, FUNKIOSIDE A, ALLIUMOSIDE A
Diosgenin 3-O-β-D-glucopyranoside

Source : *Dioscorea sativa*[1] (Dioscoreaceae), *Dioscorea zingiberensis* Wright[2] (Dioscoraceae), *Solanum melongena* L.[3] (Solanaceae), *Melilotus tauricus* (Bieb.) Ser.[4] (Leguminosae), *Dioscorea floribunda* Mart. et Gal.[5] (Dioscoreaceae) etc.
Mol. Formula : $C_{33}H_{52}O_8$
Mol. Wt. : 576
M.P. : 276-277°C[6]
[α]$_D^{25}$: -105° (c=1.1, dioxan)[2]
Registry No. : [14144-06-0]

IR (KBr)[5] : 3400 (br), 1240, 1055, 980, 960, 920, 898, 865 and 840 cm^{-1}.

IR (KBr)[6] : 3500-3200, 1630, 975, 910 < 885 (25R-spiroketal), 850, 840, 820 cm^{-1}.

PMR (CDCl$_3$, 360 MHz)[6] : 0.70 (s, CH_3), 0.97 (d, J=7.0 Hz, sec. CH_3), 0.99 (s, CH_3), 1.02 (d, J=6.0 Hz, sec. CH_3), 3.45 (m), 3.60 (dd, J=14.5, 7.0 Hz), 3.75 (dd, J=14.5, 3.0 Hz), 4.20 (d, J=8.0 Hz), 4.45 (m), 5.32 (m, J=5.5 Hz).

PMR (C$_5$D$_5$N)[5] : δ 0.70 (d, J=8.0 Hz, 3xH-27), 0.83 (s, 3xH-18), 0.92 (s, 3xH-19), 1.13 (d, J=6.0 Hz, 3xH-21), 3.52 (m, J=11.0 Hz, 2xH-26), 4.95 (d, J=7.0 Hz, H-1 of Glc), 5.28 (m, J=9.0 Hz, H-6).

CMR (C$_5$D$_5$N, 25 MHz)[7] : δ C-1) 37.5 (2) 32.2 (3) 78.5 (4) 39.3 (5) 141.0 (6) 121.7 (7 32.2 (8) 31.7 (9) 50.3 (10) 37.1 (11) 21.1 (12) 39.9 (13) 40.5 (14) 56.7 (15) 31.8 (16) 81.1 (17) 62.9 (18) 16.4 (19) 19.4 (20) 42.0 (21) 15.0 (22) 109.3 (23) 30.2 (24) 29.3 (25) 30.6 (26) 66.9 (27) 17.3 **Glc** (1) 102.6 (2) 75.3 (3) 78.4 (4) 71.7 (5) 78.1 (6) 62.9.

Mass (DCI, ammonia)[8] : *m/z* (rel.intens.) 594 [(M+H+NH$_3$)$^+$, 20], 577 [(M+H)$^+$, 28], 415 [(M+H+H$_2$O-Glc)$^+$, 100], 397 [(M+H-Glc)$^+$, 5].

Biological Activity : Hemolytic (HD$_{50}$ (200 ± 20mm) and antifungal GID (11+ 2 μm) activity,[9] used as a drug for treatment of menorrhagia. Inhibits platelet aggregation.

References

1. W.-Y. Huang, Y.-C. Chen and J.-H. Chu, *Chem. Abstr.* **52**, 11098 (1958).

2. C. Liu, Y. Chen, Y. Tang and B. Li, *Zhiwu Xuebao*, **26**, 337 (1984).

3. P.K. Kintya and S.R. Shvets, *Khim. Prir. Soedin.*, 610 (1984); *Chem. Nat. Comp.*, **20**, 575 (1984).

4. G.V. Khodakov, Yu A. Akimov and P.K. Kintya, *Khim. Prir. Soedin.*, 770 (1994); *Chem. Nat. Comp.*, **30**, 717 (1994).

5. G.-A. Hoyer, W. Sucrow and D. Winkler, *Phytochemistry*, **14**, 539 (1975).

1168

6. K.S. Reddy, M.S. Shekhani, D.E. Berry, D.G. Lynn and S.M. Hecht, *J. Chem. Soc. Perkin Trans I*, 987 (1984).

7. O. Espejo, J.C. Llavot, H. Jung and F. Giral, *Phytochemistry*, **21**, 413 (1982).

8. J.C. N. Ma and F.W. Lau, *Phytochemistry*, **24**, 1561 (1985).

9. M. Takechi, S. Shimada and Y. Tanaka, *Phytochemistry*, **30**, 3943 (1991).

ASPARAGOSIDE A
Sarsasapogenin 3-O-β-D-glucopyranoside

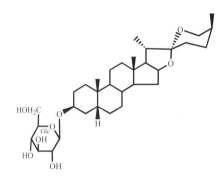

Source : *Asparagus officinalis* L.[1,2] (Liliaceae)
Mol. Formula : $C_{33}H_{54}O_8$
Mol. Wt. : 578
M.P. : 243-245°C[1]
$[\alpha]_D^{20}$: -62° (c=0.8, MeOH)[1]
Registry No.: [14835-43-9]

References

1. G.M. Goryanu, V.V. Krokhmalyuk and P.K. Kintya, *Khim. Prir. Soedin.*, 400 (1976), *Chem. Nat. Comp.*, **12**, 353 (1976).

2. G.V. Lazur'evskii, G.M. Goryanu and P.K. Kintya, *Dokl. Aka. Nauk SSSR*, **231**, 1479 (1976).

3. T. Konishi and J. Shoji, *Chem. Pharm. Bull.*, **27**, 3086 (1979).

CAPSICOSIDE A₂, ATROPOSIDE A

Tigogenin 3-O-β-D-galactopyranoside

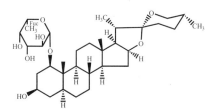

Source : *Capsicum annum* L.[1] var. *canoides* (Solanaceae)
Mol. Formula : $C_{33}H_{54}O_8$
Mol. Wt. : 578
M.P. : 290-291°C[1]
[α]ᴅ : -28° (c=0.8, MeOH)[1]
Registry No. : [35959-24-1]

IR : 900>920 cm⁻¹ (25R-spirostanol).

Mass (FAB, Positive ion)[2] : *m/z* 579 [M+H]⁺, 417 [M+H-Glc]⁺.

Biological Activity : The compound shows strong antifungal activity.[3]

References

1. E.V. Gutsu, P.K. Kintya and G.V. Lozur'evskii, *Khim. Prir. Soedin.*, **2**, 242 (1987); *Chem. Nat. Comp.*, **23**, 202 (1987).

2. Y. Ding, Y.-Y. Chen, D.-Z. Wang and C.-R. Yang, *Phytochemistry*, **28**, 2787 (1989).

3. M. Takechi and Y. Tanaka, *Phytochemistry*, **34**, 1241 (1993).

CHAMADOREA LINEARIS SAPONIN 2

Brisbagenin 1-O-β-L-fucopyranoside

Source : *Chamaedorea linearis* (Araceae)
Mol. Formula : $C_{33}H_{54}O_8$
Mol. Wt. : 578
Registry No. : [151589-17-2]

IR (KBr) : 3500, 2995, 1640, 1225, 1060, 892 cm^{-1}.

PMR (C$_5$D$_5$N, 400 MHz) : δ 0.69 (d, J=5.0 Hz, 3xH-27), 0.84 (s, 3xH-18), 1.00 (s, 3xH-19), 1.05 (d, J=7.0 Hz, 3xH-21), 1.57, 1.74 (m, H-4), 1.58 (d, J=6.4 Hz, 3xH-6 of Fuc), 1.76 (m, H-2A), 2.74 (m, H-2B), 1.80 (m, H-5), 1.83 (H-17), 1.96 (m, H-20), 3.48 (m, H-26A), 3.54 (m, H-26B), 3.76 (m, H-5 of Fuc), 3.81 (m, H-3), 3.85 (m, H-1), 4.05 (H-3 of Fuc), 4.08 (H-4 of Fuc), 4.33 (t, J=7.1 Hz, H-2 of Fuc), 4.51 (dd, J=2.6, 7.7 Hz, H-16), 4.80 (d, J=7.7 Hz, H-1 of Fuc).

CMR (C$_5$D$_5$N, 100 MHz) : δ C-1) 70.9 (2) 38.3 (3) 67.4 (4) 32.2 (5) 42.8 (6) 29.6 (7) 32.9 (8) 36.2 (9) 54.8 (10) 39.4 (11) 23.3 (12) 39.5 (13) 41.3 (14) 56.5 (15) 38.6 (16) 81.9 (17) 63.9 (18) 17.1 (19) 17.0 (20) 42.8 (21) 14.6 (22) 109.0 (23) 31.5 (24) 30.3 (25) 30.1 (26) 66.5 (27) 8.0 **Fuc** (1) 101.4 (2) 72.3 (3) 74.0 (4) 80.8 (5) 81.9 (6) 16.8.

Mass (FAB, I.R.) : m/z 579 [M+H]$^+$, 433 [M-Deoxyhexose]$^+$.

Mass (FAB, H.R.) : m/z 579.3891 [M+H]$^+$.

Reference

1. A.D. Patil, P.W. Baures, D.S. Eggleston, L. Faucette, M.E. Hemling, J.W. Westley and R.K. Johnson, *J. Nat. Prod.*, **56**, 1451 (1993).

MELONGOSIDE A
Tigogenin-3-O-β-D-glucopyranosyl

Source : *Solonum melongena* L.[1] (Solanaceae), *Agave kerchovei*[2] (Agavaceae)
Mol. Formula : C$_{33}$H$_{54}$O$_8$
Mol. Wt. : 578
M.P. : 273°C[1]
[α]$_D^{20}$: -62.0° (c=1.0, CH$_3$OH)[1]
Registry No. : [35068-81-1]

PMR (CDCl$_3$, 360 MHz)[2] : δ 0.87 (3xH-18, 3xH-27), 0.9 (3xH-19), 1.20 (3xH-21), 1.49 (s, 5H, H-23, H-24, H-25), 3.17 (m, H-3), 3.45 (d, 2xH-26).

References

1. P.K. Kintya and S.A. Shvets, *Khim. Prir. Soedin.*, 610 (1984); *Chem. Nat. Comp.*, **20**, 575 (1984).

2. M.A. El-Hashash, M.S. Amine, M.A. Shoib and L.A. Rafahy, *Indian J. Chem.*, **34B**, 836 (1995).

NICOTIANOSIDE A
Neotigogenin 3-O-β-D-glucopyranoside

Source : *Nicotiana tabacum* L.[1] (Solanaceae), *Smilax nipponica* Miq.[2] (Liliaceae)
Mol. Formula : $C_{33}H_{54}O_8$
Mol. Wt. : 578
M.P. : 276°C[1]; 265-267°C[2]
[α]$_D$: -65.0° (c=1.0, MeOH)[2]
Registry No. : [70954-49-3]

IR (KBr)[2] : 3409 (OH), 1057, 1035 (glycosidic C-O), 958, 920, 905, 850 cm^{-1} (920 > 705).

PMR (C$_5$D$_5$N, 300 MHz)[2] : δ 0.64 (s, 3xH-18), 0.78 (s, 3xH-19), 1.04 (d, *J*=7.0 Hz, 3xH-21), 1.11 (d, *J*=6.8 Hz, 3xH-27), 4.99 (d, *J*=7.7 Hz, anomeric H).

CMR (C$_5$D$_5$N, 75.5 MHz)[2] : δ C-1) 37.11 (2) 30.17 (3) 78.57 (4) 35.75 (5) 44.52 (6) 28.86 (7) 32.32 (8) 34.78 (9) 54.24 (10) 35.18 (11) 21.29 (12) 40.09 (13) 40.69 (14) 56.38 (15) 32.02 (16) 81.15 (17) 62.87 (18) 19.06 (19) 12.25 (20) 42.40 (21) 14.79 (22) 109.66 (23) 27.46 (24) 26.11 (25) 26.30 (26) 65.03 (27) 16.22 **Glc** (1) 102.07 (2) 75.29 (3) 78.43 (4) 71.76 (5) 77.17 (6) 62.77.

Mass[2] : *m/z* (rel.intens.) 599 [(M+Na)$^+$, 18.7], 577 [(M-H)$^+$, 36.2], 417 [(Neotigogenin+H)$^+$, 39.5], 399 [(Neotigogenin+H)$^+$-H$_2$O, 60.3].

References

1.   S.A. Shvets, P.K. Kintya and O.N. Gustu, *Khim. Prir. Soedin.*, **30**, 737 (1994); *Chem. Nat. Comp.*, **30**, 684 (1994).

2.   K.Y. Cho, M.H. Woo and S.O. Chung, *Yakhak Hoeji*, **39**, 141 (1995).

SOLANUM CHRYSOTRICHUM SAPONIN SC-4

(25*R*)-5α-Spirostan-3β-ol 6α-O-β-D-quinovopyranoside

Source : *Solanum chrysotrichum* Schldh. (Solanaceae)
Mol. Formula : $C_{33}H_{54}O_8$
Mol. Wt. : 578
M.P. : 194-196°C
Registry No. : [478298-14-5]

IR : 3500-3300 (OH), 2920 (CH) cm^{-1}.

Tetra-acetate :

PMR (CDCl$_3$, 400 MHz) **:** δ 0.66 (H-9 and H-24A), 0.68 (H-9, H-24A), 0.75 (s, 3xH-18), 0.78 (d, *J*=6.4 Hz, 3xH-27), 0.85 (s, 3xH-19), 0.95 (d, *J*=6.4 Hz, 3xH-21), 1.06 (H-1A and H-14A), 1.1 (H-7A), 1.15 (H-24B), 1.19 (H-4A and H-14B), 1.2 (H-11A), 1.20 (H-15A), 1.20 (d, *J*=6.0 Hz, 3xH-6 of Qui), 1.25 (H-5 and H-25), 1.4 (H-11B), 1.44 (H-2A), 1.61 (H-23A), 1.63 (H-1B and H-8), 1.68 (H-23B), 1.69 (H-12A), 1.77 (dd, *J*=8.4, 6.8 Hz, H-17), 1.82 (H-2B), 1.85 (H-20), 1.99 (H-15B), 2.07 (H-4B), 2.08 (H-12B), 2.1 (H-7B), 3.2 (ddd, *J*=11.0, 10.0, 4.5 Hz, H-6), 3.26 (dd, *J*=11.0, 1.0, Hz, H-26α), 3.47 (dd, *J*=11.0, 2.4 Hz, H-26β), 3.52 (m, H-5 of Qui), 4.39 (ddd, *J*=7.5, 7.4, 7.0 Hz, H-16), 4.47 (d, *J*=8.0 Hz, H-1 of Qui), 4.62 (dddd, *J*=11.0, 10.5, 6.0, 5.0 Hz, H-3), 4.81 (t, *J*=9.6 Hz, H-4 of Qui), 4.93 (dd, *J*=9.6, 8.0 Hz, H-2 of Qui), 5.12 (t, *J*=9.6 Hz, H-3 of Qui).

CMR (CDCl$_3$, 100 MHz) **:** δ C-1) 36.69 (2) 27.05 (3) 73.20 (4) 28.14 (5) 49.55 (6) 80.68 (7) 39.98 (8) 33.7 (9) 53.1 (10) 36.30 (11) 20.81 (12) 39.60 (13) 40.50 (14) 55.76 (15) 31.30 (16) 80.62 (17) 62.04 (18) 17.09 (19) 13.32 (20) 41.60 (21) 14.46 (22) 109.20 (23) 31.65 (24) 28.7 (25) 30.22 (26) 66.82 (27) 16.41 **Qui** (1) 101.92 (2) 72.11 (3) 72.97 (4) 73.23 (5) 69.72 (6) 17.39.

Mass (FAB, Positive ion) **:** *m/z* 769 [M+Na]$^+$, 747 [M+H]$^+$, 457 [M-Qui]$^+$, 397 [M-Qui-CH$_3$COO]$^+$, 273 [M-C$_7$H$_8$O$_2$]$^+$.

Mass (FAB, Positive ion, H.R.) **:** *m/z* 747.4320 [(M+H)$^+$, calcd. for 747.4315].

Reference

1.   A. Zamilpa, J. Tortoriello, V. Navarro, G. Delgado and L. Alvarez, *J. Nat. Prod.*, **65**, 1815 (2002).

SOLANUM HISPIDUM SAPONIN 1
(5α,25S)-3β,6α-Dihydroxyspirostane 6-O-β-D-quinovopyranoside

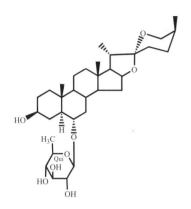

Source : *Solanum hispidum* Pers. (Solanaceae)
Mol. Formula : $C_{33}H_{54}O_8$
Mol. Wt. : 578
M.P. : 190-194°C
$[\alpha]_D^{25}$: -18.0° (c=0.002, Pyridine)

IR (KBr) : 3500-3300 (OH), 2925 (CH) cm^{-1}.

PMR (C$_5$D$_5$N, 500 MHz) : δ 0.82 (s, 3xH-18), 0.86 (s, 3xH-19), 1.06 (d, *J*=7.0 Hz, 3xH-27), 1.13 (d, *J*=6.7 Hz, 3xH-21), 1.57 (d, *J*=6.4 Hz, 3xH-6 of Qui), 3.32 (d, *J*=10.5 Hz, H-26β), 3.58 (m, H-4 of Qui), 3.70 (m, H-5 of Qui), 3.70-3.74 (m, H-6), 3.75 (m, H-3), 3.91 (m, H-26α), 4.05 (m, H-2 and H-3 of Qui), 4.46 (dd, *J*=7.6, 7.4 Hz, H-16), 4.81 (d, *J*=7.4 Hz, H-1 of Qui).

CMR (C$_5$D$_5$N, 125 MHz) : δ C-1) 37.70 (2) 32.08 (3) 70.58 (4) 33.10 (5) 51.25 (6) 78.12 (7) 41.37 (8) 34.15 (9) 53.86 (10) 36.75 (11) 21.22 (12) 40.38 (13) 40.04 (14) 56.36 (15) 32.05 (16) 81.11 (17) 62.80 (18) 16.58 (19) 13.55 (20) 42.47 (21) 14.79 (22) 109.63 (23) 26.34 (24) 26.16 (25) 27.50 (26) 67.07 (72) 16.25 **Qui** (1) 105.12 (2) 75.80 (3) 78.20 (4) 76.67 (5) 72.25 (6) 18.56.

Mass (FAB, Positive ion) : *m/z* 601 [M+Na]$^+$, 579 [M+H]$^+$, 431 [M-Qui]$^+$.

Mass (FAB, Positive ion, H.R.) : *m/z* 579.7982 [(M+H)$^+$, calc for 579.7933].

Biological Activity : Antimycotic.

Reference

1.   M. Gonzalez, A. Zamilpa, S. Marquina, V. Navarro and L. Alvareza, *J. Nat. Prod.*, **67**, 938 (2004).

TIMOSAPONIN A-1, SARSASAPOGENIN 3-O-β-D-GALACTOPYRANOSIDE

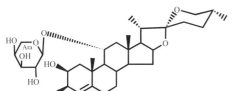

Source : *Cornus florida* L. (Cornaeacea)
Mol. Formula : $C_{33}H_{54}O_8$
Mol. Wt. : 578
M.P. : 240-245°C (decomp.)[2]
[α]$_D^{31}$: -68.5° (c=0.72, dioxane)[2]
Registry No. : [68422-00-4]

Mass (F.D.)[1] : *m/z* 601 [(M+Na)$^+$, 69.2], 579 [(M+H)$^+$, 100], 578 [(M)$^+$, 44], 417 [(M+H-Gal)$^+$, 15.4], 416 [(M$^+$-Gal)$^+$, 10.1].

References

1. K. Hostettmann, M. Hostettmann Kaldas and K. Nakonishi, *Helv. Chim. Acta*, **61**, 1990 (1978).

2. T. Kawasaki and T. Yamauchi, *Chem. Pharm. Bull.*, **11**, 1221 (1963).

PROTOMETEOGENIN 11-O-α-L-ARABINOPYRANOSIDE

Source : Soil bacterial hydrolysis product of the saponin from *Metanarthecium luteo-viride* Maxim. (Liliaceae)
Mol. Formula : $C_{32}H_{50}O_9$
Mol. Wt. : 578
M.P. : 265.5-267°C
Registry No. : [60506-77-6]

IR : 3385 cm^{-1}.

Mass : *m/z* 455 ($C_{27}H_{41}O_5$), 429 ($C_{27}H_{41}O_4$), 139 ($C_9H_{15}O$), 133 ($C_5H_9O_4$).

Penta-Acetate:

IR (CS$_2$) **:** 981, 920, 899, 846 cm^{-1} (Intensity 899 > 920 (25R-Spirostane). No OH absorption.

Mass : m/z 259 (C$_{11}$H$_{15}$O$_7$), 183 (C$_9$H$_{11}$O$_4$), 157 (C$_7$H$_9$O$_4$), 139 (100%).

CD [θ] : -70700 (200 nm).

Reference

1.   I. Kitagawa, T. Nakanishi, Y. Morii and I. Yosioka, *Tetrahedron Lett.*, 1885 (1976).

CONVALLASAPONIN A
Convallagenin A 3-O-α-L-arabinopyranoside

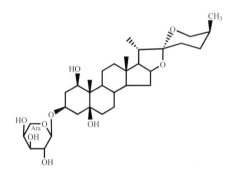

Source : *Convallaria keiskei* Miq. (Liliaceae)
Mol. Formula : C$_{32}$H$_{52}$O$_9$
Mol. Wt. : 580
M.P. : 238~240°C (decomp.)
[α]$_D^{22}$: -39.7° (c=0.527, CHCl$_3$)
Registry No. : [19316-94-0]

IR (KBr) : 3600~3200 (br, OH), 980, 921>898, 850 cm^{-1} (25S-spiroketal).

References

1.   M. Kimura, M. Tohma and I. Yoshizawa, *Chem. Pharm. Bull.*, **14**, 50 (1966).

22-*EPI*-TOKOROGENIN 1-O-α-L-ARABINOPYRANOSIDE

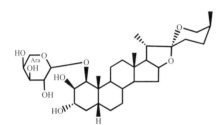

Source : *Dioscorea tenuipes* French and Savat.
(Dioscoreaceae)
Mol. Formula : $C_{32}H_{52}O_9$
Mol. Wt. : 580
M.P. : 287-290°C[2]
[α]$_D$: +19.4° (Pyridine)[2]
Registry No. : [561055-00-3]

CMR $(C_5D_5N, 25 MHz)^2$: δ C-20) 42.1 (21) 17.0 (22) 110.5 (23) 28.3[a] (24) 28.4[a] (25) 30.7 (26) 69.5 (27) 17.3.

Reference

1.   K. Kudo, K. Miyahara, N. Marubayashi and T. Kawasaki, *Chem. Pharm. Bull.*, **32**, 4229 (1984).

NEOTOKORONIN
Neotokorogenin 1-O-α-L-arabinopyranoside

Source : *Dioscorea tenuipes* French and Savat
(Dioscoreaceae)
Mol. Formula : $C_{32}H_{52}O_9$
Mol. Wt. : 580
M.P. : 284-286°C[2]
[α]$_D^{20}$: -17.2° (Pyridine)[2]
Registry No. : [50939-10-1]

IR $(CHCl_3)^1$: 3450-3250 br (OH), 988, 916 > 896 (25S-Spiroketal), 848 cm^{-1}.

CMR $(C_5D_5N, 25 MHz)^2$: δ C-20) 42.5 (21) 14.9 (22) 109.7 (23) 26.4 (24) 26.2 (25) 27.6 (26) 65.2 (27) 16.3.

Mass (E.I.)[1] : *m/z* 580 [M]$^+$, 459, 449, 432, 414, 317, 305, 299, 287, 269, 251, 230, 139, 115.

References

1.   A. Akahori, F. Yasuda, K. Kagawa and T. Iwao, *Chem. Pharm. Bull.*, **21**, 1799 (1973).

2.   K. Kudo, K. Miyahara, N. Marubayashi and T. Kawasaki, *Chem. Pharm. Bull.*, **32**, 4229 (1984).

TOKORONIN
Tokorogenin-1-O-α-L-arabinopyranoside

Source : *Dioscorea tokoro* Makino[1,2] (Dioscoreaceae)
Mol. Formula : $C_{32}H_{52}O_9$
Mol. Wt. : 580
M.P. : 275-277°C[3]
$[\alpha]_D$ **:** -13.0° (Pyridine)[3]
Registry No. : [27530-69-4]

CMR $(C_5D_5N, 25 MHz)^3$: δ C-20) 42.0 (21) 15.1 (22) 109.2 (23) 31.9 (24) 29.2 (25) 30.6 (26) 66.9 (27) 17.3.

CMR $(C_5D_5N, 80°C, 25.16/50.18 MHz)^4$: δ C-1) 88.9 (2) 74.4 (3) 71.6 (4) 35.0 (5) 36.5 (6) 26.2 (7) 26.5 (8) 35.8 (9) 42.4 (10) 41.7 (11) 21.2 (12) 40.2 (13) 40.7 (14) 56.4 (15) 32.2 (16) 81.1 (17) 63.3 (18) 16.5 (19) 19.1 (20) 42.1 (21) 14.9 (22) 109.2 (23) 32.0 (24) 29.3 (25) 30.6 (26) 67.0 (27) 17.2 **Ara** (1) 107.7 (2) 73.9 (3) 75.0 (4) 69.6 (5) 67.3.

References

1.   K. Miyahara and T. Kawasaki, *Chem. Pharm. Bull.*, **17**, 1369 (1969).

2.   K. Miyahara, F. Isozaki and T. Kawasaki, *Chem. Pharm. Bull.*, **17**, 1735 (1969).

3.   K. Kudo, K. Miyahara, N. Marubayashi and T. Kawasaki, *Chem. Pharm. Bull.*, **32**, 4229 (1984).

4.   S. Seo, A. Uomori, Y. Yoshimura and K. Tori, *J. Chem. Soc. Perkin Trans. 1*, 869 (1984).

NOLINOSPIROSIDE B
6β-Methoxy-1β-hydroxy-3β,5-cyclo-(25S)-spirostane 1-O-β-D-fucopyranoside

Source : *Nolina microcarpa* S. Wats. (Dracaenaceae)
Mol. Formula : $C_{34}H_{54}O_8$
Mol. Wt. : 590
$[\alpha]_D^{24}$ **:** -21.7±2.0° (c=0.85, Pyridine)
Registry No. : [402732-24-5]

IR (KBr) : 860, 910<930, 990 (spiroketal chain of the 25*S* series), 3200-3600 (OH) cm^{-1}.

PMR (C$_5$D$_5$N-D$_2$O, 250 MHz) : δ 0.49 (ddd, $J_{4,4}$=4.5, $J_{4,3}$=8.0 Hz, H-4), 0.86 (s, 3xH-18), 0.93 (m, H-3), 1.04 (d, $J_{27,25}$=7.0 Hz, 3xH-27), 1.13 (d, $J_{21,20}$=7.0 Hz), 1.40 (s, 3xH-19), 1.46 (m, H-14), 1.53 (d, J=, H-6 of Fuc), 1.83 (H-17), 2.08 (m, H-15'), 2.68 (t, $J_{6,7}$=2.5 Hz, H-6), 3.23 (s, OCH_3), 3.34 (br d, $J_{26,26'}$=11.0 Hz, H-26eq), 3.73 (dq, $J_{5,6}$=6.5 Hz, H-5 of Fuc), 4.02 (m, H-4 Fuc), 4.04 (br d, H-26ax), 4.04 (dd, $J_{3,4}$=3.5 Hz, H-3), 4.25 (t, $J_{2,3}$=7.5 Hz, H-2 of Fuc), 4.29 (m, H-1), 4.53 (d, $J_{1,2}$=7.5 Hz, H-1 of Fuc), 4.56 (dt, $J_{16,15}$=7.5 Hz, $J_{16,15}$=6.5 Hz, H-16).

CMR (C$_5$D$_5$N-D$_2$O, 62.5 MHz) : δ C-1) 84.51 (2) 32.72 (3) 20.89 (4) 16.79 (5) 49.51 (6) 82.52 (7) 35.78 (8) 30.57 (9) 50.24 (10) 36.43 (11) 23.37 (12) 40.58 (13) 41.09 (14) 56.48 (15) 32.19 (16) 81.36 (17) 63.07 (18) 16.79 (19) 15.94 (20) 42.58 (21) 14.89 (22) 109.81 (23) 26.47 (24) 26.24 (25) 27.60 (26) 65.21 (27) 16.34 (OCH$_3$) 56.48 **Fuc** (1) 103.27 (2) 72.13 (3) 75.72 (4) 72.84 (5) 71.49 (6) 17.34.

Mass (L.S.I.) : *m/z* 590 [M]$^+$.

Reference

1.   G.V. Shevchuk, Yu.S. Vollerner, A.S. Shashkov, M.B. Gorovits and V.Ya. Chirva, *Khim. Prir. Soedin.*, 218 (1992); *Chem. Nat. Comp.*, **28**, 187 (1992).

3-*EPI*-NEORUSCOGENIN 3-O-β-D-GLUCOPYRANOSIDE

Source : Chinese crude drug *"Senshokushichikon"*
Mol. Formula : C$_{33}$H$_{50}$O$_9$
Mol. Wt. : 590
M.P. : 130-132°C
[α]$_D$: -96° (c=0.4, CHCl$_3$)
Registry No. : [71335-76-7]

IR (CHCl$_3$) : 3350 (OH), 985, 965, 925, 885 (spirostane ring), 835 (C=CH-) cm^{-1}.

Mass (E.I.) : *m/z* (rel.intens.) 590 [(M)$^+$, 2.0], 572 [(M-H$_2$O)$^+$, 0.8], 427 (16.8), 410 [(M-Glc)$^+$, 57.4], 392 [(M-Glc-H$_2$O)$^+$, 27.7], 298 (45.5), 269 (15.8), 162 (5.9), 137 (100).

Penta-acetate :

M.P. : 225-228°C; **[α]$_D^{15}$:** -50° (c=0.2, CHCl$_3$)

IR (KBr) : 1745, 1240 (OCOCH$_3$), 985, 965, 925, 885 (spirostane), 1651, 880 (C=CH$_2$). 835 (C=CH-) cm^{-1}.

PMR (CDCl₃, 60 MHz) : δ 0.81 (s, 3xH-18), 0.97 (d, J=7.0 Hz, 3xH-21), 1.14 (s, 3xH-19), 2.02 (s, OCOCH_3), 2.05 (s, 2xOCOCH_3), 2.12 (s, 2xOCOCH_3), 3.86 (d, J=12.0 Hz, H-26A), 4.34 (d, J=12.0 Hz, H-26), 4.68 (d, J=8.0 Hz, H-1 of Glc), 4.80 (H-1), 4.80 (br s, $W\frac{1}{2}$=8.0 Hz, 2xH-27), 4.12 (br d, H-6A of Glc), 4.34 (dd, J=12.0, 6.0 Hz, H-6A of Glc), 4.34 (dd, J=12.0, 6.0 Hz, H-6B of Glc), 5.00 (dd, J=8.0, 8.0 Hz, H-3 of Glc), 5.08 (dd, J=8.0, 8.0 Hz, H-4 of Glc), 5.27 (dd, J=8.0, 8.0 Hz, H-2 of Glc), 5.51 (br d, $W\frac{1}{2}$=10.0 Hz, H-6).

Mass (E.I.) : m/z (rel.intens.) 800 [(M)⁺, 2.4], 740 [(M-AcOH)⁺, 6.5], 331 (37.1), 271 (8.1), 229 (11.3), 221 (7.2) 169 (61.3), 137 [(C₆H₁₃O), 100], 109 (13.9).

Reference

1.   M. Tanaka, Y. Kondo, G. Kusano and S. Nozoe, *Yakugaku Zasshi*, **99**, 528 (1979).

AGAVOSIDE A
Hecogenin 3-O-β-D-galactopyranoside

Source : *Agave americana* L. (Amaryllidaceae)
Mol. Formula : C₃₃H₅₂O₉
Mol. Wt. : 592
M.P. : 220-222°C
$[α]_D^{20}$ **:** -113° (c=1.01, DMF)
Registry No. : [56857-65-9]

References

1.   P.K. Kintya, V.A. Bobeiko, V.V. Krochmalyuk and V.Y. Tschirra, *Pharmazie,* **30**, 396 (1975).

2.   G.V. Lazur'evskii, V.A. Bobeiko and P.K. Kintya, *Dokl. Akad. Nauk. SSSR,* **224**, 1442 (1975).

ALLIUM NUTANS SAPONIN 2

25R-Spirostan-1β,3β-diol 1-O-β-D-galactopyranoside

Source : *Allium nutans* L. (Liliaceae)
Mol. Formula : $C_{33}H_{52}O_9$
Mol. Wt. : 592
Registry No. : [241486-81-7]

PMR (CD₃OD, 250 MHz) : δ 0.85 (s, 3xH-18), 0.85 (d, *J*=6.5 Hz, 3xH-27), 1.09 (d, *J*=6.5 Hz, 3xH-21), 1.12 (s, 3xH-19), 3.45 (m, H-3), 3.47 (ddd, *J*=2.0, 3.5, 5.0 Hz, H-5 of Gal), 3.47 (dd, *J*=8.0, 12.0 Hz, H-6A of Gal), 3.48 (dd, *J*=4.2, 9.0 Hz, H-3 of Gal), 3.52 (dd, *J*=7.8, 9.0 Hz, H-2 of Gal), 3.53 (dd, *J*=3.5, 11.5 Hz, H-1), 3.78 (d, *J*=3.5, 12.0 Hz, H-6B of Gal), 3.87 (dd, *J*=4.2, 2.0 Hz), 4.30 (d, *J*=7.8 Hz, H-1 of Gal), 4.42 (m, H-16), 5.59 (br d, *J*=5.4 Hz, H-6).

CMR (CD₃OD, 62.5 MHz) : δ C-1) 83.4 (2) 36.9 (3) 68.6 (4) 42.9 (5) 139.2 (6) 125.7 (7) 32.4 (8) 33.7 (9) 51.3 (10) 42.9 (11) 24.3 (12) 41.1 (13) 40.5 (14) 57.7 (15) 32.6 (16) 81.9 (17) 63.5 (18) 16.7 (19) 14.6 (20) 42.9 (21) 14.7 (22) 110.9 (23) 32.0 (24) 28.5 (25) 31.1 (26) 65.6 (27) 16.7 **Gal** (1) 101.9 (2) 72.9 (3) 74.7 (4) 69.6 (5) 76.0 (6) 62.0.

Mass (L.S.I., Negative ion) : *m/z* 591 [M-H]⁻, 413 [M-H-Glc-OH]⁻.

Reference

1. L.S. Akhov, M.M. Musienko, S. Piacente, C. Pizza and W. Oleszek, *J. Agric. Food Chem.*, **47**, 3193 (1999).

CORDYLINE STRICTA SAPONIN 2

5α-Spirost-25(27)-ene-1β,3α-diol 3-O-β-D-glucopyranoside

Source : *Cordyline stricta* (Agavaceae)
Mol. Formula : $C_{33}H_{52}O_9$
Mol. Wt. : 592
$[\alpha]_D^{27}$: -45.5° (c=0.44, MeOH)
Registry No. : [194146-05-9]

IR (KBr) : 3430 (OH), 2910 (CH), 1440, 1365, 1220, 1165, 1150, 1065, 1035, 985, 950, 935, 915, 890, 865 cm^{-1}.

PMR (C_5D_5N, 400 MHz) : δ 0.89 (s, 3xH-18), 1.06 (d, *J*=7.1 Hz, 3xH-21), 1.07 (s, 3xH-19), 3.91 (ddd, *J*=8.8, 4.9, 2.5 Hz, H-5 of Glc), 4.04 (br d, *J*=12.2 Hz, H-26A), 4.08 (dd, *J*=8.8, 7.7 Hz, H-2 of Glc), 4.25 (dd, *J*=8.8, 8.8 Hz, H-4 of Glc), 4.29 (overlapping, H-1), 4.30 (dd, *J*=8.8, 8.8 Hz, H-3 of Glc), 4.35 (br s, H-3), 4.42 (dd, *J*=11.8, 4.9 Hz, H-6A of Glc), 4.47 (br d, *J*=12.2 Hz, H-26B), 4.55 (2H, overlapping, H-16 and H-6B of Glc), 4.78 & 4.82 (each br s, 2xH-27), 4.97 (d, *J*=7.7 Hz, H-1 of Glc).

CMR (C_5D_5N, 100 MHz) : δ C-1) 73.7 (2) 37.1 (3) 73.9 (4) 35.1 (5) 39.1 (6) 28.7 (7) 32.5 (8) 36.0 (9) 55.1 (10) 42.6 (11) 24.8 (12) 40.9 (13) 40.5 (14) 56.8 (15) 32.3 (16) 81.4 (17) 63.3 (18) 16.7 (19) 6.5 (20) 41.9 (21) 15.0 (22) 109.4 (23) 33.2 (24) 29.0 (25) 144.5 (26) 65.0 (27) 108.6 **Glc** (1) 102.7 (2) 75.4 (3) 78.7 (4) 71.8 (5) 78.4 (6) 62.9.

Mass (FAB, Negative ion) : *m/z* 591 [M-H]$^-$.

Reference

1.   Y. Mimaki, Y. Takaashi, M. Kuroda and Y. Sashida, *Phytochemistry*, **45**, 1229 (1997).

CORDYLINE STRICTA SAPONIN 3
(25S)-5α-Spirost-20-ene-1β,3α-diol 3-O-β-D-glucopyranoside

Source : *Cordyline stricta* (Agavaceae)
Mol. Formula : $C_{33}H_{52}O_9$
Mol. Wt. : 592
$[\alpha]_D^{27}$ **:** -32.2° (c=0.21, MeOH)
Registry No. : [194143-95-8]

IR (KBr) : 3420 (OH), 2920 (CH), 1445, 1365, 1345, 1270, 1205, 1155, 1120, 1070, 1035, 1015, 980, 960, 935, 920, 890, 840 cm^{-1}.

PMR (C_5D_5N, 400 MHz) : δ 0.83 (s, 3xH-18), 1.05 (s, 3xH-19), 1.10 (d, J=7.0 Hz, 3xH-27), 2.67 (ddd, J=7.2, 1.9, 1.4 Hz, H-17), 3.39 (br d, J=11.0 Hz, H-26A), 3.91 (ddd, J=8.8, 5.1, 2.5 Hz, H-5 of Glc), 4.08 (dd, J=8.8, 7.7 Hz, H-2 of Glc), 4.11 (dd, J=11.0, 2.9 Hz, H-26B), 4.25 (dd, J=8.8, 8.8 Hz, H-4 of Glc), 4.29 (overlapping, H-1), 4.30 (dd, J=8.8, 8.8 Hz, H-3 of Glc), 4.34 (br s, H-3), 4.42 (dd, J=11.3, 5.1 Hz, H-6A of Glc), 4.55 (dd, J=11.3, 2.5 Hz, H-6B of Glc), 4.70 (ddd, J=7.2, 7.2, 4.6 Hz, H-16), 4.97 (d, J=7.7 Hz, H-1 of Glc), 5.03 (d, J=1.4 Hz, H-21A), 5.28 (d, J=1.9 Hz, H-21B).

CMR (C_5D_5N, 100 MHz) : δ C-1) 73.7 (2) 37.1 (3) 73.8 (4) 35.1 (5) 39.1 (6) 28.6 (7) 32.4 (8) 36.4 (9) 55.0 (10) 42.6 (11) 24.5 (12) 39.8 (13) 42.6 (14) 57.0 (15) 33.0 (16) 81.2 (17) 60.6 (18) 15.2 (19) 6.5 (20) 153.2 (21) 107.5 (22) 107.5 (23) 26.3 (24) 26.6 (25) 27.8 (26) 66.1 (27) 16.5 **Glc** (1) 102.6 (2) 75.4 (3) 78.7 (4) 71.8 (5) 78.4 (6) 62.9.

Mass (FAB, Negative ion) : *m/z* 591 [M-H]⁻.

Reference

1.   Y. Mimaki, Y. Takaashi, M. Kuroda and Y. Sashida, *Phytochemistry*, **45**, 1229 (1997).

DIOSCOREA PRAZERI SAPONIN 1
Prazerigenin A-3-O-β-D-glucopyranoside

Source : *Dioscorea prazeri* Prain et Burk.
(Dioscoreaceae)
Mol. Formula : $C_{33}H_{52}O_9$
Mol. Wt. : 592
M.P. : 260°C
[α]$_D$: -79.0° (c=1.20, Pyridine)
Registry No. : [63358-77-0]

Mass : *m/z* 592 [(M)$^+$, 0.6], 574 (M$^+$-H$_2$O, 2), 430 [genin, 1.5], 412 [genin-H$_2$O, 37], 394 (100), 379 (44), 371 (2), 361 (2), 353 (7), 343 (3), 340 (6), 335 (2) 325 (12), 322 (24), 298 (9), 280 (19), 269 (12), 265 (20), 251 (28), 139 (18), 126 (16), 115 (5).

Reference

1.   K. Rajaraman, V. Seshadri and S. Rangaswami, *Ind. J. Chem.*, **14B**, 735 (1976).

3-*EPI*-RUSCOGENIN 3-O-β-D-GLUCOPYRANOSIDE

Source : Chinese crude drug *"Senshokushichikon"*
Mol. Formula : $C_{33}H_{52}O_9$
Mol. Wt. : 592
M.P. : 122-125°C
[α]$_D$: -93° (c=0.44, CHCl$_3$)
Registry No. : [71335-75-6]

IR (CHCl₃) : 3350 (OH), 980, 960, 920, 895 (spirostane ring), 835 (C=CH-) cm⁻¹.

Mass (E.I.) : *m/z* (rel.intens.) 592 [(M)⁺, 1.2], 412 [(M-Glc)⁺, 100], 394 [(Agl-2xH₂O)⁺, 25.3], 429 (20.6), 413 (22.4), 298 (32.1), 213 (6.9), 139 (58.8).

Penta-acetate :

M.P. : 247-248°C; [α]$_D^{20}$: -20° (c=0.2, CHCl₃)

IR (KBr) : 3030 (CH=C), 1740, 1235 (OCOCH₃), 980, 960, 920, 895 (spirostane ring), 840 (C=CH-) cm⁻¹.

PMR (CDCl₃, 100 MHz) : δ 0.80 (s, 3xH-18), 0.81 (d, *J*=6.0 Hz, 3xH-27), 0.97 (d, *J*=6.0 Hz, 3xH-21), 1.14 (s, 3xH-19), 2.02-2.12 (5xOCOC*H₃*), 3.48 (m, 2xH-26), 3.70 (m, H-5 of Glc), 4.14 (dd, *J*=12.0, 2.0 Hz, H-6A of Glc), 4.35 (dd, *J*=12.0, 7.0 Hz, H-6B of Glc), 4.45 (m, H-15), 4.68 (d, *J*=8.0 Hz, H-1 of Glc), 4.75 (dd, *J*=10.0, 4.0 Hz, H-1), 5.00 (dd, *J*=8.0, 8.0 Hz, H-3 of Glc), 5.07 (dd, *J*=8.0, 8.0 Hz, H-4 of Glc), 5.27 (dd, *J*=8.0, 8.0 Hz, H-2 of Glc), 5.52 (br m, *W*½=10.0 Hz, H-6).

Mass (E.I.) : *m/z* (rel.intens.) 802 [(M)⁺, 2.8], 742 [(M-AcOH)⁺, 8.1], 331 (40.2), 271 (9.3), 229 (10.6), 221 (3.8), 169 (65.5) 139 (C₉H₁₅O, 100).

Reference

1.   M. Tanaka, Y. Kondo, G. Kusano and S. Nozoe, *Yakugaku Zasshi*, **99**, 528 (1979).

FLORIBUNDA SAPONIN A
Pennogenin 3-O-β-D-glucopyranoside

Source : *Dioscorea olfersiana* Klotzsch ex Griseb.[1] (Dioscoreaceae)*, *Majanthemum dilatatum* Nelsons et Macbr[2] (Liliaceae), *Dioscorea floribunda* Mart. & Gal[3]
Mol. Formula : C₃₃H₅₂O₉
Mol. Wt. : 592
M.P. : 275-278°C (decomp.)[3]
[α]$_D$: -117.4° (Pyridine)[3]
Registry No. : [37341-36-9]

CMR (CDCl₃+CD₃OD, 50 MHz)[1] : δ C-1) 37.0 (2) 29.3 (3) 78.8 (4) 38.4 (5) 140.0 (6) 121.5 (7) 31.8 (8) 31.4 (9) 49.8 (10) 36.6 (11) 20.4 (12) 36.9 (13) 43.7 (14) 52.5 (15) 31.4 (16) 90.3 (17) 89.9 (18) 16.0 (19) 19.1 (20) 44.3 (21) 7.9 (22) 109.9 (23) 30.9 (24) 29.4 (25) 29.7 (26) 66.6 (27) 16.8 **Glc** (1) 100.9 (2) 73.2 (3) 76.2 (4) 69.7 (5) 75.5 (6) 61.4.

Mass (E.I.)[1] : *m/z* (rel.intens.) 574 (12), 412 (48), 396 (66), 380 (10), 356 (14), 342 (11), 316 (3) 282 (24), 253 (12), 214 (14), 180 (30), 171 (10), 145 (35), 139 (73), 126 (55), 115 (53), 83 (100).

Mass (L.S.I., glycerol)[4] : *m/z* 593 [M+H]+, 575 [M+H-H2O]+, 441.

Mass (L.S.I., glycerol+NaCl)[4] : *m/z* 615 [M+Na]+, 481, 453 [M+Na-Glc]+, 435 575, 413 [Agl]+, 395 [Agl-H2O]+, 281, 269, 251, 213.

*Note : The compound was isolated admixed with diosgenin 3-O-β-D-glucopyranoside.

Biological Activity : The compound shows hemolytic (HD50=230±30 µM) and antifungal (GI.D50=17±2 µM) activities.[5]

References

1. M. Haraguchi, A.P.Z. D. Santos, M.C. M.Young and E.P. Chu, *Phytochemistry*, **36**, 1005 (1994).

2. E.-R. Woo, J.M. Kim, H.J. Kim, S.H. Yoon and H. Park, Planta Med., 64, 466 (1998).

3. S.B. Mahato, N.P. Sahu and A.N. Ganguly, *Phytochemistry*, **20**, 1943 (1981).

4. Yu. M. Milgrom, Ya.V. Rashkes, L.I. Sturigna and V.L. Sadovskaya, *Khim. Prir. Soedin.*, **27**, 523 (1991); *Chem. Nat. Comp.*, **27**, 455 (1991).

5. M. Takechi, S. Shimada and Y. Tanaka, *Phytochemistry*, **30**, 3943 (1991).

NEOHECOGENIN 3-O-β-D-GLUCOPYRANOSIDE

Source : *Tribulus terrestris* Linn. (Zygophyllaceae)
Mol. Formula : C33H52O9
Mol. Wt. : 592
M.P. : 280-282°C
[α]D : -14.0° (Pyridine)
Registry No. : [79974-44-0]

IR : 3400 (OH), 1700, 987, 920, 900 and 866 (spiroketal absorptions) cm^{-1}.

Reference

1. S.B. Mahato, N.P. Sahu, A.N. Ganguly, K. Miyahara and T. Kawasaki, *J. Chem. Soc. Perkin Trans. I*, 2405 (1981).

NOLINOSPIROSIDE D
(25*S*)-Spirost-5-ene-1β,3β-diol 1-O-β-D-galactopyranoside

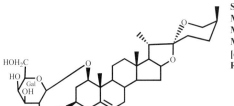

Source : *Nolina microcarpa* S. Wats. (Dracaenaceae)
Mol. Formula : $C_{33}H_{52}O_9$
Mol. Wt. : 592
M.P. : 198-200°C (MeOH)
$[\alpha]_D^{20}$: -48.0±2° (c=1.50, C_5D_5N)
Registry No. : [144028-91-1]

IR (KBr) : 860, 910<930, 990 [spiroketal chain of the (25*S*) series] 3200-3600 (OH) cm^{-1}.

PMR (C_5D_5N, 250/300 MHz) : δ 0.83 (s, 3xH-18), 1.04 (d, $J_{27,25}$=7.0 Hz, 3xH-27), 1.05 (d, $J_{21,20}$=7.0 Hz, 3xH-21), 1.22 (s, 3xH-19), 3.33 (br d, $J_{26,26}$=11.5 Hz, H-26A), 3.84 (m, H-3α), 3.86 (dd, $J_{1,2a}$=11.5 Hz, $J_{1,2e}$=3.5 Hz, H-1α), 4.00 (m, H-5 of Gal), 4.03 (dd, $J_{26',25}$=3.5 Hz, H-26B), 4.13 (dd, $J_{3,4}$=4.5 Hz, H-3 of Gal), 4.40 (dd, $J_{6',5}$=5.5 Hz, H-6A of Gal), 4.42 (dd, $J_{2,3}$=9.0 Hz, H-2 of Gal), 4.42 (dd, $J_{2,3}$=9.0 Hz, H-2 of Gal), 4.47 (m, H-16), 4.53 (dd, $J_{6,6}$=11.0 Hz, $J_{6,5}$=7.0 Hz, H-6B of Gal), 4.59 (br d, $J_{4,5}$=4.5 Hz, H-4 of Gal), 4.83 (d, $J_{1,2}$=7.5 Hz, H-1 of Gal), 5.56 (br d, J=5.5 Hz, H-6).

CMR (C_5D_5N, 62.2/75 MHz) : δ C-1) 83.73 (2) 38.02 (3) 68.17 (4) 43.78 (5) 139.81 (6) 124.70 (7) 32.07 (8) 33.14 (9) 50.55 (10) 42.98 (11) 23.88 (12) 40.46 (13) 40.32 (14) 57.11 (15) 32.47 (16) 81.31 (17) 63.04 (18) 17.91 (19) 14.91 (20) 42.61 (21) 14.91 (22) 109.81 (23) 26.52 (24) 26.30 (25) 27.64 (26) 65.18 (27) 16.37 **Gal** (1) 102.59 (2) 72.71 (3) 75.41 (4) 69.91 (5) 76.66 (6) 62.31.

Mass : *m/z* 592 [M]$^+$.

Reference

1.   G.V. Shevchuk, Y.S. Vollerner, A.S. Shashkov and V.Y. Chirva, *Khim. Prir. Soedin.*, 678 (1991), *Chem. Nat. Comp.* **27**, 597 (1991).

RHODEA JAPONICA SAPONIN 5
(5β)-Spirost-25(27)-en-1β,3β-diol 3-O-β-D-glucopyranoside

Source : *Rhodea japonica* (Thunb.) Roth (Liliaceae)
Mol. Formula : $C_{33}H_{52}O_9$
Mol. Wt. : 592
Registry No. : [86462-28-4]

Reference

1.   K. Kudo, K. Miyahara, N. Marubayashi and T. Kawasaki, *Chem. Pharm. Bull.*, **32**, 4229 (1984).

CAMASSIA CUSICKII SAPONIN 1
Chlorogenin 6-O-β-D-glucopyranoside

Source : *Camassia cusickii* S.Wats. (Liliaceae)
Mol. Formula : $C_{33}H_{54}O_9$
Mol. Wt. : 594
$[\alpha]_D^{25}$: -26.1° (c=0.36, MeOH)
Registry No. : [138877-13-1]

IR (KBr) : 3420 (OH), 2950, 2930, 2875 (CH), 1450, 1375, 1340, 1300, 1240, 1170, 1160, 1075, 1045, 980, 955, 915, 895, 860 (intensity 915<895, 25R-Spiroketal) cm^{-1}.

PMR (C$_5$D$_5$N, 400 MHz) : δ 0.61 (ddd, J=11.8, 11.8, 3.8 Hz, H-9), 0.72 (d, J=5.4 Hz, 3xH-27), 0.80 (s, H-18)[a], 0.85 (s, H-19)[a], 1.13 (d, J=6.9 Hz, 3xH-21), 1.94 (m, H-20), 2.62 (ddd, J=12.5, 4.1 Hz, H-7eq), 3.26 (br d, J=12.3 Hz, H-4eq), 3.48 (dd, J=10.5, 10.5 Hz, H-26A), 3.57 (dd, J=10.5, 3.0 Hz, H-26B), 3.73 (ddd, J=10.7, 10.7, 4.6 Hz, H-6), 3.78 (m, H-3), 3.93 (m, H-5 of Glc), 4.25 (2H, H-3 and H-4 of Glc), 4.35-4.53 (3H, 2xH-6 of Glc and H-16), 4.54 (dd, J=8.0, 8.0 Hz, H-2 of Glc), 4.94 (H-1 of Glc, overlapping with H$_2$O signal).

CMR (C$_5$D$_5$N, 100.6 MHz) : δ C-1) 37.8 (2) 32.3[a] (3) 70.7 (4) 33.2[a] (5) 51.3 (6) 79.7 (7) 41.5 (8) 34.2 (9) 54.0 (10) 36.7 (11) 21.3 (12) 40.1 (13) 40.8 (14) 56.5 (15) 32.1 (16) 81.1 (17) 63.0 (18) 16.7 (19) 13.6 (20) 42.0 (21) 15.0 (22) 109.2 (23) 31.9 (24) 29.3 (25) 30.6 (26) 66.9 (27) 17.4 **Glc** (1) 106.0 (2) 75.8 (3) 78.7[b] (4) 71.9 (5) 78.0[b] (6) 63.0.

Mass (E.I.) : m/z 594 [M]$^+$ (0.6), 535 (0.8), 522 (0.8), 433 (1.4), 432 (1.4), 415 (13), 301 (7), 271 (7), 139 (100), 115 (20), 107 (11).

Reference

1.   Y. Mimaki, Y. Sashida and K. Kawashima, *Phytochemistry*, **30**, 3721 (1991).

CAPSICOSIDE A₁, PETUNIOSIDE B'
Gitogenin 3-O-β-D-galactopyranoside

Source : *Capsicum annuum* L.[1] (Solanaceae), *Petunia hybrida* L.[2] (Solanaceae), *Hosta sieboldii*[3] (Liliaceae)
Mol. Formula : C$_{33}$H$_{54}$O$_9$
Mol. Wt. : 594
M.P. : 258°C[2]
[α]$_D^{20}$: -40° (c=1.2, MeOH)[2]
Registry No. : [60117-36-4]

IR (KBr)[1] : 3500-3400, 987, 920, 900, 850 (900>920) Intensity 900>920 cm^{-1} (25R spiorstanl).

IR (KBr)[3] : 3410 (OH), 2935, 2870 (C-H), 1445, 1375, 1235, 1205, 1170, 1150, 1120, 1085, 1065, 1045, 1005, 985, 980, 950, 920, 895, 860 cm^{-1}

PMR (C$_5$D$_5$N, 300 MHz)[2] : δ 0.70 (d, $J_{25.27}$=6.0 Hz, H-27), 1.20 ($J_{1,1}$=13.0 Hz, H-1A), 1.40 ($J_{15,16}$=8.2 Hz, H-15A), 1.55 ($J_{25,26}$=4.5 Hz, H-25), 1.62 ($J_{3,4}$=5.2 Hz, H-4A), 1.81 (H-17), 1.82 ($J_{4,4}$=12.6 Hz, H-4B), 2.03 ($J_{16,16}$=6.2 Hz, H-

15B), 2.19 ($J_{1,2}$=4.7 Hz, H-1B), 3.47 ($J_{25,26}$=10.0 Hz, H-26A), 3.56 ($J_{26,26}$=10.1 Hz, H-26B), 3.76 ($J_{3,4}$=8.9 Hz, H-3), 3.94 (H-5 of Gal), 3.95 ($J_{2,1}$=9.1 Hz, $J_{2,3}$=11.5 Hz, H-2), 4.03 ($J_{4,3}$=3.0 Hz, H-3 of Gal), 4.29 ($J_{5,6A}$=5.5 Hz, $J_{6A,6B}$=12.5 Hz, H-6A of Gal), 4.37 ($J_{4,5}$=3.5 Hz, H-4 of Gal), 4.41 ($J_{2,3}$=7.5 Hz, H-2 of Gal), 4.41 ($J_{5,6B}$=5.3 Hz, H-6B of Gal), 4.50 ($J_{16,17}$=8.2 Hz, H-16), 5.39 (d, $J_{1,2}$=7.8 Hz, H-1 of Gal).

CMR (C_5D_5N, 75.0 MHz)[2] : δ C-1) 45.7 (2) 71.9 (3) 85.15 (4) 34.3 (5) 45.05 (6) 28.3 (7) 32.45 (8) 34.9 (9) 54.8 (10) 37.1 (11) 21.6 (12) 40.3 (13) 41.0 (14) 56.6 (15) 32.25 (16) 81.3 (17) 63.3 (18) 16.65 (19) 13.5 (20) 42.2 (21) 14.9 (22) 109.4 (23) 32.1 (24) 29.4 (25) 30.7 (26) 67.1 (27) 17.3 **Gal** (1) 103.8 (2) 73.1 (3) 75.9 (4) 70.3 (5) 75.9 (6) 61.2.

Mass (FAB, Negative ion)[3] : *m/z* 593 [M-H]⁻.

References

1.   E.V. Gutsu, P.K. Kintya, S.A. Shvets and G.V. Larur'evskii, *Khim. Prir. Soedin.*, **22**, 708 (1986); *Chem. Nat. Comp.*, **22**, 661 (1986).

2.   S.A. Shvets, A.M. Naibi, P.K. Kintya, and A.S. Shashkov, *Khim. Prir. Soedin.*, **31**, 391 (1995); *Chem. Nat. Comp.*, **31**, 328 (1995).

3.   Y. Mimaki, M. Kuroda, A. Kameyama, A. Yokosuka and Y. Sashida, *Phytochemistry*, **48**, 1361 (1998).

CORDYLINE STRICTA SAPONIN 1
(25S)-5α-Spirostane-1β,3α-diol 3-O-β-D-glucopyranoside

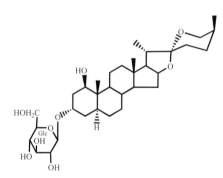

Source : *Cordyline stricta* (Agavaceae)
Mol. Formula : $C_{33}H_{54}O_9$
Mol. Wt. : 594
[α]$_D$^{27} : -55.2° (c=0.11, MeOH)
Registry No. : [194146-04-8]

IR (KBr) : 3410 (OH), 2920 (CH), 1445, 1365, 1345, 1305, 1270, 1220, 1160, 1060, 1020, 980, 935, 915, 890, 840 (intensity 915 > 890, 25S-spiroacetal) cm⁻¹.

PMR (C_5D_5N, 400 MHz) : δ 0.88 (s, 3xH-18), 1.05 (s, 3xH-19), 1.07 (d, *J*=7.1 Hz, 3xH-21), 1.11 (d, *J*=6.9 Hz, 3xH-27), 3.36 (br d, *J*=10.8 Hz, H-26A), 3.89 (ddd, *J*=8.8, 5.0, 2.3 Hz, H-5 of Glc), 4.05 (dd, *J*=8.8, 7.8 Hz, H-2 of Glc), 4.06 (dd, *J*=10.8, 2.8 Hz, H-26B), 4.22 (dd, *J*=8.8, 8.8 Hz, H-4 of Glc), 4.26 (dd, *J*=8.8, 8.8 Hz, H-3 of Glc), 4.27 (dd,

J=12.0, 3.7 Hz, H-1), 4.33 (br s, H-3), 4.38 (dd, *J*=11.6, 5.0 Hz, H-6A of Glc), 4.52 (2H, overlapping, H-16 and H-6B of Glc), 4.93 (d, *J*=7.8 Hz, H-1 of Glc).

CMR (C_5D_5N, 100 MHz) : δ C-1) 73.7 (2) 37.1 (3) 73.9 (4) 35.1 (5) 39.1 (6) 28.7 (7) 32.5 (8) 36.0 (9) 55.1 (10) 42.6 (11) 24.8 (12) 40.9 (13) 40.4 (14) 56.8 (15) 32.3 (16) 81.2 (17) 63.2 (18) 16.8 (19) 6.5 (20) 42.5 (21) 14.8 (22) 109.7 (23) 26.2 (24) 26.4 (25) 27.5 (26) 65.1 (27) 16.3 **Glc** (1) 102.6 (2) 75.3 (3) 78.7 (4) 71.7 (5) 78.3 (6) 62.9.

Mass (FAB, Negative ion) : *m/z* 593 [M-H]⁻.

Reference

1.   Y. Mimaki, Y. Takaashi, M. Kuroda and Y. Sashida, *Phytochemistry*, **45**, 1229 (1997).

NOGIRAGENIN 11-O-β-D-GALACTOPYRANOSIDE

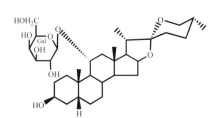

Source : *Metanarthecium luteo-viride* Maxim. (after soil-bactecial hydrolysis)
Mol. Formula : $C_{33}H_{54}O_9$
Mol. Wt. : 594
$[\alpha]_D^{10}$: -71.6° (Pyridine)
Registry No. : [50938-68-6]

Penta-Acetate :

M.P. : 229-232°C; $[\alpha]_D^{18}$: -63.0° (CHCl₃)

IR : 1762, 1741 (sh), 1738, 1250, 1239, 1220 (acetate), 980, 919 < 898, 861 (25*R*-Spiroketal) cm⁻¹.

PMR (CDCl₃, 90 MHz) **:** δ 1.95 (s, OCOC*H₃*), 1.98 (s, OCOC*H₃*), 2.00 (s, 2xOCOC*H₃*), 2.12 (s, OCOC*H₃*).

Mass : *m/z* 331 (73%), 139.

Reference

1.   I. Yosioka, Y. Morii and I. Kitagawa, *Chem. Pharm. Bull.*, **21**, 2092 (1973).

ORNITHOGALUM THYRSOIDES SAPONIN 10

(5α,25R)-Spirostan-1β,3β-diol 1-O-β-D-glucopyranoside

Source : *Ornithogalum thyrsoides* (Liliaceae)
Mol. Formula : $C_{33}H_{54}O_9$
Mol. Wt. : 594
[α]$_D^{20}$: -68.0° (c=0.10, MeOH)
Registry No. : [790661-46-0]

IR (film) : 3363 (OH), 2950 and 2926 (CH), 2875 (CH), 1073, 1055, 1025 cm^{-1}.

PMR (C_5D_5N, 500 MHz) : δ 0.70 (d, *J*=5.3 Hz, 3xH-27), 0.85 (s, 3xH-18), 1.03 (s, 3xH-19), 1.10 (d, *J*=6.9 Hz, 3xH-21), 3.51 (dd, *J*=10.6, 10.6 Hz, H-26ax), 3.58 (dd, *J*=10.6, 2.9 Hz, H-26eq), 3.88 (dd, *J*=8.9, 5.8, 3.3 Hz, H-5 of Glc), 3.90 (br m, *W*½=22.3 Hz, H-3), 3.98 (dd, *J*=11.5, 4.2 Hz, H-1), 4.03 (dd, *J*=8.9, 7.7 Hz, H-2 of Glc), 4.15 (dd, *J*=8.9, 8.9 Hz, H-4 of Glc), 4.24 (dd, *J*=8.9, 8.9 Hz, H-3 of Glc), 4.36 (dd, *J*=11.5, 5.8 Hz, H-6A of Glc), 4.54 (m, H-16), 4.59 (dd, *J*=11.5, 3.3 Hz, H-6B of Glc), 5.05 (d, *J*=7.7 Hz, H-1 of Glc).

CMR (C_5D_5N, 125 MHz) : δ C-1) 81.2 (2) 37.7 (3) 67.6 (4) 39.7 (5) 42.9 (6) 28.9 (7) 32.4 (8) 36.4 (9) 54.8 (10) 40.4 (11) 23.7 (12) 40.7 (13) 41.5 (14) 56.7 (15) 32.4 (16) 81.1 (17) 63.2 (18) 16.9 (19) 8.2 (20) 42.1 (21) 14.9 (22) 109.2 (23) 31.8 (24) 29.3 (25) 30.6 (26) 66.8 (27) 17.3 **Glc** (1) 100.6 (2) 75.6 (3) 78.6 (4) 72.3 (5) 78.4 (6) 63.6.

Mass (FAB, Positive ion) : *m/z* 595.3836 [(M+H)$^+$, requires 595.3846].

Biological Activity : Moderate cytotoxic against HL-60 leukemia cells and HSC-2 cells.

Reference

1.   M. Kuroda, Y. Mimaki, K. Ori, H. Sakagami and Y. Sashida, *J. Nat. Prod.*, **67**, 1690 (2004).

RHODEA JAPONICA SAPONIN 2
5β,22β,25(S)-Spirostane-1β,3β-diol 3-O-β-D-glucopyranoside

Source : *Rhodea japonica* (Thunb.) Roth (Liliaceae)
Mol. Formula : $C_{33}H_{54}O_9$
Mol. Wt. : 594
M.P. : 257-259°C
[α]$_D$: -31.2°
Registry No. : [94482-41-4]

IR (KBr) : 987, 946, 917 < 895, 832 cm^{-1}.

PMR (C_5D_5N, 100 MHz) : δ 0.69 (m, 3xH-27), 0.97 (s, 3xH-18), 1.01 (d, *J*=7.0 Hz, 3xH-21), 1.24 (s, 3xH-19), 3.68 (m, 2xH-26).

CMR (C_5D_5N, 25 MHz) : δ C-20) 42.1 (21) 17.0 (22) 110.3 (23) 28.2^a (24) 28.1^a (25) 30.6 (26) 69.4 (27) 17.3.

Mass (F.D.) : *m/z* 595 [M+H]$^+$.

Reference

1.   K. Kudo, K. Miyahara, N. Marubayashi and T. Kawasaki, *Chem. Pharm. Bull.*, **32**, 4229 (1984).

RHODEA JAPONICA SAPONIN 6
(5β,25R)-Spirostan-1β,3β-diol 3-O-β-D-glucopyranoside

Source : *Rhodea japonica* (Thunb.) Roth (Liliaceae)
Mol. Formula : $C_{33}H_{54}O_9$
Mol. Wt. : 594
Registry No. : [86462-30-8]

IR (KBr) : 980, 945, 918 < 900, 864 cm^{-1}.

CMR (C_5D_5N, 25 MHz) : δ C-20) 41.9 (21) 15.0 (22) 109.0 (23) 31.8 (24) 29.2 (25) 30.5 (26) 66.7 (27) 17.3.

Reference

1.   K. Kudo, K. Miyahara, N. Marubayashi and T. Kawasaki, *Chem. Pharm. Bull.*, **32**, 4229 (1984).

RHODEA JAPONICA SAPONIN R-8
Rhodeasapogenin 3-O-β-D-glucopyranoside]

Source : *Rhodea japonica*[1,2] (Thunb.) Roth (Liliaceae)
Mol. Formula : $C_{33}H_{54}O_9$
Mol. Wt. : 594
M.P. : 260-264°C[1]
[α]$_D$: -61.7° (Pyridine)[1]
Registry No. : [86462-31-9]

1194

IR (KBr)[2] : 988, 945, 920 > 898, 850 cm^{-1}.

CMR (C$_5$D$_5$N, 25.0 MHz)[2] : δ C-20) 42.4 (21) 14.8 (22) 109.5 (23) 26.1 (24) 26.1 (25) 27.4 (26) 64.9 (27) 16.2.

References

1.   K. Miyahara, K. Kudo and T. Kawasaki, *Chem. Pharm. Bull.,* **31**, 348 (1983).

2.   K. Kudo, K. Miyahara, N. Marubayashi and T. Kawasaki, *Chem. Pharm. Bull.,* **32**, 4229 (1984).

CONVALLASAPONIN B
Convallagenin B 5-O-α-L-arabinopyranoside

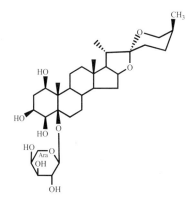

Source : *Convallaria keiskei* Miq. (Liliaceae)
Mol. Formula : C$_{32}$H$_{52}$O$_{10}$
Mol. Wt. : 596
M.P. : 237~274°C (decomp.)[1]
[α]$_D^{22}$: -55.7° (c=0.435, CHCl$_3$-MeOH)[1]
Registry No. : [19317-00-1]

IR (KBr)[1] : 3600 (br, OH), 980, 918>897, 854 cm^{-1} (25S-spiroketal).

References

1.   M. Kimura, M. Tohma and I. Yoshizawa, *Chem. Pharm. Bull.,* **14**, 50 (1966).

2.   M. Kimura, M. Tohma and I. Yoshizawa, *Chem. Phram. Bull.,* **16**, 2191 (1968).

DIOSCOREA TENUIPES SAPONIN 1

5β,25S-Spirostan-1β,2β,3α,4β-tetrol 1-O-α-L-arabinopyranoside

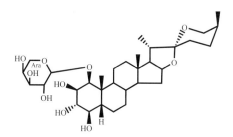

Source : *Dioscorea tenuipes* Franch. et Sav.
(Dioscoreaceae)
Mol. Formula : $C_{32}H_{52}O_{10}$
Mol. Wt. : 596
M.P. : 248-249°C (decomp.)
[α]$_D$: -12.7° (c=7.5, MeOH)
Registry No. : [87441-80-3]

IR (KBr) : 3360 (OH), 987, 915>895, 852 (25*S*-spirostane) cm^{-1}.

Hexa-acetate :

M.P. : 146-148°C; **[α]$_D$:** 1.3° (c=0.4, CHCl$_3$)

PMR (CDCl$_3$, 60/100 MHz) : δ 0.74 (s, 3xH-18), 0.97 (d, *J*=6.0 Hz, 3xH-27), 1.18 (s, 3xH-19), 1.20 (d, *J*=6.0 Hz, 3xH-21), 1.95-2.18 (6xOCOC*H*$_3$), 4.57 (d, *J*=6.0 Hz, H-1 of Ara).

Reference

1.   S. Kiyosawa, K. Goto, K. Sakamoto, and T. Kawasaki, *Phytochemistry*, **21**, 2913 (1982).

SMILAX MENISPERMOIDEA SAPONIN 2
(25*S*)-Spirost-5-en-3β,17α,27-triol 3-O-β-D-galactopyranoside

Source : *Smilax menispermoidea* (Liliaceae)
Mol. Formula : $C_{33}H_{52}O_{10}$
Mol. Wt. : 608
M.P. : 240-244°C
Registry No. : [160625-68-3]

IR (KBr) : 3475 (OH), 1660 (C=CH), 1050 (C-O-C), 839, 813 (Δ^5) cm^{-1}.

PMR (C_5D_5N, 400 MHz) : δ 0.90 (s, 3xH-18), 0.94 (s, 3xH-19), 1.23 (d, *J*=7.2 Hz, 3xH-21), 2.30 (q, *J*=7.2 Hz, H-20), 2.70 (m, H-25), 3.55 (dd, *J*=11.0, 7.0 Hz, H-27A), 3.65 (dd, *J*=11.0, 1.0 Hz, H-27B), 3.85 (d, $J_{26ax,26eq}$=-10.9 Hz, $J_{26ax,25ax}$=11.0 Hz, H-26ax), 4.07 (dd, $J_{26eq,26ax}$=-10.9 Hz, $J_{26eq,25ax}$=3.5 Hz, H-26eq), 5.29 (m, H-6), 5.02 (d, *J*=7.8 Hz, H-1 of Gal).

CMR (C_5D_5N, 100 MHz) : δ C-1) 37.5 (2) 30.2 (3) 78.1 (4) 39.0 (5) 140.9 (6) 121.7 (7) 32.1 (8) 32.2 (9) 50.2 (10) 37.1 (11) 21.0 (12) 31.9 (13) 44.9 (14) 53.1 (15) 32.0 (16) 90.1 (17) 90.1 (18) 17.1 (19) 19.4 (20) 44.9 (21) 9.7 (22) 110.2 (23) 31.9 (24) 23.7 (25) 39.1 (26) 63.9 (27) 64.4 Gal (1) 101.5 (2) 71.1 (3) 73.4 (4) 69.1 (5) 75.3 (6) 62.9.

Mass (FAB-MS) : *m/z* 631 [M+Na]$^+$.

Reference

1.   Y. Ju, Z. Jia and X. Sun , *Phytochemistry,* **37**, 1433 (1994).

AGAMENOSIDE I

(5α,22S,23S,24R,25S)-Spirostane-3β,23,24-triol 24-O-[β-D-glucopyranoside]

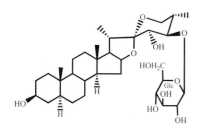

Source : *Agave americana* L. (Agavaceae)
Mol. Formula : $C_{33}H_{54}O_{10}$
Mol. Wt. : 610
$[\alpha]_D^{21}$ **:** -39.9° (c=0.41, Pyridine)
Registry No. : [738584-20-8]

IR (KBr) : 3432, 2927, 1457, 1713, 1671, 1452, 1380, 1166, 963, 941, 875, 976 cm^{-1}.

PMR (C_5D_5N, 500 MHz) : δ 0.56 (H-9), 0.75 (s, 3xH-19), 0.79 (H-7α), 0.92 (H-1α), 1.01 (s, 3xH-18), 1.09 (H-5), 1.10 (H-14), 1.11 (H-12α), 1.15 (d, J=7.1 Hz, 3xH-21), 1.19 (H-6), 1.19 (d, J=6.5 Hz, 3xH-27), 1.21 (H-11α), 1.43 (H-8), 1.46 (H-11β), 1.52 (H-4β), 1.55 (H-2β), 1.64 (H-1β, H-7β), 1.70 (H-15β), 1.76 (H-4α), 1.84 (H-12β), 1.85 (dd, J=6.5, 7.0 Hz, H-17), 2.04 (H-2α, H-15β), 2.05 (H-25), 3.01 (H-20), 3.60 (dd, J=8.9, 5.9 Hz, H-26), 3.83 (H-3), 3.84 (d, J=9.0 Hz, H-23), 3.90 (H-3 of Glc), 3.96 (dd, J=9.5, 9.0 Hz, H-24), 4.04 (H-2 of Glc), 4.18 (H-5 of Glc), 4.21 (H-4 of Glc), 4.30 (H-6A of Glc), 4.47 (H-6B of Glc), 4.62 (q-like, J=7.0 Hz, H-16), 4.88 (d, J=7.8 Hz, H-1 of Glc).

CMR (C_5D_5N, 125 MHz) : δ C-1) 37.5 (2) 32.1 (3) 70.6 (4) 39.3 (5) 45.6 (6) 29.1 (7) 32.5 (8) 35.3 (9) 54.6 (10) 35.9 (11) 21.4 (12) 40.7 (13) 41.4 (14) 56.6 (15) 32.5 (16) 82.0 (17) 62.0 (18) 17.0 (19) 12.5 (20) 36.4 (21) 14.5 (22) 112.7 (23) 71.7 (24) 87.9 (25) 37.9 (26) 64.1 (27) 13.2 **Glc** (1) 105.4 (2) 75.4 (3) 78.3 (4) 71.4 (5) 78.5 (6) 62.5.

Mass (FAB, Negative ion) : *m/z* 609 [M-H]⁻, 447 [M-H-Glc]⁻.

Mass (FAB, Negative ion, H.R.) : *m/z* 669.3676 [(M-H)⁻, calcd. for 669.3639].

Reference

1.   J.M. Jin, Y.-J. Zhang and C.R. Yang, *Chem. Pharm. Bull.*, **52**, 654 (2004).

AGAMENOSIDE C

(25R)-5α-Spirostan-3β,6α,23-triol 6-O-β-D-glucopyranoside

Source : *Agave americana* L. (Agavaceae)
Mol. Formula : $C_{33}H_{54}O_{10}$
Mol. Wt. : 610
Registry No. : [403650-97-5]

IR : 960, 945, 921, 897, 862 (intenity 897 > 921) cm^{-1}.

PMR (C_5D_5N, 500 MHz) : δ 0.73 (s, CH_3), 0.96 (s, CH_3), 0.69 (d, J=5.8 Hz, sec. CH_3), 1.17 (d, J=6.5 Hz, sec. CH_3), 4.91 (d, J=7.6 Hz, H-1 of Glc).

CMR (C_5D_5N, 125 MHz) : δ C-1) 37.8 (2) 32.1 (3) 70.8 (4) 33.2 (5) 51.3 (6) 79.8 (7) 41.4 (8) 34.1 (9) 54.0 (10) 36.8 (11) 21.4 (12) 40.5 (13) 41.5 (14) 56.5 (15) 32.3 (16) 81.7 (17) 62.6 (18) 16.9 (19) 13.6 (20) 35.9 (21) 14.7 (22) 111.7 (23) 67.5 (24) 38.8 (25) 31.8 (26) 66.0 (27) 17.0 **Glc** (1) 106.1 (2) 75.9 (3) 78.0 (4) 72.0 (5) 78.7 (6) 63.2.

Mass (FAB, Negative ion) : *m/z* 609 [M-H]$^-$.

Mass (FAB, Negative ion, H.R.) : m/z 609.3612 [(M-H)$^-$, requires 609.3639].

Reference

1.   J.-M. Jin, X.-K. Liu and C.-R. Yang, *China Journal of Chinese Materia Medica*, **27**, 431 (2002).

CORDYLINE STRICTA SAPONIN 4

(25*R*)-5α-Spirostane-1β,3α,25-triol 3-O-β-D-glucopyranoside

Source : *Cordyline stricta* (Agavaceae)
Mol. Formula : $C_{33}H_{54}O_{10}$
Mol. Wt. : 610
$[\alpha]_D^{27}$: -25.3° (c=0.30, MeOH)
Registry No. : [194143-96-9]

IR (KBr) : 3420 (OH), 2910 (CH), 1440, 1370, 1240, 1190, 1175, 1060, 1020, 985, 965, 920, 840 cm^{-1}.

PMR (C_5D_5N, 400 MHz) : δ 0.90 (s, 3xH-18), 1.07 (s, 3xH-19), 1.14 (d, *J*=6.9 Hz, 3xH-21), 1.56 (s, 3xH-27), 3.91 (ddd, *J*=8.8, 4.8, 2.4 Hz, H-5 of Glc), 4.08 (dd, *J*=8.8, 7.7 Hz, H-2 of Glc), 3.62 and 4.15 (each d, *J*=10.3 Hz, 2xH-26), 4.25 (dd, *J*=8.8, 8.8 Hz, H-4 of Glc), 4.29 (overlapping, H-1), 4.30 (dd, *J*=8.8, 8.8 Hz, H-3 of Glc), 4.35 (br s, H-3), 4.41 (dd, *J*=11.5, 4.8 Hz, H-6A of Glc), 4.55 (2H, overlapping, H-16 and H-6B of Glc), 4.97 (d, *J*=7.7 Hz, H-1 of Glc).

CMR (C_5D_5N, 100 MHz) : δ C-1) 73.7 (2) 37.1 (3) 73.8 (4) 35.1 (5) 39.1 (6) 28.7 (7) 32.4 (8) 36.0 (9) 55.1 (10) 42.6 (11) 24.8 (12) 40.9 (13) 40.5 (14) 56.8 (15) 32.3 (16) 81.3 (17) 63.2 (18) 16.8 (19) 6.5 (20) 41.5 (21) 14.9 (22) 109.3 (23) 30.2 (24) 35.6 (25) 66.4 (26) 70.0 (27) 25.0 **Glc** (1) 102.6 (2) 75.4 (3) 78.7 (4) 71.8 (5) 78.4 (6) 62.9.

Mass (FAB, Negative ion) : *m/z* 610 [M]$^-$.

Reference

1.   Y. Mimaki, Y. Takaashi, M. Kuroda and Y. Sashida, *Phytochemistry*, **45**, 1229 (1997).

DIOSCOREA TOKORO SAPONIN 1
Tokorogenin 1-O-β-D-glucopyranoside

Source : *Dioscorea tokoro* Makino (Dioscoreaceae)
Mol. Formula : $C_{33}H_{54}O_{10}$
Mol. Wt. : 610
M.P. : 275-284°C (decomp.)
$[\alpha]_D^{22}$ **:** -43.0° (c=0.66, CHCl₃)
Registry No. : [23952-48-9]

IR (KBr) : 855, 900>921, 984 (25R-spiroketal), 3250-3450 (hydroxyl) cm⁻¹.

Hexa-acetate :

M.P. : 181-184°C; $[\alpha]_D^{22}$: -27° (c=0.18, CHCl₃)

PMR (CDCl₃, 60 MHz) : δ 0.79 (s, 3xH-18), 1.09 (s, 3xH-19), 2.01, 2.03, 2.08, 2.11, 2.13 (each s, total 6xOCOCH₃), 4.62 (d, J=6.0 Hz, H-1 of Glc).

Reference

1.   K. Miyahara, F. Isozaki and T. Kawasaki, *Chem. Pharm. Bull.*, **17**, 1735 (1969).

TUBEROSIDE O
(25S)-Spirostan-2β,3β,5β-triol 3-O-β-D-glucopyranoside

Source : *Allium tuberosum* L. (Liliaceae)
Mol. Formula : $C_{33}H_{54}O_{10}$
Mol. Wt. : 610
$[\alpha]_D^{24}$ **:** -37.8° (c=0.28, MeOH]
Registry No. : [651306-80-8]

IR (KBr) : 3400, 1450, 1068, 1041, 987, 921, 900, 852 cm^{-1}.

PMR (C$_5$D$_5$N, 400 MHz) : δ 0.79 (s, 3xH-18), 1.03 (d, J=7.0 Hz, 3xH-27), 1.12 (d, J=6.3 Hz, 3xH-21), 1.13 (s, 3xH-19), 1.79 (m, H-17), 3.36 (d, J=11.8 Hz, H-26A), 3.99 (m, H-2 and H-5 of Glc), 4.02 (m, H-16), 4.04 (m, H-26B), 4.05 (m, H-2 of Glc), 4.23 (H-3 of Glc, H-4 of Glc), 4.37 (dd, J=11.8, 5.4 Hz, H-6A of Glc), 4.51 (m, H-6B of Glc), 4.69 (m, H-3), 5.08 (d, J=7.8 Hz, H-1 of Glc).

CMR (C$_5$D$_5$N, 100 MHz) : δ C-1) 35.4 (2) 66.0 (3) 78.9 (4) 35.7 (5) 72.9 (6) 35.0 (7) 28.9 (8) 34.5 (9) 44.4 (10) 42.9 (11) 21.6 (12) 39.9 (13) 40.4 (14) 56.2 (15) 32.0 (16) 81.0 (17) 62.6 (18) 16.1 (19) 17.4 (20) 42.3 (21) 14.7 (22) 109.5 (23) 26.2 (24) 26.0 (25) 27.3 (26) 64.9 (27) 16.3 **Glc** (1) 101.9 (2) 74.6 (3) 78.4 (4) 71.4 (5) 78.7 (6) 62.3.

Mass (FAB, Positive ion) : *m/z* 611 [M+H]$^+$, 449 [M+H-Glc]$^+$.

Reference

1.	S. Sang, S. Mao, A. Lao, Z. Chen and C.-T. Ho, *Food Chemistry*, **83**, 499 (2003).

REINECKEA CARNEA SAPONIN 6
(25*R*)-5β-Spirostane-1β,2β,3β,4 β,5β-pentol 1-O-β-D-xylopyranoside

Source : *Reineckea carnea* Kunth (Liliaceae)
Mol. Formula : C$_{32}$H$_{52}$O$_{11}$
Mol. Wt. : 612
[α]$_D$^{25} : -29.0° (c=0.10, CHCl$_3$-MeOH)
Registry No. : [156665-33-7]

IR (KBr) : 3400 (OH), 2940 (CH), 1410, 1240, 1045, 975, 955, 915, 895, 860 cm^{-1} (intensity 915<895, 25*R*-spiroketal).

PMR (C$_5$D$_5$N, 400 MHz) : δ 0.70 (d, J=5.3 Hz, 3xH-27), 0.87 (s, 3xH-18), 1.12 (d, J=6.9 Hz, 3xH-21), 1.68 (s, 3xH-19), 3.51 (dd, J=10.4, 10.4 Hz, H-26A), 3.59 (dd, J=10.4, 3.0 Hz, H-26B), 3.73 (dd, J=10.6, 10.6 Hz, H-5A of Xyl), 4.02 (dd, J=8.4, 7.7 Hz, H-2 of Xyl), 4.22-4.13 (overlapping H-3 and H-4 of Xyl), 4.29 (d, J=3.2 Hz, H-4), 4.34 (dd, J=3.2, 3.2 Hz, H-2), 4.39 (dd, J=10.6, 3.9 Hz, H-5B of Xyl), 4.53 (d, J=3.2 Hz, H-1), 4.58 (q-like, J=7.5 Hz, H-16), 4.81 (dd, J=3.2, 3.2 Hz, H-3), 5.44 (d, J=7.7 Hz, H-1 of Xyl).

CMR (C$_5$D$_5$N, 100 MHz) : δ C-1) 87.3 (2) 68.3 (3) 76.0 (4) 68.2 (5) 77.7 (6) 29.8 (7) 28.3 (8) 35.0 (9) 45.8 (10) 45.9 (11) 22.0 (12) 40.1 (13) 40.7 (14) 56.2 (15) 32.2 (16) 81.1 (17) 63.1 (18) 16.6 (19) 14.2 (20) 42.1 (21) 15.0 (22) 109.3 (23) 31.9 (24) 29.3 (25) 30.6 (26) 66.9 (27) 17.3 **Xyl** (1) 107.1 (2) 76.0 (3) 78.5 (4) 71.1 (5) 67.6.

Mass (FAB, Negative ion) : *m/z* 612 [M]$^-$.

Biological Activity : The compound exhibited inhibitory activity on CAMP phosphodiesterase with $IC_{50}=2.7\times10^{-5}$ M.

Reference

1.   T. Kanmoto, Y. Mimaki, Y. Sashida, T. Nikaido, K. Koike and T. Ohmoto, *Chem. Pharm. Bull.*, **42**, 926 (1994).

WATTOSIDE G
(25*R*)-1β,2β,3β,4β,5β-Pentahydroxyspirostane 5-O-β-D-xylopyranoside

Source : *Tupistra wattii* Hook. f. (Liliaceae)
Mol. Formula : $C_{32}H_{52}O_{11}$
Mol. Wt. : 612
M.P. : 214-216°C
$[α]_D^{20}$: -65.5° (c=0.03, MeOH)
Registry No. : [619319-15-2]

IR (KBr) : 3412 (OH), 2942, 1452, 1378, 1240,, 1157, 1045, 982, 920, 899, 851 cm^{-1}.

PMR (C_5D_5N, 500 MHz) : δ 0.73 (d, *J*=6.83 Hz, 3xH-27), 0.85 (s, 3xH-18), 1.18 (d, *J*=6.54 Hz, 3xH-21), 1.58 (s, 3xH-19), 5.18 (d, *J*=7.30 Hz, H-1 of Xyl).

CMR (C_5D_5N, 125 MHz) : δ C-1) 77.9 (2) 68.6 (3) 72.7 (4) 74.1 (5) 78.5 (6) 30.5 (7) 28.7 (8) 35.2 (9) 45.8 (10) 45.6 (11) 21.9 (12) 40.2 (13) 41.0 (14) 56.3 (15) 32.5 (16) 81.6 (17) 63.2 (18) 16.9 (19) 13.9 (20) 42.4 (21) 15.4 (22) 109.9 (23) 32.1 (24) 29.6 (25) 30.9 (26) 67.3 (27) 17.7 **Xyl** (1) 103.2 (2) 75.4 (3) 78.3 (4) 71.4 (5) 67.5.

Mass (FAB, Negative ion, H.R.) : *m/z* 611.3466 [(M-H)$^-$, requires 611.3431].

Mass (FAB, Negative ion) : *m/z* 611 [M-H]$^-$, 479 [M-H-Xyl]$^-$.

Biological Activity : Cytotoxic against cancer cell line K 562 with IC_{50} 35.67 µmol/l.

Reference

1.   P. Shen, S.-L. Wang, X.-K. Liu, C.-R. Yang, B. Cai and X.-S. Yao, *Chem. Pharm. Bull.*, **51**, 305 (2003).

CESTRUM SENDTENERIANUM SAPONIN 1
Spirosta-5,25(27)-diene-1β,2α,3β,12β-tetrol 3-O-β-D-galactopyranoside

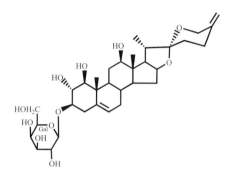

Source : *Cestrum sendtenerianum* (Solanaceae)
Mol. Formula : $C_{33}H_{50}O_{11}$
Mol. Wt. : 622
$[\alpha]_D^{25}$: -47.6° (c=0.25, MeOH)
Registry No. : [227317-05-5]

IR (KBr) : 3390 (OH), 2950 and 2905 (CH), 1445, 1365, 1225, 1065, 1035, 975, 955, 915, 890, 870, 830 cm⁻¹.

PMR (C_5D_5N, 400/500 MHz) : δ 1.14 (s, 3xH-18), 1.23 (s, 3xH-19), 1.38 (d, *J*=6.5 Hz, 3xH-21), 3.56 (d, *J*=8.8 Hz, H-1), 3.66 (dd, *J*=11.0, 4.2 Hz, H-12), 3.96 (m, H-3), 3.98 (dd, *J*=8.8, 8.8 Hz, H-2), 4.14 (dd, *J*=6.7, 5.2 Hz, H-5 of Gal), 4.21 (dd, *J*=9.5, 3.0 Hz, H-3 of Gal), 4.42 (dd, *J*=11.1, 5.2 Hz, H-6A of Gal), 4.48 (dd, *J*=11.1, 6.7 Hz, H-6B of Gal), 4.50 (dd, *J*=9.5, 7.7 Hz, H-2 of Gal), 4.51 and 4.06 (br d, *J*=12.3 Hz, 2xH-26), 4.55 (d, *J*=3.0 Hz, H-4 of Gal), 4.62 (q-like, *J*=7.3 Hz, H-16), 4.82 and 4.77 (br s, 2xH-27), 5.01 (d, *J*=7.7 Hz, H-1 of Gal), 5.51 (br d, *J*=5.2 Hz, H-6).

CMR (C_5D_5N, 100/125 MHz) : δ C-1) 82.1 (2) 75.5 (3) 81.3 (4) 37.9 (5) 138.0 (6) 125.1 (7) 32.0 (8) 31.5 (9) 50.7 (10) 43.2 (11) 34.2 (12) 79.3 (13) 45.9 (14) 55.2 (15) 32.0 (16) 81.4 (17) 63.0 (18) 11.2 (19) 14.8 (20) 42.9 (21) 14.3 (22) 109.7 (23) 33.3 (24) 29.0 (25) 144.6 (26) 65.0 (27) 108.5 **Gal** (1) 103.8 (2) 72.3 (3) 75.2 (4) 70.2 (5) 77.2 (6) 62.4.

Mass (FAB, Positive ion, H.R.) : *m/z* 645.3245 [M+Na]⁺.

Reference

1.   M. Haraguchi, M. Motidome, H. Morita, K. Takeya, H. Itokawa, Y. Mimaki and Y. Sashida, *Chem. Pharm. Bull.*, **27**, 582 (1999).

ALLIOGENIN 3-O-β-D-GLUCOPYRANOSIDE
5α,25R-Spirostane-2α,3β,5α,6β-tetraol 3-O-β-D-glucopyranoside

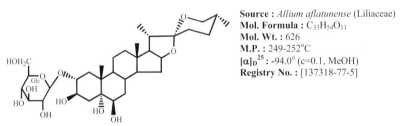

Source : *Allium giganteum* Rgl.[1] (Alliaceae),
A. karataviense Rgl.[2] (Liliaceae)
Mol. Formula : $C_{33}H_{54}O_{11}$
Mol. Wt. : 626
M.P. : 263-265°C
$[\alpha]_D^{18}$: -76.2° (c=1.39, C_5H_5N)[1]
Registry No. : [34340-23-0]

IR (KBr)[1] : 3200-3500 (OH), 870, 905 > 930, 985 cm^{-1} (spiroketal chain of the *R* series).

References

1. M.B. Gorovits, F.S. Khristulas and N.K. Abubakirov, *Khim. Prir. Soedin.*, 434 (1971); *Chem. Nat. Comp.*, **7**, 412 (1971).

2. M.B. Gorovits, F.S. Khristulas and N.K. Abubakirov, *Khim. Prir. Soedin.*, 747 (1973); *Chem. Nat. Comp.*, **9**, 715 (1973).

ALLIUM AFLATUNENSE SAPONIN 1
(25R)-5α-Spirostan-2α,3β,5α,6α-tetraol 2-O-β-D-glucopyranoside

Source : *Allium aflatunense* (Liliaceae)
Mol. Formula : $C_{33}H_{54}O_{11}$
Mol. Wt. : 626
M.P. : 249-252°C
$[\alpha]_D^{25}$: -94.0° (c=0.1, MeOH)
Registry No. : [137318-77-5]

IR (KBr) : 3400 (OH), 2950 (CH), 1450, 1370, 1070, 1030, 980, 940, 920, 890 cm^{-1}.

PMR (C$_5$D$_5$N, 400 MHz) : δ 0.69 (d, *J*=5.7 Hz, 3xH-27), 0.85 (s, 3xH-18), 1.01 (s, 3xH-19), 1.11 (d, *J*=6.9 Hz, 3xH-21), 1.98 (dd, *J*=13.7, 11.6 Hz, H-4ax), 2.09 (dd, *J*=12.0, 5.4 Hz, H-1eq), 2.24 (dd, *J*=12.0, 12.0 Hz, H-1ax), 3.01 (dd, *J*=13.7, 5.7 Hz, H-4eq), 3.47 (dd, *J*=10.8, 10.5 Hz, H-26A), 3.56 (dd, *J*=10.8, 3.5 Hz, H-26B), 4.25 (overlapped, H-2), 4.51 (ddd, *J*=7.6, 7.3, 6.9 Hz, H-16), 4.64 (m, H-3), 5.11 (d, *J*=7.8 Hz, H-1 of Glc).

CMR (C$_5$D$_5$N, 100.6 MHz) : δ C-1) 38.4 (2) 84.6 (3) 71.9 (4) 37.8 (5) 76.1 (6) 70.1 (7) 35.9 (8) 33.8 (9) 44.9 (10) 41.6[a] (11) 21.6 (12) 40.3 (13) 41.0[a] (14) 56.3 (15) 32.2[b] (16) 81.2 (17) 63.1 (18) 16.7 (19) 16.7 (20) 42.0 (21) 15.0 (22) 109.2 (23) 31.9[b] (24) 29.3 (25) 30.6 (26) 66.9 (27) 17.3 Glc (1) 104.6 (2) 75.2 (3) 78.5 (4) 71.2 (5) 78.6[c] (6) 62.8.

Mass (SIMS) : *m/z* (rel.intens.) 627 [M+H]$^+$ (66), 591 (8) 547 (9) 503 (15), 465 [M-Glc]$^+$ (37), 429 (100), 398 (8), 371 (39), 327 (32), 303 (31).

Reference

1.   K. Kawashima, Y. Mimaki and Y. Sashida, *Phytochemistry*, **30**, 3063 (1991).

ALLIUM GIGANTEUM SAPONIN 1
Alliogenin 3-O-β-D-glucopyranoside

Source : *Allium giganteum* Rgl.[1] (Alliaceae), *A. karataviense* Rgl.[2] (Liliaceae)
Mol. Formula : C$_{33}$H$_{54}$O$_{11}$
Mol. Wt. : 626
M.P. : 263-265°C[1]
[α]$_D^{18}$: -76.2° (c=1.39, C$_5$D$_5$N)[1]
Registry No. : [34340-23-3]

IR (KBr)[1] : 3200-3500 (OH), 870, 905>930, 985 cm^{-1} (25*R*-spioketal chain).

References

1.   M.B. Goravits, F.S. Khristulas and N.K. Abubakirov, *Khim. Prir. Soedin.*, 434 (1971); *Chem. Nat. Comp.*, **7**, 412 (1971).

2.   M.B. Gorovits, F.S. Khristulas and N.K. Abubakirov, *Khim. Prir. Soedin.*, 747 (1973); *Chem. Nat. Comp.*, **9**, 715 (1973).

ALLIUM GIGANTEUM SAPONIN 2

Alliogenin 2-O-β-D-glucopyranoside

Source : *Allium giganteum* Regel (Liliaceae)
Mol. Formula : $C_{33}H_{54}O_{11}$
Mol. Wt. : 626
M.P. : 253-256°C
$[\alpha]_D^{23}$: -97.0° (c=0.1, MeOH)
Registry No. : [134955-77-4]

IR (KBr) **:** 3400 (OH), 2950 (CH), 2870, 1450, 1380, 1240, 1080, 1060, 980, 920, 900, 860 cm^{-1} [(25R)-spiroketal intensity 920 < 900].

PMR (C_5D_5N, 400/500 MHz) **:** δ 0.67 (d, *J*=5.8 Hz, 3xH-27), 0.89 (s, 3xH-18), 1.12 (d, *J*=6.9 Hz, 3xH-21), 1.58 (s, 3xH-19), 2.97 (dd, *J*=13.4, 11.5 Hz, H-4ax), 3.47 (dd, *J*=10.7, 10.7 Hz, H-26A), 3.56 (dd, *J*=10.7, 3.6 Hz, H-26B), 4.17 (br s, $W\frac{1}{2}$=6.5 Hz, H-6), 4.40 (ddd, *J*=11.4, 8.6, 5.4 Hz, H-2), 4.61 (dd, *J*=11.6, 2.4 Hz, H-6A of Glc), 4.83 (ddd, *J*=11.5, 8.6, 5.9 Hz, H-3), 5.17 (d, *J*=7.8 Hz, H-1 of Glc).

CMR (C_5D_5N, 100 or 125 MHz) **:** δ C-1) 40.4^a (2) 85.2 (3) 71.5 (4) 40.7^a (5) 74.9 (6) 75.2 (7) 35.8 (8) 30.2 (9) 45.8 (10) 40.7 (11) 21.6 (12) 40.3 (13) 41.0 (14) 56.3 (15) 32.3 (16) 81.2 (17) 63.1 (18) 18.1 (19) 16.3 (20) 42.0 (21) 15.1 (22) 109.2 (23) 31.8 (24) 29.3 (25) 30.6 (26) 66.9 (27) 17.3 **Glc** (1) 104.7 (2) 75.3 (3) 78.5^b (4) 71.9 (5) 78.5^b (6) 62.0.

Mass (S.I.) **:** *m/z* 664 [M+K-H]$^+$, 627 [M+H]$^+$, 465 [M+H-Glc]$^+$.

Reference

1. Y. Sashida, K. Kawashima, Y. Mimaki, *Chem. Pharm. Bull.*, **39**, 698 (1991).

CONVALLAGENIN B 5-O-β-D-GLUCOPYRANOSIDE
(25*S*) Spirostane-1β,3β,4β,5β-tetrol 5-O-β-D-glucopyranoside

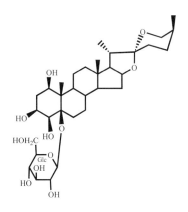

Source : *Aspidistra leshanensis* K. Lang et Z.Y. Zhu[1]
(Liliaceae), *Tupistra watti* Hook. f.[2] (Liliaceae)
Mol. Formula : $C_{33}H_{54}O_{11}$
Mol. Wt. : 626
Registry No. : [160789-42-4]

CMR (C_5D_5N, 125 MHz) : δ C-1) 73.5 (2) 33.9 (3) 71.9 (4) 68.1 (5) 87.9 (6) 24.8 (7) 28.6 (8) 34.9 (9) 46.7 (10) 47.3 (11) 27.6 (12) 40.1 (13) 40.6 (14) 56.1 (15) 32.2 (16) 81.3 (17) 63.0 (18) 18.6 (19) 13.8 (20) 42.6 (21) 14.9 (22) 109.8 (23) 26.3 (24) 26.5 (25) 27.6 (26) 65.2 (27) 16.4 **Glc** (1) 97.5 (2) 75.9 (3) 78.7 (4) 71.9 (5) 78.8 (6) 62.8.

Biological Activity : Cytotoxic against cancer cell line K 562 with IC_{50} 44.52 μmol/l.[2]

References

1.   M.Q. Chen, *Zhiwu Xuebao* (*Acta Bot. Sin.*) **36**, 568 (1994).

2.   P. Shen, S.-L. Wang, X.-K. Liu, C.-R. Yang, B. Cai and X.-S. Yao, *Chem. Pharm. Bull.*, **51**, 305 (2003).

REINECKEA CARNEA SAPONIN 4
Kitigenin 5-O-β-D-glucopyranoside

Source : *Reineckea carnea* Kunth[1] (Liliaceae), *Aspidistra leshanensis* K.Y. Lang et Z.Y. Zhu[2] (Liliaceae)
Mol. Formula : $C_{33}H_{54}O_{11}$
Mol. Wt. : 626
$[\alpha]_D^{25}$: -38.0° (c=0.10, MeOH)[1]
Registry No. : [156665-30-4]

IR (KBr)[1] **:** 3380 (OH), 2940 (CH), 1445, 1375, 1240, 1200, 1145, 1100, 1070, 1040, 980, 955, 915, 895, 875, 860, 800 cm^{-1} [intensity 915 < 895, (25R)-spiroacetal].

PMR (C_5D_5N, 400 MHz)[1] **:** δ 0.73 (d, J=5.7 Hz, 3xH-27), 0.84 (s, 3xH-18), 1.11 (d, J=6.9 Hz, 3xH-21), 1.61 (s, 3xH-19), 3.47 (dd, J=10.6, 10.6 Hz, H-26A), 3.56 (dd, J=10.6, 3.2 Hz, H-26B), 3.84 (dd, J=8.6, 7.7 Hz, H-2 of Glc), 3.90 (ddd, J=8.6, 5.4, 2.2 Hz, H-5 of Glc), 3.94 (dd, J=8.6, 8.6 Hz, H-4 of Glc), 4.07 (dd, J=3.8, 3.8 Hz, H-1), 4.09 (dd, J=8.6, 8.6 Hz, H-3 of Glc), 4.14 (dd, J=11.7, 5.4 Hz, H-6A of Glc), 4.26 (d, J=3.8 Hz, H-4), 4.37 (dd, J=11.7, 2.2 Hz, H-6B of Glc), 4.42 (q-like, J=3.8 Hz, H-3), 4.54 (q-like, J=7.3 Hz, H-16), 5.17 (d, J=7.7 Hz, H-1 of Glc).

CMR (C_5D_5N, 100 MHz)[1] **:** δ C-1) 73.4 (2) 33.9 (3) 71.9 (4) 68.0 (5) 87.9 (6) 24.8 (7) 28.5 (8) 34.9 (9) 46.7 (10) 47.3 (11) 21.6 (12) 40.1 (13) 40.6 (14) 56.1 (15) 32.2 (16) 81.1 (17) 63.0 (18) 16.6 (19) 13.7 (20) 42.1 (21) 15.0 (22) 109.3 (23) 31.9 (24) 29.3 (25) 30.6 (26) 66.9 (27) 17.3 **Glc** (1) 97.5 (2) 75.9 (3) 78.6^a (4) 72.0 (5) 78.8^a (6) 62.9.

Mass (FAB, Negative ion)[1] **:** *m/z* 625 [M-H]$^-$, 463 [M-Glc]$^-$.

Biological Activity : The compound exhibited inhibitory activity on CAMP phosphodiesterase with IC_{50}=32.5×10^{-5} M.[1]

References

1. T. Kanmoto, Y. Mimaki, Y. Sashida, T. Nikaido, K. Koike and T. Ohmoto, *Chem. Pharm. Bull.*, **42**, 926 (1994).

2. M.-J. Chen, *Zhiwu Xuebao* (*Acta Bot. Sin.*), **36**, 568 (1994).

ANZUROSIDE
(24S,25S)-2α,3β,5β,24-Tetrahydroxyspirostan-6-one 24-O-β-D-glucopyranoside

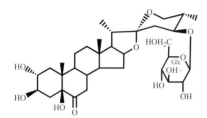

Source : *Allium suvorovii* Rgl., *A. stipitatum* Rgl. (Liliaceae)
Mol. Formula : $C_{32}H_{52}O_{12}$
Mol. Wt. : 628
M.P. : 242-244°C (decomp.)
[α]$_D^{20}$: -62±2° (c=0.96, EtOH)
Registry No. : [125445-36-5]

IR (KBr) : 820, 850, 910, 1715 (>C=O) 3300-3500 (OH) cm^{-1}.

PMR (C_5D_5N) : δ 0.71 (s, 3xH-18), 0.94 (s, 3xH-19), 1.01 (d, 3xH-27), 1.10 (d, 3xH-21), 1.88 (dd, H-4eq), 1.90 (dd,H-23ax), 2.07 (d, 2xH-1), 2.60 (dd, H-23eq), 2.98 (dd, H-4ax), 3.52 (t, H-26ax), 3.61 (dd, H-26eq), 3.95 (dt, H-24ax), 4.32 (m, H-3eq), 4.43 (m, H-2eq), 4.51 (dt, H-16), 6.90 (s, 5-OH), 4.85 (d, J=7.7 Hz, H-1 of Glc), 3.97 (dd, J=7.7, 9.0 Hz, H-2 of Glc), 4.16 (t, J=9.0 Hz, H-3 of Glc), 4.20 (t, J=9.0 Hz, H-4 of Glc), 3.81 (ddd, J=9.0, 5.0, 2.5 Hz, H-5 of Glc), 4.32 (dd, J=11.5, 5.0 Hz, H-6A), 4.44 (dd, J=11.5, 2.5 Hz, H-6B).

CMR (C_5D_5N) : δ C-1) 32.53 (2) 71.68 (3) 71.68 (4) 42.83 (5) 83.47 (6) 212.35 (7) 31.92 (8) 37.20 (9) 45.60 (10) 45.24 (11) 22.73 (12) 39.75 (13) 41.31 (14) 56.67 (15) 33.59 (16) 81.50 (17) 62.44 (18) 16.59 (19) 18.18 (20) 42.26 (21) 14.96 (22) 111.81 (23) 40.95 (24) 81.50 (25) 38.27 (26) 65.35 (27) 13.58 **Glc** (1) 106.36 (2) 75.68 (3) 78.06 (4) 71.92 (5) 78.64 (6) 63.04.

Reference

1.   Y.S. Vollerner, S.D. Kravets, A.S. Shaskkov, M.B. Gorovits and N.K. Abubakirov, *Khim. Prir. Soedin.*, 505 (1989); *Chem. Nat. Comp.*, **25**, 431 (1989).

ASPIDOSIDE A

1β,2β,3β,4β,5β-Pentahydroxyspirost-25(27)-ene 5-O-β-D-glucopyranoside

Source : *Aspidistra elatior* Blume (Liliaceae)
Mol. Formula : $C_{33}H_{52}O_{12}$
Mol. Wt. : 640
$[\alpha]_D^{23}$: -46.5° (c=0.3, C_5D_5N)
Registry No. : [288255-57-2]

IR (KBr) : 3394 (br, OH), 2951 (C–H), 1653, 1453, 1105, 967, 920 cm^{-1}.

PMR (400/500 MHz) : δ 0.81 (s, H-18), 1.05 (d, J=6.9 Hz, H-21), 1.69 (s, H-19), 3.35 (d, J=10.9 Hz, H-26), 3.97 (t, J=8.4 Hz, H-2 of Glc), 4.08 (br s, H-3), 4.21 (d, J=5.3 Hz, H-6 of Glc), 4.26 (br s, H-1), 4.36 (br s, H-2), 4.45 (d, J=12.0 Hz, H-26), 4.54 (br s, H-6 of Glc), 4.56 (m, H-16), 4.78 (br s, H-27A), 4.81 (br s, H-27B), 5.26 (d, J=7.7 Hz, H-1 of Glc).

CMR (100/125 MHz) : δ C-1) 77.85 (2) 68.13 (3) 76.25 (4) 67.64 (5) 87.53 (6) 24.96 (7) 28.50 (8) 34.72 (9) 46.74 (10) 46.38 (11) 21.76 (12) 39.97 (13) 40.62 (14) 56.05 (15) 32.19 (16) 81.48 (17) 63.06 (18) 16.36 (19) 13.78 (20) 41.98 (21) 14.98 (22) 109.51 (23) 29.02 (24) 33.31 (25) 144.46 (26) 65.12 (27) 108.76 **Glc** (1) 97.47 (2) 75.87 (3) 78.81 (4) 71.87 (5) 78.59 (6) 62.85.

Mass (FAB, Positive ion) : m/z 641 [M+H]$^+$, 479 [M-Glc+H]$^+$, 443 [M-Glc-2xH$_2$O+H]$^+$.

Mass (FAB, Negative ion, H.R.) : m/z 639.3292 [(M-H)$^-$, requires 639.3380].

Reference

1. Q.X. Yang and C.R. Yang, *Yunnan Zhiwu Yanjiu* (*Acta Botanica Yunnanica*), **22**, 109 (2000).

ALLIUM GIGANTEUM SAPONIN 5
(24S,25R)-5α-Spirostan-2α,3β,5α,6β,24-pentaol 24-O-β-D-glucopyranoside

Source : *Allium giganteum* Rgl. (Liliaceae)
Mol. Formula : $C_{33}H_{54}O_{12}$
Mol. Wt. : 642
M.P. : 200-209°C
[α]$_D^{28}$: -98° (c=0.1, MeOH)
Registry No. : [137304-82-6]

IR (KBr) : 3450 (OH), 2925 (CH), 1650, 1460, 1370, 1070, 1030, 980, 960, 890 cm^{-1}.

PMR (C$_5$D$_5$N, 400 MHz) : δ 0.82 (s, 3xH-18), 1.05 (d, J=6.9 Hz, 3xH-27), 1.12 (d, J=6.5 Hz, 3xH-21), 1.65 (s, 3xH-19), 2.43 (dd, J=11.8, 11.7 Hz, H-1ax), 2.62 (dd, J=12.6, 4.8 Hz, H-4eq), 3.03 (dd, J=12.6, 12.1 Hz, H-4ax), 3.54 (dd, J=11.2, 11.0 Hz, H-26A), 3.61 (dd, J=11.0, 4.8 Hz, H-26B), 4.82 (m, H-2 or H-3), 5.14 (d, J=7.7 Hz, H-1 of Glc).

CMR (C_5D_5N, 100.6 MHz) : δ C-1) 42.2[a] (2) 73.7[b] (3) 73.8[b] (4) 41.1[a] (5) 75.5 (6) 75.6 (7) 35.8 (8) 30.2 (9) 45.9 (10) 41.0[c] (11) 21.6 (12) 40.5 (13) 40.9[c] (14) 56.4 (15) 32.2 (16) 81.4 (17) 62.7 (18) 16.6 (19) 18.5 (20) 42.2 (21) 14.8 (22) 111.5 (23) 40.9 (24) 81.6 (25) 37.2 (26) 65.2 (27) 13.4 **Glc** (1) 106.3 (2) 75.7 (3) 77.9[d] (4) 71.8 (5) 78.6[d] (6) 62.9.

Mass (SIMS) : m/z (rel.intens.) 665 [M+Na]$^+$ (17), 643 [M+H]$^+$ (100), 627 [M-Me]$^+$ (75), 608 (14), 591 (15), 579 (14), 563 (18), 547 (23), 525 (17), 503 (36).

Reference

1. K. Kawashima, Y. Mimaki and Y. Sashida, *Phytochemistry*, **30**, 3063 (1991).

NEOPENTROGENIN 5-O-β-D-GLUCOPYRANOSIDE
25(*S*)-Spirostan-1β,2β,3β,4β,5β-pentol 5-O-β-D-glucopyranoside

Source : *Aspidistra elatior* Blume (Liliaceae)
Mol. Formula : $C_{33}H_{54}O_{12}$
Mol. Wt. : 642
[α]$_D$^{23} : -40.5° (c=0.2, C_5D_5N)
Registry No. : [91291-48-4]

IR (KBr) : 3399 (br, OH), 2951 (C–H), 1453, 1106, 1055, 988, 918 cm^{-1}.

PMR (400/500 MHz) : δ 0.80 (s, 3xH-18), 1.05 (d, *J*=7.0 Hz, 3xH-27), 1.11 (d, *J*=6.9 Hz, 3xH-21), 1.68 (s, 3xH-19), 3.35 (br d, *J*=10.9 Hz, 2xH-26), 5.24 (d, *J*=7.7 Hz, H-1 of Glc).

CMR (100/125 MHz) : δ C-1) 77.86 (2) 68.12 (3) 76.25 (4) 67.64 (5) 87.54 (6) 24.96 (7) 28.50 (8) 34.72 (9) 46.74 (10) 46.38 (11) 21.77 (12) 40.59 (13) 40.02 (14) 56.05 (15) 32.21 (16) 81.24 (17) 62.83 (18) 16.57 (19) 13.76 (20) 42.59 (21) 14.86 (22) 109.82 (23) 26.49 (24) 26.33 (25) 27.60 (26) 65.23 (27) 16.37 **Glc** (1) 97.46 (2) 75.85 (3) 78.58 (4) 71.84 (5) 78.79 (6) 62.80.

Mass (FAB) : *m/z* 641 [M-H]⁻, 479 [M-Glc-H]⁻.

Reference

1.   Q.X. Yang and C.R. Yang, *Yunnan Zhiwu Yanjiu* (*Acta Botanica Yunnanica*), **22**, 109 (2000).

REINECKEA CARNEA SAPONIN Rg-1
Pentologenin 5-O-β-D-glucopyranoside

Source : *Reineckea carnea* Kunth[1,2] (Liliaceae)
Mol. Formula : $C_{33}H_{54}O_{12}$
Mol. Wt. : 642
M.P. : 292-295°C (decomp.)[1]
$[\alpha]_D^{25}$: -46.1°
Registry No. : [108400-86-8]

IR (KBr)[1] : 3500-3300 (OH), 980, 915 < 895, 850 (25*R*-Spiroketal) cm⁻¹.

PMR (C_5D_5N, 80 MHz)[1] : δ 0.81 (s, 3xH-18), 1.07 (d, *J*=6.0 Hz, 3xH-27), 1.12 (d, *J*=6.0 Hz, 3xH-21), 1.66 (s, 3xH-19), 5.09 (d, *J*=7.0 Hz, H-1 of Glc).

CMR (C_5D_5N, 100/125 MHz)[2] : δ C-1) 77.8 (2) 68.1 (3) 76.3 (4) 67.6 (5) 87.5 (6) 24.9 (7) 28.5 (8) 34.7 (9) 46.3 (10) 46.7 (11) 21.7 (12) 40.0 (13) 40.6 (14) 56.0 (15) 32.2 (16) 81.1 (17) 63.0 (18) 16.5 (19) 13.7 (20) 42.0 (21) 15.0 (22) 109.3 (23) 31.9 (24) 29.3 (25) 30.6 (26) 66.9 (27) 17.3 **Glc** (1) 97.5 (2) 75.9 (3) 78.7[a] (4) 72.0 (5) 78.8[a] (6) 62.9.

Biological Activity : The compound exhibited inhibitory activity on CAMP phosphodiesterase with $IC_{50}=(500<) \times 10^{-5}$ M.[2]

References

1.   K. Iwagoe, T. Konishi and S. Kiyosawa, *Yakugaku Zasshi*, **107**, 140 (1987).

2.   T. Kanmoto, Y. Mimaki, Y. Sashida, T. Nikaido, K. Koike and T. Ohmoto, *Chem. Pharm. Bull.*, **42**, 926 (1994).

ALLIUM GIGANTEUM SAPONIN 3

3-O-Acetyalliogenin 2-O-β-D-glucopyranoside

Source : *Allium giganteum* Regel. (Liliaceae)
Mol. Formula : $C_{35}H_{56}O_{12}$
Mol. Wt. : 668
$[\alpha]_D^{27}$: -99.0° (c=0.1, MeOH)
Registry No. : [134981-82-1]

IR (KBr) : 3450 (OH), 2950 (C–H), 1720 (ester carbonyl), 1375, 1260, 1070, 1040, 980, 920, 890, 860. Intensity 920 < 890 (25*R*-spiroketal) cm^{-1}.

PMR (C_5D_5N, 400/500 MHz) : δ 0.68 (d, J=5.8 Hz, 3xH-27), 0.88 (s, 3xH-18), 1.12 (d, J=6.9 Hz, 3xH-21), 1.54 (s, 3xH-19), 2.12 (s, OCOCH_3), 2.33 (dd, J=12.1, 6.0 Hz, H-1 eq), 2.34 (dd, J=12.3, 6.2 Hz, H-4 eq), 2.40 (dd, J=12.1, 10.1 Hz, H-1 ax), 2.81 (dd, J=12.3, 10.4 Hz, H-4 ax), 3.48 (dd, J=10.5, 10.1 Hz, H-26A), 3.57 (dd, J=10.5, 3.5 Hz, H-26B), 4.16 (br s, $W\frac{1}{2}$=7.5 Hz, H-6), 4.43 (dd, J=16.6, 4.8 Hz, H-6A of Glc), 4.71 (ddd, J=10.2, 10.1, 6.0 Hz, H-2), 5.19 (d, J=7.7 Hz, H-1 of Glc), 6.15 (ddd, J=10.4, 10.2, 6.2 Hz, H-3).

CMR (C_5D_5N, 100 or 125 MHz) : δ C-1) 38.8^a (2) 77.7 (3) 75.0 (4) 37.8^a (5) 74.9 (6) 75.1 (7) 35.7 (8) 31.1 (9) 45.7 (10) 40.3 (11) 21.5 (12) 40.4 (13) 41.0 (14) 56.3 (15) 32.3 (16) 81.2 (17) 63.1 (18) 17.9 (19) 16.7 (20) 42.0 (21) 15.0 (22) 109.2 (23) 31.8 (24) 29.3 (25) 30.6 (26) 66.9 (27) 17.3 **Glc** (1) 103.1 (2) 75.3 (3) 78.3^b (4) 71.8 (5) 78.5^b (6) 63.0 (OCOCH₃) 170.9 (OCOCH₃) 21.6.

Mass (S.I.) : m/z 668 [M]$^+$.

Reference

1.   Y. Sashida, K. Kawashima, Y. Mimaki, *Chem. Pharm. Bull.*, **39**, 698 (1991).

LIRIOPE PLATYPHYLLA SAPONIN G
Ruscogenin-1-sulfate 3-O-α-L-rhamnopyranoside monosodium salt

Source : *Liriope platyphylla* Wang. et Tang. (Liliaceae)
Mol. Formula : $C_{33}H_{51}O_{11}SNa$
Mol. Wt. : 678*
M.P. : 300°C (decomp.)
$[\alpha]_D^{19}$: -60.4° (c=1.01, C_5D_5N)
Registry No. : [87425-37-4]

IR (KBr) : 3600-3200 (OH), 1210 (S-O), 982, 920, 902, 865 cm^{-1} (intensity 920>902, 25(R)-spiroketal).

PMR (C_5D_5N, 100 MHz) : δ 0.69 (br d, sec. CH_3), 0.84 (s, tert. CH_3), 1.18 (d, J=6.0 Hz, sec. CH_3), 1.24 (s, tert. CH_3), 1.55 (d, J=6.0 Hz, sec. CH_3).

CMR (C₅D₅N, 25.0 MHz) : δ C-1) 83.9 (2) 37.2 (3) 73.5 (4) 39.5 (5) 138.1 (6) 125.9 (7) 33.2 (8) 32.1 (9) 50.0 (10) 43.4 (11) 23.7 (12) 40.6 (13) 40.3 (14) 56.8 (15) 32.5 (16) 81.2 (17) 63.3 (18) 16.7 (19) 14.7ᵃ (20) 42.1 (21) 15.0ᵃ (22) 109.3 (23) 32.1 (24) 29.4 (25) 30.7 (26) 67.0 (27) 17.3 **Rha** (1) 99.6 (2) 72.7ᵇ (3) 72.7ᵇ (4) 74.1 (5) 69.9 (6) 18.4.

*The molecular weight refers to sodium salt. But the isolated material was a mixture of sodium potassium and calcium salts in a ratio of 6:3:1.

Reference

1.   Y. Watanabe, S. Sanada, Y. Ida and J. Shoji, *Chem. Pharm. Bull.*, **31**, 1980 (1983).

CHAMADOREA LINEARIS SAPONIN 1

Brisbagenin 1-O-[β-L-fucopyranoside-4'-sulfate]-monosodium salt

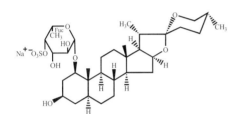

Source : *Chamaedorea linearis* (Araceae)
Mol. Formula : C₃₃H₅₃O₁₁SNa
Mol. Wt. : 680
Registry No. : [151589-15-0]

IR (KBr) : 3561, 3008, 1636, 1385, 1219, 1053, 980, 920, 900, 892 cm⁻¹.

PMR (C₅D₅N, 400 MHz) : δ 0.66 (d, *J*=5.1 Hz, 3xH-27), 0.79 (s, 3xH-18), 0.88 (s, 3xH-19), 1.04 (d, *J*=6.9 Hz, 3xH-21), 1.25 (m, H-25), 1.43 (H-15A), 1.58 (H-4A), 1.66 (d, *J*=6.1 Hz, 3xH-6 of Fuc), 1.72 (m, H-4B), 1.79 (H-2A), 1.81 (m, H-17), 1.82 (m, H-5), 1.95 (m, H-20), 2.04 (m, H-15B), 2.75 (m, H-2B), 3.48 (H-26A), 3.52 (m, H-26B), 3.78 (m, H-1), 3.78 (m, H-5 of Fuc), 3.89 (m, H-3), 4.20 (H-3 of Fuc), 4.29 (t, *J*=7.3 Hz, H-2 of Fuc), 4.52 (*J*=2.6, 7.7 Hz, H-16), 4.75 (d, *J*=7.6 Hz, H-1 of Fuc), 5.18 (d, *J*=2.9 Hz, H-4 of Fuc).

CMR (C₅D₅N, 100 MHz) : δ C-1) 70.7 (2) 38.0 (3) 67.7 (4) 32.6 (5) 41.9 (6) 28.9 (7) 32.5 (8) 36.0 (9) 54.7 (10) 36.4 (11) 23.7 (12) 39.9 (13) 41.5 (14) 55.1 (15) 38.3 (16) 81.1 (17) 63.2 (18) 17.3 (19) 16.8 (20) 42.6 (21) 14.9 (22) 109.2 (23) 32.0 (24) 30.7 (25) 30.2 (26) 66.8 (27) 8.1 **Fuc** (1) 101.5 (2) 72.2 (3) 74.1 (4) 78.9 (5) 82.7 (6) 17.5.

Mass (FAB) : m/z 703.3109 [(M+Na)⁺, requires 703.3104].

Biological Activity : The compound for inhibits recombinational DNA repair cytotoxic activity was also demonstrated. The IC₁₂ values were 3.5 μg per well on ACA Agar and 12 μg per well on control agar (3.4-fold differential).

Reference

1.   A.D. Patil, P.W. Baures, D.S. Eggleston, L. Faucette, M.E. Hemling, J.W. Westley and R.K. Johnson, *J. Nat. Prod.*, **56**, 1451 (1993).

ALLIUM GIGANTEUM SAPONIN 6

(5α,24S,25S)-Spirostane-2α,3β,5α,6β,24-pentol-3-O-acetate 2-O-β-D-glucopyranoside

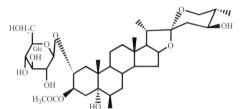

Source : *Allium giganteum* Regel. (Liliaceae)
Mol. Formula : C₃₅H₅₆O₁₃
Mol. Wt. : 684
[α]ᴅ²⁵ : -50° (c=0.10, MeOH)
Registry No. : [156006-37-0]

IR (KBr) : 3400 (OH), 2930 (CH), 1715 (C=O), 1450, 1375, 1260, 1155, 1030, 985, 950, 885, 800 cm⁻¹.

PMR (C$_5$D$_5$N, 400 MHz) : δ 0.88 (s, 3xH-18), 1.07 (d, J=6.5 Hz, 3xH-27), 1.15 (d, J=6.8 Hz, 3xH-21), 1.54 (s, 3xH-19), 1.82 (m, H-25), 2.12 (s, OCOCH_3), 2.34 (dd, J=11.9, 6.4 Hz, H-4eq), 2.39 (dd, J=10.5, 10.5 Hz, H-1), 2.80 (dd, J=11.9, 11.0 Hz, H-4ax), 3.58 (dd, J=11.1, 11.1 Hz, H-26ax), 3.69 (dd, J=11.1, 4.8 Hz, H-26eq), 4.01 (dd, J=8.8, 7.7 Hz, H-2 of Glc), 4.07-3.95 (2H, overlapping, H-24 and H-5 of Glc), 4.15 (br s, $W^{1/2}$=8.1 Hz, H-6), 4.25 (dd, J=8.8, 8.8 Hz, H-4 of Glc), 4.30 (dd, J=8.8, 8.8 Hz, H-3 of Glc), 4.42 (dd, J=11.7, 4.9 Hz, H-6A of Glc), 4.55 (dd, J=11.7, 2.4 Hz, H-6B of Glc), 4.59 (q-like, J=7.3 Hz, H-16), 4.70 (ddd, J=10.5, 9.8, 5.9 Hz, H-2), 5.18 (d, J=7.7 Hz, H-1 of Glc), 6.14 (ddd, J=11.0, 9.8, 6.4 Hz, H-3). Signals of C-25 isomer: 0.87 (s, 3xH-18), 1.17 (d, J=6.7 Hz, 3xH-21), 1.30 (d, J=6.9 Hz, 3xH-27).

CMR (C$_5$D$_5$N, 100 MHz) : δ C-1) 38.8 (2) 77.7 (3) 75.0 (4) 37.8 (5) 74.9 (6) 75.1 (7) 35.7 (8) 30.1 (9) 45.6 (10) 40.4 (11) 21.5 (12) 40.3 (13) 41.0 (14) 56.2 (15) 32.3 (16) 81.6 (17) 62.8 (62.6) (18) 16.6 (19) 17.9 (20) 42.3 (42.6) (21) 15.0 (14.8) (22) 111.8 (111.4) (23) 41.8 (36.1) (24) 70.6 (66.5) (25) 40.0 (35.9) (26) 65.3 (64.6) (27) 13.6 (9.7) (OCOCH$_3$) 170.9, 21.6 **Glc** (1) 103.1 (2) 75.4 (3) 78.3 (4) 71.8 (5) 78.5 (6) 63.0. Values in paranthesis are of the C-25 isomer.

Mass (FAB, Negative ion) : m/z 683 [M-H]$^-$, 641 [M-acetyl]$^-$.

Biological Activity : The compound shows inhibitory activity on CAMP phosphodiesterase with IC$_{50}$ 4.1x10^{-5} M.

Reference

1.   Y. Mimaki, T. Nikaido, K. Matsumoto, Y. Sashida and T. Ohmoto, *Chem. Pharm. Bull.*, **42**, 710 (1994).

WATTOSIDE H

(25S) 1β,2β,3β,4β,5β,24β-Hexahydroxyspirostane 24-O-β-D-glucopyranoside

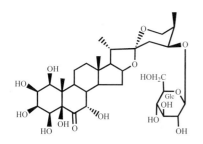

Source : *Tupistra wattii* Hook. f. (Liliaceae)
Mol. Formula : $C_{33}H_{52}O_{15}$
Mol. Wt. : 688
M.P. : 200-203°C
[α]$_D^{20}$: -78.0° (c=0.14, MeOH)
Registry No. : [619319-16-3]

PMR (C_5D_5N, 500 MHz) : δ 0.80 (s, 3xH-18), 1.04 (d, *J*=6.8 Hz, 3xH-21), 1.12 (d, *J*=6.4 Hz, 3xH-27), 1.44 (s, 3xH-19), 1.88 (m, H-25), 1.95 (dd, *J*=13.1, 9.6 Hz, H-23ax), 2.16 (dd, *J*=13.1 4.4 Hz, H-23eq), 2.44 (br s, H-9), 2.67 (dd, *J*=13.6, 4.4 Hz, H-23eq), 3.52 (br t, *J*=12.0 Hz, H-26ax), 4.02 (m, H-24), 4.52 (br s, H-7), 4.92 (d, *J*=7.7 Hz, H-1 of Glc).

CMR (C_5D_5N, 125 MHz) : δ C-1) 76.5 (2) 67.9 (3) 75.6 (4) 71.2 (5) 86.3 (6) 211.2 (7) 75.3 (8) 40.9 (9) 37.9 (10) 50.2 (11) 21.9 (12) 39.4 (13) 40.7 (14) 49.3 (15) 31.4 (16) 81.5 (17) 62.4 (18) 16.4 (19) 13.1 (20) 42.4 (21) 14.9 (22) 111.8 (23) 40.9 (24) 81.5 (25) 38.3 (26) 65.2 (27) 13.6 **Glc** (1) 106.5 (2) 75.7 (3) 78.2 (4) 71.9 (5) 78.7 (6) 62.9.

Mass (FAB, Negative ion, H.R.) : *m/z* 687.3278 [(M-H)$^-$, requires 687.3228].

Mass (FAB, Negative ion) : *m/z* 687 [M-H]$^-$, 525 [MH-Glc]$^-$.

Biological Activity : Cytotoxic against cancer cell line K 562 with IC$_{50}$ 76.16 µmol/l.

Reference

1.   P. Shen, S.-L. Wang, X.-K. Liu, C.-R. Yang, B. Cai and X.-S. Yao, *Chem. Pharm. Bull.*, **51**, 305 (2003).

RUSCOPONTICOSIDE C, RUSCUS ACULEATUS SAPONIN 2, DESGLUCORUSCIN

Neoruscogenin 1-O-[α-L-rhammnopyranosyl-(1→2)-α-L-arabinopyranoside]

Source : *Ruscus aculeatus* L.[1,2] (Liliaceae),
R. ponticus[3] (Liliaceae), *Nolina recurvata*[4] (Agavaceae)
Mol. Formula : $C_{38}H_{58}O_{12}$
Mol. Wt. : 706
$[\alpha]_D^{20}$ **:** -75.0° (c=0.10, MeOH)[5]
Registry No. : [39491-37-7]

IR (KBr)[4] : 3425 (OH), 2930 and 2850 (CH), 1450, 1375, 1340, 1255, 1230, 1195, 1135, 1040, 980, 960, 920, 875, 835, 815, 780, 700 cm^{-1}.

PMR (C_5D_5N, 400 MHz)[4] : δ 0.87 (s, 3xH-18), 1.06 (d, J=6.9 Hz, 3xH-21), 1.46 (s, 3xH-19), 1.76 (d, J=6.2 Hz, 3xH-6 of Rha), 3.67 (br d, J=12.1 Hz, H-5A of Ara), 3.85 (dd, J=11.9, 3.8 Hz, H-1), 3.89 (m, H-3), 4.16 (2H, overlapping, H-3 and H-4 of Ara), 4.28 (dd, J=12.1, 2.2 Hz, H-5B of Ara), 4.33 (dd, J=9.4, 9.4 Hz, H-4 of Rha), 4.46 and 4.02 (br d, J=12.1 Hz, 2xH-26), 4.52 (q-like, J=7.1 Hz, H-16), 4.61 (dd, J=8.4, 7.4 Hz, H-2 of Ara), 4.66 (dd, J=9.4, 2.9 Hz, H-3 of Rha), 4.74 (d, J=7.4 Hz, H-1 of Ara), 4.75 (br d, J=2.9 Hz, H-2 of Rha), 4.78 and 4.81 (br s, 2xH-27), 4.87 (dq, J=9.4, 6.2 Hz, H-5 of Rha), 5.60 (br d, J=5.5 Hz, H-6), 6.36 (br s, H-1 of Rha).

CMR (C_5D_5N, 100 MHz)[4] : δ C-1) 83.5 (2) 37.4 (3) 68.3 (4) 43.9 (5) 139.7 (6) 124.7 (7) 32.0 (8) 33.2 (9) 50.4 (10) 42.9 (11) 24.1 (12) 40.3 (12) 40.3 (14) 56.9 (15) 32.4 (16) 81.5 (17) 63.0 (18) 16.7 (19) 15.1 (20) 41.9 (21) 15.0 (22) 109.5 (23) 33.3 (24) 29.0 (25) 144.6 (26) 65.0 (27) 108.6 **Ara** (1) 100.3 (2) 75.2 (3) 75.9 (4) 70.1 (5) 67.3 **Rha** (1) 101.7 (2) 72.6 (3) 72.7 (4) 74.3 (5) 69.4 (6) 19.0.

Mass (FAB, Negative ion)[4] : m/z 706 [M]$^-$, 559 [M-Rha]$^-$.

Mass (ESI, Positive ion)[6] : m/z 729 [M+Na]$^+$, 707 [M+H]$^+$., 561 [M+H-Rha]$^+$, 429 [Agl+H]$^+$, 411 [Agl+H-H_2O]$^+$.

Mass (ESI, Negative ion)[6] : m/z 705 [M-H]$^-$.

Biological Activity : The compound shows inhibitory activity on cyclic AMP phosphodiesterase with IC_{50} (8.4x10^{-5} M)[4].

References

1.   E. Bombardelli, A. Bonati, B. Gabetta and G. Mustich, *Fitoterapia*, **42**, 127 (1971).

2.   H. Pourrat, J.L. Lamaison, J.C. Gramain, R. Remuson, *Ann. Pharm. Fr.*, **40**, 451 (1982).

3. T. Sh. Korkashvili, O.D. Dzhikiya, M.M. Vugalter, T.A. Pkheide and E.P. Kamertelidze, Soobsch. *Akad. Nauk Gruz. SSR* **120**, 561 (1985); *Chem. Abstr.* **105**, 3499s.

4. Y. Mimaki, Y. Takaashi, M. Kuroda, Y. Sashida and T. Nikaido, *Phytochemistry*, **42**, 1609 (1996).

5. Y. Mimaki, Y. Takaashi, M. Kuroda, A. Kameyama, A. Yokosuka and Y. Sashida, *Chem. Pharm. Bull.*, **46**, 298 (1998).

6. E. de Combarien, M. Falzoni, N. Fuzzati, F. Gattesco, A. Giori, M. Lovati and R. Pace, *Fitoterapia*, **73**, 583 (2002).

AFEROSIDE C

Diosgenin 3-O-[β-D-apiofuranosyl-(1→4)-β-D-glucopyranoside]

Source : *Costus afer* Ker. Gawl. (Zingiberacae)
Mol. Formula : $C_{38}H_{60}O_{12}$
Mol. Wt. : 708
$[\alpha]_D^{26}$: -72.9° (c=1.58, MeOH)
Registry No. : [197151-33-0]

IR (KBr) : 3400 (OH), 2940 (CH), 1450 1370, 1230, 1040, 970 cm^{-1}.

PMR (C$_5$D$_5$N, 500 MHz) : δ 0.68 (d, J=3.7 Hz, 3xH-27), 0.82 (s, 3xH-18), 0.88 (m, H-9), 0.89 (s, 3xH-19), 0.95 (m, H-1A), 0.99 (m, H-12A), 1.05 (m, H-14), 1.16 (d, J=6.7 Hz, 3xH-21), 1.27 (m, H-2A), 1.48 (m, H-11A), 1.55 (m, H-25), 1.56 (m, H-24A, and H-24B), 1.68 (m, H-1B and H-12B), 1.80 (m, H-17), 1.85 (m, H-7A), 1.95 (m, H-20), 2.00 (m, H-7B), 2.01 (m, H-11B), 2.08 (m, H-2B), 2.41 (m, H-4A), 2.71 (m, H-4B), 3.43 (H-26A), 3.60 (m, H-26B), 3.82 (m, H-5 of Glc), 3.90 (m, H-3), 4.02 (m, H-2 of Glc), 4.16 (m, H-5A, and H-5B of Api), 4.28 (m, H-3 of Glc), 4.29 (H-6A of Glc), 4.30 (m, H-4 of Glc), 4.33 (H-4A of Api), 4.35 (m, H-6B of Glc), 4.55 (m, H-16), 4.77 (d, J=8.0 Hz, H-4B of Api), 4.80 (m, H-2 of Api), 4.98 (d, J=8.0 Hz, H-1 of Glc), 5.35 (m, H-6), 6.03 (d, J=3.4 Hz, H-1 of Api).

CMR (C$_5$D$_5$N, 125 MHz) : δ C-1) 37.6 (2) 30.4 (3) 78.5 (4) 39.4 (5) 141.9 (6) 121.8 (7) 32.6 (8) 31.8 (9) 50.5 (10) 37.4 (11) 21.3 (12) 40.1 (13) 40.8 (14) 56.8 (15) 32.5 (16) 81.3 (17) 63.0 (18) 16.6 (19) 19.6 (20) 42.2 (21) 15.2 (22) 110.2 (23) 32.0 (24) 29.4 (25) 30.8 (26) 67.1 (27) 17.5 **Glc** (1) 99.7 (2) 75.2 (3) 76.8 (4) 79.4 (5) 76.9 (6) 61.6 **Api** (1) 111.1 (2) 77.6 (3) 80.8 (4) 75.2 (5) 64.9.

Mass (FAB, Negative ion) : *m/z* 707 [M-H]⁻.

Reference

1.   R.C. Lin, M-A. Lacaille-Dubois, B. Hanquet, M. Correia and B. Chauffert, *J. Nat. Prod.*, **60**, 1165 (1997).

ALLIOSPIROSIDE A
(25*S*)-Ruscogenin 1-O-[α-L-rhamnopyranosyl-(1→2)-α-L-arabinopyranoside]

Source : *Allium cepa* L. (Liliceae)
Mol. Formula : $C_{38}H_{60}O_{12}$
Mol. Wt. : 708
M.P. : 186-189°C
$[\alpha]_D^{25}$: -106.8±2° (c=100, Pyridine)
Registry No. : [105880-06-6]

IR (KBr) : 850, 905<930, 990 (spiroketal chain of the 25*S* series), 3000-3500 (OH) cm⁻¹.

PMR (C_5D_5N, 250 MHz) : δ 0.84 (s, 3xH-18), 1.04 (d, *J*=6.0 Hz, 3xH-27), 1.07 (d, *J*=6.0 Hz, 3xH-21), 1.41 (s, 3xH-19), 1.71 (d, *J*=Hz, 3xH-6 of Rha), 3.0 (dd, $J_{ax,ax}$=11.4 Hz, $J_{ax,eq}$=4.0 Hz, H-1), 3.33 (br d, *J*=11.8 Hz, *J*=2.6 Hz, H-26A), 3.52 (tt, $J_{a,a}$=10.6 Hz, J_{ax}=5.4 Hz, H-3), 3.64 (dd, $J_{4,5A}$=1.1 Hz, $J_{5,5A}$=11.6 Hz, H-5 of Ara), 4.02 (dd, *J*=11.0 Hz, *J*=2.6 Hz, H-26B), 4.1 (m, H-3 and H-4 of Ara), 4.23 (dd, $J_{4,5}$=2.1 Hz, $J_{5,5}$=11.6 Hz, H-5 of Ara), 4.26 (t, $J_{5,5}$=9.0 Hz, H-4 of Rha), 4.4 (dt, *J*=7.6 Hz, H-16), 4.53 (dd, $J_{2,3}$=8.9 Hz, $J_{2,3}$=6.5 Hz, H-2 of Ara), 4.57 (dd, $J_{3,4}$=9.0 Hz, $J_{2,3}$=3.1 Hz, H-3 of Rha), 4.67 (dd, $J_{2,3}$=3.1 Hz, $J_{1,2}$=1.6 Hz, H-2 of Rha), 4.68 (d, $J_{1,2}$=6.5 Hz, H-1 of Ara), 4.79 (d q, $J_{5,6}$=6.0 Hz, $J_{4,5}$=9.0 Hz, H-5 of Ara), 5.56 (d, *J*=5.6 Hz, H-6), 6.28 (d, $J_{1,2}$=1.6 Hz, H-1 of Rha).

CMR (C_5D_5N, 62.5 MHz) : δ C-1) 83.65 (2) 37.45 (3) 68.28 (4) 43.91 (5) 139.64 (6) 124.76 (7) 33.22 (8) 32.08 (9) 50.49 (10) 42.98 (11) 24.12 (12) 40.42 (13) 40.25 (14) 56.90 (15) 32.46 (16) 81.28 (17) 62.91 (18) 16.78 (19) 14.91 (20) 42.52 (21) 15.10 (22) 109.78 (23) 26.47 (24) 26.16 (25) 27.61 (26) 65.11 (27) 16.36 **Ara** (1) 100.43 (2) 75.26 (3) 75.90 (4) 70.10 (5) 67.35 **Rha** (1) 101.70 (2) 72.67 (3) 72.58 (4) 74.30 (5) 69.46 (6) 19.04.

Reference

1.   S.D. Kravets, Yu.S. Vollerner, M.B. Gorovits, A.S. Shashkov and N.K. Abubakirov, *Khim. Prir. Soedin.*, 188 (1986); *Chem. Nat. Comp.*, **22**, 174 (1986).

LIRIOPE SPICATA SAPONIN Ls-3

(25*S*)-Ruscogenin 1-O-[β-D-xylopyranoside-3-O-α-L-rhamnopyranoside]

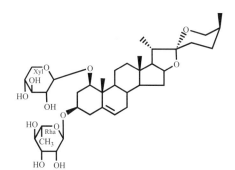

Source : *Liriope spicata* Thunb. Lour. var. *prolifera* Y.T. Ma (Liliaceae)
Mol. Formula : $C_{38}H_{60}O_{12}$
Mol. Wt. : 708
M.P. : 265-267°C (decomp.)
[α]$_D$: -98.0° (c=0.45, C_5D_5N)
Registry No. : [125225-63-0]

IR (KBr) : 3600-3200 (OH), 982, 920, 902, 865 cm^{-1} (intensity 920>902, 25*S*-Spiroketal).

CMR (C_5D_5N, 100 MHz) : δ C-1) 83.2 (2) 35.8 (3) 73.9 (4) 39.9 (5) 138.3 (6) 125.6 (7) 33.7 (8) 32.5 (9) 51.0 (10) 43.5 (11) 24.6 (12) 41.0 (13) 40.8 (14) 57.3 (15) 32.9 (16) 81.5 (17) 63.5 (18) 17.3 (19) 15.1 (20) 43.1 (21) 15.4 (22) 109.9 (23) 27.1 (24) 26.8 (25) 28.2 (26) 65.5 (27) 16.9 **Xyl** (1) 102.4 (2) 75.6 (3) 79.0 (4) 71.6 (5) 67.8 **Rha** (1) 100.1 (2) 73.14 (3) 73.10 (4) 74.5 (5) 70.3 (6) 19.1.

Reference

1. B-Y. Yu, Y. Hirai, J. Shoji and G-J. Xu, *Chem. Pharm. Bull.*, **38**, 1931 (1990).

LIRIOPE SPICATA SAPONIN Ls-4

(25*S*)-Ruscogenin 1-O-[α-L-rhamnopyranosyl-(1→2)-β-D-xylopyranoside]

Source : *Liriope spicata* Thunb. Lour. var. *prolifera* Y.T. Ma (Liliaceae)
Mol. Formula : $C_{38}H_{60}O_{12}$
Mol. Wt. : 708
M.P. : 200-202°C (decomp.)
[α]$_D$: -84.2° (c=0.71, C_5D_5N)
Registry No. : [130431-10-6]

IR (KBr) : 3600-3200 (OH), 980, 920, 902, 865 cm^{-1} (intensity 920>902, (25S)-spiroketal).

CMR (C_5D_5N, 100 MHz) : δ **Xyl** C-1) 101.7 (2) 80.3 (3) 76.9 (4) 71.8 (5) 67.6 **Rha** (1) 100.7 (2) 72.9 (3) 72.8 (4) 74.5 (5) 69.8 (6) 19.5.

Reference

1.   B-Y. Yu, Y. Hirai, J. Shoji and G-J. Xu, *Chem. Pharm. Bull.*, **38**, 1931 (1990).

PARIS POLYPHYLLA SAPONIN 4
Diosgenin 3-O-[α-L-arabinofuranosyl-(1→4)-β-D-glucopyranoside]

Source : *Paris polyphylla* Sm. (Liliaceae)
Mol. Formula : $C_{38}H_{60}O_{12}$
Mol. Wt. : 708
M.P. : 239-242°C
Registry No. : [81917-52-4]

IR (KBr) : 3600-3100 (OH), 980, 918, 898, 860 (intensity 898>918), 25R-spiroketal side chain) cm^{-1}.

Mass : m/z 396 $[C_{27}H_{40}O_2]^+$, 282 $[C_{21}H_{30}]^+$, 253, 139 $[C_9H_{15}O]^+$.

Reference

1.   M. Miyamura, K. Nakano, T. Nohara, T. Tomimatsu and T. Kawasaki, *Chem. Pharm. Bull.,* **30**, 712 (1982).

RHODEA JAPONICA SAPONIN 3
(5β)-Spirost-25(27)-en-1β,3β-diol 1-O-[α-L-rhamnopyranosyl-(1→2)-β-D-xylopyranoside]

Source : *Rhodea japonica* (Thunb.) Roth (Liliaceae)
Mol. Formula : $C_{38}H_{60}O_{12}$
Mol. Wt. : 708
Registry No. : [86462-27-3]

Reference

1.   K. Kudo, K. Miyahara, N. Marubayashi and T. Kawasaki, *Chem. Pharm. Bull.*, **32**, 4229 (1984).

SPRENGERININ A
(25*R*)-Spirost-5-en-3β-ol 3-O-[β-D-xylopyranosyl-(1→4)-β-D-glucopyranoside]

Source : *Asparagus sprengeri* Regel. (Liliaceae)
Mol. Formula : $C_{38}H_{60}O_{12}$
Mol. Wt. : 708
M.P. : 240-242°C (MeOH)
[α]$_D^{20}$: -90.4° (c=1.0, Pyridine)
Registry No. : [88866-99-3]

IR (KBr) : 3400 (OH), 2850 (C=CH), 1040, 981, 918, 898 (intensity 898>918, 25*R*-spiroketal), 855 and 810 cm^{-1}.

Reference

1.   S.C. Sharma, R. Sharma and R. Kumar, *Phytochemistry*, **22**, 2259 (1983).

YAMOGENIN 3-O-[β-D-GLUCOPYRANOSYL-(1→4)-α-D-XYLOPYRANOSIDE]

Source : *Trigonella foenum-graecum* (Leguminosae)
Mol. Formula : $C_{38}H_{60}O_{12}$
Mol. Wt. : 708
M.P. : 242-244°C

IR (KBr) : 3400 (OH), 2928 (CH$_3$ stretching), 1620 (C=C stretching), 1040 (C-O-C), 982, 914, 892, 850 cm^{-1} (intensity 915 > 896, 25S-spiroketal).

Mass : *m/z* 708 [M]$^+$.

Reference

1.   V.K. Saxena and A. Shalem, *J. Chem. Sci.*, **116**, 79 (2004).

ASPAFILIOSIDE A

Sarsasapogenin 3-O-[β-D-xylopyranosyl-(1→4)-β-D-glucopyranoside]

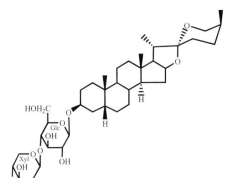

Source : *Asparagus filicinus* Buch.-Ham.[1] (Liliaceae)
Mol. Formula : $C_{38}H_{62}O_{12}$
Mol. Wt. : 710
M.P. : 210~212°C (MeOH)[1]
$[\alpha]_D^{14}$ **:** -59.2° (c=0.07, CHCl$_3$-MeOH)[1]
Registry No. : [72947-73-0]

IR (KBr)[1] **:** 3420, 1045, 985, 915 > 895, 849 cm^{-1} (25S).

PMR (C$_5$D$_5$N, 400 MHz)[1] **:** δ 0.83 (s, 3xH-19), 0.86 (s, 3xH-18), 1.09 (d, *J*=6.8 Hz, 3xH-27), 1.17 (d, *J*=6.7 Hz, 3xH-21), 4.93 (d, *J*=7.6 Hz), 5.18 (d, *J*=7.5 Hz).

CMR (C$_5$D$_5$N, 100 MHz)[1] **:** δ C-1) 30.6 (2) 27.0 (3) 74.7 (4) 31.0 (5) 37.0 (6) 27.0 (7) 26.8 (8) 35.6 (9) 40.4 (10) 35.3 (11) 21.2 (12) 40.4 (13) 40.9 (14) 56.6 (15) 32.2 (16) 81.4 (17) 63.0 (18) 16.6 (19) 24.4 (20) 42.6 (21) 15.2 (22) 109.7 (23) 26.5 (24) 26.2 (25) 27.6 (26) 65.3 (27) 16.3 **Glc** (1) 103.0 (2) 74.7 (3) 76.6 (4) 81.2 (5) 76.5 (6) 62.1 **Xyl** (1) 105.5 (2) 75.1 (3) 78.3 (4) 70.8 (5) 67.4.

Mass (F.D.)[1] **:** *m/z* 733 [M+Na]$^+$.

Biological Activity : The compound shows inhibitory effect on human spermatozoa *in vitro* at a concentration of 2 mg/ml.[2]

References

1.   Y. Ding and C.R. Yang, *Acta Pharm. Sin.*, **25**, 509 (1990).

2.   Y.-F. Wang, X.C. Li, H.-Y. Yang, J.-J. Wang and C.R. Yang, *Plant Med.*, **62**, 130 (1996).

ASPARAGUS CURILLUS SAPONIN 1

25(*S*),5β-Spirostan-3β-ol 3-O-[α-L-arabinopyranosyl-(1→4)-β-D-glucopyranoside]

Source : *Asparagus curillus* Buch.-Ham (Liliaceae)
Mol. Formula : $C_{38}H_{62}O_{12}$
Mol. Wt. : 710
M.P. : 221-223°C
[α]$_D^{20}$: -50.0° (c=0.8, CHCl$_3$-MeOH)
Registry No. : [346617-77-4]

IR (KBr) : 3450 (OH), 990, 920, 898, 850 cm^{-1}, intensity 920 < 898 (25*S*-spiroketal).

Reference

1.   S.C. Sharma, O.P. Sati and R. Chand, *Phytochemistry*, **21**, 1711 (1982).

FILIFERIN A
Sarsasapogenin 3-O-[β-D-xylopyranosyl-(1→2)-β-D-galactopyranoside]

Source : *Yucca filifera*[1] (Agavaceae), *Cornus excelsion* H.B.[2] *Cornus florida* L.[2] (Corncaceas)
Mol. Formula : $C_{38}H_{62}O_{12}$
Mol. Wt. : 710
M.P. : 286-288°C[2]
$[\alpha]_D^{31}$: -68.5° (c=0.72, dioxane)
Registry No. : [63631-39-0]

PMR (C_5D_5N, 25.02 MHz)[2] : δ **Gal** (1) 101.6 (2) 81.4 (3) 76.8 (4) 69.7 (5) 76.5 (6) 62.2 **Xyl** (1) 106.7 (2) 72.3 (3) 74.6 (4) 71.7 (5) 66.3.

Mass (F.D.)[2] : *m/z* (rel.intens.) 749 [(M+K)$^+$, 31.5], 733 [(M+Na)$^+$, 100], 711 [(M+H)$^+$, 53.1], 710 [(M)$^+$, 38.5)], 601 [(M+Na-Xyl)$^+$, 14.5], 579 [(M+H-Xyl)$^+$, 16.6], 578 [(M-Xyl)$^+$, 578], 417 [(M+H-Gal-Xyl)$^+$, 13.8], 416 [(M-Gal-Xyl)$^+$, 20.8].

Biological Activity: Molluscicidal.

References

1. R.U. Lemieux, R.M. Ratcliffe, B. Arregnin, A. Romo de Vivar, M.J. Castillo, *Carbohydr. Res.*, **55**, 113 (1977).

2. K. Hostettmann, M. Hostettmann Kaldas and K. Nakonishi, *Helv. Chim. Acta*, **61**, 1990 (1978).

3. X.A. Dominguez, R. Franco, G. Cano, S. Garcia, P. Grecian and S.G. Sanchez, *Rev. Latinomer. Quim.*, **12**, 35 (1981).

RHODEA JAPONICA SAPONIN 1, 22-*EPI*-RHODEASAPOGENIN

5β,22β,25(*S*)-Spirostane-1β,3β-diol 1-O-[α-L-rhamnopyranosyl-(1→2)-β-D-xylopyranoside]

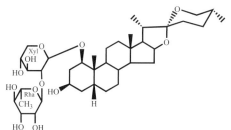

Source : *Rhodea japonica* (Thunb.) Roth (Liliaceae)
Mol. Formula : $C_{38}H_{62}O_{12}$
Mol. Wt. : 710
M.P. : 275-277°C
[α]$_D$: -51.4° (c=0.2 ~ 1.4, C_5H_5N)
Registry No. : [94535-24-7]

IR (KBr) : 983, 946, 916 < 895, 872 cm^{-1}.

PMR (C_5D_5N, 100 MHz) : δ 0.69 (m, 3xH-27), 0.97 (s, 3xH-18), 1.00 (d, *J*=6.0 Hz, 3xH-21), 1.30 (s, 3xH-19), 3.67 (m, 2xH-26).

CMR (C_5D_5N, 25 MHz) : δ C-20) 42.1 (21) 17.0 (22) 110.6 (23) 28.3^a (24) 28.1^a (25) 30.7 (26) 69.6 (27) 17.3.

Mass (F.D.) : *m/z* 711 [M+H]$^+$, 710 [M]$^+$.

Reference

1.   K. Kudo, K. Miyahara, N. Marubayashi and T. Kawasaki, *Chem. Pharm. Bull.*, **32**, 4229 (1984).

RHODEA JAPONICA SAPONIN 4

(5β,25*R*)-Spirostan-1β,3β-diol 1-O-[α-L-rhamnopyranosyl-(1→2)-β-D-glucopyranoside]

Source : *Rhodea japonica* (Thunb.) Roth (Liliaceae)
Mol. Formula : $C_{38}H_{62}O_{12}$
Mol. Wt. : 710
Registry No. : [86462-29-5]

IR (KBr) : 981, 945, 919 < 900, 863 cm^{-1}.

PMR (C$_5$D$_5$N, 100 MHz) : δ 0.69 (d, J=6 Hz, 3xH-27), 0.85 (s, 3xH-18), 1.13 (d, J=6.0 Hz, 3xH-21), 1.33 (s, 3xH-19), 3.57 (m, 2xH-26).

CMR (C$_5$D$_5$N, 25 MHz) : δ C-20) 42.0 (21) 15.0 (22) 109.2 (23) 31.7 (24) 29.2 (25) 30.6 (26) 66.9 (27) 17.3.

Reference

1.   K. Kudo, K. Miyahara, N. Marubayashi and T. Kawasaki, *Chem. Pharm. Bull.*, **32**, 4229 (1984).

RHODEA JAPONICA SAPONIN R-3
Rhodeasapogenin 1-O-[α-L-rhamnopyranosyl-(1→2)-β-D-xylopyranoside]

Source : *Rhodea japonica* (Thunb.) Roth[1,2] (Liliaceae)
Mol. Formula : C$_{38}$H$_{62}$O$_{12}$
Mol. Wt. : 710
M.P. : 243-244°C[1]
[α]$_D$: -99.2° (Pyridine)[1]
Registry No. : [86496-15-3]

IR (KBr)[2] : 986, 945, 919 > 894, 850 cm^{-1}.

CMR (C$_5$D$_5$N, 25.0 MHz) : δ C-20) 42.5 (21) 14.9 (22) 109.7 (23) 26.3 (24) 26.3 (25) 27.4 (26) 65.1 (27) 16.3.

References

1.   K. Miyahara, K. Kudo and T. Kawasaki, *Chem. Pharm. Bull.,* **31**, 348 (1983).

2.   K. Kudo, K. Miyahara, N. Marubayashi and T. Kawasaki, *Chem. Pharm. Bull.*, **32**, 4229 (1984).

SOLANUM CHRYSOTRICHUM SAPONIN SC-2

(25R)-5α-Spirostan-3β,6α-diol 6-O-[β-D-xylopyranosyl-(1→3)-β-D-quinovopyranoside]

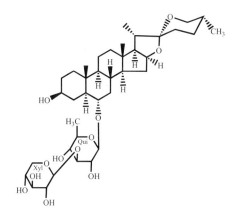

Source : *Solanum chrysotrichum* Schldh. (Solanaceae)
Mol. Formula : $C_{38}H_{62}O_{12}$
Mol. Wt. : 710
M.P. : 239-241°C
[α]$_D^{25}$: -49° (c=1.08, MeOH)
Registry No. : [478298-12-3]

IR (CHCl₃) : 3500-3300 (OH), 2925 (CH) cm⁻¹.

Hexa-acetate :

PMR (CDCl₃, 400 MHz) : δ 0.66 (H-9), 0.68 (H-24A), 0.7 (H-23A), 0.74 (s, 3xH-18), 0.79 (d, J=6.4 Hz, 3xH-27), 0.84 (s, 3xH-19), 0.95 (d, J=6.4 Hz, 3xH-21), 0.99 (H-1A), 1.05 (H-7A), 1.15 (H-14A), 1.18 (d, J=6.4 Hz, 3xH-6 of Qui), 1.2 (H-4A, H-11A, H-14B, H-15A, H-23B and H-24B), 1.21 (H-5), 1.4 (H-11B), 1.44 (H-2A), 1.63 (H-1B, H-8), 1.68 (dd, J=8.4, 6.8 Hz, H-17), 1.69 (H-12A), 1.77 (H-25), 1.82 (H-2B), 1.85 (H-20), 1.99 (H-15B), 2.07 (H-4B), 2.08 (H-12B), 2.1 (H-7B), 3.21 (ddd, J=11.0, 10.0, 4.5 Hz, H-6), 3.35 (dd, J=7.2, 1.2 Hz, H-5α of Xyl), 3.37 (dd, J=11.0, 10.0 Hz, H-26α), 3.44 (m, H-5 of Qui), 3.45 (dd, J=11.0, 2.4 Hz, H-26β), 3.72 (t, J=9.6 Hz, H-3 of Qui), 4.09 (dd, J=4.4, 1.2 Hz, H-5β of Xyl), 4.33 (d, J=8.0 Hz, H-1 of Qui), 4.39 (ddd, J=7.5, 7.4, 7.0 Hz, H-16), 4.54 (d, J=5.6 Hz, H-1 of Xyl), 4.61 (dddd, J=11.0, 10.5, 6.0, 5.5 Hz, H-3), 4.70 (t, J=9.6 Hz, H-4 of Qui), 4.74 (dd, J=7.2, 5.6 Hz, H-2 of Xyl), 4.85 (ddd, J=7.3, 7.2, 4.4 Hz, H-4 of Xyl), 4.94 (dd, J=9.6, 8.0 Hz, H-2 of Qui), 5.01 (t, J=7.2 Hz, H-3 of Xyl).

CMR (CDCl₃, 100 MHz) : δ C-1) 36.67 (2) 27.27 (3) 73.43 (4) 28.41 (5) 49.73 (6) 80.69 (7) 39.85 (8) 34.02 (9) 53.4 (10) 36.9 (11) 20.8 (12) 40.15 (13) 40.72 (14) 56.01 (15) 31.56 (16) 80.82 (17) 61.33 (18) 17.22 (19) 13.55 (20) 41.85 (21) 14.59 (22) 109.36 (23) 30.98 (24) 28.99 (25) 30.43 (26) 65.2 (27) 16.55 **Qui** (1) 102.33 (2) 73.43 (3) 80.8 (4) 73.98 (5) 69.9 (6) 17.66 **Xyl** (1) 101.07 (2) 70.09 (3) 70.54 (4) 68.75 (5) 61.63.

Mass (FAB, Positive ion) : *m/z* 1001 [M+K]⁺, 963 [M+H]⁺, 962 [M]⁺, 457 [M-Xyl-Qui]⁺, 397 [M-Xyl-Qui-60]⁺.

Mass (FAB, Positive ion, H.R.) : *m/z* 962.4901 [(M+H)⁺, calcd. for 962.4875].

Biological Activity : The compound showed strong antimycotic activity against *Trichophyton mentagrophytes T. rubrum, Aspergillus niger* and *Candida albicans* with MIC values of 12.5, 12.5, 100 and 200 μg/ml respectively.

Reference

1.   A. Zamilpa, J. Tortoriello, V. Navarro, G. Delgado and L. Alvarez, *J. Nat. Prod.*, **65**, 1815 (2002).

SOLANUM HISPIDUM SAPONIN 2

(5α,25S)-Spirostan-3β,6α-diol-6-O-[β-D-xylopyranosyl-(1→3)-β-D-quinovopyranside]

Source : *Solanum hispidum* Pers.[1] (Solanaceae) also obtained by enzymic hydrolysis of Torvoside B (qv) from *Solanum torvum* Swartz[2] (Solanaceae)

Mol. Formula : $C_{38}H_{62}O_{12}$

Mol. Wt. : 710

M.P. : 264-265°C

[α]$_D^{27}$: -18.9° (c=0.20, MeOH)[2]

Registry No. : [184684-02-0]

IR (KBr)[1] : 3423 (OH), 2929 (CH), 1077-1047 (C-O-C) cm^{-1}.

PMR (C$_5$D$_5$N, 400 MHz)[2] : δ 0.82 (s, 3xH-18), 0.88 (s, 3xH-19), 1.07 (d, *J*=7.0 Hz, 3xH-27), 1.14 (d, *J*=7.0 Hz, 3xH-21), 1.56 (d, *J*=5.9 Hz, 3xH-6 of Qui), 2.53 (br d, *J*=12.3 Hz, H-7), 3.22 (d, *J*=12.1 Hz, H-4), 3.35 (d, *J*=11.0 Hz, H-26), 4.48 (dd, *J*=7.3, 7.3 Hz, H-16), 4.84 (d, *J*=7.0 Hz, H-1 of Qui), 5.25 (d, *J*=7.7 Hz, H-1 of Xyl).

CMR (C$_5$D$_5$N, 100 MHz)[2] : δ C-1) 37.8 (2) 32.1 (3) 70.6 (4) 33.2 (5) 51.3 (6) 79.4 (7) 41.4 (8) 34.2 (9) 53.9 (10) 36.7 (11) 21.3 (12) 39.9 (13) 40.7 (14) 56.3 (15) 32.1 (16) 81.1 (17) 62.8 (18) 16.6 (19) 13.6 (20) 42.5 (21) 14.9 (22) 109.7 (23) 26.4^a (24) 26.2^a (25) 27.5 (26) 65.1 (27) 16.3 **Qui** (1) 105.3 (2) 74.9 (3) 87.6 (4) 74.7 (5) 72.3 (6) 18.6 **Xyl** (1) 106.5 (2) 75.3 (3) 78.2 (4) 70.9 (5) 67.4.

Mass (FAB, Positive ion)[2] : *m/z* 711 [M+H]$^+$, 579 [M+H-Xyl]$^+$, 433 [M+H-Xyl-Qui]$^+$.

References

1. M. Gonzalez, A. Zamilpa, S. Marquina, V. Navarro and L. Alvarez, *J. Nat. Prod.*, **67**, 938 (2004).

2. S. Yahata, T. Yamashita, N. Nozawg (nee Fujimura) and T. Nohara, *Phytochemistry*, **43**, 1069 (1996).

ALLIUM KARATAVIENSE SAPONIN 2

(25*R*)-3-O-(3-Hydroxybutyryl)-5α-spirostane-2α,3β,5,6β-tetrol 2-O-β-D-glucopyranoside

Source : *Allium karataviense* Regel (Liliaceae)
Mol. Formula : $C_{37}H_{60}O_{13}$
Mol. Wt. : 712
$[\alpha]_D^{27}$: -92.0° (c=0.10, MeOH)
Registry No. : [238398-01-1]

IR (KBr) : 3430 (OH), 2950 (CH), 2885 (CH), 1720 (C=O), 1455, 1375, 1255, 1240, 1170, 1150, 1060, 1040, 975, 955, 915, 895, 860 cm^{-1}.

PMR (C$_5$D$_5$N, 400/500 MHz) : δ 0.67 (d, *J*=5.9 Hz, 3xH-27), 0.88 (s, 3xH-18), 1.12 (d, *J*=7.0 Hz, 3xH-21), 1.54 (s, 3xH-19), 1.33 (d, *J*=6.3 Hz, H-4 of Acyl group), 2.35 (dd, *J*=12.3, 5.9 Hz, H-1eq), 2.40 (dd, *J*=12.3, 10.9 Hz, H-1ax), 2.41 (dd, *J*=12.3, 10.9 Hz, H-1ax), 2.76 (dd, *J*=15.3, 7.4 Hz, H-2A of Acyl group), 2.80 (dd, *J*=15.3, 5.2 Hz, H-2B of Acyl group), 2.86 (dd, *J*=13.2, 11.2 Hz, H-4ax), 3.47 (dd, *J*=10.7, 10.7 Hz, H-26ax), 3.56 (dd, *J*=10.7, 3.7 Hz, H-26eq), 4.00 (ddd, *J*=8.8, 3.9, 1.9 Hz, H-5 of Glc), 4.01 (dd, *J*=7.7 Hz, H-2 of Glc), 4.15 (br s, H-6), 4.30 (dd, *J*=8.8, 8.8 Hz, H-3 of Glc), 4.31 (dd, *J*=8.8, 8.8 Hz, H-4 of Glc), 4.44 (dd, *J*=11.8, 3.9 Hz, H-6A of Glc), 4.49 (dd, *J*=11.8, 1.9 Hz, H-6B of Glc), 4.57 (q-like *J*=7.3 Hz, H-16), 4.73 (ddd, *J*=10.9, 9.5, 5.9 Hz, H-2), 4.62 (m, H-3 of Acyl group), 5.20 (d, *J*=7.7 Hz, H-1 of Glc), 6.18 (ddd, *J*=11.2, 9.9, 6.1 Hz, H-3).

CMR (C$_5$D$_5$N, 100/125 MHz) : δ C-1) 38.5 (2) 77.1 (3) 74.9 (4) 37.8 (5) 74.9 (6) 75.0 (7) 35.7 (8) 30.1 (9) 45.6 (10) 40.5 (11) 21.5 (12) 40.3 (13) 41.0 (14) 56.1 (15) 32.3 (16) 81.2 (17) 63.1 (18) 16.6 (19) 17.9 (20) 42.0 (21) 15.0 (22) 109.2 (23) 31.8 (24) 29.2 (25) 30.6 (26) 66.8 (27) 17.3 **Glc** (1) 102.5 (2) 75.2 (3) 78.3 (4) 71.2 (5) 78.5 (6) 62.5 **Acyl group** (1) 171.6 (2) 45.5 (3) 64.4 (4) 23.2.

Mass (FAB, Positive ion, H.R.) : *m/z* 735.3959 [M+Na]$^+$.

1236

Reference

1.   Y. Mimaki, M. Kuroda, T. Fukasawa and Y. Sashida, *Chem. Pharm. Bull.,* **47**, 738 (1999).

ADSCENDIN A
(25*S*)-Spirostan-5-en-3β-ol 3-O-[α-L-rhamnopyranosyl-(1→6)-β-D-glucopyranoside]

Source : *Asparagus adscendens* Roxb. (Liliaceae)
Mol. Formula : $C_{39}H_{62}O_{12}$
Mol. Wt. : 722
M.P. : 219-221°C
$[\alpha]_D^{20}$: -102° (c=1.0, C_5D_5N)
Registry No. : [91095-74-8]

IR (KBr) : 3400 (OH), 980, 915, 896, 850 cm^{-1} (intensity 915>986, 25*S*-spiroketal).

Reference

1.   S.C. Sharma and H.C. Sharma, *Phytochemistry*, **23**, 645 (1984).

ASPARAGUS PLUMOSUS SAPONIN 1

Yamogenin 3-O-[α-L-rhamnopyrnosyl-(1→3)-β-D-glucopyranoside]

Source : *Asparagus plumosus* Baker (Liliaceae)
Mol. Formula : $C_{39}H_{62}O_{12}$
Mol. Wt. : 722
M.P. : 230-231°C
$[\alpha]_D^{25}$: -103° (c=0.5, CHCl$_3$-MeOH)
Registry No. : [94271-04-2]

IR (KBr) **:** 3400 (OH), 1650, 980, 920, 899, 850 cm^{-1} (intensity 920>899, 25*S*-spiroketal).

Mass (F.D.) : *m/z* (rel.intens.) 746 [(M+Na)$^+$, 5.0], 723 [(M+H)$^+$, 100], 577 [(M+H-Rha)$^+$, 34.1], 415 [(M+H-2xRha-Glc)$^+$, 15.3].

Reference

1.   O.P. Sati and G. Pant, *Phytochemistry*, **24**, 123 (1985).

COLLETTINSIDE II

Yamogenin 3-O-[α-L-rhamnopyranosyl-(1→4)-β-D-glucopyranoside]

Source : *Dioscorea colletti* Hook. f. (Dioscoreaceae)
Mol. Formula : $C_{39}H_{62}O_{12}$
Mol. Wt. : 722
M.P. : 214-216°C
$[\alpha]_D^{20}$: -97.9° (c=0.48)
Registry No. : [88668-52-4]

IR (KBr) : 3400 (OH), 1030-1060, 982, 920 > 890, 840 (25S-spiroketal) cm^{-1}.

CMR (C_5D_5N, 22.5 MHz) : δ C-1) 37.1 (2) 30.0 (3) 77.5 (4) 40.5 (5) 140.0 (6) 120.4 (7) 34.5 (8) 29.0 (9) 49.1 (10) 35.0 (11) 19.4 (12) 38.0 (13) 39.0 (14) 56.0 (15) 30.2 (16) 80.1 (17) 62.0 (18) 15.8 (19) 13.5 (20) 40.0 (21) 14.0 (22) 108.8 (23) 28.6 (24) 25.1 (25) 27.0 (26) 64.5 (27) 16.5 **Glc** (1) 102.2 (2) 74.5 (3) 77.9 (4) 78.1 (5) 77.3 (6) 61.5 **Rha** (1) 101.0 (2) 71.8 (3) 72.8 (4) 73.9 (5) 69.3 (6) 18.1.

Mass (E.I.) : m/z 723 [M+H]$^+$, 577, 396, 115.

Reference

1.   C.-L. Liu, Y.-Y. Chen, S.-B. Ge and B.-G. Li, *Yaoxue Xuebao* (*Acta Pharm. Sin.*), **18**, 597 (1983).

HISPININ B

(25S,5α)-6α-Hydroxy-spirostan 6-O-[α-L-rhamnopyranosyl-(1→3)-α-L-rhamnopyranoside]

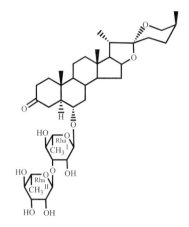

Source : *Solanum hispidum* Pers. (Solanaceae)
Mol. Formula : $C_{39}H_{62}O_{12}$
Mol. Wt. : 722
M.P. : 258-260°C
[α]$_D$: -68.3° (C_5D_5N)
Registry No. : [72176-16-0]

Reference

1.   A.K. Chakravanty, C.R. Saha and S.C. Pakrashi, *Phytochemistry*, **18**, 902 (1979).

KALLSTROEMIN E, DIOSCOREA PAZERI GLYCOSIDE A, SPRENGERININ B

Diosgenin 3-O-[α-L-rhamnopyranosyl (1→6)-β-D-glucopyranoside]

Source : *Kallstroemia pubescens* (G. Don) Dandy
(Zygophyllaceae), *Dioscorea prazeri* Prain et Burk.[2]
(Dioscoreaceae), *Asparagus sprengeri* Regel[3]
(Liliaceae)
Mol. Formula : $C_{39}H_{62}O_{12}$
Mol. Wt. : 722
M.P. : 218-219°C[1]
[α]$_D$: 101° (Pyridine)[1]
Registry No. : [64652-16-0]

IR (KBr)[1] : 3320, 980, , 919, 900, 867, 840, 812 cm^{-1}.

IR (KBr)[3] : 3395 (OH), 2845 (Δ^5), 980, 915, 896 (intensity 896>915 cm^{-1}, 25R-spiroketal).

References

1.   S.B. Mahato, N.P. Sahu, B.C. Pal and R.N. Chakravarti, *Indian J. Chem.*, **15B**, 445 (1977).

2.   M. Wij, K. Rajaraman and S. Rangaswami, *Indian J. Chem.*, **15B**, 451 (1977).

3.   S.C. Sharma, R. Sharma and R. Kumar, *Phytochemistry*, **22**, 2259 (1983).

LIRIOPE PLATYPHYLLA GLYCOSIDE B, LIRIOPE SPICATA SAPONIN LS-2

Ruscogenin 1-O-β-D-fucopyranoside-3-O-α-L-rhamnopyranoside

Source : *Liriope platyphylla* Wang. et Tang.[1] (Liliaceae), *Liriope spicata* Thunb. Lour.,[2] *Ophiopogon japonicus*[3] (Liliaceae)
Mol. Formula : $C_{39}H_{62}O_{12}$
Mol. Wt. : 722
M.P. : 225-230°C (decomp.)[3]
$[\alpha]_D^{20}$ **:** -12° (c=0.04, C_5H_5N)[3]
Registry No. : [87425-34-1]

UV (MeOH)[3] : λ_{max} 207 nm.

IR (KBr)[3] : 3500-3300 (OH), 988, 919, 897, 848 cm^{-1} [intensity 919 > 897, (25S)-spirostanol].

PMR (CD₃OD, 500.13 MHz)[3] : δ 0.85 (s, 3xH-18), 1.03 (s, 3xH-21), 1.12 (s, 3xH-19), 1.13 (s, 3xH-27), 1.17 (H-14ax), 1.24 (H-12ax), 1.29 (3xH-6 of Fuc), 1.29 (3xH-6 of Rha), 1.30 (H-9ax), 1.39 (H-23eq), 1.46 (H-24eq), 1.47 (H-11ax), 1.58 (H-7ax), 1.58 (H-15ax), 1.61 (H-8ax), 1.7 (H-2ax), 1.71 (H-12eq), 1.72 (H-25eq), 1.75 (H-17α), 1.91 (H-20β), 1.96 (H-23ax), 2.01 (H-15β), 2.02 (H-7eq), 2.07 (H-24ax), 2.23 (H-2eq), 2.25 (H-4ax), 2.37 (H-4eq), 2.58 (H-11eq), 3.31 (H-26eq), 3.41 (H-4 of Rha), 3.46 (H-1ax), 3.47 (H-3ax), 3.47 (H-2 of Fuc), 3.49 (H-3 of Fuc), 3.61 (H-5 of Fuc), 3.63 (H-4 of Fuc), 3.68 (H-3 of Rha), 3.69 (H-5 of Rha), 3.79 (H-2 of Rha), 3.97 (H-26ax), 4.25 (H-1 of Fuc), 4.43 (H-16), 4.87 (H-1 of Rha), 5.62 (H-6).

PMR (CD₃OD/C₅D₅N, 500.13 MHz)[3] : δ 0.68 (s, 3xH-18), 0.90 (s, 3xH-21), 0.95 (s, 3xH-27), 0.98 (H-14ax), 0.99 (s, 3xH-19), 1.12 (H-12ax), 1.20 (H-24eq), 1.23 (H-15α), 1.25 (H-9ax), 1.26 (H-11ax), 1.31 (3xH-6 of Fuc), 1.33 (H-23eq), 1.38 (H-7ax), 1.38 (H-8ax), 1.42 (3xH-6 of Rha), 1.46 (H-12eq), 1.46 (H-25eq), 1.61 (H-17α), 1.71 (H-20β), 1.72 (H-23ax), 1.74 (H-7eq), 1.77 (H-2ax), 1.85 (H-15β), 1.93 (H-24ax), 2.16 (H-4ax), 2.28 (H-4eq), 2.47 (H-2eq), 2.65 (H-11eq), 3.18 (2xH-26eq), 3.50 ($J_{5,6}$=6.5 Hz, H-5 of Fuc), 3.52 (H-1ax), 3.57 (H-3ax), 3.70 ($J_{3,4}$=3.2 Hz, H-3 of Fuc), 3.73 ($J_{4,5}$=1.0 Hz, H-4 of Fuc), 3.83 ($J_{4,5}$=9.2 Hz, H-4 of Rha), 3.88 ($J_{2,3}$=9.7 Hz, H-2 of Fuc), 3.88 (H-26ax), 3.99 ($J_{5,6}$=6.5 Hz, H-5 of Rha), 4.08 ($J_{3,4}$=9.2 Hz, H-3 of Rha), 4.13 ($J_{2,3}$=3.0 Hz, H-2 of Rha), 4.32 (H-16α), 4.41 ($J_{1,2}$=6.5 Hz, H-1 of Fuc), 5.17 (J= 1.5 Hz, H-1 of Rha), 5.40 (H-6).

CMR (CD₃OD, 125.76 MHz)[3] : δ C-1) 84.3 (2) 35.8 (3) 74.5 (4) 39.8 (5) 139.1 (6) 126.4 (7) 32.8 (8) 34.0 (9) 51.7 (10) 43.8 (11) 24.6 (12) 41.4 (13) 41.1 (14) 58.1 (15) 32.9 (16) 82.4 (17) 63.9 (18) 17.1 (19) 14.9 (20) 43.5 (21) 14.7 (22) 111.2 (23) 27.0 (24) 26.8 (25) 28.5 (26) 66.1 (27) 16.4 **Fuc** (1) 102.2 (2) 72.3 (3) 75.3 (4) 73.0 (5) 71.7 (6) 16.9 **Rha** (1) 99.8 (2) 72.7 (3) 72.4 (4) 74.1 (5) 70.0 (6) 17.9.

Mass (FAB, Negative ion)[1] : *m/z* 721 [(M-H)⁻, 100 (base peak)].

References

1.   Y. Watanabe, S. Sanada, Y. Ida and J. Shoji, *Chem. Pharm. Bull.*, **31**, 1980 (1983).

2.   D. Ling, L. Liu, Y. Zhu, *Zhongcaoyao*, **22**, 489 (1991).

3.   J.T. Branke and E. Haslinger, *Liebigs Ann. Chem.*, 587 (1995).

LIRIOPE PLATYPHYLLA GLYCOSIDE C,
OPHIOPOGON JAPONICUS SAPONIN OJV-IV
(25*S*)-Ruscogenin 1-O-α-L-rhamnopyranosyl-(1→2)-β-D-fucopyranoside

Source : *Liriope platyphylla* Wang. et Tang.[1] (Liliaceae),
Ophiopogon japonicus Ker-Gawler cv. Nanus[2] (Liliaceae)
Mol. Formula : $C_{39}H_{62}O_{12}$
Mol. Wt. : 722
M.P. : 201-203°C[1]
[α]$_D^{18}$: -89.6° (c=0.48, Pyridine)[1]
Registry No. : [87425-35-2]

IR (KBr)[2] : 3600-3400, 985, 920, 900, 855 cm⁻¹ (Intensity 920 > 900, 25*S*-Spiroketal).

PMR (C₅D₅N, 100 MHz)[1] : δ 0.89 (s, C*H₃*), 1.08 (d, *J*=6.0 Hz, sec. C*H₃*), 1.14 (d, *J*=6.0 Hz, sec. C*H₃*), 1.44 (s, C*H₃*), 1.52 (d, *J*=6.0 Hz, sec. C*H₃*), 1.75 (d, *J*=6.0 Hz, sec. C*H₃*).

CMR (C₅D₅N, 100 MHz)[2] : δ C-1) 84.3 (2) 38.4 (3) 68.8 (4) 44.3 (5) 139.8 (6) 124.5 (7) 33.7 (8) 32.6 (9) 51.3 (10) 43.4 (11) 24.5 (12) 41.0 (13) 40.8 (14) 57.7 (15) 33.0 (16) 81.5 (17) 63.6 (18) 17.4 (19) 15.5 (20) 43.0 (21) 15.3 (22) 109.9 (23) 27.0 (24) 26.8 (25) 28.1 (26) 65.2 (27) 16.9 **Fuc** (1) 100.4 (2) 75.0 (3) 77.0 (4) 73.5 (5) 71.3 (6) 17.7 **Rha** (1) 101.6 (2) 72.8 (3) 72.9 (4) 74.6 (5) 69.3 (6) 19.4.

References

1.   Y. Watanabe, S. Sanada, Y. Ida and J. Shoji, *Chem. Pharm. Bull.*, **31**, 1980 (1983).

2.   T. Asano, T. Murayama, Y. Hirai and J. Shoji, *Chem. Pharm. Bull.*, **41**, 566 (1993).

OPHIOPOGONIN B

Ruscogenin 1-O-[α-L-rhamnopyranosyl-(1→2)-β-D-fucopyranoside]

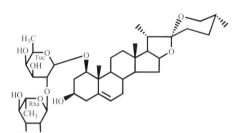

Source : *Ophiopogon japonicus* Ker-Gawler var. *genuinus* Maxim.[1] (Liliaceae), *O. ohwii* Okuyama[2] (Liliaceae), *O. japonicus* Ker-Gawler cv. Nanus[3]
Mol. Formula : $C_{39}H_{62}O_{12}$
Mol. Wt. : 722
M.P. : 270-272°C (decomp.)[3]
$[α]_D^{15}$: -105.5° (c=0.31, Pyridine)[1]
Registry No. : [38971-41-4]

IR (KBr)[2] : 3600-3200 (OH), 985, 925, 905, 870 cm⁻¹, intensity 925 < 905 (25*R*-spiroketal) cm⁻¹.

IR (KBr)[3] : 3500-3400 (OH), 990, 920, 900, 850 cm⁻¹, intensity 920 < 900 (25*R*-spiroketal) cm⁻¹.

CMR (C_5D_5N, 100 MHz)[3] : δ C-1) 84.3 (2) 38.4 (3) 68.6 (4) 44.3 (5) 139.7 (6) 124.7 (7) 33.7 (8) 32.6 (9) 51.2 (10) 43.4 (11) 24.5 (12) 41.0 (13) 40.8 (14) 57.7 (15) 33.3 (16) 81.4 (17) 63.5 (18) 17.4 (19) 15.4 (20) 42.5 (21) 15.6 (22) 109.4 (23) 32.4 (24) 29.9 (25) 31.1 (26) 67.2 (27) 17.8 **Fuc** (1) 100.4 (2) 75.0 (3) 77.0 (4) 73.5 (5) 71.3 (6) 17.7 **Rha** (1) 101.6 (2) 72.8 (3) 72.9 (4) 74.6 (5) 69.3 (6) 19.4.

References

1.   A. Tada and J. Shoji, *Chem. Pharm. Bull.*, **20**, 1729 (1972).

2.   Y. Watanabe, S. Sanada, Y. Ida and J. Shoji, *Chem. Pharm. Bull.*, **32**, 3994 (1984).

3.   T. Asano, T. Murayama, Y. Hirai and J. Shoji, *Chem. Pharm. Bull.*, **41**, 566 (1993).

POLYPHYLLIN C

Diosgenin 3-O-[α-L-rhamnopyranosyl-(1→3)-β-D-glucopyranoside]

Source : *Paris polyphylla* Sm. (Liliaceae)
Mol. Formula : $C_{39}H_{62}O_{12}$
Mol. Wt. : 722
M.P. : 185-190°C
$[α]_D^{27}$: -99.0° (c=0.5, Pyridine)
Registry No. : [76296-71-4]

IR (KBr) **:** 3500-3200, 1110-1000, 982, 921 (w), 901 (s), 860, 840 cm^{-1}.

Reference

1.   S.B. Singh, R.S. Thakur and H.-R. Schulten, *Phytochemistry*, **21**, 2925 (1982).

TRILLIUM KAMTSCHATICUM SAPONIN Ta

Diosgenin 3-O-[α-L-rhamnopyranosyl-(1→2)-β-D-glucopyranoside]

Source : *Trillium kamtschaticum* Pall. (Liliaceae)[1], *Ophiopogan planiscapus* Nakai (Liliaceae)[2], *Dioscorea collettii* var. *hypoglauca* (Dioscoraceae)[3], *Allium vineale* (Liliaceae)[4], *Smilax china* L. (Liliaceae)[5] etc.

Mol. Formula : $C_{39}H_{62}O_{12}$

Mol. Wt. : 722

M.P. : 242-246°C (decomp.)[1]

[α]$_D$: -108.3° (c=0.97, Pyridine)[1]; -121.6° (c=0.66, MeOH)[1]; -101.0° (c=0.97, dioxane)[1]

Registry No. : [19057-67-1]

IR (KBr)[1] : 3600-3200 (OH), 982, 920, 900, 865 cm^{-1}. Intensity 900 > 920 (25R-spiroketal) cm^{-1}.

IR (KBr)[3] : 3400, 2930, 1630, 1451, 1040, 980, 912 > 897 (25R-spiroketal) cm^{-1}.

PMR (C_5D_5N, 500 MHz)[3] : δ 0.69 (d, J=6.5 Hz, 3xH-27), 0.83 (s, 3xH-18), 1.05 (s, 3xH-190, 1.13 (d, J=6.9 Hz, 3xH-21), 1.76 (d, J=6.5 Hz, 3xH-6 of Rha), 3.94 (m, H-3), 5.02 (d, J=7.2 Hz, H-1 of Glc), 5.31 (br d, J=6.0 Hz, H-6), 6.35 (d, J=1.0 Hz, H-1 of Rha).

CMR (C_5D_5N-D_2O, 100 MHz)[4] : δ (C-1) 37.4 (2) 30.1 (3) 78.2 (4) 38.9 (5) 140.7 (6) 121.7 (7) 32.2 (8) 31.7 (9) 50.2 (10) 37.0 (11) 21.0 (12) 39.8 (13) 40.4 (14) 56.5 (15) 32.0 (16) 81.0 (17) 62.7 (18) 16.3 (19) 19.4 (20) 41.9 (21) 15.0 (22) 109.3 (23) 31.6 (24) 29.2 (25) 30.4 (26) 66.8 (27) 17.3 **Glc** (1) 100.4 (2) 79.7 (3) 78.0 (4) 71.9 (5) 77.9 (6) 62.7 **Rha** (1) 102.1 (2) 72.7 (3) 72.9 (4) 74.3 (5) 69.6 (6) 18.9.

Mass (FAB, Positive ion)[3] : m/z 723 [M+H]$^+$, 577 [M+H-Rha]$^+$, 415 [M+H-Rha-Glc]$^+$.

Biological Activity : Causes morphological abnormality of *Pyricularia oryzae* mycelia.[3] Exhibits cytotoxic activity against the cancer cell line K562 *in vitro*.[3] Inhibits human spermatozoa *in vitro* at a concentration of 1 mg/ml.[6]

References

1. T. Nohara, K. Miyahara and T. Kawasaki, *Chem. Pharm. Bull.*, **23**, 872 (1975).

2. Y. Watanabe, S. Sanada, Y. Ida and J. Shoji, *Chem. Pharm. Bull.*, **31**, 3486 (1983).

3. K. Hu, A. Dong, X. Yao, H. Kobayashi and S. Iwasaki, *Planta Med.*, **62**, 573 (1996).

4. S. Chen and J.K. Snyder, *J. Org. Chem.*, **54**, 3679 (1989).

5. S.W. Kim, K.C. Chung, K.H. Son and S.S. Kang, *Kor. J. Pharmacogn.*, **20**, 76 (1989).

6. Y.-F. Wang, X.-C. Li, H.-Y. Yang, J.-J. Wang and C.-R. Yang, *Planta Med.*, **62**, 130 (1996).

TUBEROSIDE C (SOLANUM)

Yamogenin 3-O-[α-L-rhamnopyranosyl-(1→2)-β-D-galactopyranoside]

Source : *Solanum tuberosum* L. (Solanaceae)
Mol. Formula : $C_{39}H_{62}O_{12}$
Mol. Wt. : 722
M.P. : 236-238°C
$[\alpha]_D^{20}$ **:** -63° (c=1.7, C_5D_5N)
Registry No. : [145680-56-4]

IR : 845, 890 < 920 and 980 cm^{-1}.

Reference

1.   P.K. Kintya and T.I. Prasol, *Khim. Prir. Soedin.*, 586 (1991); *Chem. Nat. Comp.*, **27**, 515 (1991).

UTTRONIN B
Diosgenin 3-O-[β-D-glucopyranosyl-(1→4)-α-L-rhamnopyranoside]

Source : *Solanum nigrum* L. (Solanaceae)
Mol. Formula : $C_{39}H_{62}O_{12}$
Mol. Wt. : 722
M.P. : 218-221°C
$[\alpha]_D^{20}$: -102.0°
Registry No. : [84955-03-3]

IR (KBr) **:** 3400 (OH), 3020, 2845 (Δ^5), 965, 910, 895, 845 Intensity 895 > 910 (25*R*-Spiroketal) cm^{-1}.

Permethylate :

M.P. : 103-104°C

Mass (E.I.) **:** *m/z* 806 [M]$^+$, 571 [M-tetra-O-methyl glucose + H]$^+$, 219, 139, 101, 88, 75, 71, 55, 45.

Reference

1.   S.C. Sharma, R. Chand and O.P. Sati, *Pharmazie*, **37**, 870 (1982).

ZINGIBEROSIDE A₁

Yamogenin 3-O-[α-L-rhamnopyranosyl-(1→2)-β-D-glucopyranoside

Source : *Dioscorea zinigberensis* Wright (Dioscoreaceae)
Mol. Formula : $C_{39}H_{62}O_{12}$
Mol. Wt. : 722
M.P. : 231-235°C
Registry No. : [110996-52-6]

IR (KBr) : 3350, 2900, 1640, 1440, 1360, 1210, 1040, 980, 920, 890, 845 cm⁻¹. Intensity 920>890 (25R-spiroketal).

CMR (C_5D_5N) : δ C-1) 37.3 (2) 30.0 (3) 77.7 (4) 38.8 (5) 140.4 (6) 121.3 (7) 31.6 (8) 32.1 (9) 50.1 (10) 37.0 (11) 21.0 (12) 39.8 (13) 40.1 (14) 56.5 (15) 31.6 (16) 80.9 (17) 62.5 (18) 16.2 (19) 19.3 (20) 42.3 (21) 14.3 (22) 109.4 (23) 26.1 (24) 27.3 (25) 26.1 (26) 64.8 (27) 14.3 **Glc** (1) 100.0 (2) 79.1 (3) 77.7 (4) 71.5 (5) 77.7 (6) 62.5 **Rha** (1) 101.6 (2) 72.1 (3) 72.4 (4) 73.7 (5) 69.1 (6) 18.5

Mass (E.I.) : *m/z* 561, 397, 273, 139, 115.

Reference

1.   S. Tang, and Z. Jiang, *Yunnan Zhiwu Yanjiu (Acta Botanica Yunnanica)*, **9**, 233 (1987).

HISPININ C
(25S,5α)-3β,6α-Dihydroxy-spirostan 6-O-[α-L-rhamnopyranosyl-(1→3)-α-L-rhamnopyranoside]

Source : *Solanum hispidum* Pers. (Solanaceae)
Mol. Formula : $C_{39}H_{64}O_{12}$
Mol. Wt. : 724
M.P. : 285-288°C
[α]$_D$: -59.0° (C_5D_5N)
Registry No. : [72165-40-9]

Reference

1.   A.K. Chakravanty, C.R. Saha and S.C. Pakrashi, *Phytochemistry*, **18**, 902 (1979).

NICOTIANOSIDE B
Neotigogenin 3-O-[α-L-rhamnopyranosyl-(1→2)-β-D-glucopyranoside]

Source : *Nicotiana tabacum* L. seeds (Solanceae)
Mol. Formula : $C_{39}H_{64}O_{12}$
Mol. Wt. : 724
M.P. : 241-242°C
$[\alpha]_D^{20}$ **:** -56.0° (c=1.0, MeOH)
Registry No. : [95535-06-1]

PMR (C_5D_5N, 300/500 MHz) : δ 0.74 (s, 3xH-18), 0.78 (s, 3xH-19), 1.02 (d, J=7.0 Hz, 3xH-21), 1.07 (d, J=7.0 Hz, 3xH-27), 1.64 (d, 3xH-6 of Rha), 3.62 (m, H-5 of Glc), 3.84 (m, H-3), 3.98 (dd, H-6A of Glc), 4.05 (t, $J_{1,2}=J_{2,3}$=7.5 Hz, H-2 of Glc), 4.09 (dd, $J_{2,3}$=7.5 Hz, $J_{3,4}$=10.0 Hz, H-3 of Glc), 4.18 (dd, H-6B of Glc), 4.21 (t, $J_{3,4}=J_{4,5}$=9.5 Hz, H-4 of Rha), 4.42 (t, $J_{3,4}=J_{4,5}$=10.0 Hz, H-4 of Glc), 4.45 (dd, $J_{2,3}$=3.5 Hz, $J_{3,4}$=9.5 Hz, H-3 of Rha), 4.68 (d, $J_{2,3}$=3.5 Hz, H-2 of Rha), 4.78 (dq, H-5 of Rha), 4.87 (d, $J_{1,2}$=7.5 Hz, H-1 of Glc), 6.16 (s, H-1 of Rha).

CMR (C_5D_5N, 75/125 MHz) : δ C-1) 37.4 (2) 30.0 (3) 78.0 (4) 34.6 (5) 44.8 (6) 29.1 (7) 32.6 (8) 35.4 (9) 54.7 (10) 36.1 (11) 21.5 (12) 40.3 (13) 40.9 (14) 56.6 (15) 32.3 (16) 81.4 (17) 63.0 (18) 16.8 (19) 12.6 (20) 42.6 (21) 15.0 (22) 109.9 (23) 27.7 (24) 26.5 (25) 26.3 (26) 65.3 (27) not given **Glc** (1) 100.3 (2) 78.7 (3) 78.5 (4) 71.6 (5) 76.8 (6) 61.3 **Rha** (1) 102.8 (2) 72.5 (3) 72.9 (4) 73.9 (5) 69.4 (6) 18.4.

Reference

1   S.A. Shvets, P.K. Kintya and O.N. Gutsu, *Khim. Prir. Soedin.*, 737 (1994); *Chem. Nat. Comp.*, **30**, 684 (1994).

SARSASAPOGENIN 3-O-[α-L-RHAMNOPYRANOSYL-(1→4)-β-D-GLUCOPYRANOSIDE]

Source : *Asparagus curillus* Buch.-Ham. (Liliaceae)
Mol. Formula : $C_{39}H_{64}O_{12}$
Mol. Wt. : 724
M.P. : 252-257°C
[α]$_D^{20}$: -61.5° (c=0.80, CHCl$_3$/MeOH)
Registry No. : [58881-26-8]

IR (KBr) : 3450 (OH), 980, 915, 896, 850 cm^{-1} (intensity 915 > 896, 25*S*-spiroketal).

Reference

1.   S.C. Sharma, O.M. Sati and R. Chand, *Planta Med.,* **47**, 117 (1983).

SMILANIPPIN A
Neotigogenin 3-O-[β-D-fucopyranosyl-(1→6)-β-D-glucopyranoside]

Source : *Smilax nipponica* Miq. (Liliaceae)
Mol. Formula : $C_{39}H_{64}O_{12}$
Mol. Wt. : 724
M.P. : 281-284°C
[α]$_D$: -25.7° (c=0.6, MeOH)
Registry No. : [166736-13-6]

IR (KBr) : 3435 (OH), 1062, 1043 (glycosidic C-O), 956, 920, 905, 848 cm^{-1} (920 > 905) cm^{-1}.

PMR (C_5D_5N, 300/500 MHz) : δ 0.73 (s, 3xH-18), 0.88 (s, 3xH-19), 1.15 (d, *J*=7.1 Hz, 3xH-21), 1.22 (*J*=6.9 Hz, 3xH-27), 5.08 (d, *J*=7.8 Hz, anomeric H), 5.63 (d, *J*=8.6 Hz, anomeric H).

CMR (C_5D_5N, 75/125 MHz) : δ C-1) 37.31 (2) 30.23 (3) 78.80 (4) 35.96 (5) 44.69 (6) 29.11 (7) 32.50 (8) 35.09 (9) 54.45 (10) 35.40 (11) 21.39 (12) 40.27 (13) 40.88 (14) 56.54 (15) 32.23 (16) 81.34 (17) 63.03 (18) 18.80 (19) 12.44 (20) 42.60 (21) 15.00 (22) 109.83 (23) 27.67 (24) 26.33 (25) 26.52 (26) 65.22 (27) 16.43 **Glc** (1) 102.46 (2) 74.20 (3) 77.18 (4) 72.16 (5) 77.76 (6) 69.89 **Fuc** (1) 102.80 (2) 72.51 (3) 75.42 (4) 72.95 (5) 72.45 (6) 16.73.

Mass : *m/z* (rel.intens.) 747 [(M+Na)$^+$, 20.5], 725 [(M-H)$^+$, 39.1], 579 [(M-H)$^+$-Hexose, 6.5], 417 [(Neotigogenin+H)$^+$, 32.5], 399 [(Neotigogenin+H)$^+$-H_2O, 57.1].

Reference

1.   K.Y. Cho, M.H. Woo and S.O. Chung, *Yakhak Hoeji*, **39**, 141 (1995).

SMILAX RIPARIA SAPONIN 1

Neotigogenin 3-O-[α-L-rhamnopyranosyl-(1→6)-β-D-glucopyranoside]

Source : *Smilax riparia* (Liliaceae)
Mol. Formula : $C_{39}H_{64}O_{12}$
Mol. Wt. : 724
$[\alpha]_D^{24}$: -53.4° (c=0.16, EtOH)
Registry No. : [143329-41-3]

IR (KBr) : 3450 (OH), 2950 (CH), 1470, 1455, 1385, 1340, 1275, 1265, 1220, 1155, 1135, 1100, 1060, 1045, 990, 975, 920, 900, 850, 840, 810 cm^{-1} ((25S)-spiroacetal, intensity 920 > 900).

PMR (C_5D_5N, 400 MHz) : δ 0.67 (s, 3xH-18), 0.81 (s, 3xH-19), 1.08 (d, J=7.1 Hz, 3xH-27), 1.15 (d, J=6.9 Hz, 3xH-21), 1.64 (d, J=6.2 Hz, 3xH-6 of Rha), 5.00 (d, J=7.6 Hz, H-1 of Glc), 5.53 (br s, H-1 of Rha).

CMR (C_5D_5N, 100.6 MHz) : δ C-1) 37.2 (2) 30.1 (3) 77.7 (4) 35.0 (5) 44.6 (6) 29.0 (7) 32.4^a (8) 35.3 (9) 54.4 (10) 35.9 (11) 21.3 (12) 40.2 (13) 40.8 (14) 56.4 (15) 32.1^a (16) 81.2 (17) 62.9 (18) 16.6 (19) 12.3 (20) 42.5 (21) 14.9 (22) 109.7 (23) 26.2^b (24) 26.4^b (25) 27.6 (26) 65.1 (27) 16.3 **Glc** (1) 102.4 (2) 75.3 (3) 78.7 (4) 72.1 (5) 77.1 (6) 68.6 **Rha** (1) 102.7 (2) 72.4 (3) 72.8 (4) 74.1 (5) 69.8 (6) 18.7.

Mass (S.I., Positive ion) : *m/z* 725 [M+H]$^+$.

Biological Activity : The compound shows inhibitory activity on cyclic AMP phosphodiesterase with IC$_{50}$=10.2x10^{-5} M.

Reference

1.   Y. Sashida, S. Kubo, Y. Mimaki, T. Nikaido and T. Ohmoto, *Phytochemistry*, **31**, 2439 (1992).

SOLANUM CHRYSOTRICHUM SAPONIN SC-5

(25*R*)-5α-Spirostan-3β-6α-diol 6-O-[α-L-rhamnopyranosyl-(1→3)-β-D-quinovopyranoside]

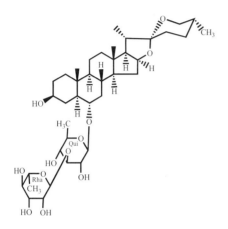

Source : *Solanum chrysotrichum* Schldh. (Solanaceae)
Mol. Formula : $C_{39}H_{64}O_{12}$
Mol. Wt. : 724
M.P. : 247-248°C
[α]$_D^{25}$: -33.89° (c=1.18, CHCl$_3$)
Registry No. : [478298-15-6]

Hexa-acetate :

IR (CHCl$_3$) **:** 2920, 1760 (ester), 1470, 1050 (C–O–C) cm^{-1}.

PMR (CDCl$_3$, 400 MHz) **:** δ 0.65 (H-24A), 0.68 (H-9), 0.75 (s, 3xH-18), 0.79 (d, *J*=6.4 Hz, 3xH-27), 0.84 (s, 3xH-19), 0.95 (d, *J*=6.4 Hz, 3xH-21), 0.99 (H-1A), 1.13 (d, *J*=6.4 Hz, 3xH-6 of Rha), 1.14 (H-14A), 1.17 (d, *J*=6.4 Hz, H-6 of Qui), 1.2 (H-4A, H-11A, H-14B, H-15A, H-23A and H-24B), 1.21 (H-2A), 1.25 (H-5), 1.27 (H-7A), 1.45 (H-11B), 1.60 (H-8), 1.62 (H-1B), 1.65 (H-2B), 1.68 (H-12A), 1.68 (dd, *J*=8.4, 6.8 Hz, H-17), 1.7 (H-23B), 1.77 (H-25), 1.80 (H-20), 2.01 (H-12B and H-15B), 2.05 (H-4B), 2.51 (H-7B), 3.22 (ddd, *J*=11.0, 10.0, 4.5 Hz, H-6), 3.36 (dd, *J*=11.0, 10.0 Hz, H-26α), 3.41 (m, H-5 of Qui), 3.45 (dd, *J*=11.0, 2.4 Hz, H-26β), 3.6 (t, *J*=9.6 Hz, H-3 of Qui), 3.87 (dd, *J*=10.0, 6.4 Hz, H-5 of Rha), 4.36 (d, *J*=8.0 Hz, H-1 of Qui), 4.39 (ddd, *J*=7.5, 7.4, 7.0 Hz, H-16), 4.62 (dddd, *J*=11.0, 10.5, 6.5 Hz, H-3), 4.77 (d, *J*=1.2 Hz, H-1 of Rha), 4.81 (t, *J*=9.6 Hz, H-4 of Qui), 5.05 (dd, *J*=9.6, 8.0 Hz, H-2 of Qui), 5.08 (m, H-2 of Rha), 5.1 (m, H-3 of Rha), 5.13 (dd, *J*=10.0, 9.6 Hz, H-4 of Rha).

CMR (CDCl$_3$, 100 MHz) **:** δ C-1) 36.67 (2) 27.07 (3) 73.25 (4) 28.14 (5) 49.43 (6) 80.78 (7) 39.62 (8) 33.8 (9) 53.12 (10) 36.39 (11) 21.2 (12) 39.91 (13) 41.51 (14) 55.76 (15) 31.66 (16) 81.74 (17) 61.98 (18) 17.12 (19) 13.36 (20) 41.6 (21) 14.49 (22) 109.24 (23) 31.3 (24) 28.76 (25) 30.24 (26) 66.82 (27) 16.44 **Qui** (1) 102.1 (2) 72.12 (3) 81.93 (4) 74.41 (5) 69.8 (6) 17.49 **Rha** (1) 99.45 (2) 69.8 (3) 68.89 (4) 70.54 (5) 67.36 (6) 17.17.

Mass (FAB, Positive ion) **:** *m/z* 999 [M+Na]$^+$, 977 [M+H]$^+$, 976 [M]$^+$, 703 [M-Rha]$^+$, 457 [M-Rha-Qui]$^+$, 397 [M-Rha-Qui-CH$_3$COO]$^+$.

Mass (FAB, Positive ion, H.R.) **:** *m/z* 977.5139 [(M+H)$^+$, calcd. for 977.5110].

Reference

1.   A. Zamilpa, J. Tortoriello, V. Navarro, G. Delgado and L. Alvarez, *J. Nat. Prod.*, **65**, 1815 (2002).

TORVONIN-A
Neochlorogenin-3-O-[β-L-rhamnopyranosyl-(1→2)-β-L-rhamnopyranoside]

Source : *Solanum torvum* Swartz (Solanaceae)
Mol. Formula : $C_{39}H_{64}O_{12}$
Mol. Wt. : 724
M.P. : 285°C
[α]$_D$: -37.3° (c=0.30, C_5D_5N)
Registry No. : [99956-65-7]

IR (KBr) : 3500-3300 (broad, OH), 1380, 1220, 1170, 1150-1000 (C-O-C), 982, 918, 892 and 850 cm^{-1} (918>892, 25S-Spiroketal).

PMR (C_5D_5N, 400 MHz) : δ 0.80 (s, 3xH-18), 0.83 (s, 3xH-19), 1.04 (d, J=7.0 Hz, 3xH-21), 1.16 (d, J=7.0 Hz, 3xH-27), 1.60 (d, J=6.0 Hz, 3xH-6 of Rha), 1.69 (d, J=6.0 Hz, 3xH-6 of Rha), 3.14 (d, J=12.0 Hz, H-4α), 3.32 (d, J=11.0 Hz, H-26β), 3.56-3.75 (6H, m, H-3α, H-6β, 4xRha-H), 4.00 (2H, m, H-26α, H-Rha), 4.28 (t, J=9.0 Hz, H-Rha), 4.33 (t, J=9.0 Hz, H-Rha), 4.49 (q, J=8.0 Hz, H-16α), 4.59 (dd, J=9.0, 4.0 Hz, H-Rha), 4.75 (d, J=8.0 Hz, H-Rha), 4.81 (br s, W½=4.0 Hz, H-1 of Rha).

CMR (C_5D_5N, 100 MHz) : δ C-1) 37.76 (2) 32.12* (3) 79.44 (4) 32.24* (5) 51.31 (6) 69.91 (7) 41.43 (8) 34.18 (9) 53.88 (10) 36.73 (11) 21.27 (12) 40.07 (13) 40.78 (14) 56.39 (15) 33.16* (16) 81.11 (17) 40.78 (18) 16.62 (19) 13.58 (20) 42.49 (21) 14.86 (22) 109.64 (23) 26.41 (24) 26.21 (25) 27.54 (26) 65.09 (27) 16.30 **Rha I** (1) 103.04 (2) 83.41 (3) 70.63 (4) 74.16 (5) 72.72 (6) 18.64 **Rha II** (1) 105.59 (2) 72.59 (3) 76.21 (4) 75.23 (5) 72.76 (6) 18.78.

Reference

1.   U. Mahmood, P.K. Agrawal and R.S. Thakur, *Phytochemistry*, **24**, 2456 (1985).

ALLIOSPIROSIDE C
Cepagenin l-O-[α-L-rhamnopyranosyl-(1→2)-α-L-arabinopyranoside]

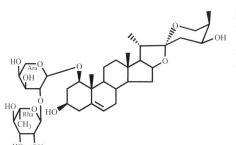

Source : *Allium cepa* L. (Liliaceae)
Mol. Formula : $C_{38}H_{60}O_{13}$
Mol. Wt. : 724
M.P. : 223-225°C (MeOH)
[α]$_D^{22}$: -105.7± 2° (c=0.94, C_5D_5N)
Registry No. : [114317-57-6]

IR (KBr) : 825, 845, 870, 880, 908, 920, 965, 1000, 3200-3600 cm^{-1}.

PMR (C_5D_5N, 250 MHz) : δ 0.77 (s, 3xH-18), 1.04 (d, 3xH-21), 1.23 (d, 3xH-27), 1.35 (s, 3xH-19), 3.49 (dd, H-26eq), 3.62 (dd, $J_{5ax,5eq}$=11.5 Hz, H-5ax of Ara), 3.72 (dd, H-1ax), 3.78 (m, H-3), 3.93 (dd, H-26ax), 4.07 (dd, $J_{3,4}$=3.5 Hz, H-3 of Ara), 4.10 (ddd, $J_{4,5ax}$=1.0 Hz, H-4 of Ara), 4.20 (dd, $J_{5eq,4}$=2.0 Hz, H-5eq of Ara), 4.22 (t, $J_{4,5}$=9.2 Hz, H-4 of Ara), 4.43 (m, H-16), 4.43 (dd, $J_{2,3}$=8.1 Hz, H-2 of Ara), 4.48 (dd, $J_{2,3}$=8.1 Hz, H-2 of Ara), 4.53 (m, H-24ax), 4.53 (dd, $J_{3,4}$=9.2 Hz, H-3 of Rha), 4.62 (d, $J_{1,2}$=7.5 Hz, H-1 of Ara), 4.64 (dd, $J_{2,3}$=3.1 Hz, H-2 of Rha), 4.75 (dq, $J_{5,6}$=6.1 Hz, H-5ax of Rha), 54.84 (dd, $J_{2,3}$=3.1 Hz, H-2 of Rha),.52 (br d, H-6), 6.21 (q, $J_{1,2}$=1.7 Hz, H-1 of Rha).

CMR (C_5D_5N, 300 MHz) : δ C-1) 83.94 (2) 37.54 (3) 68.36 (4) 43.95 (5) 139.69 (6) 124.89 (7) 32.15 (8) 33.31 (9) 50.61 (10) 43.08 (11) 24.19 (12) 40.36 (13) 40.48 (14) 57.02 (15) 32.51 (16) 81.74 (17) 62.65 (18) 16.83 (19) 15.06 (20) 42.71 (21) 15.18 (22) 111.63 (23) 36.45 (24) 66.69 (25) 35.97 (26) 64.72 (27) 9.76 **Rha** (1) 101.76 (2) 72.67 (3) 72.67 (4) 74.39 (5) 60.54 (6) 19.11 **Ara** (1) 100.63 (2) 75.39 (3) 75.97 (4) 70.19 (5) 67.44.

Reference

1.   S.D. Kravets, Yu. S. Vollerner, A.S. Shashkov, M.B. Gorovits and N.K. Abubakirov, *Khim. Prir. Soedin.*, **6**, 843 (1987); *Chem. Nat. Comp.*, **6**, 700 (1987).

OPHIOPOGON INTERMEDIUS SAPONIN 1

Ruscogenin 1-O-[α-L-arbinopyranosyl-(1→2)-β-D-glucopyranoside]

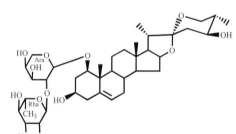

Source : *Ophiopogon intermedius* (Haemodoraceae)
Mol. Formula : $C_{38}H_{60}O_{13}$
Mol. Wt. : 724
M.P. : 203-205°C (decomp.)
$[\alpha]_D^{20}$: -71.0° (c=1.0, CHCl$_3$-MeOH)
Registry No. : [118855-42-8]

IR (KBr) : 3500-3300 (OH), 985, 950, 900, 865 (Intensity 900>920, 25*R*-spiroketal) cm^{-1}.

Reference

1.   M.S.M. Rawat, D.S. Negi, M.S. Panwar and G. Pant, *Phytochemistry*, **27**, 3326 (1988).

ORNITHOGALUM THYRSOIDES SAPONIN 12

(24*S*,25*S*)-Spirost-5-en-1β,3β,24-triol 1-O-[α-L-rhamnopyranosyl-(1→2)-α-L-arabinopyranoside]

Source : *Ornithogalum thyrsoides* (Liliaceae)
Mol. Formula : $C_{38}H_{10}O_{13}$
Mol. Wt. : 724
$[\alpha]_D^{23}$: -72.0° (c=0.10, MeOH)
Registry No. : [790661-47-1]

IR (film) : 3386 (OH), 2951, 2927 and 2906 (CH), 1084, 1054, 1037 cm^{-1}.

PMR (C₅D₅N, 500 MHz) : δ 0.86 (s, 3xH-18), 1.08 (d, *J*=6.5 Hz, 3xH-27), 1.12 (d, *J*=7.0 Hz, 3xH-21), 1.44 (s, 3xH-19), 1.75 (d, *J*=6.1 Hz, 3xH-6 of Rha), 1.82 (m, H-25), 1.98 (dd, *J*=12.7, 12.2 Hz, H-23ax), 2.31 (dd, *J*=12.7, 4.7 Hz, H-23eq), 3.59 (dd, *J*=11.3, 11.3 Hz, H-26ax), 3.69 (dd, *J*=11.3, 4.8 Hz, H-26eq), 3.84 (dd, *J*=12.1, 3.9 Hz, H-1), 3.84 (m, *W*½=19.5 Hz, H-3), 4.01 (ddd, *J*=12.2, 10.5, 4.7 Hz, H-24), 4.54 (q-like, *J*=7.4 Hz, H-16), 4.73 (d, H-1 of Ara), 5.58 (br d, *J*=5.7 Hz, H-6), 6.34 (br s, H-1 of Rha).

CMR (C₅D₅N, 125 MHz) : δ C-1) 83.5 (2) 37.4 (3) 68.2 (4) 43.9 (5) 139.6 (6) 124.7 (7) 32.4 (8) 33.1 (9) 50.4 (10) 42.9 (11) 24.0 (12) 40.3 (13) 40.2 (14) 56.8 (15) 32.0 (16) 81.5 (17) 62.6 (18) 16.7 (19) 15.0 (20) 42.2 (21) 15.0 (22) 111.8 (23) 41.9 (24) 70.6 (25) 40.0 (26) 65.3 (27) 13.6 **Ara** (1) 100.3 (2) 75.1 (3) 75.9 (4) 70.1 (5) 67.3 **Rha** (1) 101.7 (2) 72.6 (3) 72.7 (4) 74.2 (5) 69.4 (6) 19.0.

Mass (FAB, Positive ion, H.R.) : *m/z* 747.3935 [(M+Na)⁺ requires 747.3932].

Biological Activity : Moderate cytotoxicity against HL-60 leukemia cells.

Reference

1.   M. Kuroda, Y. Mimaki, K. Ori, H. Sakagami and Y. Sashida, *J. Nat. Prod.*, **67**, 1690 (2004).

PARIS POLYPHYLLA SAPONIN-3
Pennogenin 3-O-[α-L-arabinofuranosyl-(1→4)-β-D-glucopyranoside]

Source : *Paris polyphylla* Sm. (Liliaceae)
Mol. Formula : C₃₈H₆₀O₁₃
Mol. Wt. : 724
M.P. : 220-223°C
[α]ᴅ : -98.0° (c=0.50, MeOH)
Registry No. : [81941 09 5]

IR (KBr) : 3600-3100 (OH), 980, 915, 898, 890 (intensity 898>915, 25*R*-spiroketal side chain) cm⁻¹.

Mass : *m/z* 430, 412, 394, 155, 153, 127, 126.

Reference

1.   M. Miyamura, K. Nakano, T. Nohara, T. Tomimatsu and T. Kawasaki, *Chem. Pharm. Bull.,* **30**, 712 (1982).

SMILAX LEBRUNII SAPONIN 1

(25R)-5α-Spirostan-3β-ol-6-one 3-O-[α-L-arabinopyranosyl-(1→4)-β-D-glucopyranoside]

Source : *Smilax lebrunii* Levl. (Liliaceae)
Mol. Formula : $C_{38}H_{60}O_{13}$
Mol. Wt. : 724
M.P. : 175-178°C
Registry No. : [143722-96-7]

IR (KBr) : 3429 (OH), 1710 (C=O), 1053 (C–O), 954, 919, 900 and 865 cm^{-1} (Intensity 919 <900, 25R-spirostanol).

PMR (C_5D_5N, 400 MHz) : δ 0.65 (s, 3xH-19), 0.70 (d, *J*=6.7 Hz, 3xH-27), 0.78 (s, 3xH-18), 1.15 (d, *J*=6.9 Hz, 3xH-21), 3.51 (dd, *J*=10.6, 106 Hz, H-26ax), 3.61 (dd, *J*=10.6, 3.5 Hz, H-26eq), 4.98 (d, *J*=6.9 Hz, H-1 of Glc), 5.00 (d, *J*=7.5 Hz, H-1 of Ara).

CMR (C_5D_5N, 100.16 MHz) : δ C-1) 36.4 (2) 29.2 (3) 76.7 (4) 26.9 (5) 56.4 (6) 209.5 (7) 46.7 (8) 37.4 (9) 53.6 (10) 41.1 (11) 21.5 (12) 39.5 (13) 40.8 (14) 56.4 (15) 32.1 (16) 80.7 (17) 62.8 (18) 16.5 (19) 13.0 (20) 41.9 (21) 14.9 (22) 109.2 (23) 32.1 (24) 29.2 (25) 30.5 (26) 66.3 (27) 17.2 **Glc** (1) 102.1 (2) 75.1 (3) 76.7 (4) 78.4 (5) 76.9 (6) 61.9 **Ara** (1) 105.3 (2) 72.2 (3) 74.3 (4) 69.0 (5) 66.8.

Mass (FAB) : *m/z* 747 [M+Na]$^+$, 731 [M+Li]$^+$.

Reference

1.   Y. Ju and Z.-J. Jia, *Phytochemistry*, **33**, 1193 (1993).

SMILAX LEBRUNII SAPONIN 2

(25S)-Spirost-5-en-3β,27-diol 3-O-[α-L-arabinopyranosyl-(1→6)-β-D-glucopyranoside]

Source : *Smilax lebrunii* Levl. (Liliaceae)
Mol. Formula : $C_{38}H_{60}O_{13}$
Mol. Wt. : 724
M.P. : 260-262°C
Registry No. : [143601-08-5]

IR (KBr) **:** 3480 (OH), 1649 (C=CH), 1052 (C–O), 915, 856, 781 cm^{-1}.

PMR (C_5D_5N, 400 MHz) **:** δ 0.83 (s, 3xH-18), 0.91 (s, 3xH-19), 1.15 (d, *J*=7.2 Hz, 3xH-21), 3.47 (dd, *J*=10.8, 10.8 Hz, H-26ax), 3.61 (dd, *J*=10.8, 3.5 Hz, H-26eq), 4.91 (d, *J*=6.7 Hz, H-1 of Glc), 4.94 (d, *J*=7.3 Hz, H-1 of Ara), 5.31 (m, H-6).

CMR (C_5D_5N, 100.16 MHz) **:** δ C-1) 37.5 (2) 30.3 (3) 76.8 (4) 39.2 (5) 141.0 (6) 121.6 (7) 32.2 (8) 31.7 (9) 50.2 (10) 37.0 (11) 21.1 (12) 39.2 (13) 40.5 (14) 56.6 (15) 32.2 (16) 81.1 (17) 62.9 (18) 16.3 (19) 19.4 (20) 42.1 (21) 15.0 (22) 109.6 (23) 31.6 (24) 24.0 (25) 37.6 (26) 64.1 (27) 64.4 **Glc** (1) 102.8 (2) 75.1 (3) 78.6 (4) 71.8 (5) 78.4 (6) 68.9 **Ara** (1) 105.1 (2) 72.2 (3) 74.2 (4) 69.4 (5) 66.2.

Mass (FAB) **:** *m/z* 747 [M+Na]$^+$, 731 [M+Li]$^+$.

Reference

1.   Y. Ju and Z.-J. Jia, *Phytochemistry*, **33**, 1193 (1993).

SMILAX SIEBOLDII SAPONIN 2, SMILAXIN A, LAXOSIDE

Laxogenin 3-O-[α-L-arabinopyranosyl-(1→6)-β-D-glucopyranoside]

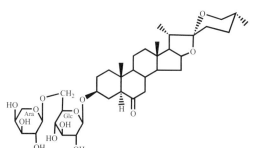

Source : *Smilax sieboldii* (Liliaceae)[1,2], *Smilax lebrunii* Lev.[3] (Liliaceae), *Allium chinense* G.Don[4] (Liliaceae)

Mol. Formula : $C_{38}H_{60}O_{13}$

Mol. Wt. : 724

M.P. : 263-265°C[2]

$[\alpha]_D^{30}$ **:** -98.7° (c=0.25, MeOH)[1]

Registry No. : [123941-70-8]

UV (MeOH)[1] : λ_{max} 282 (log ε, 108) nm.

IR (KBr)[1] : 3425 (OH), 2950 and 2890 (CH), 1710 (C=O), 1455, 1430, 1380, 1265, 1245, 1230, 1175, 1050, 1005, 980, 955, 920, 900, 865, 780 cm^{-1} [(25R)-spiroacetal, intensity 920 < 900].

PMR (C$_5$D$_5$N, 400/500 MHz)[1] : δ 0.65 (s, 3xH-19), 0.70 (d, J=5.7 Hz, 3xH-27), 0.78 (s, 3xH-18), 1.15 (d, J=6.9 Hz, 3xH-21), 2.00 (dd, J=12.4, 12.4 Hz, H-7ax), 2.19 (dd, J=12.0, 2.9 Hz, H-5), 2.35 (dd, J=12.4, 4.2 Hz, H-7eq), 3.49 (dd, J=10.5, 10.5 Hz, H-26A), 3.59 (dd, J=10.5, 3.7 Hz, H-26B), 4.53 (q-like, J=7.0 Hz, H-16), 4.94 (overlapping with H$_2$O signal, H-1 of Glc), 4.98 (d, J=6.6 Hz, H-1 of Ara).

CMR (C$_5$D$_5$N, 100/125 MHz)[1] : δ C-1) 36.8 (2) 29.6 (3) 76.8 (4) 27.0 (5) 56.4 (6) 209.1 (7) 46.7 (8) 37.3 (9) 53.6 (10) 40.8 (11) 21.5 (12) 39.6 (13) 41.1 (14) 56.4 (15) 31.8 (16) 80.8 (17) 62.8 (18) 16.4 (19) 13.1 (20) 42.0 (21) 15.0 (22) 109.2 (23) 31.8 (24) 29.2 (25) 30.6 (26) 66.9 (27) 17.3 **Glc** (1) 102.1 (2) 75.2 (3) 78.5 (4) 71.9 (5) 77.0 (6) 69.7 **Ara** (1) 105.5 (2) 72.3 (3) 74.5 (4) 69.1 (5) 66.5.

Mass (SIMS)[1] m/z : 769 [M+2Na-H]$^+$, 763 [M+K]$^+$, 747 [M+Na]$^+$.

Biological Activity : The compound shows inhibitory activity on cyclic AMP phosphodiesterase with IC$_{50}$=3.4x10^{-5} M.[1]

CD (MeOH; c=1.63x10^{-3})[1] : 293 nm (θ = -2515).

References

1.   S. Kubo, Y. Mimaki, Y. Sashida, T. Nikaido and T. Ohmoto, *Phytochemistry*, **31**, 2445 (1992).

2.   M.H. Woo, J.C. Do and K.H. Son, *J. Nat. Prod.*, **55**, 1129 (1992).

3.   Z. Jian Jia and Y. Ju, *Phytochemistry*, **31**, 3173 (1992).

4.   M. Kuroda, Y. Mimaki, A. Kameyama, Y. Sashida and T. Nikaido, *Phytochemistry*, **40,** 1071 (1995).

CAMASSIA CUSICKII SAPONIN 7

Chlorogenin 6-O-β-D-xylopyranosyl-(1→3)-β-D-glucopyranoside

Source : *Camassia cusickii* S. Wats. (Liliaceae)
Mol. Formula : $C_{38}H_{62}O_{13}$
Mol. Wt. : 726
$[\alpha]_D^{25}$: -15.6° (c=0.27, MeOH)
Registry No. : [141360-77-2]

IR (KBr) : 3410 (OH), 2930, 2860 (CH), 1445, 1370, 1235, 1165, 1150, 1075, 1035, 975, 950, 915, 895, 860 cm^{-1}; (25R)-spiroacetal, intensity 915 < 895).

PMR (C_5D_5N, 400 MHz) : δ 0.72 (d, J=5.3 Hz, 3xH-27), 0.80 (s, 3xH-18), 0.84 (s, 3xH-19), 1.13 (d, J=6.9 Hz, 3xH-21), 3.47 (dd, J=10.5, 10.5 Hz, H-26B), 3.57 (dd, J=10.5, 2.7 Hz, H-26A), 3.72 (ddd, J=9.8, 9.8, 3.5 Hz, H-6), 3.80 (m, H-3), 4.92 (d, J=7.9 Hz, H-1 of Glc), 5.27 (d, J=7.6 Hz, H-1 of Xyl).

CMR (C_5D_5N, 100 MHz) : δ C-1) 37.8 (2) 32.2^b (3) 71.0^c (4) 33.3 (5) 51.3 (6) 79.9 (7) 41.4 (8) 34.2 (9) 53.9 (10) 36.7 (11) 21.3 (12) 40.1 (13) 40.8 (14) 56.4 (15) 32.1^b (16) 81.0 (17) 63.0 (18) 16.7 (19) 13.6 (20) 42.0 (21) 15.0 (22) 109.1 (23) 31.9 (24) 29.3 (25) 30.6 (26) 66.9 (26) 17.4 **Glc** (1) 105.6 (2) 74.8 (3) 87.9 (4) 69.6 (5) 77.8 (6) 62.5 **Xyl** (1) 106.5 (2) 75.3 (3) 78.2 (4) 70.7^c (5) 67.4.

Mass (SIMS) : m/z 749 [M+Na]$^+$, 727 [M+H]$^+$, 595 [M-Pentose+H$_2$O+H]$^+$, 433 [Agl+H]$^+$, 415 [Agl-OH]$^+$.

Reference

1.   Y. Mimaki, Y. Sashida and K. Kawashima, *Chem. Pharm. Bull.*, **40**, 148 (1992).

CAMASSIA LEICHTLINII SAPONIN 7

(25R)-3β,6α-Dihydroxy-5α-spirostan 6-O-[β-D-xylopranosyl-(1→2)-β-D-glucopyranoside]

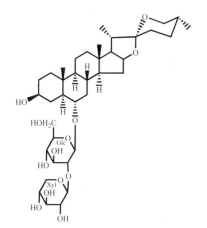

Source : *Camassia leichtlinii* (Liliaceae)
Mol. Formula : $C_{38}H_{62}O_{13}$
Mol. Wt. : 726
$[\alpha]_D^{25}$: -28.0° (c=0.10, MeOH)
Registry No. : [354575-47-6]

IR (film) : 3331 (OH), 2930 and 2874 (CH), 1451, 1376, 1243, 1171, 1074, 1052, 983, 956, 919, 897, 865 cm^{-1}.

PMR (C_5D_5N, 500 MHz) : δ 0.71 (d, *J*=5.5 Hz, 3xH-27), 0.81 (s, 3xH-18), 0.84 (s, 3xH-19), 1.13 (d, *J*=6.9 Hz, 3xH-21), 3.46 (dd, *J*=10.6, 10.6 Hz, H-26ax), 3.55 (dd, *J*=10.6, 2.9 Hz, H-26eq), 3.75 (ddd, *J*=10.9, 10.9, 4.5 Hz, H-6), 3.93 (br m, *W*½=22.3 Hz, H-3), 4.98 (d, *J*=7.7 Hz, H-1 of Glc), 5.34 (d, *J*=6.7 Hz, H-1 of Xyl).

CMR (C_5D_5N, 125 MHz) : δ C-1) 37.9 (2) 32.5 (3) 71.0 (4) 33.0 (5) 51.2 (6) 79.6 (7) 40.8 (8) 34.0 (9) 54.0 (10) 36.6 (11) 21.3 (12) 40.1 (13) 40.8 (14) 56.5 (15) 32.0 (16) 81.0 (17) 63.0 (18) 16.6 (19) 13.7 (20) 42.0 (21) 15.0 (22) 109.1 (23) 31.8 (24) 29.3 (25) 30.6 (26) 66.8 (27) 17.3 **Glc** (1) 103.4 (2) 83.7 (3) 78.5 (4) 71.5 (5) 77.9 (6) 62.7 **Xyl** (1) 106.6 (2) 76.2 (3) 77.8 (4) 71.0 (5) 67.5.

Mass (FAB, Positive ion) : *m/z* 749 [M+Na]$^+$.

Mass (FAB, Negative ion) : *m/z* 725 [M-H]$^-$, 593 [M-H-Xyl]$^-$.

Biological Activity : The compound showed cytotoxic activity against human oral squamous cell carcinoma (HSC-2) with LD_{50} value 120 µg/ml and normal human gingival fibroblasts (HGF) with LD_{50} value of 135 µg/ml.

Reference

1.   M. Kuroda, Y. Mimaki, F. Hasegawa, A. Yoksuka, Y. Sashida and H. Sakagami, *Chem. Pharm. Bull.,* **49**, 726 (2001).

PANICULONIN A

(5α,25S)-Spirostan-3β,6α,23α-triol 6-O-[β-D-xylopyranosyl-(1→3)-β-D-quinovopyranoside]

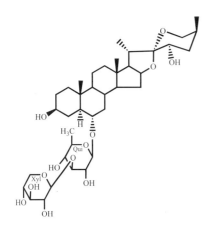

Source : *Solanum pariculatum* L.[1] , *S. hispidum* Pers.[2]
(Solanaceae)
Mol. Formula : $C_{38}H_{62}O_{13}$
Mol. Wt. : 726
M.P. : 264-265°C[2]
[α]$_D^{27}$: -61.2° (c=1.1, Pyridine)[1]
Registry No. : [20735-79-9]

PMR (C$_5$D$_5$N, 300 MHz)[2] : δ 0.81 (s, 3xH-18), 0.98 (s, 3xH-19), 1.12 (d, J=7.2 Hz, 3xH-21), 1.18 (d, J=6.9 Hz, 3xH-27), 1.58 (d, J=6.0 Hz, 3xH-6 of Qui), 3.29 (d, J=11.1 Hz, H-26β), 3.61 (m, H-4 of Qui), 3.71-3.75 (m, H-6, H-5 of Qui, H-5α of Xyl), 3.78 (m, H-3), 3.96-4.06 (m, H-23, H-2 of Qui, H-3 of Qui, H-2 of Xyl), 4.08-4.18 (m, H-26α, H-3 of Xyl, H-4 of Xyl), 4.29 (dd, J=11.1, 4.5 Hz, H-5β of Xyl), 4.57 (dd, J=6.9, 6.6 Hz, H-16), 4.77 (d, J=7.2 Hz, H-1 of Qui), 5.18 (d, J=7.8 Hz, H-1 of Xyl).

CMR (C$_5$D$_5$N, 125 MHz)[2] : δ C-1) 37.69 (2) 32.08 (3) 70.62 (4) 33.19 (5) 51.29 (6) 79.55 (7) 41.47 (8) 34.14 (9) 53.97 (10) 36.23 (11) 21.34 (12) 40.47 (13) 41.38 (14) 56.40 (15) 32.17 (16) 81.72 (17) 62.45 (18) 16.87 (19) 13.61 (20) 37.81 (21) 14.54 (22) 112.37 (23) 63.37 (24) 36.11 (25) 30.00 (26) 64.29 (27) 17.57 **Qui** (1) 105.26 (2) 74.75 (3) 87.74 (4) 75.30 (5) 72.31 (6) 18.59 **Xyl** (1) 106.49 (2) 74.82 (3) 78.22 (4) 70.91 (5) 67.43.

Mass (E.I.)[2] : *m/z* (rel.intens.) 726.3 [M$^+$, 1.2], 593.2 [(M-Xyl)$^+$, 0.5], 429 [(M-Xyl-Qui-H$_2$O)$^+$, 4], 396.3 [14], 283.2 [40], 161.1 [51], 133.1 [100], 107.1 [63].

Biological Activity : Antimycotic activity,2 weak hemolytic activity.

References

1.   H. Ripperger and K. Schreiber, *Chem. Ber.*, **101**, 2450 (1968).

2.   M. Gonzalez, A. Zamilpa, S. Marquina, V. Navarro and L. Alvarez, *J. Nat. Prod.*, **67**, 938 (2004).

TORVOSIDE D

Neosolaspigenin 6-O-[β-D-xylopyranosyl-(1→3)-β-D-quinovopyranoside]

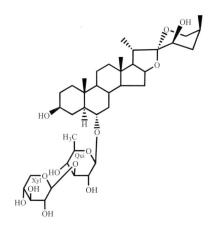

Source : *Solanum torvum* Swartz (Solanaceae)
Mol. Formula : $C_{38}H_{62}O_{13}$
Mol. Wt. : 726
M.P. : 262-264°C
$[\alpha]_D^{27}$: -30.7° (c=2.70, MeOH)
Registry No. : [20735-79-9]*

PMR (C_5D_5N, 400 MHz)[1] **:** δ 0.61 (br dd, *J*=9.3, 9.3 Hz, H-9), 0.81 (s, 3xH-18), 0.85 (s, 3xH-19), 0.96 (m, H-1), 1.06 (m, H-12), 1.06 (m, H-14), 1.20 (m, H-7), 1.36 (m, H-5), 1.42 (m, H-15), 1.48 (m, H-11), 1.52 (d, *J*=6.8 Hz, 3xH-21), 1.54 (d, *J*=7.3 Hz, 3xH-27), 1.58 (d, *J*=5.9 Hz, 3xH-6 of Qui), 1.60 (m, H-1), 1.62 (m, H-4), 1.64 (m, H-12), 1.69 (2H, m, H-2, H-25), 1.78 (2H, m, H-17, H-24), 2.02 (m, H-2), 2.07 (m, H-15), 2.33 (ddd, *J*=4.3, 4.3, 14.0 Hz, H-24), 2.53 (br d, *J*=12.7 Hz, H-7), 2.62 (m, H-20), 3.20 (br d, *J*=11.6 Hz, H-4), 3.54 (br d, *J*=9.2 Hz, H-26), 3.59 (dd, *J*=9.8, 9.8 Hz, H-4 of Qui), 3.69 (m, 3xH-5 of Qui, 3xH-6 of Qui, 3xH-5 of Xyl), 3.77 (m, H-3), 4.04 (4H, m, H-23, H-2, H-3 of Qui, H-2 of Xyl), 4.16 (m, 3xH-26, 3xH-3, 3xH-4 of Xyl), 4.30 (dd, *J*=4.9, 11.6 Hz, H-5 of Xyl), 4.63 (m, H-16), 4.82 (d, *J*=7.3 Hz, H-1 of Qui), 5.23 (d, *J*=7.3 Hz, H-1 of Xyl).

CMR (C_5D_5N, 100 MHz)[1] **:** δ C-1) 37.6 (2) 31.9 (3) 70.4 (4) 33.0 (5) 51.1 (6) 79.3 (7) 41.2 (8) 34.1 (9) 53.7 (10) 36.6 (11) 21.1 (12) 39.8 (13) 41.0 (14) 56.3 (15) 32.2 (16) 81.4 (17) 64.5 (18) 16.4 (19) 13.5 (20) 40.7 (21) 16.6 (22) 110.3 (23) 69.9 (24) 34.3 (25) 27.1 (26) 65.2 (27) 20.4 **Qui** (1) 105.1 (2) 74.7 (3) 87.4 (4) 75.4 (5) 72.1 (6) 18.5 **Xyl** (1) 106.2 (2) 75.1 (3) 78.0 (4) 70.7 (5) 67.2.

Mass (FAB, Positive ion)[1] **:** *m/z* 727 [M+H]⁺, 595 [M+H-Xyl]⁺, 449 [M+H-Xyl-Qui]⁺.

Reference

1.   S. Yahara, T. Yamashita, N. Nozawa (Nee Fujimura) and T. Nohara, *Phytochemistry*, **43**, 1069 (1996).

* This compound has been given the same registry number as Paniculonin A (qv) although it is a different compound.

ALLIUM GIGANTEUM SAPONIN 4
3-O-Benzoylalliogenin 2-O-β-D-glucopyranoside

Source : *Allium giganteum* Regel (Liliaceae)
Mol. Formula : $C_{40}H_{58}O_{12}$
Mol. Wt. : 730
$[\alpha]_D^{26}$ **:** -95.0° (c=0.1, MeOH)
Registry No. : [134955-78-5]

UV (MeOH) : λ_{max} 230 (ε, 12460), 266 sh (ε, 1800), 273 (ε, 2000), 280 (sh, ε, 1660) nm.

IR (KBr) : 3450 (OH), 2930 (CH), 1700 (carbonyl), 1600 (aromatic ring), 970, 920, 890, 860 cm^{-1} (25R-spiroketal, intensity 920 < 890), 710 (aromatic ring).

PMR (C_5D_5N, 400 or 500 MHz) : δ 0.68 (d, *J*=5.5 Hz, 3xH-27), 0.90 (s, 3xH-18), 1.13 (d, *J*=6.8 Hz, 3xH-21), 1.59 (s, 1.59 (s, 3xH-19), 2.90 (dd, *J*=12.2, 12.2 Hz, H-4ax), 3.48 (dd, *J*=10.7, 10.7 Hz, H-26A), 3.57 (dd, *J*=10.7, 3.2 Hz, H-26B), 4.41 (dd, *J*=11.4, 2.0 Hz, H-6A of Glc), 4.58 (m, H-16), 5.22 (d, *J*=7.7 Hz, H-1 of Glc), 6.31 (m, H-3), 7.44 (H-3, H-4 and H-5 of Benz), 8.44 (H-2 and H-6 of Benz).

CMR (C_5D_5N, 100 or 125 MHz) : δ C-1) 37.9^a (2) 77.7 (3) 76.3 (4) 38.9^a (5) 75.0 (6) 75.0 (7) 35.8 (8) 30.1 (9) 45.7 (10) 40.5 (11) 21.6 (12) 40.4 (13) 41.0 (14) 56.2 (15) 32.3 (16) 81.2 (17) 63.2 (18) 18.0 (19) 16.7 (20) 42.1 (21) 15.0 (22) 109.2 (23) 31.8 (24) 29.3 (25) 30.6 (26) 66.9 (27) 17.3 **Glc** (1) 103.3 (2) 75.3 (3) 78.2^b (4) 71.8 (5) 78.4^b (6) 63.0 **Benz** (1) 31.9 (2) 130.3 (3) 128.7 (4) 132.9 (5) 128.7 (6) 130.3 (7) 166.8.

Mass (S.I.) : *m/z* 731 [M+H]$^+$.

Reference

1.   Y. Shashida, K. Kawashima and Y. Mimaki, *Chem. Pharm. Bull.*, **39**, 698 (1991).

ALLIUM KARATAVIENSE SAPONIN 1
Karatavigenin 3-O-β-D-glucopyranoside

Source : *Allium karataviense* Rgl. (Liliaceae, Alliaceae)
Mol. Formula : $C_{40}H_{58}O_{12}$
Mol. Wt. : 730
M.P. : 294-296°C
[α]$_D^{20}$: -98.1±2° (c=1.64, C_5D_5N)
Registry No. : [54526-91-9]

UV (MeOH) : λ_{max} 230 (log ε, 4.20) nm.

IR (KBr) : 3400-3580 (OH), 1710, 1280 (ester group), 870, 900>930, 995 (spiroskatal chain, *R*-series) [2,3], 1605, 1585 and 720 (benzene ring) cm^{-1}.

Reference

1.   F.S. Khristulas, M.B. Gorovits and N.K. Abubakirov, *Khim. Prir. Soedin.*, 530 (1974); *Chem. Nat. Comp.*, **10**, 544 (1974).

KINGIANOSIDE B

Gentrogenin 3-O-[β-D-glucopyranosyl (1→4)-β-D-fucopyranoside]

Source : *Polygonatum kingianum* Coll. et Hemsl. (Liliaceae)
Mol. Formula : $C_{39}H_{60}O_{13}$
Mol. Wt. : 736
M.P. : 271-274°C
[α]$_D^{20}$: -46.0° (c=0.87, MeOH)
Registry No. : [145867-18-1]

PMR (C$_5$D$_5$N, 270 MHz) : δ 0.70 (d, *J*=5.5 Hz, 3xH-27), 0.94 (s, 3xH-19), 1.11 (s, 3xH-18), 1.36 (d, *J*=7.0 Hz, 3xH-21), 1.63 (d, *J*=6.6 Hz, 3xH-6 of Fuc), 4.81 (d, *J*=7.7 Hz, H-1 of Fuc), 5.23 (d, *J*=7.7 Hz, H-1 of Glc), 5.33 (br s, H-6).

CMR (C$_5$D$_5$N, 75.5 MHz) : δ C-1) 37.0 (2) 30.0 (3) 77.7 (4) 39.1 (5) 140.9 (6) 121.4 (7) 31.8^a (8) 30.9 (9) 52.3 (10) 37.6 (11) 37.5 (12) 212.6 (13) 55.0 (14) 56.0 (15) 31.6^a (16) 79.7 (17) 54.1 (18) 15.9 (19) 18.8 (20) 42.7 (21) 13.9 (22) 109.4 (23) 31.8^a (24) 29.2 (25) 30.6 (26) 67.0 (27) 17.3 **Fuc** (1) 102.7 (2) 73.0 (3) 76.2^b (4) 83.3 (5) 70.6 (6) 17.6 **Glc** (1) 107.0 (2) 75.6^b (3) 78.6 (4) 71.7 (5) 78.6 (6) 62.9.

Mass (FAB, Negative ion, H.R.) : *m/z* 735.4000 [(M-H)⁻, calcd. 735.3956].

Reference

1.   X.-C. Li, C.-R. Yang, M. Ichikawa, H. Matsuura, R. Kasai and K. Yamasaki, *Phytochemistry,* **31**, 3559 (1992).

SOLANUM LYRATUM SAPONIN 1

(25R,S)-3β-Hydroxyspirost-5-ene 3-O-[α-L-rhamnopyranosyl-(1→2)-β-D-glucuronopyranoside]

25R-epimer

Source : *Solanum lyratum* Thunb. (Solanaceae)
Mol. Formula : $C_{39}H_{60}O_{13}$
Mol. Wt. : 736
M.P. : 283-284°C
[α]$_D^{20}$: -96.8° (c=1.00, Pyridine)
Registry No. : [107783-52-8]

IR (KBr) : 3400, 1600, 980, 917, 895 cm^{-1}.

CMR (C_5D_5N, 68.0 MHz) : δ C-1) 37.5 (2) 30.0 (3) 76.4 (4) 38.8 (5) 140.9 (6) 121.8 (7) 32.5 (8) 32.1 (9) 50.3 (10) 37.1 (11) 21.1 (12) 39.8 (13) 40.5 (14) 56.7 (15) 32.3 (16) 81.2 (17) 62.9 (62.7) (18) 16.4 (19) 19.5 (20) 42.0 (42.5) (21) 15.1 (22) 109.3 (109.7) (23) 32.1 (26.4) (24) 29.3 (26.2) (25) 30.6 (27.5) (26) 66.9 (65.1) (27) 17.4 (16.4) **Glc** (1) 99.8 (2) 79.0 (3) 77.5 (4) 73.7 (5) 78.0 (6) 176.2 **Rha** (1) 102.1 (2) 72.7 (3) 72.3 (4) 74.1 (5) 69.5 (6) 18.7. The values in parenthesis are of 25S-epimer.

Mass (F.D.) : *m/e* 775 [M+K]$^+$, 414, 147.

Reference

1.   S. Yahara, M. Morooka, M. Ikeda, M. Yarrasaki and T. Nohara, *Planta Med.*, **52,** 496 (1986).

ALLIUM VINEALE SAPONIN 1

Isonuatigenin 3-O-[α-L-rhamnopyranosyl-(1→2)-β-L-glucopyranoside]

Source : *Allium vineale* (Liliaceae)
Mol. Formula : $C_{39}H_{62}O_{13}$
Mol. Wt. : 738
[α]$_D$: -64.3° (c=0.25, Pyridine)
Registry No. : [113567-55-8]

PMR (C_5D_5N-D_2O, 400 MHz) : δ 0.82 (CH_3), 1.03 (CH_3), 1.17 (d, *J*=6.8 Hz, 3xH-27), 1.22 (CH_3), 1.75 (d, *J*=6.2 Hz, 3xH-6 of Rha), 3.90* (H-5 of Glc), 4.13 (t, *J*=9.4 Hz, H-4 of Glc), 4.25* (s, H-2 of Glc), 4.25* (s, H-2 and H-3 of Glc), 4.29* (s, H-6A of Glc), 4.34 (t, *J*=9.2 Hz, H-4 of Rha), 4.48 (d, *J*=11.4 Hz, H-6B of Glc), 4.60 (dd, *J*=9.2, 3.2 Hz, H-3 of Rha), 4.79 (br, H-2 of Rha), 4.96* (H-5 of Rha), 5.01 (d, *J*=7.3 Hz, H-1 of Glc), 5.13 (d, *J*=3.2 Hz, H-6), 6.37 (H-1 of Rha). * overlapped signal.

CMR (C_5D_5N-D_2O, 100 MHz) : δ C-1) 37.8 (2) 30.2 (3) 78.3 (4) 39.0 (5) 140.8 (6) 121.7 (7) 32.3 (8) 31.7 (9) 50.4 (10) 37.0 (11) 21.2 (12) 40.0 (13) 40.5 (14) 56.8 (15) 32.2 (16) 81.4 (17) 63.0 (18) 16.4 (19) 19.6 (20) 42.0 (21) 15.2 (22) 109.6 (23) 27.9 (24) 33.8 (25) 65.9 (26) 69.8 (27) 27.0 **Glc** (1) 100.3 (2) 79.6 (3) 77.9 (4) 71.8 (5) 77.8 (6) 62.6 **Rha** (1) 102.1 (2) 72.6 (3) 72.8 (4) 74.2 (5) 69.5 (6) 18.7.

Mass (FAB, Negative ion) : *m/z* 737.41185 [(M-1)$^-$, calcd. 737.41122].

Reference

1.   S. Chen and J.K. Snyder, *J. Org. Chem.*, **54**, 3679 (1989).

DEACYLBROWNIOSIDE

(25*S*)-Spirost-5-en-3β,27-diol 3-O-α-L-rhamnopyranosyl-(1→2)-β-D-glucopyranoside

Source : *Lilium brownii* var. *colchesteri* (Liliaceae)[1]
Mol. Formula : $C_{39}H_{62}O_{13}$
Mol. Wt. : 738
[α]$_D^{26}$: -95.7° (c=0.28, MeOH)[2]
M.P. : 232-238°C[1]
Registry No. : [129744-09-8]

IR (KBr)[2] : 3410 (OH), 2430, 2895 (CH), 1445, 1370, 1160, 1130, 1040, 995, 965, 950, 905, 805 cm^{-1}.

PMR (C$_5$D$_5$N, 400 MHz)[2] : δ 0.84 (s, 3xH-18), 1.07 (s, 3xH-19), 1.17 (s, d, *J*=6.9 Hz, 3xH-21), 1.78 (d, *J*=6.2 Hz, 3xH-6 of Rha), 3.66 (dd, *J*=10.5, 7.4 Hz, H-27A), 3.74 (dd, *J*=10.5, 5.0 Hz, H-27B), 4.57 (m, H-16), 4.64 (dd, *J*=9.3, 3.2 Hz, H-3 of Rha), 4.81 (br d, *J*=3.2 Hz, H-2 of Rha), 5.01 (dq, *J*=9.5, 6.2 Hz, H-5 of Rha), 5.05 (d, *J*=7.3 Hz, H-1 of Glc), 5.31 (br d, *J*=4.4 Hz, H-6), 6.39 (br s, H-1 of Rha).

CMR (C$_5$D$_5$N, 100/125 MHz)[1,2] : δ C-1) 37.5 (2) 30.2 (3) 78.3 (4) 39.0 (5) 140.9 (6) 121.7 (7) 32.3 (8) 31.7 (9) 50.3 (10) 37.2 (11) 21.1 (12) 39.9 (13) 40.5 (14) 56.7 (15) 32.3 (16) 81.2 (17) 63.0 (18) 16.3 (19) 19.4 (20) 42.1 (21) 15.1 (22) 109.7 (23) 31.6 (24) 24.1 (25) 39.2 (26) 64.1^a (27) 64.4^a **Glc** (1) 100.4 (2) 79.7 (3) 77.9^b (4) 71.9 (5) 78.0^b (6) 62.7 **Rha** (1) 102.1 (2) 72.6 (3) 72.9 (4) 74.2 (5) 69.5 (6) 18.7.

Mass (S.I.)[2] : *m/z* 739 [M+H]$^+$, 429 [Agl-H]$^+$, 411 [Agl-H-H$_2$O]$^+$.

References

1.   Y. Mimaki and Y. Sashida, *Chem. Pharm. Bull.,* **38**, 3055 (1990).

2.   Y. Mimaki and Y. Sashida, *Phytochemistry*, **29**, 2267 (1990).

DIOSCOREA PRAZERI GLYCOSIDE B

Prazerigenin A 3-O-α-L-rhamnopyranosyl-(1→6)-β-D-glucopyranoside

Source : *Dioscorea prazeri* Prain et Burk.
(Dioscoreaceae)
Mol. Formula : $C_{39}H_{62}O_{13}$
Mol. Wt. : 738
M.P. : 273-276°C[1]; 270-271°C[2]
[α]$_D$: -87.1° (c=0.85, Pyridine)[1]; -95° (c=0.5, Pyridine)[2]
Registry No. : [63347-41-1]

IR (KBr)[2] : 3509, 1054, 980, 917, 899, 868, 837 cm^{-1}.

References

1.   K. Rajaraman, V. Seshadri and S. Rangaswami, *Ind. J. Chem.*, **14B**, 735 (1976).

2.   M. Wij, K. Rajaraman and S. Rangaswami, *Ind. J. Chem.*, **15B**, 451 (1977).

FISTULOSIDE A

Yuccagenin 3-O-[α-L-rhmnopyranosyl-(1→2)-β-D-galactopyranoside]

Source : *Allium fistulosum* L. (Liliaceae)
Mol. Formula : $C_{39}H_{62}O_{13}$
Mol. Wt. : 738
M.P. : 238-240°C
$[\alpha]_D^{21}$: -64.0° (c=0.08, MeOH)
Registry No. : [139606-31-8]

IR (KBr) : 3416, 1052, 982, 920, 900, 866 cm^{-1} (intensity of absorption 900>920, 25(*R*)-spiroketal).

PMR (C$_5$D$_5$N, 300 MHz) : δ 0.71 (d, *J*=5.5 Hz, 3xH-27), 0.82 (s, 3xH-18), 0.96 (s, 3xH-19), 1.13 (d, *J*=6.8 Hz, 3xH-21),1.61 (d, *J*=6.2 Hz, 3xH-6 of Rha), 4.97 (d, *J*=7.8 Hz, anomeric H), 5.32 (br d, *J*=4.8 Hz, H-6), 6.30 (s, anomeric H).

CMR (C$_5$D$_5$N, 75.5 MHz) : δ C-1) 45.9 (2) 70.3 (3) 85.7 (4) 37.4 (5) 140.1 (6) 121.9 (7) 32.1 (8) 31.2 (9) 50.2 (10) 37.9 (11) 21.2 (12) 39.8 (13) 40.5 (14) 56.5 (15) 32.2 (16) 81.1 (17) 62.9 (18) 16.3 (19) 20.4 (20) 42.0 (21) 15.0 (22) 109.2 (23) 31.6 (24) 29.3 (25) 30.6 (26) 66.7 (27) 17.3 **Gal** (1) 102.0 (2) 76.5 (3) 75.7 (4) 70.7 (5) 76.9 (6) 62.1 **Rha** (1) 102.0 (2) 72.5^a (3) 72.8^a (4) 74.1 (5) 69.4 (6) 18.5.

Mass (FAB, Positive ion) : *m/z* (rel.intens.) 761 [(M+Na)$^+$, 5.2], 431 [(genin+H)$^+$, 2.5].

Reference

1.   J.C. Do, K.Y. Jung and K.H. Son, *J. Nat. Prod.,* **55**, 168 (1992).

FLORIBUNDASAPONIN B
Pennogenin 3-O-[α-L-rhamnopyranosyl-(1→4)-β-D-glucopyranoside]

Source : *Dioscorea floribunda* Mart & Gal.
(Dioscoreaceae)
Mol. Formula : $C_{39}H_{62}O_{13}$
Mol. Wt. : 738
M.P. : 251-253°C (decomp.)
[α]$_D$: -86.5° (Pyridine)
Registry No. : [80666-66-6]

IR (KBr) : 3600-3200 (br, OH), 980, 920, 900, 800 (900>920) cm^{-1}.

Reference

1.   S.B. Mahato, N. P. Sahu and A.N. Ganguly, *Phytochemistry*, **20**, 1943 (1981).

FUNKIOSIDE C, CAPSICOSIDE B₃

Diosgenin 3-O-[β-D-glucopyranosyl-(1→4)-β-D-galactopyranoside]

Source : *Funkia ovata* Spr.[1] Syn. *Hosta caerulea* (Liliaceae)[1], *Polygonatum kingianum* Coll. et. Hemsl.[2] (Liliaceae), *Solanum unguiculatum* (A.) Rich.[3] (Solanaceae)

Mol. Formula : $C_{39}H_{62}O_{13}$

Mol. Wt. : 738

M.P. : 247-249° (decomp.)[3]

$[\alpha]_D^{20}$: -60.2° (c=0.59, MeOH)

Registry No. : [60454-77-5]

PMR $(C_5D_5N, 270 \text{ MHz})^2$: δ 0.70 (d, J=5.5 Hz, 3xH-27), 0.84 (s, 3xH-18), 0.89 (s, 3xH-19), 1.15 (d, J=6.9 Hz, 3xH-21), 4.92 (J=7.7 Hz, H-1 of Gal) 5.31 (br s, H-6), 5.33 (d, J=7.7 Hz, H-1 of Glc).

CMR $(C_5D_5N, 75 \text{ MHz})^2$: δ C-1) 37.5 (2) 30.3 (3) 78.1 (4) 39.3 (5) 141.0 (6) 121.7 (7) 32.3 (8) 31.7 (9) 50.3 (10) 37.1 (11) 21.1 (12) 39.9 (13) 40.5 (14) 56.7 (15) 32.2 (16) 81.1 (17) 62.9 (18) 16.4 (19) 19.4 (20) 42.0 (21) 15.0 (22) 109.3 (23) 31.8 (24) 29.3 (25) 30.6 (26) 66.9 (27) 17.3 47.3 **Gal** (1) 102.9 (2) 73.4 (3) 75.4[a] (4) 79.9 (5) 75.9[a] (6) 61.0 **Glc** (1) 107.0 (2) 75.2 (3) 78.7 (4) 72.3 (5) 78.4 (6) 63.1.

Mass (FAB, Negative ion)[2] : *m/z* 737 [M-H]⁻, 575 [M-Glc-H]⁻.

Biological Activity : The compound shows inhibitory effect on human spermatozoa *in vitro* at a concentration of 1 mg/ml.[4]

References

1. P.K. Kintya, N.E. Mashchaenko, V.I. Kononova and G.V. Lazurevskii, *Khim. Prir. Soedin.* **267** (1976), *Chem. Nat. Comp.*, **12**, 241 (1976).

2. X.-C. Li, C.-R. Yang, M. Ichikawa, H. Matsuura, R. Kasai and K. Yamasaki, *Phytochemistry*, **31**, 3559 (1992).

3. F.A. Abbas, *Sci. Pharm.*, **69**, 219 (2001).

4. Y.-F. Wang, X.-C. Li, H.-Y. Yang, J.-J. Wang and C.-R. Yang, *Planta Med.*, **62**, 130 (1996).

LIRIOPE MUSCARI SAPONIN Lm-2

Ruscogenin 1-O-[β-D-glucopyranosyl-(1→2)-β-D-fucopyranoside]

Source : *Liriope muscari* (Liliaceae)
Mol. Formula : $C_{39}H_{62}O_{13}$
Mol. Wt. : 738
M.P. : 194-196°C (decomp.)
[α]$_D$: -93.6° (c=0.59, C_5D_5N)
Registry No. : [130431-13-9]

IR (KBr) : 3600-3200 (OH), 982, 920, 902, 865 cm^{-1} (intensity 920>902, (25R)-spiroketal).

CMR (C_5D_5N, 100 MHz) : δ C-1) 83.5 (2) 37.9 (3) 68.7 (4) 44.0 (5) 139.9 (6) 124.4 (7) 33.6 (8) 32.6 (9) 51.0 (10) 43.5 (11) 24.3 (12) 41.0 (13) 40.8 (14) 57.5 (15) 33.0 (16) 81.5 (17) 63.6 (18) 17.3 (19) 15.5 (20) 42.6 (21) 15.7 (22) 109.4 (23) 32.4 (24) 29.9 (25) 31.2 (26) 67.3 (27) 17.7 **Fuc** (1) 100.1 (2) 82.6 (3) 75.3 (4) 72.4 (5) 71.3 (6) 17.7 **Glc** (1) 106.5 (2) 77.0 (3) 78.8 (4) 72.1 (5) 78.3 (6) 63.3.

Reference

1. B-Y. Yu, Y. Hirai, J. Shoji and G-J. Xu, *Chem. Pharm. Bull.*, **38**, 1931 (1990).

LIRIOPE PLATYPHYLLA SAPONIN D
Ruscogenin 3-O-[β-D-glucopyranosyl-(1→3)-α-L-rhamnopyranoside]

Source : *Liriope platyphylla* Wang. et Tang. (Liliaceae)
Mol. Formula : $C_{39}H_{62}O_{13}$
Mol. Wt. : 738
M.P. : 293-295°C (decomp.)
$[\alpha]_D^{19}$: -92.0° (c=0.87, C_5D_5N)
Registry No. : [87425-36-3]

IR (KBr) : 3600-3200 (OH), 982, 920, 902, 865 cm^{-1} (intensity 920>902, 25(*R*)-spiroketal).

PMR (C_5D_5N, 100 MHz) : δ 0.70 (br d, sec. C*H₃*), 0.91 (s, tert. C*H₃*), 1.10 (d, *J*=6.0 Hz, sec. C*H₃*), 1.26 (s, tert. C*H₃*), 1.54 (d, *J*=6.0 Hz, sec. C*H₃*).

CMR (C_5D_5N, 25.0 MHz) : δ C-1) 77.9 (2) 41.0 (3) 73.9 (4) 39.6 (5) 139.3 (6) 123.1 (7) 33.1 (8) 32.4 (9) 51.4 (10) 43.8 (11) 24.3 (12) 40.7 (13) 40.4 (14) 57.0 (15) 32.5 (16) 81.2 (17) 63.4 (18) 16.7 (19) 13.8 (20) 42.2 (21) 15.0 (22) 109.4 (23) 32.0 (24) 29.4 (25) 30.7 (26) 67.0 (27) 17.4 **Rha** (1) 100.0 (2) 71.7 (3) 83.7 (4) 73.0 (5) 69.7 (6) 18.4 **Glc** (1) 106.4 (2) 75.9 (3) 78.4^a (4) 72.1 (5) 78.3^a (6) 62.7.

Reference

1.   Y. Watanabe, S. Sanada, Y. Ida and J. Shoji, *Chem. Pharm. Bull.*, **31**, 1980 (1983).

MELONGOSIDE F
Diosgenin 3-O-[β-D-glucopyranosyl-(1→2)-β-D-glucopyranoside]

Source : *Solanum melongena* L. (Solanaceae)
Mol. Formula : $C_{39}H_{62}O_{13}$
Mol. Wt. : 738
M.P. : 233°C
$[\alpha]_D^{20}$ **:** -55.0° (c=1.0, CH_3OH)
Registry No. : [94805-85-3]

Reference

1.   P.K. Kintya, and S.A. Shvets, *Khim. Prir. Soedin.*, 610 (1984); *Chem. Nat. Comp.*, **20**, 575 (1984).

PINGPEISAPONIN
24α-Hydroxydiosgenin 3-O-[α-L-rhamnopyranopy-(1→2)-β-D-glucopyranoside]

Source : *Fritillaria ussuriensis* Maxim. (Liliaceae)
Mol. Formula : $C_{39}H_{62}O_{13}$
Mol. Wt. : 738
M.P. : 239.5-240.5°C
$[\alpha]_D^{25}$: -72.6° (c=0.15, MeOH)
Registry No. : [126453-84-7]

IR (KBr) : 3370, 2930, 2840, 1448, 1375, 1335, 1128, 1045, 1032, 920, 900, 867, 840, 810 cm^{-1}. Intensity 920<900 (25R-spiroketal).

CMR (C$_5$D$_5$N) : δ C-1) 37.5 (2) 30.1 (3) 77.9 (4) 39.1 (5) 140.2 (6) 121.6 (7) 32.2 (8) 31.6 (9) 50.2 (10) 37.1 (11) 21.1 (12) 39.8 (13) 40.4 (14) 56.5 (15) 32.3 (16) 81.1 (17) 62.8 (18) 16.3 (19) 19.4 (20) 42.0 (21) 15.0 (22) 109.6 (23) 38.9 (24) 64.3 (25) 31.7 (26) 64.0 (27) 24.0 **Glc** (1) 100.3 (2) 79.5 (3) 78.1 (4) 71.7 (5) 77.9 (6) 62.6 **Rha** (1) 101.9 (2) 72.4 (3) 72.7 (4) 74.0 (5) 69.4 (6) 18.3.

Mass (F.D.) : *m/z* 722, 430, 414, 412, 282, 271, 253, 150 (100%) 131, 85, 69, 43.

Note : molecular ion should have been at *m/z* 738.

Reference

1.   D-Xu, S.Q. Wang, E.X. Huang and M. L. Xu, *Acta. Bot. Sin.*, **31**, 285 (1989).

RUSCUS ACULEATUS SAPONIN 14

Ruscogenin 1-O-{α-L-rhamnopyranosyl-(1→2)-β-D-galactopyranoside}

Source : *Ruscus aculeatus* L. (Liliaceae)
Mol. Formula : $C_{39}H_{62}O_{13}$
Mol. Wt. : 738
$[\alpha]_D^{25}$ **:** -74.0° (c=0.10, MeOH)
Registry No. : [211059-91-5]

IR (KBr) **:** 3410 (OH), 2935 (CH), 1445, 1375, 1235, 1130, 1050, 980, 960, 915, 895, 860, 830, 805 cm⁻¹.

PMR (C_5D_5N, 400 MHz) **:** δ 0.72 (d, *J*=5.2 Hz, 3xH-27), 0.90 (s, 3xH-18), 1.08 (d, *J*=6.9 Hz, 3xH-21), 1.48 (s, 3xH-19), 1.80 (d, *J*=6.1 Hz, 3xH-6 of Rha), 3.52 (dd, *J*=10.4, 10.4 Hz, H-26A), 3.61 (dd, *J*=10.4, 2.4 Hz, H-26B), 3.68 (m, H-3), 3.88 (dd, *J*=11.6, 4.0 Hz, H-1), 3.99 (br dd, H-5 of Gal), 4.24 (dd, *J*=9.3, 9.3 Hz, H-3 of Gal), 4.38 (dd, *J*=9.4, 9.4 Hz, H-4 of Rha), 4.40 (dd, *J*=10.3, 5.7 Hz, H-6A of Gal), 4.50 (q-like, *J*=7.5 Hz, H-16), 4.54 (br d, *J*=3.3 Hz, H-4 of Gal), 4.55 (dd, *J*=10.3, 6.9 Hz, H-6B of Gal), 4.70 (dd, *J*=9.3, 7.8 Hz, H-2 of Gal), 4.71 (dd, *J*=9.4, 3.5 Hz, H-3 of Rha), 4.81 (*J*=3.5 Hz, H-2 of Rha), 4.82 (d, *J*=7.8 Hz, H-1 of Gal), 4.97 (dq, *J*=9.4, 6.1 Hz, H-5 of Rha), 5.60 (br d, *J*=5.5 Hz, H-6), 6.45 (br s, H-1 of Rha).

CMR (C_5D_5N, 100 MHz) **:** δ C-1) 84.2 (2) 38.0 (3) 68.1 (4) 43.8 (5) 139.5 (6) 124.7 (7) 31.9 (8) 33.0 (9) 50.5 (10) 42.8 (11) 24.0 (12) 40.4 (13) 40.1 (14) 57.1 (15) 32.4 (16) 81.1 (17) 62.9 (18) 16.8 (19) 15.0 (20) 41.9 (21) 14.9 (22) 109.2 (23) 31.7 (24) 29.2 (25) 30.6 (26) 66.8 (27) 17.2 **Gal** (1) 100.6 (2) 74.9 (3) 76.8 (4) 70.4 (5) 76.3 (6) 61.9 **Rha** (1) 101.7 (2) 72.6 (3) 72.6 (4) 74.2 (5) 69.3 (6) 19.0.

Mass (FAB, Negative ion) **:** *m/z* 737 [M-H]⁻, 591 [M-Rha]⁻.

Reference

1.   Y. Mimaki, M. Kuroda, A. Kameyama, A. Yokosuka and Y. Sashida, *Phytochemistry*, **48**, 485 (1998).

SOLANUM UNGUICULATUM SAPONIN 1
Laxogenin 3-O-[α-L-rhamonopyranosyl-(1→2)-β-D-glucopyranoside]

Source : *Solanum unguiculatum* (A.) Rich. (Solanaceae)
Mol. Formula : $C_{39}H_{62}O_{13}$
Mol. Wt. : 738
[α]$_D$: -90° (c=0.4, MeOH)
Registry No. : [384343-15-1]

UV (MeOH) : λ_{max} 284 nm.

IR (KBr) : 3440 (OH), 2940, 2870 (CH), 1705 (C=O), 1455, 1375, 1260, 1175, 1045, 975, 895, 800, 700 cm^{-1} (intensity 915<895, 25R-spiroketal).

PMR (C$_5$D$_5$N, 300 MHz) : δ 0.64 (s, 3xH-19), 0.70 (d, J=5.8 Hz, 3xH-27), 0.79 (s, 3xH-18), 1.15 (d. J=6.9 Hz, 3xH-21), 2.00 (dd, J=12.7, 12.7 Hz, H-7ax), 2.16 (dd, J=12.6, 2.4 Hz, H-5), 2.37 (dd, J=12.7, 4.1 Hz, H-7eq), 3.49 (dd, J=10.6, 10.6 Hz, H-26A), 3.59 (dd, J=10.6, 3.6 Hz, H-26B), 3.88 (dd, J=9.3, 7.8 Hz, H-2 of Glc), 3.96 (m, H-3 and H-5 of Glc), 4.22 (overlapped, H-3 of Glc), 4.42 (dd, J=9.5, 9.5 Hz, H-4 of Rha), 4.45 (dd, J=9.3, 9.3 Hz, H-4 of Glc), 4.53 (q-like, J=7.0 Hz, H-16), 4.64 (dd, J=9.5, 3.4 Hz, H-3 of Rha), 4.68 (dd, J=10.8, 3.6 Hz, H-6A of Glc), 4.76 (br d, J=3.4 Hz, H-2 of Rha), 4.83 (br d, J=10.4 Hz, H-6B of Glc), 4.93 (d, J=7.8 Hz, H-1 of Glc), 4.96 (dq, J=9.5, 6.2 Hz, H-5 of Rha), 6.33 (br s, H-1 of Rha).

CMR (C$_5$D$_5$N, 75.0 MHz) : δ C-1) 36.7 (2) 37.4 (3) 77.0 (4) 27.0 (5) 56.4 (6) 209.6 (7) 46.7 (8) 37.4 (9) 53.7 (10) 40.8 (11) 21.5 (12) 39.6 (13) 41.1 (14) 56.5 (15) 31.8 (16) 80.0 (17) 62.8 (18) 16.5 (19) 13.3 (20) 42.0 (21) 15.0 (22) 109.3 (23) 31.8 (24) 29.4 (25) 30.7 (26) 66.7 (27) 17.2 **Glc** (1) 99.6 (2) 79.8 (3) 78.4 (4) 72.0 (5) 78.2 (6) 62.9 **Rha** (1) 102.2 (2) 72.6 (3) 72.9 (4) 74.2 (5) 69.5 (6) 18.7.

Mass (FAB, Positive ion) : *m/z* 777 [M+K]⁺, 762 [M+Na+H]⁺, 721 [M-OH]⁺.

Reference

1.   F.A. Abbas, *Sci. Pharm.*, **69**, 219 (2001).

TRILLIUM KAMTSCHATICUM SAPONIN Tb
Pennogenin 3-O-[α-L-rhamnopyranosyl-(1→2)-β-D-glucopyranoside]

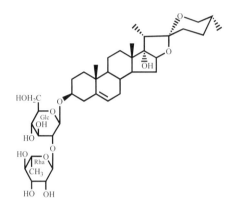

Source : *Trillium kamtschaticum* Pall.[1] (Liliaceae), *Paris polyphylla* Sm. Var. yunnanensis (Fr.) H-M.[2] (Liliaceae), *Polygonatum stenophyllum* Maxim[3] (Liliaceae) etc.
Mol. Formula : $C_{39}H_{62}O_{13}$
Mol. Wt. : 738
M.P. : 273-276°C (decomp.)[1]
[α]ᴅ : -117.8° (c=0.87, Pyridine)[1]
[M]ₓ : -870°
Registry No. : [55916-51-3]

IR (KBr)[1] **:** 3600-3200 (OH), 980, 920, 900, 890 cm⁻¹. Intensity 900 > 920 (25*R*-spiroketal) cm⁻¹.

CMR (C_5D_5N, 22.5 MHz)[4] : δ C-1) 37.5 (2) 30.0 (3) 77.9 (4) 38.9 (5) 140.8 (6) 121.6 **Glc** (1) 100.3 (2) 79.3 (3) 77.9 (4) 71.7 (5) 77.9 (6) 62.6 **Rha** (1) 101.6 (2) 72.7ᵃ (3) 72.3ᵃ (4) 73.9 (5) 69.1 (6) 18.5.

Mass (E.I.)[5] (rel.intens.) : *m/z* 720 [(M-H₂O)⁺, 0.08], 705 [(M-H₂O-CH₃)⁺, 0.06], 648 (0.05), 647 (0.06), 616 (0.08), 603 (0.21), 601 (0.04), 592 [(M-Rha+H)⁺, 0.16], 577 (592-CH₃, 0.5), 574 [(720-Rha+H)⁺, 0.4], 559 [(574-CH₃)⁺, 0.33], 533 (0.02), 502 (0.15), 467 (0.04), 463 (0.03), 430 (0.2), 412 (21), 395 (23.4), 379 (1.5), 377 (2.1), 371 (0.06), 358 (0.6), 353 (0.3), 340 (4.1), 316 (3.2), 298 (13.8), 287 (0.5), 281 (8.3), 277 [C₁₈H₂₉O₂, 3.6], 275 [C₁₈H₂₇O₂, 2.7], 269 (0.07), 259 [C₁₈H₂₇O, 10.5], 257 [C₁₈H₂₅O), 10.0], 241 (259-H₂O, 6.5), 239 (257-H₂O, 8.2), 214 [(C₁₆H₂₂), 18.7], 213 [(C₁₆H₂₁), 16.4], 199 [(C₁₅H₁₉ and C₁₁H₁₉O₃), 15.8], 197 [(C₁₅H₁₇ and C₁₁H₁₇O₃), 16.4], 155 [(C₉H₁₅O₂), 29.5], 153 [(C₉H₁₂O₂), 64.0], 126 [(C₈H₁₄O), 100]

Mass (F.D.)[6] : *m/z* 739 [M+H]⁺, 721 [M+H-H₂O]⁺, 593, 575, 412.

Mass (FAB, Negative ion)[7] : *m/z* 737 [(M-H)⁻, 20], 590 [M-H-Rha]⁻, 325 [Agl-H]⁻, 247, 205, 163, 119.

Peracetate :

M.P. : 195-197°C[1]; **[α]_D :** -51.8° (c=1.21, CHCl₃)[1]

PMR (CDCl₃, 100 MHz)[1] : δ 0.80 (s, 3xH-18), 1.01 (s, 3xH-19), 1.20 (d, J=7.0 Hz, 3xH-6 of Rha), 1.98-2.10 (6xOCOCH_3), 3.94 (t, J=7.0 Hz, H-16), 4.56 (d, J=7.0 Hz, H-1 of Glc), 4.96 (d, J=1.5 Hz, H-1 of Rha).

Mass (E.I.)[1] : m/z 990 [M⁺, $C_{51}H_{74}O_{19}$], 972 [M-H₂O]⁺, 561 [$C_{24}H_{33}O_{16}$]⁺, 412 [$C_{27}H_{40}O_3$]⁺, 394 [$C_{27}H_{38}O_2$]⁺, 273 [$C_{12}H_{17}O_7$]⁺.

References

1. T. Nohara, K. Miyahara and T. Kawasaki, *Chem. Pharm. Bull.*, **23**, 872 (1975).

2. C. Chen, Y. Zhang and J. Zhou, *Yunnan Zhiwu Yanjiu*, **5**, 91 (1983).

3. L.I. Strigina, *Khim. Prir. Soedin*, 654 (1983); *Chem. Nat. Comp.*, **19**, 623 (1983).

4. L.I. Strigina and V.V. Isakov, *Khim. Prir. Soedin.*, 474 (1982); *Chem. Nat. Comp.*, **18**, 440 (1982).

5. Y.M. Milgrom, Y.V. Rashkes and L.I. Strigina, *Khim. Prir. Soedin.*, **22**, 337 (1986); *Chem. Nat. Comp.*, **22**, 310 (1986).

6. H.R. Schulten, T. Komori, R. Higuchi and T. Kawasaki, *Tetrahedron*, **34**, 1003 (1978).

7. N.-Y. Chen, N. Chen, H. Li, Y.-Z. Chen, F.-Z. Zhao, C.-X. Chen and C.-R. Yang, *Huaxue Xuebao*, **45**, 682 (1987).

ZINGIBEROSIDE A₂

Zingiberogenin 3-O-|α-L-rhamnopyranosyl-(1→2)-β-D-glucopyranoside

Source : *Dioscorea zinigberensis* Wright (Dioscoreaceae)
Mol. Formula : $C_{39}H_{62}O_{13}$
Mol. Wt. : 738
M.P. : 231-232°C
Registry No. : [110996-53-7]

IR (KBr) : 3300, 2900, 1630, 1440, 1360, 1160, 1110, 1030, 905, 860, 830 cm^{-1}.

CMR (C_5D_5N) : δ C-1) 37.4 (2) 30.1 (3) 77.7 (4) 39.0 (5) 140.5 (6) 121.3 (7) 31.6 (8) 32.2 (9) 50.2 (10) 37.0 (11) 21.0 (12) 39.8 (13) 40.4 (14) 56.5 (15) 31.6 (16) 80.9 (17) 62.7 (18) 16.2 (19) 19.3 (20) 41.9 (21) 14.8 (22) 109.3 (23) 32.2 (24) 64.2 (25) 29.8 (26) 63.8 (27) 14.0 **Glc** (1) 100.1 (2) 79.3 (3) 77.7 (4) 71.7 (5) 77.9 (6) 62.6 **Rha** (1) 101.5 (2) 72.1 (3) 72.5 (4) 73.9 (5) 69.0 (6) 18.5.

Mass (E.I.) : *m/z* 561, 455, 273, 193.

Reference

1.   S. Tang, and Z. Jiang, *Yunnan Zhiwu Yanjiu (Acta Botanica Yunnanica)*, **9**, 233 (1987).

ALLIUM TUBEROSUM SAPONIN 3

(25S)-5β-Spirostane-3β,6α-diol 3-O-[α-L-rhamnopyranosyl-(1→4)-β-D-glucopyranoside]

Source : *Allium tuberosum* Rottler ex Spreng. (Liliaceae)
Mol. Formula : $C_{39}H_{64}O_{13}$
Mol. Wt. : 740
$[\alpha]_D^{30}$: -60.3° (c=0.27, MeOH)
Registry No. : [265092-99-7]

PMR (C_5D_5N, 500 MHz) : δ 0.81 (s, 3xH-18), 0.85 (s, 3xH-19), 1.08 (d, *J*=6.7 Hz, 3xH-27), 1.15 (d, *J*=7.3 Hz, 3xH-21), 1.73 (d, *J*=6.7 Hz, 3xH-6 of Rha), 1.94 (br d, *J*=6.7 Hz, H-5), 3.86 (m, *W*½=14.8 Hz, H-6), 4.45 (m, *W*½=7.5 Hz, H-3), 4.94 (d, *J*=7.3 Hz, H-1 of Glc), 5.89 (br s, H-1 of Rha).

CMR (C_5D_5N, 125 MHz) : δ C-1) 30.0 (2) 26.7 (3) 80.1 (4) 31.7 (5) 36.4 (6) 66.9 (7) 40.5 (8) 35.6 (9) 41.4 (10) 36.9 (11) 21.4 (12) 40.2 (13) 40.8 (14) 56.4 (15) 32.1 (16) 81.3 (17) 63.0 (18) 16.6 (19) 23.8 (20) 42.5 (21) 14.9 (22) 109.7 (23) 26.2 (24) 26.4 (25) 27.6 (26) 65.1 (27) 16.3 **Glc** (1) 104.0 (2) 75.0 (3) 76.6 (4) 78.3 (5) 77.4 (6) 61.4 **Rha** (1) 103.0 (2) 72.6 (3) 72.8 (4) 74.0 (5) 70.5 (6) 18.6.

Mass (FAB, Negative ion, H.R.) : *m/z* 739 [M-H]⁻, 593 [M-H-Rha]⁻, 431 [M-H-Rha-Glc]⁻.

Mass (FAB, Positive ion) : *m/z* 763.4249 [M+Na]⁺.

Reference

1.   T. Ikeda, H. Tsumagari and T. Nohara, *Chem. Pharm. Bull.,* **48,** 362 (2000).

ASPARAGOSIDE C
Sarsasapogenin 3-O-[β-D-glucopyranosyl-(1→3)-β-D-glucopyranoside]

Source : *Asparagus officinalis* L. (Liliaceae)
Mol. Formula : $C_{39}H_{64}O_{13}$
Mol. Wt. : 740
M.P. : 287-290°C
[α]$_D^{20}$: -13.0° (c=0.43, MeOH)
Registry No. : [60267-23-4]

Reference

1.   G.M. Goryamu, V.V. Krokhmalyuk and P.K. Kintya, *Khim. Prir. Soedin.*, 823 (1976); *Chem. Nat. Comp.*, **12**, 743 (1976).

ASPARANIN A
(25S)-5β-Spirostan-3β-ol 3-O-[β-D-glucopyranosyl-(1→2)-β-D-glucopyranoside]

Source : *Asparagus adscendens* (Liliaceae)
Mol. Formula : $C_{39}H_{64}O_{13}$
Mol. Wt. : 740
M.P. : 276-280°C (decomp.)
[α]$_D^{18}$: -61.5° (c=1.0, C_5D_5N)
Registry No. : [84633-33-0]

IR (KBr) : 3400 (OH), 980, 918, 898, 850 cm^{-1} (intensity 918 > 898, 25S spiroketal).

Reference

1.   S.C. Sharma, R. Chand and O.P. Sati, *Phytochemistry*, **21**, 2075 (1982).

CAPSICOSIDE B₂, ATROPOSIDE C

Tigogenin 3-O-[β-D-glucopyranosyl-(1→4)-β-D-galactopyranoside]

Source : *Capsicum annum* L. (Solanaceae)
Mol. Formula : $C_{39}H_{64}O_{13}$
Mol. Wt. : 740
M.P. : 293°C
[α]$_D$: -37° (c=1.0, MeOH)
Registry No. : [110124-76-0]

Reference

1.   E.V. Gutsu, P.K. Kintya and G.V. Lozur'evskii, *Khim. Prir. Soedin.*, **2**, 242 (1987); *Chem. Nat. Comp.*, **23**, 202 (1987).

CURILLIN G

(25S)-5β-Spirostain-3β-ol 3-O-[β-D-glucopyranosyl-(1→4)-β-D-glucopyranoside]

Source : *Asparagus curillus* (Liliaceae)
Mol. Formula : $C_{39}H_{64}O_{13}$
Mol. Wt. : 740
M.P. : 276-280°C
[α]$_D^{18}$: -61.7° (Pyridine)
Registry No. : [150677-83-1]

IR : 3400 (OH), 980, 917, 900, 855 cm^{-1} (intensity 917>900, 25S-spiroketal).

Reference

1.   S.C. Sharma and H.C. Sharma, *Phytochemistry*, **33**, 683 (1993).

DESGLUCODESRHAMNOPARILLIN
Sarsasapogenin 3-O-[β-D-glucopyranosyl-(1→6)-β-D-glucopyranoside]

Source : *Smilax aristolochiaefolia* Mill. (Liliaceae)
Mol. Formula : $C_{39}H_{64}O_{13}$
Mol. Wt. : 740
M.P. : *ca.* 250-265°C
[α]$_D^{20}$: -65.5° (c=0.8, CHCl$_3$-EtOH, 1:1)
Registry No. : [11025-85-7]

Reference

1.   R. Tschesche, R. Kottler and G. Wulff, *Justus Liebigs Ann. Chem.*, **699**, 212 (1966).

DRACAENA OMBET SAPONIN 1

Smilagenin 3-O-[β-D-galactopyranosyl-(1→2)-β-D-glucopyranside]

Source : *Dracaena ombet* (Agavaceae)
Mol. Formula : $C_{39}H_{64}O_{13}$
Mol. Wt. : 740
M.P. : 320°C
Registry No. : [675606-16-3]

IR : 3450 (OH), 1073, 980, 920, 900, 860 cm^{-1} (intensity 920>900, 25R-spirostan).

PMR (CD$_3$OD, 500 MHz) : δ 0.83 (s, 3xH-18)^a, 0.98 (s, 3xH-19)^a, 0.70 (d, 3xH-21)^b, 1.14 (d, 3xH-27)^b, 4.9 (anomric H), 5.3 (anomeric H).

CMR (C$_5$D$_5$N, 125 MHz) : δ C-1) 30.95 (2) 29.0 (3) 75.23 (4) 31.85 (5) 36.49 (6) 27.08 (7) 26.71 (8) 35.54 (9) 40.25 (10) 35.80 (11) 21.15 (12) 40.33 (13) 40.09 (14) 56.50 (15) 32.00 (16) 81.10 (17) 63.20 (18) 16.60 (19) 23.90 (20) 42.00 (21) 14.50 (22) 109.20 (23) 32.10 (24) 29.27 (25) 30.30 (26) 66.89 (27) 17.30 **Glc** (1) 102.55 (2) 81.80 (3) 76.90 (4) 69.85 (5) 76.60 (6) 62.81 **Gal** (1) 106.10 (2) 75.10 (3) 78.40 (4) 71.70 (5) 78.03 (6) 62.81.

Mass (FAB, Positive ion) : *m/z* 914 [M+NBA+Na]$^+$, 739, 577 [M+H-Glc]$^+$, 550, 491, 463, 415 [M-Glc-Gal]$^+$, 399, 343, 3339, 329, 302, 115, 139.

Reference

1.   S.M. El-Amin, M.A. Yousef, L.A. Refahy and M. Abdel-Montagally, *J. Drug Res. Egypt*, **24**, 109 (2002).

22-*EPI*-TIMOSAPONIN A-III
22-*Epi*sarsasapogenin 3-O-[β-D-glucopyranosyl-(1→2)-β-D-galactopyranoside]

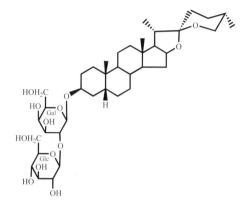

Source : *Anemarrhena asphodeloides* Bunge (Liliaceae)
Mol. Formula : $C_{39}H_{64}O_{13}$
Mol. Wt. : 740
M.P. : 295-300°C
[α]$_D$: -0.1°
Registry No. : [894959-82-5]

IR (KBr) **:** 994, 925, 910 < 896, 850 cm^{-1}.

PMR (C_5D_5N, 100 MHz) **:** δ 0.70 (m, 3xH-27), 0.96 (s, 3xH-18 and 3xH-19), 1.02 (d, *J*=7.0 Hz, 3xH-21), 3.70 (m, 2xH-26).

CMR (C_5D_5N, 25 MHz) **:** δ C-20) 42.2 (21) 17.0 (22) 110.5 (23) 28.2 (24) 28.2 (25) 30.9 (26) 69.6 (27) 17.4.

Mass (F.D.) **:** *m/z* 763 [M+Na]$^+$, 741 [M+H]$^+$.

Reference

1.   K. Kudo, K. Miyahara, N. Marubayashi and T. Kawasaki, *Chem. Pharm. Bull.*, **32**, 4229 (1984).

HOSTA LONGIPES SAPONIN 7
Gitogenin 3-O-[α-L-rhamnopyranosyl-(1→2)-β-D-galactopyranoside]

Source : *Hosta longipes*[1], *H. sieboldii*[2] (Liliaceae)
Mol. Formula : $C_{39}H_{64}O_{13}$
Mol. Wt. : 740
M.P. : 150-152°C[1]
[α]$_D^{25}$: -70.0° (c=0.10 CHCl$_3$-MeOH)
Registry No. : [178494-78-5]

IR (KBr)[1] : 3410 (OH), 2935 (CH), 1450, 1375, 1240, 1175, 1125, 1045, 980, 950, 920, 895, 865, 815, 780, 700 cm^{-1}.

PMR (C$_5$D$_5$N, 400 MHz)[1] : δ 0.70 (d, J=5.4 Hz, 3xH-27), 0.81 (s, 3xH-18), 0.90 (s, 3xH-19), 1.13 (d, J=7.0 Hz, 3xH-21), 1.61 (d, J=6.2 Hz, 3xH-6 of Rha), 3.50 (dd, J=10.4, 10.4, Hz, H-26A), 3.59 (dd, J=10.4, 3.1 Hz, H-26B), 3.89 (ddd, J=11.2, 8.6, 5.3 Hz, H-3), 4.07-4.15 (overlapping, H-2, H-5 of Gal), 4.27 (dd, J=9.2, 3.4 Hz, H-3 of Gal), 4.28 (dd, J=9.3, 9.3 Hz, H-4 of Rha), 4.38 (dd, J=10.8, 5.1 Hz, H-6A of Gal), 4.47 (dd, J=10.8, 5.9 Hz, H-6B of Gal), 4.48 (br d, J=3.4 Hz, H-4 of Gal), 4.55 (q-like, J=7.0 Hz, H-16), 4.61 (dd, J=9.3, 3.3 Hz, H-3 of Rha), 4.66 (dd, J=9.2, 7.7 Hz, H-2 of Gal), 4.81 (br d, J=3.3 Hz, H-2 of Rha), 4.86 (dq, J=9.3, 6.2 Hz, H-5 of Rha), 5.01 (d, J=7.7 Hz, H-1 of Gal), 6.29 (br s, H-1 of Rha).

CMR (C$_5$D$_5$N, 00 MHz)[1] : δ C-1) 45.8 (2) 70.7^a (3) 85.6 (4) 33.7 (5) 44.7 (6) 28.2 (7) 32.2 (8) 34.7 (9) 54.5 (10) 36.9 (11) 21.5 (12) 40.1 (13) 40.8 (14) 56.4 (15) 32.3 (16) 81.2 (17) 63.1 (18) 16.6 (19) 13.5 (20) 42.0 (21) 15.0 (22) 109.2 (23) 31.9 (24) 29.3 (25) 30.6 (26) 66.9 (27) 17.3 **Gal** (1) 101.8 (20) 77.0 (3) 76.5 (4) 70.8^a (5) 76.2 (6) 62.2 **Rha** (1) 102.2 (2) 72.5 (3) 72.8 (4) 74.2 (5) 69.4 (6) 18.5.

Mass (FAB, Negative ion)[1] : m/z 739 [M-H]$^-$, 595 [M-Rha]$^-$.

Biological Activity : The compound inhibited the phospholipid metabolism of HeLa cells with percentage value of 77.8 at a sample concentration of 50 μg/ml^{-1}. It exhibited cytotoxic activity on leukemia HL-60 cells with the IC$_{50}$=3.0 μg/ml cause 94.6% cell growth inhibition at the sample concentration of 10 μg/ml.[2]

References

1. Y. Mimaki, T. Kanmoto, M. Kuroda, Y. Sashida, Y. Satomi, A. Nishino and H. Nishino, *Phytochemistry*, **42**, 1065 (1996).

2. Y. Mimaki, M. Kuroda, A. Kameyama, A. Yokosuka and Y. Sashida, *Phytochemistry*, **48**, 1361 (1998).

MELONGOSIDE E, YUCCA ALOIFOLIA SAPONIN 1

Tigogenin 3-O-[β-D-glucopyranosyl (1→2)-β-D-glucopyranoside

Source : *Solanum melongena* L.[1] (Solanaceae), *Yucca aloifolia* L.[2] (Liliaceae)

Mol. Formula : $C_{39}H_{64}O_{13}$

Mol. Wt. : 740

M.P. : 265-268 °C

[α]$_D$: -52.3 (MeOH)

Registry No. : [94805-84-2]

IR[2] : 3400 (OH), 980, 928, 900 and 875 (intensity 875>900 cm^{-1}, 25R-stereochemistry) cm^{-1}.

Mass (E.I., rel.inten. %)[2] : *m/z* 416 [M]$^+$, 139.

Mass (F.D.)[2] : *m/z* 763 [M+Na]$^+$, 741 [M+H]$^+$, 601 [M+H-162]$^+$, 439 [M+Na-2x162]$^+$, 415 [M+Na-(2x162+23)]$^+$, 399, 347, 344, 272, 139 (base peak).

References

1.   P.K. Kintya and S.A. Shvets, *Khim. Prir. Soedin.*, **20**, 610 (1984); *Chem. Nat Comp.*, **20**, 575 (1984).

2.   S. Bahuguna and O.P. Sati, *Phytochemistry*, **29**, 342 (1990).

PANICULONIN B

Paniculogenin 6-O-[α-L-rhamnopyranosyl-(1→3)-β-D-quinovopyranoside]

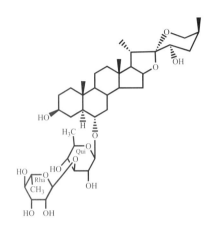

Source : *Solanum paniculatum* L.[1,2] (Solanaceae)
Mol. Formula : $C_{39}H_{64}O_{13}$
Mol. Wt. : 740
M.P. : 237-238°C[2]
[α]$_D^{19}$: -78.9° (c=1.12, Pyridine)[2]
Registry No. : [20735-80-2]

Biological Activity : weak hemolytic activity.

References

1.   H. Ripperger, K. Schreiber and H. Budzikewicz, *Chem. Ber.*, **100**, 741 (1967).

2.   H. Ripperger and K. Schreiber, *Chem. Ber.*, **101**, 2450 (1968).

(25*R* and *S*)-SCHIDIGERA-SAPONIN D5

(25R,S)-5β-Spirostan-3β-ol 3-O-[β-D-glucopyranosyl-(1→2)-β-D-glucopyranoside]

Source : *Yucca schidigera* Roezl ex Ortgies (Agavaceae)
Mol. Formula : $C_{39}H_{64}O_{13}$
Mol. Wt. : 740
[α]$_D^{24}$: -44.4° (c=0.65, MeOH)
Registry No. : [266998-04-3]

PMR (C$_5$D$_5$N, 500 MHz) : δ **25R** 0.64 (d, J=5.6 Hz, 3xH-27), 0.77 (s, 3xH-18), 0.93 (s, 3xH-19), 1.09 (d, J=6.8 Hz, 3xH-21); **25S** 0.76 (s, 3xH-18), 0.93 (s, 3xH-19), 1.02 (d, J=6.8 Hz, 3xH-27), 1.08 (d, J=7.1 Hz 3xH-21), 4.91 (overlapped with HDO signal, H-1 of Glc I), 5.34 (d, J=7.6 Hz, H-1 of Glc II).

CMR (C$_5$D$_5$N, 125 MHz) : δ C-1) 30.9 (2) 26.8 (3) 75.3 (4) 30.9 (5) 37.0 (6) 26.8 (7) 26.8 (8) 35.6 (9) 40.4 (10) 35.3 (11) 21.2 (12) 40.4 (13) 40.9 (14) 56.6 (15) 32.1 (16) 81.3 (17) 63.1 (18) 16.5 (19) 24.0 **25R** C-20) 42.0 (21) 14.8 (22) 109.0 (23) 32.1 (24) 29.0 (25) 30.9 (26) 66.9 (27) 17.2 **25S** C-20) 42.5 (21) 14.8 (22) 109.7 (23) 26.5 (24) 26.2 (25) 27.5 (26) 65.1 (27) 16.5 **Glc I** (1) 101.9 (2) 83.2 (3) 77.9 (4) 78.2 (5) 71.8 (6) 63.1 **Glc II** (1) 105.8 (2) 76.8 (3) 77.9 (4) 72.1 (5) 78.2 (6) 62.9.

Mass (FAB, Negative ion) : m/z 739 [M-H]$^-$, 577 [M-Glc-H]$^-$, 421 [M-Glc-Glc-H]$^-$.

Mass (FAB, Negative ion, H.R.) : m/z 739.4254 [(M-H)$^-$, calcd. for 739.4269].

Biological Activity : It exhibited relatively low antiyeast activity against *Saccharomyces cerevisiae* IFO 203 (12.5 μg/ml), *Candida albicans* TIMM 0134 (12.5 μg/ml), *Hansenula anomala* HUT 7083 (6.25 μg/ml), *Pichia nakazawae* HUT 1688 (3.13 μg/ml), *Kloeckera apiculata* IFO 154 (>100 μg/ml), *Debaryomyces hansenii* IF018 (>100 μg/ml).

Reference

1.   M. Miyakoshi, Y. Tamura, H. Masuda, K. Mizutani, O. Tanaka, T. Ikeda, K. Ohtani, R. Kasai and K. Yamasaki, *J. Nat. Prod.*, **63**, 332 (2000).

SMILAGENINOSIDE
Smilagenin 3-O-[β-D-glucopyranosyl-(1→2)-β-D-mannopyranoside]

Source : *Anemarrhena asphodeloides* Bge. (Liliaceae)
Mol. Formula : $C_{39}H_{64}O_{13}$
Mol. Wt. : 740
M.P. : 265-267°C
$[\alpha]_D^{12}$: -189.3°
Registry No. : [138831-68-2]

IR (KBr) : 3450 (OH), 2900, 2860, 1450, 1380 (C–H), 1042, 1035 (C–O), 987, 922, 900, 850 cm^{-1}.

PMR (C_5D_5N, 100 MHz) : δ 0.72 (s, 3xH-18), 0.91 (3xH-27), 0.92 (s, 3xH-19), 1.20 (3xH-21), 2.80-4.50 (m),3.30 (2xH-26), 4.25 (H-16).

CMR (C_5D_5N, 25 MHz) : δ C-1) 29.95 (2) 27.20 (3) 79.07 (4) 32.40 (5) 36.75 (6) 26.45 (7) 26.45 (8) 35.40 (9) 40.07 (10) 35.78 (11) 21.35 (12) 39.64 (13) 40.55 (14) 56.64 (15) 31.45 (16) 81.26 (17) 62.81 (18) 16.79 (19) 24.41 (20) 41.01 (21) 14.39 (22) 109.67 (23) 30.90 (24) 27.20 (25) 30.09 (26) 65.22 (27) 17.01 **Man** (1) 101.6 (2) 77.8 (3) 74.1 (4) 68.4 (5) 75.6 (6) 61.2 **Glc** (1) 104.6 (2) 74.7 (3) 76.0 (4) 71.0 (5) 76.0 (6) 62.1.

Mass (FAB, Positive ion) : m/z 779 [M+K]$^+$, 741 [M+H]$^+$, 579 [M-Glc]$^+$, 399 [M-Glc-Man]$^+$.

Reference

1.   D. Gou, S. Li, Q. Chi, W.G. Sun and Z.F. Sha, *Yaoxue Xuebao*, **26**, 619 (1991).

SOLANUM CHRYSOTRICHUM SAPONIN SC-6

(25*R*)-5α-Spirostan-3β,23α-diol 6-O-[α-L-rhamnopyranosyl-(1→3)-β-D-quinovopyranoside]

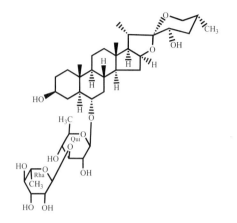

Source : *Solanum chrysotrichum* Schldh. (Solanaceae)
Mol. Formula : C₃₉H₆₄O₁₃
Mol. Wt. : 740
M.P. : 198-199°C
Registry No. : [478298-16-7]

IR : 3500-3300 (OH), 2925 (CH) cm⁻¹.

Hepta-acetate :

PMR (CDCl₃, 400 MHz) : δ 0.69 (H-9), 0.69 (dd, *J*=8.4, 6.4 Hz, H-17), 0.79 (s, 3xH-18), 0.83 (d, *J*=6.4 Hz, 3xH-27), 0.84 (s, 3xH-19), 0.94 (d, *J*=6.4 Hz, 3xH-21), 1.05 (H-1A), 1.14 (H-14A), 1.14 (d, *J*=6.4 Hz, 3xH-6 of Rha), 1.16 (H-11A), 1.18 (d, *J*=6.4 Hz, 3xH-6 of Qui, 1.2 (H-11B), 1.22 (H-5), 1.23 (H-14B), 1.25 (H-4A), 1.46 (H-2A), 1.61 (H-8 and H-24A), 1.68 (H-12A), 1.7 (H-1B), 1.70 (H-24B), 1.72 (H-7A), 1.88 (H-25), 1.92 (H-2B), 2.01 (H-7B), 2.05 (H-15), 2.08 (H-12B), 2.09 (H-4B and H-20), 3.22 (ddd, *J*=10.0, 10.0, 5.0 Hz, H-6), 3.36 (dd, *J*=11.0, 10.0 Hz, H-26α), 3.44 (m, H-5 of Qui), 3.46 (dd, *J*=11.0, 3.0 Hz, H-26β), 3.70 (t, *J*=9.6 Hz, H-3 of Qui), 3.88 (dd, *J*=10.0, 6.4 Hz, H-5 of Rha), 4.37 (d, *J*=8.0 Hz, H-1 of Qui), 4.46 (ddd, *J*=7.5, 7.4, 7.0 Hz, H-16), 4.63 (dddd, *J*=10.5, 10.0, 6.0, 5.0 Hz, H-3), 4.79 (d, *J*=1.6 Hz, H-1 of Rha), 4.81 (m, H-23), 5.00 (t, *J*=9.6 Hz, H-4 of Qui), 5.02 (dd, *J*=10.0, 9.6 Hz, H-4 of Rha), 5.06 (dd, *J*=9.6, 8.0 Hz, H-2 of Qui), 5.07 (m, H-2 of Rha), 5.08 (m, H-3 of Rha).

CMR (CDCl₃, 100 MHz) : δ C-1) 36.71 (2) 27.09 (3) 73.19 (4) 28.15 (5) 49.5 (6) 80.52 (7) 39.55 (8) 33.8 (9) 53.1 (10) 36.39 (11) 20.78 (12) 39.91 (13) 41.02 (14) 55.78 (15) 31.65 (16) 81.1 (17) 61.42 (18) 16.09 (19) 13.35 (20) 36.03 (21) 14.09 (22) 108.58 (23) 68.6 (24) 34.0 (25) 30.65 (26) 65.68 (27) 16.4 **Qui** (1) 102.17 (2) 72.16 (3) 81.88 (4) 74.43 (5) 69.87 (6) 17.44 **Rha** (1) 99.46 (2) 69.87 (3) 68.89 (4) 70.56 (5) 67.38 (6) 17.18.

Mass (FAB, Positive ion) : *m/z* 1073 [M+K]⁺, 1057 [M+Na]⁺, 1034 [M]⁺, 999 [(M+Na)-CH₃COO]⁺, 710 [(M+Na)-Rha-CH₃COO]⁺, 709 [(M+Na)-Rha-CH₃COO]⁺, 515 [M-Rha-Qui]⁺, 457 [M-Rha-Qui-CH₃COO]⁺, 395 [M-Rha-Qui-2xCH₃COO]⁺.

Mass (FAB, Positive ion, H.R.) : m/z 1034.5070 [(M+H)$^+$, calcd. for 1034.5086].

Reference

1.   A. Zamilpa, J. Tortoriello, V. Navarro, G. Delgado and L. Alvarez, *J. Nat. Prod.*, **65**, 1815 (2002).

TIMOSAPONIN A-III, FILIFERIN B

Sarsasapogenin 3-O-[β-D-glucopyranosyl-(1→2)-β-D-galactopyrnoside]

Source : *Cornus florida* L.[1] (Cornaceae), *Anemarrahena asphodeloides* Bunge[2] (Liliaceae), *Yucca macrocarpa* Engelm.[3] (Agavaceae)
Mol. Formula : $C_{39}H_{64}O_{13}$
Mol. Wt. : 740
M.P. : 317-322°C[1]
[α]$_D$: -41.3° (c=0.68, Pyridine)[1]
Registry No. : [41059-79-4]

IR (KCl)[4] : 3380 (OH), 2970, 2900, 2875, 2865, 1450, 1380, 1370, 1335, 1300, 1225, 1215, 1175, 1115, 1070, 1045, 1000, 985, 930, 920, 895, 876, 870, 850 cm^{-1}.

PMR (C$_5$D$_5$N, 400 MHz)[2] : δ 0.82 (s, 3xH-18), 0.95 (s, 3xH-19), 1.08 (d, *J*=8.0 Hz, 3xH-21), 1.15 (d, *J*=7.5 Hz, 3xH-27), 3.33 (d, *J*=12.0 Hz, H-26β), 4.83 (d, *J*=7.5 Hz, H-1 of Gal), 5.15 (d, *J*=8.0 Hz, H-1 of Glc).

CMR (C$_5$D$_5$N, 100 MHz)[5] : δ C-1) 30.9 (2) 26.8 (3) 75.2 (4) 30.9 (5) 36.9 (6) 26.4 (7) 26.4 (8) 35.3 (9) 40.3 (10) 35.3 (11) 21.2 (12) 40.3 (13) 40.9 (14) 56.5 (15) 32.2 (16) 81.4 (17) 63.0 (18) 16.6 (19) 24.0 (20) 42.5 (21) 14.9 (22) 109.7 (23) 26.2 (24) 26.2 (25) 27.6 (26) 65.1 (27) 16.3 **Gal** (1) 106.1 (2) 75.5 (3) 78.0 (4) 71.8 (5) 78.4 (6) 62.8 **Glc** (1) 102.6 (2) 81.3 (3) 76.9 (4) 69.8 (5) 76.6 (6) 62.2.

Mass (PD, Positive ion)[1] : m/z 786 [M+2Na-H]$^+$, 780 [M+Na]$^+$, 763 [M+Na]$^+$, 748, 733, 714, 691, 663, 645, 624, 603, 581, 552, 307, 301.

Mass (F.D., Positive ion)[1] : m/z 779 [(M+K)$^+$, 18.2], 763 [(M+Na)$^+$, 100], 741 [(M+H)$^+$, 60.2], 740 [(M)$^+$, 53.7], 601 [(M+Na-Glc)$^+$, 8.9], 579 [(M+H-Glc)$^+$, 11.8], 578 [(M$^+$-Glc), 12.9], 417 [(M+H-Glc-Gal)$^+$, 10.7], 416 [(M-Glc-Gal)$^+$, 12].

Biological Activity : The compound shows weak hypoglycemic activity.[5] It inhibits ADP-induced aggregation as well on 5-HT or arachidonic acid-induced aggegation of human platelets,[2] molluscidal.[6,7]

References

1. K. Hostettman, M. Hostettman-Kaldas and K. Nakanishi, *Helv. Chim. Acta*, **61**, 1990 (1978).

2. A. Niwa, O. Takeda, M. Ishimaru, Y. Nakamoto, K. Yamasaki, H. Kohda, H. Nishio, T. Segawa, K. Fujimura and A. Kuramoto, *Yakugaku, Zasshi*, **108**, 555 (1988).

3. V.I. Grishkovets, P.A. Karpov, S.V. Iksanova and V.Ya Chirva, *Khim. Prir. Soedin.*, 828 (1998), *Chem. Nat. Comp.*, **34**, 738 (1998).

4. Y.F. Hong, G.M. Zhang, L.N. Sun, G.Y. Han and G.Z. Ji, *Yaoxue Xuebao* (*Acta Pharm. Sin.*), **34**, 518 (1999).

5. S. Nagumo, S.-I. Kishi, T. Inoue and M. Nagai, *Yakugaku Zasshi*, **111**, 306 (1991).

6. N. Nakashima, I. Kimura, M. Kimura and H. Matsuura, *J. Nat. Prod.*, **56**, 345 (1993).

7. O. Takeda, S. Tanaka, K. Yamasaki, H. Kohda, Y. Iwanaga and M. Tsuji, *Chem. Pharm. Bull.*, **37**, 1090 (1989).

TORVONIN-B

Neosolaspigenin 3-O-[β-D-fucopyranosyl-(1→2)-β-D-quinovopyranoside]

Source : *Solanum torvum* Swartz (Solanaceae)
Mol. Formula : $C_{39}H_{64}O_{13}$
Mol. Wt. : 740
M.P. : 274°C
[α]$_D$: -4.5° (c=0.30, C_5D_5N)
Registry No. : [125850-42-2]

1300

IR (KBr) : 3700-3200 (br OH), 1380, 1215, 1170, 1160-1000 (br C–O–C), 950, 930, 900 and 830 cm⁻¹.

PMR (C₅D₅N, 400 MHz) : δ 0.82 (s, 3xH-18), 0.84 (s, 3xH-19), 1.53 (d, *J*=6.5 Hz, 3xH-21), 1.55 (d, *J*=6.5 Hz, 3xH-27), 1.64 (d, *J*=6.5 Hz, 3xH-6 of sugar), 1.71 (d, *J*=6.5 Hz, 3xH-6 of sugar), 3.55 (d, *J*=11.0 Hz, H-26α), 3.65 (td, *J*=10.5 Hz, H-6β), 3.73 (m, *W½*=22.0 Hz, H-3α), 3.80 (m, sugar-H), 4.00 (m, sugar-H), 4.06 (t, *J*=7.0 Hz, H-23α), 4.20 (dd, *J*=11.0, 2.0 Hz, H-26β), 4.28 (t, *J*=9.0 Hz, sugar-H), 4.35 (t, *J*=11.0 Hz, sugar-H), 4.61 (dd, *J*=9.0 Hz, sugar-H), 4.63 (q, *J*=7.5 Hz, H-16), 4.75 (d, *J*=8.0 Hz, anomeric H), 4.81 (br s, *W½*=4.0 Hz, anomeric H).

CMR (C₅D₅N, 100 MHz) : δ C-1) 37.79 (2) 32.23 (3) 79.45 (4) 32.38 (5) 51.33 (6) 69.92 (7) 41.43 (8) 34.30 (9) 53.91 (10) 36.75 (11) 21.22 (12) 39.95 (13) 40.86 (14) 66.48 (15) 33.17 (16) 81.56 (17) 64.57 (18) 17.20 (19) 13.59 (20) 41.17 (21) 16.56 (22) 110.44 (23) 65.35 (24) 34.49 (25) 27.28 (26) 65.35 (27) 20.52 **Qui** (1) 103.06 (2) 83.49 (3) 76.19 (4) 75.24 (5) 72.77 (6) 18.64 **Fuc** (1) 105.59 (2) 72.68 (3) 74.16 (4) 72.58 (5) 70.63 (6) 16.79.

Reference

1.   P.K. Agrawal, U. Mahmood and R.S. Thakur, *Heterocycles*, **29**, 1895 (1989).

TORVOSIDE-C
Neosolaspigenin 6-O-[α-L-rhamnopyranosyl-(1→3)-β-D-quinovopyranoside]

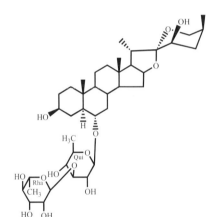

Source : *Solanum torvum* Swartz (Solanaceae)
Mol. Formula : C₃₉H₆₄O₁₃
Mol. Wt. : 740
[α]ᴅ²⁷ : -47.9° (c=2.90, MeOH)
Registry No. : [185012-38-8]

PMR (C₅D₅N, 400 MHz) : δ 0.59 (br dd, *J*=9.5, 9.5 Hz, H-9), 0.83 (s, 3xH-18), 0.84 (s, 3xH-19), 0.95 (m, H-1), 1.09 (m, H-12), 1.13 (m, H-14), 1.19 (m, H-7), 1.32 (m, H-5), 1.42 (m, H-15), 1.50 (m, H-11), 1.52 (d, *J*=7.3 Hz, 3xH-27), 1.54 (d, *J*=7.0 Hz, 3xH-21), 1.63 (d, *J*=6.2 Hz, 3xH-6 of Rha), 1.65 (4H, m, H-1, 4, 8, 12), 1.70 (d, *J*=6.2 Hz, 3xH-6 of Qui), 1.70 (2H, m, H-2, 25), 1.81 (2H, m, H-17, 24), 2.04 (m, H-2), 2.09 (m, H-15), 2.36 (br ddd, *J*=3.6, 3.6, 14.0

Hz, H-24), 2.52 (br dd, *J*=4.0, 12.2 Hz, H-7), 2.63 (m, H-20), 3.15 (br d, *J*=12.1 Hz, H-4), 3.55 (br d, *J*=11.0 Hz, H-26), 3.64 (2H, m, H-6, H-4 of Qui), 3.73 (2H, m, H-3, H-5 of Qui), 4.03 (br dd, *J*=9.2, 9.2 Hz, H-2 of Qui), 4.07 (dd, *J*=3.6, 3.6 Hz, H-23), 4.20 (dd, *J*=3.7, 11.0 Hz, H-26), 4.26 (dd, *J*=9.2, 9.2 Hz, 3xH-3 of Qui), 4.35 (dd, *J*=9.2, 9.2 Hz, H-4 of Rha), 4.60 (dd, *J*=3.1, 9.2 Hz, H-3 of Rha), 4.64 (m, H-16), 4.75 (d, *J*=8.1 Hz, H-1 of Qui), 4.82 (br s, H-2 of Rha), 5.00 (m, H-5 of Rha), 6.28 (s, H-1 of Rha).

CMR (C_5D_5N, 100 MHz) : δ C-1) 37.7 (2) 32.1 (3) 70.6 (4) 33.1 (5) 51.1 (6) 79.4 (7) 41.4 (8) 34.3 (9) 53.9 (10) 36.7 (11) 21.2 (12) 39.9 (13) 41.1 (14) 56.4 (15) 32.4 (16) 81.4 (17) 64.5 (18) 16.5 (19) 13.6 (20) 40.8 (21) 16.6 (22) 110.4 (23) 69.9 (24) 34.3 (25) 27.3 (26) 65.3 (27) 20.5 **Qui** (1) 105.6 (2) 76.2 (3) 83.2 (4) 75.2 (5) 72.6 (6) 18.8 **Rha** (1) 103.0 (2) 72.7 (3) 72.6 (4) 74.1 (5) 69.9 (6) 18.6.

Mass (FAB, Positive ion) : *m/z* 741 [M+H]$^+$, 595 [M+H-Rha]$^+$, 449 [M+H-Rha-Qui]$^+$.

Reference

1.	S. Yahara, T. Yamashita, N. Nozawa (nee Fujimura) and T. Nohara, *Phytochemistry*, **43**, 1069 (1996).

TUBEROSIDE M

(25*S*,5β)-Spirostane-1β,3β-diol 3-O-[α-L-rhamnopyranosyl-(1→4)-β-D-glucopyranoside]

Source : *Allium tuberosum* Rottl. (Liliaceae)
Mol. Formula : $C_{39}H_{64}O_{13}$
Mol. Wt. : 740
$[\alpha]_D^{24}$: -47.9° (c=0.20, MeOH)
Registry No. : [477217-05-3]

IR (KBr) : 3406, 1450, 1051, 987 916, 900, 850 cm^{-1}, intensity 916 > 900 (24*S*-Spirostanol).

PMR (C_5D_5N, 400 MHz) : δ 0.82 (s, 3xH-18), 1.16 (d, *J*=6.7 Hz, 3xH-27), 1.23 (d, *J*=6.5 Hz, 3xH-21), 1.33 (s, 3xH-19), 1.79 (d, *J*=6.2 Hz, 3xH-6 of Rha), 1.87 (m, H-17), 3.45 (d, *J*=10.9 Hz, H-26A), 3.82 (m, H-5 of Glc), 3.94 (m, H-1), 3.97 (m, H-2 of Glc), 4.14 (m, H-26B), 4.18 (m, H-6A of Glc), 4.26 (t, *J*=9.0 Hz, H-3 of Glc), 4.35 (m, H-6B of

Glc), 4.42 (t, *J*=9.3 Hz, H-4 of Rha), 4.50 (t, *J*=9.3 Hz, H-4 of Glc), 4.60 (m, H-3), 4.62 (m, H-16 and H-3 of Rha), 4.77 (m, H-2), 4.99 (d, *J*=7.8 Hz, H-1 of Glc), 5.03 (m, H-5), 5.96 (s, H-1 of Rha).

CMR (C_5D_5N, 100 MHz) : δ C-1) 72.5 (2) 29.2 (3) 74.8 (4) 32.3 (5) 31.1 (6) 26.6 (7) 26.6 (8) 35.8 (9) 42.6 (10) 40.5 (11) 21.7 (12) 40.4 (13) 40.7 (14) 56.5 (15) 32.0 (16) 81.3 (17) 63.1 (18) 16.7 (19) 19.2 (20) 42.3 (21) 15.0 (22) 109.8 (23) 26.5 (24) 26.3 (25) 27.6 (26) 65.2 (27) 16.5 **Glc** (1) 101.2 (2) 74.8 (3) 76.9 (4) 78.4 (5) 77.4 (6) 61.6 **Rha** (1) 102.8 (2) 72.7 (3) 72.8 (4) 74.1 (5) 70.5 (6) 18.7.

Mass (MALDI, Positive ion, H.R.) : *m/z* 741 [M+H]$^+$, 595 [M+H-Rha]$^+$, 433 [M+H-Rha-Glc]$^+$.

Biological Activity : The compound shows significant inhibitory effect on the growth of the human promyelocytic leukemia cell line HL-60 with IC_{50} value of 6.8 µg/ml.

Reference

1.   S.-M. Sang, M.-L. Zou, X.-W. Zhang, A.-N. Lao and Z.-L. Chen, *J. Asian Nat. Prod. Res.*, **4**, 69 (2002).

YUCCA GLORIOSA SAPONIN YS-I
Smilagenin 3-O-β-D-glucopyranosyl-(1→2)-β-D-glucopyranoside

Source : *Yucca gloriosa* L. (Agavaceae)
Mol. Formula : $C_{39}H_{64}O_{13}$
Mol. Wt. : 740
M.P. : 247-249°C
[α]$_D^{26}$: -0.2° (c=0.1, C_5D_5N)
Registry No. : [122566-56-7]

IR (KBr) : 3200-3500 (OH), 980, 920, 900, 860 cm^{-1} (intensity 900 > 920, 25*R*-spiroketal).

CMR (C_5D_5N) : δ C-1) 30.7 (2) 26.8 (3) 76.8 (4) 30.9 (5) 36.8 (6) 26.8 (7) 27.0 (8) 35.5 (9) 40.3 (10) 35.2 (11) 21.1 (12) 40.3 (13) 40.9 (14) 56.5 (15) 31.9^a (16) 81.2 (17) 63.1 (18) 16.6 (19) 24.0 (20) 42.0 (21) 15.0 (22) 109.1 (23)

32.1^a (24) 29.3 (25) 30.6 (26) 66.9 (27) 17.3 **Glc I** (1) 101.8 (2) 83.0 (3) 77.8 (4) 71.6 (5) 78.3^b (6) 62.7^c **Glc II** (1) 105.8 (2) 75.2 (3) 78.1^b (4) 71.8 (5) 78.0^b (6) 62.9^c.

Mass (FAB, Positive ion) : *m/z* 740 [M]$^+$.

Reference

1. K. Nakano, T. Yamasaki, Y. Imamura, K. Murakami, Y. Takaishi and T. Tomimatsu, *Phytochemistry*, **28**, 1215 (1989).

YUCCA GLORIOSA SAPONIN YS-II
Smilagenin 3-O-[β-D-glucopyranosyl-(1→2)-β-D-galactopyranoside]

Source : *Yucca aloifolia* L.[1], *Y. gloriosa* L.[2] (Agavaceae)
Mol. Formula : C$_{39}$H$_{64}$O$_{13}$
Mol. Wt. : 740
M.P. : 252-255°C[2]
[α]$_D^{26}$: -51.0° (c=1.0, CHCl$_3$-MeOH, 1:1)[2]
Registry No. : [95722-22-8]

IR (KBr)[2] : 3200-3500 (OH), 980, 920, 900, 860 cm^{-1} (intensity 900 > 920, 25*R*-spiroketal).

CMR (C$_5$D$_5$N)[2] : δ C-1) 30.9 (2) 26.7 (3) 76.4 (4) 30.9 (5) 36.9 (6) 26.7 (7) 27.0 (8) 35.5 (9) 40.3 (10) 35.2 (11) 21.1 (12) 40.3 (13) 40.9 (14) 56.5 (15) 31.9^a (16) 81.2 (17) 63.2 (18) 16.6 (19) 24.0 (20) 42.0 (21) 15.0 (22) 109.2 (23) 32.1^a (24) 29.3 (25) 30.6 (26) 66.9 (27) 17.3 **Gal** (1) 102.4 (2) 81.7 (3) 75.4 (4) 69.8 (5) 76.7 (6) 62.1^b **Glc** (1) 105.9 (2) 75.1 (3) 77.9 (4) 71.7 (5) 78.2 (6) 62.9^b.

Mass (FAB, Positive ion) : *m/z* 763 [M+Na]$^+$.

References

1. M.M. Benidze, O.D. Dzhikiya, M.M. Vugal'ter, T.A. Pkheidze and E.P. Kemertelidze, *Khim. Prir. Soedin.*, 744 (1984), *Chem. Nat. Comp.*, **20**, 703 (1984).
2. K. Nakano, T. Yamasaki, Y. Imamura, K. Murakami, Y. Takaishi and T. Tomimatsu, *Phytochemistry*, **28**, 1215 (1989).

YUCCOSIDE B
Tigogenin 3-O-[β-D-galactopyranosyl-(1→4) β-D-glucopyranoside]

Source : *Yucca filamentosa* L. (Liliaceae)
Mol. Formula : $C_{39}H_{64}O_{13}$
Mol. Wt. : 740
M.P. : 285-286°C
$[\alpha]_D^{20}$: -19.5° (c=1.6, C_5D_5N)
Registry No. : [41679-10-1]

Reference

1. P.K. Kintya, I.P. Dragalin and V.Y. Chirva, *Khim. Prir. Soedin.*, 615 (1972); *Chem. Nat. Comp.*, **8**, 584 (1972).

SIEBOLDIIN B
Sieboldogein 3-O-[α-L-arabinopyranosyl-(1→6)-β-D-glucopyranoside]

Source : *Smilax sieboldii* Miq. (Liliaceae)
Mol. Formula : $C_{38}H_{60}O_{14}$
Mol. Wt. : 740
M.P. : 266-267°
$[\alpha]_D^{21}$: 40.0° (c=0.07, Pyridine)
Registry No. : [145385-66-6]

IR (KBr) : 3447, 1710, 1073 cm^{-1}.

PMR (C$_5$D$_5$N, 300 MHz) : δ 0.62 (s, 3xH-18), 0.76 (s, 3xH-19), 1.14 (d, J=7.0 Hz, 3xH-21), 4.81 (d, J=10.6 Hz, anomeric H), 5.06 (d, J=7.6 Hz, anomeric H).

CMR (C$_5$D$_5$N, 75.5 MHz) : δ C-1) 36.7 (2) 29.5 (3) 77.0 (4) 27.0 (5) 56.4 (6) 209.7 (7) 46.7 (8) 37.3 (9) 53.6 (10) 41.0 (11)21.5 (12) 39.6 (13) 40.8 (14) 56.4 (15) 31.8 (16) 80.9 (17) 62.8 (18) 16.4 (19) 13.1 (20) 42.0 (21) 15.0 (22) 109.7 (23) 31.5 (24) 24.0 (25) 39.1 (26) 64.4^a (27) 64.1^a **Glc** (1) 102.1 (2) 75.2 (3) 78.5 (4) 71.8 (5) 76.7 (6) 69.7 **Ara** (1) 105.5 (2) 72.3 (3) 74.4 (4) 69.1 (5) 66.5.

Mass (FABMS, rel. intens. %) : m/z 763 [(M+Na)$^+$, 20.1], 447 [(genin + H}$^+$, 18.7].

Reference

1.   M.H. Woo, J.C. Do and K.H. Son, *J. Nat. Prod.*, **55**, 1129 (1992).

SMILAX LEBRUNII SAPONIN 3
(25*S*)-Spirost-5-en-3β,17α-27-triol 3-O-[α-L-arabinopyranosyl-(1→6)-β-D-glucopyranoside]

Source : *Smilax lebrunii* Levl. (Liliaceae)
Mol. Formula : C$_{38}$H$_{60}$O$_{14}$
Mol. Wt. : 740
M.P. : 280-282°C
Registry No. : [143722-97-8]

IR (KBr) : 3472, 1639, 1057, 993 cm^{-1}.

PMR (C$_5$D$_5$N, 400 MHz) : δ 0.94 (s, 3xH-18), 0.98 (s, 3xH-19), 1.28 (d, J=7.2 Hz, 3xH-21), 2.33 (q, J=7.2 Hz, H-20), 3.66 (dd, J=11.2, 11.2 Hz, H-26A), 3.76 (dd, J=11.2, 3.3 Hz, H-26B), 4.96 (d, J=7.0 Hz, H-1 of Glc), 5.00 (d, J=7.2 Hz, H-1 of Ara), 5.31 (m, H-6).

CMR (C$_5$D$_5$N, 100.16 MHz) : δ C-1) 37.5 (2) 30.3 (3) 77.0 (4) 39.0 (5) 141.0 (6) 121.6 (7) 32.0 (8) 32.4 (9) 50.1 (10) 37.0 (11) 21.5 (12) 31.8 (13) 44.1 (14) 53.0 (15) 31.8 (16) 90.1 (17) 90.1 (18) 17.2 (19) 19.4 (20) 45.1 (21) 9.7 (22) 110.3 (23) 31.8 (24) 23.6 (25) 39.4 (26) 63.9 (27) 64.4 **Glc** (1) 102.9 (2) 75.1 (3) 78.6 (4) 71.8 (5) 78.5 (6) 69.0 **Ara** (1) 105.3 (2) 72.3 (3) 74.4 (4) 69.5 (5) 66.4.

Mass (FAB) : m/z 763 [M+Na]$^+$, 747 [M+Li]$^+$.

Reference

1.   Y. Ju and Z.-J. Jia, *Phytochemistry,* **33**, 1193 (1993).

GLUCOCONVALLASAPONIN A
Convallagenin A 3-O-[β-D-glucopyranosyl-(1→2)-α-L-arabinopyranoside]

Source : *Convallaria keiskei* Miq. (Liliaceae)
Mol. Formula : C$_{38}$H$_{62}$O$_{14}$
Mol. Wt. : 742
M.P. : 213°C
$[\alpha]_D^{18}$: -41° (c=0.87, MeOH)
Registry No. : [19316-99-5]

IR (Nujol) : 3500-3200 (br, OH), 981, 921>897, 850 cm^{-1} (25S-spiroketal).

Reference

1.   M. Kimura, M. Tohma and I. Yoshizawa, *Chem. Pharm. Bull.*, **16**, 25 (1968).

ALLIUM KARATAVIENSE SAPONIN 3
(24S,25S)-3-O-Benzoyl-5α-spirostane-2α,3β,5,6β,24-pentol 2-O-β-D-glucopyranoside

Source : *Allium karataviense* Regel (Liliaceae)
Mol. Formula : C$_{40}$H$_{58}$O$_{13}$
Mol. Wt. : 746
[α]$_D^{27}$: -106.0° (c=0.10, MeOH)
Registry No. : [238075-79-1]

IR (KBr) : 3425 (OH), 2935 and 2890 (CH), 1700 (C=O), 1605 and 1585 (aromatic ring), 1455, 1380, 1320, 1275, 1210, 1175, 1120, 1060, 1025, 990, 950, 890 cm^{-1}.

PMR (C_5D_5N, 400/500 MHz) : δ 0.89 (s, 3xH-18), 1.08 (d, J=6.5 Hz, 3xH-27), 1.15 (d, J=7.0 Hz, 3xH-21), 1.58 (s, 3xH-19), 1.82 (m, H-25), 2.00 (dd, J=12.6, 10.5 Hz, H-23ax), 2.31 (dd, J=12.6, 4.8 Hz, H-23eq), 2.38 (dd, J=12.3, 5.6 Hz, H-1eq), 2.48 (dd, J=12.3, 11.1 Hz, H-1ax), 2.57 (dd, J=13.2, 6.1 Hz, H-4eq), 2.90 (dd, J=13.2, 11.0 Hz, H-4ax), 3.58 (dd, J=11.2, 11.2 Hz, H-26ax), 3.69 (dd, J=11.2, 4.8 Hz, H-26eq), 3.91 (ddd, J=8.9, 4.6, 2.5 Hz, H-5 of Glc), 3.96 (dd, J=8.9, 7.7 Hz, H-2 of Glc), 4.00 (ddd, J=10.5, 10.5, 4.8 Hz, H-24), 4.19 (br s, H-6), 4.19 (dd, J=8.9, 8.9 Hz, H-4 of Glc), 4.29 (dd, J=8.9, 8.9 Hz, H-3 of Glc), 4.30 (dd, J=11.5, 4.6 Hz, H-6A of Glc), 4.41 (dd, J=11.5, 2.5 Hz, H-6B of Glc), 4.59 (q-like J=7.3 Hz, H-16), 4.90 (ddd, J=11.1, 9.8, 5.6 Hz, H-2), 5.23 (d, J=7.7 Hz, H-1 of Glc), 6.30 (ddd, J=13.2, 11.0 Hz, H-4ax), 7.45 (H-3, H-4 and H-5 of Benz), 8.44 (dd, J=7.8, 2.1 Hz, H-2 of Benz), 8.44 (dd, J=7.8, 2.1 Hz, H-6 of Benz).

CMR (C_5D_5N, 100/125 MHz) : δ C-1) 38.9 (2) 77.7 (3) 76.4 (4) 37.9 (5) 75.0 (6) 75.0 (7) 35.7 (8) 30.1 (9) 45.6 (10) 40.4 (11) 21.5 (12) 40.3 (13) 41.0 (14) 56.2 (15) 32.3 (16) 81.6 (17) 62.8 (18) 16.6 (19) 18.0 (20) 42.3 (21) 15.0 (22) 111.8 (23) 41.8 (24) 70.6 (25) 39.9 (26) 65.3 (27) 13.6 **Glc** (1) 103.4 (2) 75.3 (3) 78.4 (4) 71.7 (5) 78.3 (6) 62.9 **Benz** (1) 131.9 (2) 130.3 (3) 128.7 (4) 132.9 (5) 128.7 (6) 130.3 (7) 166.8.

Mass (FAB, Positive ion, H.R.) : m/z 747.3400 [M+H]$^+$.

Reference

1.   Y. Mimaki, M. Kuroda, T. Fukasawa and Y. Sashida, *Chem. Pharm. Bull.,* **47**, 738 (1999).

DICHELOSTEMMA MULTIFLORUM SAPONIN 1

Brisbagenin 1-O-{α-L-rhamnopyranosyl-(1→3)-4-O-acetyl-α-L-arabinopyranoside}

Source : *Dichelostemma multiflorum* (Liliaceae)
Mol. Formula : $C_{40}H_{64}O_{13}$
Mol. Wt. : 752
$[α]_D^{28}$: -36.7° (c=0.10, CHCl₃-MeOH 1:1)
Registry No. : [167960-60-3]

IR (KBr) : 3430 (OH), 2930 (CH), 1740 (C=O), 1455, 1380, 1245, 1045, 1020, 980, 920, 900, 865, 835, 810, 755, 700 cm^{-1}.

PMR (C_5D_5N, 400/500 MHz) : δ 0.69 (d, *J*=4.9 Hz, 3xH-27), 0.87 (s, 3xH-18), 0.97 (s, 3xH-19), 1.11 (d, *J*=6.9 Hz, 3xH-21), 1.69 (d, *J*=6.1 Hz, 3xH-6 of Rha), 1.99 (s, OCOC*H₃*), 3.50 (dd, *J*=10.3, 10.3 Hz, H-26ax), 3.58 (dd, *J*=10.3, 2.6 Hz, H-26eq), 3.73 (br d, *J*=12.8 Hz, H-5A of Ara), 3.87 (dd, *J*=11.8, 4.2 Hz, H-1), 3.92 (m, H-3), 4.35-4.23 (4H, overlapping H-2, H-3, H-5B of Ara, H-4 of Rha), 4.38 (dd, *J*=9.2, 6.1 Hz, H-5 of Rha), 4.52 (dd, *J*=9.2, 3.0 Hz, H-3 of Rha), 4.54 (q-like, *J*=7.5 Hz, H-16), 4.76 (br d, *J*=3.0 Hz, H-2 of Rha), 4.78 (d, *J*=7.4 Hz, H-1 of Ara), 5.63 (br s, H-4 of Ara), 5.96 (br s, H-1 of Rha).

CMR (C_5D_5N, 100/125 MHz) : δ C-1) 81.6 (2) 37.7 (3) 67.6 (4) 39.6 (5) 43.0 (6) 28.8 (7) 32.4 (8) 36.5 (9) 54.9 (10) 41.4 (11) 23.8 (12) 40.9 (13) 40.4 (14) 56.8 (15) 32.4 (16) 81.1 (17) 63.3 (18) 16.9 (19) 8.2 (20) 42.0 (21) 15.0 (22) 109.2 (23) 31.8 (24) 29.3 (25) 30.6 (26) 66.8 (27) 17.3 **Ara** (1) 101.3 (2) 71.9 (3) 78.5 (4) 72.2 (5) 64.5 **Rha** (1) 104.2 (2) 72.3 (3) 72.7 (4) 73.9 (5) 70.5 (6) 18.6 (OCH₃) 170.6, 20.9.

Mass (FAB, Negative ion) : *m/z* 752 [M]⁻, 709 [M-acetyl]⁻, 563 [M-Rha-acetyl]⁻.

Biological Activity : The compound exhibited inhibitory activity on cAMP phosphodiesterase with IC_{50} value of 20.6x10^{-5} M.

Reference

1.   T. Inoue, Y. Mimaki, Y. Shashida, T. Nikaido and T. Ohmoto, *Phytochemistry*, **39**, 1103 (1995).

LILIUM SPECIOSUM SAPONIN 1

(25*R*,26*R*)-26-methoxyspirost-5-en-3β-ol 3-O-[α-L-rhamnopyranosyl-(1→2)-β-D-glucopyranoside]

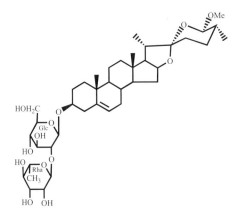

Source : *Lilium speciosum*[1], *Lilium dauricum*[2] (Liliaceae)
Mol. Formula : $C_{40}H_{64}O_{13}$
Mol. Wt. : 752
$[\alpha]_D^{23}$: -95.9° (c=0.74, MeOH)[1]
Registry No.: [133632-71-0]

IR (KBr)[1] : 3425 (OH), 2940 (CH), 1455, 1375, 1300, 1245, 1220, 1200, 1170, 1135, 1060, 1030, 980, 955, 910, 890, 835, 10 cm⁻¹.

PMR (C_5D_5N, 400 MHz)[1]: δ 0.87 (s, 3xH-18), 0.96 (d, J=5.7 Hz, 3xH-27), 1.08 (s, 3xH-19), 1.14 (d, J=6.7 Hz, 3xH-21), 1.79 (d, J=6.2 Hz, 3xH-6 of Rha), 3.53 (s, OCH₃), 3.91 (m, H-5 of Glc), 3.97 (m, H-3), 4.72-4.13 (9H, overlapped m, H-16, H-26, H-2, H-3, H-4 and 2xH-6 of Glc, H-3 and H-4 of Rha), 4.81 (br s, H-2 of Rha), 5.02 (overlapped, H-5 of Rha), 5.06 (d, J=7.3 Hz, H-1 of Glc), 5.34 (br d, J=5.1 Hz, H-6), 6.39 (br s, H-1 of Rha).

CMR (C_5D_5N, 100.6 MHz)[2] : δ C-1) 37.5 (2) 30.2 (3) 78.2 (4) 39.0 (5) 140.9 (6) 121.7 (7) 32.2ᵃ (8) 31.8 (9) 50.4 (10) 37.2 (11) 21.1 (12) 39.8 (13) 40.5 (14) 56.7 (15) 32.4ᵃ (16) 81.4 (17) 62.9 (18) 16.3 (19) 19.4 (20) 42.0 (21) 15.0 (22) 111.8 (23) 31.4 (24) 28.4 (25) 35.5 (26) 103.1 (27) 16.7 (OCH₃) 55.6 **Glc** (1) 100.4 (2) 79.6 (3) 77.9ᵇ (4) 71.8 (5) 77.8ᵇ (6) 62.7 **Rha** (1) 102.0 (2) 72.5 (3) 72.8 (4) 74.2 (5) 69.4 (6) 18.6.

Mass (SI, Positive ion)[1] : *m/z* 790 [M+K-H]⁺, 775 [M+Na]⁺, 735 [M-OH]⁺.

Mass (CI)[1] (rel.intens.) : *m/z* 441 (5), 409 (73), 342 (25), 338 (43), 267 (45), 229 (57), 180 (21), 147 (80), 145 (47), 129 (100), 127 (33), 113 (41).

Biological Activity : The compound showed 11.8% inhibitory effect on TPA-enhanced ³²P-incorporation into phospholipids of HeLa cells at 50 μg ml⁻¹ concentration.[3]

References

1.   Y. Mimaki and Y. Sashida, *Phytochemistry*, **30**, 937 (1991).

2.   Y. Mimaki, N. Ishibashi, K. Ori and Y. Sashida, *Phytochemistry*, **31**, 1753 (1992).

3.   O. Nakamura, Y. Mimaki, H. Nishino and Y. Sashida, *Phytochemistry*, **36**, 463 (1994).

CESTRUM SENDTENERIANUM SAPONIN 2

1β,2α,3β-Trihydroxyspirosta-5,25(27)-diene 3-O-[α-L-rhamnopyranosyl-(1→2)-β-D-galactopyranoside]

Source : *Cestrum sendtenerianum* (Solanaceae)
Mol. Formula : $C_{39}H_{60}O_{14}$
Mol. Wt. : 752
$[\alpha]_D^{25}$ **:** -70.7° (c=0.20, MeOH)
Registry No. : [329267-25-6]

IR (film) :3371 (OH), 2903 and 2848 (CH), 1449, 1373, 1230, 1130, 1043, 983, 920, 876 cm^{-1}.

PMR (C_5D_5N, 400/500 MHz) : δ 0.90 (s, 3xH-18), 1.07 (d, *J*=7.0 Hz, 3xH-21), 1.12 (H-14), 1.22 (ddd, *J*=12.8, 12.8, 3.2 Hz, H-12ax), 1.33 (ddd, *J*=10.7, 10.7, 3.5 Hz, H-9), 1.36 (s, 3xH-19), 1.49 (H-15β), 1.52 (H-7α), 1.55 (H-8), 1.55 (d, *J*=6.2 Hz, 3xH-6 of Rha), 1.70 (H-11ax), 1.74 (ddd, *J*=12.8, 3.1, 3.1 Hz, H-12 eq), 1.80 (2xH-23), 1.82 (dd, *J*=7.3, 6.7 Hz, H-17), 1.87 (br dd, *J*=13.8, 5.1 Hz, H-7β), 1.97 (H-20), 2.05 (H-15α), 2.27 (br d, *J*=12.7 Hz, H-24 eq), 2.74 (ddd, *J*=12.7, 12.7, 5.7 Hz, H-24ax), 2.87 (H-11eq), 2.89 (dd, *J*=12.3, 6.0 Hz, H-4eq), 2.95 (dd, *J*=12.3, 12.3 Hz, H-4ax), 3.58 (d, *J*=8.9 Hz, H-1), 3.96 (ddd, *J*=12.3, 8.9, 6.0 Hz, H-3), 4.04 (d, *J*=12.1 Hz, H-26eq), 4.09 (br dd, *J*=6.7, 5.2 Hz, H-5 of Gal), 4.12 (dd, *J*=8.9, 8.9 Hz, H-2), 4.27 (dd, *J*=9.5, 3.3 Hz, H-3 of Gal), 4.28 (dd, *J*=9.6, 9.6 Hz, H-4 of Rha), 4.35 (dd, *J*=11.1, 5.2 Hz, H-6A of Gal), 4.42 (dd, *J*=11.1, 5.2 Hz, H-6A of Gal), 4.42 (dd, *J*=11.1, 6.7 Hz, H-6B of Gal), 4.46 (br d, *J*=3.3 Hz, H-4 of Gal), 4.48 (d, *J*=12.1 Hz, H-26ax), 4.55 (q-like, *J*=7.3 Hz, H-16), 4.64 (dd, *J*=9.6, 3.7 Hz, H-3 of Rha), 4.66 (dd, *J*=9.5, 7.8 Hz, H-2 of Gal), 4.80 (br s, H-27A), 4.80 (br d, *J*=3.7 Hz, H-2 of Rha), 4.83 (br s, H-27B), 4.89 (dq, *J*=9.6, 6.2 Hz, H-5 of Rha), 5.02 (d, *J*=7.8 Hz, H-1 of Gal), 5.53 (br d, *J*=5.1 Hz, H-6), 6.31 (br s, H-1 of Rha).

CMR (C$_5$D$_5$N, 100/125 MHz) : δ C-1) 82.3 (2) 75.7 (3) 81.9 (4) 37.7 (5) 138.0 (6) 125.1 (7) 32.2 (8) 32.4 (9) 51.2 (10) 43.1 (11) 23.9 (12) 40.4 (13) 40.3 (14) 56.7 (15) 32.4 (16) 81.4 (17) 63.2 (18) 16.5 (19) 14.8 (20) 41.9 (21) 15.0 (22) 109.4 (23) 33.2 (24) 29.0 (25) 144.5 (26) 65.0 (27) 108.6 **Gal** (1) 101.7 (2) 75.7 (3) 76.6 (4) 70.8 (5) 76.9 (6) 62.2 **Rha** (1) 101.9 (2) 72.5 (3) 72.8 (4) 74.2 (5) 69.3 (6) 18.4.

Mass (FAB, Positive ion, H.R.) : *m/z* 753.4081 [(M+H)$^+$, requires 753.4061].

Reference

1.   M. Haraguchi, Y. Mimaki, M. Motidome, H. Morita, K. Takeya, H. Itokawa, A. Yakosuka and Y. Sashida, *Phytochemistry*, **55**, 715 (2000).

KINGIANOSIDE A

Gentrogenin 3-O-[β-D-glucopyranosyl-(1→4)-β-D-galactopyranoside]

Source : *Polygonatum kingianum* Coll. et Hemsl. (Liliaceae)
Mol. Formula : C$_{39}$H$_{60}$O$_{14}$
Mol. Wt. : 752
M.P. : 250-252°C
[α]$_D^{20}$: -42.4° (c=0.51, MeOH)
Registry No. : [145854-03-1]

PMR (C$_5$D$_5$N, 270 MHz) : δ 0.70 (d, *J*=5.5 Hz, 3xH-27), 0.93 (s, 3xH-19), 1.11 (s, 3xH-18), 1.35 (d, *J*=6.6 Hz, 3xH-21), 4.87 (d, *J*=7.3 Hz, H-1 of Gal), 5.28 (d, *J*=8.1 Hz, H-1 of Glc), 5.29 (br s, H-6).

CMR (C$_5$D$_5$N, 75.5 MHz) : δ C-1) 37.0 (2) 30.0 (3) 77.9 (4) 39.1 (5) 140.8 (6) 121.4 (7) 31.8^a (8) 30.9 (9) 52.3 (10) 37.6 (11) 37.6 (12) 212.7 (13) 55.0 (14) 56.0 (15) 31.7^a (16) 79.7 (17) 54.1 (18) 15.9 (19) 18.8 (20) 42.7 (21) 13.9 (22) 109.4 (23) 31.8^a (24) 29.2 (25) 30.6 (26) 67.0 (27) 17.3 **Gal** (1) 102.9 (2) 73.4 (3) 75.4^b (4) 80.0 (5) 75.9^b (6) 61.0 **Glc** (1) 107.1 (2) 75.2 (3) 78.7 (4) 72.3 (5) 78.4 (6) 63.2.

Mass (FAB, Negative ion) : m/z 751 [M-H]$^-$, 589 [M-Glc-H]$^-$.

Reference

1. X.-C. Li, C.-R. Yang, M. Ichikawa, H. Matsuura, R. Kasai and K. Yamasaki, *Phytochemistry*, **31**, 3559 (1992).

AGAVOSIDE B
Hecogenin 3-O-[β-D-glucopyranosyl-(1→4)-β-D-galactopyranoside]

Source : *Agave americana* L.[1] (Agavaceae), *Tribulus terrestris* L.[2] (Zygophyllaceae)
Mol. Formula : $C_{39}H_{62}O_{14}$
Mol. Wt. : 754
M.P. : 268-270°C[3]
[α]$_D$: -150.72° (c=0.04, Pyridine)[2]
Registry No. : [56857-66-0]

UV [2] **:** λ_{max} 199, 255 nm.

IR (KBr)[2] **:** 3410 (br) cm^{-1}.

PMR (C$_5$D$_5$N, 400/600 MHz)[2] : δ 0.62 (s, 3xH-19), 0.65 (d, J=6.8 Hz, 3xH-27), 0.70 (H-1A), 0.80 (H-5), 0.85 (H-9), 1.05 (s, 3xH-18), 1.10 (H-6), 1.25 (H-1B), 1.25 (H-4A), 1.34 (d, J=6.9 Hz, 3xH-21), 1.35 (H-7), 1.45 (H-2), 1.50 (H-15A), 1.55 (H-25), 1.60 (H-8), 1.65 (H-23), 1.76 (H-4B), 1.86 (H-20), 1.95 (H-24), 2.06 (H-15B), 2.20 (H-11A), 2.35 (H-11B), 2.75 (H-17), 3.45 (H-26A), 3.55 (H-26B), 3.84 (H-3), 4.05 (H-4 of Glc and H-5 of Gal), 4.06 (H-5 of Glc), 4.10 (H-6A of Glc), 4.22 (H-3 of Glc), 4.25 (H-3 of Gal), 4.25 (H-6A of Gal), 4.35 (H-2 of Gal), 4.45 (H-2 of Glc), 4.60 (H-6B of Glc), 4.65 (H-6B of Gal), 4.78 (H-4 of Gal), 4.86 (d, J=7.6 Hz, H-1 of Gal), 5.29 (d, J=7.8 Hz, H-1 of Glc).

CMR (C_5D_5N, 100/150 MHz)[2] : δ C-1) 36.55 (2) 29.69 (3) 76.78 (4) 34.26 (5) 44.31 (6) 28.53 (7) 31.73 (8) 34.54 (9) 55.43 (10) 36.21 (11) 37.93 (12) 210.8 (13) 55.30 (14) 55.48 (15) 31.64 (16) 79.64 (17) 54.23 (18) 16.04 (19) 11.50 (20) 42.58 (21) 13.88 (22) 109.27 (23) 31.73 (24) 29.16 (25) 29.95 (26) 66.89 (27) 17.26 **Glc** (1) 107.10 (2) 75.94 (3) 78.71 (4) 72.26 (5) 78.50 (6) 63.09 **Gal** (1) 102.48 (2) 73.06 (3) 75.22 (4) 80.80 (5) 75.45 (6) 61.03.

Mass (FAB, Positive ion)[2] : *m/z* (rel.intens.) 754 [(M)$^+$, 28], 777 [(M+Na)$^+$, 7].

References

1.   a) P.K. Kintya, V.A. Bobeiko Tezisy Dokl.-Vses. *Simp. Bioorg. Khim.*, 20 (1975); *Chem. Abstr.*, **85**, 160461j.

       b) G.V. Lazur'evskii, V.A. Bobeiko and P. Kintya, *Dokl. Akad. Nauk. SSSR*, 224, 1442 (1975).

       c) P.K. Kintya, V.A. Bobeiko, V.V. Krochmalyuk and V.Ya. Chirva, *Pharmazie*, **30**, 396 (1975).

2.   G. Wu, S. Jiang, F. Jiang, D. Zhu, H. Wu and S. Jiang, *Phytochemistry*, **42**, 1677 (1996).

3.   X.C. Li, D.Z. Wang and C.R. Yang, *Phytochemistry*, **29**, 3893 (1990).

ALLIOSPIROSIDE D

Cepagenin l-O[α-L-rhamnopyranosyl-(1→2)-β-galactopyranoside]

Source : *Allium cepa* L. (Liliaceae)
Mol. Formula : $C_{39}H_{62}O_{14}$
Mol. Wt. : 754
M.P. : 242-243°C (MeOH)
$[\alpha]_D^{22}$: -89.9± 2° (c=0.75, C_5D_5N)
Registry No. : [114317-58-7]

IR (KBr) : 830, 850, 870, 880, 910, 925, 965, 990, 1000, 3200-3600 cm^{-1}.

PMR (C_5D_5N, 250 MHz) : δ 0.77 (s, 3xH-18), 1.11 (d, 3xH-21), 1.27 (d, 3xH-27), 1.29 (s, 3xH-19), 1.63 (d, 3xH-6 of Rha), 3.47 (dd, H-26eq), 3.64 (m, H-3ax), 3.70 (m, H-1ax), 3.84 (dt, $J_{5,6}$=6.6 Hz, H-5 of Gal), 3.95 (dd, H-26ax), 4.07 (dd, $J_{3,4}$=3.5 Hz, H-3 of Gal), 4.18 (t, $J_{4,5}$=9.2 Hz, H-4 of Rha), 4.23 (dd, $J_{6,6}$=10.5 Hz, H-6B of Gal), 4.36 (dd, $J_{6,5}$=5.6 Hz, H-6A of Gal), 4.37 (dd, $J_{4,5}$=1.0 Hz, H-4 of Gal), 4.48 (m, H-16), 4.48 (dd, $J_{2,3}$=9.0 Hz, H-2 of Gal), 4.50 (dd, $J_{3,4}$=9.2 Hz, H-3 of Rha), 4.50 (m, H-24ax), 4.62 (dd, $J_{2,3}$=3.5 Hz, H-2 of Rha), 4.64 (dd, $J_{1,2}$=8.0 Hz, H-1 of Gal), 4.76 (dq, $J_{5,6}$=6.0 Hz, H-5 of Rha), 5.50 (br d, H-6), 6.21 (d, $J_{1,2}$=1.7 Hz, H-1 of Rha).

CMR (C$_5$D$_5$N, 300 MHz) : δ C-1) 84.67 (2) 36.12 (3) 68.39 (4) 43.97 (5) 139.74 (6) 125.06 (7) 32.21 (8) 33.30 (9) 50.84 (10) 43.05 (11) 24.23 (12) 40.40 (13) 40.60 (14) 57.32 (15) 32.56 (16) 81.76 (17) 62.73 (18) 17.62 (19) 15.04 (20) 42.77 (21) 15.21 (22) 111.70 (23) 36.00 (24) 66.77 (25) 36.00 (26) 64.76 (27) 9.96 **Rha** (1) 101.80 (2) 72.71 (3) 72.71 (4) 74.49 (5) 69.48 (6) 19.18 **Gal** (1) 100.97 (2) 76.96 (3) 75.21 (4) 70.56 (5) 76.46 (6) 62.13.

Reference

1. S.D. Kravets, Yu. S. Vollerner, A.S. Shashkov, M.B. Gorovits and N.K. Abubakirov, *Khim. Prir. Soedin.*, **6**, 843 (1987); *Chem. Nat. Comp.*, **6**, 700 (1987).

CAMASSIA CUSICKII SAPONIN 5

(25R)-6α-Hydroxy-5α-spirostan-3-one 6-O-[β-D-glucopyranosyl-(1→3)-β-D-glucopyranoside]

Source : *Camassia cusickii* S. Wats. (Liliaceae)
Mol. Formula : C$_{39}$H$_{62}$O$_{14}$
Mol. Wt. : 754
[α]$_D^{25}$: -22.3° (c=0.26, MeOH)
Registry No. : [138867-29-5]

UV (MeOH) : λ$_{max}$ 289 (log ε, 102).

IR (KBr) : 3420 (OH), 2950, 2875 (CH), 1695 (C=O), 1450, 1375, 1255, 1235, 1170, 1155, 1075, 1040, 980, 960, 915, 895, 860 (intensity 915 < 895, 25R-Spiroketal) cm^{-1}.

PMR (C$_5$D$_5$N, 400 MHz) : δ 0.61 (ddd, J=11.0, 11.0, 3.4 Hz, H-9), 0.73 (d, J=5.3 Hz, 3xH-27), 0.82 (s, 3xH-18), 0.97 (s, 3xH-19), 1.13 (d, J=6.8 Hz, 3xH-21), 1.94 (m, H-20), 2.31 (br d, J=14.8 Hz, H-2eq), 2.42 (ddd, J=14.8, 14.8, 6.4 Hz, H-2ax), 2.46 (dd, J=14.5, 14.5 Hz, H-4ax), 2.57 (ddd, J=12.7, 3.9, 3.9 Hz, H-7eq), 3.48 (dd, J=10.5, 10.5 Hz, H-26A), 3.57 (overlapping, H-4eq and H-26B), 3.71 (ddd, J=10.6, 10.6, 4.4 Hz, H-6), 4.83 (d, J=7.8 Hz, H-1 of Glc I), 5.32 (d, J=7.8 Hz, H-1 of Glc II).

1316

CMR (C$_5$D$_5$N, 100.4 MHz) : δ C-1) 39.9^a (2) 38.7^a (3) 210.7 (4) 38.1^a (5) 52.3 (6) 80.4 (7) 40.7 (8) 34.0 (9) 53.3 (10) 36.8 (11) 21.3 (12) 39.9 (13) 40.9 (14) 56.2 (15) 32.0 (16) 81.0 (17) 63.0 (18) 16.6 (19) 12.6 (20) 42.0 (21) 15.0 (22) 109.2 (23) 31.8 (24) 29.3 (25) 30.6 (26) 66.9 (27) 17.4 **Glc I** (1) 105.5 (2) 74.4 (3) 88.9 (4) 69.8 (5) 77.8 (6) 62.5 **Glc II** (1) 106.0 (2) 75.7 (3) 78.7^b (4) 71.7 (5) 78.3^b (6) 62.5.

Mass (S.I.) : *m/z* 777 [M+Na]$^+$, 755 [M+H]$^+$, 595, 431.

CD MeOH (c=9.55 x 10^{-4}) (θ) : 289 (+1099) nm.

Reference

1.   Y. Mimaki, Y. Sashida and K. Kawashima, *Phytochemistry*, **30**, 3721 (1991).

CANTALANIN-A
Hecogenin 3-O-[α-D-glucopyranosyl-(1→6)-α-D-glucopyranoside]

Source : *Agave cantala* Roxb. (Agavaceae)
Mol. Formula : C$_{39}$H$_{62}$O$_{14}$
Mol. Wt. : 754
M.P. : 210-213°C
[α]$_D$: +75.21° (c=1.0, MeOH)
Registry No. : [80938-28-9]

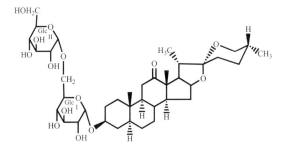

IR : 860, 900, 918 and 980 cm^{-1}.

Reference

1.   I.P. Varshney, D.C. Jain and H.C. Srivastava, *J. Nat. Prod.*, **44**, 662 (1981).

CESTRUM SENDTENERIANUM SAPONIN 3

(25R)-1β,2α,3β-Trihydroxyspirost-5-ene 3-O-[α-L-rhamnopyranosyl-(1→2)-β-D-galactopyranoside]

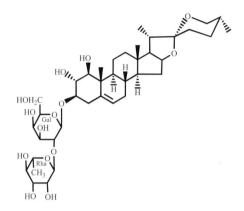

Source : *Cestrum sendtenerianum* (Solanaceae)
Mol. Formula : $C_{39}H_{62}O_{14}$
Mol. Wt. : 754
[α]$_D^{25}$: -57.1° (c=0.14, MeOH)
Registry No. : [329267-26-7]

IR (film) :3377 (OH), 2951 and 2926 (CH), 1454, 1375, 1260, 1242, 1130, 1054, 982, 920, 899, 867 cm^{-1}.

PMR (C$_5$D$_5$N, 400/500 MHz) : δ 0.70 (d, J=5.3 Hz, 3xH-27), 0.90 (s, 3xH-18), 1.12 (d, J=6.9 Hz, 3xH-21), 1.36 (s, 3xH-19), 1.55 (d, J=6.0 Hz, 3xH-6 of Rha), 3.51 (dd, J=10.0, 10.0 Hz, H-26ax), 3.58 (br d, J=10.0 Hz, H-25eq), 3.58 (d, J=8.8 Hz, H-1), 3.97 (ddd, J=12.1, 8.8, 6.0 Hz, H-3), 4.13 (dd, J=8.8, 8.8 Hz, H-2), 4.56 (q-like, J=7.1 Hz, H-16), 5.02 (d, J=7.7 Hz, H-1 of Gal), 5.52 (br d, J=4.8 Hz, H-6), 6.32 (d, J=1.3 Hz, H-1 of Rha).

CMR (C$_5$D$_5$N, 100/125 MHz) : δ C-1) 82.3 (2) 75.7 (3) 81.9 (4) 37.7 (5) 138.0 (6) 125.2 (7) 32.2 (8) 32.4 (9) 51.2 (10) 43.1 (11) 23.9 (12) 40.4 (13) 40.2 (14) 56.7 (15) 32.4 (16) 81.1 (17) 63.2 (18) 16.5 (19) 14.8 (20) 42.0 (21) 15.0 (22) 109.2 (23) 31.9 (24) 29.3 (25) 30.6 (26) 66.9 (27) 17.3 **Gal** (1) 101.7 (2) 75.7 (3) 76.6 (4) 70.8 (5) 76.9 (6) 62.2 **Rha** (1) 101.9 (2) 72.5 (3) 72.8 (4) 74.2 (5) 69.3 (6) 18.4.

Mass (FAB, Positive ion) : *m/z* 777.4000 [(M+Na)$^+$, requires 777.4037].

Reference

1.   M. Haraguchi, Y. Mimaki, M. Motidome, H. Morita, K. Takeya, H. Itokawa, A. Yakosuka and Y. Sashida, *Phytochemistry*, **55**, 715 (2000).

CESTRUM SENDTENERIANUM SAPONIN 4

1β,2α,3β-Trihydroxy-5α-spirost-25(27)-ene 3-O-[α-L-rhamnopyranosyl-(1→2)-β-D-galactopyranoside]

Source : *Cestrum sendtenerianum* (Solanaceae)
Mol. Formula : $C_{39}H_{62}O_{14}$
Mol. Wt. : 754
$[\alpha]_D^{25}$: -54.2° (c=0.43, EtOH)
Registry No. : [329267-27-8]

IR (film) : 3379 (OH), 2925 and 2948 (CH), 1451, 1374, 1258, 1232, 1127, 1047, 921, 878 cm^{-1}.

PMR (C_5D_5N, 400/500 MHz) : δ 0.88 (s, 3xH-18), 1.07 (d, J=6.9 Hz, 3xH-21), 1.18 (s, 3xH-19), 1.54 (d, J=6.1 Hz, 3xH-6 of Rha), 3.51 (d, J=9.0 Hz, H-1), 3.99 (dd, J=9.0, 9.0 Hz, H-2), 4.02 (overlapping, H-3), 4.04 and 4.48 (d, J=12.2 Hz, 2xH-26), 4.55 (q-like, J=7.6 Hz, H-16), 4.79 and 4.82 (br s, 2xH-27), 5.02 (d, J=7.7 Hz, H-1 of Gal), 6.25 (br s, H-1 of Rha).

CMR (C_5D_5N, 100/125 MHz) : δ C-1) 81.7 (2) 76.4 (3) 81.7 (4) 33.5 (5) 41.9 (6) 28.4 (7) 32.3 (8) 35.5 (9) 55.6 (10) 41.7 (11) 24.5 (12) 40.7 (13) 40.5 (14) 56.6 (15) 32.3 (16) 81.4 (17) 63.2 (18) 16.6 (19) 8.7 (20) 41.9 (21) 14.9 (22) 109.4 (23) 33.2 (24) 28.9 (25) 144.5 (26) 65.0 (27) 108.6 **Gal** (1) 101.4 (2) 761 (3) 76.5 (4) 70.8 (5) 76.9 (6) 62.3 **Rha** (1) 102.1 (2) 72.4 (3) 72.7 (4) 74.1 (5) 69.3 (6) 62.3 **Rha** (1) 102.1 (2) 72.4 (3) 72.7 (4) 74.1 (5) 69.3 (6) 18.4.

Mass (FAB, Positive ion, H.R.) : *m/z* 777.4076 [(M+Na)$^+$, requires 777.4037].

Reference

1.   M. Haraguchi, Y. Mimaki, M. Motidome, H. Morita, K. Takeya, H. Itokawa, A. Yakosuka and Y. Sashida, *Phytochemistry*, **55**, 715 (2000).

DRACAENA SURCULOSA SAPONIN 2

(24S,25R)-1β,6β,24β-Trihydroxy-3α,5α-cyclospirostane-1-O-β-D-fucopyranoside-24-O-β-D-glucopyranoside

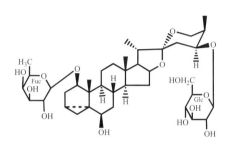

Source : *Dracaena surculosa* Lindle. (Agavaceae)
Mol. Formula : $C_{39}H_{62}O_{14}$
Mol. Wt. : 754
$[\alpha]_D^{26}$: -90.0° (c=0.10, MeOH)
Registry No. : [463962-98-3]

IR (film) : 3388 (OH), 2928 (CH), 1074 cm^{-1}.

CMR (C$_5$D$_5$N, 500 MHz) : δ 0.56 (dd, *J*=8.0, 4.0 Hz, H-4A), 0.87 (m, H-9), 0.88 (s, 3xH-18), 1.12 (d, *J*=6.9 Hz, 3xH-21), 1.14 (m, H-3), 1.18 (m, H-14), 1.21 (m, H-12ax), 1.26 (dd, *J*=13.4, 12.0, 2.4 Hz, H-7ax), 1.32 (d, *J*=6.9 Hz, 3xH-27), 1.44 (m, H-15A), 1.51 (t-like, *J*=4.0 Hz, H-4B), 1.56 (d, *J*=6.4 Hz, H-6 of Fuc), 1.64 (s, 3xH-19), 1.68 (m, H-11ax), 1.77 (m, H-12eq), 1.81 (m, H-11eq), 1.85 (dd, *J*=8.5, 6.7 Hz, H-17), 1.97 (m, H-20), 2.05 (ddd, J=13.4, 3.3, 2.4 Hz, H-7eq), 2.05 (d-like, *J*=10.8 Hz, H-23 (2H)), 2.08 (m, H-15B), 2.26 (m, H-25), 2.28 (br m, H-2 (2H)), 2.48 (m, H-8), 3.47 (br s, H-6), 3.52 (br d, *J*=10.2 Hz, H-26ax), 3.77 (q-like, *J*=6.4 Hz, H-5 of Fuc), 3.93 (br d, *J*=10.2 Hz, H-26eq), 3.93 (ddd, *J*=9.2, 4.7, 2.2 Hz, H-5 of Glc), 4.04 (overlapping, H-4 of Fuc), 4.06 (dd, *J*=8.6, 7.7 Hz, H-2 of Glc), 4.07 (dd, *J*=9.3, 3.5 Hz, H-3 of Fuc), 4.27 (dd, *J*=9.3, 7.8 Hz, H-2 of Fuc), 4.27 (dd, *J*=9.0, 8.6 Hz, H-3 of Glc), 4.32 (dd, *J*=9.2, 9.0 Hz, H-4 of Glc), 4.35 (br d, *J*=4.0 Hz, H-1), 4.40 (dd, *J*=11.9, 4.7 Hz, H-6A of Glc), 4.50 (dd, *J*=11.9, 2.2 Hz, H-6B of Glc), 4.54 (m, H-16), 4.57 (d, *J*=7.8 Hz, H-1 of Fuc), 4.80 (ddd, *J*=10.8, 10.8, 6.1 Hz, H-24), 5.03 (d, *J*=7.7 Hz, H-1 of Glc).

CMR (C$_5$D$_5$N, 125 MHz) : δ C-1) 84.8 (2) 33.1 (3) 23.3 (4) 15.8 (5) 40.0 (6) 73.0 (7) 38.6 (8) 30.1 (9) 50.2 (10) 49.2 (11) 23.3 (12) 40.5 (13) 41.0 (14) 56.6 (15) 32.1 (16) 81.6 (17) 62.5 (18) 16.8 (19) 16.9 (20) 42.5 (21) 14.7 (22) 111.2 (23) 34.1 (24) 72.8 (25) 31.7 (26) 64.2 (27) 9.9 **Fuc** (1) 103.4 (2) 72.0 (3) 75.7 (4) 72.8 (5) 71.4 (6) 17.3 **Glc** (1) 101.2 (2) 75.3 (3) 78.7 (4) 71.5 (5) 78.4 (6) 62.5.

Mass (FAB, Positive ion) : *m/z* 777 [M+Na]$^+$.

Reference

1.   A. Yokosuka, Y. Mimaki and Y. Sashida, *Chem. Pharm. Bull.*, **50**, 992 (2002).

LILIUM DAURICUM SAPONIN 4

(25*R*)-3β,17α-Dihydroxy-5α-spirostan-6-one 3-O-α-L-rhamnopyranosyl-(1→2)-β-D- glucopyranoside

Source : *Lilium dauricum* (Liliaceae)
Mol. Formula : $C_{39}H_{62}O_{14}$
Mol. Wt. : 754
$[\alpha]_D^{25}$: -97.7° (c=0.04, MeOH)
Registry No. : [143051-94-9]

UV (MeOH) : λ_{max} 280 (log ε, 175) nm.

IR (KBr): 3440 (OH), 2940, 2870 (CH), 1705 (C=O), 1455, 1375, 1260, 1175, 1045, 975, 915, 895, 860, 800, 700 cm^{-1} [intensity 915 < 895 (25*R*)-spiroacetal].

PMR (C_5D_5N, 400 MHz) : δ 0.69 (d, *J*=5.5 Hz, 3xH-27), 0.81 (s, 3xH-18), 0.90 (s, 3xH-19), 1.22 (d, *J*=7.2 Hz, 3xH-21), 1.78 (d, *J*=6.2 Hz, 3xH-6 of Rha), 2.25 (q, *J*=7.2 Hz, H-20), 3.50 (br m, 2xH-26), 3.90 (m, H-5 of Glc), 3.98 (m, H-3), 4.28-4.10 (H-2, H-3, H-4 of Glc), 4.32 (dd, *J*=9.5, 9.5 Hz, H-4 of Rha), 4.33 (dd, *J*=10.4, 5.3 Hz, H-6A of Glc), 4.42 (dd, *J*=7.5, 6.3 Hz, H-16), 4.55 (br d, *J*=10.4 Hz, H-6B of Glc), 4.64 (dd, *J*=9.5, 3.4 Hz, H-3 of Rha), 4.76 (br d, *J*=3.4 Hz, H-2 of Rha), 4.96 (dq, *J*=9.5, 6.2 Hz, H-5 of Rha), 5.06 (d, *J*=7.5 Hz, H-1 of Glc), 6.33 (br s, H-1 of Rha).

CMR (C_5D_5N, 100.6 MHz) : δ C-1) 36.9 (2) 29.4 (3) 76.2 (4) 26.6 (5) 56.4 (6) 209.4 (7) 46.9 (8) 38.0 (9) 53.7 (10) 41.0 (11) 21.3 (12) 32.0^a (13) 45.8 (14) 53.0 (15) 31.9^a (16) 89.8 (17) 89.9 (18) 17.3^b (19) 13.2 (20) 44.8 (21) 9.7 (22) 109.8 (23) 31.3^a (24) 28.8 (25) 30.4 (26) 66.7 (27) 17.2^b **Glc** (1) 99.6 (2) 79.6 (3) 78.4^c (4) 72.0 (5) 78.2^c (6) 62.9 **Rha** (1) 102.2 (2) 72.6 (3) 72.9 (4) 74.2 (5) 69.5 (6) 18.7.

Mass (SIMS) : *m/z* 793 [M+K]$^+$, 778 [M+Na+H]$^+$, 737 [M-OH]$^+$.

CD (MeOH; c=5.70x10^{-4}) : 293 nm (θ = -2870).

Reference

1. Y. Mimaki, N. Ishibashi, K. Ori and Y. Sashida, *Phytochemistry*, **31**, 1753 (1992).

OPHIOPOGON JAPONICUS SAPONIN 1
Ophiogenin 3-O-α-L-rhamnopyranosyl (1→2)-β-D-glucopyranoside

Source : *Ophiopogon japonicus* Ker.-Gawl. (Liliaceae)
Mol. Formula : $C_{39}H_{62}O_{14}$
Mol. Wt. : 754
Registry No. : [128502-94-3]

PMR (C_5D_5N, 400 MHz)[1] : δ 0.66 (d, *J*=5.8 Hz, 3xH-27), 1.09[b] (s, 3xH-18), 1.12[b] (s, 3xH-19), 1.26 (d, *J*=7.3 Hz, 3x H-21), 1.60 (H-25), 1.77 (d, *J*=6.3 Hz, 3xH-6 of Rha), 1.85 (ddd, *J*=5.8, 8.3, 13.4 Hz, H-15B), 1.91 (H-2A), 2.10 (H-2B), 2.39 (q, *J*=7.3 Hz, H-20), 2.54 (ddd, *J*=5.8, 8.3, 13.4 Hz, H-15A), 2.81 (d, *J*=12.8 Hz, H-4), 3.88 (m, H-5 of Glc), 3.90 (m, H-3), 3.50 (m, H-26), 4.16 (m, H-4 of Glc), 4.26 (H-3 of Glc), 4.28 (H-2 of Glc), 4.35 (t, *J*=5.86, 11.7 Hz, H-6A of Glc), 4.35 (t, *J*=9.3 Hz, H-4 of Rha), 4.49 (dd, *J*=2.4, 11.7 Hz, H-6B of Glc), 4.63 (dd, *J*=3.4, 9.3 Hz, H-3 of Rha), 4.80 (H-2 of Rha), 4.8 (H-16), 5.0 (H-5 of Rha), 5.02 (d, *J*=7.9 Hz, H-1 of Glc), 5.38 (m, *W*½=5.0 Hz, H-6), 6.37 (d, *J*=1.5 Hz, H-1 of Rha).

CMR (C_5D_5N, 100.62 MHz)[1] : δ C-1) 37.8 (2) 32.2 (3) 79.7 (4) 40.4 (5) 140.3 (6) 122.3 (7) 26.2 (8) 36.3 (9) 43.6 (10) 37.4 (11) 20.1 (12) 26.6 (13) 48.4 (14) 87.8 (15) 39.0 (16) 90.5 (17) 91.1 (18) 20.7 (19) 19.4 (20) 45.1 (21) 9.8 (22) 109.9 (23) 30.3 (24) 28.9 (25) 30.4 (26) 66.8 (27) 17.3 **Glc** (1) 102.1 (2) 77.9 (3) 77.9 (4) 71.5 (5) 78.3 (6) 62.7 **Rha** (1) 110.3 (2) 72.6 (3) 72.9 (4) 74.2 (5) 69.5 (6) 18.7.

Mass (FAB, Positive ion)[1] : *m/z* 793 [M+K]⁺, 777 [M+Na]⁺, 719 [M+Na-58 (CH₃CHCH₂O)]⁺, 573 [M+Na-58-146]⁺, 411 [M+Na-58-146-162]⁺, 393 [M+Na-58-146-162-18]⁺.

References

1.   M. Adinolfi, M. Parrilli and Y. Zhu, *Phytochemistry*, **29**, 1696 (1990).

RUSCUS ACULEATUS SAPONIN 27

(23*S*,25*R*)-Spirost-5-ene-3β,23-diol 23-O-{β-D-glucopyranosyl-(1→6)-β-D-glucopyranoside}

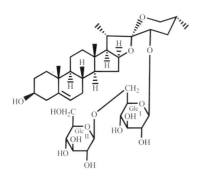

Source : *Ruscus aculeatus* L. (Liliaceae)
Mol. Formula : $C_{39}H_{62}O_{14}$
Mol. Wt. : 754
$[\alpha]_D^{25}$ **:** -44.0° (c=0.10, MeOH)
Registry No. : [239105-68-1]

IR (KBr) **:** 3420 (OH), 2930 (CH), 1450, 1370, 1270, 1250, 1155, 1055, 960, 940, 900 cm^{-1}.

PMR (C_5D_5N, 500 MHz) **:** δ 0.78 (d, *J*=6.5 Hz, 3xH-27), 0.95 (H-9), 0.99 (s, 3xH-19), 1.10 (H-1ax), 1.11 (H-14), 1.19 (H-12ax), 1.19 (s, 3xH-18), 1.22 (d, *J*=7.0 Hz, 3xH-21), 1.40 (H-11ax), 1.47 (H-11eq), 1.48 (H-15β), 1.52 (H-7β), 1.57 (H-8), 1.78 (H-2ax), 1.79 (H-12eq), 1.80 (H-1eq), 1.85 (H-7α), 1.94 (dd, *J*=8.5, 7.7 Hz, H-17), 1.99 (H-15α), 2.01 (q-like, *J*=11.6 Hz, H-24ax), 2.08 (H-2eq), 2.14 (m, H-25), 2.57 (dd, *J*=13.2, 13.2 Hz, H-4ax), 2.61 (dd, *J*=13.2, 4.6 Hz, H-4eq), 2.64 (H-24eq), 3.19 (m, H-20), 3.45 (dd, *J*=11.0, 4.2 Hz, H-26eq), 3.51 (dd, *J*=11.0, 11.0 Hz, H-26ax), 3.83 (m, H-3), 3.93 (m, H-5 of Glc II), 3.95 (dd, *J*=9.2, 7.7 Hz, H-2 of Glc I), 4.02 (dd, *J*=9.2, 9.2 Hz, H-4 of Glc I), 4.04 (dd, *J*=8.6, 7.8 Hz, H-2 of Glc II), 4.15 (m, H-5 of Glc I), 4.16 (dd, *J*=11.6, 5.1 Hz, H-23), 4.17 (dd, *J*=9.2, 9.2 Hz, H-3 of Glc I), 4.22 (dd, *J*=8.6, 8.6 Hz, H-3 of Glc II), 4.25 (dd, *J*=8.6, 8.6 Hz, H-4 of Glc I), 4.33 (dd, *J*=11.8, 6.8 Hz, H-6A of Glc I), 4.39 (dd, *J*=11.7, 5.1 Hz, H-6A of Glc II), 4.53 (dd, *J*=11.7, 2.0 Hz, H-6B of Glc II), 4.61 (q-like, *J*=7.7 Hz, H-16), 4.85 (dd, *J*=11.8, 1.1 Hz, H-6B of Glc I), 4.94 (d, *J*=7.7 Hz, H-1 of Glc I), 5.17 (d, *J*=7.8 Hz, H-1 of Glc II), 5.35 (br d, *J*=4.9 Hz, H-6).

CMR (C_5D_5N, 125 MHz) **:** δ C-1) 37.8 (2) 32.6 (3) 71.2 (4) 43.5 (5) 141.9 (6) 121.1 (7) 32.4 (8) 31.6 (9) 50.4 (10) 37.0 (11) 21.2 (12) 40.4 (13) 41.0 (14) 56.7 (15) 32.0 (16) 81.3 (17) 62.1 (18) 17.3 (19) 19.5 (20) 35.7 (21) 14.7 (22) 110.8 (23) 76.2 (24) 37.6 (25) 31.4 (26) 65.8 (27) 16.8 **Glc I** (1) 106.4 (2) 75.1 (3) 78.7 (4) 71.8 (5) 77.7 (6) 70.1 **Glc II** (1) 105.6 (2) 75.3 (3) 78.4 (4) 71.6 (5) 78.5 (6) 62.8.

Mass (FAB, Negative ion) **:** *m/z* 753 [M-H]$^-$, 591 [M-Glc]$^-$.

Reference

1.   Y. Mimaki, M. Kuroda, A. Yokosuka and Y. Sashida, *Phytochemistry*, **51**, 689 (1999).

SCHIDIGERA-SAPONIN C2

5β-Spirost-25(27)-ene-2β,3β-diol 3-O-[β-D-glucopyranosyl-(1→2)-β-D-galactopyranoside]

Source : *Yucca schidigera* Roezl ex Ortgies (Agavaceae)
Mol. Formula : $C_{39}H_{62}O_{14}$
Mol. Wt. : 754
$[\alpha]_D^{24}$ **:** -38.2° (c=0.55, MeOH)
Registry No. : [266997-36-8]

PMR (C_5D_5N, 500 MHz) **:** δ 0.74 (s, 3xH-18), 0.90 (s, 3xH-19), 1.02 (d, *J*=6.8 Hz, 3xH-21), 4.72, 4.75 (each 1H, s, H-27), 4.92 (overlapped with HDO signal, H-1 of Gal), 5.22 (d, *J*=7.8 Hz, H-1 of Glc).

CMR (C_5D_5N, 125 MHz) **:** δ C-1) 40.3 (2) 67.2 (3) 81.6 (4) 31.8 (5) 36.5 (6) 26.3 (7) 26.8 (8) 35.7 (9) 41.6 (10) 37.2 (11) 21.4 (12) 40.3 (13) 41.0 (14) 56.5 (15) 32.1 (16) 81.3 (17) 63.1 (18) 16.5 (19) 23.9 (20) 42.0 (21) 14.9 (22) 109.5 (23) 28.9ᵃ (24) 33.3ᵃ (25) 144.5 (26) 65.1 (27) 108.4 **Gal** (1) 102.9 (2) 81.3 (3) 76.7 (4) 69.9 (5) 76.7 (6) 63.1 **Glc** (1) 105.7 (2) 75.2 (3) 78.0 (4) 72.1 (5) 78.0 (6) 62.1.

Mass (FAB, Negative ion) **:** *m/z* 753 [M-H]⁻, 591 [M-Glc-H]⁻.

Mass (FAB, Negative ion, H.R.) **:** *m/z* 753.4068 [(M-H)⁻, calcd. for 753.4061].

Reference

1. M. Miyakoshi, Y. Tamura, H. Masuda, K. Mizutani, O. Tanaka, T. Ikeda, K. Ohtani, R. Kasai and K. Yamasaki, *J. Nat. Prod.*, **63**, 332 (2000).

SMILAX MENISPERMOIDEA SAPONIN 3

(25*S*)-Spirost-5-en-3β,17α,27-triol 3-O-[α-L-rhamnopyranosyl-(1→4)]-β-D-glucopyranoside

Source : *Smilax menispermoidea* (Liliaceae)
Mol. Formula : $C_{39}H_{62}O_{14}$
Mol. Wt. : 754
M.P. : 156-160°C
Registry No. : [155297-07-7]

IR (KBr) : 3424, 1653, 1037, 998 cm^{-1}.

PMR (C_5D_5N, 400 MHz) : δ 0.87 (3H, s), 0.95 (3H, s), 1.21 (3H, d, J=7.1 Hz), 2.25 (q, J=7.1 Hz), 2.71 (m), 3.60 (dd, J=11.0, 7.0 Hz), 3.75 (dd, J=11.0, 1.0 Hz), 3.90 (d, $J_{26ax,26eq}$=11.0 Hz), 4.05 (dd, $J_{26ax,26eq}$=11.0 Hz, $J_{26eq,25ax}$=3.5 Hz), 5.28 (m), 4.94 (d, J=7.0 Hz, H-1 of Glc), 5.80 (br s, H-1 of Rha).

CMR (C_5D_5N, 100 MHz) : δ C-1) 37.2 (2) 29.9 (3) 77.7 (4) 38.8 (5) 140.6 (6) 121.5 (7) 31.8 (8) 32.1 (9) 49.9 (10) 37.0 (11) 20.7 (12) 31.6 (13) 44.8 (14) 52.8 (15) 31.8 (16) 89.9 (17) 89.8 (18) 17.1 (19) 19.2 (20) 44.6 (21) 9.6 (22) 110.0 (23) 31.6 (24) 23.3 (25) 39.0 (26) 63.6 (27) 64.1 **Glc** (1) 102.3 (2) 75.0 (3) 76.3 (4) 78.2 (5) 76.7 (6) 61.5 **Rha** (1) 102.0 (2) 72.7 (3) 72.5 (4) 73.5 (5) 69.1 (6) 18.5.

Mass (FAB, Positive ion) : *m/z* 777 [M+Na]$^+$.

Reference

1.   Y. Ju, Z. Jia and X. Sun , *Phytochemistry* **37**, 1433 (1994).

SURCULOSIDE B

(25R)-1β,3β,24S-Trihydroxyspirost-5-ene 1-O-[β-D-fucopyranoside]-24-O-[β-D-glucopyranoside]

Source : *Dracaena surculosa* Lindle. (Agavaceae)
Mol. Formula : $C_{39}H_{62}O_{14}$
Mol. Wt. : 754
$[\alpha]_D^{25}$: -114.0° (c=0.10, CHCl$_3$-MeOH)
Registry No. : [295313-06-3]

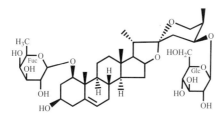

IR (KBr) **:** 3380 (OH), 2905 (CH), 1450, 1375, 1160, 1070, 995, 955, 900, 865, 835 cm⁻¹.

PMR (C$_5$D$_5$N, 400/500 MHz) **:** δ 0.82 (s, 3xH-18), 1.05 (d, J=6.9 Hz, 3xH-21), 1.24 (s, 3xH-19), 1.32 (d, J=7.0 Hz, 3xH-27), 1.57 (d, J=6.3 Hz, H-6 of Fuc), 2.10 (d-like, J=8.8 Hz, 2xH-23), 3.75 (br q, J=6.3 Hz, H-5 of Fuc), 3.83 (dd, J=11.5, 3.9 Hz, H-1), 3.89 (H-4 of Fuc), 3.91 (br m, overlapping, H-3), 3.92 (m, H-5 of Glc), 4.06 (H-3 of Fuc), 4.06 (dd, J=8.8, 7.7 Hz, H-2 of Glc), 4.26 (dd, J=9.1, 8.8 Hz, H-3 of Glc), 4.31 (dd, J=9.1, 9.1 Hz, H-4 of Glc), 4.32 (dd, J=9.3, 7.7 Hz, H-2 of Fuc), 4.38 (H-6A of Glc), 4.48 (br d, J=11.6 Hz, H-6B of Glc), 4.74 (d, J=7.7 Hz, H-1 of Fuc), 4.81 (td-like, J=8.5, 5.0 Hz, H-24), 5.04 (d, J=7.7 Hz, H-1 of Glc), 5.60 (br d, J=5.7 Hz, H-6).

CMR (C$_5$D$_5$N, 100/125 MHz) **:** δ C-1) 83.9 (2) 38.1 (3) 68.1 (4) 43.8 (5) 139.6 (6) 124.7 (7) 32.0 (8) 33.0 (9) 50.4 (10) 42.8 (11) 23.8 (12) 40.4 (13) 40.1 (14) 57.0 (15) 32.2 (16) 81.5 (17) 62.4 (18) 16.8 (19) 14.8 (20) 42.4 (21) 14.6 (22) 111.2 (23) 34.1 (24) 72.9 (25) 31.7 (26) 64.2 (27) 9.9 **Fuc** (1) 102.5 (2) 72.1 (3) 75.4 (4) 72.5 (5) 71.2 (6) 17.4 **Glc** (1) 101.2 (2) 75.3 (3) 78.6 (4) 71.5 (5) 78.4 (6) 62.5.

Mass (FAB, Positive ion, H.R.) **:** *m/z* 777.4054 [M+Na]⁺ (calcd. for C$_{39}$H$_{62}$O$_{14}$Na, 777.4037).

Reference

1.   A. Yokosuka, Y. Mimaki, Y. Sashida, *J. Nat. Prod.*, **63**, 1239 (2000).

TRICHOSANTHES KIRILOWII SAPONIN 1

(5β)-Spirost-25(27)-ene-2β,3β-diol 3-O-[β-D-glucopyranosyl-(1→2)-β-D-galactopyranoside]

Source : *Trichosanthes kirilowii* Maxim. (Cucurbitaceae)
Mol. Formula : $C_{39}H_{62}O_{14}$
Mol. Wt. : 754
M.P. : 265-266°C
Registry No. : [266997-36-8]

IR (KBr) : 3425.4, 2927.3, 1629.0, 1384.1, 1075.5 cm^{-1}.

PMR (C$_5$D$_5$N, 500 MHz) : δ 0.78 (s, 3xH-18)[a], 0.92 (s, 3xH-19)[a], 1.06 (d, J=8.0 Hz, 3xH-21), 2.24 (H-24ax), 2.72 (H-24eq), 4.02 (d, J=11.6 Hz, H-26ax), 4.30 (m, H-2), 4.45 (d, J=11.6 Hz, H-26eq), 4.46 (m, H-3), 4.60 (m, H-16), 4.66 (t, J=8.0 Hz, H-2 of Gal), 4.78 (s, H-27A), 4.82 (s, H-27B), 4.98 (d, J=7.8 Hz, H-1 of Gal), 5.28 (d, J=7.6 Hz, H-1 of Glc).

CMR (C$_5$D$_5$N, 125 MHz) : δ C-1) 40.2 (2) 67.2 (3) 82.0 (4) 31.8 (5) 36.6 (6) 26.3 (7) 26.8 (8) 35.6 (9) 40.6 (10) 37.1 (11) 21.3 (12) 40.8 (13) 41.4 (14) 56.3 (15) 32.1 (16) 81.6 (17) 63.1 (18) 16.6 (19) 23.9 (20) 41.9 (21) 15.0 (22) 109.4 (23) 33.3 (24) 29.0 (25) 144.5 (26) 65.0 (27) 108.7 **Gal** (1) 103.4 (2) 81.8 (3) 75.2 (4) 69.8 (5) 77.0 (6) 62.0 **Glc** (1) 106.3 (2) 75.2 (3) 78.6 (4) 71.7 (5) 78.0 (6) 62.8.

Mass (FAB, Positive ion) : *m/z* 755 [M+H]$^+$, 593 [M+H-Glc]$^+$, 431 [M+H-Glc-Gal]$^+$, 413 [aglycone+H-H$_2$O]$^+$.

Reference

1.   X. He, F. Qiu, Y. Shoyama, H. Tanaka and X. Yao, *Chem. Pharm. Bull.*, **50**, 653 (2002).

Note : This compound is identical to Schidigera-Saponin C2 (qv).

TRILLENOSIDE C

Trillenogenin 1-O-[α-L-rhamnopyranosyl-(1→2)-α-L-arabinopyranoside]

Source : *Trillium kamtschaticum* Pall. (Liliaceae)
Mol. Formula : $C_{37}H_{54}O_{16}$
Mol. Wt. : 754
Registry No. : [58808-76-7]

PMR (C_5D_5N, 500 MHz) : δ 1.01 (dddd, *J*=4.5, 12.0, 12.0, 12.0 Hz, H-11A), .1.03 (d, *J*=6.5 Hz, 3xH-27), 1.24 (s, 3xH-19), 1.66 (dd, *J*=11.5, 17.0 Hz, H-7A), 1.71 (d, *J*=6.0 Hz, 3xH-6 of Rha), *ca.* 1.72 (H-9), 2.02 (m, H-25), 2.23 (br dd, *J*=1.5, 11.5 Hz, H-8), 2.40 (ddd, *J*=12.0, 12.0, 12.0 Hz, H-2A), *ca.* 2.53 (H-4A), *ca.* 2.58 (H-12A), *ca.* 2.60 (H-4B), *ca.* 2.65 (H-2B), 2.78 (br d, *J*=17.0 Hz, H-12B), 3.14 (ddd, *J*=6.0, 7.5, 7.5 Hz, H-20), *ca.* 3.20 (H-11B), 3.25 (ddd, *J*=5.5, 5.5, 17.0 Hz, H-7B), 3.32 (dd, *J*=6.5, 7.5 Hz, H-17), 3.61 (dd, *J*=5.0, 11.5 Hz, H-26A), 3.65 (d, *J*=12.0 Hz, H-5A of Ara), 3.73 (dd, *J*=11.5, 11.5 Hz, H-26B), 3.77 (dd, *J*=3.5, 12.0 Hz, H-1), 3.83 (dddd, *J*=4.5, 4.5, 10.0, 10.0 Hz, H-3), 4.01 (dd, *J*=9.5, 9.5 Hz, H-24), *ca.* 4.14 (H-3 of Ara), *ca.* 4.15 (H-4 of Ara), 4.18 (dd, *J*=7.5, 11.0 Hz, H-21A), 4.22 (d, *J*=12.0 Hz, H-5B of Ara), 4.26 (d, *J*=9.5 Hz, H-23), *ca.* 4.32 (H-21B), 4.33 (dd, *J*=9.5, 9.5 Hz, H-4 of Rha), 4.56 (dd, *J*=7.5, 8.0 Hz, H-2 of Ara), 4.62 (d, *J*=7.5 Hz, H-1 of Ara), 4.63 (br d, *J*=9.5 Hz, H-3 of Rha), 4.75 (d, *J*=6.5 Hz, H-16), 4.75 (br s, H-2 of Rha), 4.85 (dq, *J*=9.5, 6.0 Hz, H-5 of Rha), 5.59 (br d, *J*=6.0 Hz, H-6), 6.34 (s, H-1 of Rha I).

CMR (C_5D_5N, 125 MHz) : δ 84.2 (2) 37.5 (3) 68.3 (4) 43.3 (5) 139.5 (6) 124.9 (7) 29.6 (8) 32.0 (9) 47.9 (10) 42.6 (11) 25.3 (12) 28.4 (13) 176.4 (14) 138.9 (15) 204.2 (16) 81.5 (17) 48.9 (18) - (19) 13.9 (20) 49.5 (21) 61.6 (22) 114.5 (23) 74.6 (24) 75.5 (25) 39.0 (26) 65.1 (27) 13.3 **Ara** (1) 100.7 (2) 75.0 (3) 75.9 (4) 70.2 (5) 67.5 **Rha** (1) 101.6 (2) 72.5 (3) 72.7 (4) 74.2 (5) 69.3 (6) 19.0.

Mass (FAB, Positive ion) : *m/z* 777 [M+Na]⁺, 755 [M+H]⁺.

Mass (FAB, Negative ion) : *m/z* 754 [M-H]⁻.

Reference

1.   M. Ono, Y. Yanai, T. Ikda, M. Okawa and T. Nohara, *Chem. Pharm. Bull.*, **51**, 1328 (2003).

AGAVE CANTALA SAPONIN 1

Gitogenin 3-O-[β-D-glucopyranosyl-(1→3)-β-D-glucopyranoside]

Source : *Agve cantala* Roxb. (Agavaceae)
Mol. Formula : $C_{39}H_{64}O_{14}$
Mol. Wt. : 756
M.P. : 235-238°C
[α]$_D$: -62.03° (c=1.0, MeOH)
Registry No. : [110219-90-4]

IR (KBr) : 3400 (OH), 981, 955, 920, 898, 865 cm^{-1} (intensity 898>920 (25R) spiroketal).

CMR (C$_5$D$_5$N, 80 MHz) : δ C-1) 45.7 (2) 72.7 (3) 84.6 (4) 34.1 (5) 44.9 (6) 28.6 (7) 32.5^a (8) 34.9 (9) 54.7 (10) 37.2 (11) 21.7 (12) 40.3 (13) 41.0 (14) 56.6 (15) 32.3^a (16) 81.4 (17) 62.8 (18) 16.8 (19) 13.6 (20) 42.2 (21) 15.2 (22) 109.5 (23) 32.1^a (24) 29.5 (25) 30.8 (26) 67.2 (27) 17.6 **Glc I** (1) 103.4 (2) 72.7 (3) 87.4 (4) 69.7 (5) 77.5 (6) 62.5 **Glc II** (1) 104.9 (2) 75.0 (3) 78.4^b (4) 70.5 (5) 78.2^b (6) 63.0.

Mass (E.I) : *m/z* 432 [M]$^+$, 414, 373, 363, 360, 318, 303, 300, 289, 271, 139 (base peak), 115.

Reference

1.   D.C. Jain, *Phytochemistry*, **26**, 1789 (1987).

AGAVE CANTALA SAPONIN 2
Chlorogenin 3,6-*bis*-O-β-D-glucopyranoside

Source : *Agave cantala* Roxb. (Agavaceae)
Mol. Formula : $C_{39}H_{64}O_{14}$
Mol. Wt. : 756
M.P. : 245-246°C
$[\alpha]_D^{20}$ **:** -78° (c=0.5, $CHCl_3$-MeOH)
Registry No. : [84321-98-2]

IR (KBr) : 965, 930, 907, 870 cm^{-1} (intensity 907>930, 25*R*-spiroketal).

Octamethyl derivative :

Mass (E.I.) : *m/z* 868 [M$^+$, 6], 633 (12), 414 (7), 400 (10), 399 (15), 398 (75), 253 (6), 219 (20), 187 (100).

Reference

1.   S.C. Sharma and O.P. Sati, *Phytochemistry*, **21**, 1820 (1982).

ALLIUM TUBEROSUM SAPONIN 2
(25*S*)-Spirostane-3β,5β,6α-triol 3-O-[α-L-rhamnopyranosyl-(1→4)-β-D-glucopyranoside]

Source : *Allium tuberosum* Rottler ex Spreng. (Liliaceae)
Mol. Formula : $C_{39}H_{64}O_{14}$
Mol. Wt. : 756
$[\alpha]_D^{29}$: -61.1° (c=0.14, MeOH)
Registry No. : [265092-98-6]

PMR (C$_5$D$_5$N, 500 MHz) : δ 0.83 (s, 3xH-18), 1.08 (d, *J*=7.3 Hz, 3xH-27), 1.15 (s, 3xH-19), 1.16 (d, *J*=6.7 Hz, 3xH-21), 1.71 (d, *J*=6.1 Hz, 3xH-6 of Rha), 1.90 (m, H-9), 3.38 (br d, *J*=11.0 Hz, H-26), 4.04 (br d, *J*=12.2 Hz, H-6), 4.07 (m, H-26), 4.63 (br s, *W*½=7.6 Hz, H-3), 5.00 (d, *J*=8.0 Hz, H-1 of Glc), 5.88 (br s, of Rha I).

CMR (C$_5$D$_5$N, 125 MHz) : δ C-1) 35.9 (2) 29.1 (3) 79.2 (4) 35.1 (5) 73.0 (6) 66.1 (7) 35.6 (8) 34.7 (9) 42.5 (10) 43.1 (11) 21.8 (12) 40.1 (13) 40.6 (14) 56.4 (15) 32.2 (16) 81.2 (17) 62.9 (18) 16.5 (19) 17.6 (20) 44.7 (21) 14.9 (22) 109.8 (23) 26.4 (24) 26.2 (25) 27.6 (26) 65.1 (27) 16.3 **Glc** (1) 101.9 (2) 74.9 (3) 76.8 (4) 78.5 (5) 77.5 (6) 61.2 **Rha** (1) 102.8 (2) 72.6 (3) 72.8 (4) 74.0 (5) 70.4 (6) 18.5.

Mass (FAB, Positive ion, H.R.) : *m/z* 779.4178 [M+Na]$^+$.

Mass (FAB, Negative ion, H.R.) : *m/z* 755 [M-H]$^-$, 609 [M-H-Rha]$^-$, 447 [M-H-Rha-Glc]$^-$.

Reference

1.   T. Ikeda, H. Tsumagari and T. Nohara, *Chem. Pharm. Bull.,* **48**, 362 (2000).

ANEMARRHENA ASPHODELOIDES SAPONIN 1
Markogenin-3-O-β-D-glucopyranosyl-(1→2)-β-D-galactopyranoside

Source : *Anemarrarrhena asphedeloides* Bunge (Liliaceae)
Mol. Formula : C$_{39}$H$_{64}$O$_{14}$
Mol. Wt. : 756
M.P. : 302°C[1]
[α]$_D$: -38.4° (c=0.42, MeOH)[1]
Registry No. : [108027-19-6]

PMR (C$_5$D$_5$N, 400 MHz)[1] : δ 0.80 (s, 3xH-18), 0.95 (s, 3xH-19), 1.02 (d, J=8.0 Hz, 3xH-21), 1.08 (d, J=8.0 Hz, 3xH-27), 3.37 (d, J=12.0 Hz, H-26β), 4.90 (d, J=8.0 Hz, H-1 of Gal), 5.20 (d, J=7.0 Hz, H-1 of Glc).

CMR (C$_5$D$_5$N, 100 MHz) : δ C-1) 40.6 (2) 67.2 (3) 81.9[a] (4) 32.0 (5) 36.6 (6) 26.8 (7) 26.5 (8) 35.7 (9) 41.5 (10) 37.1 (11) 21.4 (12) 40.3 (13) 40.9 (14) 56.4 (15) 32.2 (16) 81.3[a] (17) 67.9 (18) 16.6 (19) 23.9 (20) 42.5 (21) 14.9 (22) 109.7 (23) 26.3[b] (24) 26.2[b] (25) 27.6 (26) 65.5 (27) 16.3 **Gal** (1) 106.2 (2) 75.2 (3) 78.1 (4) 71.8 (5) 78.5 (6) 62.9 **Glc** (1) 103.4 (2) 82.0[a] (3) 77.0 (4) 69.8 (5) 76.9 (6) 62.0.

Biological Activity : Inhibits ADP-induced as well as SH-T or arachidonic acid-induced aggregation of human platelets.[1]

References

1.   A. Niwa, O. Takeda, M. Ishimaru, Y. Nakamoto, K. Yamasaki, H. Kohda, H. Nishio, T. Segawa, K. Fujimura and A. Kuramoto, *Yakugaku, Zasshi*, **108**, 555 (1988).

2.   S. Nagumo, S.I. Kishi, T. Inoue, and M. Nagai, *Yakugaku Zasshi*, **111**, 306 (1991).

ANEMARRHENASAPONIN III
(25S)-3β,15α-Dihydroxyspirostane 3-O-[β-D-glucopyranosyl-(1→2)-β-D-galactopyranoside]

Source : *Anemarrhena asphodeloides* Bunge (Liliaceae)
Mol. Formula : $C_{39}H_{64}O_{14}$
Mol. Wt. : 756
M.P. : 260-262°C
[α]$_D$: -45.1° (c=1.12, C_5D_5N)
Registry No. : [163047-23-2]

PMR (C_5D_5N, 500 MHz) : δ 0.87 (s, 3xH-18), 0.99 (s, 3xH-19), 1.04 (d, *J*=6.7 Hz, 3xH-27), 1.12 (d, *J*=6.7 Hz, 3xH-21), 2.00-3.00 (H-17), 3.27 (d, *J*=11.0 Hz, H-26A), 3.75 (ddd, *J*=9.2, 4.0, 2.4 Hz, H-5 of Glc), 3.91 (dd, *J*=11.0, 2.5 Hz, H-26B), 3.99 (dd, *J*=8.8 7.7 Hz, H-2 of Glc), 4.00-4.30 (H-15), 4.00-4.32 (H-5 of Gal), 4.00-4.32 (H-6A and H-6B of Gal), 4.10 (dd, *J*=9.2, 8.8 Hz, H-3 of Glc), 4.20 (dd, *J*=9.2, 3.0 Hz, H-3 of Gal), 4.22 (dd, *J*=9.2, 9.2 Hz, H-4 of Glc), 4.25 (br s, H-3), 4.30-4.62 (H-6A and H-6B of Glc), 4.49 (d, *J*=3.0 Hz, H-4 of Gal), 4.56 (dd, *J*=9.2, 7.6 Hz, H-2 of Gal), 4.83 (d, *J*=7.6 Hz, H-1 of Gal), 4.91 (dd, *J*=8.9, 3.7 Hz, H-16), 5.18 (d, *J*=7.7 Hz, H-1 of Glc).

CMR (C_5D_5N, 125 MHz) : δ C-1) 30.8^a (2) 27.4^b (3) 75.3 (4) 31.0^a (5) 36.3 (6) 27.1^b (7) 26.8 (8) 36.9 (9) 40.3 (10) 35.3 (11) 21.1 (12) 40.9 (13) 41.0 (14) 60.8 (15) 78.7 (16) 91.3 (17) 60.3 (18) 17.9 (19) 24.1 (20) 42.4 (21) 14.9 (22) 109.3 (23) 26.3 (24) 26.2 (25) 27.4 (26) 65.0 (27) 16.3 **Gal** (1) 102.3 (2) 81.7 (3) 75.0 (4) 69.7 (5) 76.6 (6) 62.0 **Glc** (1) 105.9 (2) 76.4 (3) 77.8 (4) 71.5 (5) 78.2 (6) 62.6.

Mass (FAB, Positive ion) : *m/z* 779 [M+Na]$^+$.

Reference

1.   S. Saito, S. Nagase and K. Ichinose, *Chem. Pharm. Bull.*, **42**, 2342 (1994).

CAMASSIA CUSICKII SAPONIN 2

Chlorogenin 6-O-[β-D-glucopyranoside-(1→2)-β-D-glucopyranoside]

Source : *Camassia cusickii* S. Wats. (Liliaceae)
Mol. Formula : $C_{39}H_{64}O_{14}$
Mol. Wt. : 756
$[\alpha]_D^{25}$: -20.0° (c=0.26, MeOH)
Registry No. : [138867-26-2]

IR (KBr) : 3400 (OH), 2920, 2865 (CH), 1445, 1370, 1235, 1165, 1150, 1065, 1045, 975, 945, 910, 890, 855 (intensity 910 < 890, 25R-spiroketal) cm^{-1}.

PMR (C_5D_5N, 400 MHz) : δ 0.62 (ddd, J=10.6, 10.6, 3.6 Hz, H-9), 0.73 (d, J=5.0 Hz, 3xH-27), 0.80 (s, 3xH-18)[a], 0.82 (s, 3xH-19)[a], 1.13 (d, J=6.9 Hz, 3xH-21), 1.95 (m, H-20), 2.03 (br d, J=11.6 Hz, H-2eq), 2.63 (d, J=12.7, 3.6, 3.6 Hz, H-7eq), 3.42 (br d, J=13.3 Hz, H-4eq), 3.47 (dd, J=10.4, 10.4 Hz, H-26A), 3.58 (dd, J=10.4, 1.8 Hz, H-26B), 3.63 (ddd, J=10.6, 10.6, 4.1 Hz, H-6), 3.87 (m, H-3), 4.91 (H-1 of Glc I, overlapping with H_2O signal), 5.41 (d, J=7.4 Hz, H-1 of Glc II).

CMR (C_5D_5N, 100.6 MHz) : δ C-1) 37.9 (2) 32.2[a] (3) 70.9 (4) 32.4[a] (5) 51.0 (6) 80.7 (7) 41.0 (8) 34.1 (9) 54.0 (10) 36.7 (11) 21.3 (12) 40.1 (13) 40.8 (14) 56.5 (15) 32.1 (16) 81.1 (17) 63.0 (18) 16.6 (19) 13.7 (20) 42.0 (21) 15.0 (22) 109.2 (23) 31.8 (24) 29.3 (25) 30.6 (26) 66.9 (27) 17.3 **Glc I** (1) 103.7 (2) 84.6 (3) 77.9 (4) 71.4[b] (5) 79.0 (6) 62.2 **Glc II** (1) 106.3 (2) 76.6 (3) 78.5[c] (4) 71.3[b] (5) 78.4[c] (6) 62.8.

Mass (SI) : *m/z* 779 [M+Na]$^+$, 757 [M+H]$^+$, 595.

Reference

1.   Y. Mimaki, Y. Sashida and K. Kawashima, *Phytochemistry*, **30**, 3721 (1991).

CAMASSIA CUSICKII SAPONIN 3

Chlorogenin 6-O-[β-D-glucopyranosyl-(1→3)-β-D-glucopyranoside]

Source : *Camassia cusickii* S. Wats. (Liliaceae)
Mol. Formula : $C_{39}H_{64}O_{14}$
Mol. Wt. : 756
[α]$_D^{25}$: -20° (c=0.21, MeOH)
Registry No. : [138867-27-3]

IR (KBr) **:** 3410 (OH), 2935, 2875 (CH), 1450, 1375, 1340, 1300, 1260, 1170, 1155, 1075, 1050, 980, 965, 920, 895, 865 (intensity 920 < 895, 25R-Spiroketal) cm^{-1}.

PMR (C_5D_5N, 400 MHz) **:** δ 0.60 (ddd, J=11.8, 11.8, 3.4 Hz, H-9), 0.72 (d, J=5.3 Hz, 3xH-27), 0.80 (s, 3xH-18)[a], 0.84 (s, 3xH-19)[a], 1.13 (d, J=6.9 Hz, 3xH-21), 1.94 (m, H-20), 2.04 (br d, J=11.7 Hz, H-2eq), 2.58 (ddd, J=12.7, 3.6, 3.6 Hz, H-7eq), 3.22 (br d, J=12.2 Hz, H-4eq), 3.47 (dd, J=10.5 Hz, H-26A), 3.57 (dd, J=10.5, 2.6 Hz, H-26B), 3.72 (ddd, J=10.8, 10.8, 4.5 Hz, H-6), 3.80 (m, H-3), 4.89 (overlapped with H_2O signal, H-1 of Glc I), 5.30 (d, J=7.8 Hz, H-1 of Glc II).

CMR (C_5D_5N, 100.6 MHz) **:** δ C-1) 37.8 (2) 32.2[a] (3) 70.7 (4) 33.3[a] (5) 51.3 (6) 79.8 (7) 41.4 (8) 34.2 (9) 53.9 (10) 36.7 (11) 21.3 (12) 40.1 (13) 40.8 (14) 56.4 (15) 32.1 (16) 81.0 (17) 63.0 (18) 16.7 (19) 13.6 (20) 42.0 (21) 15.0 (22) 109.2 (23) 31.9 (24) 29.3 (25) 30.6 (26) 66.9 (27) 17.4 **Glc I** (1) 105.5 (2) 74.5 (3) 89.1 (4) 69.8 (5) 77.7 (6) 62.6[b] **Glc II** (1) 106.1 (2) 75.7 (3) 78.7[c] (4) 71.7 (5) 78.3[c] (6) 62.5[b].

Mass (SI) **:** m/z 779 [M+Na]$^+$, 757 [M+H]$^+$, 595.

Reference

1.   Y. Mimaki, Y. Sashida and K. Kawashima, *Phytochemistry*, **30**, 3721 (1991).

CESTRUM SENDTENERIANUM SAPONIN 5

(25R)-1β,2α,3β-Trihydroxy-5α-spirostane 3-O-[α-L-rhamnopyranosyl-(1→2)-β-D-galactopyranoside]

Source : *Cestrum sendtenerianum* (Solanaceae)
Mol. Formula : $C_{39}H_{64}O_{14}$
Mol. Wt. : 756
[α]$_D^{25}$: -56.4° (c=0.11, EtOH)
Registry No. : [329267-28-9]

IR (film) :3377 (OH), 2950, 2926 and 2857 (CH), 1453, 1375, 1241, 1129, 1072, 1053, 982, 921, 899, 866 cm^{-1}.

PMR (C_5D_5N, 400/500 MHz) : δ 0.70 (d, J=5.4 Hz, 3xH-27), 0.88 (s, 3xH-18), 1.11 (d, J=7.0 Hz, 3xH-21), 1.19 (s, 3xH-19), 1.54 (d, J=6.2 Hz, 3xH-6 of Rha), 3.51 (dd, J=10.3, 10.3 Hz, H-26ax), 3.52 (d, J=8.9 Hz, H-1), 3.58 (br d, J=10.3 Hz, H-26eq), 4.01 (dd, J=8.9, 8.9 Hz, H-2), 4.03 (overlapping, H-3), 4.55 (q-like, J=7.4 Hz, H-16), 5.04 (d, J=7.8 Hz, H-1 of Gal), 6.28 (d, J=1.4 Hz, H-1 of Rha).

CMR (C_5D_5N, 100/125 MHz) : δ C-1) 81.8 (2) 76.4 (3) 81.8 (4) 33.5 (5) 41.9 (6) 28.5 (7) 32.4 (8) 35.6 (9) 55.6 (10) 41.7 (11) 24.6 (12) 40.8 (13) 40.5 (14) 56.7 (15) 32.5 (16) 81.1 (17) 63.3 (18) 16.7 (19) 8.8 (20) 42.0 (21) 15.0 (22) 109.2 (23) 31.9 (24) 29.3 (25) 30.6 (26) 66.9 (27) 17.3 **Gal** (1) 101.4 (2) 76.1 (3) 76.6 (4) 70.9 (5) 77.0 (6) 62.4 **Rha** (1) 102.1 (2) 72.5 (3) 72.8 (4) 74.2 (5) 69.4 (6) 18.5.

Mass (FAB, Positive ion, H.R.) : *m/z* 779.4198 [(M+Na)$^+$, requires 779.4194].

Reference

1. M. Haraguchi, Y. Mimaki, M. Motidome, H. Morita, K. Takeya, H. Itokawa, A. Yakosuka and Y. Sashida, *Phytochemistry*, **55**, 715 (2000).

HOSTA SIEBOLDII SAPONIN 6

(25R)-5α-Spirostan-3α,3β,12β-triol 3-O-{α-L-rhamnopyranosyl-(1→2)-β-D-galactopyranoside}

Source : *Hosta sieboldii* (Liliaceae)
Mol. Formula : $C_{39}H_{64}O_{14}$
Mol. Wt. : 756
$[\alpha]_D^{25}$: -84° (c=0.10, MeOH)
Registry No. : [213771-54-1]

IR (KBr) : 3410 (OH), 2930 (CH), 1450, 1375, 1340, 1255, 1235, 1065, 1045, 975, 955, 915, 895, 860 cm^{-1}.

PMR (C$_5$D$_5$N, 400/500 MHz) : δ 0.71 (d, J=5.2 Hz, 3xH-27), 0.93 (s, 3xH-19), 1.08 (s, 3xH-18), 1.43 (d, J=6.6 Hz, 3xH-21), 1.60 (d, J=6.1 Hz, 3xH-6 of Rha), 3.51 (dd, J=12.0, 5.4 Hz, H-12), 3.55 (dd, J=10.5, 10.5 Hz, H-26A), 3.61 (dd, J=10.5, 2.3 Hz, H-26B), 3.89 (ddd, J=11.1, 8.8, 5.2 Hz, H-3), 4.09 (ddd, J=11.7, 8.8 Hz, H-2), 4.12 (br dd, J=6.6, 5.3 Hz, H-5 of Gal), 4.27 (dd, J=9.2, 3.7 Hz, H-3 of Gal), 4.28 (dd, J=9.5, 9.5 Hz, H-4 of Rha), 4.37 (dd, J=11.1, 5.3 Hz, H-6A of Gal), 4.45 (dd, J=11.1, 6.6 Hz, H-6B of Gal), 4.46 (br d, J=3.7 Hz, H-4 of Gal), 4.61 (dd, J=9.5, 3.4 Hz, H-3 of Rha), 4.62 (overlapped H-16), 4.65 (dd, J=9.2, 7.8 Hz, H-2 of Gal), 4.82 (dd, J=3.4, 1.2 Hz, H-2 of Rha), 4.85 (dq, J=9.5, 6.1 Hz, H-5 of Rha), 5.00 (d, J=7.8 Hz, H-1 of Gal), 6.29 (d, J=1.2 Hz, H-1 of Rha).

CMR (C$_5$D$_5$N, 100/125 MHz) : δ C-1) 45.7 (2) 70.6 (3) 85.5 (4) 33.6 (5) 44.7 (6) 28.2 (7) 31.9 (8) 33.8 (9) 53.5 (10) 36.9 (11) 31.8 (12) 79.1 (13) 46.5 (14) 55.0 (15) 32.1 (16) 81.2 (17) 62.9 (18) 11.2 (19) 13.5 (20) 43.0 (21) 14.3 (22) 109.5 (23) 31.9 (24) 29.3 (25) 30.6 (26) 66.9 (27) 17.4 **Gal** (1) 102.1 (2) 76.1 (3) 76.5 (4) 70.8 (5) 76.9 (6) 62.2 **Rha** (1) 101.7 (2) 72.5 (3) 72.8 (4) 74.1 (5) 69.3 (6) 18.5.

Mass (FAB, Negative ion) : *m/z* 755 [M-H]$^-$.

Reference

1.   Y. Mimaki, M. Kuroda, A. Kameyama, A. Yokosuka and Y. Sashida, *Phytochemistry*, **48**, 1361 (1998).

MARKOGENIN DIGLYCOSIDE
Markogenin 3-O-[β-D-glucopyranosyl-(1→2)-β-D-galactopyranoside]

Source : *Anemarrhena asphodeloides* Bge. (Liliaceae)
Mol. Formula : $C_{39}H_{64}O_{14}$
Mol. Wt. : 756
M.P. : 302°C
[α]$_D$: -38.4° (c=0.42, MeOH)
Registry No. : [117210-12-5]

PMR (C_5D_5N) : δ 0.80 (s, 3xH-18), 0.95 (s, 3xH-19), 1.02 (d, *J*=8.0 Hz, 3xH-21), 1.08 (d, *J*=8.0 Hz, 3xH-27), 3.37 (d, *J*=12.0 Hz, H-26β), 4.90 (d, *J*=8.0 Hz, H-1 of Gal), 5.20 (d, *J*=7.0 Hz, H-1 of Glc).

CMR (C_5D_5N) : δ C-1) 40.5 (2) 67.2 (3) 81.8 (4) 31.8 (5) 36.5 (6) 27.6 (7) 27.6 (8) 35.4 (9) 41.4 (10) 37.1 (11) 21.3 (12) 40.2 (13) 40.8 (14) 56.3 (15) 32.1 (16) 81.3 (17) 62.9 (18) 16.6 (19) 23.9 (20) 42.5 (21) 14.9 (22) 109.6 (23) 26.3 (24) 26.3 (25) 27.6 (26) 65.1 (27) 16.3 Gal (1) 106.1 (2) 75.1 (3) 77.9 (4) 71.7 (5) 78.4 (6) 62.9 **Glc** (1) 103.3 (2) 81.8 (3) 76.9 (4) 69.7 (5) 76.9 (6) 62.0.

Biological Activity : Inhibits ADP-induced aggregation as well as 5-HT or arachidonic acid induced aggregation of human platelet.

Reference

1. A. Niwa, O. Takeda, M. Ishimaru, Y. Nakamoto, K. Yamasaki, H. Koda, H. Nishio, T. Segawa, K. Fujimura and A. Kuramoto, *Yakugaku Zasshi*, **108**, 555 (1988).

PETUNIOSIDE D', CAPSICOSIDE B$_1$

Gitogenin 3-O-[β-D-glucopyranosyl-(1→4)-β-D-galactopyranoside]

Source : *Petunia hybrida* L.(Solanaceae)
Mol. Formula : C$_{39}$H$_{64}$O$_{14}$
Mol. Wt. : 756
M.P. : 263°C
[α]$_D^{20}$: -48° (c=1.0, CH$_3$OH)
Registry No. : [160067-95-8]

IR (KBr) : 3500-3400, 987, 920, 900, 850, Intensity 900>920 cm^{-1} (25R-spirostane).

PMR (C$_5$D$_5$N, 300 MHz) : δ 0.70 (d, $J_{25,27}$=6.0 Hz, H-27), 1.20 ($J_{1,1}$=13.0 Hz, H-1A), 1.40 ($J_{15,16}$=8.2 Hz, H-15A), 1.55 ($J_{25,26}$=4.5 Hz, H-25), 1.62 ($J_{3,4}$=5.2 Hz, H-4A), 1.81 (H-17), 1.82 ($J_{4,4}$=12.6 Hz, H-4B), 2.03 ($J_{15,16}$=6.2 Hz, H-15B), 2.19 ($J_{1,2}$=4.7 Hz, H-1B), 3.47 ($J_{25,26}$=10.1 Hz, H-26A), 3.56 ($J_{26,26}$=10.1 Hz, H-26B), 3.76 ($J_{3,4}$=8.9 Hz, H-3), 3.93 ($J_{4,5}$ =9.2 Hz, H-4 of Glc), 3.95 ($J_{2,1}$=9.1 Hz, $J_{2,3}$=11.5 Hz, H-2), 3.97 ($J_{5,6}$=2.0 Hz, H-5 of Glc), 3.98 ($J_{2,3}$=9.2 Hz, H-2 of Glc), 4.05, ($J_{5,6}$=8.0 Hz, H-5 of Gal), 4.10 ($J_{3,4}$=9.2 Hz, H-3 of Glc), 4.10 ($J_{5,6}$=7.0 Hz, H-6A of Glc), 4.14 ($J_{3,4}$=3.1 Hz, H-3 of Gal), 4.18 ($J_{5,6}$=6.0 Hz, H-6B of Gal), 4.33 ($J_{2,3}$=10.0 Hz, H-2 of Gal), 4.45 ($J_{6,6}$=12.6 Hz, H-6B of Glc), 4.49 ($J_{6,6}$=11.8 Hz, H-6B of Gal), 4.50 ($J_{16,17}$=8.2 Hz, H-16), 4.56 ($J_{4,5}$=2.0 Hz, H-4 of Gal), 4.82 (d, $J_{1,2}$=7.8 Hz, H-1 of Gal), 5.13 (d, $J_{1,2}$=8.0 Hz, H-1 of Glc).

CMR (C$_5$D$_5$N, 62.0 MHz) : δ C-1) 45.2 (2) 72.0 (3) 86.0 (4) 34.5 (5) 45.1 (6) 28.3 (7) 32.4 (8) 34.85 (9) 54.85 (10) 37.2 (11) 21.6 (12) 40.45 (13) 41.1 (14) 56.6 (15) 32.2 (16) 81.4 (17) 63.4 (18) 16.55 (19) 13.7 (20) 42.3 (21) 14.9 (22) 109.0 (23) 32.1 (24) 29.3 (25) 30.6 (26) 67.15 (27) 17.1 **Gal** (1) 103.65 (2) 72.9 (3) 75.7 (4) 79.85 (5) 75.4 (6) Not reported **Glc** (1) 106.95 (2) 75.3 (3) 78.4 (4) 70.7 (5) 78.7 (6) 63.3.

Reference

1.   S.A. Shvets, A.M. Naibi, P.K. Kintya, and A.Shashkov, *Khim. Prir. Soedin.*, **31**, 391 (1995).

(25*R* and *S*)-SCHIDIGERA-SAPONIN F2

25(*R*,*S*)-5β-Spirostane-2β,3β-diol 3-*O*-[β-D-glucopyranosyl-(1→2)]-β-D-galactopyranoside

Source : *Yucca schidigera* Roezl ex Ortgies (Agavaceae)
Mol. Formula : $C_{39}H_{64}O_{14}$
Mol. Wt. : 756
[α]$_D^{24}$: -57.3° (c=1.07, MeOH)
Registry No. : [267003-05-4]

PMR (C$_5$D$_5$N, 500 MHz) : δ **25*R*** 0.64 (d, *J*=5.3 Hz, 3xH-27), 0.73 (s, 3xH-18), 0.89 (s, 3xH-19), 1.07 (d, *J*=7.1 Hz, 3xH-21); **25*S*** 0.73 (s, 3xH-18), 0.89 (s, 3xH-19), 1.01 (d, *J*=7.0 Hz, 3xH-27), 1.07 (d, *J*=7.1 Hz, 3xH-21), 4.92 (d, *J*=8.3 Hz, H-1 of Gal), 5.22 (d, *J*=7.8 Hz, H-1 of Glc).

CMR (C$_5$D$_5$N, 125 MHz) : δ C-1) 40.5 (2) 67.1 (3) 81.5 (4) 31.8 (5) 36.5 (6) 26.2 (7) 26.8 (8) 35.7 (9) 41.5 (10) 37.1 (11) 21.4 (12) 40.3 (13) 40.9 (14) 56.4 (15) 32.1 (16) 80.3 (17) 63.0 (18) 16.4 (19) 23.8 **25*R*** C-20) 42.0 (21) 14.7 (22) 109.0 (23) 32.1 (24) 29.2 (25) 30.8 (26) 67.1 (27) 17.2 **25*S*** C-20) 42.5 (21) 14.7 (22) 109.6 (23) 26.5 (24) 26.2 (25) 27.5 (26) 65.1 (27) 16.3 **Gal** (1) 103.0 (2) 81.5 (3) 76.7 (4) 69.8 (5) 76.7 (6) 63.0 **Glc** (1) 105.9 (2) 75.2 (3) 78.2 (4) 72.0 (5) 78.0 (6) 62.0.

Mass (FAB, Negative ion) : *m/z* 755 [M-H]⁻, 593 [M-Glc-H]⁻.

Mass (FAB, Negative ion, H.R.) : *m/z* 755.4193 [(M-H)⁻, calcd. for 755.4218].

Reference

1.   M. Miyakoshi, Y. Tamura, H. Masuda, K. Mizutani, O. Tanaka, T. Ikeda, K. Ohtani, R. Kasai and K. Yamasaki, *J. Nat. Prod.*, **63**, 332 (2000).

TIMOSAPONIN G'

(5β,25S)-Spirostan-3β,23α-diol 3-O-[β-D-glucopyranosyl-(1→2)-β-D-galactopyranoside]

Source : *Anemarrhena asphodeloides* Bge. (Liliaceae)
Mol. Formula : $C_{39}H_{64}O_{14}$
Mol. Wt. : 756
M.P. : >210°C (decomp.)
[α]$_D^{25}$: -42.2° (MeOH)
Registry No. : [254751-21-8]

IR (KBr) **:** 986, 923, 900, 845 cm^{-1} (intensity 923 > 900).

PMR (C_5D_5N, 300 MHz) **:** δ 0.90 (3xH-19), 0.99 (3xH-18), 1.07 (3xH-27), 1.20 (3xH-21), 4.90 (d, *J*=7.5 Hz), 5.27 (d, *J*=7.5 Hz).

CMR (C_5D_5N, 75 MHz) **:** δ C-1) 31.0 (2) 27.0 (3) 75.2 (4) 31.0 (5) 36.9 (6) 26.8 (7) 26.8 (8) 35.4 (9) 40.3 (10) 35.3 (11) 21.2 (12) 40.7 (13) 41.5 (14) 56.5 (15) 32.1 (16) 81.9 (17) 62.6 (18) 18.2 (19) 24.0 (20) 36.1 (21) 14.6 (22) 112.5 (23) 63.3 (24) 36.0 (25) 30.5 (26) 64.3 (27) 17.6 **Gal** (1) 102.6 (2) 81.9 (3) 77.0 (4) 69.9 (5) 76.6 (6) 62.2 **Glc** (1) 106.2 (2) 75.5 (3) 78.0 (4) 71.8 (5) 78.4 (6) 62.8.

Mass (FAB) **:** *m/z* 779 [M+Na]$^+$, 757 [M+H]$^+$, 739 [M-H₂O+H]$^+$, 595 [M-162+H]$^+$, 415 [M-2x162-H₂O+H]$^+$, 397 [M-2x162xH₂O+H]$^+$.

Reference

1.   Z.Y. Meng, W. Li, S. Xu, X. Qi and Y. Sha, *Yaoxue Xuebao (Acta Pharm. Sin.)*, **34**, 451 (1999).

TUBEROSIDE J

(25S,5α)-Spirostan-2α,3β,27-triol 3-O-[α-L-rhamnopyranosyl-(1→2)-β-D-glucopyranoside]

Source : *Allium tuberosum* Rottl. (Liliaceae)
Mol. Formula : $C_{39}H_{64}O_{14}$
Mol. Wt. : 756
$[\alpha]_D^{24}$: -35.1° (c=0.28, MeOH)
Registry No. : [373647-08-6]

IR (KBr) : 3400, 1452, 1381, 1045, 989, 912, 816 cm⁻¹.

PMR (C_5D_5N, 400 MHz) **:** δ 0.87 (s, 3xH-18), 0.97 (s, 3xH-19), 1.25 (d, *J*=7.1 Hz, 3xH-21), 1.82 (d, *J*=6.1 Hz, 3xH-6 of Rha), 1.90 (m, H-17), 3.37 (dd, *J*=7.3, 0.6 Hz, H-26A), 3.81 (dd, *J*=5.2, 10.6 Hz, H-26B), 3.95 (m, H-27A), 3.97 (m, H-3), 4.08 (m, H-5 of Glc), 4.16 (m, H-4 of Glc), 4.20 (m, H-2), 4.21 (m, H-27B), 4.29 (dd, *J*=7.7, 9.0 Hz, H-2 of Glc), 4.34 (m, H-3 of Glc), 4.38 (m, H-4 of Rha), 4.46 (m, H-6A of Glc), 4.61 (m, H-6B of Glc), 4.62 (m, H-16), 4.64 (m, H-3 of Rha), 4.88 (m, H-2 of Rha), 4.96 (m, H-5 of Rha), 5.14 (d, *J*=7.5 Hz, H-1 of Glc), 6.41 (s, H-1 of Rha).

CMR (C_5D_5N, 100 MHz) **:** δ C-1) 45.6 (2) 70.5 (3) 85.2 (4) 33.5 (5) 44.5 (6) 27.9 (7) 32.1 (8) 34.4 (9) 54.2 (10) 36.7 (11) 21.3 (12) 39.9 (13) 40.6 (14) 56.2 (15) 32.0 (16) 81.0 (17) 62.8 (18) 16.4 (19) 13.3 (20) 41.9 (21) 14.8 (22) 109.5 (23) 31.4 (24) 23.9 (25) 39.0 (26) 64.2 (27) 63.9 **Glc** (1) 101.1 (2) 77.9 (3) 79.3 (4) 31.3 (5) 78.1 (6) 62.3 **Rha** (1) 102.0 (2) 72.3 (3) 72.6 (4) 73.9 (5) 69.3 (6) 18.4.

Reference

1.   S. Sang, M. Zou, Z. Xia, A. Lao, Z. Chen and C.-T. Ho, *J. Agric. Food Chem.*, **49**, 4780 (2001).

TUBEROSIDE P

(25S)-Spirostan-2β,3β,5β-triol 3-O-[α-L-rhamnopyranosyl-(1→4)-β-D-glucopyranoside]

Source : *Allium tuberosum* L. (Liliaceae)
Mol. Formula : $C_{39}H_{64}O_{14}$
Mol. Wt. : 756
Registry No. : [651306-81-9]

IR (KBr) : 3400, 1452, 1041, 986, 922, 900, 850 cm^{-1}.

PMR (C_5D_5N, 400 MHz) : δ 0.91 (s, 3xH-18), 1.16 (d, J=7.0 Hz, 3xH-27), 1.21 (s, 3xH-19), 1.22 (d, J=6.4 Hz, 3xH-21), 1.88 (d, J=6.2 Hz, 3xH-6 of Rha), 1.90 (m, H-17), 3.45 (d, J=11.8 Hz, H-26A), 3.81 (br d, J=8.5 Hz, H-5 of Glc), 3.98 (t, J=8.4 Hz, H-2 of Glc), 4.10 (m, H-2), 4.15 (m, H-6A of Glc), 4.25 (t, J=9.0 Hz, H-3 of Glc), 4.30 (m, H-6B of Glc), 4.40 (H-4 of Rha), 4.45 (t, J=9.3 Hz, H-4 of Glc), 4.60 (m, H-3 of Rha), 4.72 (m, H-2 of Rha), 5.00 (m, H-5 of Rha), 5.07 (d, J=7.8 Hz, H-1 of Glc I), 5.90 (br s, H-1 of Rha).

CMR (C_5D_5N, 100 MHz) : δ C-1) 35.4 (2) 65.9 (3) 78.9 (4) 35.6 (5) 72.9 (6) 34.9 (7) 28.9 (8) 34.5 (9) 44.4 (10) 42.9 (11) 21.6 (12) 39.9 (13) 40.4 (14) 56.2 (15) 32.0 (16) 81.0 (17) 62.6 (18) 16.1 (19) 17.4 (20) 42.3 (21) 14.7 (22) 109.5 (23) 26.2 (24) 26.0 (25) 27.3 (26) 64.9 (27) 16.3 **Glc** (1) 101.7 (2) 74.7 (3) 76.5 (4) 78.1 (5) 77.3 (6) 60.9 **Rha** (1) 102.5 (2) 72.3 (3) 72.5 (4) 73.7 (5) 70.2 (6) 18.3.

Mass (FAB, Positive ion) : m/z 757 [M+H]$^+$. 611 [M+H-Rha]$^+$, 449 [M+H-Rha-Glc]$^+$.

Reference

1.   S. Sang, S. Mao, A. Lao, Z. Chen and C.-T. Ho, *Food Chemistry*, **83**, 499 (2003).

YUCCA GLORIOSA SAPONIN YS-V

25*R*,5β-Spirostan-2β,3β-diol 3-O-β-D-glucopyranosyl-(1→2)-β-D-glucopyranoside

Source : *Yucca gloriosa* L. (Agavaceae)
Mol. Formula : $C_{39}H_{64}O_{14}$
Mol. Wt. : 756
M.P. : 258-261°C
[α]$_D^{26}$: -75.0° (c=1.0, CHCl$_3$-MeOH, 1:1)
Registry No. : [122537-18-2]

IR (KBr) : 3200-3500 (OH), 990, 930, 905, 870 cm^{-1} (905 > 930, 25*R*-spiroketal).

CMR (C$_5$D$_5$N) : δ C-1) 40.5 (2) 67.2 (3) 81.7 (4) 31.9 (5) 36.5 (6) 26.3 (7) 26.8 (8) 35.6 (9) 41.1 (10) 37.1 (11) 21.3 (12) 40.3 (13) 40.8 (14) 56.4 (15) 32.1^a (16) 81.2 (17) 63.2 (18) 16.5 (19) 23.9 (20) 42.0 (21) 15.0 (22) 109.1 (23) 31.9^a (24) 29.3 (25) 30.6 (26) 66.9 (27) 17.3 **Gal** (1) 103.2 (2) 81.8 (3) 76.8 (4) 69.7 (5) 76.8 (6) 62.9^b **Glc** (1) 106.1 (2) 75.1 (3) 78.3^c (4) 71.7 (5) 78.0^c (6) 62.0^b.

Mass (FAB, Positive ion) : *m/z* 757 [M+H]$^+$.

Reference

1. K. Nakano, T. Yamasaki, Y. Imamura, K. Murakami, Y. Takaishi and T. Tomimatsu, *Phytochemistry*, **28**, 1215 (1989).

YUCCA GLORIOSA SAPONIN YS-XI

12β-Hydroxysmilagenin 3-O-β-D-glucopyranosyl-(1→2)-β-D-galactopyranoside

Source : *Yucca gloriosa* L. (Agavaceae)
Mol. Formula : $C_{39}H_{64}O_{14}$
Mol. Wt. : 756
M.P. : 281-282°C
$[\alpha]_D^{26}$: -26.6° (c=1.05, MeOH)
Registry No. : [135707-33-4]

IR (KBr) : 3200-3500, 980, 920, 900, 860 cm^{-1} (900 > 920).

CMR (C$_5$D$_5$N, 100 MHz) : δ C-1) 30.9 (2) 26.7 (3) 76.5 (4) 31.4 (5) 36.8 (6) 26.7 (7) 27.1 (8) 34.1 (9) 39.5 (10) 35.3 (11) 31.9^a (12) 79.5 (13) 46.7 (14) 55.3 (15) 31.0^a (16) 81.3 (17) 63.0 (18) 11.2 (19) 23.9 (20) 43.0 (21) 14.3 (22) 109.5 (23) 31.9 (24) 29.4 (25) 30.7 (26) 66.9 (27) 17.4 **Gal** (1) 102.4 (2) 81.7 (3) 75.5 (4) 69.8 (5) 76.7 (6) 62.9 **Glc** (1) 105.9 (2) 75.2 (3) 78.0^b 94) 71.8 (5) 78.2^b (6) 62.0.

Mass (FAB, Positive ion) : *m/z* 779 [M+Na]$^+$.

Reference

1. K. Nakano, Y. Hara, K. Murakami, Y. Takaishi and T. Tomimatsu, *Phytochemistry*, **30**, 1993 (1991).

GLUCOCONVALLASAPONIN B
Convallagenin B 3-O-[β-D-glucopyranoside-5-O-α-L-arbainopyranoside]

Source : *Convallaria keiskei* Miq. (Liliaceae)
Mol. Formula : $C_{38}H_{62}O_{15}$
Mol. Wt. : 758
M.P. : 221-222.5°C
[α]$_D$: -35°
Registry No. : [16939-88-1]

IR (Nujol) : 3500-3200 (OH), 980, 920>895, 850 cm^{-1} (25S-spirostane).

Reference

1.   M. Kimura, M. Tohma and I. Yoshizawa, *Chem. Pharm. Bull.*, **15**, 129 (1967).

HEMEROSIDE A
(24S,25R)-5β-Spirostan-1β,2β,3α,24-tetraol 1-O-α-L-arabinopyranoside 24-O-β-D-glucopyranoside

Source : *Hemerocallis fulva* var. *kwanso* Regel (Liliaceae)
Mol. Formula : $C_{38}H_{62}O_{15}$
Mol. Wt. : 758
M.P. : 120-125°C
[α]$_D^{26}$: -17.6° (c=0.9, MeOH)
Registry No. : [337515-94-3]

IR (KBr) : 3400 (OH), 997, 949 > 898, 843 cm^{-1} (*S*-spiroketal).

PMR (C$_5$D$_5$N, 600 MHz) : δ 0.77 (s, 3xH-18), 0.95 (m, H-12ax), 0.95 (m, H-14), 1.06 (m, H-7ax), 1.10 (d, *J*=7.0 Hz, 3xH-21), 1.26 (m, H-6ax), 1.26 (m, H-7eq), 1.31 (ddd, *J*=4.0, 6.0, 11.0 Hz, H-15α), 1.32 (d, *J*=6.0 Hz, H-27), 1.39 (m, H-11), 1.39 (s, 3xH-19), 1.44 (br, dd, *J*=4.0, 10.0 Hz, H-9), 1.58 (br, dd, *J*=4.0, 10.0 Hz, H-8), 1.62 (br, dt, *J*=2.0, 10.0 Hz, H-12eq), 1.71 (m, H-6eq), 1.76 (dd, *J*=6.5, 8.6 Hz, H-17), 1.86 (ddd, *J*=4.0, 4.0, 11.5 Hz, H-4eq), 1.95 (dd, *J*=6.5, 7.0 Hz, H-20), 1.96 (ddd, *J*=5.0, 6.0, 11.0 Hz, H-15b), 2.06 (m, H-5), 2.11 (dd, *J*=5.0, 11.0 Hz, H-23eq), 2.12 (dd, *J*=9.0, 11.0 Hz, H-23ax), 2.22 (ddd, *J*=10.0, 11.5, 11.5 Hz, H-4ax), 2.27 (ddt, *J*=2.0, 4.0, 6.0 Hz, H-25), 3.52 (dd, *J*=2.0, 11.0 Hz, H-26ax), 3.81 (dd, *J*=1.5, 12.1 Hz, H-5A of Ara), 3.92 (ddd, *J*=2.3, 5.0, 8.5 Hz, H-5 of Glc), 3.95 (dd, *J*=2.0, 11.0 Hz, H-26eq), 4.05 (dd, *J*=7.5, 8.3 Hz, H-2 of Glc), 4.11 (dd, *J*=2.0, 9.2 Hz, H-2ax), 4.15 (dd, *J*=3.6, 8.9 Hz, H-3 of Ara), 4.23 (d, *J*=2.0 Hz, H-1), 4.25 (ddd, *J*=1.5, 2.2, 3.6 Hz, H-4 of Ara), 4.26 (dd, *J*=8.3, 9.3 Hz, H-3 of Glc), 4.30 (dd, *J*=8.5, 9.3 Hz, H-4 of Glc), 4.35 (dd, *J*=2.2, 12.1 Hz, H-5B of Ara), 4.38 (dd, *J*=5.0, 12.0 Hz, H-6A of Glc), 4.47 (dd, *J*=2.3, 12.0 Hz, H-6B of Glc), 4.52 (dd, *J*=7.3, 8.9 Hz, H-2 of Ara), 4.53 (ddd, *J*=6.0, 6.0, 8.6 Hz, H-16), 4.60 (ddd, *J*=4.0, 9.2, 11.5 Hz, H-3), 4.81 (ddd, *J*=4.0, 5.0, 9.0 Hz, H-24), 5.03 (d, *J*=7.5 Hz, H-1 of Glc), 5.12 (d, *J*=7.3 Hz, H-1 of Ara).

CMR (C$_5$D$_5$N, 150 MHz) : δ C-1) 89.7 (2) 75.1 (3) 71.8 (4) 35.1 (5) 36.4 (6) 36.1 (7) 26.4 (8) 35.6 (9) 42.1 (10) 41.6 (11) 21.1 (12) 40.0 (13) 40.6 (14) 56.2 (15) 32.0 (16) 81.6 (17) 62.4 (18) 16.6 (19) 19.2 (20) 42.5 (21) 14.7 (22) 111.3 (23) 34.2 (24) 73.0 (25) 31.8 (26) 64.3 (27) 9.9 Ara (1) 108.1 (2) 74.1 (3) 75.4 (4) 69.9 (5) 67.7.

Mass (FAB, Positive ion, H.R.) : *m/z* 781.4001 [M+Na]$^+$.

Reference

1.   T. Konishi, Y. Fujiwara, T. Konoshima, S. Kiyosawa, M. Nishi and K. Miyahara, *Chem. Pharm. Bull.*, **49**, 318 (2001).

LIRIOPROLIOSIDE B

25(*S*)-Ruscogenin 1-O-[(3-O-acetyl)-α-L-rhamnopyranosyl-(1→2)-β-D- fucopyranoside]

Source : *Liriope spicata* (Thunb.) Lour. var. Y.T. Ma. *prolifera* (Liliaceae)
Mol. Formula : C$_{41}$H$_{64}$O$_{13}$
Mol. Wt. : 764
M.P. : 185-188°C
[α]$_D$^{19} : -64.4° (c=0.51, Pyridine.)
Registry No. : [182284-68-0]

Isolated admixed with Ophiopogonin A (qv).

IR : 3600-3200 (OH), 1734 (CO), 982, 922, 900, 860 (spiroketal) cm^{-1}.

PMR (C$_5$D$_5$N, 400/300 MHz) : δ 0.88 (s, 3xH-18), 1.06 (d, J=6.7 Hz, 3xH-27), 1.08 (d, J=6.2 Hz, 3xH-21), 1.44 (s, 3xH-19), 1.66 (d, J=6.4 Hz, 3xH-6 of Fuc), 1.78 (d, J=5.9 Hz, 3xH-6 of Rha), 2.51 (s, OCOCH_3), 4.67 (d, J=7.8 Hz, H-1 of Fuc), 5.59 (d, J=5.4 Hz, H-6), 5.91 (dd, J=3.1, 9.8 Hz, H-3 of Rha), 6.38 (br s, H-1 of Rha).

CMR (C$_5$D$_5$N, 100/75 MHz) : δ C-1) 84.6 (2) 38.5 (3) 68.7 (4) 44.2 (5) 139.7 (6) 124.7 (7) 33.7 (8) 32.6 (9) 51.2 (10) 43.4 (11) 24.6 (12) 41.0 (13) 40.8 (14) 57.6 (15) 33.0 (16) 81.5 (17) 63.7 (18) 17.4 (19) 15.5 (20) 43.0 (21) 15.3 (22) 109.9 (23) 27.0 (24) 26.8 (25) 28.1 (26) 65.5 (27) 16.9 **Fuc** (1) 100.5 (2) 75.3 (3) 76.9 (4) 73.5 (5) 71.4 (6) 17.7 **Rha** (1) 101.7 (2) 70.4 (3) 76.9 (4) 71.3 (5) 69.9 (6) 19.4 (OCOCH$_3$) 170.8 (OCOCH$_3$) 21.7.

Mass (FAB, Positive ion) : m/z 765 [M+H]$^+$, 723 [M+H-Ac]$^+$, 577 [M+H-Ac-146]$^+$, 431 [M+H-Ac-146-146, Agl+H]$^+$, 413 [Agl+H-H$_2$O]$^+$.

Mass (FAB, Negative ion) : m/z 763 [M-H]$^-$, 721 [M-H-Ac]$^-$.

Reference

1.   B.-Y. Yu, S.-X. Qiu, K. Zaw, G.-J. Xu, Y. Hirai, J. Shoji, H.H.S. Fong and A.D. Kinghorn, *Phytochemistry*, **43**, 201 (1996).

LIRIOPROLIOSIDE C
25(S)-Ruscogenin 1-O-[(2-O-acetyl)-α-L-rhamnopyranosyl-(1→2)-β-D-fucopyranoside]

Source : *Liriope spicata* (Thunb.) Lour. var. *prolifera*
Y.T. Ma. (Liliaceae)
Mol. Formula : C$_{41}$H$_{64}$O$_{13}$
Mol. Wt. : 764
M.P. : 203-207°C
$[\alpha]_D^{19}$: -101.6° (c=0.50, Pyridine)
Registry No. : [182284-69-1]

Isolated admixed with Lirioprolioside D (qv).

IR : 3600-3200 (OH), 1734 (CO), 982, 922, 900, 860 (spiroketal) cm^{-1}.

PMR (C$_5$D$_5$N, 400/300 MHz) : δ 0.89 (s, 3xH-18), 1.07 (d, *J*=6.7 Hz, 3xH-27), 1.08 (d, *J*=6.2 Hz, 3xH-21), 1.45 (s, 3xH-19), 1.58 (d, *J*=6.4 Hz, 3xH-6 of Fuc), 1.78 (d, *J*=5.9 Hz, 3xH-6 of Rha), 2.51 (s, C*H*$_3$-CO-), 4.68 (d, *J*=7.8 Hz, H-1 of Fuc), 5.60 (d, *J*=4.7 Hz, H-6), 6.05 (t, *J*=2.1 Hz, H-2 of Rha), 6.27 (br s, H-1 of Rha).

CMR (C$_5$D$_5$N, 100/75 MHz) : δ C-1) 84.2 (2) 38.4 (3) 68.6 (4) 44.4 (5) 139.7 (6) 124.8 (7) 33.7 (8) 32.7 (9) 51.2 (10) 43.4 (11) 24.5 (12) 41.0 (13) 40.8 (14) 57.7 (15) 32.9 (16) 81.5 (17) 63.6 (18) 17.4 (19) 15.4 (20) 43.1 (21) 15.3 (22) 109.9 (23) 27.1 (24) 26.8 (25) 28.2 (26) 65.5 (27) 16.9 **Fuc** (1) 100.2 (2) 75.0 (3) 76.6 (4) 73.6 (5) 71.4 (6) 17.7 **Rha** (1) 98.8 (2) 74.6 (3) 70.8 (4) 74.4 (5) 69.7 (6) 19.4 (OCOCH$_3$) 170.3 (OCOCH$_3$) 21.6.

Mass (FAB, Positive ion) : *m/z* 765 [M+H]$^+$, 723 [M+H-Ac]$^+$, 577 [M+H-Ac-146]$^+$, 432 [M+2H-Ac-146-146, Agl+2H]$^+$, 414 [Agl+H-H$_2$O]$^+$.

Mass: (FAB, Negative ion) : *m/z* 763 [M-H]$^-$, 721 [M-H-Ac]$^-$.

Reference

1.   B.-Y. Yu, S.-X. Qiu, K. Zaw, G.-J. Xu, Y. Hirai, J. Shoji, H.H.S. Fong and A.D. Kinghorn, *Phytochemistry*, **43**, 201 (1996).

LIRIOPROLIOSIDE D
Ruscogenin 1-O-[2-O-acetyl-α-L-rhamnopyranosyl (1→2)]-β-D-fucopyranoside}

Source : *Liriope spicata* (Thunb.) Lour. var. *prolifera*
Y.T. Ma. (Liliaceae)
Mol. Formula : C$_{41}$H$_{64}$O$_{13}$
Mol. Wt. : 764
M.P. : 203-207°C
[α]$_D^{19}$ **:** -101.6° (c=0.50, Pyridine)
Registry No. : [182284-70-4]

Isolated admixed with Lirioprolioside C (qv).

IR : 3600-3200 (OH), 1734 (CO), 982, 922, 900, 860 (spiroketal) cm^{-1}.

PMR (C$_5$D$_5$N, 400/300 MHz) : δ 0.69 (s, 3xH-27), 0.89 (s, 3xH-18), 1.08 (d, *J*=6.2 Hz, 3xH-21), 1.45 (s, 3xH-19), 1.58 (d, *J*=6.4 Hz, 3xH-6 of Fuc), 1.78 (d, *J*=5.9 Hz, 3xH-6 of Rha), 2.51 (s, C*H*$_3$-CO-), 4.68 (d, *J*=7.8 Hz, H-1 of Fuc), 5.60 (d, *J*=4.7 Hz, H-6), 6.05 (t, *J*=2.1 Hz, H-2 of Rha), 6.27 (br s, H-1 of Rha).

CMR (C$_5$D$_5$N, 100/75 MHz) : δ C-1) 84.2 (2) 38.4 (3) 68.6 (4) 44.4 (5) 139.7 (6) 124.8 (7) 33.7 (8) 32.7 (9) 51.2 (10) 43.4 (11) 24.5 (12) 41.0 (13) 40.8 (14) 57.7 (15) 32.9 (16) 81.5 (17) 63.6 (18) 17.4 (19) 15.4 (20) 42.6 (21) 15.6 (22) 109.4 (23) 32.4 (24) 29.9 (25) 31.2 (26) 67.3 (27) 17.9 **Fuc** (1) 100.2 (2) 75.0 (3) 76.6 (4) 73.6 (5) 71.4 (6) 17.7 **Rha** (1) 98.8 (2) 74.6 (3) 70.8 (4) 74.4 (5) 69.7 (6) 19.4 (OCOCH$_3$) 170.3 (OCOCH$_3$) 21.6.

Mass (FAB, Negative ion) : *m/z* 763 [M-H]$^-$, 721 [M-H-Ac]$^-$.

Mass (FAB, Positive ion) : *m/z* 765 [M+H]$^+$, 723 [M+H-Ac]$^+$, 577 [M+H-Ac-146]$^+$, 432 [M+2H-Ac-146-146, Aglycone+2H]$^+$, 414 [Aglycone+H-H$_2$O]$^+$.

Reference

1.   B.-Y. Yu, S.-X. Qiu, K. Zaw, G.-J. Xu, Y. Hirai, J. Shoji, H.H.S. Fong and A.D. Kinghorn, *Phytochemistry* **43**, 201 (1996).

OPHIOPOGONIN A
Ruscogenin 1-O-[(3-O-acetyl)-α-L-rhamnopyranosyl-(1→2)-β-D-fucopyranoside]

Source : *Ophiopogon japonicus* Ker-Gawl. (Liliaceae)[1], *Liriope spicata* (Thunb.) Lour. var. *prolifera* Y.T. Ma. (Liliaceae)[2]
Mol. Formula : C$_{41}$H$_{64}$O$_{13}$
Mol. Wt. : 764
M.P. : 185-188°C
[α]$_D^{19}$: -64.4° (c=0.51, Pyridine)
Registry No. : [11054-24-3]

IR[2] : 3600-3200 (OH), 1734 (CO), 982, 922, 900, 860 (spiroketal) cm^{-1}.

PMR (C$_5$D$_5$N, 400/300 MHz)[2] : δ 0.69 (s, 3xH-27), 0.88 (s, 3xH-18), 1.08 (d, *J*=6.2 Hz, 3xH-21), 1.44 (s, 3xH-19), 1.66 (d, *J*=6.4 Hz, 3xH-6 of Fuc), 1.78 (d, *J*=5.9 Hz, 3xH-6 of Rha), 2.51 (C*H*$_3$CO), 4.67 (d, *J*=7.8 Hz, H-1 of Fuc), 5.59 (d, *J*=5.4 Hz, H-6), 5.91 (dd, *J*=3.1, 9.8 Hz, H-3 of Rha), 6.38 (br s, H-1 of Rha).

CMR (C$_5$D$_5$N, 100/75 MHz)[2] : δ C-1) 84.6 (2) 38.5 (3) 68.7 (4) 44.2 (5) 139.7 (6) 124.7 (7) 33.7 (8) 32.6 (9) 51.2 (10) 43.4 (11) 24.6 (12) 41.0 (13) 40.8 (14) 57.6 (15) 33.0 (16) 81.5 (17) 63.7 (18) 17.4 (19) 15.5 (20) 42.5 (21) 15.6 (22) 109.4 (23) 32.4 (24) 29.8 (25) 31.1 (26) 67.2 (27) 17.8 **Fuc** (1) 100.5 (2) 75.3 (3) 76.9 (4) 73.5 (5) 71.4 (6) 17.7 **Rha** (1) 101.7 (2) 70.4 (3) 76.9 (4) 71.3 (5) 69.9 (6) 19.4 (OCOCH$_3$) 170.8 (OCOCH$_3$) 21.7.

Mass (FAB, Positive ion)[2] : *m/z* 765 [M+H]$^+$, 723 [M+H-Ac]$^+$, 577 [M+H-Ac-146]$^+$, 431 [M+H-Ac-146-146, Agl+H]$^+$, 413 [Agl+H-H$_2$O]$^+$.

Mass (FAB, Negative ion)^2 : m/z 763 [M-H]$^-$, 721 [M-H-Ac]$^-$.

References

1.   Y. Watanabe, S. Sanada, A. Tada and J. Shoji, *Chem. Pharm. Bull.*, **25**, 3049 (1977).

2.   B.-Y. Yu, S.-X. Qiu, K. Zaw, G.-J. Xu, Y. Hirai, J. Shoji, H.H.S. Fong and A.D. Kinghorn, *Phytochemistry*, **43**, 201 (1996).

ALLIUM CHINENSE SAPONIN 1
Laxogenin 3-O-{(2-O-acetyl)-α-L-arabinopyranosyl-(1→6)-β-D-glucopyranoside}

Source : *Allium chinense* (Liliaceae)
Mol. Formula : $C_{40}H_{62}O_{14}$
Mol. Wt. : 766
$[\alpha]_D^{30}$ **:** -62.9° (c=0.10, MeOH)
Registry No. : [170473-65-1]

IR (KBr) : 3380 (OH), 2920 and 2850 (CH), 1725 and 1705 (C=O), 1455, 1375, 1235, 1170, 1045, 1010, 980, 960, 916, 895, 865, 770 cm^{-1} (intensity 915 < 895, 25R-spiroacetal).

PMR (C_5D_5N, 400 MHz) : δ 0.68 (s, 3xH-19), 0.70 (d, J=5.5 Hz, 3xH-27), 0.79 (s, 3xH-18), 1.15 (d, J=6.9 Hz, 3xH-21), 2.12 (s, OCOCH_3), 3.49 (dd, J=10.6, 10.6 Hz, H-26A), 3.59 (dd, J=10.6, 3.4 Hz, H-26B), 4.54 (q-like, J=6.7 Hz, H-16), 4.98 (d, J=8.5 Hz, H-1 of Glc), 5.00 (d, J=6.7 Hz, H-1 of Ara), 5.88 (dd, J=7.9, 6.7 Hz, H-2 of Ara).

CMR (C_5D_5N, 100 MHz) : δ C-1) 36.8 (2) 29.5 (3) 76.8 (4) 27.0 (5) 56.4 (6) 209.6 (7) 46.7 (8) 37.4 (9) 53.7 (10) 40.9 (11) 21.5 (12) 39.6 (13) 41.1 (14) 56.4 (15) 31.8 (16) 80.8 (17) 62.8 (18) 16.4 (19) 13.1 (20) 41.9 (21) 14.9 (22) 109.2 (23) 31.7 (24) 29.2 (25) 30.6 (26) 66.9 (27) 17.3 **Glc** (1) 102.1 (2) 75.2 (3) 78.5 (4) 72.1 (5) 76.9 (6) 69.7 **Ara** (1) 102.1 (2) 73.7 (3) 72.1 (4) 69.0 (5) 66.1 (OCOCH_3) 170.1 (OCOCH_3) 21.2.

Mass (FAB, Negative ion) : m/z 765 [M-H]$^-$, 723 [M-acetyl]$^-$.

Biological Activity : The compound exhibited strong inhibitory activity on CAMP phosphodiesterase showing an IC50 value of 3.3 x 10^{-5} M.

Reference

1.   M. Kuroda, Y. Mimaki, A. Kameyama, Y. Sashida and T. Nikaido, *Phytochemistry*, **40**, 1071 (1995).

FOLIUMIN

(22*S*,23*R*,25*R*)-3β,15α,23-Trihydroxyspirost-5-ene-26-one 3-O-[α-L-rhamnopyranosyl-(1→2)-
β-D-glucopyranoside]

Source : *Solanum amygdalifolium* Steud. (Solanaceae)
Mol. Formula : $C_{39}H_{60}O_{15}$
Mol. Wt. : 768
$[\alpha]_D^{22}$: -42° (c=0.2, H_2O)
Registry No. : [159126-18-8]

PMR (CD$_3$OD, 600, 400 or 270 MHz) : δ 0.82 (s, H-18), 1.00 (s, 3xH-9), 1.05 (3xH-21), 1.07 (s, 3xH-19), 1.08 (H-1α), 1.20 (s, 3xH-27), 1.22 (H-14α), 1.24 (H-12α), 1.24 (H-6 of Rha), 1.49 (H-11α), 1.56 (H-11β), 1.59 (H-2β), 1.72 (H-12β), 1.74 (H-7α), 1.84 (H-8β), 1.86 (H-1β), 1.88 (H-24B), 1.90 (H-17α), 1.92 (H-2α), 2.30 (H-4β), 2.35 (H-7β), 2.44 (H-4α), 2.45 (H-20β), 2.73 (H-24A), 2.88 (H-25α), 3.24 (H-5 of Glc), 3.28 (H-4 of Glc), 3.35 (H-2 of Glc), 3.39 (H-4 of Rha), 3.47 (H-3 of Glc), 3.59 (H-3α), 3.65 (H-6 of Glc), 3.66 (H-3 of Rha), 3.82 (H-15β), 3.84 (H-6 of Glc), 3.92 (H-2 of Rha), 4.12 (H-5 of Rha), 4.32 (H-16α), 4.38 (H-23β), 4.48 (H-1 of Glc), 5.19 (H-1 of Rha), 5.39 (H-6).

CMR (CD$_3$OD, 150, 100 or 67.8 MHz) : δ C-1) 38.6 (2) 30.7 (3) 79.2 (4) 39.4 (5) 141.1 (6) 123.1 (7) 33.3 (8) 33.0 (9) 51.5 (10) 37.9 (11) 21.7 (12) 41.4 (13) 41.8 (14) 61.2 (15) 80.4 (16) 91.6 (17) 60.5 (18) 18.0 (19) 19.9 (20) 37.8 (21) 15.2 (22) 110.5 (23) 78.7 (24) 31.5 (25) 35.1 (26) 183.2 (27) 16.4 **Glc** (1) 100.5 (2) 79.0 (3) 79.3 (4) 71.8 (5) 77.7 (6) 62.8 **Rha** (1) 102.2 (2) 72.2 (3) 72.4 (4) 73.9 (5) 69.7 (6) 18.0.

Mass: (FAB, Positive ion) : *m/z* 791.6 [M+Na]$^+$, 807.6 [M+K]$^+$.

Mass (FAB, Negative ion) : *m/z* 767.5 [M-H]$^-$, 621.2 [M-Rha]$^-$, 459.2 [M-H-Rha-Glc]$^-$.

Reference

1.   F. Ferreira, A. Vazquez, P. Moyna and L. Kenne, *Phytochemistry*, **36**, 1473 (1994).

AGAVE AMERICANA SAPONIN 1

(25*R*)-3β,6α-Dihroxy-5α-spirostan-12-one 3,6-di-O-[β-D-glucopyranoside]

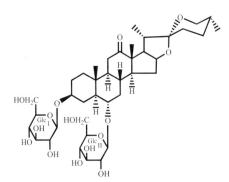

Source : *Agave americana* L. (Agavceae)
Mol. Formula : $C_{39}H_{62}O_{15}$
Mol. Wt. : 770
[α]$_D^{25}$: -57.1° (c=0.11, CHCl$_3$-MeOH)
Registry No. : [284677-34-5]

IR (film) : 3381 (OH), 2925 and 2865 (CH), 1686 (C=O), 1077, 1041 cm^{-1}.

PMR (C$_5$D$_5$N) : δ 0.72 (d, *J*=5.8 Hz, 3xH-27), 0.76 (s, 3xH-19), 0.78 (H-1ax), 0.95 (H-9), 1.05 (s, 3xH-18), 1.16 (H-7ax), 1.23 (H-5), 1.32 (H-1eq), 1.34 (d, *J*=6.9 Hz, 3xH-21), 1.37 (H-14), 1.45 (q-like, *J*=12.1 Hz, H-4ax), 1.54 (H-15β), 1.56 (2xH-24), 1.57 (H-25), 1.61 (H-23ax), 1.63 (H-2ax), 1.68 (H-23eq), 1.90 (H-20), 1.92 (H-8), 2.01 (H-2eq), 2.03 (H-15α), 2.22 (dd, *J*=13.9, 5.0 Hz, H-11eq), 2.37 (dd, *J*=13.9, 13.9 Hz, H-11ax), 2.65 (H-7eq), 2.70 (dd, *J*=8.5, 6.7 Hz, H-17), 3.40 (br d, *J*=12.1 Hz, H-4eq), 3.46 (dd, *J*=10.8, 3.3 Hz, H-26ax), 3.58 (dd, *J*=10.8, 10.8 Hz, H-26ax), 3.69 (ddd, *J*=10.8, 10.8, 4.7 Hz, H-6), 3.84 (ddd, *J*=8.8, 5.3, 2.4 Hz, H-5 of Glc I), 3.94 (H-3), 3.95 (ddd, *J*=8.8, 5.3, 2.5 Hz, H-5 of Glc II), 4.04 (dd, *J*=8.1, 7.6 Hz, H-2 of Glc II), 4.05 (dd, *J*=8.3, 7.6 Hz, H-2 of Glc I), 4.23 (dd, *J*=8.8, 8.3 Hz, H-3 of Glc I), 4.24 (dd, *J*=8.8, 8.8 Hz, H-4 of Glc II), 4.26 (dd, *J*=8.8, 8.8 Hz, H-4 of Glc I), 4.26 (dd, *J*=8.8, 8.8 Hz,, H-3 of Glc II), 4.32 (dd, *J*=11.8, 5.3 Hz, H-6A of Glc I), 4.36 (q-like, *J*=6.7 Hz, H-16), 4.41 (dd, *J*=11.6, 5.3 Hz, H-6A Glc II), 4.42 (dd, *J*=11.8, 2.4 Hz, H-6B of Glc I), 4.90 (d, *J*=7.8 Hz, H-1 of Glc II), 5.10 (d, *J*=7.6 Hz, H-1 of Glc I), 5.54 (dd, *J*=11.6, 2.5 Hz, H-6B of Glc II).

CMR (C$_5$D$_5$N) : δ C-1) 36.9 (2) 29.6 (3) 76.7 (4) 28.6 (5) 50.7 (6) 79.5 (7) 40.7 (8) 33.1 (9) 54.8 (10) 36.8 (11) 37.8 (12) 212.6 (13) 55.2 (14) 55.7 (15) 31.3 (16) 79.5 (17) 54.3 (18) 16.1 (19) 12.8 (20) 42.6 (21) 13.9 (22) 109.2 (23) 31.8 (24) 29.2 (25) 30.5 (26) 66.9 (27) 17.3 **Glc I** (1) 101.6 (2) 75.6 (3) 78.5 (4) 71.8 (5) 78.1 (6) 62.6 **Glc II** (1) 106.3 (2) 75.4 (3) 78.6 (4) 71.7 (5) 78.1 (6) 63.0.

Mass (FAB, Positive ion, H.R.) : *m/z* 771.4185 [(M+H)$^+$, calcd. for 771.4167].

Reference

1.   A. Yokosuka, Y. Mimaki, M. Kuroda, Y.Sashida, *Planta Med.*, **66,** 393 (2000).

DRACAENA SURCULOSA SAPONIN 3

(24*S*,25*R*)-1β,6β,24β-Trihydroxy-3α,5α-cyclospirostane-1,24-*bis*-O-β-D-glucopyranoside

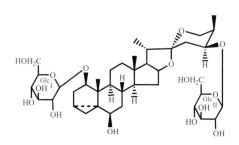

Source : *Dracaena surculosa* Lindle. (Agavaceae)
Mol. Formula : $C_{39}H_{62}O_{15}$
Mol. Wt. : 770
$[\alpha]_D^{26}$ **:** -42.0° (c=0.10, MeOH)
Registry No. : [463962-99-4]

IR (film) : 3388 (OH), 2926 (CH), 1077 cm⁻¹.

CMR (C_5D_5N, 500 MHz) : δ 0.56 (dd, *J*=7.9, 3.9 Hz, H-4A), 0.83 (s, 3xH-18), 1.11 (d, *J*=7.0 Hz, 3xH-21), 1.14 (m, H-3), 1.32 (d, *J*=6.9 Hz, 3xH-27), 1.58 (t-like, *J*=3.9 Hz, H-4B), 1.65 (s, 3xH-19), 3.47 (br s, H-6), 4.41 (br d, *J*=5.3 Hz, H-1), 4.52 (m, H-16), 4.78 (d, *J*=7.9 Hz, H-1 of Glc I), 4.79 (ddd, *J*=10.9, 10.8, 6.4 Hz, H-24), 5.04 (d, *J*=7.8 Hz, H-1 of Glc II).

CMR (C_5D_5N, 125 MHz) : δ C-1) 85.0 (2) 33.0 (3) 23.2 (4) 15.9 (5) 39.9 (6) 73.0 (7) 38.6 (8) 30.0 (9) 50.0 (10) 49.1 (11) 23.3 (12) 40.5 (13) 40.9 (14) 56.5 (15) 32.0 (16) 81.6 (17) 62.4 (18) 16.7 (19) 16.9 (20) 42.5 (21) 14.7 (22) 111.2 (23) 34.1 (24) 72.8 (25) 31.7 (26) 64.2 (27) 9.9 **Glc I** (1) 103.1 (2) 74.9 (3) 78.9 (4) 72.0 (5) 78.2 (6) 63.1 **Glc II** (1) 101.2 (2) 75.3 (3) 78.7 (4) 71.5 (5) 78.4 (6) 62.5.

Mass (FAB, Positive ion) : *m/z* 793 [M+Na]⁺.

Mass (MADI-TOF, H.R.) : *m/z* 793.3990 [calcd. for $C_{39}H_{62}O_{15}Na$: 793.3986].

Reference

1.   A. Yokosuka, Y. Mimaki and Y. Sashida, *Chem. Pharm. Bull.*, **50**, 992 (2002).

SCOPOLOSIDE II
Scopologenin 3-O-[β-D-glucopyranosyl-(1→4)-β-D-galactopyranoside]

Source : *Scopolia japonica* Maxim (Solanaceae)
Mol. Formula : $C_{39}H_{62}O_{15}$
Mol. Wt. : 770
M.P. : 229-234°C
$[\alpha]_D^{22}$ **:** -58.3 (c=0.49, MeOH)
Registry No. : [148332-58-5]

IR (KBr) : 3426 (OH) cm^{-1}.

PMR (C_5D_5N, 400 MHz) : δ 0.68 (d, *J*=5.9 Hz, 3xH-27), 0.88 (s, 3xH-18), 1.12 (s, 3xH-19), 1.20 (d, *J*=7.0 Hz, 3xH-21), 3.36 (t, *J*=10.7 Hz, 3xH-26ax), 3.46 (m, H-26eq) 4.88 (d, *J*=7.7 Hz, H-1 of Gal), 5.29 (d, *J*=8.0 Hz, H-1 of Glc), 5.40 (br s, H-1).

CMR (C_5D_5N, 100 MHz) : δ C-1) 37.5 (2) 30.2 (3) 77.9 (4) 39.1 (5) 140.1 (6) 122.5 (7) 33.0 (8) 32.2 (9) 50.2 (10) 36.9 (11) 20.9 (12) 40.7 (13) 40.7 (14) 60.0 (15) 79.0 (16) 91.3 (17) 60.7 (18) 14.7 (19) 19.3 (20) 35.7 (21) 16.8 (22) 111.4 (23) 67.3 (24) 38.7 (25) 31.6 (26) 65.8 (27) 17.8 **Gal** (1) 102.7 (2) 73.4 (3) 75.1 (4) 79.9 (5) 75.3 (6) 60.8 **Glc** (1) 107.0 (2) 75.8 (3) 78.6 (4) 72.2 (5) 78.4 (6) 63.0.

Mass (FAB, Positive ion, H.R.) : *m/z* 793.4033 [(M+Na)$^+$ requires 793.3986].

Mass (FAB, Negative ion) : *m/z* 769 [M-H]$^-$, 607 [M-H-Hexose]$^-$.

Reference

1. S. Okamura, K. Shingu, S. Yahara, H. Kohoda and T. Nohara, *Chem. Pharm. Bull.*, **40**, 2981 (1992).

SOLADULCOSIDE A

(22R,25R)-3β,15α,23α-Trihydroxy-5α-spirostan-26-one 3-O-α-L-rhamnopyranosyl-(1→2)-β-D-glucopyranoside

Source : *Solanum dulcamara* L. (Solanaceae)
Mol. Formula : $C_{39}H_{62}O_{15}$
Mol. Wt. : 770
[α]$_D$: -73.8°
Registry No. : [137031-53-9]

IR : 1764 cm^{-1}.

PMR (C_5D_5N) : δ 0.91 (s, CH_3), 0.99 (s, CH_3), 1.16 (d, *J*=6.2 Hz, sec. CH_3), 1.29 (d, *J*=6.6 Hz, sec. CH_3), 1.77 (d, *J*=5.7 Hz, sec. CH_3), 5.07 (d, *J*=7.3 Hz, H-1 of Glc), 6.37 (s, H-1 o Rha).

CMR (C_5D_5N) : δ C-1) 37.4 (2) 30.0 (3) 76.8 (4) 34.4 (5) 44.6 (6) 29.2 (7) 32.8 (8) 35.9 (9) 54.6 (10) 36.1 (11) 21.3 (12) 40.9 (13) 41.4 (14) 60.7 (15) 79.0 (16) 92.0 (17) 60.3 (18) 18.1 (19) 12.5 (20) 37.3 (21) 15.6 (22) 109.9 (23) 77.7 (24) 31.1 (25) 34.1 (26) 180.6 (27) 16.3 **Glc** (1) 99.8 (2) 79.6 (3) 78.1 (4) 71.9 (5) 78.3 (6) 62.8 **Rha** (1) 102.2 (2) 72.5 (3) 72.8 (4) 74.1 (5) 69.5 (6) 18.7.

Mass (FAB, Positive ion) : *m/z* (rel.intens.) 793 [M+Na]$^+$.

Reference

1.   T. Yamashita, T. Matsumoto, S. Yahara, N. Yoshida and T. Nohara, *Chem. Pharm. Bull.*, **39**, 1626 (1991).

YUCCA GLORIOSA SAPONIN YS-VI

Mexogenin 3-O-[β-D-glucopyranosyl-(1→2)-β-D-galactopyranoside]

Source : *Yucca gloriosa* L. (Agavaceae)
Mol. Formula : $C_{39}H_{62}O_{15}$
Mol.Wt. : 770
M.P. : 245-247°C
[α]$_D^{26}$: -13.8° (c=0.58, MeOH)
Registry No. : [137853-59-9]

IR (KBr) : 3200-3500 (OH), 1708, 980, 915, 900, 860 (915 < 900) cm^{-1}.

CMR (C_5D_5N 100 MHz) : δ C-1) 40.2 (2) 66.8 (3) 81.6 (4) 31.8 (5) 36.1 (6) 26.1 (7) 26.5 (8) 34.7 (9) 42.7 (10) 37.5 (11) 37.9 (12) 212.7 (13) 55.6 (14) 55.8 (15) 31.8 (16) 79.4 (17) 54.3 (18) 16.0 (19) 23.1 (20) 42.9 (21) 13.9 (22) 109.3 (23) 31.5 (24) 29.2 (25) 30.5 (26) 67.0 (27) 17.3 **Gal** (1) 103.1 (2) 81.6 (3) 76.9 (4) 69.8 (5) 77.0 (6) 62.9 **Glc** (1) 106.1 (2) 75.2 (3) 78.1 (4) 71.8 (5) 78.5 (6) 62.0.

Mass (FAB, Positive ion) : *m/z* 809 [M+K]$^+$, 793 [M+Na]$^+$.

Reference

1.   K. Nakano, Y. Midzuta, Y. Hara, K. Murakami, Y. Takaishi and T. Tomimatsu, *Phytochemistry*, **30**, 633 (1991).

CANTALASAPONIN 1
3β,6α,23α-Trihydroxyspirostane 3,6-*bis*-O-β-D-glucopyranoside

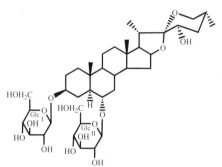

Source : *Agave cantala* Roxb.[1] (Agavaceae), *A. americana*[2]
Mol. Formula : $C_{39}H_{64}O_{15}$
Mol. Wt. : 772
M.P. : 243-245°C[1]
$[\alpha]_D^{15}$: -51.5 (c=2.2, Pyridine)[1]
Registry No. : [98569-61-0]

IR (KBr)[1] : 3400 (OH), 977, 928, 900, 870 cm^{-1}. Intensity 900>928, 25R-spiroketal.

PMR (C_5D_5N, 100 MHz)[1] : δ 0.64 (s, 3xH-18), 0.74 (d, 3xH-27), 0.96 (s, 3xH-19), 1.18 (d, 3xH-21), 4.84 (d, J=7.8 Hz, H-1 of Glc), 5.08 (d, J=7.8 Hz, H-1 of Glc).

CMR (C_5D_5N, 25.0 MHz)[1] : δ C-1) 37.5 (2) 29.8 (3) 79.9 (4) 28.5 (5) 50.8 (6) 77.0 (7) 41.4^a (8) 33.9 (9) 53.8 (10) 36.7 (11) 21.2 (12) 40.4 (13) 41.3^a (14) 56.3 (15) 31.7 (16) 81.6 (17) 62.5 (18) 16.9 (19) 13.3 (20) 38.8 (21) 14.7 (22) 111.6 (23) 67.4 (24) 35.7 (25) 32.0 (26) 65.9 (27) 16.9 **Glc I** (1) 106.2^b (2) 75.4^c (3) 77.9^d (4) 71.7 (5) 78.5^e (6) 63.1^f **Glc II** (1) 101.6^b (2) 75.6^c (3) 78.0^d (4) 71.7 (5) 78.3^e (6) 62.5^f.

Mass (FAB, Positive ion)[1] : *m/z* (rel.intens.) 811 [(M+K)$^+$, 0.34], 795 [(M+Na)$^+$, 2.71], 773 [(M+H)$^+$, 0.58], 772 [(M)$^+$, 0.68], 755 [(M+H-H$_2$O)$^+$, 2.08], 611 [(M+H-Glc)$^+$, 1.34], 593 [(M+H-Glc-H$_2$O)$^+$, 3.44], 575 [(M+H-Glc-2xH$_2$O)$^+$, 1.76], 449 (M+H-2xGlc)$^+$, 2.07], 431 [(M+H-2xGlc-H$_2$O)$^+$, 6.42], 4.13 [(M+H-2x H$_2$O-2x Glc)$^+$, 16.6], 395 (12.25), 363 (1.86), 345 (5.3), 327 (13.3), 271 (33.36), 253 (33.92), 155 (21.82), 142 (23.25), 131 (63.50), 113 (23.14), 105 (100).

Biological Activity : Cytotoxic against JTC-26 cells.[3]

References

1.   O.P. Sati, G. Pant, K. Miyahara and T. Kawasaki, *J. Nat. Prod.*, **48**, 395 (1985).

2.   Y. Yokusaka, Y. Mimaki, M. Kurido and Y. Sashida, *Planta Med.*, **66**, 393 (2000).

3.   O.P. Sati, G. Pant, T. Nohara and A. Sato, *Pharmazie*, **40**, 586 (1985).

TIMOSAPONIN F
(5β,25S)-Spirostan-3β,15α,23α-triol 3-O-[β-glucopyranosyl-(1→2)-β-D-galactopyranoside]

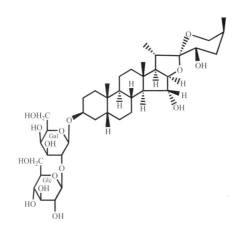

Source : *Anemarrhena asphodeloides*[1,2] Bunge (Liliaceae)
Mol. Formula : $C_{39}H_{64}O_{15}$
Mol. Wt. : 772
M.P. : >200°C[1]
[α]$_D^{25}$: -47.8° (c=0.5, MeOH)[1]
Registry No. : [249729-36-0]

IR (KBr)[1] : 988, 922, 900, 850 (intensity 922>900, 25S-spirostane) cm^{-1}.

PMR[1] : δ 0.95 (s, 3xH-19), 1.09 (s, 3xH-18), 1.09 (s, 3xH-27), 1.20 (3xH-21), 1.54 (H-14), 1.84 (H-24eq), 2.13 (H-17), 2.22 (d, *J*=4.3 Hz, H-24ax), 3.05 (q, H-20), 3.23 (d, *J*=10.1 Hz, H-26ax), 3.82 (H-5 of Glc), 3.90 (d, *J*=10.1 Hz, H-26eq), 4.00 (H-5 of Gal), 4.07 (dd, *J*=7.6, 9.1 Hz, H-2 of Glc), 4.19 (H-3 of Glc), 4.26 (H-3 of Gal), 4.33 (H-4 of Glc), 4.41 (H-6 of Gal), 4.45 (H-6 of Glc), 4.50 (H-15), 4.55 (H-4 of Gal), 4.65 (dd, *J*=8.0, 8.8 Hz, H-2 of Gal), 4.77 (dd, H-16), 4.88 (d, *J*=7.2 Hz, H-1 of Gal), 5.26 (d, *J*=7.2 Hz, H-1 of Glc).

CMR (C$_5$D$_5$N)[1] : δ C-1) 31.0^a (2) 27.2^b (3) 75.5 (4) 31.1^a (5) 36.4 (6) 27.0^b (7) 26.8^b (8) 37.1 (9) 40.4 (10) 35.4 (11) 21.2 (12) 41.4 (13) 41.5 (14) 60.9 (15) 78.8 (16) 91.9 (17) 60.1 (18) 18.2 (19) 24.2 (20) 36.2 (21) 14.7 (22) 112.2 (23) 63.4 (24) 36.0 (25) 30.5 (26) 64.3 (27) 17.6 **Gal** (1) 102.5 (2) 81.9 (3) 75.2 (4) 69.9 (5) 76.6 (6) 62.2 **Glc** (1) 106.2 (2) 77.0 (3) 78.0 (4) 71.7 (5) 78.4 (6) 62.8.

Mass (E.S.I.)[1] : *m/z* 790.4 [M+H$_2$O]$^+$, 609.5 [M-Glc+H$_2$O]$^+$, 430.4 [M-Glc-Gal-H$_2$O]$^+$.

References

1.   Z.-Y. Meng, J.-Y. Zhang, S.-X. Xu and K. Sugahara, *Planta Med.*, **65**, 661 (1999).

2.   Z-Y. Meng, W. Li, S.-X. Xu, X.-G. Qi and Y. Sha, *Huaxue Xuebao*, **34**, 451 (1999).

TUBEROSIDE Q

(24S,25S)-Spirostan-2β,3β,5β-24-tetrol 3-O-[α-L-rhamnopyranosyl-(1→4)-β-D-glucopyranoside]

Source : *Allium tuberosum* L. (Liliaceae)
Mol. Formula : $C_{39}H_{64}O_{15}$
Mol. Wt. : 772
[α]$_D^{24}$: -53.2° (c=0.18, MeOH)
Registry No. : [651306-82-0]

IR (KBr) : 3400, 1452, 1379, 1041, 993, 895 cm^{-1}.

PMR (C$_5$D$_5$N, 400 MHz) : δ 1.21 (s, 3xH-19), 1.26 (d, J=7.0 Hz, 3xH-21), 1.39 (d, J=7.0 Hz, 3xH-27), 1.78 (d, J=6.1 Hz, 3xH-6 of Rha), 3.64 (d, J=10.3 Hz, H-26A), 3.82 (br d, J=8.3, H-5 of Glc), 3.98 (dd, J=8.3, 8.7 Hz, H-2 of Glc), 4.10 (m, H-2), 4.13 (m, H-26B), 4.20 (m, H-6A of Glc), 4.27 (t, J=9.0 Hz, H-3 of Glc), 4.33 (d, J=11.2 Hz, H-6B of Glc), 4.41 (t, J=9.4 Hz, H-4 of Rha), 4.51 (t, J=9.2 Hz, H-4 of Glc), 4.62 (m, H-3 of Rha), 4.65 (m, H-16), 4.68 (m, H-24), 4.73 (m, H-3), 4.77 (m, H-2 of Rha), 5.03 (m, H-5 of Rha), 5.10 (d, J=7.9 Hz, H-1 of Glc), 5.96 (br s, H-1 of Rha).

CMR (C$_5$D$_5$N, 100 MHz) : δ C-1) 35.7 (2) 66.3 (3) 79.3 (4) 36.0 (5) 73.1 (6) 35.2 (7) 29.2 (8) 34.8 (9) 44.7 (10) 43.2 (11) 21.9 (12) 40.1 (13) 40.7 (14) 56.5 (15) 32.2 (16) 81.7 (17) 62.6 (18) 16.6 (19) 17.7 (20) 42.7 (21) 14.9 (22) 111.6 (23) 36.2 (24) 66.6 (25) 36.0 (26) 64.7 (27) 9.9 **Glc** (1) 102.0 (2) 75.0 (3) 76.9 (4) 78.4 (5) 77.7 (6) 61.3 **Rha** (1) 102.9 (2) 72.7 (3) 72.9 (4) 74.1 (5) 70.5 (6) 18.7.

Mass (FAB, Positive ion) : *m/z* 773 [M+H]$^+$. 627 [M+H-Rha]$^+$, 465 [M+H-Rha-Glc]$^+$.

Reference

1.   S. Sang, S. Mao, A. Lao, Z. Chen and C.-T. Ho, *Food Chemistry*, **83**, 499 (2003).

WATTOSIDE I

(5β,25S)-1β,3β,24β-Trihydroxyspirostane 24-O-[β-D-glucopyranosyl-(1→6)-β-D-glucopyranoside]

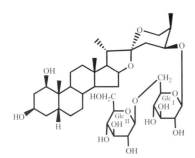

Source : *Tupistra wattii* Hook. f. (Liliaceae)
Mol. Formula : $C_{39}H_{64}O_{15}$
Mol. Wt. : 772
M.P. : 205-207°C
[α]$_D^{20}$: -76.2° (c=0.027, MeOH)
Registry No. : [619319-17-4]

IR (KBr) : 3396 (OH), 2929, 1569, 1381, 1165, 1055, 897 cm^{-1}.

PMR (C_5D_5N, 500 MHz) : δ 0.81 (s, 3xH-18), 1.10 (d, *J*=6.4 Hz, 3xH-21), 1.16 (d, *J*=6.8 Hz, 3xH-27)m, 1.26 (s, 3xH-19), 1.89 (m, H-25), 2.00 (dd, *J*=13.0, 9.8 Hz, H-23ax), 2.73 (dd, *J*=13.0, 4.4 Hz, H-23eq), 3.57 (br t, *J*=11.7 Hz, H-26ax), 3.65 (dd, *J*=11.7, 3.5 Hz, H-26eq), 4.05 (m, H-24), 4.86 (d, *J*=7.7 Hz, H-1 of Glc I), 5.02 (d, *J*=7.6 Hz, H-1 of Glc II).

CMR (C_5D_5N, 125 MHz) : δ C-1) 73.4 (2) 33.0 (3) 68.2 (4) 34.4 (5) 31.3 (6) 26.7 (7) 26.6 (8) 35.7 (9) 42.2 (10) 40.4 (11) 21.1 (12) 40.7 (13) 40.4 (14) 56.5 (15) 32.1 (16) 81.6 (17) 62.2 (18) 16.8 (19) 19.4 (20) 42.2 (21) 14.9 (22) 111.8 (23) 40.7 (24) 81.6 (25) 38.2 (26) 65.1 (27) 13.6 **Glc I** (1) 106.1 (2) 75.1 (3) 78.5 (4) 71.3 (5) 76.7 (6) 70.0 **Glc II** (1) 105.6 (2) 75.5 (3) 75.4 (4) 71.5 (5) 78.5 (6) 62.8.

Mass (FAB, Negative ion, H.R.) : *m/z* 771.4153 [(M-H)⁻, calcd. for 771.4167].

Biological Activity : Cytotoxic against cancer cell line K 562 with IC$_{50}$ 76.96 μmol/l.

Reference

1.   P. Shen, S.-L. Wang, X.-K. Liu, C.-R. Yang, B. Cai and X.-S. Yao, *Chem. Pharm. Bull.*, **51**, 305 (2003).

YUCCA GLORIOSA SAPONIN YS-XIII
12b-Hydroxysamogenin 3-O-[β-D-glucopyranosyl-(1→2)-β-D-galactopyranoside]

Source : *Yucca gloriosa* L. (Agavaceae)
Mol. Formula : $C_{39}H_{64}O_{15}$
Mol.Wt. : 772
$[\alpha]_D^{26}$: -50.0° (c=0.9, CHCl$_3$-MeOH, 1:1)
Registry No. : [135688-76-5]

IR (KBr) : 3200-3500, 980, 920, 900, 860 (900 < 920) cm^{-1}.

CMR (C$_5$D$_5$N, 100 MHz) : δ C-1) 40.5 (2) 67.2 (3) 81.6 (4) 31.6 (5) 36.3 (6) 26.3 (7) 26.7 (8) 34.7 (9) 40.5 (10) 37.1 (11) 31.8^a (12) 79.3 (13) 46.6 (14) 55.2 (15) 31.9^a (16) 81.3 (17) 63.0 (18) 11.2 (19) 23.8 (20) 43.0 (21) 14.4 (22) 109.5 (23) 32.0 (24) 29.3 (25) 30.6 (26) 66.9 (27) 17.4 **Gal** (1) 103.3 (2) 81.8 (3) 76.8 (4) 69.8 (5) 76.9 (6) 62.8 **Glc** (1) 106.1 (2) 75.1 (3) 78.0^b (4) 71.7 (5) 78.5^b (6) 62.0.

Mass (FAB, Positive ion) : *m/z* 795 [M+Na]$^+$.

Reference

1.   K. Nakano, Y. Hara, K. Murakami, Y. Takaishi and T. Tomimatsu, *Phytochemistry*, **30**, 1993 (1991).

RUSCUS ACULEATUS SAPONIN 15

Ruscogenin 1-O-{α-L-rhamnopyranosyl-(1→2)-6-O-acetyl-β-D-galactopyranoside}

Source : *Ruscus aculeatus* L. (Liliaceae)
Mol. Formula : $C_{41}H_{64}O_{14}$
Mol. Wt. : 780
[α]$_D^{25}$: -76.0° (c=0.10, MeOH)
Registry No. : [211036-47-4]

IR (KBr) : 3420 (OH), 2940 (CH), 1730 (C=O), 1445, 1370, 1235, 1135, 1050, 980, 960, 915, 895, 860, 830, 805 cm^{-1}.

PMR (C_5D_5N, 400 MHz) : δ 0.68 (d, *J*=5.1 Hz, 3xH-27), 0.89 (s, 3xH-18), 1.07 (d, *J*=6.4 Hz, 3xH-21), 1.42 (s, 3xH-19), 1.74 (d, *J*=6.2 Hz, 3xH-6 of Rha), 2.02 (s, OCOC*H₃*), 3.49 (dd, *J*=10.3, 10.3 Hz, H-26A), 3.56 (dd, *J*=10.3, 2.6 Hz, II-26B), 3.77 (dd, *J*=11.9, 3.9 Hz, H-1), 3.82 (m, H-3), 3.96 (br dd, *J*=7.7, 4.8 Hz, H-5 of Gal), 4.17 (H-3 of Gal), 4.17 (H-4 of Gal), 4.30 (dd, *J*=9.4, 9.4 H,z H-4 of Rha), 4.53 (dd, *J*=11.2, 4.8 Hz, H-6A of Gal), 4.58 (dd, *J*=8.8, 7.5 Hz, H-2 of Gal), 4.59 (1H, H-16), 4.63 (dd, *J*=9.4, 3.4 Hz, H-3 of Rha), 4.72 (br d, *J*=3.4 Hz, H-2 of Rha), 4.73 (d, *J*=7.5 Hz, H-1 of Gal), 4.90 (dq, *J*=9.4, 6.2 Hz, H-5 of Rha), 4.95 (dd, *J*=11.2, 7.7 Hz, H-6B of Gal), 5.61 (br d, *J*=5.3 Hz, H-6), 6.34 (br s, H-1 of Rha).

CMR (C_5D_5N, 100 MHz) : δ C-1) 84.8 (2) 38.1 (3) 68.3 (4) 43.8 (5) 139.6 (6) 124.8 (7) 32.2 (8) 33.2 (9) 50.6 (10) 42.9 (11) 24.1 (12) 40.3 (13) 40.3 (14) 57.2 (15) 32.5 (16) 81.2 (17) 63.1 (18) 16.9 (19) 15.0 (20) 42.0 (21) 15.0 (22) 109.3 (23) 31.9 (24) 29.3 (25) 30.6 (26) 66.9 (27) 17.3 **Gal** (1) 100.7 (2) 74.6 (3) 76.4 (4) 70.6 (5) 73.2 (6) 64.6 **Rha** (1) 101.7 (2) 72.5 (3) 72.7 (4) 74.3 (5) 69.3 (6) 19.0 (Ac) 170.5 (2) 20.9.

Mass (FAB, Negative ion) : *m/z* 779 [M-H]⁻, 737 [M-acetyl]⁻, 616 [M-Rha]⁻, 591 [M-Rha-acetyl]⁻.

Reference

1.   Y. Mimaki, M. Kuroda, A. Kameyama, A. Yokosuka and Y. Sashida, *Phytochemistry*, **48**, 485 (1998).

DRACONIN C

(23*S*,24*S*)-Spirosta-5,25(27)-diene-1β,3β,23,24-tetrol 1-O-[(2-O-acetyl-α-L-rhamnopyranosyl-(1→2)-α-L-arabinopyranoside]

Source : *Dracaena draco* (Dracaenacea)
Mol. Formula : $C_{40}H_{60}O_{15}$
Mol. Wt. : 780
$[\alpha]_D^{20}$: -85° (c=11.5, EtOH)
Registry No. : [565205-17-6]

IR (KBr) : 3393 (OH), 2975, 2904, 1730 (C=O), 1251, 1052 cm^{-1}.

PMR (C$_5$D$_5$N, 250 MHz) : δ 0.98 (s, 3xH-18), 1.06 (d, *J*=7.0 Hz, 3xH-21), 1.07 (H-14), 1.25 (H-4A), 1.26 (m, H-12A), 1.35 (H-15A), 1.36 (s, 3xH-19), 1.49 (H-9), 1.50 (H-8), 1.51 (H-7), 1.55 (m, H-4B, H-12B), 1.60 (H-11),, 1.69 (m, H-17), 1.70 (d, *J*=6.1 Hz, 3xH-6 o Rha), 1.90 (H-15B), 1.91 (s, OCOC*H$_3$*), 2.35 (H-2ax) 2.65 (H-2eq), 2.66 (m, H-20), 3.58 (dd, *J*=12.0, 0.5 Hz, H-5A of Ara), 3.78 (m, H-1), 3.80 (m, H-3), 3.88 (d, *J*=4.9 Hz, H-23), 3.95 (s, H-26A), 3.98 (s, 3xH-26B), 4.06 (H-4 of Ara), 4.07 (H-5 of Ara), 4.16 (m, H-24), 4.19 (m, H-4 of Rha), 4.20 (m, H-5B of Ara), 4.51 (m, H-2 of Ara), 4.52 (m, H-16), 4.62 (d, *J*=8.0 Hz, H-2 of Ara), 4.73 (dd, *J*=8.6, 3.5 Hz, H-3 of Rha), 4.81 (m, H-5 of Rha), 5.04 (d, *J*=1.4 Hz, H-27), 5.40 (d, *J*=5.5 Hz, H-6), 5.96 (dd, *J*=1.5, 3.5 Hz, H-2 of Rha), 6.17 (d, *J*=1.2 Hz, H-1 of Rha).

CMR (C$_5$D$_5$N, 125 MHz) : δ C-1) 84.32 (2) 38.27 (3) 69.00 (4) 43.75 (5) 140.4 (6) 125.6 (7) 32.87 (8) 33.85 (9) 51.24 (10) 44.71 (11) 24.89 (12) 40.40 (13) 41.43 (14) 57.71 (15) 33.15 (16) 84.10 (17) 62.26 (18) 17.71 (19) 15.87 (20) 37.93 (21) 15.48 (22) 113.2 (23) 70.51 (24) 75.00 (25) 147.3 (26) 61.64 (27) 113.5 **Ara** (1) 101.0 (2) 75.79 (3) 76.51 (4) 71.08 (5) 68.27 **Rha** (1) 99.5 (2) 75.10 (3) 71.25 (4) 75.00 (5) 70.28 (6) 19.7 (OCOC*H$_3$*) 171.6 (OCOC*H$_3$*) Not reported.

Reference

1. A.G. Gonzalez, J.C. Hernandez, F. Leon, J.I. Padron, F. Estever, J. Quintana and J. Bermejo, *J. Nat. Prod.*, **66**, 793 (2003).

DIOSPOLYSAPONIN A

(23S,25R)-Spirost-5-en-3β,12α,14α,17α,23-pentol 3-O-[α-L-rhamnopyranosyl-(1→2)-β-D-glucopyranoside]

Source : *Dioscorea polygonoides* Humb. et Bonpl.
(Dioscoreaceae)
Mol. Formula : $C_{39}H_{62}O_{16}$
Mol. Wt. : 786
M.P. : 260-265°C
[α]$_D$: -61.3° (c=0.10, CHCl$_3$-MeOH 1:1)
Registry No. : [623939-65-1]

IR (film) : 3376 (OH), 2955, 2928, 2870 (CH), 1053 cm^{-1}.

PMR (C$_5$D$_5$N, 500 MHz) : δ 0.69 (d, J=5.8 Hz, 3xH-27), 1.05 (s, 3xH-19), 1.13 (s, 3xH-18), 1.38 (d, J=7.1 Hz, 3xH-21), 1.78 (d, J=6.2 Hz, 3xH-6 of Rha), 3.43 (br s, 2xH-26), 3.47 (q, J=7.1 Hz, H-20), 3.90 (H-3 of aglycone, H-5 of Glc), 3.92 (dd, J=10.9, 4.6 Hz, H-23), 4.17 (t, J=9.0 Hz, H-4 of Glc), 4.27 (dd, J=9.0, 7.3 Hz, H-2 of Gc), 4.30 (t, J=9.0 Hz, H-3 of Glc), 4.35 (dd, J=11.9, 5.4 Hz, H-6A of Glc), 4.36 (t, J=9.3 Hz, H-4 of Rha), 4.49 (br s, H-12), 4.52 (dd, J=11.9, 2.3 Hz, H-6B of Glc), 4.81 (dd, J=3.4, 1.4 Hz, H-2 of Rha), 5.00 (dq, J=9.3, 6.2 Hz, H-5 of Rha), 5.03 (d, J=7.3 Hz, H-1 of Glc), 5.15 (t, J=7.2 Hz, H-16), 5.38 (br d, J=4.8 Hz, H-6), 6.38 (d, J=1.4 Hz, H-1 of Rha).

CMR (CDCl$_3$, 125 MHz) : δ C-1) 37.6 (2) 30.2 (3) 77.7 (4) 39.0 (5) 139.8 (6) 122.6 (7) 26.0 (8) 36.1 (9) 39.3 (10) 37.1 (11) 29.2 (12) 76.5 (13) 49.1 (14) 88.3 (15) 39.9 (16) 91.8 (17) 94.2 (18) 21.9 (19) 19.3 (20) 40.4 (21) 9.2 (22) 112.2 (23) 67.7 (24) 38.3 (25) 31.6 (26) 66.0 (27) 16.9 **Glc** (1) 100.3 (2) 77.9 (3) 79.7 (4) 71.9 (5) 78.4 (6) 62.7 **Rha** (1) 102.1 (2) 72.6 (3) 72.9 (4) 74.2 (5) 69.5 (6) 18.7.

Mass (E.S.I., Positive ion, H.R.) : *m/z* 809.3969 [(M+Na)$^+$, requires 809.3930].

Reference

1. J.N. Osorio, O.M.M. Martinez, Y.M.C. Navarro, Y. Mimaki, H. Sakagami and Y. Sashida, *Heterocycles*, 60, 1709 (2003).

AGAMENOSIDE H

(5α,22*S*,23*S*,24*R*,25*S*)-Spirostane-3β,6α,23,24-tetrol 6,24-*bis*-O-β-D-glucopyranoside

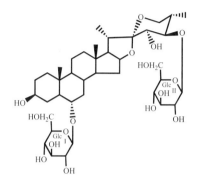

Source : *Agave americana* L. (Agavaceae)
Mol. Formula : $C_{39}H_{64}O_{16}$
Mol. Wt. : 788
[α]$_D^{14}$: -39.9° (c=0.41, Pyridine)
Registry No. : [738584-19-5]

IR (KBr) : 3439 (OH), 2925 (CH), 1457, 1380, 1165, 1036, 895, 868 cm^{-1}.

PMR (C$_5$D$_5$N, 500 MHz) : δ 0.58 (H-9), 0.76 (s, 3xH-19), 0.92 (H-1α), 0.93 (s, 3xH-18), 1.04 (H-14), 1.12 (d, *J*=7.0 Hz, 3xH-21), 1.15 (H-12α), 1.18 (H-7α, H-11α), 1.20 (d, *J*=6.4 Hz, 3xH-27), 1.35 (H-5), 1.41 (H-11β), 1.48 (H-2β), 1.56 (H-8), 1.62 (H-4β), 1.58 (H-1β), 1.70 (H-12β, H-15β), 1.78 (dd, *J*=6.8, 7.5 Hz, H-17), 1.97 (H-15α), 2.02 (H-2α), 2.08 (H-25), 2.56 (H-7β), 2.98 (H-20), 3.22 (H-4α), 3.56 (dd, *J*=6.1, 8.6 Hz, H-26), 3.67 (H-6), 3.74 (H-3), 3.87 (H-5 of Glc I), 3.90 (d, *J*=8.6 Hz, H-23), 3.96 (dd, *J*=9.2, 8.6 Hz, H-24), 3.96 (H-5 of Glc II), 4.04 (H-2 of Glc I and of Glc II), 4.18 (H-4 of Glc II), 4.22 (H-4 of Glc I), 4.24 (H-3 of Glc I), 4.25 (H-3 of Glc II), 4.27 (H-6A of Glc II), 4.33 (H-6A of Glc I), 4.43 (H-6B of Glc I), 4.49 (q-like, *J*=7.5 Hz, H-16), 4.49 (H-6B of Glc II), 4.91 (d, *J*=7.8 Hz, H-1 of Glc I), 4.94 (d, *J*=7.9 Hz, H-1 of Glc II).

CMR (C$_5$D$_5$N, 125 MHz) : δ C-1) 37.9 (2) 32.3 (3) 70.8 (4) 33.2 (5) 51.4 (6) 79.6 (7) 41.4 (8) 34.1 (9) 54.1 (10) 36.8 (11) 21.4 (12) 40.5 (13) 41.4 (14) 56.6 (15) 32.1 (16) 81.9 (17) 62.0 (18) 17.0 (19) 13.6 (20) 36.5 (21) 14.5 (22) 112.7 (23) 71.4 (24) 87.7 (25) 38.0 (26) 64.2 (27) 13.2 **Glc I** (1) 106.0 (2) 75.9 (3) 78.7 (4) 72.0 (5) 78.2 (6) 63.0 **Glc II** (1) 105.4 (2) 75.4 (3) 78.5 (4) 71.7 (5) 78.4 (6) 62.6.

Mass (FAB, Negative ion) : *m/z* 787 [M-H]$^-$.

Mass (FAB, Negative ion, H.R.) : *m/z* 787.4177 [(M-H)]$^-$, calcd. for 787.4177].

Reference

1.   J.M. Jin, Y.-J. Zhang and C.R. Yang, *Chem. Pharm. Bull.*, **52**, 654 (2004).

OPHIOPOGON PLANISCAPUS SAPONIN E

Ruscogenin 1-O-[α-L-rhamnopyranosyl-(1→2)-4-O-sulfo-α-L-arabinopyranoside]

Source : *Ophiopogon planiscapus* Nakai (Liliaceae)[1], *O. ohwii* Okuyama (Liliaceae)[2]
Mol. Formula : $C_{38}H_{60}O_{15}S$
Mol. Wt. : 788
M.P. : 220-221°C (decomp.)[1]
$[\alpha]_D^{23}$: -82.5° (c=0.77, Pyridine)[1]
Registry No. : [88623-84-1]

IR (KBr)[1] : 3600-3200 (OH), 1215 (S–O), 980, 920, 900, 865 cm^{-1}. Intensity 920 < 900 (25*R*-spiroketal) cm^{-1}.

CMR (C$_5$D$_5$N, 100 MHz)[1] : δ **Ara** (C-1) 100.2 (2) 75.9 (3) 74.5 (4) 76.0 (5) 65.5 **Rha** (1) 101.2 (2) 72.1 (3) 72.1 (4) 74.0 (5) 69.3 (6) 18.6. C$_1$-H *J* values (Ara) 159 Hz, (Rha) 172 Hz.

Disulfated Product :

Mol. Formula : $C_{38}H_{60}O_{11}$; **M.P. :** 266-267°C (decomp.).[1]

IR (KBr)[1] : 3600-3200 (OH), 982, 920, 900, 865 cm^{-1}. Intensity 920 < 900 (25*R*-spiroketal) cm^{-1}.

CMR (C$_5$D$_5$N, 100 MHz)[1] : δ **Ara** C-1) 100.2 (2) 75.6^a (3) 75.5^a (4) 69.9 (5) 67.1 **Rha** (1) 101.6 (2) 75.5^b (3) 72.7^b (4) 74.3 (5) 69.4 (6) 18.9.

References

1.   Y. Watanabe, S. Sanada, Y. Ida and J. Shoji, *Chem. Pharm. Bull.*, **31**, 3486 (1983).

2.   Y. Watanabe, S. Sanada, Y. Ida and J. Shoji, *Chem. Pharm. Bull.*, **32**, 3994 (1984).

CAMASSIA CUSICKII SAPONIN 6

(25R)-3,3-Dimethoxy-5α-spirostan-6α-ol 6-O-β-D-glucopyranosyl-(1→3)-β-D-glucopyranoside

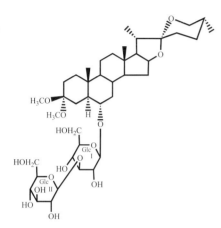

Source : *Camassia cusickii* (Liliaceae)
Mol. Formula : $C_{41}H_{68}O_{15}$
Mol. Wt. : 800
$[\alpha]_D^{25}$: -29° (c=0.20, MeOH)
Registry No. : [138867-30-8]

IR (KBr) : 3430 (OH), 2945 (CH), 1445, 1370, 1335, 1235, 1170, 1150, 1070, 1035, 975, 945, 915, 890, 870, 860 (intensity 915 < 890, 25R-spiroketal) cm^{-1}.

PMR (C_5D_5N, 400 MHz) : δ 0.64 (ddd, J=11.4, 11.4, 3.7 Hz, H-9), 0.72 (d, J=5.5 Hz, 3xH-27), 0.79 (s, 3xH-18)^a, 0.80 (s, 3xH-19)^a, 1.12 (d, J=6.9 Hz, 3xH-21), 2.53 (ddd, J=12.5, 4.0, 4.0 Hz, H-7eq), 3.20 (s, OCH$_3$), 3.24 (s, OCH$_3$), 3.47 (dd, J=10.6, 10.6 Hz, H-26A), 3.57 (dd, J=10.6, 2.8 Hz, H-26B), 3.65 (ddd, J=10.9, 10.9, 4.4 Hz, H-6), 4.86 (d, J=7.8 Hz, H-1 of Glc I), 5.30 (d, J=7.8 Hz, H-1 of Glc II).

CMR (C_5D_5N, 100.6 MHz) : δ C-1) 35.7 (2) 30.1^a (3) 100.8 (4) 28.9^a (5) 48.9 (6) 79.8 (7) 41.2 (8) 34.1 (9) 53.6 (10) 36.8 (11) 21.2 (12) 40.1 (13) 40.8 (14) 56.4 (15) 32.0 (16) 81.0 (17) 63.0 (18) 16.7 (19) 12.9 (20) 42.0 (21) 15.0 (22) 109.1 (23) 31.8 (24) 29.3 (25) 30.6 (26) 66.9 (27) 17.4 **Glc I** (1) 105.9^b (2) 74.3 (3) 88.9 (4) 69.9 (5) 77.7 (6) 62.6 **Glc II** (1) 105.8^b (2) 75.6 (3) 78.6^c (4) 71.7 (5) 78.3^c (6) 62.6. (OCH$_3$) 47.5 (OCH$_3$) 47.5.

Mass (S.I.) : *m/z* 823 [M+Na]$^+$, 799 [M-H]$^+$.

Reference

1. Y. Mimaki, Y. Sashida and K. Kawashima, *Phytochemistry*, **30**, 3721 (1991).

ALLIUM KARATAVIENSE SAPONIN 4

(24*S*,25*S*)-5α-Spirostane-2α,3β,5,6β,24-pentol 2,24-di-O-β-D-glucopyranoside

Source : *Allium karataviense* Regel (Liliaceae)
Mol. Formula : $C_{39}H_{64}O_{17}$
Mol. Wt. : 804
$[α]_D^{27}$: -78.0° (c=0.10, MeOH)
Registry No. : [238075-80-4]

IR (KBr) : 3400 (OH), 2920 (CH), 1455, 1370, 1255, 1155, 1060, 1020, 885 cm^{-1}.

PMR (C_5D_5N, 400/500 MHz) : δ 0.80 (s, 3xH-18), 1.04 (d, *J*=6.9 Hz, 3xH-21), 1.12 (d, *J*=6.4 Hz, 3xH-27), 1.55 (s, 3xH-19), 1.88 (m, H-25), 1.94 (dd, *J*=12.9, 10.7 Hz, H-23ax), 2.15 (dd, *J*=12.3, 5.0 Hz, H-1eq), 2.36 (dd, *J*=12.3, 11.5 Hz, H-1ax), 2.38 (dd, *J*=13.1, 6.2 Hz, H-4eq), 2.63 (dd, *J*=12.9, 4.8 Hz, H-23eq), 2.97 (dd, *J*=13.1, 11.2 Hz, H-4ax), 3.54 (dd, *J*=11.4, 11.4 Hz, H-26 ax), 3.61 (dd, *J*=11.4, 4.2 Hz, H-26eq), 3.85 (ddd, *J*=8.9, 5.0, 2.2 Hz, H-5 of Glc II), 3.99 (ddd, *J*=10.7, 10.7, 4.8 Hz, H-24), 4.04 (dd, *J*=8.9, 7.7 Hz, H-2 of Glc II), 4.09 (ddd, *J*=8.9, 5.9, 2.0 Hz, H-5 of Glc I), 4.11 (dd, *J*=8.9, 7.8 Hz, H-2 of Glc I), 4.15 (br s, H-6), 4.20 (dd, *J*=8.9, 8.9 Hz, H-3 of Glc II), 4.23 (dd, *J*=8.9, 8.9 Hz, H-4 of Glc I), 4.27 (dd, *J*=8.9, 8.9 Hz, H-4 of Glc II), 4.30 (dd, *J*=8.9, 8.9 Hz, H-3 of Glc II), 4.33 (dd, *J*=11.5, 5.9 Hz, H-6A of Glc I), 4.38 (dd, *J*=11.5, 5.0 Hz, H-6A of Glc II), 4.39 (ddd, *J*=11.5, 8.8, 5.0 Hz, H-2), 4.50 (dd, *J*=11.5, 2.2 Hz, H-6B of Glc II), 4.56 (q-like *J*=6.9 Hz, H-16), 4.61 (dd, *J*=11.5, 2.0 Hz, H-6B of Glc II), 4.82 (ddd, *J*=11.5, 8.8, 5.0 Hz, H-3), 4.89 (d, *J*=7.7 Hz, H-1 of Glc II), 5.16 (d, *J*=7.8 Hz, H-1 of Glc I).

CMR (C_5D_5N, 100/125 MHz) : δ C-1) 39.6 (2) 85.0 (3) 71.4 (4) 40.2 (5) 74.9 (6) 75.3 (7) 35.7 (8) 30.1 (9) 45.7 (10) 40.6 (11) 21.5 (12) 40.3 (13) 40.9 (14) 56.3 (15) 32.1 (16) 81.6 (17) 62.6 (18) 16.6 (19) 18.1 (20) 42.1 (21) 14.9 (22) 111.5 (23) 40.8 (24) 81.4 (25) 38.2 (26) 65.1 (27) 13.4 **Glc I** (1) 104.6 (2) 75.2 (3) 78.5 (4) 71.8 (5) 78.5 (6) 62.7 **Glc II** (1) 106.4 (2) 75.6 (3) 78.6 (4) 71.7 (5) 78.0 (6) 62.8.

Mass (FAB, Positive ion, H.R.) : *m/z* 827.4111 [M+Na]$^+$.

Reference

1.   Y. Mimaki, M. Kuroda, T. Fukasawa and Y. Sashida, *Chem. Pharm. Bull.,* **47**, 738 (1999).

RUSCUS ACULEATUS SAPONIN 7

Neoruscogenin 1-O-{α-L-rhamnopyranosyl-(1→2)-(4-O-sulfo)-α-L-arabinopyranoside}-monosodium salt

Source : *Ruscus aculeatus* L. (Liliaceae)
Mol. Formula : $C_{38}H_{57}O_{15}NaS$
Mol. Wt. : 808
$[\alpha]_D^{26}$: -84.0° (c=0.10, MeOH)
Registry No. : [205191-08-8]

IR (KBr) : 3420 (OH), 2930 (CH), 1255, 1225, 1040 cm^{-1}.

PMR (C$_5$D$_5$N+CD$_3$OD, 400 MHz) : δ 0.82 (s, 3xH-18), 1.02 (d, *J*=7.0 Hz, 3xH-21), 1.38 (s, 3xH-19), 1.68 (d, *J*=6.2 Hz, 3xH-6 of Rha), 3.73 (br d, *J*=11.8 Hz, H-5A of Ara), 3.74 (overlapping, H-1), 3.77 (m, H-3), 3.98 (br d, *J*=12.0 Hz, H-26A), 4.18 (dd, *J*=9.4, 9.4 Hz, H-4 of Rha), 4.19 (dd, *J*=9.1, 2.8 Hz, H-3 of Ara), 4.41 (dd, *J*=9.1, 7.4 Hz, H-2 of Ara), 4.42 (br d, *J*=12.0 Hz, H-26B), 4.48 (overlapping, H-16), 4.49 (dd, *J*=9.4, 3.3 Hz, H-3 of Rha), 4.61 (dd, *J*=3.3, 0.9 Hz, H-2 of Rha), 4.62 (br d, *J*=11.8 Hz, H-5B of Ara), 4.65 (d, *J*=7.4 Hz, H-1 of Ara), 4.70 (dq, *J*=9.4, 6.2 Hz, H-5 of Rha), 4.76 and 4.80 (each 1H, br s, 2xH-27), 5.19 (br d, *J*=2.8 Hz, H-4 of Ara), 5.57 (overlaping with H$_2$O signal, H-6), 6.13 (d, *J*=0.9 Hz, H-1 of Rha).

CMR (C$_5$D$_5$N+CD$_3$OD, 100 MHz) : δ C-1) 83.5 (2) 36.9 (3) 67.9 (4) 43.5 (5) 139.2 (6) 124.8 (7) 31.9 (8) 33.0 (9) 50.2 (10) 42.8 (11) 23.8 (12) 39.7 (13) 40.0 (14) 56.5 (15) 32.2 (16) 81.3 (17) 62.7 (18) 16.4 (19) 14.6 (20) 41.7 (21) 14.8 (22) 109.2 (23) 33.1 (24) 28.8 (25) 144.4 (26) 64.8 (27) 108.3 **Ara** (1) 100.0 (2) 75.7 (3) 74.5 (4) 75.8 (5) 65.5 **Rha** (1) 101.3 (2) 71.9 (3) 72.0 (4) 73.8 (5) 69.2 (6) 18.6.

Mass (FAB, Negative ion) : *m/z* 807 [M-H]$^-$, 785 [M-Na]$^-$.

Biological Activity : It possesses cytostatic activity on growth of Leukemia HL 60 cells and shows 16.8% inhibition at 10 µg/ml sample concentration.

Reference

1.   Y. Mimaki, M. Kuroda, A. Kameyama, A. Yokosuka and Y. Sashida, *Chem. Pharm. Bull.*, **46**, 298 (1998).

1370

CONVALLASAPONIN E
Diosgenin-3-O-[α-L-arabinopyranosyl-(1→2)-α-l-arabinopyranosyl-(1→2)-α-L-arabinopyranoside]

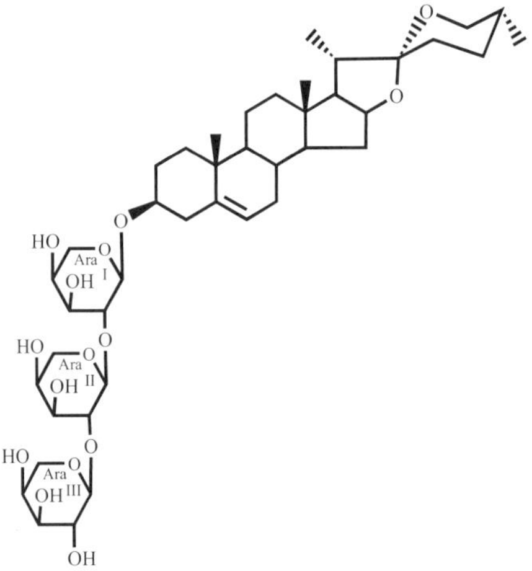

Source : *Convallaria keiskei* Miq. (Liliaceae)
Mol. Formula : $C_{42}H_{66}O_{15}$
Mol. Wt. : 810
M.P. : 213-217°C
$[\alpha]_D^{18}$: -149.0° (c=0.44, MeOH)
Registry No. : [23558-73-8]

IR (Nujol) : 3600-3200 (br, OH) cm^{-1}.

Reference

1.   M. Kimura, M. Tohma, I. Yoshizawa and A. Fujino, *Chem. Pharm. Bull.*, **16**, 2191 (1968).

OPHIOPOGONIN B'
Diosgenin 3-O-{(4-O-acetyl)-α-L-rhamnopyranosyl-(1→2)-[β-D-xylopyranosyl-(1→3)]-β-D-glucopyranoside}

Source : *Ophiopogon japonicus* Ker-Gawler var. *genuinus* Maxim (Liliaceae)
Mol. Formula : $C_{46}H_{72}O_{12}$
Mol. Wt. : 816
M.P. : 245-248°C (decomp.)
$[\alpha]_D$: -86.65° (c=0.67, Pyridine)
Registry No. : [65604-79-7]

IR (KBr) : 3400 (br, OH), 1735 (ester), 980, 920, 900, 865 (intensity 900>920, 25*R*-spiroketal) cm^{-1}.

PMR (C$_5$D$_5$N, 90.0 MHz) : δ 0.77 (m, C*H$_3$*), 0.87 (s, C*H$_3$*), 1.08 (s, C*H$_3$*), 1.18 (d, *J*=6.0 Hz, sec. C*H$_3$*), 1.78 (d, *J*=6.0 Hz, sec. C*H$_3$*), 2.06 (s, OCOC*H$_3$*).

Reference

1.   Y. Watanabe, S. Sanada, A. Tada and J. Shoji, *Chem. Phram. Bull.*, **25**, 3049 (1977).

RUSCUS ACULEATUS SAPONIN 16
Ruscogenin 1-O-{α-L-rhamnopyranosyl-(1→2)-4,6-di-O-acetyl-β-D-galactopyranoside}

Source : *Ruscus aculeatus* L. (Liliaceae)
Mol. Formula : C$_{43}$H$_{66}$O$_{15}$
Mol. Wt. : 822
[α]$_D^{25}$: -62.0° (c=0.10, MeOH)
Registry No. : [211036-48-5]

IR (KBr) : 3425 (OH), 2945 (CH), 1745 (C=O), 1450, 1375, 1240, 1180, 1130, 1065, 1055, 980, 965, 915, 900, 865, 835, 810 cm^{-1}.

PMR (C$_5$D$_5$N, 400 MHz) : δ 0.69 (d, J=5.3 Hz, 3xH-27), 0.98 (s, 3xH-18), 1.13 (d, J=6.6 Hz, 3xH-21), 1.44 (s, 3xH-19), 1.78 (d, J=6.2 Hz, 3xH-6 of Rha), 1.99 (s, OCOCH_3), 2.03 (s, OCOCH_3), 3.50 (dd, J=10.4, 10.4 Hz, H-26A), 3.58 (dd, J=10.4, 2.6 Hz, H-26B), 3.78 (dd, J=12.0, 4.1 Hz, H-1), 3.81 (m, H-3), 4.12 (br dd, J=7.6, 5.2 Hz, H-5 of Gal), 4.31 (dd, J=9.5, 9.5 Hz, H-4 of Rha), 4.33 (dd, J=9.5, 3.4 Hz, H-3 of Gal), 4.34 (dd, J=11.4, 5.2 Hz, H-6A of Gal), 4.45 (dd, J=9.5, 7.6 Hz, H-2 of Gal), 4.57 (dd, J=11.4, 7.6 Hz, H-6B of Gal), 4.58 (dd, J=9.5, 3.3 Hz, H-3 of Rha), 4.59 (H-16), 4.69 (dd, J=3.3, 0.9 Hz, H-2 of Rha), 4.78 (d, J=7.6 Hz, H-1 of Gal), 4.89 (dq, J=9.5, 6.2 Hz, H-5 of Rha), 5.62 (br d, J=5.5 Hz, H-6), 5.68 (br d, J=3.4 Hz, H-4 of Gal), 6.27 (d, J=0.9 Hz, H-1 of Rha).

CMR (C$_5$D$_5$N, 100 MHz) : δ C-1) 84.2 (2) 37.9 (3) 68.3 (4) 43.7 (5) 139.4 (6) 124.7 (7) 32.0 (8) 33.2 (9) 50.4 (10) 42.7 (11) 24.4 (12) 40.6 (13) 40.5 (14) 57.2 (15) 32.4 (16) 81.2 (17) 63.3 (18) 17.0 (19) 15.0 (20) 42.0 (21) 15.0 (22) 109.3 (23) 31.9 (24) 29.3 (25) 30.6 (26) 66.9 (27) 17.3 **Gal** (1) 100.1 (2) 75.0 (3) 74.1 (4) 72.0 (5) 71.3 (6) 63.0 **Rha** (1) 102.0 (2) 72.4 (3) 72.6 (4) 74.2 (5) 69.6 (6) 19.1 (OCOCH_3) 170.9, 20.8 (OCOCH_3) 170.2, 20.6.

Mass (FAB, Negative ion) : m/z 821 [M-H]$^-$, 779 [M-Acetyl]$^-$, 737 [M-Acetylx2]$^-$, 675 [M-Rha]$^-$, 616 [M-Rha-Acetyl]$^-$, 591 [M-Rha-Acetylx2]$^-$.

Biological Activity : Weak cytotoxic activity on leukemia HL-60 cells, show 43.7% inhibition at 10 μg/ml.

Reference

1.   Y. Mimaki, M. Kuroda, A. Kameyama, A. Yokosuka and Y. Sashida, *Phytochemistry*, **48**, 485 (1998).

DRACONIN B

(23S,24S)-Spirosta-5,25(27)-diene-1β,3β,23,24-tetrol 1-O-[(2,3,4-di-O-acetyl-α-L-rhamnopyranosyl-(1→2)-α-L-arabinopyranoside]

Source : *Dracaena draco* (Dracaenacea)
Mol. Formula : $C_{42}H_{62}O_{16}$
Mol. Wt. : 822
M.P. :
$[\alpha]_D^{20}$: $-100°$ (c =2.6, EtOH)
Registry No. : [565205-16-5]

IR (KBr) : 3400 (OH), 1740, 1250, 1100 cm^{-1}.

PMR (C$_5$D$_5$N, 500 MHz) : δ 0.85 (s, 3xH-18), 0.94 (d, J=7.0 Hz, 3xH-21), 1.06 (s, 3xH-19), 1.15 (m, H-14), 1.17 (m, H-12A), 1.18 (t, J=7.0 Hz, H-9), 1.30 (H-11). 1.33 (d, J=6.1 Hz, 3xH-6 of Rha), 1.52 (H-7A, H-15A), 1.55 (m, H-8), 1.72 (m, H-12B), 1.73 (br d, J=9.6 Hz, H-17), 1.96 (H-15B), 1.97 (H-7B), 2.06 (s, OCOCH$_3$), 2.11 (s, OCOCH$_3$), 2.16 (H-2ax), 2.21 (br s, H-4), 2.62 (H-2 eq), 2.65 (m, H-26), 3.34 (dd, J=3., 11.5 Hz, H-1), 3.45 (ddd, J=2.3, 2.9, 9.5 Hz, H-3), 3.51 (d, J=3.1 Hz, H-23), 3.61 (d, J=7.7 Hz H-2 of Ara), 3.68 (dd, J=8.8, 3.6 Hz, H-3 of Ara), 3.79 (br s, H-4 of Ara), 3.85 (d, J=12.8 Hz, H-26 eq), 3.92 (dd, J=12., 1.9 Hz, H-5A of Ara), 4.15 (m, H-4 and H-5 of Rha), 4.24 (m, H-24), 4.24 (d, J=7.3 Hz, H-1 of Ara), 4.32 (m, H-5B of Ara), 4.4 (d, J=12.9, H-26 ax), 4.49 (dd, J=7.2, 15.3 Hz, H-16), 5.02 (s, H-27A), 5.10 (br s, H-27B), 5.19 (d, J=3.3 Hz, H-2 of Rha), 5.21 (s, H-4 of Rha), 5.36 (d, J=1.9 Hz, H-1 of Rha), 5.57 (d, J=5.8 Hz, H-6).

CMR (C$_5$D$_5$N, 125 MHz) : δ C-1) 84.44 (2) 37.24 (3) 68.03 (4) 42.13 (5) 137.72 (6) 125.37 (7) 31.47 (8) 32.51 (9) 49.97 (10) 42.13 (11) 23.41 (12) 39.86 (13) 40.53 (14) 56.70 (15) 31.85 (16) 83.32 (17) 60.80 (18) 16.54 (19) 14.37 (20) 36.06 (21) 14.05 (22) 112.40 (23) 68.45 (24) 72.24 (25) 143.07 (26) 60.13 (27) 114.13 **Ara** (1) 99.77 (2) 74.11 (3) 74.87 (4) 70.13 (5) 66.03 **Rha** (1) 96.75 (2) 71.81 (3) 71.00 (4) 69.27 (5) 68.63 (6) 17.56 (OCOCH$_3$) 170.24, 171.59 (OCOCH$_3$) 20.93, 21.04.

Mass (FAB, Positive ion, H.R.) : *m/z* 846.3954 [(M+Na+H)$^+$, requires 846.4013].

Biological Activity : Inhibits the growth of HL-60 cells with IC$_{50}$: 39.0±13.8 μMol..

Reference

1. A.G. Gonzalez, J.C. Hernandez, F. Leon, J.I. Padron, F. Estever, J. Quintana and J. Bermejo, *J. Nat. Prod.*, **66**, 793 (2003).

OPHIOPOGON OHWII SAPONIN O-4

Ruscogenin 1-O-[α-L-rhamnopyranosyl-(1→2)-(4-O-sulfo)-β-D-fucopyranoside] sodium salt

Source : *Ophiopogon ohwii* Okuyama (Liliaceae)
Mol. Formula : $C_{39}H_{61}O_{15}SNa$
Mol. Wt. : 824
M.P. : 228-231°C (decomp.)
$[\alpha]_D^{21}$ **:** -98.2° (c=0.33, Pyridine)
Registry No. : [94898-64-3]

IR (KBr) **:** 3600-3200 (OH), 1215 (S–O), 985, 925, 905, 870 cm^{-1}. Intensity 925 < 905 (25R-spiroketal) cm^{-1}.

CMR (C_5D_5N, 25 MHz) **:** δ **Fuc** C-1) 100.3 (2) 76.0^a (3) 75.7^a (4) 79.1 (5) 70.6 (6) 17.3 **Rha** (1) 101.5 (2) 72.3 (3) 72.3 (4) 74.3 (5) 69.3 (6) 18.9.

Desulfated Product : see Ophiopogonin B.

Reference

1.   Y. Watanabe, S. Sanada, Y. Ida and J. Shoji, *Chem. Pharm. Bull.*, **32**, 3994 (1984).

TRIBULUS TERRESTRIS SAPONIN 1

Diosgenin 3-O-[α-L-rhamnopyranosyl-(1→2)-(4-O-sulfo)-β-D-glucopyranoside]-monosodium salt

Source : *Tribulus terrestris* L. (Zygopullaceae)
Mol. Formula : $C_{39}H_{61}O_{15}SNa$
Mol. Wt. : 824
M.P. : 310-312°C
[α]$_D^{20}$: -120°
Registry No. : [134461-54-4]

PMR (C_5D_5N, 300 MHz) : δ 1.71 (d, *J*=6.1 Hz, 3xH-6 of Rha), 1.86 (d, *J*=9.2, 6.1 Hz, H-5 of Rha), 3.82 (ddd, *J*=2.8, 10.0, 2.8 Hz, H-5 of Glc), 4.19 (dd, *J*=8.9, 8.1 Hz, H-2 of Glc), 4.31 (t, *J*=9.2 Hz, H-4 of Rha), 4.40 (m, $J_{6A,6B}$=6.1 Hz, H-6A and H-6B of Glc), 4.43 (t, *J*=8.9 Hz, H-3 of Glc), 4.56 (dd, $J_{3,6}$=9.2 Hz, H-3 of Rha), 4.88 (d, *J*=8.1 Hz, H-1 of Glc), 4.75 (dd, *J*=1.6, 3.6 Hz, H-2 of Rha), 5.1 (dd, *J*=10.0, 8.0 Hz, H-4 of Glc), 6.20 (d, *J*=1.6 Hz, H-1 of Rha).

CMR (C_5D_5N, 75.0 MHz) : δ C-1) 37.67 (2) 30.77 (3) 78.44 (4) 39.12 (5) 141.04 (6) 121.98 (7) 32.00 (8) 31.87 (9) 50.45 (10) 37.67 (11) 21.26 (12) 40.04 (13) 40.64 (14) 56.81 (15) 32.84 (16) 81.35 (17) 63.05 (18) 16.51 (19) 19.56 (20) 42.16 (21) 15.20 (22) 109.51 (23) 32.60 (24) 29.43 (25) 30.77 (26) 67.06 (27) 17.58 **Rha** (1) 102.13 (2) 72.54 (3) 72.88 (4) 74.27 (5) 69.68 (6) 18.80 **Glc** (1) 100.19 (2) 78.35 (3) 76.73 (4) 81.35 (5) 76.34 (6) 62.35.

Reference

1.   N.E. Mashchenko, R. Gyulemetova, P.K Kintya and A.S.Shashkov, *Khim. Prir. Soedin.*, **26**, 649 (1990); *Chem. Nat. Comp.*, **26**, 552 (1990).

NOLINA RECURVATA SAPONIN 4

Neoruscogenin 1-O-{α-L-rhamnopyranosyl-(1→2)-[β-D-xylopyranosyl-(1→3)]-α-L-arabinopyranoside}

Source : *Nolina recurvata* (Agavaceae)
Mol. Formula : $C_{43}H_{66}O_{16}$
Mol.Wt. : 838
$[\alpha]_D^{27}$: -63.8° (c=0.26, MeOH)
Registry No. : [180161-85-7]

IR (KBr) : 3410 (OH), 2940 (CH), 1440, 1370, 1225, 1135, 1085, 1035, 975, 960, 915, 870, 830, 810, 775, 695 cm^{-1}.

PMR (C_5D_5N, 400 MHz) : δ 0.85 (s, 3xH-18), 1.05 (d, *J*=6.8 Hz, 3xH-21), 1.43 (s, 3xH-19), 1.74 (d, *J*=6.1 Hz, 3xH-6 of Rha), 4.73 (d, *J*=7.3 Hz, H-1 of Ara), 4.77 and 4.80 (br s, 2xH-27), 4.98 (d, *J*=7.5 Hz, H-1 of Xyl), 5.59 (br d, *J*=5.3 Hz, H-6), 6.33 (br s, H-1 of Rha).

CMR (C_5D_5N, 100 MHz) : δ C-1) 83.8 (2) 37.5 (3) 68.2 (4) 43.9 (5) 139.6 (6) 124.7 (7) 32.0 (8) 33.1 (9) 50.3 (10) 42.9 (11) 24.1 (12) 40.2 (13) 40.2 (14) 56.8 (15) 32.4 (16) 81.5 (17) 63.0 (18) 16.7 (19) 15.0 (20) 41.8 (21) 14.9 (22) 109.4 (23) 33.2 (24) 29.0 (25) 144.5 (26) 65.0 (27) 108.6 **Ara** (1) 100.5 (2) 74.2 (3) 84.5 (4) 69.6 (5) 67.1 **Rha** (1) 101.8 (2) 72.5 (3) 72.5 (4) 74.1 (5) 69.5 (6) 19.1 **Xyl** (1) 106.5 (2) 74.6 (3) 78.2 (4) 71.0 (5) 67.0.

Mass (FAB, Negative ion) : *m/z* 838 [M]$^-$, 706 [M-Xyl]$^-$, 692 [M-Rha]$^-$, 559 [M-Xyl-Rha]$^-$.

Biological Activity : The compound shows inhibitory activity on cyclic AMP phosphodiesterase with IC$_{50}$ (9.2x10^{-5} M).

Reference

1.   Y. Mimaki, Y. Takaashi, M. Kuroda, Y. Sashida and T. Nikaido, *Phytochemistry*, **42**, 1609 (1996).

CORDYLINE STRICTA SAPONIN 11

5α-Spirost-25(27)-ene-1β-3α-diol 1-O-{α-L-rhamnopyranosyl-(1→2)-[β-D-xylopyranosyl-(1→3)]-β-D-xylopyranoside}

Source : *Cordyline stricta* (Agavaceae)
Mol. Formula : $C_{43}H_{68}O_{16}$
Mol. Wt. : 840
$[\alpha]_D^{27}$: -72.6° (c=0.52, MeOH)
Registry No. : [202185-87-3]

IR (KBr) : 3410 (OH), 2915 (CH), 1445, 1365, 1225, 1155, 1060, 1035, 975, 935, 915, 865, 825, 805 cm^{-1}.

PMR (C_5D_5N, 400/500 MHz) : δ 0.93 (s, 3xH-18), 1.10 (d, *J*=6.9 Hz, 3xH-21), 1.23 (s, 3xH-19), 1.79 (d, *J*=6.1 Hz, 3xH-6 of Rha), 4.73 (d, *J*=7.7 Hz., H-1 of Xyl I), 4.77 and 4.80 (each 1H, br s, 2xH-27), 4.95 (d, *J*=7.6 Hz, H-1 of Xyl II), 6.48 (br s, H-1 of Rha).

CMR (C_5D_5N, 100/125 MHz) : δ C-1) 80.9 (2) 35.0 (3) 65.8 (4) 37.3 (5) 39.5 (6) 28.7 (7) 32.5 (8) 36.7 (9) 54.9 (10) 42.5 (11) 23.9 (12) 40.9 (13) 40.5 (14) 57.0 (15) 32.3 (16) 81.5 (17) 63.3 (18) 17.0 (19) 7.8 (20) 41.9 (21) 14.9 (22) 109.4 (23) 33.2 (24) 29.0 (25) 144.5 (26) 65.0 (27) 108.6 **Xyl I** (1) 100.5 (2) 75.8 (3) 88.9 (4) 69.5 (5) 66.6 **Rha** (1) 101.5 (2) 72.5 (3) 72.5 (4) 74.2 (5) 69.6 (6) 19.3 **Xyl II** (1) 105.2 (2) 74.8 (3) 78.5 (4) 70.6 (5) 67.3.

Mass (FAB, Positive ion) : *m/z* 879 [M+K]$^+$.

Reference

1.   Y. Mimaki, M. Kuroda, Y. Takaashi and Y. Sashida, *Phytochemistry*, **47**, 79 (1998).

NOLINA RECURVATA SAPONIN 5

(25S)-Ruscogenin 1-O-{α-L-rhamnopyranosyl-(1→2)-[β-D-xylopyranosyl-(1→3)]-α-L-arabinopyranoside}

Source : *Nolina recurvata* (Agavaceae)
Mol. Formula : $C_{43}H_{68}O_{16}$
Mol. Wt. : 840
$[\alpha]_D^{27}$: -63.5° (c=0.57, MeOH)
Registry No. : [180161-86-8]

IR (KBr) : 3425 (OH), 2950 (CH), 1450, 1375, 1225, 1135, 1090, 1040, 980, 915, 890, 870, 835, 810, 780, 700 cm⁻¹.

PMR (C_5D_5N, 400 MHz) : δ 0.85 (s, 3xH-18), 1.06 (d, J=6.9 Hz, 3xH-21), 1.09 (d, J=6.9 Hz, 3xH-27), 1.43 (s, 3xH-19), 1.74 (d, J=5.8 Hz, 3xH-6 of Rha), 4.73 (d, J=7.3 Hz, H-1 of Ara), 4.98 (overlapping with H_2O signal, H-1 of Xyl), 5.59 (br d, J=5.1 Hz, H-6), 6.33 (br s, H-1 of Rha).

CMR (C_5D_5N, 100 MHz) : δ : C-1) 83.8 (2) 37.5 (3) 68.2 (4) 43.9 (5) 139.5 (6) 124.7 (7) 32.0 (8) 33.1 (9) 50.4 (10) 42.9 (11) 24.1 (12) 40.3 (13) 40.2 (14) 56.8 (15) 32.4 (16) 81.2 (17) 62.8 (18) 16.7 (19) 15.0 (20) 42.4 (21) 14.8 (22) 109.7 (23) 26.4 (24) 26.2 (25) 27.5 (26) 65.0 (27) 16.3 **Ara** (1) 100.5 (2) 74.2 (3) 84.5 (4) 69.6 (5) 67.1 **Rha** (1) 101.8 (2) 72.5 (3) 72.5 (4) 74.1 (5) 69.5 (6) 19.1 **Xyl** (1) 106.5 (2) 74.6 (3) 78.3 (4) 71.0 (5) 67.0.

Mass (FAB, Negative ion) : *m/z* 840 [M]⁻, 708 [M-Xyl]⁻, 694 [M-Rha]⁻.

Biological Activity : The compound shows inhibitory activity on cyclic AMP phosphodiesterase with IC_{50} (8.7x10⁻⁵ M).

Reference

1.   Y. Mimaki, Y. Takaashi, M. Kuroda, Y. Sashida and T. Nikaido, *Phytochemistry*, **42**, 1609 (1996).

ORNITHOGALUM THYRSOIDES SAPONIN 11

(25*R*)-Spirostan-5-en-1β,3β-diol 1-O-{α-L-rhamnopyranosyl-(1→2)-[β-D-xylopyranosyl-(1→3)]-
α-L-arabinopyranoside}

Source : *Ornithogalum thyrsoides* (Liliaceae)
Mol. Formula : $C_{43}H_{68}O_{16}$
Mol. Wt. : 840
$[\alpha]_D^{25}$: -54.0° (c=0.10, MeOH)

IR (film) : 3377 (OH), 2950, 2928 and 2905 (CH), 1090, 1050 cm^{-1}.

PMR (C$_5$D$_5$N) : δ 0.68 (d, *J*=5.3 Hz, 3xH-27), 0.85 (s, 3xH-18), 1.09 (d, *J*=7.0 Hz, 3xH-21), 1.43 (s, 3xH-19), 1.74 (d, *J*=6.2 Hz, 3xH-6 of Rha) 3.48 (dd, *J*=10.5, 10.5 Hz, H-26ax), 3.56 (dd, *J*=10.5, 2.5 Hz, H-26eq), 3.82 (dd, *J*=12.0, 3.9 Hz, H-1), 3.89 (m, *W*½=21.2 Hz, H-3), 4.52 (q-like *J*=7.4 Hz, H-16), 4.73 (d, *J*=7.3 Hz, H-1 of Ara), 4.98 (d, *J*=7.5 Hz, H-1 of Xyl), 5.58 (br d, *J*=5.7 Hz, H-6), 6.34 (br s, H-1 of Rha).

CMR (C$_5$D$_5$N, 125 MHz) : δ C-1) 83.8 (2) 37.5 (3) 68.2 (4) 43.9 (5) 139.5 (6) 124.7 (7) 32.4 (8) 33.1 (9) 50.3 (10) 42.9 (11) 24.1 (12) 40.3 (13) 40.2 (14) 56.8 (15) 32.0 (16) 81.1 (17) 63.0 (18) 16.7 (19) 15.0 (20) 41.9 (21) 15.0 (22) 109.2 (23) 31.8 (24) 29.3 (25) 30.6 (26) 66.8 (27) 17.3 **Ara** (1) 100.5 (2) 74.2 (3) 84.6 (4) 69.7 (5) 67.1 **Rha** (1) 101.8 (2) 72.6 (3) 72.6 (4) 74.1 (5) 69.5 (6) 19.2 **Xyl** (1) 106.5 (2) 74.6 (3) 78.3 (4) 71.0 (5) 67.0.

Mass (FAB, Positive ion, H.R.) : *m/z* 863.4404 [(M+Na)$^+$, calcd. for 863.4405].

Biological Activity : Moderate cytotoxic against HL-60 leukemia cells and HS C-2 cells.

Reference

1.   M. Kuroda, Y. Mimaki, K. Ori, H. Sakagami and Y. Sashida, *J. Nat. Prod.*, **67**, 1690 (2004).

ASPAFILIOSIDE B

Sarsasapogenin 3-O-{β-D-xylopyranosyl-(1→4)-[α-L-arabinopyranosyl-(1→6)]-β-D-glucopyranoside}

Source : *Asparagus filicinus* Buch.-Ham. (Liliaceae)
Mol. Formula : $C_{43}H_{70}O_{16}$
Mol. Wt. : 842
M.P. : 180~182°C (MeOH)
$[\alpha]_D^{14}$ **:** -47.3° (c=0.07, CHCl₃-MeOH)
Registry No. : [131123-73-4]

IR (KBr) : 3420, 1045, 985, 915 > 895, 850 cm⁻¹ (25S).

PMR (C_5D_5N, 400 MHz) : δ 0.82 (s, 3xH-18), 0.84 (s, 3xH-19), 1.08 (d, *J*=6.6 Hz, 3xH-27), 1.16 (d, *J*=6.7 Hz, 3xH-21), 4.80 (d, *J*=7.6 Hz), 5.00 (d, *J*=7.3 Hz), 5.39 (d, *J*=7.7 Hz).

CMR (C_5D_5N, 100 MHz) : δ C-1) 30.8 (2) 27.2 (3) 74.7 (4) 31.1 (5) 37.2 (6) 27.2 (7) 27.0 (8) 35.8 (9) 40.6 (10) 35.4 (11) 21.4 (12) 40.6 (13) 41.1 (14) 56.7 (15) 32.4 (16) 81.5 (17) 63.2 (18) 16.8 (19) 24.1 (20) 42.7 (21) 15.0 (22) 109.9 (23) 26.7 (24) 26.4 (25) 27.7 (26) 65.4 (27) 16.5 **Glc** (1) 103.1 (2) 75.0 (3) 76.5 (4) 80.2 (5) 74.9 (6) 68.3 **Xyl** (1) 105.2 (2) 75.1 (3) 78.5 (4) 71.1 (5) 67.2 **Ara** (1) 105.6 (2) 72.6 (3) 75.0 (4) 69.8 (5) 67.2.

Mass (F.D.) : *m/z* 865 [M+Na]⁺, 733 [M-132]⁺.

Reference

1.   Y. Ding and C.R. Yang, *Acta Pharm. Sin.*, **25**, 509 (1990).

CORDYLINE STRICTA SAPONIN 10

(25S)-5α-Spirostane-1β-3α-diol 1-O-{α-L-rhamnopyranosyl-(1→2)-[β-D-xylopyranosyl-(1→3)]-β-D-xylopyranoside}

Source : *Cordyline stricta* (Agavaceae)
Mol. Formula : $C_{43}H_{70}O_{16}$
Mol. Wt. : 842
$[\alpha]_D^{28}$: -46.5° (c=0.16, MeOH)
Registry No. : [202185-86-2]

IR (KBr) : 3420 (OH), 2930 (CH), 1445, 1370, 1215, 1165, 1125, 1060, 1035, 985, 915, 890, 860, 805 cm^{-1}.

PMR (C_5D_5N, 400/500 MHz) : δ 0.92 (s, 3xH-18), 1.07 (J=7.0 Hz, 3xH-21). 1.14 (d, J=6.9 Hz, 3xH-27), 1.23 (s, 3xH-19), 1.78 (d, J=6.1 Hz, 3xH-6 of Rha), 4.72 (d, J=7.7 Hz., H-1 of Xyl I), 4.94 (d, J=7.6 Hz, H-1 of Xyl II), 6.47 (br s, H-1 of Rha).

PMR (C_5D_5N-CD$_3$OD, 400/500 MHz) : δ 1.70 (d, J=6.0 Hz, 3xH-6 of Rha), 3.36 (dd, J=10.1, 10.1 Hz, H-5A of Xyl I), 3.61 (dd, J=10.9, 10.5 Hz, H-5 A of Xyl II), 3.86 (dd, J=8.7, 7.6 Hz, H-2 of Xyl II), 3.88 (H-3 of Xyl I), 3.89 (H-4 of Xyl I), 3.95 (dd, J=8.7, 8.7 Hz, H-3 of Xyl II), 4.02 (H-4 of Xyl II), 4.03 (dd, J=8.8, 7.7 Hz, H-2 of Xyl I), 4.18 (dd, J=9.5, 9.5 Hz, H-4 of Rha), 4.19 (H-5B of Xyl I and Xyl II), 4.41 (dd, J=9.5, 2.6 Hz, H-3 of Rha), 4.65 (br d, J=2.6 Hz, H-2 of Rha), 4.67 (dq, J=9.5, 6.0 Hz, H-5 of Rha), 4.69 (d, J=7.7 Hz, H-1 of Xyl), 4.86 (d, J=7.6 Hz, H-1 of Xyl II), 6.36 (br s, H-1 of Rha).

CMR (C_5D_5N, 100/125 MHz) : δ C-1) 80.8 (2) 34.9 (3) 65.8 (4) 37.3 (5) 39.4 (6) 28.7 (7) 32.5 (8) 36.7 (9) 54.9 (10) 42.5 (11) 23.9 (12) 40.9 (13) 40.4 (14) 56.9 (15) 32.3 (16) 81.2 (17) 63.1 (18) 17.0 (19) 7.7 (20) 42.5 (21) 14.8 (22) 109.7 (23) 26.2 (24) 26.4 (25) 27.5 (26) 65.0 (27) 16.3 **Xyl I** (1) 100.5 (2) 75.8 (3) 88.9 (4) 69.5 (5) 66.6 **Rha** (1) 101.5 (2) 72.5 (3) 72.5 (4) 74.2 (5) 69.6 (6) 19.3 **Xyl II** (1) 105.2 (2) 74.7 (3) 78.4 (4) 70.6 (5) 67.2.

Mass (FAB, Positive ion) : *m/z* 881 [M+K]$^+$.

Reference

1.   Y. Mimaki, M. Kuroda, Y. Takaashi and Y. Sashida, *Phytochemistry*, **47**, 79 (1998).

NOLINA RECURVATA SAPONIN 6

Neoruscogenin 1-O-{α-L-rhamnopyranosyl-(1→2)-[β-D-xylopyranosyl-(1→3)]-β-D-fucopyranoside}

Source : *Nolina recurvata.* (Agavaceae)
Mol. Formula : $C_{44}H_{68}O_{16}$
Mol. Wt. : 852
$[\alpha]_D^{27}$: -45.0° (c=0.44, MeOH)
Registry No. : [180161-87-9]

IR (KBr) : 3430 (OH), 2930 (CH), 1440, 1370, 1225, 1150, 1060, 1040, 980, 915, 875, 835, 700 cm^{-1}.

PMR (C_5D_5N, 400 MHz) : δ 0.87 (s, 3xH-18), 1.03 (d, J=6.9 Hz, 3xH-21), 1.42 (s, 3xH-19), 1.52 (d, J=6.3 Hz, 3xH-6 of Fuc), 1.75 (d, J=6.1 Hz, 3xH-6 of Rha), 4.69 (d, J=7.7 Hz, H-1 of Fuc), 4.77 and 4.81 (each br s, 2xH-27), 5.00 (d, J=7.5 Hz, H-1 of Xyl), 5.61 (br d, J=5.0 Hz, H-6), 6.37 (br s, H-1 of Rha).

CMR (C_5D_5N, 100 MHz) : δ C-1) 84.4 (2) 38.1 (3) 68.3 (4) 43.9 (5) 139.7 (6) 124.7 (7) 32.1 (8) 33.1 (9) 50.6 (10) 42.9 (11) 24.0 (12) 40.4 (13) 40.3 (14) 57.2 (15) 32.4 (16) 81.5 (17) 63.1 (18) 16.8 (19) 15.0 (20) 41.9 (21) 14.9 (22) 109.5 (23) 33.2 (24) 29.0 (25) 144.6 (26) 65.0 (27) 108.6 **Fuc** (1) 100.5 (2) 73.5 (3) 85.6 (4) 72.7 (5) 70.9 (6) 17.1 **Rha** (1) 101.8 (2) 72.6 (3) 72.6 (4) 74.3 (5) 69.4 (6) 19.2 **Xyl** (1) 106.7 (2) 74.7 (3) 78.4 (4) 71.0 (5) 67.1.

Mass (FAB, Negative ion) : *m/z* 852 [M]$^-$, 720 [M-Xyl]$^-$, 706 [M-Rha]$^-$, 573 [M-Xyl-Rha]$^-$.

Biological Activity : The compound shows inhibitory activity on cyclic AMP phosphodiesterase with IC$_{50}$ (16.1x10^{-5} M).

Reference

1. Y. Mimaki, Y. Takaashi, M. Kuroda, Y. Sashida and T. Nikaido, *Phytochemistry*, **42**, 1609 (1996).

AFEROSIDE A

Diosgenin 3-O-{β-D-apiofuranosyl-(1→4)-[α-L-rhamnopyranosyl-(1→2)]-β-D-glucopyranoside}

Source : *Costus afer* Ker-Gawl. (Zingiberacae)
Mol. Formula : $C_{44}H_{70}O_{16}$
Mol. Wt. : 854
$[\alpha]_D^{20}$: -108° (c=0.16, MeOH)[1]
Registry No. : [182067-83-0]

PMR (C_5D_5N, 500 MHz)[2] : δ 0.70 (d, *J*=3.7 Hz, 3xH-27), 0.82 (s, 3xH-18), 0.97 (m, H-1A), 1.04 (s, 3xH-19), 1.08 (m, H-12A), 1.12 (d, *J*=6.9 Hz, H-21), 1.39 (m, H-11A and H-11B), 1.50 (m, 2xH-13), 1.55 (m, H-8 and H-25), 1.56 (H-24A), 1.57 (m, H-24B), 1.62 (m, H-23A and H-23B), 1.71 (H-12B), 1.72 (m, H-1B), 1.73 (d, *J*=6.0 Hz, 3xH-6 of Rha), 1.75 (m, H-17), 1.83 (m, H-2A and H-2B), 1.86 (m, H-7A and H-7B), 1.95 (m, H-20), 2.73 (H-4A), 2.76 (m, H-4B), 3.50 (H-26A), 3.58 (m, H-26B), 3.73 (m, H-5 of Glc), 3.89 (m, H-3), 4.12 (H-5A of Api), 4.13 (m, H-2 of Glc), 4.16 (m, H-4 of Glc), 4.16 (m, H-5B of Api), 4.18 (m, H-3 of Glc), 4.20 (H-6A of Glc), 4.30 (m, H-6B of Glc), 4.32 (d, *J*=8.0 Hz, H-4A of Api), 4.33 (m, H-4 of Rha), 4.54 (m, H-16), 4.58 (dd, *J*=9.5, 3.5 Hz, H-3 of Rha), 4.72 (d, *J*=8.0 Hz, H-4B of Api), 4.73 (m, H-2 of Api), 4.78 (m, H-2 of Rha), 4.90 (m, H-5 of Rha), 4.92 (d, *J*=8.0 Hz, H-1 of Glc), 5.35 (m, H-6), 5.88 (d, *J*=3.0 Hz, H-1 of Api), 6.20 (s, H-1 of Rha).

CMR (C_5D_5N, 125 MHz)[2] : δ C-1) 37.6 (2) 30.3 (3) 78.3 (4) 39.1 (5) 140.4 (6) 122.0 (7) 32.5 (8) 31.0 (9) 50.5 (10) 37.8 (11) 21.3 (12) 40.1 (13) 40.2 (14) 56.8 (15) 32.4 (16) 81.3 (17) 63.0 (18) 16.5 (19) 19.6 (20) 42.1 (21) 15.2 (22) 109.1 (23) 32.0 (24) 29.4 (25) 30.7 (26) 67.0 (27) 17.5 **Glc** (1) 100.1(2) 77.8 (3) 77.7 (4) 79.6 (5) 76.6 (6) 61.4 **Api** (1) 111.3 (2) 77.5 (3) 80.0 (4) 75.1 (5) 64.7 **Rha** (1) 102.1 (2) 72.6 (3) 72.9 (4) 74.2 (5) 69.6 (6) 18.8.

Mass (FAB, Negative ion)[1] : *m/z* 853 [M-H]⁻, 721 [M-H-Api]⁻, 707 [M-H-Rha]⁻.

References

1. R.-C. Lin, B. Hanquet and M.-A. Lacaille-Dubois, *Phytochemistry*, **43**, 665 (1996).

2. R.-C. Lin, M.-A. Lacaille-Dubois, B. Hanquet, M. Correia and B. Chauffert, *J. Nat. Prod.*, **60**, 1165 (1997).

CORDYLINE STRICTA SAPONIN 9

5α-Spirost-25(27)-ene-1β,3α-diol 1-O-{α-L-rhamnopyranosyl-(1→2)-[β-D-xylopyranosyl-(1→3)]-
β-D-fucopyranoside}

Source : *Cordyline stricta* (Agavaceae)
Mol. Formula : $C_{44}H_{70}O_{16}$
Mol. Wt. : 854
$[\alpha]_D^{27}$: -54.5° (c=0.29, MeOH)
Registry No. : [202185-85-1]

IR (KBr) : 3420 (OH), 2905 and 2835 (CH), 1440, 1365, 1295, 1220, 1150, 1120, 1055, 1030, 965, 935, 910, 865, 825, 800 cm^{-1}.

PMR (C$_5$D$_5$N, 400/500 MHz) : δ 0.86 (s, 3xH-18), 1.03 (d, *J*=6.9 Hz, 3xH-21). 1.24 (s, 3xH-19), 1.45 (d, *J*=6.3 Hz, 3xH-6 of Fuc), 1.75 (d, *J*=6.1 Hz, 3xH-6 of Rha), 4.74 (d, *J*=8.0 Hz., H-1 of Fuc), 4.76 and 4.80 (each 1H, br s, 2xH-27), 4.99 (d, *J*=7.5 Hz, H-27), 4.99 (d, *J*=7.5 Hz, H-1 of Xyl), 6.40 (br s, H-1 of Rha).

PMR (C$_5$D$_5$N-CD$_3$OD, 11:1, 400/500 MHz) : δ 0.83 (s, 3xH-18), 0.92 (dddd, *J*=13.1, 13.1, 13.1, 3.5 Hz, H-7ax), 1.01 (d, *J*=7.0 Hz, 3xH-21), 1.14 (H-14), 1.19 (H-4), 1.20 (s, 3xH-19), 1.25 (H-6ax), 1.23 (H-12ax), 1.33 (H-6eq), 1.36 (H-11ax), 1.40 (H-15β), 1.43 (d, *J*=6.3 Hz, 3xH-6 of Fuc), 1.52 (br d, *J*=13.5 Hz, H-4eq), 1.57 (H-12eq), 1.59 (H-7eq), 1.60 (H-8), 1.61 (br dd, *J*=13.5, 13.5 Hz, H-4ax), 1.70 (d, *J*=6.2 Hz, H-6 of Rha), 1.74 (H-23A), 1.78 (H-23B), 1.79 (dd, *J*=8.6, 6.8 Hz, H-17), 1.93 (H-20), 1.98 (H-5), 2.00 (H-15α), 2.20 (ddd, *J*=12.5, 12.5, 2.7 Hz, H-2ax), 2.23 (br d, *J*=13.0 Hz, H-24eq), 2.45 (br dd, *J*=12.5, 4.3 Hz, H-2eq), 2.68 (ddd, *J*=13.0, 13.0, 5.4 Hz, H-24ax), 3.20 (br dd,

J=13.8, 3.2 Hz, H-11eq), 3.60 (br q, *J*=6.3 Hz, H-5 of Fuc), 3.62 (dd, *J*=11.3, 10.1 Hz, H-5A), 3.82 (dd, *J*=8.6, 7.5 Hz, H-2 of Xyl), 3.98 (dd, *J*=8.6, 8.6 Hz, H-3 of Xyl), 3.99 (br d, *J*=12.1 Hz, H-26A), 4.01 (dd, *J*=9.5, 2.9 Hz, H-3 of Fuc), 4.03 (ddd, *J*=10.1, 8.6, 5.3 Hz, H-4 of Xyl), 4.16 (br d, *J*=2.9 Hz, H-4 of Fuc), 4.19 (dd, *J*=9.4, 9.4 Hz, H-4 of Rha), 4.23 (dd, *J*=11.3, 5.3 Hz, H-5B of Xyl), 4.25 (br d, *J*=2.7 Hz, H-3), 4.35 (dd, *J*=11.5, 4.3 Hz, H-1α), 4.43 (br d, *J*=12.1 H-26B), 4.48 (dd, *J*=9.4 Hz, 3.5 H-3 of Rha), 4.50 (q-like, *J*=6.8 Hz, H-16), 4.53 (dd, *J*=9.5, 8.0 Hz, H-2 of Fuc), 4.70 (dd, *J*=3.5, 1.1 Hz, H-2 of Rha), 4.71 (d, *J*=8.0 Hz, H-1 of Fuc), 4.75 (dq, *J*=9.4, 6.2 Hz, H-5 of Rha), 4.77 (br s, H-27A), 4.80 (br s, H-27B), 4.92 (d, *J*=7.5 Hz, H-1 of Xyl), 6.27 (H-1 of Rha).

CMR (C_5D_5N, 100/125 MHz) : δ C-1) 81.3 (2) 35.5 (3) 65.9 (4) 37.4 (5) 39.5 (6) 28.8 (7) 32.6 (8) 36.7 (9) 55.3 (10) 42.6 (11) 23.8 (12) 40.7 (13) 40.4 (14) 57.1 (15) 32.3 (16) 81.6 (17) 63.1 (18) 17.0 (19) 7.7 (20) 41.9 (21) 14.8 (22) 109.4 (23) 33.2 (24) 28.9 (25) 144.5 (26) 64.9 (27) 108.6 **Fuc** (1) 100.4 (2) 73.3 (3) 85.8 (4) 72.8 (5) 70.8 (6) 17.0 **Rha** (1) 101.6 (2) 72.6 (3) 72.6 (4) 74.3 (5) 69.3 (6) 19.2 **Xyl** (1) 106.6 (2) 74.7 (3) 78.3 (4) 71.0 (5) 67.1.

Mass (FAB, Negative ion) : *m/z* 853 [M-H]⁻, 720 [M-Xyl]⁻.

Reference

1.   Y. Mimaki, M. Kuroda, Y. Takaashi and Y. Sashida, *Phytochemistry*, **47**, 79 (1998).

INDIOSIDE E
Diosgenin 3-O-{α-L-rhamnopyranosyl-(1→2)-[β-D-xylopyranosyl-(1→3)]-β-D-galactopyranoside}

Source : *Solanum indicum* L. (Solanaceae)
Mol. Formula : $C_{44}H_{70}O_{16}$
Mol. Wt. : 854
[α]$_D^{26}$: -78.9° (c=0.50, Pyridine)
Registry No. : [146388-18-3]

PMR (C$_5$D$_5$N, 400 MHz) : δ 0.70 (d, *J*=5.5 Hz, 3xH-27), 0.83 (s, 3xH-18), 1.07 (s, 3xH-19), 1.15 (d, *J*=7.0 Hz, 3xH-27), 1.69 (d, *J*=7.2 Hz, 3xH-6 of Rha), 5.00 (d, *J*=7.7 Hz, H-1 of Gal), 5.05 (d, *J*=7.3 Hz, H-1 of Xyl), 5.32 (br s, H-6), 6.23 (s, H-1 of Rha).

CMR (C$_5$D$_5$N, 100 MHz) : δ C-1) 37.4 (2) 30.1 (3) 78.2 (4) 38.7 (5) 140.8 (6) 121.7 (7) 32.1 (8) 31.6 (9) 50.2 (10) 37.1 (11) 21.0 (12) 39.8 (13) 40.4 (14) 56.6 (15) 32.2 (16) 81.6 (17) 62.8 (18) 16.2 (19) 19.3 (20) 41.9 (21) 15.0 (22) 109.2 (23) 31.1 (24) 29.2 (25) 30.5 (26) 66.9 (27) 17.2 **Gal** (1) 100.4 (2) 76.4 (3) 84.9 (4) 70.2 (5) 75.0 (6) 62.1 **Rha** (1) 102.2 (2) 72.5 (3) 72.8 (4) 74.6 (5) 69.4 (6) 18.5 **Xyl** (1) 106.7 (2) 74.1 (3) 77.4 (4) 70.9 (5) 67.0.

Mass (FAB, Negative ion) : *m/z* 853 [M-H]⁻.

Reference

1.   S. Yahara, T. Nakamura, Y. Someya, T. Matsumoto, T. Yamashita and T. Nohara, *Phytochemistry,* **43**, 1319 (1996).

LILIUM DAURICUM SAPONIN 3, LILILANCIFOLOSIDE A
Diosgenin 3-O-{α-L-rhamnopyranosyl-(1→2)-[α-L-arabinopyranosyl-(1→3)]-β-D-glucopyranoside}

Source : *Lilium dauricum*[1], *L. longiflorum*[2], *L. lancifolium* Thumb.[3] (Liliaceae)
Mol. Formula : C$_{44}$H$_{70}$O$_{16}$
Mol. Wt. : 854
Registry No. : [143051-93-8]

PMR (C$_5$D$_5$N, 500 MHz)^3 : δ 0.67 (d, J=5.0 Hz, 3xH-27), 0.81 (s, 3xH-18), 0.86 (dd, J=5.0, 10.0 Hz, H-9α), 0.91 (dd, J=5.5, 10.0 Hz, H-1β), 1.05 (s, 3xH-19), 1.06 (dd, J=7.5, 10.8 Hz, H-14α), 1.10 (dd, J=5.0, 9.5 Hz, H-12α), 1.12 (d, J=7.0 Hz, 3xH-21), 1.42 (dd, J=6.0, 9.5 Hz, H-11β), 1.44 (dd, J=6.0, 9.5 Hz, H-11α), 1.47 (dd, J=7.5, 14.5 Hz, H-15β), 1.49 (dd, J=7.5, 14.5 Hz, H-15α), 1.50 (m, 2xH-23), 1.52 (dd, J=4.5, 10.5 Hz, 2xH-24), 1.55 (dd, J=4.5, 10.5 Hz, H-25β), 1.64 (m, H-8β), 1.68 (dd, J=6.0, 9.5 Hz, H-12β), 1.72 (dd, J=5.5, 10.0 Hz, H-1α), 1.75 (d, J=6.5 Hz, 3xH-6 of Rha), 1.78 (dd, J=7.0, 10.0 Hz, H-17α), 1.84 (dd, J=5.0, 10.0 Hz, H-2α), 1.87 (dt, J=7.5, 11.0 Hz, H-7β), 1.93 (dd, J=7.0, 7.0 Hz, H-20β), 2.02 (dt, J=7.5, 11.0 Hz, H-7α), 2.08 (dd, J=5.0, 10.0 Hz, H-2β), 2.75 (d, J=14.0 Hz, H-4β), 2.79 (dd, J=10.0, 14.0 Hz, H-4α), 3.48 (t, J=10.5 Hz, H-26β), 3.56 (dd, J=5.0, 10.5 Hz, H-25α), 3.66 (dd, J=5.0, 11.5 Hz, H-5A of Ara), 3.83 (m, H-3α), 3.94 (dd, J=7.5, 7.0 Hz, H-2 of Ara), 3.97 (t, J=7.5 Hz, H-5 of Glc), 4.03 (dd, J=8.0, 16.8 Hz, H-4 of Glc), 4.08 (d, J=8.5 Hz, H-2 of Glc), 4.11 (dd, J=4.5, 7.0 Hz, H-4 of Ara), 4.17 (t, J=8.5 Hz, H-3 of Glc), 4.18 (dd, J=4.5, 7.0 Hz, H-3 of Ara), 4.24 (dd, J=5.0, 10.8 Hz, H-5B of Ara), 4.25 (dd, J=5.0, 11.5 Hz, H-6A of Glc), 4.33 (t, J=9.5 Hz, H-4 of Rha), 4.45 (d, J=11.5 Hz, H-6B of Glc), 4.52 (dd, J=7.5, 14.5 Hz, H-16), 4.58 (dd, J=3.0, 9.0 Hz, H-3 of Rha), 4.87 (br s, H-2 of Rha), 4.92 (dd, J=3.5, 11.5 Hz, H-5 of Rha), 4.97 (d, J=7.5 Hz, H-1 of Ara), 4.98 (d, J=7.0 Hz, H-1 of Glc), 5.29 (d-like H-6), 6.33 (br s, H-1 of Rha).

CMR (C$_5$D$_5$N, 125 MHz)^3 : δ C-1) 37.4 (2) 30.0 (3) 78.0 (4) 38.6 (5) 140.7 (6) 121.9 (7) 32.3 (8) 31.6 (9) 50.2 (10) 37.1 (11) 21.1 (12) 39.8 (13) 40.4 (14) 56.6 (15) 32.2 (16) 81.1 (17) 62.8 (18) 16.3 (19) 19.4 (20) 41.9 (21) 15.0 (22) 109.2 (23) 31.8 (24) 29.2 (25) 30.6 (26) 66.8 (27) 17.3 **Glc** (1) 99.8 (2) 77.4 (3) 88.1 (4) 69.4 (5) 77.6 (6) 62.3 **Rha** (1) 102.4 (2) 72.4 (3) 72.8 (4) 74.1 (5) 69.7 (6) 18.7 **Ara** (1) 105.6 (2) 74.6 (3) 77.4 (4) 70.6 (5) 67.3.

Mass (FAB, Negative ion)^3 : m/z 853.459 [M-H]$^-$.

References

1.   Y. Mimaki, N. Ishibashi, K. Ori and Y. Sashida, *Phytochemistry*, **31**, 1753 (1992).

2.   Y. Mimaki, O. Nakamura, Y. Sashida, Y.Y. Satomi, A. Nishino and H. Nishino, *Phytochemistry*, **37**, 227 (1994).

3.   X.-W. Yang, Y.-X. Cui, X.-H. Liu, *Chin. J. Magn. Reson.*, **19**, 301 (2002).

LIRIOPE SPICATA SAPONIN Ls-7

Yamogenin 3-O-[α-L-rhamnopyranosyl-(1→2)-[β-D-xylopyranosyl-(1→3)]-β-D-glucopyranoside}

Source : *Liriope spicata* Thumb. Lour. var. *prolifera* Y.T. Ma. (Liliaceae)
Mol. Formula : C$_{44}$H$_{70}$O$_{16}$
Mol. Wt. : 854
M.P. : 279-282°C (decomp.)
[α]$_D$: -54.4° (c=0.23, C$_5$D$_5$N)
Registry No. : [87480-46-4]

IR (KBr) : 3600-3200 (OH), 980, 920, 902, 860 cm^{-1} (intensity 920>902, (25S)-spiroketal).

CMR (C$_5$D$_5$N, 100 MHz) : δ C-1) 38.1 (2) 30.7 (3) 78.8 (4) 39.7 (5) 141.0 (6) 121.8 (7) 32.9 (8) 32.3 (9) 50.9 (10) 37.7 (11) 21.7 (12) 40.4 (13) 41.0 (14) 57.2 (15) 32.7 (16) 81.5 (17) 63.2 (18) 16.9 (19) 20.0 (20) 43.0 (21) 15.4 (22) 109.9 (23) 28.1 (24) 27.1 (25) 26.8 (26) 65.5 (27) 16.9 **Glc** (1) 100.5 (2) 77.6 (3) 82.1 (4) 71.1 (5) 78.8 (6) 62.4 **Rha** (1) 102.0 (2) 74.5 (3) 73.1 (4) 75.2 (5) 69.7 (6) 19.1 **Xyl** (1) 105.9 (2) 76.5 (3) 78.0 (4) 72.7 (5) 67.7.

Reference

1.   B-Y. Yu, Y. Hirai, J. Shoji and G-J. Xu, *Chem. Pharm. Bull.*, **38**, 1931 (1990).

OPHIOPOGONIN D, OPHIOPOGON JAPONICUS SAPONIN OJV-V
Ruscogenin 1-O-{α-L-rhamnopyranosyl-(1→2)-[β-D-xylopyranosyl-(1→3)]-β-D-fucopyranoside}

Source : *Ophiopogon japonicus* Ker-Gawler var. *genuinus* Maxim.[1] (Liliaceae), *O. ohwii* Okuyama[2] (Liliaceae), *O. japonicus* Ker-Gawler cv. Nanus[3] (Liliaceae)
Mol. Formula : C$_{44}$H$_{70}$O$_{16}$
Mol. Wt. : 854
M.P. : 263-265°C (decomp.)[1]
[α]$_D$^{14} : -107.9° (c=0.66, Pyridine)[1]
Registry No. : [41753-55-3]

IR (KBr)[3] : 3600-3200 (OH), 980, 920, 900, 860 cm^{-1}, intensity 920 < 900 (25R-spiroketal) cm^{-1}.

PMR (C$_5$D$_5$N, 400/500 MHz)[4] : δ 0.82 (d, J=5.7 Hz, 3xH-27), 0.89 (s, 3xH-18), 1.10 (d, J=6.7 Hz, 3xH-21), 1.33 (s, 3xH-19), 1.49 (d, J=6.1 Hz, 3xH-6 of Fuc), 1.59 (d, J=5.8 Hz, 3xH-6 of Rha), 4.80 (d, J=7.4 Hz, H-1 of Xyl), 5.61 (d, J=5.2 Hz, H-1 of Fuc), 6.00 (br s, H-1 of Rha).

CMR (C$_5$D$_5$N, 100 MHz)[3] : δ C-1) 84.3 (2) 38.0 (3) 68.4 (4) 43.9 (5) 139.7 (6) 124.6 (7) 33.2 (8) 32.2 (9) 50.7 (10) 50.0 (11) 24.1 (12) 40.6 (13) 40.3 (14) 57.3 (15) 32.5 (16) 81.4 (17) 63.2 (18) 16.8 (19) 14.7 (20) 42.5 (21) 14.9 (22) 109.2 (23) 32.4 (24) 29.3 (25) 30.5 (26) 66.9 (27) 17.3 **Fuc** (1) 100.4 (2) 74.7 (3) 85.7 (4) 72.5 (5) 71.0 (6) 17.1 **Rha** (1) 101.7 (2) 72.6 (3) 72.7 (4) 74.3 (5) 69.3 (6) 19.1 **Xyl** (1) 106.5 (2) 73.6 (3) 78.2 (4) 70.9 (5) 66.9.

Mass (FAB, Negative ion)[4] : m/z 853 [M-H]$^-$.

References

1.   A. Tada, M. Kobayashi and J. Shoji, *Chem. Pharm. Bull.*, **21**, 308 (1973).

2.   Y. Watanabe, S. Sanada, Y. Ida and J. Shoji, *Chem. Pharm. Bull.*, **32**, 3994 (1984).

3.   T. Asano, T. Murayama, Y. Hirai and J. Shoji, *Chem. Pharm. Bull.*, **41**, 566 (1993).

4.   H.-F. Dai, J. Zhou, N.-H. Tau and Z.-T. Ding, *Zhiwu Xuebao (Acta Bot. Sin.)*, **43**, 97 (2001).

OPHIOPOGONIN D'
Diosgenin 3-O-{α-L-rhamnopyranosyl-(1→2)-[β-D-xylopyranosyl-(1→3)]-β-D-glucopyranoside}

Source : *Ophiopogon japonicus* Ker.-Gawl.[1,2] (Liliaceae), *Liriope platyphylla* Wang. et Tang.[3] (Liliaceae)
Mol. Formula : C$_{44}$H$_{70}$O$_{16}$
Mol. Wt. : 854
M.P. : 256-258°C
[α]$_D^{21}$: -40.4° (c=0.16, C$_5$H$_5$N)[2]
Registry No. : [65604-80-0]

IR (KBr)[2] : 3400 (OH), 1630 (C=C), 1060 (C-O-C), 980, 920, 910, 870 cm^{-1}.

PMR (C$_5$D$_5$N, 500 MHz)[4] : δ 0.67 (d, J=5.0 Hz, 3xH-27), 0.81 (s, 3xH-18), 0.85 (dd, J=5.5, 12.3 Hz, H-9), 0.92 (dd, J=5.5, 10.0 Hz, H-1β), 1.04 (s, 3xH-19), 1.06 (dd, J=6.5, 12.0 Hz, H-14α), 1.08 (dd, J=5.0, 9.5 Hz, H-12α), 1.12 (d, J=7.0 Hz, 3xH-21), 1.42 (dd, J=6.0, 9.5 Hz, H-11β), 1.44 (dd, J=6.0, 9.5 Hz, H-14α), 1.47 (dd, J=7.5, 14.5 Hz, H-15β), 1.49 (dd, J=7.5, 14.5 Hz, H-15α), 1.52 (m, 2xH-23), 1.52 (dd, J=4.5, 10.5 Hz, 2xH-24), 1.55 (dd, J=4.5, 10.5 Hz, H-25α), 1.64 (m, H-8), 1.65 (dd, J=6.0, 9.5 Hz, H-12β), 1.70 (dd, J=5.5, 10.0 Hz, H-1α), 1.73 (d, J=6.0 Hz, 3xH-6 of Rha), 1.78 (dd, J=7.5, 8.0 Hz, H-17α), 1.84 (dd, J=5.0, 10.0 Hz, H-2β), 1.89 (dt, J=6.8, 12.3 Hz, H-7β), 1.93 (dd, J=7.0, 8.0 Hz, H-20β), 2.02 (dt, J=6.8, 12.3 Hz, H-7α), 2.08 (dd, J=5.0, 10.0 Hz, H-2α), 2.73 (d, J=12.5 Hz, H-4β), 2.77 (dd, J=10.0, 12.5 Hz, H-4α), 3.48 (t, J=10.5 Hz, H-26β), 3.57 (dd, J=5.0, 10.5 Hz, H-26α), 3.77 (d, J=11.5 Hz, H-5A of Xyl), 3.83 (m, H-3), 3.93 (dd, J=5.0, 7.5 Hz, H-5 of Glc), 4.03 (dd, J=7.5, 9.0 Hz, H-4 of Glc), 4.04 (dd, J=5.0, 9.0 Hz, H-2 of Xyl), 4.12 (d, J=7.5 Hz, H-2 of Glc), 4.15 (dd, J=7.5, 8.5 Hz, H-3 of Glc), 4.25 (dd, J=3.5, 12.0 Hz, H-6A of Glc), 4.25 (dd, J=5.0, 9.0 Hz, H-3 of Xyl), 4.26 (dd, J=3.5, 11.5 Hz, H-5B of Xyl), 4.34 (dd, J=7.5, 9.0 Hz, H-4 of Rha), 4.44 (dd, J=3.5, 9.0 Hz, H-4 of Xyl), 4.45 (d, J=12.0 Hz, H-6B of Glc), 4.53 (dd, J=7.5, 15.0 Hz, H-16α), 4.58 (dd, J=3.0, 9.0 Hz, H-3 of Rha), 4.88 (br s, H-2 of Rha), 4.89 (d, J=7.5 Hz, H-1 of Xyl), 4.90 (dd, J=6.0, 9.0 Hz, H-5 of Rha), 4.96 (d, J=7.5 Hz, H-1 of Glc), 5.30 (d, J=5.0 Hz, H-6), 6.27 (br s, H-1 of Rha).

CMR (C$_5$D$_5$N, 100 MHz)[2] : δ C-1) 37.6 (2) 29.7 (3) 78.4 (4) 37.6 (5) 140.9 (6) 121.9 (7) 32.0 (8) 30.7 (9) 50.5 (10), 37.3 (11) 21.1 (12) 39.1 (13) 40.0 (14) 56.8 (15) 31.8 (16) 81.2 (17) 63.1 (18) 15.1 (19) 17.4 (20) 42.1 (21) 14.3 (22) 109.4 (23) 30.7 (24) 29.4 (25) 29.7 (26) 67.0 (27) 16.4 **Glc** (1) 98.8 (2) 77.7 (3) 81.7 (4) 70.9 (5) 78.4 (6) 61.8 **Rha** (1) 100.2 (2) 72.8 (3) 72.5 (4) 74.9 (5) 69.6 (6) 18.7 **Xyl** (1) 105.8 (2) 75.0 (3) 76.3 (4) 70.9 (5) 67.4.

Mass (FAB, Negative ion)[2] : m/z 853 [M-H]$^-$, 722 [M-Xyl]$^-$.

References

1. Y. Watanabe, S. Sanada, A. Tada and J. Shoji, *Chem. Pharm. Bull.*, **25**, 3055 (1977).

2. H.-F. Dai, J. Zhou, N.-H. Tan and Z.-T. Ding, *Zhiwu Xuebao* (*Acta Bot. Sin.*), **43**, 97 (2001).

3.   Y. Watanabe, S. Sanada, Y. Ida and J. Shoji, *Chem. Pharm. Bull.*, **31**, 1980 (1983).

4.   X.-W. Yang, Y.-X. Cui and X.-H. Liu, *Chin. J. Magn. Reson.*, **19**, 301 (2002).

OPHIOPOGON JAPONICUS SAPONIN OJV-VI
(25S)-Ruscogenin 1-O-{α-L-rhamnopyranosyl-(1→2)-[β-D-xylopyranosyl-(1→3)]-β-D-fucopyranoside

Source : *Ophiopogon japonicus* Ker-Gawler cv. Nanus[1] (Liliaceae), *Nolina recurvata*[2] (Agavaceae), *Liriope spicata* var. *prolifera*[3] (Liliaceae)
Mol. Formula : $C_{44}H_{70}O_{16}$
Mol. Wt. : 854
M.P. : 201-202°C[1]
[α]$_D^{23}$: -93.4° (c=0.41, Pyridine)[1]
Registry No. : [125150-67-6]

IR (KBr)[1] : 3600-3200, 980, 920, 902, 865 cm^{-1} (Intensity 920 > 902, 25S-spiroketal).

PMR (C_5D_5N, 400 MHz)[2] : δ 0.86 (s, 3xH-18), 1.07 (d, J=6.9 Hz, 3xH-21), 1.08 (d, J=6.9 Hz, 3xH-27), 1.41 (s, 3xH-19), 1.50 (d, J=6.2 Hz, 3xH-6 of Fuc), 1.73 (d, J=6.1 Hz, 3xH-6 of Rha), 4.67 (d, J=7.7 Hz, H-1 of Fuc), 4.98 (d, J=7.7 Hz, H-1 of Xyl), 5.59 (br d, J=5.1 Hz, H-6), 6.35 (br s, H-1 of Xyl).

CMR (C_5D_5N, 100 MHz)[1] : δ C-1) 84.3 (2) 38.1 (3) 68.4 (4) 44.0 (5) 139.7 (6) 124.6 (7) 33.2 (8) 32.2 (9) 50.7 (10) 43.5 (11) 24.1 (12) 40.6 (13) 40.6 (14) 57.3 (15) 32.5 (16) 81.2 (17) 63.0 (18) 16.8 (19) 14.8 (20) 42.8 (21) 14.8 (22) 109.8 (23) 26.5 (24) 26.2 (25) 28.1 (26) 65.2 (27) 16.3 **Fuc** (1) 100.4 (2) 74.7 (3) 85.7 (4) 72.5 (5) 71.0 (6) 17.1 **Rha** (1) 101.7 (2) 72.6 (3) 72.7 (4) 74.3 (5) 69.3 (6) 19.1 **Xyl** (1) 106.5 (2) 73.6 (3) 78.2 (4) 70.9 (5) 66.9.

Mass (FAB, Positive ion)[3] : m/z 855 [M+H]$^+$, 709 [M+H-Rha]$^+$, 577 [M+H-Xyl-Rha]$^+$, 431 [Agl+H]$^+$.

Mass (FAB, Negative ion)[3] : m/z 853 [M-H]$^-$, 721 [M-H-Xyl]$^-$.

1392

Biological Activity : The compound shows inhibitory activity on cyclic AMP-phosphodiesterases with IC_{50} (10.3×10^{-5} M).[2]

References

1. T. Asano, T. Murayama, Y. Hirai and J. Shoji, *Chem. Pharm. Bull.*, **41**, 566 (1993).

2. Y. Mimaki, Y. Takaashi, M. Kuroda, Y. Sushida and T. Nakaido, *Phytochemistry*, **42**, 1609 (1996).

3. B.-Y. Yu, S.-X. Qiu, K. Zaw, G.J. Xu, Y. Hirai, J. Shoji, H.H.S. Fong and A.D. Kinghorn, *Phytochemistry*, **43**, 201 (1996).

PARIS POLYPHYLLA SAPONIN Pa
Diosgenin 3-O-{α-L-rhamnopyranosyl-(1→2)-[α-L-arabinofuranosyl-(1→4)]-β-D-glucopyranoside}

Source : *Paris polyphylla* Sm.[1] (Liliaceae), *Paris vietnamensis* (Takht.) H.Li[2] (Liliaceae)
Mol. Formula : $C_{44}H_{70}O_{16}$
Mol. Wt. : 854
M.P. : 276-278°C (decomp.)[1]
[α]$_D$: -133.0 (c=0.56, MeOH)[1]
Registry No. : [50773-41-6]

IR (KBr)[1] : 3500-3300 (OH), 982, 920, 898, 866 cm^{-1}, intensity 898 > 920 (25R-spiroketal)

CMR $(C_5D_5N)^2$: δ C-1) 37.5 (2) 30.0 (3) 77.6 (4) 40.5 (5) 140.8 (6) 121.7 (7) 32.2 (8) 31.7 (9) 50.3 (10) 37.1 (11) 21.1 (12) 39.9 (13) 40.5 (14) 56.6 (15) 32.2 (16) 81.1 (17) 62.9 (18) 16.3 (19) 19.3 (20) 41.9 (21) 15.0 (22) 109.2 (23) 31.7 (24) 29.2 (25) 30.5 (26) 66.9 (27) 17.1 **Glc** (1) 100.1 (2) 78.1 (3) 77.5 (4) 76.5 (5) 77.1 (6) 62.3 **Rha** (1) 101.8 (2) 72.3 (3) 72.7 (4) 73.9 (5) 69.5 (6) 18.5 **Ara** (1) 109.6 (2) 82.5 (3) 77.8 (4) 86.5 (5) 61.3.

References

1.   T. Nohara, H. Yabuta, M. Suenobu, R. Hida, K. Miyahara and T. Kawasaki, *Chem. Pharm. Bull.*, **21**, 1240 (1973).

2.   T. Namba, H.L. Huang, Y. Zhongshu, S.L. Huang, M. Hottori, N. Kakiuchi, Q. Wang and G. Jun Xu, *Planta Med.*, **55**, 501 (1989).

POLYPHYLLIN D
Diosgenin 3-O-{α-L-rhmanopyranosyl-(1→2)-[α-L-arabinofuranosyl-(1→4)]-β-D-glucopyranoside}

Source : *Paris polyphylla*[1], *Paris polyphylla* Sm. var. *chinensis* Hara[2] (Liliaceae)
Mol. Formula : $C_{44}H_{70}O_{16}$
Mol. Wt. : 854
M.P. : 275-280°C[2]
[α]$_D$: -113.0° (c=0.53, MeOH)[2]
Registry No. : [76296-72-5]

IR $(KBr)^2$: 3414 (br), 901 > 920 cm^{-1} (25*R*-spirostanol).

CMR (CDCl$_3$, 62.8 MHz)[2] : δ C-1) 38.0 (2) 30.2 (3) 78.0 (4) 38.3 (5) 140.9 (6) 122.2 (7) 32.3 (8) 32.2 (9) 50.9 (10) 37.7 (11) 21.5 (12) 39.0 (13) 40.4 (14) 56.6 (15) 32.7 (16) 81.7 (17) 62.9 (18) 16.6 (19) 19.6 (20) 41.6 (21) 14.7 (22) 109.4 (23) 31.9 (24) 29.3 (25) 30.8 (26) 67.5 (27) 17.3 **Glc** (1) 101.4 (2) 71.5 (3) 71.9 (4) 73.5 (5) 69.1 (6) 17.7 **Ara** (1) 109.4 (2) 82.0 (3) 77.4 (4) 85.6 (5) 61.5.

Mass (FAB, Positive ion)[2] : *m/z* (rel.intens.) 855 [M+H]$^+$ (19), 723 [855-Ara+H$_2$O]$^+$ (9), 577 [723-Rha-H$_2$O]$^+$ (6), 415 [577-Glc+H$_2$O] (75), 397 [diosgenin-H$_2$O]$^+$ (100).

Biological Activity : Halmostatic.[2]

References

1. S.B. Singh, R.S. Thakur and H.-R. Schulten, *Phytochemistry*, **21**, 2925 (1982).

2. J.C.N. Ma and F.W. Lau, *Phytochemistry*, **24**, 1561 (1985).

REINECKIA CARNEA SAPONIN 2

Convallomarogenin 1-O-[α-L-rhamnopyrasyl-(1→2)-β-D-xylopyrnoside]-3-O-α-L-rhamnopyranoside

Source : *Reineckia carnea* Kunth (Liliaceace)
Mol. Formula : C$_{44}$H$_{70}$O$_{16}$
Mol. Wt. : 854
M.P. : 238-241°C (decomp.)
[α]$_D$^{36} : -93.3° (c=0.9, MeOH)
Registry No. : [108470-60-6]

IR (KBr) : 3500-3300 (OH), 980, 945, 918, 895, 875, 830 cm^{-1} (spiroketal).

PMR (C$_5$D$_5$N, 80.0 MHz) : δ 0.86 (s, 3xH-18), 1.10 (d, *J*=6.0 Hz, 3xH-21), 1.33 (s, 3xH-19), 1.67 (d, *J*=5.0 Hz, sec. C*H*$_3$), 4.79 (br s, 2xH-27), 5.09 (d, *J*=6.0 Hz, anomeric H), 5.41, 6.42 (each br s, anomeric H).

CMR (C₅D₅N, 20.0 MHz) : C-1) 78.0 (2) 30.2 (3) 70.3 (4) 34.4 (5) 30.9 (6) 27.2^a (7) 26.3^a (8) 34.4 (9) 45.8 (10) 39.5 (11) 22.5 (12) 40.4 (13) 40.6 (14) 57.0 (15) 32.1 (16) 81.4 (17) 63.1 (18) 16.6 (19) 17.2 (20) 41.8 (21) 15.0 (22) 109.4 (23) 33.2^b (24) 28.9^b (25) 144.4 (26) 65.0 (27) 108.6 **Xyl** (1) 99.6^c (2) 79.1 (3) 77.0 (4) 71.3 (5) 66.8 **Rha I** (1) 101.5^c (2) 72.2^d (3) 72.6^d (4) 74.0 (5) 69.4 (6) 18.9^e **Rha II** (1) 99.3 (2) 72.2 (3) 72.2 (4) 73.9 (5) 69.8 (6) 18.6^e.

Mass (F.D.) : *m/z* 878 [M+Na+H]$^+$.

Note : The compound was isolated admixed with 25,27-dihydroderivative.

Reference

1.   K. Iwagoe, T. Konishi and S. Kiyosawa, *Yakugaku Zasshi*, **107**, 140 (1987).

SPICATOSIDE C
25(*S*)-Ruscogenin 1-O-{β-D-fucopyranosyl-(1→2)-[β-D-xylopyranosyl-(1→4)]-β-D-fucopyranoside}

Source : *Liriope spicata* (Thunb.) (Liliaceae)
Mol. Formula : C₄₄H₇₀O₁₆
Mol. Wt. : 854
M.P. : 224-226°C
[α]$_D$^{23} : -55.0° (c=0.5, MeOH).
Registry No. : [166334-39-0]

IR (KBr) : 3414, 1066, 988, 919, 897, 851 [919>897, 25(*S*)-spiroketal] cm^{-1}.

PMR (C₅D₅N, 300 MHz) : δ 0.86 (s, 3xH-18), 1.06 (d, *J*=7.1 Hz, 3xH-27), 1.09 (d, *J*=7.3 Hz, 3xH-21), 1.34 (s, 3xH-19), 1.47 (d, *J*=6.4 Hz, 3xH-6 of Fuc), 1.62 (d, *J*=6.3 Hz, 3xH-6 of Fuc), 4.78 (d, *J*=7.6 Hz, anomeric H), 5.10 (d, *J*=7.5 Hz, anomeric H), 5.16 (d, *J*=7.8 Hz, anomeric H), 5.56 (br d, *J*=4.7 Hz, H-6).

CMR (C$_5$D$_5$N, 75.5 MHz) **:** δ C-1) 83.3 (2) 37.1 (3) 68.2 (4) 43.7 (5) 139.9 (6) 124.4 (7) 32.4 (8) 33.0 (9) 50.3 (10) 42.9 (11) 23.7 (12) 40.4 (13) 40.2 (14) 56.9 (15) 32.1 (16) 81.2 (17) 62.9 (18) 16.7 (19) 14.8 (20) 42.5 (21) 15.0 (22) 109.7 (23) 26.4 (24) 26.2 (25) 27.5 (26) 65.1 (27) 16.3 **Fuc I** (1) 99.8 (2) 82.0 (3) 72.9 (4) 78.7 (5) 73.5 (6) 17.1 **Fuc II** (1) 105.4 (2) 71.7 (3) 75.0 (4) 72.2 (5) 70.6 (6) 17.4 **Xyl** (1) 106.1 (2) 75.1 (30 78.5 (4) 71.0 (5) 67.2.

Mass (FAB) **:** m/z 877 [M+Na]$^+$, 856 [M+2H]$^+$, 724 [M+2H-132]$^+$, 710 [M+2H-146]$^+$, 577 [genin+H+146]$^+$, 431 [genin+H]$^+$.

Reference

1.   J.C. Do, K.Y. Jung, Y.K. Sung, J.H. Jung and K.H. Son, *J. Nat. Prod.,* **58**, 778 (1995).

SPRENGERININ C

(25*R*)-Spirost-5-en-3β-ol 3-O-{α-L-rhmnopyranosyl-(1→2)-[β-D-xylopyranosyl-(1→4)]-β-D-glucopyranosyl}

Source : *Asparagus sprengeri* Regel.[1] (Liliaceae), *Allium narcissiflorum*[2] (Liliaceae)
Mol. Formula : C$_{44}$H$_{70}$O$_{16}$
Mol. Wt. : 854
M.P. : 245-250°C (MeOH)[1]
[α]$_D^{20}$ **:** -86.4° (c=1.0, Pyridine)[1]
Registry No. : [88861-91-0]

IR (KBr)[1] : 3400 (OH), 2845 (C=CH), 1040, 980, 916, 898 (intensity 898>916, 25*R*-spiroketal), 850 cm^{-1}.

PMR (DMSO-d_6, 100 MHz)[1] : δ 4.20 (d, J=7.0 Hz, H-1 of Xyl), 4.60 (d, J=6.5 Hz, H-1 of Glc), 5.28 (d, J=1.5 Hz, H-1 of Rha).

Mass (FD)[1] (Silicone emitter 29-33 MA) : m/z (rel.intens.) 893 [M+K]$^+$ (30.0), 877 [M+Na]$^+$ (100), 855 [M+H]$^+$ (10), 854 [M]$^+$ (8), 745 [(M+Na)-132]$^+$ (18), 731 [(M+Na)-146]$^+$ (10), 599 [(M+Na)-278]$^+$ (4), 450 [M+2Na]$^{2+}$ (10), 147 [Rha-OH] (8), 133 [Xyl-OH] (6).

References

1.   S.C. Sharma, R. Sharma and R. Kumar, *Phytochemistry*, **22**, 2259 (1983).

2.   Y. Mimaki, T. Satou, M. Ohmura and Y. Sashida, *Natural Phytomedicine*, **50**, 308 (1996).

RECURVOSIDE A
(23*S*)-Spirosta-5,25(27)-diene-1β,3β,23-triol 1-O-{α-L-rhamnopyranosyl-(1→2)-[β-D-xylopyranosyl-(1→3)]-α-L-arabinopyranoside

Source : *Nolina recurvata*[1] (Agavaceae), *Sansevieria trifasciata*[2] (Agavaceae)
Mol. Formula : $C_{43}H_{66}O_{17}$
Mol. Wt. : 854
[α]$_D^{26}$: -62.9° (c=0.25, MeOH)
Registry No. : [163136-07-0]

IR (KBr)[1] : 3420 (OH), 2900 (CH), 1445, 1365, 1245, 1130, 1035, 975 cm^{-1}.

PMR (C_5D_5N, 400/500 MHz)[1] : δ 1.03 (s, 3xH-18), 1.11 (d, J=7.0 Hz, 3xH-21), 1.37 (s, 3xH-19), 1.70 (d, J=6.1 Hz, 3xH-6 of Rha), 4.73 (d, J=7.2 Hz, H-1 of Ara), 4.76 and 4.83 (br s, 2xH-27), 4.97 (d, J=7.5 Hz, H-1 of Xyl), 5.56 (br d, J=5.5 Hz, H-6), 6.30 (br s, H-1 of Rha).

PMR (C_5D_5N-CD_3OD 11:1, 400/500 MHz)[1] : δ 1.00 (s, 3xH-18), 1.08 (d, J=7.0 Hz, 3xH-21), 1.14 (H-14), 1.29 (H-12A), 1.34 (s, 3xH-19), 1.47 (H-9), 1.50 (H-7A and H-15A), 1.52 (H-8), 1.57 (H-11A, H-12B), 1.65 (d, J=6.1 Hz, 3xH-6 of Rha), 1.80 (dd, J=8.5, 7.1 Hz, H-17), 1.86 (H-7B), 2.01 (H-15B), 2.27 (q-like, J=12.0 Hz, H-2A), 2.52 (dd, J=12.7, 4.3 Hz, H-4A), 2.61 (dd, J=12.7, 12.2 Hz, H-4B), 2.63 (H-2B), 2.76 (dd, J=12.6, 5.2 Hz, H-24A), 2.86 (dd, J=12.6, 12.6 Hz, H-24B), 2.87 (H-11B), 2.94 (m, H-20), 3.61 (dd, J=11.0, 10.2 Hz, H-5A of Xyl), 3.69 (br d, J=11.5 Hz, H-5A of Ara), 3.74 (dd, J=12.0, 4.0 Hz, H-1), 3.79 (m, H-3), 3.83 (dd, J=8.4, 7.5 Hz, H-2 of Xyl), 3.96 (d, J=12.5 Hz, H-26A), 3.97 (dd, J=9.5, 8.4 Hz, H-3 of Xyl), 4.02 (ddd, J=10.2, 9.5, 5.0 Hz, H-4 of Xyl), 4.06 (dd, J=8.9, 3.2 Hz, H-3 of Ara), 4.16 (dd, J=9.5, 9.4 Hz, H-4 of Rha), 4.21 (dd, J=11.5, 2.1 Hz, H-5B of Ara), 4.22 (dd, J=11.0, 5.0 Hz, H-5B of Xyl), 4.37 (br s, H-4 of Ara), 4.38 (d, J=12.5 Hz, H-26B), 4.46 (dd, J=9.4, 3.4 Hz, H-3 of Rha), 4.54 (dd, J=8.9, 7.5 Hz, H-2 of Ara), 4.58 (q-like, J=8.5 Hz, H-6), 4.66 (br d, J=3.4 Hz, H-2 of Rha), 4.67 (d, J=7.5 Hz, H-1 of Ara), 4.69 (dq, J=9.5, 6.1 Hz, H-5 of Rha), 4.81 (br s, H-27A), 4.84 (br s, H-27B), 4.90 (d, J=7.5 Hz, H-1 of Xyl), 5.56 (br d, H-6), 6.19 (br s, H-1 of Rha).

CMR (C_5D_5N, 100/125 MHz)[1] : δ C-1) 83.7 (2) 37.4 (3) 68.3 (4) 43.9 (5) 139.6 (6) 124.7 (7) 32.1 (8) 33.1 (9) 50.4 (10) 42.9 (11) 24.1 (12) 40.6 (13) 40.8 (14) 56.9 (15) 32.4 (16) 82.0 (17) 62.5 (18) 16.9 (19) 15.0 (20) 35.8 (21) 14.6 (22) 111.8 (23) 68.6 (24) 38.9 (25) 144.4 (26) 64.3 (27) 109.3 **Ara** (1) 100.5 (2) 74.3 (3) 84.4 (4) 69.6 (5) 67.1 **Rha** (1) 101.9 (2) 72.5 (3) 72.6 (4) 74.3 (5) 69.6 (6) 19.1 **Xyl** (1) 106.5 (2) 74.7 (3) 78.3 (4) 71.0 (5) 66.9.

Mass (FAB, Negative ion)[1] : m/z 853 [M-H]⁻.

Biological Activity : Recurvoside A exhibited medium inhibitory activity on cAMP phosphodiesterase with IC_{50}=19.5x10⁻⁵M.[1]

References

1.   Y. Takaashi, Y. Mimaki, M. Kuroda, Y. Sashida, T. Nikaido and T. Ohmoto, *Tetrahedron*, **51**, 2281 (1995).

2.   Y. Mimaki, T. Inoue, M. Kuroda and Y. Sashida, *Phytochemistry*, **43**, 1325 (1996).

ASPARAGUS CURILLUS SAPONIN 2

(25S)-5β-Spirostan-3β-ol 3-O-{α-L-rhamnopyranosyl-(1→2)-[α-L-arabinopyranosyl-(1→4)-β-D-glucopyranoside}

Source : *Asparagus curillus* Buch.-Ham. (Liliaceae)
Mol. Formula : $C_{44}H_{72}O_{16}$
Mol. Wt. : 856
M.P. : 223-225°C
[α]$_D^{20}$: -57.0° (c=1.0, Pyridine)
Registry No. : [84765-73-1]

IR (KBr) : 3400 (OH), 988, 920, 898, 850 cm^{-1}, intensity 920 > 898 (25S-spiroketal).

Reference

1.   S.C. Sharma, O.P. Sati and R. Chand, *Phytochemistry*, **21**, 1711 (1982).

ASPARANIN C
Sarsasapogenin 3-O-{α-L-arabinopyranosyl-(1→4)-[α-L-rhamnopyranosyl-(1→6)]-β-D-glucopyranoside}

Source : *Asparagus adscendens* Roxb. (Liliaceae)
Mol. Formula : C$_{44}$H$_{72}$O$_{16}$
Mol. Wt. : 856
M.P. : 229-232°C
[α]$_D$^{18} : -56.6° (Pyridine)
Registry No. : [83997-41-5]

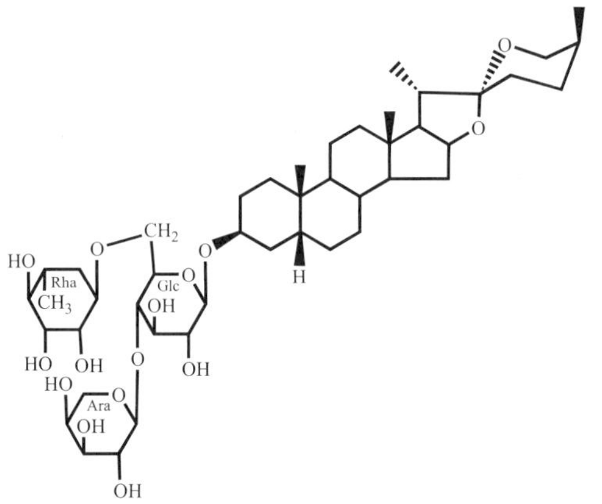

IR (KBr) : 3400 (OH), 981, 915, 900, 855, intensity 915 > 900 (25S-spiroketal) cm^{-1}.

Mass (F.D.) : m/z 895 [M+K]$^+$, 879 [M+Na]$^+$, 763, 744, 733.

Octa-O-methyl derivative :

Mass (E.I.) : m/z 968 [M]$^+$, 777 [M-tri-O-methylarabinose+H]$^+$, 763 [M-tri-O-methyl-rhamnose+H]$^+$, 189 [Tri-O-methylrhamnose]$^+$, 175 [Tri-O-methylarabinose]$^+$.

Biological Activity : Anti-ulcerogenic activity.

Reference

1. S.C. Sharma, R. Chand, B.S. Bhatti and O.P. Sati, *Planta Med.*, **46**, 48 (1982).

CONVALLASAPONIN C
Isorhodeasapogenin 3-O-[α-L-rhamnopyranosyl-(1→3)-α-L-rhamnopyranosyl-(1→2)-α-L-arabinopyranoside]

Source : *Convallaria keiskei* Miq. (Liliaceae)
Mol. Formula : $C_{44}H_{72}O_{16}$
Mol. Wt. : 856
M.P. : 218~221°C (decomp.)[2]
$[\alpha]_D^{19}$: -89.7° (c=0.78, CHCl$_3$-MeOH)[2]

IR (KBr)[2] : 3500~3200 (br, OH), 983, 918<900, 865 cm^{-1} (25R-spiroketal).

References

1.   M. Kimura, M. Tohma and I. Yoshizawa, *Chem. Pharm. Bull.*, **14**, 55 (1966).

2.   M. Kimura, M. Tohma and I. Yoshizawa, *Chem. Phram. Bull.*, **14**, 50 (1966).

ALLIUM CHINENSE SAPONIN 2, XIEBAISAPONIN I
Laxogenin 3-O-{β-D-xylopyranosyl-(1→4)-[α-L-arabinopyranosyl-(1→6)]-β-D-glucopyranoside}

Source : *Allium chinense* G. Don[1,2] (Liliaceae)
Mol. Formula : C$_{43}$H$_{68}$O$_{17}$
Mol. Wt. : 856
[α]$_D^{27}$: -71.0° (c=0.12, MeOH)
Registry No. : [123941-67-3]

IR (KBr)[1] : 3390 (OH), 1705 (C=O) cm^{-1}.

PMR (C$_5$D$_5$N, 400 MHz)[1] : δ 0.64 (s, 3xH-19), 0.70 (d, J=5.7 Hz, 3xH-27), 0.79 (s, 3xH-18), 1.15 (d, J=6.8 Hz, 3xH-21), 4.94 (d, J=7.8 Hz, H-1 of Glc), 5.06 (d, J=7.4 Hz, H-1 of Xyl), 5.46 (d, J=7.8 Hz, H-1 of Ara).

CMR (C$_5$D$_5$N, 100 MHz)[1] : δ C-1) 36.7 (2) 29.4 (3) 77.0 (4) 26.9 (5) 56.4 (6) 209.6 (7) 46.7 (8) 37.3 (9) 53.7 (10) 40.8 (11) 21.5 (12) 39.6 (13) 41.0 (14) 56.4 (15) 31.7 (16) 80.8 (17) 62.8 (18) 16.4 (19) 13.0 (20) 41.9 (21) 14.9 (22) 109.2 (23) 31.7 (24) 29.2 (25) 30.5 (26) 66.9 (27) 17.3 **Glc** (1) 102.0 (2) 74.9 (3) 78.4 (4) 79.8 (5) 74.8 (6) 68.1 **Ara** (1) 105.1 (2) 74.8 (3) 76.3 (4) 71.0 (5) 67.3 **Xyl** (1) 105.6 (2) 72.5 (3) 74.5 (4) 69.8 (5) 67.3.

Mass (FAB, Positive ion)[2] : m/z 879 [M+Na]$^+$, 857 [M+H]$^+$, 431 [Agl+H]$^+$.

Mass (FAB, Negative ion)[1] : m/z 856 [M]$^-$.

Biological Activity : The compound exhibited considerable inhibitory activity on cAMP phosphodiesterase with IC$_{50}$ value of 12.3x10^{-5} M. It showed strong cytotoxicity against HeLa cells at 25 μg/ml (14% inhibition and at 10 μg/l (8.3% inhibition).[2] Inhibit hypoxia/reoxygenation-induced protein tyrosine kinase activation in cultured human umbilical vein endothelial cells.[3]

References

1. M. Kuroda, Y. Mimaki, A. Kameyama, Y. Sashida and T. Nikaido, *Phytochemistry*, **40**, 1071 (1995).

2. M. Baba, M. Ohmura (neé Matsuda), N. Kishi, Y. Okada, S. Shibata, J. Peng, S.-S. Yao, H. Nishino and T. Okuyama, *Biol. Pharm. Bull.*, **23**, 660 (2000).

3. Y.-W. Zhang, I. Morita, G. Shao, X.-S. Yao and S.I. Murota, *Planta Med.*, **66**, 114 (2000).